Human Drug Targets

Human Drug Targets

A Compendium for Pharmaceutical Discovery

Edward D. Zanders

PharmaGuide Ltd, Cambridge, UK

WILEY Blackwell

Contents

Preface

The drug discovery scientist in the 21st century has access to a vast amount of information about the workings of living organisms and the nature of human disease. It is now possible to survey the entire human genome and proteome in a systematic way at different levels of organization (through sequence and expression analysis, epigenetic modification, etc.) in order to identify potential targets for drug development. This information comes from a rapidly expanding global scientific workforce that is producing an equally expanding output of literature, aided in part by the open-access publishing model.

My personal involvement in drug target selection started in a large pharmaceutical company, looking at molecules of the human immune system that could be useful as targets for drugs to treat allergic and autoimmune diseases. This was the time when cytokine biology was beginning to develop with the discovery of the first interleukins; there was a compelling case for making inhibitors of cytokine–receptor interactions that would interfere with immunoinflammatory processes in a highly selective manner. At the time, there was a humble tally of just three named interleukins (the latest at the time of writing is interleukin-37). The search for drug targets, then as now, involved a survey of the basic biomedical literature: the latest issue of *Nature* (or similar), delivered by post (in pre-Internet days), might contain the description of a new cytokine or adhesion molecule or something that was shown to affect the behaviour of cells, at least *in vitro*.

The landscape of target discovery changed around the turn of the millennium with the rise of genomics. The increasing availability of human genome sequence (some of it still only accessible by subscription) meant that target discovery could potentially become less hit and miss. It was much easier to identify closely related protein families using sequence analysis; families with established drug targets like the peptidases could be mined for target opportunities, a process that continues to this day.

From a pharmaceutical perspective, the ultimate aim of a systematic survey of the human genome must be the delineation of all possible drug targets. I (and many others) have often wondered how big this number is; it was while reading one of the several reviews on this subject that the idea of a compendium of drug targets came to me. The *Oxford English Dictionary* definition puts the idea into words perfectly: '(a compendium is) a collection of concise but detailed information about a particular subject, especially in a book or other publication'. Furthermore, the word is derived from the Latin *compendere*, 'what is weighed together', literally meaning 'profit, saving', something with a certain commercial appeal. This book then is a compendium of human drug targets, both established and potential, based on the roughly 19,000 protein-coding genes of the human genome plus some non-coding RNA targets. Since only human genes are covered, this excludes most infectious disease targets except those relating to host cell–pathogen interactions. The compendium is concise, in having just enough information to attract the readers' attention to a particular entry, then allowing them to access the relevant information online using the HUGO Gene Nomenclature Committee (HGNC) approved gene names and symbols. The book format presents the information in such a way as to encourage browsing by thumbing through pages rather than by scrolling down a screen and getting distracted by various hyperlinks.

There is sufficient information on potential targets to keep investigators busy for a long time. The book contains a survey of approximately 50% of the human protein-coding genome and includes established drug target classes such as enzymes, receptors and transporters. The remaining gene entries will be curated for inclusion in a future volume, eventually resulting in a significant coverage of the human genome. It is inevitable that more data will become available for each entry as the years go by, but so long as there is a fixed point of reference in the book, changes in nomenclature will be flagged online in the HGNC pages and new publications revealed through the 'related citations' feature in PubMed.

There is clearly no shortage of potential drug targets, but of course not all are created equal. Going back to my earlier days with cytokines, we discovered the hard way that the small-molecule receptors so successfully targeted by the medicinal chemists and pharmacologists are not the same as cytokine receptors because the latter generally lack suitable pockets for high-affinity binding of small molecules. This did not in any way deter us (or indeed our rivals) as we tried to find small-molecule inhibitors by random screening. To paraphrase the 18th-century English writer Samuel Johnson, it was 'the triumph of hope over experience' (although he was referring to second marriages). It takes a vast amount of effort to move from drug targets to effective medicines. Sometimes, it takes a complete re-engineering of drug development, as happened with the introduction of monoclonal antibodies or other recombinant proteins as cytokine inhibitors. Thus, many years later, after a fruitless search for small-molecule inhibitors of interleukin action for atopic and asthmatic diseases (IL-4 and IL-5), positive clinical data with antibodies are finally becoming available. Hopefully, it will not be too long before the same level of technological maturity can be achieved with RNA drugs and gene editing/therapy.

Every effort has been made to minimize errors and omissions in content and layout. If the book is likened to a large menu, for example, some desserts will appear under 'entrées' and so on. However, I like to think that I have been reasonably conscientious; perhaps my DNA contains a relevant mutation in *KATNAL2*, a gene that might show some association with this personality trait [1].

Finally, I would like to thank the organizations that have made it possible to select gene entries and annotations for this compendium. These include the HGNC and UniProt Knowledgebase, both based at the European Bioinformatics Institute in Cambridge, United Kingdom, and the US National Library of Medicine for PubMed references. In particular, I'd like to thank Dr Elspeth Bruford at the HGNC for her helpful comments as well as permission to show the HGNC web page in Chapter 2.

I am grateful to Wiley, in particular Lucy Sayer for her willingness to accept this book project and Celia Carden for helping to turn it into reality. Their enthusiasm is much appreciated.

Last, but not least, I thank my wife Rosie for her patience while I spent many hours in front of the computer sorting through lists; I dedicate this book to her and to our children.

Reference

1. De Moor MHM *et al.* (2012) Meta-analysis of genome-wide association studies for personality. *Molecular Psychiatry* **17**, 337–49.

Chapter 1

Introduction

Global sales of prescription medicines reached nearly 1 trillion dollars in 2013 and show no sign of abating [1]. At first sight, this might give the impression that all is well with the biopharmaceutical industry; however, this hides the well-documented fact that company pipelines of innovative drugs are not full enough to keep up with the escalating costs and difficulty of bringing them to market [2]. There are many points in the drug development pipeline where improvements can be made to increase the chance of success. One such point lies at the beginning of the discovery process itself, the identification of drug targets with therapeutic potential. Organized drug target discovery was once almost exclusively undertaken in the pharmaceutical industry, but this situation is changing through stronger collaboration between companies and academia. Regardless of where drug target discovery is actually undertaken, there is a need for as much scientific information as possible to guide the research; this information is provided through biology, chemistry and medicine but is overwhelming in its totality. Individual pieces of information can be readily accessed in online databases, publications and verbal communication with colleagues, but it is difficult to present this totality of target opportunities in a format that is easily browsed. This book is designed to address this issue by presenting a large list of potential human drug targets in a physical form that is easy to browse through, rather like a catalogue; each entry contains just enough information to attract interest without adding undue clutter to the text while at the same time supplying the key information required to follow up online. This is a book about potential and actual human drug targets, not the drugs themselves; microbial targets are not included in order to keep the book within manageable proportions for the sake of both the reader and the author.

This chapter sets the scene for the rest of the book by describing the drug target concept from its origins in 19th century pharmacology through to the Human Genome Project and the present day.

1.1   Magic bullets

If we picture an organism as infected by a certain species of bacterium, it will obviously be easy to effect a cure if substances have been discovered which have an exclusive affinity for these bacteria and act deleteriously or lethally on these alone, while at the same time they possess no affinity for the normal constituents of the body and can therefore have the least harmful, or other, effect on

Human Drug Targets: A Compendium for Pharmaceutical Discovery, First Edition. Edward D. Zanders.
© 2016 John Wiley & Sons, Ltd. Published 2016 by John Wiley & Sons, Ltd.

that body. Such substances would then be able to exert their full action exclusively on the parasite harboured within the organism and would represent, so to speak, magic bullets, which seek their target of their own accord.

These words were spoken in 1906 by Paul Ehrlich as part of an address to inaugurate the Georg-Speyer Haus, an institute devoted to chemotherapy research in Frankfurt, Germany [3]. His comments provide a useful summary of the concept of a drug target and are applicable to all diseases, not just those caused by infectious agents. Ehrlich's research represented a transition point between the beginnings of the modern pharmacology that emerged in the 19th century and the description and eventual isolation of defined receptors for synthetic drug molecules that occurred in the 20th.

The following sections present modern ideas about drug targets in a historical context, highlighting the relatively recent molecular characterization of receptors for drugs which, in many cases, have been used for over a century.

1.2   Background to modern pharmacology

Some of the following is taken from Prüll, Maehle and Halliwell's informative history of the development of the drug receptor concept [4].

Natural products have been isolated from living organisms to treat diseases for thousands of years, but a coherent understanding of the disease process itself and how the agents actually worked was lacking until only the last 200 years or so. Of the many examples of rational and quasi-religious theories propounded for disease and drug action, I rather enjoy that of the 18th-century Scottish physician John Brown; he suggested that illness was due to either a lack of bodily excitement or to overexcitement. The cure was a mixture of alcohol and opium for the former and a vegetable diet or bloodletting for the latter. Despite this being at odds with modern thinking, there is an air of familiarity about it, although nowadays the bloodletting is generally a side effect rather than a therapeutic intervention.

Pharmacology as a named discipline was born in France, through the work of François Magendie in Paris and later Rudolf Buchheim, who established the first laboratory for experimental pharmacology at the University of Dorpat in Estonia. Magendie and a collaborator, Alire Raffeneau-Delille, studied the toxic action in dogs of several drugs of vegetable origin, including *nux vomica*, marking the first experiments of modern pharmacology. The results suggested to Magendie that the action of natural drugs depended on the chemical substances they contain, and it should be possible to obtain these substances in a pure state. This emphasis on pure substances rather than compound remedies was a turning point in pharmaceutical research. Later in the 19th century, the first hints of structure–activity relationships between drugs and physiological responses were obtained as a result of advances in organic chemistry. For example, Sir Benjamin Ward Richardson showed that chemical modifications of amyl nitrate produced anaesthetics with varying degrees of activity in frogs. Alexander Crum Brown and Thomas Fraser presented a paper to the Royal Society of Edinburgh in 1868 entitled 'On the Connection between Chemical Constitution and Physiological Action; with Special Reference to the Physiological Action of the Salts of the Ammonium Bases Derived from Strychnia, Brucia, Thebata, Codeia, Morphia, and Nicotia'. They showed that whatever the normal effect of these alkaloids, the change of a tertiary nitrogen atom to the quaternary form invariably produced a curare-like paralysing action, thus providing the opportunity for making novel agents.

This early medicinal chemistry was not fully developed until the 20th century. In the meantime, it was necessary to develop theories of drug action that fitted the experimental observations made by pioneering pharmacologists, microbiologists and chemists. One important aspect of this related to the idea of *affinity* between a drug and the cells and tissues of the body. The title of Goethe's 1809 novel about understanding human relationships, *Wahlverwandtschaften (Elective Affinities)*, was applied to pharmacology by Friedrich Sobernheim in terms of *specific elective affinities*. Another key aspect of drug action was that disease results from alterations in cellular structure and activity, an idea published by Rudolf Virchow in 1858. This observation, coupled with data showing that dyes would selectively bind to specific cell types and structures, created the groundwork for a receptor theory of drug action.

1.2.1 The receptor theory

Ehrlich worked on antibody-mediated haemolysis of red blood cells that stimulated his theory in which a countless number of side-chains would adapt to the "constantly changing chemistry" of the body. This chemistry would be influenced by race, sex, nutrition, energy, secretion and other factors, and so there were continuous changes taking place in the blood serum'. In 1900, Ehrlich and his collaborator Julius Morgenroth introduced the term 'receptor' for the first time: 'For the sake of brevity, that combining group of the protoplasmic molecule to which the introduced group is anchored will hereafter be termed receptor'.

Independent support for the receptor theory was provided by the Cambridge (UK) physiologist John Newport Langley with his concept of *receptive substances*. He interpreted these as 'atom-groups of the protoplasm' of the cell. When compounds bonded to the receptive atom groups, they would alter the protoplasmic molecule of the cell and in this way change the cell's function. In more differentiated cells, such as those of the muscles and glands, the receptive atom groups had undergone a 'special development' which enabled them to combine with hormones or with alkaloids. Due to those cells' connection with nerve fibres, these further developed atom groups tended to concentrate in the region of the nerve endings. In contrast, *fundamental atom groups* were essential for the cell's life. If a chemical substance bound to such a group, the cell would be damaged and die [4]. These comments bring to mind the modern distinction between genes coding for drug targets and those housekeeping genes that are essential for cellular viability.

The receptor theory was not immediately accepted (despite the support of Sir Arthur Conan Doyle, formerly an ophthalmologist but better known as the creator of Sherlock Holmes; this support was reciprocated, as Ehrlich was a great fan of detective stories [4]). The most prominent alternative to a chemical receptor theory was the idea that the physical properties of molecules and target tissues dictated drug action. This viewpoint, held notably by Walther Straub in Germany, was part of a major controversy in pharmacology, amazingly until as late as the 1940s.

Pharmacology was advancing rapidly in the early 20th century despite the aforementioned controversy at the end of the previous paragraph. The neurotransmitter acetylcholine was discovered through the work of Sir Henry Dale and Otto Loewi, earning them the Nobel Prize in 1936. Dale also discovered histamine, while the Japanese chemist Jokichi Takamine, working in the United States for the Parke–Davis and Company, purified adrenaline for the first time in 1900. For this latter feat, the Emperor of Japan donated fifteen imperial cherry trees to Parke–Davis which were planted outside their administrative offices [5].

Despite the ability of pharmacologists to affect cells and tissues with these chemically defined molecules, the idea of a specific receptor was still resisted; in a practical world, they were considered to be too theoretical, at least until the point where their existence could be proven experimentally.

One approach to this was to put pharmacology on a quantitative footing, whereas previously it had been almost entirely descriptive. In 1909, Archibald Hill described the action of nicotine and curare on the contraction or relaxation of frog muscle; through analysing the concentration–effect curves and the temperature dependence of the reactions, he deduced that the drug action was due to a chemical process. This pioneering quantitative work was taken up by Alfred Clark in London in the 1920s, using isolated tissues in a similar manner to Hill. The sigmoidal dose–response curves familiar to drug discovery scientists gave him insight into receptor function as well as the phenomenon of antagonism. To quote Clark, 'atropine and acetyl choline (*sic*), therefore, appear to be attached to different receptors in the heart cells and their antagonism appears to be an antagonism of effects rather than of combination'.

By the 1950s and 1960s, concepts such as agonists, affinity and drug efficacy were well known, but little of this knowledge had been applied to pharmaceutical discovery. This all changed with the identification and exploitation of adrenaline receptor subtypes by Raymond Alquist and Sir James Black, respectively.

The American pharmacologist Alquist embarked upon a study of sympathomimetic compounds designed to relax uterine muscles in the cases of dysmenorrhoea in the 1940s. Briefly, he determined the rank order of potency of a series of compounds (including adrenaline) on the excitation or inhibition of various tissues and in the process discovered two classes of adrenergic receptors which he named α- and β-receptors. This work was aided in part by having access to using sophisticated instruments developed from technology developed during the recent World War. Alquist's seminal work was published in 1948, but he considered the idea of receptors as a theoretical tool; the later subdivisions of adrenoreceptors into α_1, α_2, β_1 and β_2 subtypes caused him anxiety (β_3 came much later). He believed that 'if there are too many receptors, something is obviously wrong' [4]. What he would have made of our current inventory of receptors and subtypes is probably best left to the imagination.

Sir James Black worked for the UK company Imperial Chemical Industries (now subsumed into AstraZeneca) in the late 1950s. His work on agents to treat angina pectoris led to the first 'beta blocker' drugs, pronethalol and propranolol, thus pioneering the exploitation of receptor subtypes that is now routine practice in biopharmaceutical companies.

By the 1960s, the receptor theory of drug action was accepted by pharmacologists, but still not understood at the molecular level. D.K. de Jongh's comments written in 1964 sum the situation up in a rather literary manner: 'To most of the modern pharmacologists the receptor is like a beautiful but remote lady. He has written her many a letter and quite often she has answered the letters. From these answers the pharmacologist has built himself an image of this fair lady. He cannot, however, truly claim ever to have seen her, although one day he may do so'. That day came soon enough after the application of cell and molecular biology to pharmacological problems.

1.2.2 Molecular pharmacology

The following is adapted from Halliwell's article published in *Trends in Pharmacological Sciences* [6]. Early work on drug receptors provided hints that they were located in specialized regions of tissues. R.P. Cook showed in 1926 that acetylcholine action on frog muscle was blocked by methylene blue dye before it stained the muscle tissue. This antagonist action was reversible, as demonstrated by washing away and reapplying the methylene blue even though the heart muscle retained the blue staining throughout. This suggested that methylene blue had reversibly bound to receptors located at the cell surface. Much later in the late 1960s, Eduardo Robertis and colleagues disrupted tissue with detergents and isolated synaptosomes by differential centrifugation. Synaptosomal membranes contain the nicotinic acetylcholine receptors that were the first receptor molecules to be purified. This purification was independently achieved in the early 1970s by Jean-Pierre Changeux and Ricardo Miledi using the electric organ of rays and eels as rich sources of receptor protein. This period saw the introduction of radioligand binding and affinity purification with potent ligands, in this case α-bungarotoxin. The receptor was then shown to be a 275,000 dalton complex formed from multiple protein subunits.

Thus, 70 years after Ehrlich's time and nearly half a century from our own, the receptor theory was no longer dealing with the abstract but with real molecular entities.

1.2.3 Receptors, signals and enzymes

The nicotinic receptor highlighted earlier is now known to be one of some 300 ion channels, many of which are of major pharmaceutical interest. However, a significant number of current medicines act through a different system, the G-protein-coupled receptors (GPCRs). The prototypic GPCR is the visual transducer rhodopsin, first characterized in the 19th century and sequenced in the 1980s [7]. The protein sequence of bovine rhodopsin revealed a serpentine structure which traversed the cell membrane seven times. Work on rhodopsin signalling revealed the action of a GTPase, initially called transducin and later shown to be a heterotrimeric protein composed of α, β and γ subunits; thus, the term G-protein coupled or 7TM receptor entered the pharmaceutical lexicon. This signalling system is of course one of the many which have been exploited as drug targets or which have the potential to be so.

So far, this historical summary has focused on cell surface receptors for therapeutic ligands, but enzymes are also excellent drug targets. The true nature of enzymes as catalytic proteins was not known until Sumner's crystallization of urease in 1926 and Northrop's studies on pepsin in 1929. Nevertheless, enzyme inhibition was understood at this time and was exploited, for example, in the development in the 1930s of cholinesterase inhibitors that could be used in glaucoma treatment; unfortunately, they had more potential as nerve agents for military use. Later in the 1940s, the antibacterial sulphonamide drugs, by now being superseded by penicillin, provided a lead for novel diuretic drugs (thiazides) based on the inhibition of carbonic anhydrase in the renal tubule of the kidney [8]. The list of enzyme targets discovered through the remainder of the 20th century includes those for highly successful drugs used in the treatment of millions of patients; these include angiotensin-converting enzyme for hypertension, cyclooxygenases as targets for anti-inflammatory drugs, HMGCoA reductase inhibitors for hypercholesterolaemia and tyrosine kinase inhibitors for oncology.

1.2.4   Recombinant DNA technology and target discovery

Results of the first molecular cloning experiment, in which ribosomal RNA from *Xenopus laevis* was transferred to *Escherichia coli* in a plasmid vector, were published in 1974 [9]. Ten years later, the first pharmacological receptor molecules were cloned (the α, β, γ and δ subunits of the nicotinic acetylcholine receptor from *Torpedo californica*; see Ref. [7]). This achievement was possible because sufficient amino acid sequence of the receptor protein was available to allow investigators to design degenerate oligonucleotides for screening cDNA libraries. This approach has been used many times since then to clone genes encoding a wide variety of human proteins, some of which are known drug targets. However, the only way to identify the full repertoire of human protein-encoding genes was to sequence the entire 3.5 gigabase genome, which of course is what happened between 1990 and 2003 [10]. At the time of publication (in 2001) of the first draft of the human genome sequence, there was some inevitable speculation about how this affected the search for drug targets [11]. In the intervening years to the present, the number of protein-coding genes has dropped from around 30,000 to around 19,000 as more proteomics data and improved bioinformatics analysis have become available [12]. This value of 19,000 genes is the one I have used to constrain the number of drug target proteins that could potentially exist. However, alternative splicing and the identification of micro and other non-coding RNAs have added a new dimension to the analysis of target numbers. A more detailed discussion about identifying drug targets from human genome and proteome data follows later in the book.

1.3   Drug and therapeutic targets in the biomedical literature

The explicit use of the phrase 'drug target' or 'therapeutic target' in the literature began around the time that molecular pharmacology was beginning to grow in the late 1970s. Figure 1.1 shows the number of papers containing each phrase taken from the PubMed database for every year from 1979 to 2014 (both phrases occurred in the same paper only 174 times). Prior to this date, there was only one use of the phrase 'drug target' (in 1975) and two for 'therapeutic target' (in 1954 and 1977). For the historical record, the 1954 paper was entitled 'The human mouth flora as a therapeutic target'.

This exercise has an important bearing on the way some of the data were gathered for this book; the results from these PubMed searches were used to annotate many of the compendium entries as described in the next chapter.

What is notable about this admittedly not very scientific analysis is that the explicit use of the phrases 'drug target' and 'therapeutic target' in the literature really only occurred in a significant way from the mid-1990s onwards. Incidentally, this exceeds the rate of growth in total PubMed citations over this period, which roughly trebled between 1979 and 2012. In a separate analysis, the Espacenet patent database was searched with the 'drug target' and 'therapeutic target' phrases, giving 548 and 1156 worldwide patent citations, respectively (at the time of writing).

1.4   How many drug targets are there?

The distinction between the number of targets for current drugs and the number of *potential* targets must be made at the outset; this book covers drug targets and not the drugs themselves; named drug molecules are listed in the entries listed from Chapter 3 onwards purely to highlight the fact that the target is of

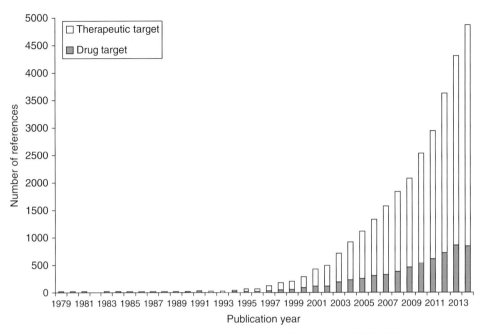

Figure 1.1 Occurrence of the phrase 'drug target' and 'therapeutic target' in papers published between 1979 and 2014. Data taken from a search of PubMed [30]

interest and has been subject to preclinical or clinical investigation. It should be noted that microbial targets are not covered since they are outside the scope of this book.

In 1997, Drews and Ryser [13] published an analysis of the number of targets for drugs listed in Goodman and Gilman's *The Pharmacological Basis of Therapeutics*. Their figure of 483 human and microbial targets was reduced to 324 by Overington *et al.* [14] and even further to 218 by Imming *et al.* [15], both in 2006. Rask-Andersen *et al.* [16] identified 435 human drug targets in the human genome which were affected by 989 unique drugs. These data were obtained by analysing the 2009 entries from DrugBank, a comprehensive database of drugs and targets curated at the University of Alberta in Canada [17]. Whatever the exact figure, the number of targets to FDA-approved drugs is of the order of hundreds rather than the thousands of targets that might be expected from the thousands of human protein-coding genes. While drug development has continued apace since 2009, there is clearly a large discrepancy between actual and potential target numbers, although how many of the latter would lead to useful therapeutics is presently just a matter of conjecture. Whatever the final size of the 'playing field', it is hardly surprising that industry and academia are investing much time and effort into the discovery and exploitation of new targets. Some of this activity is summarized in Section 1.4.1.

1.4.1 Systematic target discovery

The last decades of the 20th century saw the birth of genomics, a discipline which arose naturally on the back of DNA sequencing technology and bioinformatics. Further technological advances followed, in particular the introduction of DNA microarrays for high-throughput analysis of mRNA expression. Proteomics followed as a matter of course, based on sophisticated mass spectrometry for identifying multiple proteins in complex mixtures and affinity-based methods for purifying these proteins on defined ligands. Genomics and proteomics (along with metabolomics and other 'omics' technologies) have, for the first time, made it possible to consider pharmaceutical targets in a systematic way [18]. Given the commercial interest in finding novel targets, genome sequence became a commodity and was sold to pharmaceutical companies by companies such as Incyte and Celera. These commercial restrictions fell away once the human genome sequence was made publically available at the beginning of the new century and posted online in databases such as GenBank [19] and Ensembl [20]; this has allowed a far wider group of scientists to scrutinize the sequence data than would otherwise have been possible. This ongoing analysis of sequence data is not only revealing new members of existing protein target families (e.g. the GPCRs) but also completely new molecules with important regulatory functions such as the microRNAs. Of course, sequence alone does not reveal the biological or pharmaceutical relevance of candidate proteins (or RNAs); a range of different technologies is required in order to achieve this. Examples of these include expression analysis of genes or proteins in normal or diseased tissues or *in vitro* cell culture systems. Many of the human genes referenced in this compendium have been highlighted as potential drug targets on this basis, often because aberrant protein expression was detected in diseased tissue using immunohistochemistry. Another approach is to create transgenic organisms with the gene of interest expressed or removed in order to determine its function *in vivo*. Amgen scientists generated transgenic mice expressing potentially interesting human genes and in doing so discovered osteoprotegerin (TNFRSF11B) which enhanced bone mineralization and led to the discovery of novel targets and treatments for osteoporosis [21]. Human genetics has been revitalized with the advent of next-generation sequencing technology to decrease the cost and increase the throughput of DNA sequencing. Many potential drug targets are being identified by sequencing samples of normal and diseased tissue; this is notably the case in cancer research, as many 'driver mutations' are being identified and assigned to particular tumour types. It is even possible to sequence DNA in individual tumour cells, thereby demonstrating tumour heterogeneity, something that needs to be considered in devising therapies [22].

Lastly, the powerful CRISPR–Cas system for selective gene editing has added to the existing antisense and small interfering RNA (siRNA) technologies for assessing the functions of potential drug targets and promises to broaden opportunities for novel drug development [23].

1.5 Screening for active molecules

This introductory chapter concludes with a brief discussion about how pharmaceutically active molecules are currently being discovered, either through exploiting drug targets that are initially defined or that have to be uncovered retrospectively.

New medicines (or the precursors to them) have been discovered in several ways: by accident, by screening for phenotypic changes or by screening defined molecular targets. Accidental discovery, like that of the cisplatin drugs [24], cannot be made to order, so it will be excluded from this discussion except to remind readers of Pasteur's dictum: 'in the field of observation, chance only favours the prepared mind'.

Current drug discovery practice involves the choice between screening test molecules on phenotypic targets or on defined molecular targets. For small-molecule screening, these activities have been described in terms of chemical genomics, with forward and reverse chemical genetics for phenotypic and target-based screening, respectively (for an overview of chemical genomics and proteomics, see Ref. [25]). There are benefits to each approach: phenotypic (or whole cell/organism) screening has the advantage of selecting molecules with desired biological actions without prior knowledge of the molecular target(s). In addition, only cell-permeant compounds will be selected if the drug target is intracellular. The disadvantage of course is that the target protein(s) has to be identified for compound optimization through a process of target deconvolution [26]. Defined molecular targets may be easier to screen and to obtain reliable SAR data for medicinal chemistry, but any hits must be extensively optimized to show *in vivo* activity. It is also (currently) impossible to predict in advance precisely which other cellular targets might be affected by the test compound, although chemical proteomics strategies for affinity purification of targets on drug ligands have proven successful (e.g. [27]).

It is worth examining the relative contributions of phenotypic and target-based screening to actual drug discovery. Two 2014 reviews have covered this topic; the first, by Eder *et al.* [28], presents an analysis of all 113 first-in-class drugs approved by the FDA between 1999 and 2013. The second review, by Moffatt *et al.* [29], describes a similar analysis but is restricted to oncology drugs. However, it is noteworthy that a very significant proportion of annotations in this compendium are related to oncology, the largest therapeutic

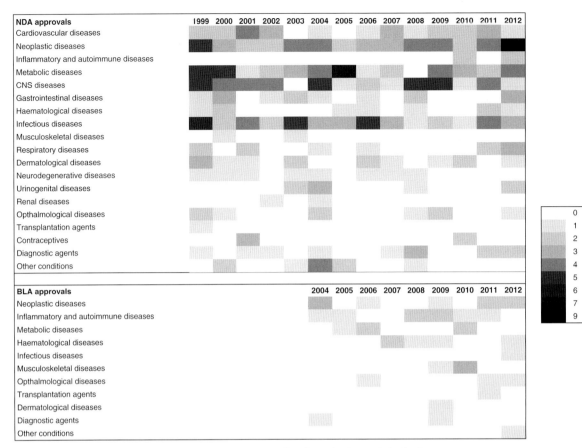

Figure 1.2   FDA approvals by therapeutic area. Data were taken from the FDA website [31]

Table 1.1   Number of approved first-in-class or oncology drugs described in Refs [25] and [26] according to means of discovery

Drug type	Target based	Chemocentric/ mechanism based	Phenotypic screen based
All small molecules from Eder *et al.* [25]	45	25	8
All biologicals from Eder *et al.* [25]	33	—	—
Oncology drugs from Moffat *et al.* [26]	31	7	10

area in 2013, with $10 billion more in sales than the runner up (pain) [1]. Figure 1.2 shows a table of all FDA approvals for small molecules and biologicals (for the periods 1999–2012 and 2004–2012, respectively), listed by therapeutic area. This again shows the importance of oncology targets in pharmaceutical development as well as other trends; for example, there has been a noticeable increase in the number of approvals for biological drugs for inflammatory and autoimmune diseases compared with their small-molecule counterparts.

The key data from Refs [27] and [28] are summarized in Table 1.1.

It can be seen that the majority of first-in-class approved drugs or oncology drugs (either on the market or in clinical development) were originally discovered using a target-centric approach. Despite this bias, there are clearly some advantages to phenotypic screening, as discussed extensively in both reviews, so it is reasonable to expect that this approach to drug discovery will continue to be used alongside target discovery depending upon the individual features of the system under investigation. Whatever the type of drug screening undertaken, I hope that this book will prove useful in providing ideas for either selecting the target in the first place, or for assisting in the identification of the targets uncovered in phenotypic screens.

References

1. Data from IMS Health (2013) *Top 20 Global Therapy Areas 2013*. Available at http://www.imshealth.com/deployedfiles/imshealth/Global/Content/Corporate/Press%20Room/Top_line_data/2014/Top_20_Global_Therapy_Classes_2014.pdf (accessed 11 May 2015).
2. Pammolli F, Magazzini L, Riccaboni M. (2011) The productivity crisis in pharmaceutical R&D. *Nature Reviews Drug Discovery* **10**, 428–38.
3. Gradmann C. (2011) Magic bullets and moving targets: antibiotic resistance and experimental chemotherapy, 1900–1940. *Dynamis* **31**, 305–21.
4. Prüll C-R, Maehle A-H, Halliwell RF. (2009) *A Short History of the Drug Receptor Concept*. Palgrave Macmillan, Basingstoke.
5. Bennett JW. (2001) Adrenaline and cherry trees. *Modern Drug Discovery* **4**, 47–8.
6. Halliwell RF. (2007) A short history of the rise of the molecular pharmacology of ionotropic drug receptors. *Trends in Pharmacological Sciences* **28**, 214–19.
7. Costanzi1 S, Siegel J, Tikhonova IG, Jacobson KA. (2009) Rhodopsin and the others: a historical perspective on structural studies of G protein-coupled receptors. *Current Pharmaceutical Design* **15**, 3994–4002.
8. Sneader W. (2005) *Drug Discovery: A History*. John Wiley & Sons, Ltd, Chichester.
9. Morrow JF *et al.* (1974) Replication and transcription of eukaryotic DNA in *Escherichia coli*. *Proceedings of the National Academy of Sciences* **71**, 1743–7.
10. Roberts L, Davenport RJ, Pennisi E, Marshall E. (2001) A history of the Human Genome Project. *Science* **291**, 1177–80.
11. Bailey D, Zanders E, Dean P. (2001) The end of the beginning for genomic medicine. *Nature Biotechnology* **19**, 207–9.
12. Ezkurdia I *et al.* (2014) Multiple evidence strands suggest that there may be as few as 19 000 human protein-coding genes. *Human Molecular Genetics* **23**, 5866–78.
13. Drews J, Ryser S. (1997) The role of innovation in drug development. *Nature Biotechnology* **15**, 1318–19.
14. Overington JP, Al-Lazikani B, Hopkins AL. (2006) How many drug targets are there? *Nature Reviews Drug Discovery* **5**, 993–6.
15. Imming P, Sinning C, Meyer A. (2006) Drugs, their targets and the nature and number of drug targets. *Nature Reviews Drug Discovery* **5**, 821–34.
16. Rask-Andersen M, Almén MS, Schiöth HB. (2011) Trends in the exploitation of novel drug targets. *Nature Reviews Drug Discovery* **10**, 579–90.
17. Knox C *et al.* (2011) DrugBank 3.0: a comprehensive resource for 'omics' research on drugs. *Nucleic Acids Research* **39**, D1035–41.
18. Yadav SP. (2007) The wholeness in suffix -omics, -omes, and the word om. *Journal of Biomolecular Techniques* **18**, 277.
19. GenBank® (2014) *GenBank Overview*. Available at http://www.ncbi.nlm.nih.gov/genbank/ (accessed 11 May 2015).
20. EMBL-EBI and Wellcome Trust Sanger Institute (2015) Available at http://www.ensembl.org/Homo_sapiens/Info/Index (accessed 11 May 2015).
21. Lacey DL *et al.* (2012) Bench to bedside: elucidation of the OPG–RANK–RANKL pathway and the development of denosumab. *Nature Reviews Drug Discovery* **11**, 401–19.
22. Wang Y *et al.* (2014) Clonal evolution in breast cancer revealed by single nucleus genome sequencing. *Nature* **512**, 155–60.
23. Kasap C *et al.* (2014) DrugTargetSeqR: a genomics- and CRISPR-Cas9-based method to analyze drug targets. *Nature Chemical Biology* **10**, 626–8.
24. Rosenberg B, VanCamp L, Trosko JE, Mansour VH. (1969) Platinum compounds: a new class of potent antitumour agents. *Nature* **222**, 385–6.
25. Zanders ED. (2012) Overview of chemical genomics and proteomics. *Methods in Molecular Biology* **800**, 3–10.
26. Lee J, Bogyo M. (2013) Target deconvolution techniques in modern phenotypic profiling. *Current Opinion in Chemical Biology* **17**, 118–26.

27. Ito T *et al.* (2010) Identification of a primary target of thalidomide teratogenicity. *Science* **327**, 1345–50.
28. Eder J, Sedrani R, Wiesmann C. (2014) The discovery of first-in-class drugs: origins and evolution. *Nature Reviews Drug Discovery* **13**, 577–87.
29. Moffat JG, Rudolph J, Bailey D. (2014) Phenotypic screening in cancer drug discovery – past, present and future. *Nature Reviews Drug Discovery* **13**, 588–602.
30. PubMed. Available at http://www.ncbi.nlm.nih.gov/pubmed (accessed 11 May 2015).
31. FDA. Available at http://www.fda.gov (accessed 11 May 2015).

Chapter 2
Overview of the drug target compendium

2.1 Introductory comments

The process of selecting targets for pharmaceutical development requires decision making based on both scientific and commercial criteria, as discussed in a review by Knowles and Gromo [1]. From a scientific perspective, initial consideration might be given to understanding fundamental biological mechanisms which could then be exploited to control specific disease processes; alternatively, a disease-centric view may be taken at the outset, with knowledge of specific pathogenic mechanisms used to guide target selection. This book is designed to be used by academic and industrial scientists who will be familiar with both of these discovery strategies. Entries are laid out in a way which allows readers to browse through the target lists and hopefully find the inspiration to start new projects or modify old ones (perhaps through employing the 'lateral thinking' techniques espoused by Edward de Bono [2]?).

The primary identifiers for the target entries are the gene name and symbol approved by the HUGO Gene Nomenclature Committee (HGNC) [3]. Some caveats are in order when describing drug targets purely on the basis of these single protein-coding genetic loci. The potential repertoire of drug targets in the human genome is greater than the number of protein-coding genes. One reason for this is the alternative splicing of mRNAs throughout the genome, thus increasing the number of individual proteins encoded by single genes [4], something that can be seen after the gene identifier is entered into a protein database such as UniProt [5]. However, as noted by Barrie *et al.* [6], this diversity may provide opportunities, rather than obstacles, for new target discovery. Secondly, many proteins are subject to post-translational modification, either constitutively or dynamically, in signalling networks. These modifications are the focus of much attention in drug discovery as some of the enzymes involved in adding or removing them (such as protein kinases and phosphatases) are clinically validated targets. Lastly, not all drug targets are proteins; some are nucleic acids. DNA is a well-known oncology target, but there are many long non-coding RNAs and microRNAs in the human genome [7]; some of these have potential as drug targets [8] and are listed in Chapter 7.

Another issue to consider is the fact that a single molecular target used to guide drug development may not play a significant part in the actual mechanism of action of that drug in the clinic, clearly a problem when using a purified target as a starting point. This is less of a problem with phenotypic screens (see Chapter 1), because the molecular target or targets of an effective compound are not known

Human Drug Targets: A Compendium for Pharmaceutical Discovery, First Edition. Edward D. Zanders.
© 2016 John Wiley & Sons, Ltd. Published 2016 by John Wiley & Sons, Ltd.

at the outset and indeed may never be fully characterized. Some examples of successful 'promiscuous' drugs are given in a review by Imming *et al.* [9].

2.2   Selection of entries

Most publicly accessible information on drugs and their targets is available online in databases, research publications and pharmaceutical industry news media. The use of computerized search tools is mandatory due to the sheer volume of information available. As is the case with bio- and chemoinformatics, the number of databases in general appears to be on the increase, providing the investigator with more choice (or more confusion) depending on one's point of view. In recognizing these realities, I have designed this book to present just enough information on human drug targets for the reader to use as a launch pad for online searches while at the same time ensuring that material will attract interest. The target nomenclature and information sources must therefore be readily accessible and fully accepted by the scientific community. The entries comprise the following:

- Gene name and symbol
- Drug/investigational compound name and therapeutic indication or literature reference indicating potential for therapeutic intervention
- Genetic association with disease, if known

2.2.1   Gene name and symbol

I have chosen the HGNC list of approved gene names and symbols for each entry. This is because the drug targets are restricted to the human genome, the nomenclature is widely used in biomedical publications and the data are readily accessible from the HGNC website [7]. The HGNC entries set an upper limit to the number of targets that it is possible to include. These numbers are (at the beginning of 2015) as follows: 19,000 protein-coding loci, 1,200 other loci (T-cell receptors, immunoglobulin genes, etc.), 2,637 long non-coding RNAs and 1,879 microRNAs. Reliable annotation is essential: international project consortia have published annotation data for the entire genome (ENCODE; [10]) and for protein-coding regions (GENCODE; [11]). The latter data support the removal of spurious electronically annotated protein-coding regions, bringing the total number down from the low 20,000s to around 19,000. Crucially, mass spectrometry data have been used to verify true protein expression and support a figure of roughly 19,000 genes [12]. Interestingly, a draft map of the human proteome includes 17,294 protein-coding genes and has revealed a small (low hundreds) number of novel loci derived from non-coding RNA, pseudogenes and upstream open reading frames [13].

2.2.2   Drug/investigational compound name

Any target entry associated with a drug that shows clinical efficacy represents, by definition, a validated target; these are annotated with the name of the drug (in bold type) and the primary disease indication. Most drug entries are entered as the World Health Organization's International Nonproprietary Name (INN) [14] and assigned to a therapeutic area, adapted from the Anatomical Therapeutic Chemical (ATC) classification system [15]. The drug molecules listed may be small molecules, therapeutic proteins, RNA molecules (antisense, small interfering RNA) or gene therapy constructs.

Some of the targets may be affected by more than one drug, in which case a representative example of the class is shown. On the other hand, some drugs have more than one named target, but here, in most, but not all cases, only the main target is annotated. There are about fifty duplicated entries where two targets reside in the same molecule, the majority being cytokine receptors with tyrosine kinase domains.

Table 2.1   Data sources for drug and experimental compound entries

Data source	Description	Reference
Databases		
DrugBank	Canadian database >7000 drug entries	[16]
DGIdb	US drug–gene interaction database >7000 entries	[17]
ChEMBL	United Kingdom-based chemoinformatics resource >1M structures	[18]
KEGG	Kyoto Encyclopedia of Genes and Genomes with drug entries	[19]
INN list	WHO list of International Nonproprietary Names	[14]
Publications		
Agarwal *et al.*	Novelty in the target landscape of the pharmaceutical industry	[20]
Imming *et al.*	Drugs, their targets and the nature and number of drug targets	[9]
Okada *et al.*	Genetics of rheumatoid arthritis contributes to biology and drug discovery	[21]
Nelson *et al.*	An abundance of rare functional variants in 202 drug target genes sequenced in 14,002 people	[22]
Rask-Andersen *et al.*	Trends in the exploitation of novel drug targets	[23]
Others		
Ad hoc browsing	Review articles with drug data, business press and lay press	
Company web pages	Pipeline searches for >40 pharmaceutical and biotech companies	

For 'publications', the titles are shown in the description field.

Therapeutic antibodies against these targets are shown in the section on cytokines and receptors in Chapter 3, and the same targets annotated with small-molecule tyrosine kinase inhibitors in the protein kinase section in Chapter 4.

While some of the drugs used to annotate target entries have been approved and marketed, many have not and are at any stage from preclinical development through to phase III candidates. Agents which are unproven (at the time of writing) are included in order to demonstrate that the associated target entry has attracted a critical level of interest from drug developers, regardless of the eventual fate of the molecule in question.

The drug names were obtained from a number of sources, including databases, research literature and company pipelines, using both systematic and *ad hoc* searches. The information acquired in this way is all publicly available, but a company may have compounds/biologicals acting on a proprietary target which would not be listed as such in this book. However, experience shows that a surprisingly large number of randomly chosen genes/proteins/RNAs have been investigated in some way and the results published; in this case, it will be included in the compendium, unless the link to drug target discovery is just too spurious.

The sources of drug/experimental compound data are listed in Table 2.1.

2.2.3   Literature reference

Compendium entries with named drugs take up a relatively small proportion of the total list, the majority being literature references and disease links. Literature references are taken from searches of the PubMed database [24] and regular browsing of journals such as *Nature, Science, Nature Reviews Drug Discovery, etc.*, which provide useful summaries of research published in other journals as well as their own articles. An obvious question is how comprehensive can the coverage be, given the fact that there are approximately 19,000 protein-coding entries alone. This can only really be answered by knowing the proportion of the human genome that is likely to provide therapeutic targets. Although the true proportion is unknown, the number of potential targets identified in this compendium still runs into the thousands, hopefully providing more than enough items to interest the reader.

Table 2.2 Annotation categories of approximately 44% of the 19,123 protein-coding genes downloaded from the HGNC

Dataset	Number of entries
HGNC protein-coding entries included in this volume	9641
Annotated entries	8504
Entries with specified drug/experimental agent	742
Literature references	5329
Disease association/note from UniProtKB[a]	3353
Annotated entries for non-coding RNA	169

Non-coding RNA entries were taken from literature references only.
[a] UniProt Knowledgebase.

The references were obtained in different phases. Firstly, a systematic search of the PubMed database was made for the literature published from 2009 onwards using the search term 'therapeutic target'. This approach has already been highlighted in Chapter 1, where the number of papers containing the phrase 'therapeutic target' or 'drug target' was analysed by year of publication. This, coupled with *ad hoc* entries taken directly from journals, has generated thousands of publications which have been manually curated and used to populate the compendium list. For all references, whatever the source, only the titles and PMID numbers are shown, so that the display of each entry on the written page is as uncluttered as possible. The title of the publication conveys the potential for interest in the associated target, and the PMID number allows rapid access to the full citation *via* PubMed. Note that the titles are taken directly from the PubMed website without alteration, so any typographical errors in the original will remain in the compendium. This is a deliberate strategy in case it is necessary to use the title as a text string in any subsequent searches.

The search strategy described in the previous paragraph resulted in a somewhat patchy coverage of the human genome entries, with a higher representation of more 'traditional' target classes with known pharmaceutical potential, that is, ligand/receptors, transporters, channels and enzymes. It should be noted that all the non-coding RNA entries listed (in Chapter 7) are assigned a literature reference on the basis of this first search strategy alone. The second search phase involved filling as many gaps in these entries as possible, given the constraints on the size of the book and the time involved. A PubMed search was made for each gene that had not already been annotated with a drug, literature reference or disease association. The gene was listed if any of the publications associated with that gene gave at least some indication of therapeutic potential, even if not explicitly stated by the authors. Many of these annotations related to the involvement of the relevant gene in basic cell biology, pointing to their potential as targets in oncology or related areas.

Taken together, both search strategies have resulted in several thousand literature references (see Table 2.2 in Section 2.4).

The distribution of the references by year is shown in Figure 2.1, emphasizing the fact that the majority were published about a decade after the completion of the human genome sequence.

2.2.4 Involvement in disease

Some of the HGNC gene entries are associated with specific diseases, and these have been annotated as such using data from the UniProt knowledgebase [5]. These data have been extracted by UniProt consortium members and include entries from the Online Mendelian Inheritance in Man (OMIM) database [25]. A 'Note:' field is used to describe the role of the gene/protein in disease pathogenesis

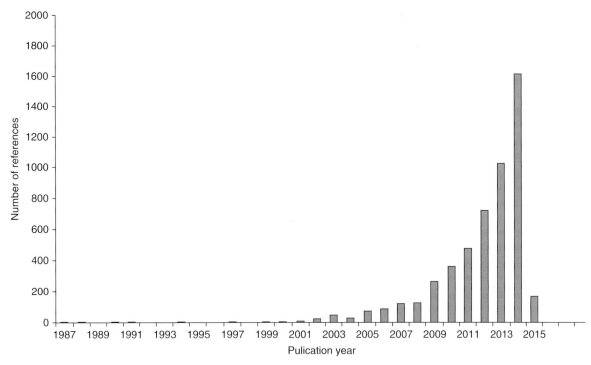

Figure 2.1   Distribution of reference annotations by year of publication. Numbers taken from PubMed website after loading PMID identifiers from Chapters 3–7

and distinguish, where possible, between causative, susceptibility and modifier genes according to literature and OMIM reports [5]. Each disease entry in the compendium is in the format disease name (disease abbreviation) [link to OMIM]: disease description. The link to OMIM is an alphanumeric identifier that can be used to search the OMIM database for detailed information about the disease and its association with the target.

2.3   Organization of entries

I confess to having experienced a certain panic when confronted with a spreadsheet of the 19,000+ genes used as the basis for this compendium. This subsided as I thought of several ways in which the entries could be organized. One would be to create a long list, but the result would be too monotonous. Another possibility was to organize all the genes according to their biological functions. This has been achieved by Stewart Scherer in his highly impressive *Guide to the Human Genome* [26], which runs to 1008 pages in the print version. The gene entries are distributed through 14 chapters, each covering an aspect of biology such as metabolism, cell cycle, signals, organs and tissues and the nervous system. Scherer's book is in effect an extremely useful textbook of genome biology; this book is different, however, in that it focuses exclusively on genes as drug targets, with any information on their biological function being conveyed in the reference associated with the gene entry. Most readers of this book will be familiar with the ways in which drug targets are organized into classes such as receptors, enzymes and transporters. For this reason, I have assigned target entries in to these categories, reflecting their importance as potential or actual drug targets. However, these entries represent a relatively small part of the human genome/proteome, so the remainder have been subdivided in different ways as described in the following sections.

2.3.1   Cell surface and secreted proteins

A master list of HGNC gene entries was used to generate subgroups corresponding to different protein families. The different family members were selected by using online databases specific for a given family (kinases, peptidases, etc.). The cell surface and secreted protein group contains a variety of subgroups whose main headings are:

G-protein-coupled receptors (GPCRs)
Nuclear hormone receptors
Cytokines and receptors
Adhesion molecules
Host defence molecules
Transporters and channels

The GPCR and nuclear hormone receptor entries were extracted from the International Union of Basic and Clinical Pharmacology (IUPHAR) database [27]. The cytokines/receptors, adhesion molecules and host defence molecules are grouped by the function implied by each title. Much use was made of Scherer's human genome guide [26], UniProt functional annotations and personal knowledge of molecular immunology. These groupings are much looser than those based on similarities of DNA or protein sequence, with the assignment in a particular category being fairly arbitrary. To give one example, several 'host defence' molecules could be classified under 'adhesion molecules', and *vice versa*. The transporters and channels entries were taken from the Transporter Classification Database [28] and the IUPHAR database for ion channel entries [27].

2.3.2   Enzymes

This large group of entries is subdivided according to the enzyme's function. The headings are:

Signalling enzymes

- Protein kinases
- Protein phosphatases
- Cyclic nucleotides and phosphodiesterases
- GTPase signalling proteins
- Other signalling enzymes

Protein metabolism

- Peptidases
- Peptidase inhibitors
- Glycosylation
- Ubiquitylation and related modifications
- Chromatin modification
- Protein synthesis and folding
- Other protein modifications

Metabolic and related enzymes

- Lipids and related
- Amino acids and related
- Nucleotides and related

- Carbohydrates and related
- Vitamins, cofactors and related
- DNA-processing enzymes
- RNA-processing enzymes
- Stress response and homeostasis
- Miscellaneous enzymes

The majority of the above were obtained by selecting HGNC names ending in 'ase' and also by searching the Gene Ontology (GO) annotations for each entry in UniProt. These annotations are controlled vocabularies from the GO consortium whose aim is to assign consistent nomenclature about biological function to every gene [29]. It was then possible to search these terms for 'ase activity', eliminate (most) proteins without catalytic activity and then search for the biological activity associated with the final classification. The following examples illustrate how this procedure was used to assign an enzyme with known catalytic activity and one whose activity is only inferred from the protein sequence. Firstly, the UniProt and GO annotations for pyruvate carboxylase (symbol: PC) include the terms 'biotin carboxylase activity', 'carbohydrate metabolic process' and 'lipid metabolic process'. The enzymatic activity is real, with biotin carboxylation being a first step in a tissue-specific process of glucose or lipid synthesis from pyruvate. This enzyme was then listed in Chapter 5 under the heading 'Carbohydrate and related', although it could equally have been included in 'Lipid and related'. The second example is adenosine deaminase-like (symbol: ADAL); the GO annotation included the terms 'adenosine deaminase activity' and 'nucleobase-containing small molecule metabolic process', but these are only inferred through sequence similarity and not biological activity. ADAL was assigned to the 'Nucleotide and related' enzymes in Chapter 5 on the basis of the main GO annotation, fully recognizing that future investigations might alter this view completely.

The peptidases and peptidase inhibitor entries were more straightforward to deal with as they have been curated in the MEROPS database [30]; similarly, the protein kinases were readily extracted from the list compiled by Manning *et al.* [31].

2.3.3 Entries classified by subcellular location

Each of the aforementioned categories of target families were prioritized over all others because they contain most of the pharmaceutically tractable targets identified thus far. The (rather inelegant) term 'druggable target' is commonly used for protein targets with structural features that allow small molecules to bind to them selectively and with high affinity [32]. The GPCRs, nuclear hormone receptors, enzymes and transporters fall into this category, but most adhesion molecules, cytokines, receptors and host defence molecules do not. The latter are often either secreted or cell surface proteins and generally lack a defined small-molecule binding site; however, they can be successfully targeted with therapeutic antibodies or receptor decoy proteins, which is the reason why these categories have been included with the more druggable targets.

Once all of the druggable and cell surface/secreted proteins have been accounted for, the majority of the human genome still remains to be organized into groups. This is where I have used a 'target first, pharmacology second' approach (to borrow a phrase from Makley and Gestwicki [33]). Each entry could have been assigned to different functional categories (by for example using the *Guide to the Human Genome* [26]), but this would make the book far too fragmented. I have taken instead only those entries that have drug/literature/disease annotations and crudely sorted them by their subcellular location as specified in the UniProt database. These locations are as follows:

Cell surface and secreted proteins
Cytoskeleton
Cytoplasm to nucleus

Nucleus
Internal membranes and organelles
Cytoplasmic proteins
No location specified in UniProt

Many different functional types are captured in the above categories: transcription factors, signalling adapters, chromatin proteins and indeed anything that is not likely to have enzymatic activity or to act as a direct receptor for ligands. For a number of reasons, there will be misassignments, including the fact that some proteins may act in more than one cellular compartment. I hope that this issue will not present any problems for readers; if a target annotation of interest has been found for a gene, it will always be included somewhere in the book, even if not correctly assigned to a chapter heading.

2.3.4   Non-coding RNAs

The majority of drug targets are proteins, but RNA molecules have potential as both targets and as therapeutic agents. Genome sequencing has revealed populations of RNAs which were absent from conventional cDNA libraries but involved in the important phenomenon of RNA interference [34]. The human genome contains thousands of non-coding RNAs, including the long non-coding RNAs (lncRNAs) and approximately 22 nucleotide microRNAs (miRNAs). Both are the subject of intense study as they are involved in diverse cellular processes and, in the case of miRNAs at least, have real pharmaceutical potential [35]. A number of lncRNAs and miRNAs have been annotated with PubMed references and these are presented in Chapter 7.

A schematic overview of the process of generating target entries from the HGNC database and their subsequent assignment into different categories is shown in Figure 2.2:

2.4   Summary of data entries

There are 19,123 unique protein-coding gene entries taken from HGNC site updated to February 2015 (http://www.genenames.org/cgi-bin/statistics). Approximately 50% of these entries were sorted into the different categories described in this chapter (the remaining 50% being reserved for a future volume). The number of entries for each annotation type is listed in Table 2.2.

The curation of drug, disease and literature information generally required subjective judgements on which items to include in the entries; despite this, there are some interesting features in the data which are highlighted as follows.

Firstly, the mostly heavily featured therapeutic area is oncology, with nearly 40% of the drug candidates developed (or under investigation) for this indication and 36% of the literature citations. The closest following areas are cardiovascular and immunoinflammatory diseases, each with approximately 12% of the drugs listed; infectious diseases are clearly an important therapeutic area but hardly feature in this book as it deals with human drug targets; there are however a few examples of host cell–virus interactions. The high proportion of potential and actual oncology targets may reflect the significant advances made in recent years in understanding basic cellular functions such as proliferation, autophagy and apoptosis.

A second point to emerge from analysing the drug data is that approximately 18% of the drug entries are biologicals rather than small molecules. The majority of these were proteins (predominantly monoclonal antibodies) plus some RNA drugs. Even if only a small number of the gene products listed in this compendium turn out to be useful drug targets in the future, a significant proportion will only be affected by large molecules or by modifying their expression using nucleic acid-based agents. It is reasonable to expect therefore that the proportion of marketed biological agents relative to small-molecule drugs will continue its upward trajectory.

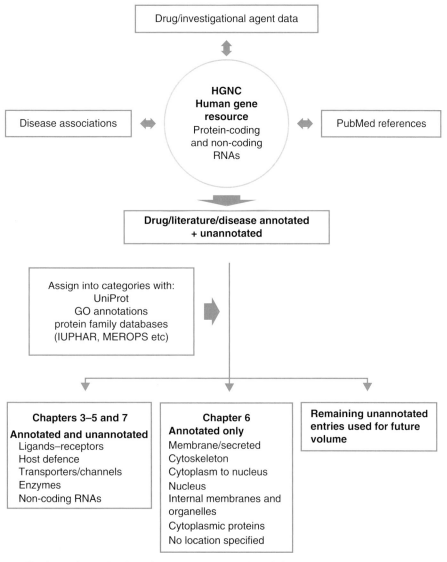

Figure 2.2   Data collection and organization schema. Gene names and symbols using the HGNC nomenclature were used as entries for a compendium of drug targets. Data on drugs and diseases, and references to target potential, were fed into the gene list to create a group of annotated entries. All gene entries, whether annotated or not, were sorted into categories and headings within chapters using data from UniProt, GO, etc. as indicated

2.5   How to use this book

My aim in writing this book is to provide the maximum amount of useful information for each entry in as little space as possible. In this case, 'useful information' means the following:

- A universally recognized gene name and symbol, without synonyms
- If available, the name of a drug molecule (of any type) and relevant disease indication to indicate that the target is clearly of pharmaceutical interest
- An association of the gene entry with a particular disease, if known
- Some brief text to indicate that the entry is a drug target or has the potential to be one

This is illustrated by the examples shown in Table 2.3.

The three entries between them show all of the annotation types to be found throughout the book.

A survey of the target information fields shows a drug entry (pimasertib (in bold type)) and the associated use (antineoplastic agents) for the MAPK2 target. It is possible to search for detailed information about pimasertib in specific drug databases at the outset, but for this and most drug names, a simple Google (or similar) search will bring up relevant entries. In this example, the National Cancer Institute Drug Dictionary (http://www.cancer.gov/drugdictionary?cdrid=653634) lists pimasertib as a drug targeting the protein kinases MEK1 and MEK2, giving summary information about the cancer pathway and links to clinical trials. The problem of nomenclature immediately becomes apparent since the drug is listed as MAP2K2 in this book. This is easily rectified by entering the symbol text into the HGNC website (http://www.genenames.org/) to extract the synonyms. This is shown in Figure 2.3, where MEK2 is shown as a synonym for MAP2K2 (MEK1 is also a target for pimasertib and is listed as MAP2K1 in Chapter 4).

Continuing with MAP2K2, information on the disease 'cardiofaciocutaneous syndrome 4 (CFC4)' can be found by opening the OMIM database directly and inputting CFC4 or else through the direct link to OMIM on the symbol report page for MAPK2. This page links the gene entry to a variety of important databases, including the UniProtKB used to retrieve the disease links and GO annotations.

Table 2.3 Example entries taken from Section 4.1.1 to illustrate the different types of annotation for target information

HGNC-approved name	Symbol	Target information
Mitogen-activated protein kinase kinase 2	MAP2K2	**Pimasertib** – antineoplastic agents. Disease: cardiofaciocutaneous syndrome 4 (CFC4)
Mitogen-activated protein kinase kinase 3	MAP2K3	Note: Defects in MAP2K3 may be involved in colon cancer
Mitogen-activated protein kinase kinase 4	MAP2K4	Mitogen-activated protein kinase kinase 4 (MAP2K4) promotes human prostate cancer metastasis. PMID: 25019290

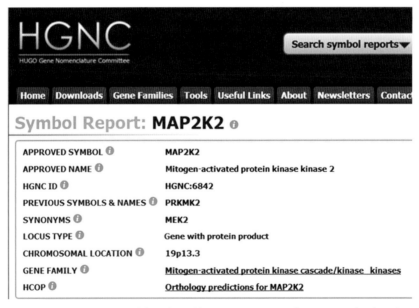

Figure 2.3 HGNC symbol report page accessed from http://www.genenames.org for MAP2K2 showing the synonym MEK2 and other information

The MAP2K3 entry has a 'Note:' annotation which is described in the UniProt entry (P46734) corresponding to this gene. Finally, MAP2K4 contains a PubMed reference which gives an indication (and no more) of why the entry could be a pharmaceutical target. The PMID number can then be entered into the PubMed database to bring up the title, abstract and links to full text. The 'related citations in PubMed' field can be particularly useful as these entries may lead in the future to more up-to-date or relevant publications that either support or refute the validity of the target.

2.6   Final comments

This book is intended to stimulate ideas for new drug discovery programmes by providing a comprehensive list of possible targets; there are however widely differing degrees of 'possible', so each annotated entry must be evaluated carefully for the following reasons:

- A few of the marketed drugs associated with particular targets have been withdrawn from sale due to toxicity or other issues. This does not necessarily cast doubt over the value of affecting the target with other molecules, but the possibility of mechanism-based toxicity must be considered.
- Some of the compounds or biologicals in preclinical or clinical development listed in the annotations will not reach the market; this is for all the usual reasons for drug development failure such as lack of efficacy, poor pharmacokinetics or safety problems. This does not necessarily invalidate the target, as there could be selectivity issues with related proteins or undesirable off-target effects which could be 'bred out' using medicinal chemistry or other types of drug molecule.
- Many of the literature entries link to laboratory data indicating that a particular molecule could be a useful therapeutic target. The strength of evidence supporting this conclusion varies widely, from correlation of expression level with disease to support from human genetics and animal studies *in vivo*. There is serious concern about the reproducibility of some published data (e.g. see commentary in Ref. [36]). However, the costs involved in running a drug discovery project over several years mean that decisions to pursue a given target will not be taken lightly. Published work will be repeated and target candidates either pushed forwards or rejected. It is of course desirable to identify failures as quickly as possible, which is more readily achievable if the primary sources of information are sound.
- A protein or RNA molecule may have impeccable credentials as a potential target but may not be chemically tractable. This relates to druggability (the problem of inhibiting protein–protein interactions (PPIs)), or delivery. This is an unfortunate situation as many potentially valuable targets lie on the surfaces of interacting protein partners such that it is impossible to produce the orally available small-molecule drugs favoured by the pharmaceutical industry. Examples of this abound in the section on cytokines and receptors, where in general only large antibody molecules are able to provide the efficacy and selectivity necessary for an effective drug. This is not to say that there is not a great deal of computational and medicinal chemistry activity in trying to produce small molecule PPIs (see Ref. [37] for a review).

Finally, in once again quoting Pasteur's dictum that 'in the field of observation, chance only favours the prepared mind', I hope that this compendium helps in that mental preparation and contributes to new and effective therapies in the future.

References

1. Knowles JR, Gromo G. (2002) Target selection in drug discovery. *Nature Reviews Drug Discovery* **2**, 63–9.
2. De Bono E. (1970) *Lateral Thinking: Creativity Step by Step*. Harper & Row, New York.
3. Daugherty LC, Seal RL, Wright MW, Bruford EA. (2012) Gene family matters: expanding the HGNC resource. *Human Genomics* **6**, 1–6.
4. Nilsen TW, Graveley BR. (2010) Expansion of the eukaryotic proteome by alternative splicing. *Nature* **463**, 457–63.
5. UniProt Knowledgebase (2015) *UniProt Consortium*. Available at http://www.uniprot.org/uniprot/ (accessed 11 May 2015).

6. Barrie ES, Smith RM, Sanford JC, Sadee W. (2012) mRNA transcript diversity creates new opportunities for pharmacological intervention. *Molecular Pharmacology* **81**, 620–30.

7. HUGO Gene Nomenclature Committee (2015). Available at http://www.genenames.org/ (accessed 11 May 2015).

8. Li Z, Rana TM. (2014) Therapeutic targeting of microRNAs: current status and future challenges. *Nature Reviews Drug Discovery* **13**, 622–38.

9. Imming P, Sinning C, Meyer A. (2006) Drugs, their targets and the nature and number of drug targets. *Nature Reviews Drug Discovery* **5**, 821–34.

10. ENCODE Project Consortium (2012) An integrated encyclopaedia of DNA elements in the human genome. *Nature* **489**, 57–74.

11. Harrow J *et al.* (2012) GENCODE: the reference human genome annotation for The ENCODE Project. *Genome Research* **22**, 760–74.

12. Ezkurdia I *et al.* (2014) Multiple evidence strands suggest that there may be as few as 19 000 human protein-coding genes. *Human Molecular Genetics* **23**, 5866–78.

13. Kim MS *et al.* (2014) A draft map of the human proteome. *Nature* **509**, 575–87.

14. WHO (2015) *International Nonproprietary Names.* Available at http://www.who.int/medicines/services/inn/en/ (accessed 11 May 2015).

15. ATC Index (2015) *ATC/DDD Index 2015.* Available at http://www.whocc.no/atc_ddd_index/ (accessed 11 May 2015).

16. Knox C *et al.* (2011) DrugBank 3.0: a comprehensive resource for 'omics' research on drugs. *Nucleic Acids Research* **39**, D1035–41.

17. Griffith M *et al.* (2013) DGIdb: mining the druggable genome. *Nature Methods* **10**, 1209–15.

18. Bento AP *et al.* (2014) The ChEMBL bioactivity database: an update. *Nucleic Acids Research* **42**, D1083–90.

19. Kanehisa M, Goto S, Sato Y, Furumichi M, Tanabe M. (2012) KEGG for integration and interpretation of large-scale molecular data sets. *Nucleic Acids Research* **40**, D109–14.

20. Agarwal P, Sanseau P, Cardon LR. (2013) Novelty in the target landscape of the pharmaceutical industry. *Nature Reviews Drug Discovery* **12**, 575–6.

21. Okada Y *et al.* (2014) Genetics of rheumatoid arthritis contributes to biology and drug discovery. *Nature* **506**, 376–81.

22. Nelson MR *et al.* (2012) An abundance of rare functional variants in 202 drug target genes sequenced in 14,002 people. *Science* **337**, 100–4.

23. Rask-Andersen M, Almén MS, Schiöth HB. (2011) Trends in the exploitation of novel drug targets. *Nature Reviews Drug Discovery* **10**, 579–90.

24. PubMed. Available at http://www.ncbi.nlm.nih.gov/pubmed (accessed 11 May 2015).

25. OMIM Database. Available at http://omim.org/ (accessed 11 May 2015).

26. Scherer S. (2010) *Guide to the Human Genome.* Cold Spring Harbor Press, New York.

27. Pawson AJ *et al.* (2014) The IUPHAR/BPS Guide to PHARMACOLOGY: an expert-driven knowledgebase of drug targets and their ligands. *Nucleic Acids Research* **42**, 1098–106.

28. Saier MH, Reddy VS, Tamang DG, Västermark A. (2014) The transporter classification database. *Nucleic Acids Research* **42**, 251–8.

29. Ashburner M *et al.* (2000) Gene ontology: tool for the unification of biology. *Nature Genetics* **25**, 25–9.

30. Rawlings ND, Barrett AJ, Bateman A. (2012) MEROPS: the database of proteolytic enzymes, their substrates and inhibitors. *Nucleic Acids Research* **40**, 343–50.

31. Manning G, Whyte DB, Martinez R, Hunter T, Sudarsanam S. (2002) The protein kinase complement of the human genome. *Science* **298**, 1912–34.

32. Hopkins AR, Groom CR. (2002) The druggable genome. *Nature Reviews Drug Discovery* **1**, 727–30.

33. Makley LN, Gestwicki JE. (2013) Expanding the number of 'druggable' targets: non-enzymes and protein–protein interactions. *Chemical Biology & Drug Design* **81**, 22–32.

34. Ambros V. (2001) microRNAs: tiny regulators with great potential. *Cell* **107**, 823–6.

35. Li Z, Rana TM. (2014) Therapeutic targeting of microRNAs: current status and future challenges. *Nature Reviews Drug Discovery* **13**, 622–38.

36. Mullard A. (2011) Reliability of 'new drug target' claims called into question. *Nature Reviews Drug Discovery* **10**, 643–4.

37. Villoutreix BO, Labbé C, Lagorce D, Laconde G, Sperandio O. (2012) A leap into the chemical space of protein-protein interaction inhibitors. *Current Pharmaceutical Design* **18**, 4648–67.

Chapter 3
Cell surface and secreted proteins

This chapter includes HGNC entries for proteins that are expressed on the cell surface, namely receptors for different ligand classes, transporters and ion channels. Targets for existing drugs are strongly over-represented in this group; for example, approximately 40% of current medicines affect GPCRs, and hundreds of these molecules are listed with annotations that indicate pharmaceutical potential. Realising this potential in the form of potent specific drug molecules is challenging however, given the nature of this difficult class of membrane proteins, although technical advances in 3D structure determination are providing the tools to greatly increase the chance of success. Over the next few years, scientific innovation will help to accelerate the production of structures in concert with organizational initiatives such as the GPCR consortium which was created from academia and pharmaceutical companies [1].

Unlike the other entries, the nuclear receptors are not extracellular proteins, but instead bind their (small-molecule) ligands intracellularly and operate as transcription factors; this family does, however, contain a significant number of targets for current drugs, so it seems reasonable to list them in this chapter.

The next three sections include targets that involve protein–protein or protein–carbohydrate interactions. These are cytokines and their receptors, adhesion molecules and host defence molecules. The allocation of entries to each section is less clear-cut than with the GPCRs and should be understood as such. For example, 'host defence molecules' include many CD antigens, which themselves can be part of a cytokine receptor complex or an adhesion molecule. Much of this is due to the vagaries of nomenclature which can at least be tamed with a single HGNC name and symbol. On looking through these entries, it will be immediately apparent that many are targeted with therapeutic antibodies or other biotherapeutics, which is not surprising given the difficulties of finding small-molecule inhibitors for the relevant protein–protein interactions. If a small-molecule inhibitor/activator approach is not going to be feasible for many of the possible targets listed in these sections, it is likely that alternatives such as monoclonal antibodies will have to be developed. Chemical inhibitors with antibody-like properties have been developed and have the advantage over antibodies both in origin (produced by chemical synthesis) and size (7 kDa as opposed to 150 kDa) [2].

The last group of entries comes under the heading of transporters and channels. The ATP-binding cassette and solute carrier families are not only primary therapeutic targets but also contain proteins that contribute to drug resistance, through acting as efflux pumps for chemotherapeutics, for example. The ion channels are well-known targets for drugs and toxins and, like the GPCRs, are complex membrane proteins that are yielding to new techniques in structural biology (e.g. Ref. [3]).

Human Drug Targets: A Compendium for Pharmaceutical Discovery, First Edition. Edward D. Zanders.
© 2016 John Wiley & Sons, Ltd. Published 2016 by John Wiley & Sons, Ltd.

References

1. The GPCR Consortium. Available at http://gpcrconsortium.org/ (accessed 16 April 2015).
2. McEnaney PJ *et al.* (2014) Chemically synthesized molecules with the targeting and effector functions of antibodies. *Journal of the American Chemical Society* **136**, 18034–43.
3. Efremov RG, Leitner A, Aebersold R, Raunser S. (2015) Architecture and conformational switch mechanism of the ryanodine receptor. *Nature* **517**, 39–43.

3.1   G-protein-coupled receptors (GPCRs)

HGNC-approved name	Symbol	Target information
Class A		
5-Hydroxytryptamine (serotonin) receptor 1A, G protein-coupled	HTR1A	**Buspirone** – anxiolytics. Disease: periodic fever, menstrual cycle-dependent (PFMC)
5-Hydroxytryptamine (serotonin) receptor 1B, G protein-coupled	HTR1B	**Sumatriptan** – antimigraine preparations
5-Hydroxytryptamine (serotonin) receptor 1D, G protein-coupled	HTR1D	**Rizatriptan** – antimigraine preparations
5-Hydroxytryptamine (serotonin) receptor 1E, G protein-coupled	HTR1E	Toward selective drug development for the human 5-hydroxytryptamine 1E receptor: A comparison of 5-hydroxytryptamine 1E and 1F receptor structure-affinity relationships. PMID: 21422162
5-Hydroxytryptamine (serotonin) receptor 1F, G protein-coupled	HTR1F	**Almotriptan** – antimigraine preparations
5-Hydroxytryptamine (serotonin) receptor 2A, G protein-coupled	HTR2A	**Pimavanserin** – antipsychotics
5-Hydroxytryptamine (serotonin) receptor 2B, G protein-coupled	HTR2B	**Triflupromazine** – antipsychotics
5-Hydroxytryptamine (serotonin) receptor 2C, G protein-coupled	HTR2C	SB 242084, a selective and brain penetrant 5-HT2C receptor antagonist. PMID: 9225286
5-Hydroxytryptamine (serotonin) receptor 4, G protein-coupled	HTR4	**Cisapride** – parasympathomimetics
5-Hydroxytryptamine (serotonin) receptor 5A, G protein-coupled	HTR5A	ASP5736, a novel 5-HT5A receptor antagonist, ameliorates positive symptoms and cognitive impairment in animal models of schizophrenia. PMID: 25108314
5-Hydroxytryptamine (serotonin) receptor 6, G protein-coupled	HTR6	**GSK 742457** – parasympathomimetics
5-Hydroxytryptamine (serotonin) receptor 7, adenylate cyclase-coupled	HTR7	Serotonin 5-HT7 receptor agents: Structure-activity relationships and potential therapeutic applications in central nervous system disorders. PMID: 20923682
Cholinergic receptor, muscarinic 1	CHRM1	Discovery and characterization of novel subtype-selective allosteric agonists for the investigation of M(1) receptor function in the central nervous system. PMID: 21961051
Cholinergic receptor, muscarinic 2	CHRM2	**Aclidinium bromide** – drugs for obstructive airway diseases. Disease: major depressive disorder (MDD)
Cholinergic receptor, muscarinic 3	CHRM3	**Cevimeline** – dry mouth in Sjögren's syndrome. Disease: Eagle–Barrett syndrome (EGBRS)
Cholinergic receptor, muscarinic 4	CHRM4	An allosteric potentiator of M4 mAChR modulates hippocampal synaptic transmission. PMID: 18059262
Cholinergic receptor, muscarinic 5	CHRM5	The M5 muscarinic receptor as possible target for treatment of drug abuse. PMID: 20175795

HGNC-approved name	Symbol	Target information
Adenosine A1 receptor	ADORA1	**Theophylline** – drugs for obstructive airway diseases
Adenosine A2a receptor	ADORA2A	**Regadenoson** – cardiac therapy
Adenosine A2b receptor	ADORA2B	**Enprofylline** – drugs for obstructive airway diseases
Adenosine A3 receptor	ADORA3	**CF101** – anti-inflammatory and antirheumatic products
Adrenoceptor alpha 1A	ADRA1A	**Silodosin** – urologicals
Adrenoceptor alpha 1B	ADRA1B	**Labetalol** – antihypertensives
Adrenoceptor alpha 1D	ADRA1D	The alpha 1D-adrenergic receptor is expressed intracellularly and coupled to increases in intracellular calcium and reactive oxygen species in human aortic smooth muscle cells. PMID: 18304336
Adrenoceptor alpha 2A	ADRA2A	**Bethanidine** – antihypertensives
Adrenoceptor alpha 2B	ADRA2B	Characterization of α(2B)-adrenoceptor ligand binding in the presence of muscarinic toxin α and delineation of structural features of receptor binding selectivity. PMID: 22465183
Adrenoceptor alpha 2C	ADRA2C	Pharmacological characterization and CNS effects of a novel highly selective alpha 2C-adrenoceptor antagonist JP-1302. PMID: 17220913
Adrenoceptor beta 1	ADRB1	**Atenolol** – antiarrhythmics
Adrenoceptor beta 2, surface	ADRB2	**Salmeterol** – drugs for obstructive airway diseases
Adrenoceptor beta 3	ADRB3	**Mirabegron** – urologicals
Angiotensin II receptor, type 1	AGTR1	**Losartan** – antihypertensives. Disease: renal tubular dysgenesis (RTD)
Angiotensin II receptor, type 2	AGTR2	AT2 receptors: Functional relevance in cardiovascular disease. PMID: 18804122
Apelin receptor	APLNR	The apelin-APJ system in heart failure: Pathophysiologic relevance and therapeutic potential. PMID: 18272138
G protein-coupled bile acid receptor 1	GPBAR1	The bile acid membrane receptor TGR5: A novel pharmacological target in metabolic, inflammatory and neoplastic disorders. PMID: 23782454
Neuromedin B receptor	NMBR	Characterization of putative GRP- and NMB-receptor antagonist's interaction with human receptors. PMID: 19463875
Gastrin-releasing peptide receptor	GRPR	Powerful inhibition of in-vivo growth of experimental hepatic cancers by bombesin/gastrin-releasing peptide antagonist RC-3940-II. PMID: 22926257
Bombesin-like receptor 3	BRS3	Antiobesity effect of MK-5046, a novel bombesin receptor subtype-3 agonist. PMID: 21036912
Bradykinin receptor B1	BDKRB1	Bradykinin B1 receptor antagonists as potential therapeutic agents for pain. PMID: 20369879
Bradykinin receptor B2	BDKRB2	**Icatibant** – anti-inflammatory and antirheumatic products
Cannabinoid receptor 1 (brain)	CNR1	**Rimonabant** – antiobesity preparations
Cannabinoid receptor 2 (macrophage)	CNR2	Targeting cannabinoid agonists for inflammatory and neuropathic pain. PMID: 17594182
Chemokine-like receptor 1	CMKLR1	Chemokine-like receptor 1 (CMKLR1) and chemokine (C—C motif) receptor-like 2 (CCRL2) two multifunctional receptors with unusual properties. PMID: 21056554
Chemokine (C—C mot if) receptor 1	CCR1	**BMS-817399** – anti-inflammatory and antirheumatic products
Chemokine (C—C motif) receptor 2	CCR2	**CCX140** – diabetic nephropathy
Chemokine (C—C motif) receptor 3	CCR3	Discovery and structure-activity relationships of urea derivatives as potent and novel CCR3 antagonists. PMID: 22749826
Chemokine (C—C motif) receptor 4	CCR4	**Mogamulizumab** – antineoplastic agents

(Continued)

HGNC-approved name	Symbol	Target information
Chemokine (C—C motif) receptor 5 (gene/pseudogene)	CCR5	**Maraviroc** – anti-HIV. Disease: diabetes mellitus, insulin-dependent, 22 (IDDM22)
Chemokine (C—C motif) receptor 6	CCR6	CCR6 as a possible therapeutic target in psoriasis. PMID: 20629596
Chemokine (C—C motif) receptor 7	CCR7	Multifaceted activities of CCR7 regulate T-cell homeostasis in health and disease. PMID: 22700449
Chemokine (C—C motif) receptor 8	CCR8	Orally bioavailable allosteric CCR8 antagonists inhibit dendritic cell, T cell and eosinophil migration. PMID: 22209712
Chemokine (C—C motif) receptor 9	CCR9	**Vercirnon** – selective immunosuppressants
Chemokine (C—C motif) receptor 10	CCR10	CCR10 and its ligands in regulation of epithelial immunity and diseases. PMID: 22684736
Chemokine (C—X—C motif) receptor 1	CXCR1	Upregulation of C—X—C chemokine receptor type 1 expression is associated with late-stage gastric adenocarcinoma. PMID: 23060922
Chemokine (C—X—C motif) receptor 2	CXCR2	Pharmacological characterization of Sch527123, a potent allosteric CXCR1/CXCR2 antagonist. PMID: 17496166
Chemokine (C—X—C motif) receptor 3	CXCR3	A selective and potent CXCR3 antagonist SCH 546738 attenuates the development of autoimmune diseases and delays graft rejection. PMID: 22233170
Chemokine (C—X—C motif) receptor 4	CXCR4	**Plerixafor** – immunostimulants. Disease: WHIM syndrome (WHIM)
Chemokine (C—X—C motif) receptor 5	CXCR5	**SAR113244** – selective immunosuppressants
Chemokine (C—X—C motif) receptor 6	CXCR6	CXCR6/CXCL16 functions as a regulator in metastasis and progression of cancer. PMID: 20122997
Chemokine (C—X3—C motif) receptor 1	CX3CR1	Pharmacological inhibition of the chemokine receptor CX3CR1, reduces atherosclerosis in mice. PMID: 23887641. Disease: macular degeneration, age-related, 12 (ARMD12)
Chemokine (C motif) receptor 1	XCR1	The XC chemokine receptor 1 is a conserved selective marker of mammalian cells homologous to mouse CD8alpha+ dendritic cells. PMID: 20479118
Chemokine (C—C motif) receptor-like 2	CCRL2	Chemokine-like receptor 1 (CMKLR1) and chemokine (C—C motif) receptor-like 2 (CCRL2) two multifunctional receptors with unusual properties. PMID: 21056554
Atypical chemokine receptor 1 (Duffy blood group)	ACKR1	Regulation of motor function and behavior by atypical chemokine receptor 1. PMID: 24997773
Atypical chemokine receptor 2	ACKR2	D6 and the atypical chemokine receptor family: Novel regulators of immune and inflammatory processes. PMID: 19130487
Atypical chemokine receptor 3	ACKR3	**CCX650** – antineoplastic agents
Atypical chemokine receptor 4	ACKR4	β-Arrestin recruitment and G protein signaling by the atypical human chemokine decoy receptor CCX—CKR. PMID: 23341447
Cholecystokinin A receptor	CCKAR	CCK1 antagonists: Are they ready for clinical use? PMID: 16699265
Cholecystokinin B receptor	CCKBR	**Netazepide** – antineoplastic agents
Complement component 3a receptor 1	C3AR1	Identification of a selective nonpeptide antagonist of the anaphylatoxin C3a receptor that demonstrates antiinflammatory activity in animal models. PMID: 11342658
Complement component 5a receptor 1	C5AR1	**NN8209** – Anti-inflammatory and antirheumatic products
Complement component 5a receptor 2	C5AR2	Fosb gene products contribute to excitotoxic microglial activation by regulating the expression of complement C5a receptors in microglia. PMID: 24771617
Dopamine receptor D1	DRD1	**Fenoldopam** – antihypertensives
Dopamine receptor D2	DRD2	**Olanzapine** – antipsychotics. Disease: dystonia 11 (DYT11)
Dopamine receptor D3	DRD3	**Pramipexole** – Anti-Parkinson drugs. Disease: tremor, hereditary essential 1 (ETM1)
Dopamine receptor D4	DRD4	**Clozapine** – antipsychotics

HGNC-approved name	Symbol	Target information
Dopamine receptor D5	DRD5	Novel insights in dopamine receptor physiology. PMID: 17413183. Disease: benign essential blepharospasm (BEB)
Endothelin receptor type A	EDNRA	**Atrasentan** – antihypertensives
Endothelin receptor type B	EDNRB	BQ-788, a selective endothelin ET(B) receptor antagonist. PMID: 12070534. Disease: Waardenburg syndrome 4A (WS4A)
G protein-coupled estrogen receptor 1	GPER1	G protein-coupled oestrogen receptor 1 (GPER1)/GPR30: A new player in cardiovascular and metabolic oestrogenic signalling. PMID: 21250980
Receptor (G protein-coupled) activity-modifying protein 3	RAMP3	G-protein-coupled receptor 30 interacts with receptor activity-modifying protein 3 and confers sex-dependent cardioprotection. PMID: 23674134
Formyl peptide receptor 1	FPR1	G protein-coupled receptor FPR1 as a pharmacologic target in inflammation and human glioblastoma. PMID: 22863814
Formyl peptide receptor 2	FPR2	Distinct signaling cascades elicited by different formyl peptide receptor 2 (FPR2) agonists. PMID: 23549262
Formyl peptide receptor 3	FPR3	*N*-Formyl peptide receptor 3 (FPR3) departs from the homologous FPR2/ALX receptor with regard to the major processes governing chemoattractant receptor regulation, expression at the cell surface, and phosphorylation. PMID: 21543323
Free fatty acid receptor 1	FFAR1	Discovery of potent and selective agonists for the free fatty acid receptor 1 (FFA(1)/GPR40), a potential target for the treatment of type II diabetes. PMID: 18947221
Free fatty acid receptor 2	FFAR2	**GLPG0974** – anti-inflammatory and antirheumatic products
Free fatty acid receptor 3	FFAR3	Short-chain free fatty acid receptors FFA2/GPR43 and FFA3/GPR41 as new potential therapeutic targets. PMID: 23060857
Free fatty acid receptor 4	FFAR4	Identification of G protein-coupled receptor 120-selective agonists derived from PPARgamma agonists. PMID: 19007110
Galanin receptor 1	GALR1	Galanin, through GalR1 but not GalR2 receptors, decreases motivation at times of high appetitive behavior. PMID: 23142608
Galanin receptor 2	GALR2	Novel systemically active galanin receptor 2 ligands in depression-like behavior. PMID: 23600864
Galanin receptor 3	GALR3	The galanin-3 receptor antagonist, SNAP 37889, reduces operant responding for ethanol in alcohol-preferring rats. PMID: 20736033
Growth hormone secretagogue receptor	GHSR	Ghrelin – a key pleiotropic hormone-regulating systemic energy metabolism. PMID: 23652395. Disease: short stature, idiopathic, autosomal (ISSA)
Luteinizing hormone/choriogonadotropin receptor	LHCGR	**Lutropin alfa** – fertility agents. Disease: familial male precocious puberty (FMPP)
Follicle-stimulating hormone receptor	FSHR	**Urofollitropin** – fertility agents. Disease: ovarian dysgenesis 1 (ODG1)
Gonadotropin-releasing hormone receptor	GNRHR	**Leuprolide** – antineoplastic agents. Disease: hypogonadotropic hypogonadism 7 with or without anosmia (HH7)
Histamine receptor H1	HRH1	**Cetirizine** – antihistamines for systemic use
Histamine receptor H2	HRH2	**Ranitidine** – drugs for acid-related disorders
Histamine receptor H3	HRH3	Histamine H(3) receptor (H(3)R) antagonists and inverse agonists in the treatment of sleep disorders. PMID: 21476958
Histamine receptor H4	HRH4	Is the H4 receptor a new drug target for allergies and asthma? PMID: 23276980
Hydroxycarboxylic acid receptor 1	HCAR1	Biological and pharmacological roles of HCA receptors. PMID: 21907911
Hydroxycarboxylic acid receptor 2	HCAR2	Discovery of a biaryl cyclohexene carboxylic acid (MK-6892): A potent and selective high affinity niacin receptor full agonist with reduced flushing profiles in animals as a preclinical candidate. PMID: 20184326

(Continued)

HGNC-approved name	Symbol	Target information
Hydroxycarboxylic acid receptor 3	HCAR3	Biological and pharmacological roles of HCA receptors. PMID: 21907911
KISS1 receptor	KISS1R	Discovery of potent kisspeptin antagonists delineate physiological mechanisms of gonadotropin regulation. PMID: 19321788. Disease: hypogonadotropic hypogonadism 8 with or without anosmia (HH8)
Leukotriene B4 receptor	LTB4R	Discovery of novel and potent leukotriene B4 receptor antagonists. PMID: 20380377
Leukotriene B4 receptor 2	LTB4R2	A leukotriene B4 receptor-2 is associated with paclitaxel resistance in MCF-7/DOX breast cancer cells. PMID: 23799854
Cysteinyl leukotriene receptor 1	CYSLTR1	**Montelukast** – drugs for obstructive airway diseases
Cysteinyl leukotriene receptor 2	CYSLTR2	A selective cysteinyl leukotriene receptor 2 antagonist blocks myocardial ischemia/reperfusion injury and vascular permeability in mice. PMID: 21903747
Oxoglutarate (alpha-ketoglutarate) receptor 1	OXGR1	Identification of GPR99 protein as a potential third cysteinyl leukotriene receptor with a preference for leukotriene E4 ligand. PMID: 23504326
Oxoeicosanoid (OXE) receptor 1	OXER1	OXER1, a G protein-coupled oxoeicosatetraenoid receptor, mediates the survival-promoting effects of arachidonate 5-lipoxygenase in prostate cancer cells. PMID: 23643940
Lysophosphatidic acid receptor 1	LPAR1	The lysophosphatidic acid receptor LPA1 links pulmonary fibrosis to lung injury by mediating fibroblast recruitment and vascular leak. PMID: 18066075
Lysophosphatidic acid receptor 2	LPAR2	Lysophosphatidic acid receptor 2 deficiency confers protection against bleomycin-induced lung injury and fibrosis in mice. PMID: 23808384
Lysophosphatidic acid receptor 3	LPAR3	Lysophosphatidic acid receptor-3 increases tumorigenicity and aggressiveness of rat hepatoma RH7777 cells. PMID: 22161812
Lysophosphatidic acid receptor 4	LPAR4	Strategy for the identification of GPR23/LPA4 receptor agonists and inverse agonists. PMID: 20482379
Lysophosphatidic acid receptor 5	LPAR5	Unique ligand selectivity of the GPR92/LPA5 lysophosphatidate receptor indicates role in human platelet activation. PMID: 19366702
Lysophosphatidic acid receptor 6	LPAR6	Disruption of P2RY5, an orphan G protein-coupled receptor, underlies autosomal recessive woolly hair. PMID: 18297072. Disease: Woolly hair autosomal recessive 1 with or without hypotrichosis (ARWH1)
Sphingosine-1-phosphate receptor 1	S1PR1	**Fingolimod** – selective immunosuppressants
Sphingosine-1-phosphate receptor 2	S1PR2	A sphingosine 1-phosphate receptor 2 selective allosteric agonist. PMID: 23849205
Sphingosine-1-phosphate receptor 3	S1PR3	Novel selective allosteric and bitopic ligands for the S1P(3) receptor. PMID: 22971058
Sphingosine-1-phosphate receptor 4	S1PR4	Identification of a novel agonist of the sphingosine 1-phosphate receptor 4 (S1P4). PMID: 23762926
Sphingosine-1-phosphate receptor 5	S1PR5	Sphingosine 1-phosphate receptor 5 mediates the immune quiescence of the human brain endothelial barrier. PMID: 22715976
Melanin-concentrating hormone receptor 1	MCHR1	Preclinical evaluation of melanin-concentrating hormone receptor 1 antagonism for the treatment of obesity and depression. PMID: 19182070
Melanin-concentrating hormone receptor 2	MCHR2	Discovery and characterization of a potent and selective antagonist of melanin-concentrating hormone receptor 2. PMID: 22123324
Melanocortin 1 receptor (alpha melanocyte-stimulating hormone receptor)	MC1R	MC1R and cAMP signaling inhibit cdc25B activity and delay cell cycle progression in melanoma cells. PMID: 23908401. Disease: melanoma, cutaneous malignant 5 (CMM5)
Melanocortin 2 receptor (adrenocorticotropic hormone)	MC2R	**Cosyntropin** – diagnostic agents. Disease: glucocorticoid deficiency 1 (GCCD1)
Melanocortin 3 receptor	MC3R	Melanocortin-3 receptors and metabolic homeostasis. PMID: 23317784

HGNC-approved name	Symbol	Target information
Melanocortin 4 receptor	MC4R	Activation of melanocortin 4 receptors reduces the inflammatory response and prevents apoptosis induced by lipopolysaccharide and interferon-gamma in astrocytes. PMID: 17595227. Disease: obesity (OBESITY)
Melanocortin 5 receptor	MC5R	α-MSH signalling via melanocortin 5 receptor promotes lipolysis and impairs re-esterification in adipocytes. PMID: 24046867
Melatonin receptor 1A	MTNR1A	**Ramelteon** – hypnotics and sedatives
Melatonin receptor 1B	MTNR1B	Design and synthesis of N-(3,3-diphenylpropenyl)alkanamides as a novel class of high-affinity MT2-selective melatonin receptor ligands. PMID: 17149869
Motilin receptor	MLNR	**Camicinal** – gastroparesis agents
Neuromedin U receptor 1	NMUR1	Neuromedin U can exert colon-specific, enteric nerve-mediated prokinetic activity, via a pathway involving NMU1 receptor activation. PMID: 17211455
Neuromedin U receptor 2	NMUR2	Discovery and pharmacological characterization of a small-molecule antagonist at neuromedin U receptor NMUR2. PMID: 19369576
Neuropeptide FF receptor 1	NPFFR1	Development of sub-nanomolar dipeptidic ligands of neuropeptide FF receptors. PMID: 23131340
Neuropeptide FF receptor 2	NPFFR2	Structure-activity studies of RFamide-related peptide-1 identify a functional receptor antagonist and novel cardiac myocyte signaling pathway involved in contractile performance. PMID: 22909119
Neuropeptide S receptor 1	NPSR1	Neuropeptide S receptor: Recent updates on nonpeptide antagonist discovery. PMID: 21452188. Disease: asthma-related traits 2 (ASRT2)
Neuropeptides B/W receptor 1	NPBWR1	Design, synthesis and SAR analysis of novel potent and selective small molecule antagonists of NPBWR1 (GPR7). PMID: 23079522
Neuropeptides B/W receptor 2	NPBWR2	Identification of natural ligands for the orphan G protein-coupled receptors GPR7 and GPR8. PMID: 12401809
Neuropeptide Y receptor Y1	NPY1R	Re-exposure and environmental enrichment reveal NPY-Y1 as a possible target for post-traumatic stress disorder. PMID: 22659471
Neuropeptide Y receptor Y2	NPY2R	Polymorphisms in the NPY2R gene show significant associations with BMI that are additive to FTO, MC4R, and NPFFR2 gene effects. PMID: 21818152
Neuropeptide Y receptor Y4	NPY4R	Functional and molecular properties of the human recombinant Y4 receptor: Resistance to agonist-promoted desensitization. PMID: 10640301
Neuropeptide Y receptor Y5	NPY5R	Identification of a novel benzimidazole derivative as a highly potent NPY Y5 receptor antagonist with an anti-obesity profile. PMID: 23206862
Neurotensin receptor 1 (high affinity)	NTSR1	Structure of the agonist-bound neurotensin receptor. PMID: 23051748
Neurotensin receptor 2	NTSR2	Spinal NTS2 receptor activation reverses signs of neuropathic pain. PMID: 23756650
Opioid receptor, delta 1	OPRD1	**Naltrindole** – analgesics
Opioid receptor, kappa 1	OPRK1	**Meperidine** – analgesics
Opioid receptor, mu 1	OPRM1	**Oxycodone** – analgesics
Opiate receptor-like 1	OPRL1	The endogenous nociceptin/orphanin FQ-NOP receptor system as a potential therapeutic target for intestinal disorders. PMID: 25307525
Hypocretin (orexin) receptor 1	HCRTR1	**Suvorexant** – hypnotics and sedatives
Hypocretin (orexin) receptor 2	HCRTR2	**Suvorexant** – hypnotics and sedatives
Purinergic receptor P2Y, G-protein coupled, 1	P2RY1	The purinergic P2Y1 receptor supports leptin secretion in adipose tissue. PMID: 20185765
Purinergic receptor P2Y, G-protein coupled, 2	P2RY2	**Diquafosol** – opthalmologicals

(Continued)

HGNC-approved name	Symbol	Target information
Pyrimidinergic receptor P2Y, G-protein coupled, 4	P2RY4	Apical targeting of the P2Y(4) receptor is directed by hydrophobic and basic residues in the cytoplasmic tail. PMID: 23054062
Pyrimidinergic receptor P2Y, G-protein coupled, 6	P2RY6	Diisothiocyanate derivatives as potent, insurmountable antagonists of P2Y6 nucleotide receptors. PMID: 15081875
Purinergic receptor P2Y, G-protein coupled, 8	P2RY8	Transforming activity of purinergic receptor P2Y, G protein coupled, 8 revealed by retroviral expression screening. PMID: 17487742
Purinergic receptor P2Y, G-protein coupled, 10	P2RY10	Identification of the orphan GPCR, P2Y(10) receptor as the sphingosine-1-phosphate and lysophosphatidic acid receptor. PMID: 18466763
Purinergic receptor P2Y, G-protein coupled, 11	P2RY11	Synthesis and structure-activity relationships of suramin-derived P2Y11 receptor antagonists with nanomolar potency. PMID: 16250663
Purinergic receptor P2Y, G-protein coupled, 12	P2RY12	**Ticagrelor** – antithrombotic agents. Disease: bleeding disorder, platelet-type 8 (BDPLT8)
Purinergic receptor P2Y, G-protein coupled, 13	P2RY13	Synthesis of pyridoxal phosphate derivatives with antagonist activity at the P2Y13 receptor. PMID: 15913566
Purinergic receptor P2Y, G-protein coupled, 14	P2RY14	Synthesis and SAR of pyrimidine-based, non-nucleotide P2Y14 receptor antagonists. PMID: 21507642
Pyroglutamylated RFamide peptide receptor	QRFPR	Roles of prolactin-releasing peptide and RFamide related peptides in the control of stress and food intake. PMID: 21126313
Platelet-activating factor receptor	PTAFR	**Israpafant** – drugs for obstructive airway diseases
Prokineticin receptor 1	PROKR1	Triazinediones as prokineticin 1 receptor antagonists. PMID: 19375913
Prokineticin receptor 2	PROKR2	Functional analysis of the distal region of the third intracellular loop of PROKR2. PMID: 23969157. Disease: hypogonadotropic hypogonadism 3 with or without anosmia (HH3)
Prolactin-releasing hormone receptor	PRLHR	Biological properties of prolactin-releasing peptide analogs with a modified aromatic ring of a C-terminal phenylalanine amide. PMID: 21872625
Prostaglandin D2 receptor (DP)	PTGDR	Does prostaglandin D2 hold the cure to male pattern baldness? PMID: 24521203. Disease: asthma-related traits 1 (ASRT1)
Prostaglandin D2 receptor 2	PTGDR2	Pharmacologic profile of OC000459, a potent, selective, and orally active D prostanoid receptor 2 antagonist that inhibits mast cell-dependent activation of T helper 2 lymphocytes and eosinophils. PMID: 22106101
Prostaglandin E receptor 1 (subtype EP1), 42 kDa	PTGER1	**Iloprost** – antihypertensives
Prostaglandin E receptor 2 (subtype EP2), 53 kDa	PTGER2	**Carboprost (tromethamine)** – uterotonics
Prostaglandin E receptor 3 (subtype EP3)	PTGER3	3-(2-Aminocarbonylphenyl)propanoic acid analogs as potent and selective EP3 receptor antagonists. PMID: 20385498
Prostaglandin E receptor 4 (subtype EP4)	PTGER4	Prostaglandin E(2) receptors as potential bone anabolic targets – selective EP4 receptor agonists. PMID: 21787288
Prostaglandin F receptor (FP)	PTGFR	**Travoprost** – opthalmologicals
Prostaglandin I2 (prostacyclin) receptor (IP)	PTGIR	Selexipag: A selective prostacyclin receptor agonist that does not affect rat gastric function. PMID: 20660124
Thromboxane A2 receptor	TBXA2R	**Ridogrel** – antithrombotic agents. Disease: bleeding disorder, platelet-type 13 (BDPLT13)
Coagulation factor II (thrombin) receptor	F2R	**Vorapaxar** – antithrombotic agents
Coagulation factor II (thrombin) receptor-like 1	F2RL1	Novel agonists and antagonists for human protease activated receptor 2. PMID: 20873792
Coagulation factor II (thrombin) receptor-like 2	F2RL2	Protease-activated receptor-3 (PAR3) regulates PAR1 signaling by receptor dimerization. PMID: 17376866

HGNC-approved name	Symbol	Target information
Coagulation factor II (thrombin) receptor-like 3	F2RL3	Synthesis of indole derived protease-activated receptor 4 antagonists and characterization in human platelets. PMID: 23776495
Relaxin/insulin-like family peptide receptor 1	RXFP1	Identification and optimization of small-molecule agonists of the human relaxin hormone receptor RXFP1. PMID: 23764525
Relaxin/insulin-like family peptide receptor 2	RXFP2	**Serelaxin** – cardiac therapy. Disease: cryptorchidism (CRYPTO)
Relaxin/insulin-like family peptide receptor 3	RXFP3	Minimization of human relaxin-3 leading to high-affinity analogues with increased selectivity for relaxin-family peptide 3 receptor (RXFP3) over RXFP1. PMID: 22257012
Relaxin/insulin-like family peptide receptor 4	RXFP4	Relaxin family peptides and their receptors. PMID: 23303914
Somatostatin receptor 1	SSTR1	Somatostatin receptor 1 selective analogues: 2 N(alpha)-Methylated scan. PMID: 15658864
Somatostatin receptor 2	SSTR2	**Octreotide** – antineoplastic agents
Somatostatin receptor 3	SSTR3	SST3-selective potent peptidic somatostatin receptor antagonists. PMID: 11095748
Somatostatin receptor 4	SSTR4	Novel sst(4)-selective somatostatin (SRIF) agonists. PMID: 14667213
Somatostatin receptor 5	SSTR5	**Pasireotide** – Cushing's disease
Succinate receptor 1	SUCNR1	The succinate receptor as a novel therapeutic target for oxidative and metabolic stress-related conditions. PMID: 22649411
Tachykinin receptor 1	TACR1	**Aprepitant** – antiemetics and antinauseants
Tachykinin receptor 2	TACR2	Pharmacology of an original and selective nonpeptide antagonist ligand for the human tachykinin NK2 receptor. PMID: 15925360
Tachykinin receptor 3	TACR3	**Talnetant** – drugs for functional gastrointestinal disorders. Disease: hypogonadotropic hypogonadism 11 with or without anosmia (HH11)
Thyrotropin-releasing hormone receptor	TRHR	**Taltirelin** – adult spinal muscular atrophy
Thyroid-stimulating hormone receptor	TSHR	**Thyrotropin alfa** – diagnostic agents. Note: Defects in TSHR are found in patients affected by hyperthyroidism with different etiologies. Somatic, constitutively activating TSHR mutations and/or constitutively activating G(s)alpha mutations have been identified
Trace amine-associated receptor 1	TAAR1	Optimisation of imidazole compounds as selective TAAR1 agonists: Discovery of RO5073012. PMID: 22795332
Trace amine-associated receptor 2	TAAR2	A second class of chemosensory receptors in the olfactory epithelium. PMID: 16878137
Trace amine-associated receptor 3 (gene/pseudogene)	TAAR3	Structural and functional evolution of the trace amine-associated receptors TAAR3, TAAR4 and TAAR5 in primates. PMID: 20559446
Trace amine-associated receptor 5	TAAR5	Human trace amine-associated receptor TAAR5 can be activated by trimethylamine. PMID: 23393561
Trace amine-associated receptor 6	TAAR6	A second class of chemosensory receptors in the olfactory epithelium. PMID: 16878137
Trace amine-associated receptor 8	TAAR8	A second class of chemosensory receptors in the olfactory epithelium. PMID: 16878137
Trace amine-associated receptor 9 (gene/pseudogene)	TAAR9	Correlation of a set of gene variants, life events and personality features on adult ADHD severity. PMID: 20006992
Urotensin 2 receptor	UTS2R	Discovery of new antagonists aimed at discriminating UII and URP-mediated biological activities: Insight into UII and URP receptor activation. PMID: 22994258
Arginine vasopressin receptor 1A	AVPR1A	**Selepressin** – cardiac therapy

(Continued)

HGNC-approved name	Symbol	Target information
Arginine vasopressin receptor 1B	AVPR1B	Characterization of a novel and selective V1B receptor antagonist. PMID: 18655906
Arginine vasopressin receptor 2	AVPR2	**Tolvaptan** – antihyponatraemic agents. Disease: nephrogenic syndrome of inappropriate antidiuresis (NSIAD)
Oxytocin receptor	OXTR	**Carbetocin** – labour-inducing agents
Class A orphans		
G protein-coupled receptor 1	GPR1	The genetic design of signaling cascades to record receptor activation. PMID: 18165312
G protein-coupled receptor 3	GPR3	The orphan G protein-coupled receptor 3 modulates amyloid-beta peptide generation in neurons. PMID: 19213921
G protein-coupled receptor 4	GPR4	Vascular abnormalities in mice deficient for the G protein-coupled receptor GPR4 that functions as a pH sensor. PMID: 17145776
G protein-coupled receptor 6	GPR6	Novel clusters of receptors for sphingosine-1-phosphate, sphingosylphosphorylcholine, and (lyso)-phosphatidic acid: New receptors for 'old' ligands. PMID: 15258916
G protein-coupled receptor 12	GPR12	Role of the G-protein-coupled receptor GPR12 as high-affinity receptor for sphingosylphosphorylcholine and its expression and function in brain development. PMID: 12574419
G protein-coupled receptor 15	GPR15	Gp120-induced Bob/GPR15 activation: A possible cause of human immunodeficiency virus enteropathy. PMID: 11696454
G protein-coupled receptor 17	GPR17	Enigmatic GPCR finds a stimulating drug. PMID: 24150253
G protein-coupled receptor 18	GPR18	Quantitative expression profiling of G-protein-coupled receptors (GPCRs) in metastatic melanoma: The constitutively active orphan GPCR GPR18 as novel drug target. PMID: 20880198
G protein-coupled receptor 19	GPR19	Expression of G protein-coupled receptor 19 in human lung cancer cells is triggered by entry into S-phase and supports G(2)-M cell-cycle progression. PMID: 22912338
G protein-coupled receptor 20	GPR20	Characterization of an orphan G protein-coupled receptor, GPR20, that constitutively activates Gi proteins. PMID: 18347022
G protein-coupled receptor 21	GPR21	G protein-coupled receptor 21 deletion improves insulin sensitivity in diet-induced obese mice. PMID: 22653059
G protein-coupled receptor 22	GPR22	Myocardial expression, signaling, and function of GPR22: A protective role for an orphan G protein-coupled receptor. PMID: 18539757
G protein-coupled receptor 25	GPR25	Discovery of a novel human G protein-coupled receptor gene (GPR25) located on chromosome 1. PMID: 9020062
G protein-coupled receptor 26	GPR26	GPR26-deficient mice display increased anxiety- and depression-like behaviors accompanied by reduced phosphorylated cyclic AMP responsive element-binding protein level in central amygdala. PMID: 21924326
G protein-coupled receptor 27	GPR27	An siRNA screen in pancreatic beta cells reveals a role for Gpr27 in insulin production. PMID: 22253604
G protein-coupled receptor 31	GPR31	Identification of the orphan G protein-coupled receptor GPR31 as a receptor for 12-(*S*)-hydroxyeicosatetraenoic acid. PMID: 21712392
G protein-coupled receptor 32	GPR32	Resolvins as new fascinating drug candidates for inflammatory diseases. PMID: 22297737
G protein-coupled receptor 34	GPR34	GPR34 as a lysophosphatidylserine receptor. PMID: 23389307
G protein-coupled receptor 35	GPR35	High-throughput identification and characterization of novel, species-selective GPR35 agonists. PMID: 23262279
G protein-coupled receptor 37	GPR37	GPR37 and GPR37L1 are receptors for the neuroprotective and glioprotective factors prosaptide and prosaposin. PMID: 23690594
G protein-coupled receptor 37-like 1	GPR37L1	GPR37 and GPR37L1 are receptors for the neuroprotective and glioprotective factors prosaptide and prosaposin. PMID: 23690594

HGNC-approved name	Symbol	Target information
G protein-coupled receptor 39	GPR39	Deficiency of the GPR39 receptor is associated with obesity and altered adipocyte metabolism. PMID: 21784784
G protein-coupled receptor 42 (gene/pseudogene)	GPR42	A family of fatty acid binding receptors. PMID: 15684720
G protein-coupled receptor 45	GPR45	Brain-specific expression of novel G-protein-coupled receptors, with homologies to Xenopus PSP24 and human GPR45. PMID: 11027574
G protein-coupled receptor 50	GPR50	A role for the melatonin-related receptor GPR50 in leptin signaling, adaptive thermogenesis, and torpor. PMID: 22197240
G protein-coupled receptor 52	GPR52	Identification and cloning of three novel human G protein-coupled receptor genes GPR52, PsiGPR53 and GPR55: GPR55 is extensively expressed in human brain. PMID: 9931487
G protein-coupled receptor 55	GPR55	2-Arachidonoyl-sn-glycero-3-phosphoinositol: A possible natural ligand for GPR55. PMID: 18845565
G protein-coupled receptor 61	GPR61	Characterization of metabolic phenotypes of mice lacking GPR61, an orphan G-protein coupled receptor. PMID: 21971119
G protein-coupled receptor 62	GPR62	Identification of four novel human G protein-coupled receptors expressed in the brain. PMID: 11165367
G protein-coupled receptor 63	GPR63	Sphingosine 1-phosphate and dioleoylphosphatidic acid are low affinity agonists for the orphan receptor GPR63. PMID: 12618218
G protein-coupled receptor 65	GPR65	The G protein-coupled receptor T-cell death-associated gene 8 (TDAG8) facilitates tumor development by serving as an extracellular pH sensor. PMID: 20855608
G protein-coupled receptor 68	GPR68	The proton-activated G protein coupled receptor OGR1 acutely regulates the activity of epithelial proton transport proteins. PMID: 22508039
G protein-coupled receptor 75	GPR75	RANTES stimulates Ca^{2+} mobilization and inositol trisphosphate (IP3) formation in cells transfected with G protein-coupled receptor 75. PMID: 17001303
G protein-coupled receptor 78	GPR78	Association analysis of the chromosome 4p-located G protein-coupled receptor 78 (GPR78) gene in bipolar affective disorder and schizophrenia. PMID: 16389273
G protein-coupled receptor 82	GPR82	Reduced food intake and body weight in mice deficient for the G protein-coupled receptor GPR82. PMID: 22216272
G protein-coupled receptor 83	GPR83	Downregulation of GPR83 in the hypothalamic preoptic area reduces core body temperature and elevates circulating levels of adiponectin. PMID: 22560055
G protein-coupled receptor 84	GPR84	Medium-chain fatty acid-sensing receptor, GPR84, is a proinflammatory receptor. PMID: 23449982
G protein-coupled receptor 85	GPR85	SREB2/GPR85, a schizophrenia risk factor, negatively regulates hippocampal adult neurogenesis and neurogenesis-dependent learning and memory. PMID: 22697179
G protein-coupled receptor 87	GPR87	G protein-coupled receptor 87 (GPR87) promotes the growth and metastasis of CD133+ cancer stem-like cells in hepatocellular carcinoma. PMID: 23593389
G protein-coupled receptor 88	GPR88	The GPR88 receptor agonist 2-PCCA does not alter the behavioral effects of methamphetamine in rats. PMID: 23123351
G protein-coupled receptor 101	GPR101	Characterization of Gpr101 expression and G-protein coupling selectivity. PMID: 16647048
G protein-coupled receptor 119	GPR119	**DS-8500** – drugs used in diabetes
G protein-coupled receptor 132	GPR132	Free fatty acid receptors: Emerging targets for treatment of diabetes and its complications. PMID: 23148161

(Continued)

HGNC-approved name	Symbol	Target information
G protein-coupled receptor 135	GPR135	Seven evolutionarily conserved human rhodopsin G protein-coupled receptors lacking close relatives. PMID: 14623098
G protein-coupled receptor 139	GPR139	Identification of surrogate agonists and antagonists for orphan G-protein-coupled receptor GPR139. PMID:19525486
G protein-coupled receptor 141	GPR141	Seven evolutionarily conserved human rhodopsin G protein-coupled receptors lacking close relatives. PMID: 14623098
G protein-coupled receptor 142	GPR142	Discovery and optimization of a novel series of GPR142 agonists for the treatment of type 2 diabetes mellitus. PMID: 22884988
G protein-coupled receptor 146	GPR146	Evidence for an interaction between proinsulin C-peptide and GPR146. PMID: 23980258
G protein-coupled receptor 148	GPR148	Nine new human Rhodopsin family G-protein coupled receptors: Identification, sequence characterisation and evolutionary relationship. PMID: 15777626
G protein-coupled receptor 149	GPR149	Deletion of the novel oocyte-enriched gene, Gpr149, leads to increased fertility in mice. PMID: 19887567
G protein-coupled receptor 150	GPR150	Identification of PRTFDC1 silencing and aberrant promoter methylation of GPR150, ITGA8 and HOXD11 in ovarian cancers. PMID: 17303177
G protein-coupled receptor 151	GPR151	Cloning and characterization of a novel G-protein-coupled receptor with homology to galanin receptors. PMID: 15111018
G protein-coupled receptor 152	GPR152	Nine new human Rhodopsin family G-protein coupled receptors: Identification, sequence characterisation and evolutionary relationship. PMID: 15777626
G protein-coupled receptor 153	GPR153	The G protein coupled receptor Gpr153 shares common evolutionary origin with Gpr162 and is highly expressed in central regions including the thalamus, cerebellum and the arcuate nucleus. PMID: 21981325
G protein-coupled receptor 155	GPR155	GPR155: Gene organization, multiple mRNA splice variants and expression in mouse central nervous system. PMID: 20537985
G protein-coupled receptor 160	GPR160	Quantitative expression profiling of G-protein-coupled receptors (GPCRs) in metastatic melanoma: The constitutively active orphan GPCR GPR18 as novel drug target. PMID: 20880198
G protein-coupled receptor 161	GPR161	G-protein-coupled receptor GPR161 is overexpressed in breast cancer and is a promoter of cell proliferation and invasion. PMID: 24599592
G protein-coupled receptor 162	GPR162	The G protein coupled receptor Gpr153 shares common evolutionary origin with Gpr162 and is highly expressed in central regions including the thalamus, cerebellum and the arcuate nucleus. PMID: 21981325
G protein-coupled receptor 171	GPR171	GPR171 is a hypothalamic G protein-coupled receptor for BigLEN, a neuropeptide involved in feeding. PMID: 24043826
G protein-coupled receptor 173	GPR173	The metabolite GnRH-(1-5) inhibits the migration of immortalized GnRH neurons. PMID: 23321696
G protein-coupled receptor 174	GPR174	Expression of orphan G-protein coupled receptor GPR174 in CHO cells induced morphological changes and proliferation delay via increasing intracellular cAMP. PMID: 23178570
G protein-coupled receptor 176	GPR176	Isoform level expression profiles provide better cancer signatures than gene level expression profiles. PMID: 23594586
G protein-coupled receptor 182	GPR182	The ADMR receptor mediates the effects of adrenomedullin on pancreatic cancer cells and on cells of the tumor microenvironment. PMID: 19847298
G protein-coupled receptor 183	GPR183	Molecular characterization of oxysterol binding to the Epstein-Barr virus-induced gene 2 (GPR183). PMID: 22875855
Leucine-rich repeat containing G protein-coupled receptor 4	LGR4	R-spondins function as ligands of the orphan receptors LGR4 and LGR5 to regulate Wnt/beta-catenin signaling. PMID: 21693646. Disease: osteoporosis (OSTEOP)

HGNC-approved name	Symbol	Target information
Leucine-rich repeat containing G protein-coupled receptor 5	LGR5	Increased expression of Lgr5 is associated with chemotherapy resistance in human gastric cancer. PMID: 24859092
Leucine-rich repeat containing G protein-coupled receptor 6	LGR6	LGR6 is a high affinity receptor of R-spondins and potentially functions as a tumor suppressor. PMID: 22615920
MAS1 oncogene	MAS1	Endothelial dysfunction through genetic deletion or inhibition of the G protein-coupled receptor Mas: A new target to improve endothelial function. PMID: 17984663. Note: The MAS oncogene has a weak focus-inducing activity in transfected NIH 3T3 cells.
MAS1 oncogene-like	MAS1L	Cloning and functional characterization of a novel mas-related gene, modulating intracellular angiotensin II actions. PMID: 1723144
MAS-related GPR, member D	MRGPRD	MRGD, a MAS-related G-protein coupled receptor, promotes tumorigenisis and is highly expressed in lung cancer. PMID: 22715397
MAS-related GPR, member E	MRGPRE	The effect of deletion of the orphan G - protein coupled receptor (GPCR) gene MrgE on pain-like behaviours in mice. PMID: 18197975
MAS-related GPR, member F	MRGPRF	Proenkephalin A gene products activate a new family of sensory neuron – specific GPCRs. PMID: 11850634
MAS-related GPR, member G	MRGPRG	Proenkephalin A gene products activate a new family of sensory neuron – specific GPCRs. PMID: 11850634
MAS-related GPR, member X1	MRGPRX1	Identification of small molecule antagonists of the human mas-related gene-X1 receptor. PMID: 16510108
MAS-related GPR, member X2	MRGPRX2	Identification of a mast-cell-specific receptor crucial for pseudo-allergic drug reactions. PMID: 25517090
MAS-related GPR, member X3	MRGPRX3	Transgenic rats overexpressing the human MrgX3 gene show cataracts and an abnormal skin phenotype. PMID: 15809047
MAS-related GPR, member X4	MRGPRX4	Identification of candidate oncogenes in human colorectal cancers with microsatellite instability. PMID: 23684749
Class B		
Calcitonin (and amylin) receptor	CALCR	**Pramlintide** – drugs used in diabetes
Calcitonin receptor-like	CALCRL	Adrenomedullin is a therapeutic target in colorectal cancer. PMID: 24519534
Receptor (G protein-coupled)-activity-modifying protein 1	RAMP1	RAMP1 is a direct NKX31 target gene up-regulated in prostate cancer that promotes tumorigenesis. PMID: 23867798
Receptor (G protein-coupled)-activity-modifying protein 2	RAMP2	The GPCR modulator protein RAMP2 is essential for angiogenesis and vascular integrity. PMID: 18097473
Corticotropin-releasing factor receptors		
Corticotropin-releasing hormone receptor 1	CRHR1	**NBI-77860** – anxiolytics
Corticotropin-releasing hormone receptor 2	CRHR2	**NBI-77860** – anxiolytics
Secretin receptor	SCTR	**Secretin** – diagnostic agents
Glucagon receptor	GCGR	**Glucagon recombinant** – drugs used in diabetes
Glucagon-like peptide 1 receptor	GLP1R	**Liraglutide recombinant** – drugs used in diabetes
Glucagon-like peptide 2 receptor	GLP2R	**Teduglutide** – short bowel syndrome
Growth hormone-releasing hormone receptor	GHRHR	**Sermorelin** – hormone replacement agents. Disease: growth hormone deficiency, isolated, 1B (IGHD1B)
Gastric inhibitory polypeptide receptor	GIPR	Two incretin hormones GLP-1 and GIP: Comparison of their actions in insulin secretion and β cell preservation. PMID: 21820006
Parathyroid hormone 1 receptor	PTH1R	**Teriparatide** – drugs affecting bone structure and mineralization. Disease: Jansen metaphyseal chondrodysplasia (JMC)

(Continued)

HGNC-approved name	Symbol	Target information
Parathyroid hormone 2 receptor	PTH2R	Parathyroid hormone 2 receptor and its endogenous ligand tuberoinfundibular peptide of 39 residues are concentrated in endocrine, viscerosensory and auditory brain regions in macaque and human. PMID: 19401215
Adenylate cyclase-activating polypeptide 1 (pituitary) receptor type I	ADCYAP1R1	Novel stable PACAP analogs with potent activity towards the PAC1 receptor. PMID: 18353507
Vasoactive intestinal peptide receptor 1	VIPR1	Identification and optimization of small molecule antagonists of vasoactive intestinal peptide receptor-1 (VIPR1). PMID: 22365758
Vasoactive intestinal peptide receptor 2	VIPR2	A potent and highly selective VPAC2 agonist enhances glucose-induced insulin release and glucose disposal: A potential therapy for type 2 diabetes. PMID: 11978642
Class C		
Calcium-sensing receptor	CASR	**Cinacalcet** – hyperparathyroidism. Disease: hypocalciuric hypercalcemia, familial 1 (HHC1)
G protein-coupled receptor, family C, group 6, member A	GPRC6A	Multiligand specificity and wide tissue expression of GPRC6A reveals new endocrine networks. PMID: 22374969
Gamma-aminobutyric acid (GABA) B receptor, 1	GABBR1	**Baclofen** – muscle relaxants
Gamma-aminobutyric acid (GABA) B receptor, 2	GABBR2	**Baclofen** – muscle relaxants
Glutamate receptor, metabotropic 1	GRM1	Unique antipsychotic activities of the selective metabotropic glutamate receptor 1 allosteric antagonist 2-cyclopropyl-5-[1-(2-fluoro-3-pyridinyl)-5-methyl-1H-1,2,3-triazol-4-yl]-2,3-dihydro-1H-isoindol-1-one. PMID: 19359526. Disease: spinocerebellar ataxia
Glutamate receptor, metabotropic 2	GRM2	**RG1578** – antidepressants
Glutamate receptor, metabotropic 3	GRM3	Anxiolytic effects of a novel group II metabotropic glutamate receptor agonist (LY354740) in the fear-potentiated startle paradigm in humans. PMID: 12709777
Glutamate receptor, metabotropic 4	GRM4	mGluR4-positive allosteric modulation as potential treatment for Parkinson's disease. PMID: 20161443
Glutamate receptor, metabotropic 5	GRM5	Metabotropic glutamate receptor 5 in the pathology and treatment of schizophrenia. PMID: 23253944
Glutamate receptor, metabotropic 6	GRM6	mGluR6 deletion renders the TRPM1 channel in retina inactive. PMID: 22131384. Disease: night blindness, congenital stationary, 1B (CSNB1B)
Glutamate receptor, metabotropic 7	GRM7	A selective mGlu7 receptor antagonist MMPIP reversed antidepressant-like effects of AMN082 in rats. PMID: 23085340
Glutamate receptor, metabotropic 8	GRM8	Metabotropic glutamate receptor subtype 8 in the amygdala modulates thermal threshold, neurotransmitter release, and rostral ventromedial medulla cell activity in inflammatory pain. PMID: 21430167
Class C orphans		
G protein-coupled receptor 156	GPR156	Molecular cloning and characterisation of a novel GABAB-related G-protein coupled receptor. PMID: 12591167
G protein-coupled receptor 158	GPR158	GPR158, an orphan member of G protein-coupled receptor Family C: Glucocorticoid-stimulated expression and novel nuclear role. PMID: 23451275
G protein-coupled receptor 179	GPR179	GPR179 is required for depolarizing bipolar cell function and is mutated in autosomal-recessive complete congenital stationary night blindness. PMID: 22325362. Disease: Night blindness, congenital stationary, 1E (CSNB1E)
G protein-coupled receptor, family C, group 5, member A	GPRC5A	Identification of the retinoic acid-inducible Gprc5a as a new lung tumor suppressor gene. PMID: 18000218

HGNC-approved name	Symbol	Target information
G protein-coupled receptor, family C, group 5, member B	GPRC5B	Molecular cloning and characterization of two novel retinoic acid-inducible orphan G-protein-coupled receptors (GPRC5B and GPRC5C). PMID: 10945465
G protein-coupled receptor, family C, group 5, member C	GPRC5C	Molecular cloning and characterization of two novel retinoic acid-inducible orphan G-protein-coupled receptors (GPRC5B and GPRC5C). PMID: 10945465
G protein-coupled receptor, family C, group 5, member D	GPRC5D	Cloning and characterization of a human orphan family C G-protein coupled receptor GPRC5D. PMID: 11311935
Class F		
Frizzled family receptor 1	FZD1	**OMP-18R5** – antineoplastic agents
Frizzled family receptor 2	FZD2	**OMP-18R5** – antineoplastic agents
Frizzled family receptor 3	FZD3	Frizzled3 is required for neurogenesis and target innervation during sympathetic nervous system development. PMID: 21325504
Frizzled family receptor 4	FZD4	Norrin/Frizzled4 signaling in retinal vascular development and blood brain barrier plasticity. PMID: 23217714. Disease: vitreoretinopathy, exudative 1 (EVR1)
Frizzled family receptor 5	FZD5	**OMP-18R5** – antineoplastic agents
Frizzled family receptor 6	FZD6	FZD6 encoding the Wnt receptor frizzled6 is mutated in autosomal-recessive nail dysplasia. PMID: 22211385. Disease: nail disorder, non-syndromic congenital, 10 (NDNC10)
Frizzled family receptor 7	FZD7	**Vantictumab** – antineoplastic agents
Frizzled family receptor 8	FZD8	**Ipafricept** – antineoplastic agents
Frizzled family receptor 9	FZD9	Control of bone formation by the serpentine receptor Frizzled-9. PMID: 21402791
Frizzled family receptor 10	FZD10	Activation of the non-canonical Dvl-Rac1-JNK pathway by Frizzled homologue 10 in human synovial sarcoma. PMID: 19137009
Smoothened, frizzled family receptor	SMO	**Vismodegib** – antineoplastic agents
G protein-coupled receptor 125	GPR125	Gpr125 modulates Dishevelled distribution and planar cell polarity signaling. PMID: 23821037
Adhesion class		
CD97 molecule	CD97	Overexpression of CD97 confers an invasive phenotype in glioblastoma cells and is associated with decreased survival of glioblastoma patients. PMID: 23658650
Brain-specific angiogenesis inhibitor 1	BAI1	Phosphatidylserine receptor BAI1 and apoptotic cells as new promoters of myoblast fusion. PMID: 23615608
Brain-specific angiogenesis inhibitor 2	BAI2	Antidepressant-like behavior in brain-specific angiogenesis inhibitor 2-deficient mice. PMID: 21110148
Brain-specific angiogenesis inhibitor 3	BAI3	The cell-adhesion G protein-coupled receptor BAI3 is a high-affinity receptor for C1q-like proteins. PMID: 21262840
Cadherin, EGF LAG seven-pass G-type receptor 1	CELSR1	Planar cell polarity genes, Celsr1-3, in neural development. PMID: 22622831. Disease: neural tube defects (NTD)
Cadherin, EGF LAG seven-pass G-type receptor 2	CELSR2	Planar cell polarity genes, Celsr1-3, in neural development. PMID: 22622831
Cadherin, EGF LAG seven-pass G-type receptor 3	CELSR3	Planar cell polarity genes, Celsr1-3, in neural development. PMID: 22622831
EGF, latrophilin and seven transmembrane domain containing 1	ELTD1	A core human primary tumor angiogenesis signature identifies the endothelial orphan receptor ELTD1 as a key regulator of angiogenesis. PMID: 23871637
EGF-like module containing, mucin-like, hormone receptor-like 1	EMR1	EMR1, the human homolog of F4/80, is an eosinophil-specific receptor. PMID: 17823986

(Continued)

HGNC-approved name	Symbol	Target information
EGF-like module containing, mucin-like, hormone receptor-like 2	EMR2	Identification of the epidermal growth factor-TM7 receptor EMR2 and its ligand dermatan sulfate in rheumatoid synovial tissue. PMID: 15693006
EGF-like module containing, mucin-like, hormone receptor-like 3	EMR3	EMR-3: A potential mediator of invasive phenotypic variation in glioblastoma and novel therapeutic target. PMID: 20827226
G protein-coupled receptor 56	GPR56	G protein-coupled receptor 56 and collagen III, a receptor-ligand pair, regulates cortical development and lamination. PMID: 21768377. Disease: polymicrogyria, bilateral frontoparietal (BFPP)
G protein-coupled receptor 64	GPR64	Adhesion-GPCRs in the male reproductive tract. PMID: 21618837
G protein-coupled receptor 97	GPR97	Signaling property study of adhesion G-protein-coupled receptors. PMID: 22575658
G protein-coupled receptor 98	GPR98	GPR98/Gpr98 gene is involved in the regulation of human and mouse bone mineral density. PMID: 22419726. Disease: Usher syndrome 2C (USH2C)
G protein-coupled receptor 110	GPR110	Orphan receptor GPR110, an oncogene overexpressed in lung and prostate cancer. PMID: 20149256
G protein-coupled receptor 111	GPR111	Characterization and functional study of a cluster of four highly conserved orphan adhesion-GPCR in mouse. PMID: 22837050
G protein-coupled receptor 112	GPR112	Novel markers for enterochromaffin cells and gastrointestinal neuroendocrine carcinomas. PMID: 18953328
G protein-coupled receptor 113	GPR113	Two novel genes, Gpr113, which encodes a family 2 G-protein-coupled receptor, and Trcg1, are selectively expressed in taste receptor cells. PMID: 15780750
G protein-coupled receptor 114	GPR114	Specific expression of GPR56 by human cytotoxic lymphocytes. PMID: 21724806
G protein-coupled receptor 115	GPR115	Chromatin immunoprecipitation (ChIP) scanning identifies primary glucocorticoid receptor target genes. PMID: 15501915
G protein-coupled receptor 116	GPR116	GPR116, an adhesion G-protein coupled receptor, promotes breast cancer metastasis via the Gαq-p63RhoGEF-Rho GTPase pathway. PMID: 24008316
G protein-coupled receptor 123	GPR123	The evolutionary history and tissue mapping of GPR123: Specific CNS expression pattern predominantly in thalamic nuclei and regions containing large pyramidal cells. PMID: 17212699
G protein-coupled receptor 124	GPR124	Angiogenic sprouting into neural tissue requires Gpr124, an orphan G protein-coupled receptor. PMID: 21282641
G protein-coupled receptor 126	GPR126	Gpr126 is essential for peripheral nerve development and myelination in mammals. PMID: 21613327
G protein-coupled receptor 128	GPR128	TFG, a target of chromosome translocations in lymphoma and soft tissue tumors, fuses to GPR128 in healthy individuals. PMID: 19797732
G protein-coupled receptor 133	GPR133	Genetic variation in GPR133 is associated with height: Genome wide association study in the self-contained population of Sorbs. PMID: 19729412
G protein-coupled receptor 144	GPR144	The secretin GPCRs descended from the family of adhesion GPCRs. PMID: 18845549
G protein-coupled receptor 157	GPR157	
Latrophilin 1	LPHN1	Latrophilin 1 and its endogenous ligand Lasso/teneurin-2 form a high-affinity transsynaptic receptor pair with signaling capabilities. PMID: 21724987
Latrophilin 2	LPHN2	The latrophilin family: Multiply spliced G protein-coupled receptors with differential tissue distribution. PMID: 10025961
Latrophilin 3	LPHN3	FLRT proteins are endogenous latrophilin ligands and regulate excitatory synapse development. PMID: 22405201

HGNC-approved name	Symbol	Target information
Other 7TM proteins		
G protein-coupled receptor 107	GPR107	Evidence for an interaction of neuronostatin with the orphan G protein-coupled receptor, GPR107. PMID: 22933024
G protein-coupled receptor 137	GPR137	Inhibition of GPR137 expression reduces the proliferation and colony formation of malignant glioma cells. PMID: 24870220
G protein-coupled receptor 137B	GPR137B	TM7SF1 (GPR137B): A novel lysosome integral membrane protein. PMID: 22729905
G protein-coupled receptor 137C	GPR137C	
G protein-coupled receptor 143	GPR143	A novel nonsense mutation of the GPR143 gene identified in a Chinese pedigree with ocular albinism. PMID: 22916221. Disease: albinism ocular 1 (OA1)
G protein-coupled receptor 180	GPR180	Inhibition of experimental intimal thickening in mice lacking a novel G-protein-coupled receptor. PMID: 12538434
Transmembrane protein, adipocyte associated 1	TPRA1	TPRA40/GPR175 regulates early mouse embryogenesis through functional membrane transport by Sjögren's syndrome-associated protein NA14. PMID: 18459117
Olfaction, taste and vision receptors		
Olfactory receptor, family 51, subfamily E, member 1	OR51E1	Olfactory receptor 51E1 as a novel target in somatostatin receptor negative lung carcinoids. PMID: 23969981
Taste receptor, type 1, member 1	TAS1R1	Expression, regulation and putative nutrient-sensing function of taste GPCRs in the heart. PMID: 23696900
Taste receptor, type 1, member 2	TAS1R2	Expression, regulation and putative nutrient-sensing function of taste GPCRs in the heart. PMID: 23696900
Taste receptor, type 1, member 3	TAS1R3	Expression, regulation and putative nutrient-sensing function of taste GPCRs in the heart. PMID: 23696900
Opsin 1 (cone pigments), long-wave-sensitive	OPN1LW	Disease: colorblindness, partial, protan series (CBP)
Opsin 1 (cone pigments), medium-wave-sensitive	OPN1MW	Disease: colorblindness, partial, deutan series (CBD)
Opsin 1 (cone pigments), medium-wave-sensitive 2	OPN1MW2	Disease: colorblindness, partial, deutan series (CBD)
Opsin 1 (cone pigments), short-wave-sensitive	OPN1SW	Disease: tritan color blindness (CBT)
Retinal G protein-coupled receptor	RGR	Disease: retinitis pigmentosa 44 (RP44)
Rhodopsin	RHO	Disease: retinitis pigmentosa 4 (RP4)

3.1.1   G-protein-coupled receptor (GPCR) ligands

HGNC-approved name	Symbol	Target information
Adrenomedullin	ADM	
Adrenomedullin 2	ADM2	
Adrenomedullin 5 (putative)	ADM5	
Agouti related protein homolog (mouse)	AGRP	Disease: obesity (OBESITY)
Apelin receptor early endogenous ligand	APELA	
Apelin	APLN	
Arginine vasopressin	AVP	Disease: diabetes insipidus, neurohypophyseal (NDI)

(Continued)

HGNC-approved name	Symbol	Target information
Calcitonin-related polypeptide alpha	CALCA	
Calcitonin-related polypeptide beta	CALCB	
CART prepropeptide	CARTPT	
Cholecystokinin	CCK	
Glycoprotein hormones, alpha polypeptide	CGA	
Chorionic gonadotropin, beta polypeptide	CGB	
Chorionic gonadotropin, beta polypeptide 1	CGB1	
Chorionic gonadotropin, beta polypeptide 2	CGB2	
Chorionic gonadotropin, beta polypeptide 5	CGB5	
Chorionic gonadotropin, beta polypeptide 7	CGB7	
Chorionic gonadotropin, beta polypeptide 8	CGB8	
Cortistatin	CORT	
Corticotropin-releasing hormone	CRH	
Endothelin 1	EDN1	Disease: question mark ears, isolated (QME)
Endothelin 2	EDN2	
Endothelin 3	EDN3	Disease: Hirschsprung disease 4 (HSCR4)
Follicle-stimulating hormone, beta polypeptide	FSHB	Disease: isolated follicle-stimulating hormone deficiency (IFSHD)
Galanin/GMAP prepropeptide	GAL	
Galanin-like peptide	GALP	
Gastrin	GAST	Progastrin a new pro-angiogenic factor in colorectal cancer. PMID: 25109333
Glucagon	GCG	
Glycoprotein hormone alpha 2	GPHA2	
Glycoprotein hormone beta 5	GPHB5	
Growth hormone-releasing hormone	GHRH	
Ghrelin/obestatin prepropeptide	GHRL	
Gastric inhibitory polypeptide	GIP	
Gonadotropin-releasing hormone 1 (luteinizing-releasing hormone)	GNRH1	Disease: hypogonadotropic hypogonadism 12 with or without anosmia (HH12)
Gonadotropin-releasing hormone 2	GNRH2	
Gastrin-releasing peptide	GRP	
Hypocretin (orexin) neuropeptide precursor	HCRT	Disease: narcolepsy 1 (NRCLP1)
Islet amyloid polypeptide	IAPP	IAPP-driven metabolic reprogramming induces regression of p53-deficient tumours in vivo. PMID: 25409149
KiSS-1 metastasis-suppressor	KISS1	KISS1 methylation and expression as predictors of disease progression in colorectal cancer patients. PMID: 25110434. Disease: hypogonadotropic hypogonadism 13 with or without anosmia (HH13)
Luteinizing hormone beta polypeptide	LHB	Disease: hypogonadism LHB-related (HGON-LHB)
Motilin	MLN	
Neuromedin B	NMB	
Neuromedin S	NMS	
Neuromedin U	NMU	Neuromedin U: A candidate biomarker and therapeutic target to predict and overcome resistance to HER-tyrosine kinase inhibitors. PMID: 24876102
Neuropeptide B	NPB	
Neuropeptide FF-amide peptide precursor	NPFF	
Neuropeptide S	NPS	

HGNC-approved name	Symbol	Target information
Neuropeptide VF precursor	NPVF	
Neuropeptide W	NPW	
Neuropeptide Y	NPY	
Neurotensin	NTS	
Oxytocin/neurophysin I prepropeptide	OXT	
Prodynorphin	PDYN	Disease: spinocerebellar ataxia 23 (SCA23)
Proenkephalin	PENK	
Prokineticin 1	PROK1	Anti-prokineticin1 (PROK1) monoclonal antibody suppresses angiogenesis and tumor growth in colorectal cancer. PMID: 24838366
Prokineticin 2	PROK2	Disease: hypogonadotropic hypogonadism 4 with or without anosmia (HH4)
Pro-melanin-concentrating hormone	PMCH	
Prepronociceptin	PNOC	
Proopiomelanocortin	POMC	Disease: obesity (OBESITY)
Pancreatic polypeptide	PPY	
Prolactin-releasing hormone	PRLH	
Parathyroid hormone	PTH	Disease: hypoparathyroidism, familial isolated (FIH)
Parathyroid hormone 2	PTH2	
Parathyroid hormone-like hormone	PTHLH	**Abaloparatide** – drugs affecting bone structure and mineralization. Disease: brachydactyly E2 (BDE2)
Peptide YY	PYY	
Pyroglutamylated RFamide peptide	QRFP	
Relaxin 1	RLN1	Up-regulating relaxin expression by G-quadruplex interactive ligand to achieve antifibrotic action. PMID: 22673230
Relaxin 2	RLN2	The plasma levels of relaxin-2 and relaxin-3 in patients with diabetes. PMID: 23994775
Relaxin 3	RLN3	The electrostatic interactions of relaxin-3 with receptor RXFP4 and the influence of its B-chain C-terminal conformation. PMID: 24802387
Retinoic acid receptor responder (tazarotene induced) 2	RARRES2	
R-spondin 1	RSPO1	Disease: keratoderma, palmoplantar, with squamous cell carcinoma of skin and sex reversal (PKKSCC)
R-spondin 2	RSPO2	
R-spondin 3	RSPO3	**Anti-RSPO3 (OncoMed)** – antineoplastic agents
R-spondin 4	RSPO4	Disease: nail disorder, non-syndromic congenital, 4 (NDNC4)
Secretin	SCT	
Spexin hormone	SPX	
Somatostatin	SST	
Tachykinin, precursor 1	TAC1	
Tachykinin 3	TAC3	Disease: hypogonadotropic hypogonadism 10 with or without anosmia (HH10)
Tachykinin 4 (hemokinin)	TAC4	
Thyroid-stimulating hormone, beta	TSHB	
Thyrotropin-releasing hormone	TRH	
Urocortin	UCN	
Urocortin 2	UCN2	
Urocortin 3	UCN3	

(*Continued*)

HGNC-approved name	Symbol	Target information
Urotensin 2	UTS2	
Urotensin 2B	UTS2B	
Vasoactive intestinal peptide	VIP	
Chemokine (C—C motif) ligand 1	CCL1	
Chemokine (C—C motif) ligand 2	CCL2	**Carlumab** – antineoplastic agents
Chemokine (C—C motif) ligand 3	CCL3	
Chemokine (C—C motif) ligand 4	CCL4	
Chemokine (C—C motif) ligand 5	CCL5	Disrupting functional interactions between platelet chemokines inhibits atherosclerosis in hyperlipidemic mice. PMID: 19122657
Chemokine (C—C motif) ligand 7	CCL7	
Chemokine (C—C motif) ligand 8	CCL8	
Chemokine (C—C motif) ligand 11	CCL11	**Bertilimumab** – anti-inflammatory and antirheumatic products
Chemokine (C—C motif) ligand 13	CCL13	
Chemokine (C—C motif) ligand 14	CCL14	Binding of the JmjC demethylase JARID1B to LSD1/NuRD suppresses angiogenesis and metastasis in breast cancer cells by repressing chemokine CCL14. PMID: 21937684
Chemokine (C—C motif) ligand 15	CCL15	
Chemokine (C—C motif) ligand 16	CCL16	
Chemokine (C—C motif) ligand 17	CCL17	
Chemokine (C—C motif) ligand 18 (pulmonary and activation-regulated)	CCL18	CC-chemokine ligand 18 induces epithelial to mesenchymal transition in lung cancer A549 cells and elevates the invasive potential. PMID: 23349697
Chemokine (C—C motif) ligand 19	CCL19	
Chemokine (C—C motif) ligand 20	CCL20	
Chemokine (C—C motif) ligand 21	CCL21	Role of the CCL21 and CCR7 pathways in rheumatoid arthritis angiogenesis. PMID: 22392503
Chemokine (C—C motif) ligand 22	CCL22	
Chemokine (C—C motif) ligand 23	CCL23	
Chemokine (C—C motif) ligand 24	CCL24	
Chemokine (C—C motif) ligand 25	CCL25	
Chemokine (C—C motif) ligand 26	CCL26	
Chemokine (C—C motif) ligand 27	CCL27	
Chemokine (C—C motif) ligand 28	CCL28	Characterising the expression and function of CCL28 and its corresponding receptor, CCR10, in RA pathogenesis. PMID: 24833787
Chemokine (C—X3—C motif) ligand 1	CX3CL1	Therapeutic potential of the chemokine-receptor duo fractalkine/CX3CR1: An update. PMID: 22530606
Chemokine (C—X—C motif) ligand 1 (melanoma growth stimulating activity, alpha)	CXCL1	Lymphatic endothelial cell-secreted CXCL1 stimulates lymphangiogenesis and metastasis of gastric cancer. PMID: 21387301
Chemokine (C—X—C motif) ligand 2	CXCL2	
Chemokine (C—X—C motif) ligand 3	CXCL3	
Platelet factor 4	PF4	Disrupting functional interactions between platelet chemokines inhibits atherosclerosis in hyperlipidemic mice. PMID: 19122657
Platelet factor 4, variant 1	PF4V1	
Chemokine (C—X—C motif) ligand 5	CXCL5	CXCL5 drives neutrophil recruitment in TH17-mediated GN. PMID: 24904089
Chemokine (C—X—C motif) ligand 6	CXCL6	

HGNC-approved name	Symbol	Target information
Chemokine (C—X—C motif) ligand 8	CXCL8	
Chemokine (C—X—C motif) ligand 9	CXCL9	
Chemokine (C—X—C motif) ligand 10	CXCL10	**Eldelumab** – drugs for functional gastrointestinal disorders
Chemokine (C—X—C motif) ligand 11	CXCL11	
Chemokine (C—X—C motif) ligand 12	CXCL12	Development and preclinical characterization of a humanized antibody targeting CXCL12. PMID: 23812669
Chemokine (C—X—C motif) ligand 13	CXCL13	Role of CXCL13 in asthma: Novel therapeutic target. PMID: 22016489
Chemokine (C—X—C motif) ligand 14	CXCL14	Expression and effect of CXCL14 in colorectal carcinoma. PMID: 24938992
Chemokine (C—X—C motif) ligand 16	CXCL16	SR-PSOX/CXCL16 plays a critical role in the progression of colonic inflammation. PMID: 21471570
Chemokine (C—X—C motif) ligand 17	CXCL17	
Pro-platelet basic protein (chemokine (C—X—C motif) ligand 7)	PPBP	
Interleukin 8	IL8	
Chemokine (C motif) ligand 1	XCL1	
Chemokine (C motif) ligand 2	XCL2	
Chemokine (C—C motif) ligand 3-like 1	CCL3L1	
Chemokine (C—C motif) ligand 3-like 3	CCL3L3	
Chemokine (C—C motif) ligand 4-like 1	CCL4L1	
Chemokine (C—C motif) ligand 4-like 2	CCL4L2	
Chemokine-like factor	CKLF	

3.2   Nuclear hormone receptors (NHRS)

HGNC-approved name	Symbol	Target information
Thyroid hormone receptor, alpha	THRA	**Levothyroxine** – hormone replacement agents. Disease: hypothyroidism, congenital, non-goitrous, 6 (CHNG6)
Thyroid hormone receptor, beta	THRB	Molecular aspects of thyroid hormone actions. PMID: 20051527. Disease: generalized thyroid hormone resistance (GTHR)
Retinoic acid receptor, alpha	RARA	**Isotretinoin** – keratolytic agents. Note: Chromosomal aberrations involving RARA are commonly found in acute promyelocytic leukemia. Translocation t(1117)(q32q21) with ZBTB16/PLZF translocation t(1517)(q21q21) with PML translocation t(517)(q32q11) with NPM
Retinoic acid receptor, beta	RARB	**Tamibarotene** – antineoplastic agents. Disease: microphthalmia, syndromic, 12 (MCOPS12)
Retinoic acid receptor, gamma	RARG	Randomised controlled trial for emphysema with a selective agonist of the γ-type retinoic acid receptor. PMID: 2282548
Peroxisome proliferator-activated receptor alpha	PPARA	**Fenofibrate** – lipid-modifying agents
Peroxisome proliferator-activated receptor delta	PPARD	Targeting PPARβ/δ for the treatment of type 2 diabetes mellitus. PMID: 22280315
Peroxisome proliferator-activated receptor gamma	PPARG	**Rosiglitazone** – drugs used in diabetes. Note: Defects in PPARG can lead to type 2 insulin-resistant diabetes and hypertension. PPARG mutations may be associated with colon cancer. Disease: obesity (OBESITY)

(Continued)

HGNC-approved name	Symbol	Target information
Nuclear receptor subfamily 1, group D, member 1	NR1D1	Optimized chemical probes for REV-ERBα. PMID: 23656296
Nuclear receptor subfamily 1, group D, member 2	NR1D2	Regulation of circadian behaviour and metabolism by REV-ERB-α and REV-ERB-β. PMID: 22460952
RAR-related orphan receptor A	RORA	Genome-wide identification of transcriptional targets of RORA reveals direct regulation of multiple genes associated with autism spectrum disorder. PMID: 23697635
RAR-related orphan receptor B	RORB	Examination of nuclear receptor expression in osteoblasts reveals Rorβ as an important regulator of osteogenesis. PMID: 22189870
RAR-related orphan receptor C	RORC	Small molecule inhibitors of RORγt: Targeting Th17 cells and other applications. PMID: 22949321
Nuclear receptor subfamily 1, group H, member 4	NR1H4	Recent advances in non-steroidal FXR antagonists development for therapeutic applications. PMID: 25388534
Nuclear receptor subfamily 1, group H, member 3	NR1H3	Liver LXRα expression is crucial for whole body cholesterol homeostasis and reverse cholesterol transport in mice. PMID: 22484817
Nuclear receptor subfamily 1, group H, member 2	NR1H2	LXRbeta is required for adipocyte growth, glucose homeostasis, and beta cell function. PMID: 15831500
Vitamin D (1,25- dihydroxyvitamin D3) receptor	VDR	**Calcitriol** – drugs affecting bone structure and mineralization. Disease: Rickets vitamin D-dependent 2A (VDDR2A)
Nuclear receptor subfamily 1, group I, member 2	NR1I2	Pregnane X receptor as a target for treatment of inflammatory bowel disorders. PMID: 22609277
Nuclear receptor subfamily 1, group I, member 3	NR1I3	The constitutive androstane receptor is a novel therapeutic target facilitating cyclophosphamide-based treatment of hematopoietic malignancies. PMID: 23160467
Hepatocyte nuclear factor 4, alpha	HNF4A	Hepatocyte nuclear factor 4 alpha suppresses the development of hepatocellular carcinoma. PMID: 20876809. Disease: maturity-onset diabetes of the young 1 (MODY1)
Hepatocyte nuclear factor 4, gamma	HNF4G	Orphan nuclear receptor HNF4G promotes bladder cancer growth and invasion through the regulation of the hyaluronan synthase 2 gene. PMID: 23896584
Retinoid X receptor, alpha	RXRA	**Tretinoin** – antineoplastic agents
Retinoid X receptor, beta	RXRB	**Bexarotene** – antineoplastic agents
Retinoid X receptor, gamma	RXRG	The retinoid X receptors and their ligands. PMID: 22020178
Nuclear receptor subfamily 2, group C, member 1	NR2C1	Roles of testicular orphan nuclear receptors 2 and 4 in early embryonic development and embryonic stem cells. PMID: 19131575
Nuclear receptor subfamily 2, group C, member 2	NR2C2	Evidence for orphan nuclear receptor TR4 in the etiology of Cushing disease. PMID: 23653479
Nuclear receptor subfamily 2, group E, member 1	NR2E1	The nuclear receptor TLX is required for gliomagenesis within the adult neurogenic niche. PMID: 23028043
Nuclear receptor subfamily 2, group E, member 3	NR2E3	Nr2e3-directed transcriptional regulation of genes involved in photoreceptor development and cell-type specific phototransduction. PMID: 19379737. Disease: enhanced S cone syndrome (ESCS)
Nuclear receptor subfamily 2, group F, member 1	NR2F1	COUP-TFII inhibits TGF-β-induced growth barrier to promote prostate tumorigenesis. PMID: 23201680
Nuclear receptor subfamily 2, group F, member 2	NR2F2	COUP-TFII orchestrates venous and lymphatic endothelial identity by homo- or hetero-dimerisation with PROX1. PMID: 23345397
Nuclear receptor subfamily 2, group F, member 6	NR2F6	Nuclear orphan receptor NR2F6 directly antagonizes NFAT and RORγt binding to the Il17a promoter. PMID: 22921335

HGNC-approved name	Symbol	Target information
Estrogen receptor 1	ESR1	**Norgestrel** – contraceptives. Disease: estrogen resistance (ESTRR)
Estrogen receptor 2 (ER beta)	ESR2	**Tamoxifen** – antineoplastic agents
Estrogen-related receptor alpha	ESRRA	Molecular pathways: The metabolic regulator estrogen-related receptor α as a therapeutic target in cancer. PMID: 23019305
Estrogen-related receptor beta	ESRRB	Pluripotency re-centered around Esrrb. PMID: 23064149. Disease: deafness, autosomal recessive, 35 (DFNB35)
Estrogen-related receptor gamma	ESRRG	ERRgamma suppresses cell proliferation and tumor growth of androgen-sensitive and androgen-insensitive prostate cancer cells and its implication as a therapeutic target for prostate cancer. PMID: 17510420
Androgen receptor	AR	**Enzalutamide** – antineoplastic agents. Disease: androgen insensitivity syndrome (AIS)
Nuclear receptor subfamily 3, group C, member 1 (glucocorticoid receptor)	NR3C1	**Methylprednisolone** – anti-inflammatory and antirheumatic products. Disease: glucocorticoid resistance (GCRES)
Nuclear receptor subfamily 3, group C, member 2	NR3C2	**Spironolactone** – diuretics. Disease: pseudohypoaldosteronism 1, autosomal dominant (PHA1A)
Progesterone receptor	PGR	**Dydrogesterone** – antidysmennorheal agents
Nuclear receptor subfamily 4, group A, member 1	NR4A1	Activation of nuclear TR3 (NR4A1) by a diindolylmethane analog induces apoptosis and proapoptotic genes in pancreatic cancer cells and tumors. PMID: 21362629
Nuclear receptor subfamily 4, group A, member 2	NR4A2	Structure-dependent activation of NR4A2 (Nurr1) by 1,1-bis(3′-indolyl)-1-(aromatic)methane analogs in pancreatic cancer cells. PMID: 22405837
Nuclear receptor subfamily 4, group A, member 3	NR4A3	6-Mercaptopurine augments glucose transport activity in skeletal muscle cells in part via a mechanism dependent upon orphan nuclear receptor NR4A3. PMID: 24022864. Disease: Ewing sarcoma (ES)
Nuclear receptor subfamily 5, group A, member 1	NR5A1	Distinct functions of steroidogenic factor-1 (NR5A1) in the nucleus and the centrosome. PMID: 23219870. Disease: 46,XY sex reversal 3 (SRXY3)
Nuclear receptor subfamily 5, group A, member 2	NR5A2	Structure-based discovery of antagonists of nuclear receptor LRH-1. PMID: 23667258
Nuclear receptor subfamily 6, group A, member 1	NR6A1	Expression screening of cancer/testis genes in prostate cancer identifies NR6A1 as a novel marker of disease progression and aggressiveness. PMID: 23532770
Nuclear receptor subfamily 0, group B, member 1	NR0B1	Tumorigenic role of orphan nuclear receptor NR0B1 in lung adenocarcinoma. PMID: 19644015. Disease: X-linked adrenal hypoplasia congenital (XL-AHC)
Nuclear receptor subfamily 0, group B, member 2	NR0B2	Adamantyl-substituted retinoid-related molecules bind small heterodimer partner and modulate the Sin3A repressor. PMID: 17210713. Disease: obesity (OBESITY)

3.3 Cytokines and receptors

HGNC-approved name	Symbol	Target information
Activin A receptor, type I[a]	ACVR1	**LDN 212854** – drugs affecting bone structure and mineralisation. Disease: fibrodysplasia ossificans progressiva (FOP)
Activin A receptor, type IB[a]	ACVR1B	Note: ACVRIB is abundantly expressed in systemic sclerosis patient fibroblasts and production of collagen is also induced by activin-A/INHBA. This suggests that the activin/ACRV1B signaling mechanism is involved in systemic sclerosis

(Continued)

HGNC-approved name	Symbol	Target information
Activin A receptor, type IC[a]	ACVR1C	Roles of activin receptor-like kinase 7 signaling and its target, peroxisome proliferator-activated receptor γ, in lean and obese adipocytes. PMID: 24052900
Activin A receptor, type IIA[a]	ACVR2A	**Sotatercept** – antianaemic preparations
Activin A receptor, type IIB[a]	ACVR2B	**Bimagrumab** – sporadic inclusion body myositis treatments. Disease: heterotaxy, visceral, 4, autosomal (HTX4)
Activin A receptor type II-like 1[a]	ACVRL1	ALK1 as an emerging target for antiangiogenic therapy of cancer. PMID: 21467543. Disease: telangiectasia, hereditary hemorrhagic, 2 (HHT2)
Adiponectin, C1Q and collagen domain containing	ADIPOQ	Adiponectin isoforms: A potential therapeutic target in rheumatoid arthritis? PMID: 22532632. Disease: adiponectin deficiency (ADPND)
Adiponectin receptor 1	ADIPOR1	A small-molecule AdipoR agonist for type 2 diabetes and short life in obesity. PMID: 24172895
Adiponectin receptor 2	ADIPOR2	A small-molecule AdipoR agonist for type 2 diabetes and short life in obesity. PMID: 24172895
Angiopoietin 1	ANGPT1	**Trebananib** – antineoplastic agents
Angiopoietin 2	ANGPT2	**Nesvacumab** – antineoplastic agents
Angiopoietin 4	ANGPT4	Angiopoietin-4 promotes glioblastoma progression by enhancing tumor cell viability and angiogenesis. PMID: 20823154
TEK tyrosine kinase, endothelial[a]	TEK	**Rebastinib** – antineoplastic agents. Disease: dominantly inherited venous malformations (VMCM)
Angiopoietin-like 1	ANGPTL1	Angiopoietin-like protein 1 suppresses SLUG to inhibit cancer cell motility. PMID: 23434592
Angiopoietin-like 2	ANGPTL2	ANGPTL2 promotes tumor metastasis in hepatocellular carcinoma. PMID: 25090954
Angiopoietin-like 3	ANGPTL3	Angiopoietin-like proteins 3, 4 and 8: Regulating lipid metabolism and providing new hope for metabolic syndrome. PMID: 24960069
Angiopoietin-like 4	ANGPTL4	Angiopoietin-like 4 protein elevates the prosurvival intracellular $O_2(-)$: H_2O_2 ratio and confers anoikis resistance to tumors. PMID: 21397862
Angiopoietin-like 5	ANGPTL5	Mesenchymal stem cells secreting angiopoietin-like-5 support efficient expansion of human hematopoietic stem cells without compromising their repopulating potential. PMID: 21142526
Angiopoietin-like 6	ANGPTL6	Evidence of a role of ANGPTL6 in resting metabolic rate and its potential application in treatment of obesity. PMID: 21460784
Angiopoietin-like 7	ANGPTL7	Angiopoietin-like 7, a novel pro-angiogenetic factor over-expressed in cancer. PMID: 24903490
Chromosome 19 open reading frame 80	C19orf80	Angiopoietin-like proteins 3, 4 and 8: Regulating lipid metabolism and providing new hope for metabolic syndrome. PMID: 24960069
Bone morphogenetic protein 2	BMP2	Bone morphogenetic protein 2: A potential new player in the pathogenesis of diabetic retinopathy. PMID: 24910902
Bone morphogenetic protein 3	BMP3	BMP3 suppresses osteoblast differentiation of bone marrow stromal cells via interaction with Acvr2b. PMID: 22074949
Bone morphogenetic protein 4	BMP4	Bone morphogenetic protein-4: A novel therapeutic target for pathological cardiac hypertrophy/heart failure. PMID: 24736806. Disease: microphthalmia, syndromic, 6 (MCOPS6)
Bone morphogenetic protein 5	BMP5	Whole-exome sequencing identifies a polymorphism in the BMP5 gene associated with SSRI treatment response in major depression. PMID: 23926243
Bone morphogenetic protein 6	BMP6	Enhanced expression of BMP6 inhibits hepatic fibrosis in non-alcoholic fatty liver disease. PMID: 25011936

HGNC-approved name	Symbol	Target information
Bone morphogenetic protein 7	BMP7	BMP4 and BMP7 induce the white-to-brown transition of primary human adipose stem cells. PMID: 24284793
Bone morphogenetic protein 8a	BMP8A	Genome-wide copy number variation analysis identifies deletion variants associated with ankylosing spondylitis. PMID: 24692264
Bone morphogenetic protein 8b	BMP8B	BMP8B mediates the survival of pancreatic cancer cells and regulates the progression of pancreatic cancer. PMID: 25176058
Bone morphogenetic protein 10	BMP10	**Dalantercept** – antineoplastic agents
Bone morphogenetic protein 15	BMP15	Disease: ovarian dysgenesis 2 (ODG2)
Bone morphogenetic protein receptor, type IA[a]	BMPR1A	DMH1, a small molecule inhibitor of BMP type i receptors, suppresses growth and invasion of lung cancer. PMID: 24603907. Disease: juvenile polyposis syndrome (JPS)
Bone morphogenetic protein receptor, type IB[a]	BMPR1B	Endogenous BMPR-IB signaling is required for early osteoblast differentiation of human bone cells. PMID: 21136190. Disease: acromesomelic chondrodysplasia, with genital anomalies (AMDGA)
Bone morphogenetic protein receptor, type II (serine/threonine kinase)[a]	BMPR2	Disease: pulmonary hypertension, primary, 1 (PPH1)
Brain-derived neurotrophic factor	BDNF	BDNF repairs podocyte damage by microRNA-mediated increase of actin polymerization. PMID: 25408545
Cardiotrophin-like cytokine factor 1	CLCF1	Inactivation of cardiotrophin-like cytokine, a second ligand for ciliary neurotrophic factor receptor, leads to cold-induced sweating syndrome in a patient. PMID: 16782820
Neurotrophin 3	NTF3	A promoter polymorphism of neurotrophin 3 gene is associated with curve severity and bracing effectiveness in adolescent idiopathic scoliosis. PMID: 22158057
Neurotrophin 4	NTF4	Neurotrophin-4 regulates the survival of gustatory neurons earlier in development using a different mechanism than brain-derived neurotrophic factor. PMID: 22353733
Ciliary neurotrophic factor	CNTF	Ciliary neurotrophic factor (CNTF) activation of astrocytes decreases spreading depolarization susceptibility and increases potassium clearance. PMID: 25092804
Ciliary neurotrophic factor receptor	CNTFR	Humanin inhibits neuronal cell death by interacting with a cytokine receptor complex or complexes involving CNTF receptor alpha/WSX-1/gp130. PMID: 19386761
Cytokine receptor-like factor 1	CRLF1	Cytokine-like factor 1 gene expression is enriched in idiopathic pulmonary fibrosis and drives the accumulation of CD4+ T cells in murine lungs: Evidence for an antifibrotic role in bleomycin injury. PMID: 22429962. Disease: cold-induced sweating syndrome 1 (CISS1)
Thymic stromal lymphopoietin	TSLP	**MEDI9929** – drugs for obstructive airway diseases
Cytokine receptor-like factor 2	CRLF2	Anti-CRLF2 antibody-armored biodegradable nanoparticles for childhood B-ALL. PMID: 23976822
Cytokine receptor-like factor 3	CRLF3	Uncovering genes required for neuronal morphology by morphology-based gene trap screening with a revertible retrovirus vector. PMID: 22874834
Colony-stimulating factor 1 (macrophage)	CSF1	Note: Aberrant expression of CSF1 or CSF1R can promote cancer cell proliferation, invasion and formation of metastases. Overexpression of CSF1 or CSF1R is observed in a significant percentage of breast, ovarian, prostate, and endometrial cancers
Colony-stimulating factor 2 (granulocyte–macrophage)	CSF2	**Namilumab** – immunomodulating agents
Colony-stimulating factor 3 (granulocyte)	CSF3	**Pegfilgrastim** – immunostimulants

(Continued)

HGNC-approved name	Symbol	Target information
Colony-stimulating factor 1 receptor[a]	CSF1R	**AMG 820** – antineoplastic agents. Note: Aberrant expression of CSF1 or CSF1R can promote cancer cell proliferation, invasion and formation of metastases. Overexpression of CSF1 or CSF1R is observed in a significant percentage of breast, ovarian, prostate, and endometrial cancers
Colony-stimulating factor 2 receptor, alpha, low-affinity (granulocyte–macrophage)	CSF2RA	**Mavrilimumab** – anti-inflammatory and antirheumatic products. Disease: pulmonary surfactant metabolism dysfunction 4 (SMDP4)
Colony-stimulating factor 2 receptor, beta, low-affinity (granulocyte–macrophage)	CSF2RB	Disease: pulmonary surfactant metabolism dysfunction 5 (SMDP5)
Colony-stimulating factor 3 receptor (granulocyte)	CSF3R	Disease: hereditary neutrophilia (NEUTROPHILIA)
Connective tissue growth factor	CTGF	**EXC 001** – dermatologicals
Cytokine-like 1	CYTL1	C17 prevents inflammatory arthritis and associated joint destruction in mice. PMID: 21799806
Ectodysplasin A	EDA	Generation and characterization of function-blocking anti-ectodysplasin A (EDA) monoclonal antibodies that induce ectodermal dysplasia. PMID: 24391090. Disease: ectodermal dysplasia 1, hypohidrotic, X-linked (XHED)
Ectodysplasin A receptor	EDAR	Molecular and therapeutic characterization of anti-ectodysplasin A receptor (EDAR) agonist monoclonal antibodies. PMID: 21730053
Ectodysplasin A2 receptor	EDA2R	Crosstalk of EDA-A2/XEDAR in the p53 signaling pathway. PMID: 20501644
EGF-like-domain, multiple 8	EGFL8	Down-regulation of EGFL8: A novel prognostic biomarker for patients with colorectal cancer. PMID: 21737648
Ephrin-A1	EFNA1	Association between ephrin-A1 mRNA expression and poor prognosis after hepatectomy to treat hepatocellular carcinoma. PMID: 24969670
Ephrin-A2	EFNA2	**siRNA-EphA2-DOPC** – antineoplastic agents
Ephrin-A3	EFNA3	MicroRNA-210 promotes proliferation and invasion of peripheral nerve sheath tumor cells targeting EFNA3. PMID: 24512729
Ephrin-A4	EFNA4	An impaired transendothelial migration potential of chronic lymphocytic leukemia (CLL) cells can be linked to ephrin-A4 expression. PMID: 19828693
Ephrin-A5	EFNA5	Hypothalamic EphA5 facilitates counterregulatory responses: Possible role for bidirectional signaling leading to bistability that enhances responsiveness to hypoglycemia. PMID: 23520275
Ephrin-B1	EFNB1	EphrinB1 expression is dysregulated and promotes oncogenic signaling in medulloblastoma. PMID: 25258252. Disease: craniofrontonasal syndrome (CFNS)
Ephrin-B2	EFNB2	EphrinB2 signaling in osteoblasts promotes bone mineralization by preventing apoptosis. PMID: 24982128
Ephrin-B3	EFNB3	EphB3 is overexpressed in non-small-cell lung cancer and promotes tumor metastasis by enhancing cell survival and migration. PMID: 21266352
EPH receptor A1[a]	EPHA1	Lack of ephrin receptor A1 is a favorable independent prognostic factor in clear cell renal cell carcinoma. PMID: 25025847
EPH receptor A2[a]	EPHA2	**MEDI-547** – antineoplastic agents. Disease: cataract 6, multiple types (CTRCT6)
EPH receptor A3[a]	EPHA3	**KB004** – antineoplastic agents. Disease: colorectal cancer (CRC)
EPH receptor A4[a]	EPHA4	EPHA4 is a disease modifier of amyotrophic lateral sclerosis in animal models and in humans. PMID: 22922411
EPH receptor A5[a]	EPHA5	After repeated division, bone marrow stromal cells express inhibitory factors with osteogenic capabilities, and EphA5 is a primary candidate. PMID: 24029132

HGNC-approved name	Symbol	Target information
EPH receptor A6[a]	EPHA6	Learning and memory impairment in Eph receptor A6 knockout mice. PMID: 18450376
EPH receptor A7[a]	EPHA7	Mining the cancer genome uncovers therapeutic activity of EphA7 against lymphoma. PMID: 22356769
EPH receptor A8[a]	EPHA8	Expression of EphA8-Fc in transgenic mouse embryos induces apoptosis of neural epithelial cells during brain development. PMID: 23696555
EPH receptor A10[a]	EPHA10	Ephrin receptor A10 is a promising drug target potentially useful for breast cancers including triple negative breast cancers. PMID: 24946238
EPH receptor B1[a]	EPHB1	Ligand-dependent EphB1 signaling suppresses glioma invasion and correlates with patient survival. PMID: 24121831
EPH receptor B2[a]	EPHB2	Overexpression of EPH receptor B2 in malignant mesothelioma correlates with oncogenic behavior. PMID: 23887168. Disease: prostate cancer (PC)
EPH receptor B3[a]	EPHB3	Silencing of the EPHB3 tumor-suppressor gene in human colorectal cancer through decommissioning of a transcriptional enhancer. PMID: 24707046
EPH receptor B4[a]	EPHB4	**EphB4-HSA fusion protein** – antineoplastic agents
EPH receptor B6[a]	EPHB6	Mutations of the EPHB6 receptor tyrosine kinase induce a pro-metastatic phenotype in non-small cell lung cancer. PMID: 23226491
Epidermal growth factor	EGF	Disease: hypomagnesemia 4 (HOMG4)
Amphiregulin	AREG	Amphiregulin induces human ovarian cancer cell invasion by down-regulating E-cadherin expression. PMID: 25261255
Epiregulin	EREG	Oncogenic KRAS-induced epiregulin overexpression contributes to aggressive phenotype and is a promising therapeutic target in non-small-cell lung cancer. PMID: 22964644
Betacellulin	BTC	The ABC of BTC: Structural properties and biological roles of betacellulin. PMID: 24440602
Epithelial mitogen	EPGN	Structure and function of epigen, the last EGFR ligand. PMID: 24374012
Heparin-binding EGF-like growth factor	HBEGF	Autocrine HBEGF expression promotes breast cancer intravasation, metastasis and macrophage-independent invasion in vivo. PMID: 24013225
Epidermal growth factor receptor (see also protein kinases)[a]	EGFR	**Cetuximab** – antineoplastic agents. Disease: lung cancer (LNCR)
Erythropoietin	EPO	**Epoetin alfa** – antianaemic preparations. Disease: microvascular complications of diabetes 2 (MVCD2)
Erythropoietin receptor	EPOR	**ARA 290** – analgesics. Disease: erythrocytosis, familial, 1 (ECYT1)
Fibroblast growth factor 1 (acidic)	FGF1	Endocrinization of FGF1 produces a neomorphic and potent insulin sensitizer. PMID: 25043058
Fibroblast growth factor 2 (basic)	FGF2	Fibroblast growth factor-2 alone as an efficient inducer for differentiation of human bone marrow mesenchymal stem cells into dopaminergic neurons. PMID: 25248378
Fibroblast growth factor 3	FGF3	Disease: deafness with labyrinthine aplasia, microtia and microdontia (LAMM)
Fibroblast growth factor 4	FGF4	**Ad5FGF-4** – cardiac therapy
Fibroblast growth factor 5	FGF5	Decapeptide with fibroblast growth factor (FGF)-5 partial sequence inhibits hair growth suppressing activity of FGF-5. PMID: 14502567
Fibroblast growth factor 6	FGF6	Regulation of osteoblast and osteoclast functions by FGF-6. PMID: 20458746
Fibroblast growth factor 7	FGF7	FGF7/KGF regulates autophagy in keratinocytes: A novel dual role in the induction of both assembly and turnover of autophagosomes. PMID: 24577098

(Continued)

HGNC-approved name	Symbol	Target information
Fibroblast growth factor 8 (androgen induced)	FGF8	FGF8 activates proliferation and migration in mouse post-natal oligodendrocyte progenitor cells. PMID: 25259688
Fibroblast growth factor 9	FGF9	Fgf9 from dermal γδ T cells induces hair follicle neogenesis after wounding. PMID: 23727932. Disease: multiple synostoses syndrome 3 (SYNS3)
Fibroblast growth factor 10	FGF10	FGF10: Type III epithelial mesenchymal transition and invasion in breast cancer cell lines. PMID: 25057305
Fibroblast growth factor 11	FGF11	Infiltrating T cells promote prostate cancer metastasis via modulation of FGF11 → miRNA-541 → androgen receptor (AR) → MMP9 signaling. PMID: 25135278
Fibroblast growth factor 12	FGF12	Involvement of intracellular expression of FGF12 in radiation-induced apoptosis in mast cells. PMID: 18525161
Fibroblast growth factor 13	FGF13	Upregulated expression of FGF13/FHF2 mediates resistance to platinum drugs in cervical cancer cells. PMID: 24113164
Fibroblast growth factor 14	FGF14	FGF14 regulates presynaptic Ca^{2+} channels and synaptic transmission. PMID: 23831029
Fibroblast growth factor 16	FGF16	FGF16 promotes invasive behavior of SKOV-3 ovarian cancer cells through activation of mitogen-activated protein kinase (MAPK) signaling pathway. PMID: 24253043
Fibroblast growth factor 17	FGF17	Abnormal social behaviors in mice lacking Fgf17. PMID: 17908176
Fibroblast growth factor 18	FGF18	Intra-articular injection of rhFGF-18 improves the healing in microfracture treated chondral defects in an ovine model. PMID: 24436147
Fibroblast growth factor 19	FGF19	FGF19 action in the brain induces insulin-independent glucose lowering. PMID: 24084738
Fibroblast growth factor 20	FGF20	Roles of FGF20 in dopaminergic neurons and Parkinson's disease. PMID: 23754977
Fibroblast growth factor 21	FGF21	**LY2405319** – drugs used in diabetes
Fibroblast growth factor 22	FGF22	Fibroblast growth factor 22 is not essential for skin development and repair but plays a role in tumorigenesis. PMID: 22737238
Fibroblast growth factor 23	FGF23	KRN23 X-linked hypophosphataemic rickets. Disease: hypophosphatemic rickets, autosomal dominant (ADHR)
Fibroblast growth factor receptor 1[a]	FGFR1	**GSK3052230** – antineoplastic agents. Disease: Pfeiffer syndrome (PS)
Fibroblast growth factor receptor 2[a]	FGFR2	**Palifermin** – antimucositis agents. Disease: Crouzon syndrome (CS)
Fibroblast growth factor receptor 3[a]	FGFR3	**Dovitinib** – antineoplastic agents. Disease: achondroplasia (ACH)
Fibroblast growth factor receptor 4[a]	FGFR4	**ISIS-FGFR4Rx** – antiobesity preparations
Fibroblast growth factor receptor-like 1	FGFRL1	The FgfrL1 receptor is required for development of slow muscle fibers. PMID: 25172430
Fibroblast growth factor binding protein 1	FGFBP1	Anti-tumor effects of fibroblast growth factor-binding protein (FGF-BP) knockdown in colon carcinoma. PMID: 22111880
Fibroblast growth factor binding protein 2	FGFBP2	POLD2 and KSP37 (FGFBP2) correlate strongly with histology, stage and outcome in ovarian carcinomas. PMID: 21079801
Fibroblast growth factor binding protein 3	FGFBP3	Inactivation of fibroblast growth factor binding protein 3 causes anxiety-related behaviors. PMID: 20851768
Gastrokine 1	GKN1	The role of gastrokine 1 in gastric cancer. PMID: 25328759
Gastrokine 2	GKN2	GKN2 contributes to the homeostasis of gastric mucosa by inhibiting GKN1 activity. PMID: 24151046
Glial cell-derived neurotrophic factor	GDNF	Disease: Hirschsprung disease 3 (HSCR3)
Cerebral dopamine neurotrophic factor	CDNF	Gene therapy with AAV2-CDNF provides functional benefits in a rat model of Parkinson's disease. PMID: 23532969

HGNC-approved name	Symbol	Target information
Artemin	ARTN	Neurotrophic factor artemin promotes invasiveness and neurotrophic function of pancreatic adenocarcinoma in vivo and *in vitro*. PMID: 25242023
Neurturin	NRTN	The neurotrophic factor neurturin contributes toward an aggressive cancer cell phenotype, neuropathic pain and neuronal plasticity in pancreatic cancer. PMID: 24067900. Note: Genetic variations in NRTN may contribute to Hirschsprung disease, in association with mutations of RET gene, and possibly mutations in other loci
Persephin	PSPN	Persephin: A potential key component in human oral cancer progression through the RET receptor tyrosine kinase-mitogen-activated protein kinase signaling pathway. PMID: 24375483
GDNF family receptor alpha 1	GFRA1	GFRα1 released by nerves enhances cancer cell perineural invasion through GDNF-RET signaling. PMID: 24778213
GDNF family receptor alpha 2	GFRA2	A genome-wide association study suggests an association of Chr8p21.3 (GFRA2) with diabetic neuropathic pain. PMID: 24974787
GDNF family receptor alpha 3	GFRA3	**Neublastin** – analgesia
GDNF family receptor alpha 4	GFRA4	A model for GFR alpha 4 function and a potential modifying role in multiple endocrine neoplasia 2. PMID: 15592530
Growth differentiation factor 1	GDF1	Disease: conotruncal heart malformations (CTHM)
Growth differentiation factor 2	GDF2	**Dalantercept** – antineoplastic agents. Disease: telangiectasia, hereditary hemorrhagic, 5 (HHT5)
Growth differentiation factor 3	GDF3	Growth and differentiation factor 3 induces expression of genes related to differentiation in a model of cancer stem cells and protects them from retinoic acid-induced apoptosis. PMID: 23950971. Disease: Klippel–Feil syndrome 3, autosomal dominant (KFS3)
Growth differentiation factor 5	GDF5	Disease: acromesomelic chondrodysplasia, Grebe type (AMDG)
Growth differentiation factor 6	GDF6	Disease: Klippel–Feil syndrome 1, autosomal dominant (KFS1)
Growth differentiation factor 7	GDF7	BMP12 induces tenogenic differentiation of adipose-derived stromal cells. PMID: 24155967
Myostatin	MSTN	**Stamulumab** – muscular dystrophy treatments. Disease: muscle hypertrophy (MSLHP)
Growth differentiation factor 9	GDF9	Note: Altered GDF9 function may be involved in ovarian disorders. Rare variants in GDF9 have been found in patients with premature ovarian failure and mothers of dizygotic twins
Growth differentiation factor 10	GDF10	Runx2 mediates epigenetic silencing of the bone morphogenetic protein-3B (BMP-3B/GDF10) in lung cancer cells. PMID: 22537242
Growth differentiation factor 11	GDF11	Restoring systemic GDF11 levels reverses age-related dysfunction in mouse skeletal muscle. PMID: 24797481
Growth differentiation factor 15	GDF15	Sustained elevation of circulating growth and differentiation factor-15 and a dynamic imbalance in mediators of muscle homeostasis are associated with the development of acute muscle wasting following cardiac surgery. PMID: 23328263
Growth hormone 1	GH1	Disease: growth hormone deficiency, isolated, 1A (IGHD1A)
Growth hormone 2	GH2	Developmental programming of growth: Genetic variant in GH2 gene encoding placental growth hormone contributes to adult height determination. PMID: 24035309
Chorionic somatomammotropin hormone-like 1	CSHL1	Differential expression profile of growth hormone/chorionic somatomammotropin genes in placenta of small- and large-for-gestational-age newborns. PMID: 20233782
Growth hormone receptor	GHR	**Pegvisomant** – pituitary and hypothalamic hormones and analogues. Disease: Laron syndrome (LARS)

(Continued)

HGNC-approved name	Symbol	Target information
Hepatocyte growth factor (hepapoietin A; scatter factor)	HGF	**Ficlatuzumab** – antineoplastic agents. Disease: deafness, autosomal recessive, 39 (DFNB39)
Met proto-oncogene[a]	MET	**Onartuzumab** – antineoplastic agents. Note: Activation of MET after rearrangement with the TPR gene produces an oncogenic protein. Note: Defects in MET may be associated with gastric cancer. Disease: hepatocellular carcinoma (HCC)
Hepatoma-derived growth factor	HDGF	Antibodies targeting hepatoma-derived growth factor as a novel strategy in treating lung cancer. PMID: 19435872
Hepatoma-derived growth factor-like 1	HDGFL1	
IK cytokine, down-regulator of HLA II	IK	The interferon inhibiting cytokine IK is overexpressed in cutaneous T cell lymphoma derived tumor cells that fail to upregulate major histocompatibility complex class II upon interferon-gamma stimulation. PMID: 11407974
Inhibin, alpha	INHA	Inhibin alpha-subunit (INHA) expression in adrenocortical cancer is linked to genetic and epigenetic INHA promoter variation. PMID: 25111790
Inhibin, beta A	INHBA	Significance of INHBA expression in human colorectal cancer. PMID: 24085226
Inhibin, beta B	INHBB	Endometrial inhibin/activin beta-B subunit expression is related to decidualization and is reduced in tubal ectopic pregnancy. PMID: 18381568
Inhibin, beta C	INHBC	The inhibin-βC subunit is down-regulated, while inhibin-βE is up-regulated by interferon-β1a in Ishikawa carcinoma cell line. PMID: 23580013
Inhibin, beta E	INHBE	The inhibin-βC subunit is down-regulated, while inhibin-βE is up-regulated by interferon-β1a in Ishikawa carcinoma cell line. PMID: 23580013
Insulin	INS	Disease: hyperproinsulinemia, familial (FHPRI)
Insulin-like 3 (Leydig cell)	INSL3	Disease: cryptorchidism (CRYPTO)
Insulin-like 4 (placenta)	INSL4	Early placental insulin-like protein (INSL4 or EPIL) in placental and fetal membrane growth. PMID: 15958731
Insulin-like 5	INSL5	Insulin-like peptide 5 is an orexigenic gastrointestinal hormone. PMID: 25028498
Insulin-like 6	INSL6	Insulin-like 6 is induced by muscle injury and functions as a regenerative factor. PMID: 20807758
Insulin receptor[a]	INSR	**Linsitinib** – antineoplastic agents. Disease: Rabson–Mendenhall syndrome (RMS)
Insulin-like growth factor 1 (somatomedin C)	IGF1	**BI 836845** – antineoplastic agents. Disease: insulin-like growth factor I deficiency (IGF1 deficiency)
Insulin-like growth factor 2 (somatomedin A)	IGF2	Insulin-like growth factor 2 silencing restores taxol sensitivity in drug resistant ovarian cancer. PMID: 24932685
Insulin-like growth factor 1 receptor[a]	IGF1R	**Ganitumab** – antineoplastic agents. Disease: insulin-like growth factor 1 resistance (IGF1RES)
Insulin-like growth factor 2 receptor	IGF2R	Knockdown of IGF2R suppresses proliferation and induces apoptosis in hemangioma cells *in vitro* and in vivo. PMID: 24968760
Insulin-like growth factor binding protein, acid-labile subunit	IGFALS	IGFALS gene dosage effects on serum IGF-I and glucose metabolism, body composition, bone growth in length and width, and the pharmacokinetics of recombinant human IGF-I administration. PMID: 24423360
Insulin-like growth factor binding protein 1	IGFBP1	Insulin-like growth factor binding protein-1 inhibits cancer cell invasion and is associated with poor prognosis in hepatocellular carcinoma. PMID: 25337205
Insulin-like growth factor binding protein 2, 36 kDa	IGFBP2	**OGX-225** – antineoplastic agents

HGNC-approved name	Symbol	Target information
Insulin-like growth factor binding protein 3	IGFBP3	Insulin-like growth factor-binding protein-3 (IGFBP-3) blocks the effects of asthma by negatively regulating NF-κB signaling through IGFBP-3R-mediated activation of caspases. PMID: 21383009
Insulin-like growth factor binding protein 4	IGFBP4	Control of adipose tissue expandability in response to high fat diet by the insulin-like growth factor-binding protein-4. PMID: 24778188
Insulin-like growth factor binding protein 5	IGFBP5	**OGX-225** – antineoplastic agents
Insulin-like growth factor binding protein 6	IGFBP6	Serum insulin-like growth factor binding protein 6 (IGFBP6) is increased in patients with type 1 diabetes and its complications. PMID: 22837797
Insulin-like growth factor binding protein 7	IGFBP7	Disease: retinal arterial macroaneurysm with supravalvular pulmonic stenosis (RAMSVPS)
Insulin-like growth factor binding protein-like 1	IGFBPL1	Epigenetic inactivation implies independent functions for insulin-like growth factor binding protein (IGFBP)-related protein 1 and the related IGFBPL1 in inhibiting breast cancer phenotypes. PMID: 17634530
IGF-like family member 1	IGFL1	IGFL: A secreted family with conserved cysteine residues and similarities to the IGF superfamily. PMID: 16890402
IGF-like family member 2	IGFL2	
IGF-like family member 3	IGFL3	
IGF-like family member 4	IGFL4	
IGF-like family receptor 1	IGFLR1	Murine insulin growth factor-like (IGFL) and human IGFL1 proteins are induced in inflammatory skin conditions and bind to a novel tumor necrosis factor receptor family member, IGFLR1. PMID: 21454693
Interferon, alpha 1	IFNA1	**Sifalimumab** – selective immunosuppressants
Interferon, alpha 2	IFNA2	**Peginterferon alfa-2b** – antineoplastic agents
Interferon, alpha 4	IFNA4	
Interferon, alpha 5	IFNA5	Genetic association of interferon-alpha subtypes 1, 2 and 5 in systemic lupus erythematosus. PMID: 19000144
Interferon, alpha 6	IFNA6	
Interferon, alpha 7	IFNA7	
Interferon, alpha 8	IFNA8	
Interferon, alpha 10	IFNA10	
Interferon, alpha 13	IFNA13	
Interferon, alpha 14	IFNA14	
Interferon, alpha 16	IFNA16	
Interferon, alpha 17	IFNA17	
Interferon, alpha 21	IFNA21	
Interferon, beta 1, fibroblast	IFNB1	**Interferon Beta-1B** – immunomodulating agents
Interferon, gamma	IFNG	**Fontolizumab** – immunomodulating agents. Disease: aplastic anemia (AA)
Interferon, epsilon	IFNE	
Interferon, kappa	IFNK	The role of genetic variation near interferon-kappa in systemic lupus erythematosus. PMID: 20706608
Interferon, lambda 1	IFNL1	Regulation of interferon lambda-1 (IFNL1/IFN-λ1/IL-29) expression in human colon epithelial cells. PMID: 24140069
Interferon, lambda 2	IFNL2	Interleukin-28A enhances autoimmune disease in a retinal autoimmunity model. PMID: 25138017
Interferon, lambda 3	IFNL3	Hydrodynamic delivery of IL-28B (IFN-λ3) gene ameliorates lung inflammation induced by cigarette smoke exposure in mice. PMID: 24732350

(Continued)

HGNC-approved name	Symbol	Target information
Interferon, lambda 4 (gene/pseudogene)	IFNL4	A variant upstream of IFNL3 (IL28B) creating a new interferon gene IFNL4 is associated with impaired clearance of hepatitis C virus. PMID: 23291588
Interferon, omega 1	IFNW1	
Interferon (alpha, beta and omega) receptor 1	IFNAR1	**Anifrolumab** – immunomodulating agents
Interferon (alpha, beta and omega) receptor 2	IFNAR2	Interferon-α acts on the S/G2/M phases to induce apoptosis in the G1 phase of an IFNAR2-expressing hepatocellular carcinoma cell line. PMID: 25012666
Interferon gamma receptor 1	IFNGR1	IFN-γ Rα is a key determinant of CD8+ T cell-mediated tumor elimination or tumor escape and relapse in FVB mouse. PMID: 24324806
Interferon gamma receptor 2 (interferon gamma transducer 1)	IFNGR2	Partial IFN-γR2 deficiency is due to protein misfolding and can be rescued by inhibitors of glycosylation. PMID: 23963039
Interferon, lambda receptor 1	IFNLR1	Role of the interleukin (IL)-28 receptor tyrosine residues for antiviral and antiproliferative activity of IL-29/interferon-lambda 1: Similarities with type I interferon signaling. PMID: 15166220
Interleukin 1, alpha	IL1A	**Rilonacept** – anti-inflammatory and antirheumatic products
Interleukin 1, beta	IL1B	**Canakinumab** – anti-inflammatory and antirheumatic products
Interleukin 1 family, member 10 (theta)	IL1F10	
Interleukin 2	IL2	Note: A chromosomal aberration involving IL2 is found in a form of T-cell acute lymphoblastic leukemia (T-ALL). Translocation t(416)(q26p13) with involves TNFRSF17
Interleukin 3 (colony-stimulating factor, multiple)	IL3	
Interleukin 4	IL4	Disease: ischemic stroke (ISCHSTR)
Interleukin 5 (colony-stimulating factor, eosinophil)	IL5	**Mepolizumab** – drugs for obstructive airway diseases
Interleukin 6 (interferon, beta 2)	IL6	Disease: rheumatoid arthritis systemic juvenile (RASJ)
Interleukin 7	IL7	
Interleukin 9	IL9	**Enokizumab** – drugs for obstructive airway diseases
Interleukin 10	IL10	
Interleukin 11	IL11	
Interleukin 12A (natural killer cell stimulatory factor 1, cytotoxic lymphocyte maturation factor 1, p35)	IL12A	**Ustekinumab** – antipsoriatics
Interleukin 12B (natural killer cell stimulatory factor 2, cytotoxic lymphocyte maturation factor 2, p40)	IL12B	Disease: Mendelian susceptibility to mycobacterial disease (MSMD)
Interleukin 13	IL13	**Tralokinumab** – drugs for obstructive airway diseases. Disease: allergic rhinitis (ALRH)
Interleukin 15	IL15	**HuMax-IL15** – selective immunosuppressants
Interleukin 16	IL16	Serum interleukin-16 levels in patients with nasal polyposis. PMID: 22447494
Interleukin 17A	IL17A	**Secukinumab** – antipsoriatics
Interleukin 17B	IL17B	
Interleukin 17C	IL17C	
Interleukin 17D	IL17D	
Interleukin 17F	IL17F	Increased plasma IL-17F levels in rheumatoid arthritis patients are responsive to methotrexate, anti-TNF, and T cell costimulatory modulation. PMID: 25240765

HGNC-approved name	Symbol	Target information
Interleukin 18 (interferon gamma-inducing factor)	IL18	**rIL18BP** – anti-inflammatory and antirheumatic products
Interleukin 19	IL19	IL-19 reduces ligation-mediated neointimal hyperplasia by reducing vascular smooth muscle cell activation. PMID: 24814101
Interleukin 20	IL20	**NN8226** – anti-inflammatory and antirheumatic products
Interleukin 21	IL21	**NN8828** – anti-inflammatory and antirheumatic products
Interleukin 22	IL22	**Fezakinumab** – antipsoriatics
Interleukin 23, alpha subunit p19	IL23A	**Tildrakizumab** – antipsoriatics
Interleukin 24	IL24	MDA-7/IL-24: Multifunctional cancer killing cytokine. PMID: 25001534
Interleukin 25	IL25	IL-25 causes apoptosis of IL-25R-expressing breast cancer cells without toxicity to nonmalignant cells. PMID: 21490275
Interleukin 26	IL26	Bringing in the cavalry: IL-26 mediates neutrophil recruitment to the lungs. PMID: 25398102
Interleukin 27	IL27	Interleukin-27 as a potential therapeutic target for rheumatoid arthritis: Has the time come? PMID: 23877489
Interleukin 31	IL31	**IL-31 mAb** – agents for dermatitis
Interleukin 32	IL32	Towards a role of interleukin-32 in atherosclerosis. PMID: 23727326
Interleukin 33	IL33	Interleukin (IL)-33: New therapeutic target for atopic diseases. PMID: 25213717
Interleukin 34	IL34	Targeting IL-34 in chronic inflammation. PMID: 24906044
Epstein–Barr virus induced 3	EBI3	IL-35-producing B cells are critical regulators of immunity during autoimmune and infectious diseases. PMID: 24572363
Interleukin 36, alpha	IL36A	
Interleukin 36, beta	IL36B	
Interleukin 36, gamma	IL36G	IL-36γ sustains a proinflammatory self-amplifying loop with IL-17C in anti-TNF-induced psoriasiform skin lesions of patients with Crohn's disease. PMID: 25299544
Interleukin 37	IL37	IL-37 inhibits the production of inflammatory cytokines in peripheral blood mononuclear cells of patients with systemic lupus erythematosus: Its correlation with disease activity. PMID: 24629023
Interleukin 1 receptor, type I	IL1R1	**Anakinra** – anti-inflammatory and antirheumatic products
Interleukin 1 receptor, type II	IL1R2	
Interleukin 1 receptor-like 1	IL1RL1	Inhibition of interleukin-33 signaling attenuates the severity of experimental arthritis. PMID: 19248109
Interleukin 1 receptor-like 2	IL1RL2	Anti-IL-36R antibodies, potentially useful for the treatment of psoriasis: A patent evaluation of WO2013074569. PMID: 24456078
Interleukin 1 receptor antagonist	IL1RN	Disease: Microvascular complications of diabetes 4 (MVCD4)
Interleukin 1 receptor accessory protein	IL1RAP	Selective killing of candidate AML stem cells by antibody targeting of IL1RAP. PMID: 23479569
Interleukin 1 receptor accessory protein-like 1	IL1RAPL1	Disease: mental retardation, X-linked 21 (MRX21)
Interleukin 1 receptor accessory protein-like 2	IL1RAPL2	
Interleukin 2 receptor, alpha	IL2RA	**Daclizumab** – selective immunosuppressants. Disease: diabetes mellitus, insulin-dependent, 10 (IDDM10)
Interleukin 2 receptor, beta	IL2RB	
Interleukin 2 receptor, gamma	IL2RG	Disease: severe combined immunodeficiency X-linked T-cell-negative/B-cell-positive/NK-cell-negative (XSCID)

(Continued)

HGNC-approved name	Symbol	Target information
Interleukin 3 receptor, alpha (low affinity)	IL3RA	**Daniplestim** – chemotherapy adjuvant
Interleukin 4 receptor	IL4R	**Dupilumab** – drugs for obstructive airway diseases
Interleukin 5 receptor, alpha	IL5RA	**Benralizumab** – drugs for obstructive airway diseases
Interleukin 6 receptor	IL6R	**Tocilizumab** – anti-inflammatory and antirheumatic products
Interleukin 6 signal transducer (gp130, oncostatin M receptor)	IL6ST	Glycoprotein 130 cytokine signal as a therapeutic target against cardiovascular diseases. PMID: 22056652
Interleukin 7 receptor	IL7R	Interleukin 7 receptor alpha as a potential therapeutic target in transplantation. PMID: 19585222. Disease: severe combined immunodeficiency autosomal recessive T-cell-negative/B-cell-positive/NK-cell-positive (T(−)B(+)NK(+) SCID)
Interleukin 9 receptor	IL9R	Expression of IL-9 receptor alpha chain on human germinal center B cells modulates IgE secretion. PMID: 17919707
Interleukin 10 receptor, alpha	IL10RA	**Dekavil** – anti-inflammatory and antirheumatic products
Interleukin 10 receptor, beta	IL10RB	Disease: inflammatory bowel disease 25 (IBD25)
Interleukin 11 receptor, alpha	IL11RA	**Oprelvekin** – antithrombotic agents
Interleukin 11 receptor, beta	IL11RB	
Interleukin 12 receptor, beta 1	IL12RB1	The dichotomous pattern of IL-12r and IL-23R expression elucidates the role of IL-12 and IL-23 in inflammation. PMID: 24586521
Interleukin 12 receptor, beta 2	IL12RB2	
Interleukin 13 receptor, alpha 1	IL13RA1	Bone marrow-derived IL-13Rα1-positive thymic progenitors are restricted to the myeloid lineage. PMID: 22351937
Interleukin 13 receptor, alpha 2	IL13RA2	Interleukin-13 receptor alpha 2 is a novel therapeutic target for human adrenocortical carcinoma. PMID: 22570059
Interleukin 15 receptor, alpha	IL15RA	IL15RA drives antagonistic mechanisms of cancer development and immune control in lymphocyte-enriched triple-negative breast cancers. PMID: 24980552
Interleukin 17 receptor A	IL17RA	**Brodalumab** – antipsoriatics. Disease: candidiasis, familial, 5 (CANDF5)
Interleukin 17 receptor B	IL17RB	Autocrine/paracrine mechanism of interleukin-17B receptor promotes breast tumorigenesis through NF-κB-mediated antiapoptotic pathway. PMID: 23851503
Interleukin 17 receptor C	IL17RC	
Interleukin 17 receptor D	IL17RD	Disease: hypogonadotropic hypogonadism 18 with or without anosmia (HH18)
Interleukin 17 receptor E	IL17RE	
Interleukin 18 binding protein	IL18BP	The IL-18 antagonist IL-18-binding protein is produced in the human ovarian cancer microenvironment. PMID: 23873689
Interleukin 18 receptor 1	IL18R1	**Iboctadekin** – antineoplastic agents
Interleukin 18 receptor accessory protein	IL18RAP	Identification of IL18RAP/IL18R1 and IL12B as leprosy risk genes demonstrates shared pathogenesis between inflammation and infectious diseases. PMID: 23103228
Interleukin 20 receptor, alpha	IL20RA	**MDA-7/IL-24** – antineoplastic agents
Interleukin 20 receptor beta	IL20RB	
Interleukin 21 receptor	IL21R	**Denenokin** – antineoplastic agents. Disease: IL21R immunodeficiency (IL21RID)
Interleukin 22 receptor, alpha 1	IL22RA1	Interleukin-22: A likely target for treatment of autoimmune diseases. PMID: 2441829
Interleukin 22 receptor, alpha 2	IL22RA2	The multiple sclerosis risk gene IL22RA2 contributes to a more severe murine autoimmune neuroinflammation. PMID: 25008863

HGNC-approved name	Symbol	Target information
Interleukin 23 receptor	IL23R	IL23R polymorphisms influence phenotype and response to therapy in patients with ulcerative colitis. PMID: 24168842
Interleukin 27 receptor, alpha	IL27RA	A soluble form of IL-27Rα is a natural IL-27 antagonist. PMID: 24771852
Interleukin 31 receptor A	IL31RA	**CIM331** – agents for dermatitis. Disease: amyloidosis, primary localized cutaneous, 2 (PLCA2)
Interleukin 36 receptor antagonist	IL36RN	Disease: psoriasis generalized pustular (PSORP)
Single immunoglobulin and toll–interleukin 1 receptor (TIR) domain	SIGIRR	Therapeutic potential of SIGIRR in systemic lupus erythematosus. PMID: 23546688
KIT ligand	KITLG	Disease: familial progressive hyperpigmentation (FPH)
v-kit Hardy–Zuckerman 4 feline sarcoma viral oncogene homolog[a]	KIT	**Imatinib** – antineoplastic agents. Disease: Piebald trait (PBT)
Leptin	LEP	**Metreleptin** – lipid-modifying agents. Disease: leptin deficiency (LEPD)
Leptin receptor	LEPR	**Metreleptin** – lipid-modifying agents. Disease: leptin receptor deficiency (LEPRD)
Leukemia inhibitory factor	LIF	LIF negatively regulates tumour-suppressor p53 through Stat3/ID1/MDM2 in colorectal cancers. PMID: 25323535
Cardiotrophin 1	CTF1	Cardiotrophin-1 (CTF1) ameliorates glucose-uptake defects and improves memory and learning deficits in a transgenic mouse model of Alzheimer's disease. PMID: 23541490
Leukemia inhibitory factor receptor alpha	LIFR	**Emfilermin** – fertility agents. Disease: Stueve–Wiedemann syndrome (STWS)
Macrophage stimulating 1 (hepatocyte growth factor-like)	MST1	Note: MST1 variant Cys-686 may be associated with inflammatory bowel disease (IBD), a chronic, relapsing inflammation of the gastrointestinal tract with a complex etiology
Macrophage stimulating 1 receptor (c-met-related tyrosine kinase)[a]	MST1R	**Narnatumab** – antineoplastic agents
Midkine (neurite growth-promoting factor 2)	MDK	Increased midkine expression correlates with desmoid tumour recurrence: A potential biomarker and therapeutic target. PMID: 21826666
Pleiotrophin	PTN	Functional receptors and intracellular signal pathways of midkine (MK) and pleiotrophin (PTN). PMID: 24694599
Macrophage migration inhibitory factor (glycosylation-inhibiting factor)	MIF	A novel allosteric inhibitor of macrophage migration inhibitory factor (MIF). PMID: 22782901. Disease: rheumatoid arthritis systemic juvenile (RASJ)
CD74 molecule, major histocompatibility complex, class II invariant chain	CD74	**Milatuzumab** – antineoplastic agents. Note: A chromosomal aberration involving CD74 is found in a non-small cell lung tumor. Results in the formation of a CD74-ROS1 chimeric protein
Myeloid-derived growth factor	MYDGF	Myeloid-derived growth factor (C19orf10) mediates cardiac repair following myocardial infarction. PMID: 25581518
Natriuretic peptide A	NPPA	Disease: atrial fibrillation, familial, 6 (ATFB6)
Natriuretic peptide B	NPPB	NPPB is a novel candidate biomarker expressed by cancer-associated fibroblasts in epithelial ovarian cancer. PMID: 25047817
Natriuretic peptide C	NPPC	C-type natriuretic peptide attenuates LPS-induced endothelial activation: Involvement of p38, Akt, and NF-κB pathways. PMID: 25096521
Nerve growth factor (beta polypeptide)	NGF	**Fulranumab** – drugs affecting bone structure and mineralization. Disease: neuropathy, hereditary sensory and autonomic, 5 (HSAN5)
Nerve growth factor receptor	NGFR	Epigenetic inactivation and tumor suppressor behavior of NGFR in human colorectal cancer. PMID: 25244921

(Continued)

HGNC-approved name	Symbol	Target information
Neurotrophic tyrosine kinase, receptor, type 1[a]	NTRK1	Gambogic amide, a selective agonist for TrkA receptor that possesses robust neurotrophic activity, prevents neuronal cell death. PMID: 17911251. Disease: congenital insensitivity to pain with anhidrosis (CIPA)
Neurotrophic tyrosine kinase, receptor, type 2[a]	NTRK2	A monoclonal antibody TrkB receptor agonist as a potential therapeutic for Huntington's disease. PMID: 24503862. Disease: obesity hyperphagia and developmental delay (OHPDD)
Neurotrophic tyrosine kinase, receptor, type 3[a]	NTRK3	ETV6-NTRK3 as a therapeutic target of small molecule inhibitor PKC412. PMID: 23131561
Nodal growth differentiation factor	NODAL	Nodal promotes invasive phenotypes via a mitogen-activated protein kinase-dependent pathway. PMID: 23334323. Disease: heterotaxy, visceral, 5, autosomal (HTX5)
Oncostatin M	OSM	**GSK315234** – anti-inflammatory and antirheumatic products
Oncostatin M receptor	OSMR	Oncostatin M receptor is a novel therapeutic target in cervical squamous cell carcinoma. PMID: 24659184. Disease: amyloidosis, primary localized cutaneous, 1 (PLCA1)
Osteocrin	OSTN	Osteocrin is a specific ligand of the natriuretic peptide clearance receptor that modulates bone growth. PMID: 17951249
Natriuretic peptide receptor A/guanylate cyclase A (atrionatriuretic peptide receptor A)	NPR1	**Erythrityl tetranitrate** – cardiac therapy
Natriuretic peptide receptor B/guanylate cyclase B (atrionatriuretic peptide receptor B)	NPR2	Emerging roles of natriuretic peptides and their receptors in pathophysiology of hypertension and cardiovascular regulation. PMID: 19746200. Disease: acromesomelic dysplasia, Maroteaux type (AMDM)
Natriuretic peptide receptor C/guanylate cyclase C (atrionatriuretic peptide receptor C)	NPR3	Endothelial C-type natriuretic peptide maintains vascular homeostasis. PMID: 25105365
Platelet-derived growth factor alpha polypeptide	PDGFA	SOX11 promotes tumor angiogenesis through transcriptional regulation of PDGFA in mantle cell lymphoma. PMID: 25092176
Platelet-derived growth factor beta polypeptide	PDGFB	Disease: basal ganglia calcification, idiopathic, 5 (IBGC5)
Platelet-derived growth factor C	PDGFC	Inverse correlation between PDGFC expression and lymphocyte infiltration in human papillary thyroid carcinomas. PMID: 19968886
Platelet-derived growth factor D	PDGFD	Platelet-derived growth factor D: A new player in the complex cross-talk between cholangiocarcinoma cells and cancer-associated fibroblasts. PMID: 23696149
Platelet-derived growth factor receptor, alpha polypeptide[a]	PDGFRA	**Olaratumab** – antineoplastic agents. Note: A chromosomal aberration involving PDGFRA is found in some cases of hypereosinophilic syndrome. Interstitial chromosomal deletion del(4)(q12q12) causes the fusion of FIP1L1 and PDGFRA (FIP1L1-PDGFRA)
Platelet-derived growth factor receptor, beta polypeptide[a]	PDGFRB	**Crenolanib** – antineoplastic agents. Note: A chromosomal aberration involving PDGFRB is found in a form of chronic myelomonocytic leukemia (CMML). Translocation t(512)(q33p13) with EVT6/TEL. It is characterized by abnormal clonal myeloid proliferation
Prolactin	PRL	**Dusigitumab** – antineoplastic agents. Disease: Silver–Russell syndrome (SRS)
Prolactin receptor	PRLR	**Somatropin recombinant** – pituitary and hypothalamic hormones and analogues. Disease: multiple fibroadenomas of the breast (MFAB)

HGNC-approved name	Symbol	Target information
Neuregulin 1	NRG1	Note: A chromosomal aberration involving NRG1 produces gamma-heregulin. Translocation t(811) with TENM4. The translocation fuses the 5′-end of TENM4 to NRG1 (isoform 8)
Neuregulin 2	NRG2	
Neuregulin 3	NRG3	Evidence from mouse and man for a role of neuregulin 3 in nicotine dependence. PMID: 23999525
Neuregulin 4	NRG4	The brown fat-enriched secreted factor Nrg4 preserves metabolic homeostasis through attenuation of hepatic lipogenesis. PMID: 25401691
v-erb-b2 avian erythroblastic leukemia viral oncogene homolog 2[a]	ERBB2	**Trastuzumab** – antineoplastic agents. Disease: hereditary diffuse gastric cancer (HDGC)
v-erb-b2 avian erythroblastic leukemia viral oncogene homolog 3[a]	ERBB3	**Patritumab** – antineoplastic agents. Disease: lethal congenital contracture syndrome 2 (LCCS2)
v-erb-b2 avian erythroblastic leukemia viral oncogene homolog 4[a]	ERBB4	**Afatinib** – antineoplastic agents. Disease: amyotrophic lateral sclerosis 19 (ALS19)
Resistin	RETN	Human resistin promotes neutrophil proinflammatory activation and neutrophil extracellular trap formation and increases severity of acute lung injury. PMID: 24719460
CAP, adenylate cyclase-associated protein 1 (yeast)	CAP1	Adenylyl cyclase-associated protein 1 is a receptor for human resistin and mediates inflammatory actions of human monocytes. PMID: 24606903
Resistin-like beta	RETNLB	Resistin-like molecule β is abundantly expressed in foam cells and is involved in atherosclerosis development. PMID: 23702657
Secretogranin II	SCG2	Differential expression and processing of secretogranin II in relation to the status of pheochromocytoma: Implications for the production of the tumoral marker EM66. PMID: 22217803
Sema domain, immunoglobulin domain (Ig), short basic domain, secreted, (semaphorin) 3A	SEMA3A	Class 3 semaphorins as a therapeutic target. PMID: 22834859. Disease: hypogonadotropic hypogonadism 16 with or without anosmia (HH16)
Sema domain, immunoglobulin domain (Ig), short basic domain, secreted, (semaphorin) 3B	SEMA3B	Semaphorin 3B inhibits the phosphatidylinositol 3-kinase/Akt pathway through neuropilin-1 in lung and breast cancer cells. PMID: 18922901
Sema domain, immunoglobulin domain (Ig), short basic domain, secreted, (semaphorin) 3C	SEMA3C	Semaphorin 3C is a novel adipokine linked to extracellular matrix composition. PMID: 23666167
Sema domain, immunoglobulin domain (Ig), short basic domain, secreted, (semaphorin) 3D	SEMA3D	Semaphorin-3D and semaphorin-3F inhibit the development of tumors from glioblastoma cells implanted in the cortex of the brain. PMID: 22936999
Sema domain, immunoglobulin domain (Ig), short basic domain, secreted, (semaphorin) 3E	SEMA3E	Semaphorin 3E secreted by damaged hepatocytes regulates the sinusoidal regeneration and liver fibrosis during liver regeneration. PMID: 24930441
Sema domain, immunoglobulin domain (Ig), short basic domain, secreted, (semaphorin) 3F	SEMA3F	Merlin/NF2 regulates angiogenesis in schwannomas through a Rac1/semaphorin 3F-dependent mechanism. PMID: 22431917
Sema domain, immunoglobulin domain (Ig), short basic domain, secreted, (semaphorin) 3G	SEMA3G	Effects of SEMA3G on migration and invasion of glioma cells. PMID: 22562223
Sema domain, immunoglobulin domain (Ig), transmembrane domain (TM) and short cytoplasmic domain, (semaphorin) 4A	SEMA4A	Stability and function of regulatory T cells is maintained by a neuropilin-1-semaphorin-4a axis. PMID: 23913274. Disease: retinitis pigmentosa 35 (RP35)
Sema domain, immunoglobulin domain (Ig), transmembrane domain (TM) and short cytoplasmic domain, (semaphorin) 4B	SEMA4B	SEMA4b inhibits MMP9 to prevent metastasis of non-small cell lung cancer. PMID: 25095981

(Continued)

HGNC-approved name	Symbol	Target information
Sema domain, immunoglobulin domain (Ig), transmembrane domain (TM) and short cytoplasmic domain, (semaphorin) 4C	SEMA4C	Sema4C expression in neural stem/progenitor cells and in adult neurogenesis induced by cerebral ischemia. PMID: 19189244
Sema domain, immunoglobulin domain (Ig), transmembrane domain (TM) and short cytoplasmic domain, (semaphorin) 4D	SEMA4D	Suppression of bone formation by osteoclastic expression of semaphorin 4D. PMID: 22019888
Sema domain, immunoglobulin domain (Ig), transmembrane domain (TM) and short cytoplasmic domain, (semaphorin) 4F	SEMA4F	Semaphorin 4F as a critical regulator of neuroepithelial interactions and a biomarker of aggressive prostate cancer. PMID: 24097862
Sema domain, immunoglobulin domain (Ig), transmembrane domain (TM) and short cytoplasmic domain, (semaphorin) 4G	SEMA4G	
Sema domain, seven thrombospondin repeats (type 1 and type 1-like), transmembrane domain (TM) and short cytoplasmic domain, (semaphorin) 5A	SEMA5A	Secreted semaphorin 5A activates immune effector cells and is a biomarker for rheumatoid arthritis. PMID: 24585544
Sema domain, seven thrombospondin repeats (type 1 and type 1-like), transmembrane domain (TM) and short cytoplasmic domain, (semaphorin) 5B	SEMA5B	Semaphorin 5B mediates synapse elimination in hippocampal neurons. PMID: 19463192
Sema domain, transmembrane domain (TM), and cytoplasmic domain, (semaphorin) 6A	SEMA6A	Semaphorin 6A regulates angiogenesis by modulating VEGF signaling. PMID: 23007403
Sema domain, transmembrane domain (TM), and cytoplasmic domain, (semaphorin) 6B	SEMA6B	The role of axon guidance factor semaphorin 6B in the invasion and metastasis of gastric cancer. PMID: 23781008
Sema domain, transmembrane domain (TM), and cytoplasmic domain, (semaphorin) 6C	SEMA6C	Semaphorin 6C expression in innervated and denervated skeletal muscle. PMID: 17605078
Sema domain, transmembrane domain (TM), and cytoplasmic domain, (semaphorin) 6D	SEMA6D	Semaphorin 6D regulates the late phase of CD4+ T cell primary immune responses. PMID: 18728195
Semaphorin 7A, GPI membrane anchor (John Milton Hagen blood group)	SEMA7A	Semaphorin7A: Branching beyond axonal guidance and into immunity. PMID: 24222277
Neuropilin 1	NRP1	**Vesencumab** – antineoplastic agents
Neuropilin 2	NRP2	Expression of neuropilin-2 in salivary adenoid cystic carcinoma: Its implication in tumor progression and angiogenesis. PMID: 20851535
Plexin A1	PLXNA1	Plexin-A1 is required for Toll-like receptor-mediated microglial activation in the development of lipopolysaccharide-induced encephalopathy. PMID: 24604454
Plexin A2	PLXNA2	Plexin-A2 limits recovery from corticospinal axotomy by mediating oligodendrocyte-derived Sema6A growth inhibition. PMID: 22564823
Plexin A3	PLXNA3	
Plexin A4	PLXNA4	Plexin-A4 promotes tumor progression and tumor angiogenesis by enhancement of VEGF and bFGF signaling. PMID: 21832283
Plexin B1	PLXNB1	Plexin-B1 silencing inhibits ovarian cancer cell migration and invasion. PMID: 21059203
Plexin B2	PLXNB2	Plexin-B2 negatively regulates macrophage motility, Rac, and Cdc42 activation. PMID: 21966369
Plexin B3	PLXNB3	Semaphorin 5A and plexin-B3 inhibit human glioma cell motility through RhoGDIalpha-mediated inactivation of Rac1 GTPase. PMID: 20696765
Plexin C1	PLXNC1	Crucial role of Plexin C1 for pulmonary inflammation and survival during lung injury. PMID: 24345803
Plexin D1	PLXND1	Tumour growth inhibition and anti-metastatic activity of a mutated furin-resistant Semaphorin 3E isoform. PMID: 22247010

HGNC-approved name	Symbol	Target information
Thrombopoietin	THPO	Disease: thrombocythemia 1 (THCYT1)
Myeloproliferative leukemia virus oncogene[a]	MPL	**Eltrombopag** – immunostimulants. Disease: congenital amegakaryocytic thrombocytopenia (CAMT)
Transforming growth factor, alpha	TGFA	Generation and activity of a humanized monoclonal antibody that selectively neutralizes the epidermal growth factor receptor ligands transforming growth factor-α and epiregulin. PMID: 24518034
Transforming growth factor, beta 1	TGFB1	**LY2382770** – diabetic nephropathy treatments. Disease: Camurati–Engelmann disease (CE)
Transforming growth factor, beta 2	TGFB2	**Trabedersen** – antineoplastic agents. Note: A chromosomal aberration involving TGFB2 is found in a family with Peters anomaly. Translocation t(1;7)(q41;p21) with HDAC9. Disease: Loeys–Dietz syndrome 4 (LDS4)
Transforming growth factor, beta 3	TGFB3	**Fresolimumab** – antineoplastic agents. Disease: arrhythmogenic right ventricular dysplasia, familial, 1 (ARVD1)
Transforming growth factor, beta receptor 1[a]	TGFBR1	**LY2157299** – antineoplastic agents. Disease: Loeys–Dietz syndrome 1A (LDS1A)
Transforming growth factor, beta receptor II (70/80 kDa)[a]	TGFBR2	**Avotermin** – dermatologicals. Disease: hereditary non-polyposis colorectal cancer 6 (HNPCC6)
Transforming growth factor, beta receptor III	TGFBR3	Type III TGF-β receptor promotes FGF2-mediated neuronal differentiation in neuroblastoma. PMID: 24216509
Tumor necrosis factor	TNF	**Infliximab** – anti-inflammatory and antirheumatic products. Disease: psoriatic arthritis (PSORAS)
Granulin	GRN	The growth factor progranulin binds to TNF receptors and is therapeutic against inflammatory arthritis in mice. PMID: 21393509. Disease: ubiquitin-positive frontotemporal dementia (UP-FTD)
Lymphotoxin alpha	LTA	**Pateclizumab** – immunomodulating agents. Disease: psoriatic arthritis (PSORAS)
Lymphotoxin beta (TNF superfamily, member 3)	LTB	**Baminercept** – anti-inflammatory and antirheumatic products
Tumor necrosis factor (ligand) superfamily, member 4	TNFSF4	**Oxelumab** – drugs for obstructive airway diseases. Disease: systemic lupus erythematosus (SLE)
CD40 ligand	CD40LG	**Ruplizumab** – selective immunosuppressants. Disease: X-linked immunodeficiency with hyper-IgM 1 (HIGM1)
Fas ligand (TNF superfamily, member 6)	FASLG	Enhancing production and cytotoxic activity of polymeric soluble FasL-based chimeric proteins by concomitant expression of soluble FasL. PMID: 23991192
CD70 molecule	CD70	**Vorsetuzumab mafodotin** – antineoplastic agents
Tumor necrosis factor (ligand) superfamily, member 8	TNFSF8	Targeting CD30/CD30L in oncology and autoimmune and inflammatory diseases. PMID: 19760074
Tumor necrosis factor (ligand) superfamily, member 9	TNFSF9	CD137 ligand is expressed in primary and secondary lymphoid follicles and in B-cell lymphomas: Diagnostic and therapeutic implications. PMID: 23095505
Tumor necrosis factor (ligand) superfamily, member 10	TNFSF10	Mesenchymal progenitors expressing TRAIL induce apoptosis in sarcomas. PMID: 25420617
Tumor necrosis factor (ligand) superfamily, member 11	TNFSF11	**Denosumab** – drugs affecting bone structure and mineralization. Disease: osteopetrosis, autosomal recessive 2 (OPTB2)
Tumor necrosis factor (ligand) superfamily, member 12	TNFSF12	**RO5458640** – antineoplastic agents
Tumor necrosis factor (ligand) superfamily, member 13	TNFSF13	Selective APRIL blockade delays systemic lupus erythematosus in mouse. PMID: 22355399
Tumor necrosis factor (ligand) superfamily, member 13b	TNFSF13B	**Belimumab** – selective immunosuppressants

(Continued)

HGNC-approved name	Symbol	Target information
Tumor necrosis factor (ligand) superfamily, member 14	TNFSF14	**SAR252067** – selective immunosuppressants
Tumor necrosis factor (ligand) superfamily, member 15	TNFSF15	The TNF-family cytokine TL1A promotes allergic immunsopathology through group 2 innate lymphoid cells. PMID: 24368564
Tumor necrosis factor (ligand) superfamily, member 18	TNFSF18	**TRX518** – antineoplastic agents
Tumor necrosis factor receptor superfamily, member 1A	TNFRSF1A	**GSK1995057** – acute lung injury treatments. Disease: familial Hibernian fever (FHF)
Tumor necrosis factor receptor superfamily, member 1B	TNFRSF1B	TNF receptor 2 pathway: Drug target for autoimmune diseases. PMID: 20489699
Lymphotoxin beta receptor (TNFR superfamily, member 3)	LTBR	The unexpected role of lymphotoxin beta receptor signaling in carcinogenesis: From lymphoid tissue formation to liver and prostate cancer development. PMID: 20603617
Tumor necrosis factor receptor superfamily, member 4	TNFRSF4	**Nivolumab** – antineoplastic agents. Disease: immunodeficiency 16 (IMD16)
CD40 molecule, TNF receptor superfamily member 5	CD40	**Lucatumumab** – antineoplastic agents. Disease: immunodeficiency with hyper-IgM 3 (HIGM3)
Fas cell surface death receptor	FAS	**APO010** – antineoplastic agents. Disease: autoimmune lymphoproliferative syndrome 1A (ALPS1A)
Tumor necrosis factor receptor superfamily, member 6b, decoy	TNFRSF6B	'Decoy' and 'non-decoy' functions of DcR3 promote malignant potential in human malignant fibrous histiocytoma cells. PMID: 23817777
Tumor necrosis factor receptor superfamily, member 8	TNFRSF8	**Brentuximab vedotin** – antineoplastic agents
Tumor necrosis factor receptor superfamily, member 9	TNFRSF9	**Urelumab** – antineoplastic agents
Tumor necrosis factor receptor superfamily, member 10a	TNFRSF10A	**Mapatumumab** – antineoplastic agents
Tumor necrosis factor receptor superfamily, member 10b	TNFRSF10B	**Drozitumab** – antineoplastic agents. Disease: squamous cell carcinoma of the head and neck (HNSCC)
Tumor necrosis factor receptor superfamily, member 10c, decoy without an intracellular domain	TNFRSF10C	High TRAIL-R3 expression on leukemic blasts is associated with poor outcome and induces apoptosis-resistance which can be overcome by targeting TRAIL-R2. PMID: 21281967
Tumor necrosis factor receptor superfamily, member 10d, decoy with truncated death domain	TNFRSF10D	TRAIL-R4 promotes tumor growth and resistance to apoptosis in cervical carcinoma HeLa cells through AKT. PMID: 21625476
Tumor necrosis factor receptor superfamily, member 11a, NFKB activator	TNFRSF11A	Brief report: Involvement of TNFRSF11A molecular defects in autoinflammatory disorders. PMID: 24891336
Tumor necrosis factor receptor superfamily, member 11b	TNFRSF11B	**CEP-37251** – Drugs affecting bone structure and mineralization. Disease: juvenile Paget disease (JPD)
Tumor necrosis factor receptor superfamily, member 12A	TNFRSF12A	**Enavatuzumab** – antineoplastic agents
Tumor necrosis factor receptor superfamily, member 13B	TNFRSF13B	**Atacicept** – anti-inflammatory and antirheumatic products. Disease: immunodeficiency, common variable, 2 (CVID2)
Tumor necrosis factor receptor superfamily, member 13C	TNFRSF13C	**VAY736** – anti-inflammatory and antirheumatic products. Disease: immunodeficiency, common variable, 4 (CVID4)
Tumor necrosis factor receptor superfamily, member 14	TNFRSF14	TNFRSF14 deficiency protects against ovariectomy-induced adipose tissue inflammation. PMID: 24287621
Tumor necrosis factor receptor superfamily, member 17	TNFRSF17	B-cell maturation antigen (BCMA) activation exerts specific proinflammatory effects in normal human keratinocytes and is preferentially expressed in inflammatory skin pathologies. PMID: 22166983

HGNC-approved name	Symbol	Target information
Tumor necrosis factor receptor superfamily, member 18	TNFRSF18	**TRX518** – antineoplastic agents
Tumor necrosis factor receptor superfamily, member 19	TNFRSF19	β-catenin regulates NF-κB activity via TNFRSF19 in colorectal cancer cells. PMID: 24623448
Tumor necrosis factor receptor superfamily, member 21	TNFRSF21	Death receptor 6 (DR6) antagonist antibody is neuroprotective in the mouse SOD1G93A model of amyotrophic lateral sclerosis. PMID: 24113175
Tumor necrosis factor receptor superfamily, member 25	TNFRSF25	TL1A and DR3, a TNF family ligand-receptor pair that promotes lymphocyte costimulation, mucosal hyperplasia, and autoimmune inflammation. PMID: 22017439
RELT tumor necrosis factor receptor	RELT	RELT induces cellular death in HEK 293 epithelial cells. PMID: 19969290
RELT-like 1	RELL1	Identification of RELT homologues that associate with RELT and are phosphorylated by OSR1. PMID: 16389068
RELT-like 2	RELL2	Identification of RELT homologues that associate with RELT and are phosphorylated by OSR1. PMID: 16389068
Vascular endothelial growth factor A	VEGFA	**Bevacizumab** – antineoplastic agents. Disease: microvascular complications of diabetes 1 (MVCD1)
Vascular endothelial growth factor B	VEGFB	Targeting VEGF-B as a novel treatment for insulin resistance and type 2 diabetes. PMID: 23023133
Vascular endothelial growth factor C	VEGFC	Vascular endothelial growth factor-C promotes breast cancer progression via a novel anti-oxidant mechanism that involves regulation of superoxide dismutase 3. PMID: 25358638
c-Fos-induced growth factor (vascular endothelial growth factor D)	FIGF	Nuclear localization of vascular endothelial growth factor-D and regulation of c-Myc-dependent transcripts in human lung fibroblasts. PMID: 24450584
Fms-related tyrosine kinase 3 ligand	FLT3LG	Fms-related tyrosine kinase 3 ligand promotes proliferation of placenta amnion and chorion mesenchymal stem cells *in vitro*. PMID: 24820950
Placental growth factor	PGF	**Aflibercept** – antineovascularisation agents
Kinase insert domain receptor (a type III receptor tyrosine kinase)[a]	KDR	**Ramucirumab** – antineoplastic agents. Disease: hemangioma, capillary infantile (HCI)
Fms-related tyrosine kinase 1[a]	FLT1	**Icrucumab** – Antineoplastic agents. Note: Can contribute to cancer cell survival, proliferation, migration, and invasion, and tumor angiogenesis and metastasis.
Fms-related tyrosine kinase 3[a]	FLT3	**Quizartinib** – antineoplastic agents. Disease: leukemia, acute myelogenous (AML)
Fms-related tyrosine kinase 4[a]	FLT4	**Axitinib** – antineoplastic agents. Disease: lymphedema, hereditary, 1A (LMPH1A)
Vasohibin 1	VASH1	Vasohibin-1 increases the malignant potential of colorectal cancer and is a biomarker of poor prognosis. PMID: 25275025
Vasohibin 2	VASH2	Vasohibin 2 is transcriptionally activated and promotes angiogenesis in hepatocellular carcinoma. PMID: 22614011
WNT inhibitory factor 1	WIF1	Wnt inhibitory factor-1 functions as a tumor suppressor through modulating Wnt/β-catenin signaling in neuroblastoma. PMID: 24561119
WNT1-inducible-signaling pathway protein 1	WISP1	Clinical significance of Wnt-induced secreted protein-1 (WISP-1/CCN4) in esophageal squamous cell carcinoma. PMID: 21498727
WNT1-inducible-signaling pathway protein 2	WISP2	The novel secreted adipokine WNT1-inducible signaling pathway protein 2 (WISP2) is a mesenchymal cell activator of canonical WNT. PMID: 24451367
WNT1-inducible-signaling pathway protein 3	WISP3	Silencing of WISP3 suppresses gastric cancer cell proliferation and metastasis and inhibits Wnt/β-catenin signaling. PMID: 25400723

[a] Also listed under protein kinases (Chapter 4).

3.4   Adhesion molecules

HGNC-approved name	Symbol	Target information
Asialoglycoprotein receptor 1	ASGR1	Expression of asialoglycoprotein receptor 1 in human hepatocellular carcinoma. PMID: 23979840
Asialoglycoprotein receptor 2	ASGR2	
Cadherin 1, type 1, E-cadherin (epithelial)	CDH1	Soluble E-cadherin: A critical oncogene modulating receptor tyrosine kinases, MAPK and PI3K/Akt/mTOR signaling. PMID: 23318419. Disease: hereditary diffuse gastric cancer (HDGC)
Cadherin 2, type 1, N-cadherin (neuronal)	CDH2	N-cadherin expression is associated with acquisition of EMT phenotype and with enhanced invasion in erlotinib-resistant lung cancer cell lines. PMID: 23520479
Cadherin 3, type 1, P-cadherin (placental)	CDH3	Disease: hypotrichosis congenital with juvenile macular dystrophy (HJMD)
Cadherin 4, type 1, R-cadherin (retinal)	CDH4	CDH4 as a novel putative tumor suppressor gene epigenetically silenced by promoter hypermethylation in nasopharyngeal carcinoma. PMID: 21665361
Cadherin 5, type 2 (vascular endothelium)	CDH5	Cadherin 5 is regulated by corticosteroids and associated with central serous chorioretinopathy. PMID: 24665005
Cadherin 6, type 2, K-cadherin (fetal kidney)	CDH6	Cadherin 6 is a new RUNX2 target in TGF-β signalling pathway. PMID: 24069422
Cadherin 7, type 2	CDH7	Cadherin-7 interacts with melanoma inhibitory activity protein and negatively modulates melanoma cell migration. PMID: 19200257
Cadherin 8, type 2	CDH8	CHD8 is an independent prognostic indicator that regulates Wnt/β-catenin signaling and the cell cycle in gastric cancer. PMID: 23835524
Cadherin 9, type 2 (T1-cadherin)	CDH9	Cadherin-9 regulates synapse-specific differentiation in the developing hippocampus. PMID: 21867881
Cadherin 10, type 2 (T2-cadherin)	CDH10	Cadherin-10 is a novel blood-brain barrier adhesion molecule in human and mouse. PMID: 16181616
Cadherin 11, type 2, OB-cadherin (osteoblast)	CDH11	Cadherin-11 contributes to pulmonary fibrosis: Potential role in TGF-β production and epithelial to mesenchymal transition. PMID: 21990376. Note: A chromosomal aberration involving CDH11 is a common genetic feature of aneurysmal bone cyst
Cadherin 12, type 2 (N-cadherin 2)	CDH12	Cadherin-12 contributes to tumorigenicity in colorectal cancer by promoting migration, invasion, adhesion and angiogenesis. PMID: 24237488
Cadherin 13	CDH13	Involvement of members of the cadherin superfamily in cancer. PMID: 20457567
Cadherin 15, type 1, M-cadherin (myotubule)	CDH15	Note: A chromosomal aberration involving CDH15 and KIRREL3 is found in a patient with severe mental retardation and dysmorphic facial features. Translocation t(11;16)(q24.2;q24). Disease: mental retardation, autosomal dominant 3 (MRD3)
Cadherin 16, KSP-cadherin	CDH16	CDH16/Ksp-cadherin is expressed in the developing thyroid gland and is strongly down-regulated in thyroid carcinomas. PMID: 22028439
Cadherin 17, LI cadherin (liver-intestine)	CDH17	Role of cadherin-17 in oncogenesis and potential therapeutic implications in hepatocellular carcinoma. PMID: 20580775
Cadherin 18, type 2	CDH18	Cadherins and neuropsychiatric disorders. PMID: 22765916
Cadherin 19, type 2	CDH19	Myelin-forming cell-specific cadherin-19 is a marker for minimally infiltrative glioblastoma stem-like cells. PMID: 25361488
Cadherin 20, type 2	CDH20	
Cadherin 22, type 2	CDH22	CDH22 expression is reduced in metastatic melanoma. PMID: 21969168
Cadherin-related 23	CDH23	Disease: Usher syndrome 1D (USH1D)

HGNC-approved name	Symbol	Target information
Cadherin 24, type 2	CDH24	
Cadherin 26	CDH26	
Cadherin-related family member 1	CDHR1	Disease: cone-rod dystrophy 15 (CORD15)
Cadherin-related family member 2	CDHR2	Protocadherin LKC, a new candidate for a tumor suppressor of colon and liver cancers, its association with contact inhibition of cell proliferation. PMID: 12117771
Cadherin-related family member 3	CDHR3	A genome-wide association study identifies CDHR3 as a susceptibility locus for early childhood asthma with severe exacerbations. PMID: 24241537
Cadherin-related family member 4	CDHR4	
Cadherin-related family member 5	CDHR5	Cdx2 controls expression of the protocadherin Mucdhl, an inhibitor of growth and β-catenin activity in colon cancer cells. PMID: 22202456
Calcium and integrin binding 1 (calmyrin)	CIB1	A novel splice variant of calcium and integrin-binding protein 1 mediates protein kinase D2-stimulated tumour growth by regulating angiogenesis. PMID: 23503467
Calcium and integrin binding family member 2	CIB2	Disease: deafness, autosomal recessive, 48 (DFNB48)
Calcium and integrin binding family member 3	CIB3	
Calcium and integrin binding family member 4	CIB4	
Calsyntenin 1	CLSTN1	Calsyntenins function as synaptogenic adhesion molecules in concert with neurexins. PMID: 24613359
Calsyntenin 2	CLSTN2	Calsyntenins function as synaptogenic adhesion molecules in concert with neurexins. PMID: 24613359
Calsyntenin 3	CLSTN3	The specific α-neurexin interactor calsyntenin-3 promotes excitatory and inhibitory synapse development. PMID: 24094106
Carcinoembryonic antigen-related cell adhesion molecule 1 (biliary glycoprotein)	CEACAM1	**Arcitumomab** – diagnostic agents
Carcinoembryonic antigen-related cell adhesion molecule 3	CEACAM3	HemITAM signaling by CEACAM3, a human granulocyte receptor recognizing bacterial pathogens. PMID: 22469950
Carcinoembryonic antigen-related cell adhesion molecule 4	CEACAM4	Carcinoembryonic antigen-related cell adhesion molecule 4 (CEACAM4) is specifically expressed in medullary thyroid carcinoma cells. PMID: 25152032
Carcinoembryonic antigen-related cell adhesion molecule 5	CEACAM5	**MEDI-565** – antineoplastic agents
Carcinoembryonic antigen-related cell adhesion molecule 6 (non-specific cross-reacting antigen)	CEACAM6	CD66c is a novel marker for colorectal cancer stem cell isolation, and its silencing halts tumor growth in vivo. PMID: 23027178
Carcinoembryonic antigen-related cell adhesion molecule 8	CEACAM8	Soluble CEACAM8 interacts with CEACAM1 inhibiting TLR2-triggered immune responses. PMID: 24743304
Carcinoembryonic antigen-related cell adhesion molecule 19	CEACAM19	The expression of the CEACAM19 gene, a novel member of the CEA family, is associated with breast cancer progression. PMID: 23525470
CD207 molecule, langerin	CD207	Disease: Birbeck granule deficiency (BIRGD)
CD209 molecule	CD209	Depletion of inflammatory dendritic cells with anti-CD209 conjugated to saporin toxin. PMID: 24781193
CD302 molecule	CD302	The novel endocytic and phagocytic C-Type lectin receptor DCL-1/CD302 on macrophages is colocalized with F-actin, suggesting a role in cell adhesion and migration. PMID: 17947679

(Continued)

HGNC-approved name	Symbol	Target information
Cell adhesion molecule 1	CADM1	CADM1/TSLC1 inhibits melanoma cell line A375 invasion through the suppression of matrix metalloproteinases. PMID: 25215547
Cell adhesion molecule 2	CADM2	Aberrant methylation and loss of CADM2 tumor suppressor expression is associated with human renal cell carcinoma tumor progression. PMID: 23643812
Cell adhesion molecule 3	CADM3	Loss of NECL1, a novel tumor suppressor, can be restored in glioma by HDAC inhibitor-Trichostatin A through Sp1 binding site. PMID: 19062177
Cell adhesion molecule 4	CADM4	Clinicopathological significance of CADM4 expression in invasive ductal carcinoma of the breast. PMID: 23559354
Chondroitin sulfate proteoglycan 4	CSPG4	Combining NK cells and mAb9.2.27 to combat NG2-dependent and anti-inflammatory signals in glioblastoma. PMID: 24575382
Chondroitin sulfate proteoglycan 5 (neuroglycan C)	CSPG5	Neuroglycan C, a brain-specific chondroitin sulfate proteoglycan, interacts with pleiotrophin, a heparin-binding growth factor. PMID: 20369290
Collectin subfamily member 10 (C-type lectin)	COLEC10	Structure and function of collectin liver 1 (CL-L1) and collectin 11 (CL-11, CL-K1). PMID: 22475410
Collectin subfamily member 11	COLEC11	Disease: 3MC syndrome 2 (3MC2)
Collectin subfamily member 12	COLEC12	Possible role of scavenger receptor SRCL in the clearance of amyloid-beta in Alzheimer's disease. PMID: 16868960
C-type lectin domain family 1, member A	CLEC1A	
C-type lectin domain family 1, member B	CLEC1B	CLEC-2 is required for development and maintenance of lymph nodes. PMID: 24532804
C-type lectin domain family 2, member A	CLEC2A	Interaction of C-type lectin-like receptors NKp65 and KACL facilitates dedicated immune recognition of human keratinocytes. PMID: 20194751
C-type lectin domain family 2, member B	CLEC2B	Peptide mimicry of AICL inhibits cytolysis of NK cells by blocking NKp80-AICL recognition. PMID: 20653426
C-type lectin domain family 2, member D	CLEC2D	Osteoclast inhibitory lectin (OCIL) inhibits osteoblast differentiation and function *in vitro*. PMID: 17049328
C-type lectin domain family 2, member L	CLEC2L	BACL is a novel brain-associated, non-NKC-encoded mammalian C-type lectin-like receptor of the CLEC2 family. PMID: 23776472
C-type lectin domain family 3, member A	CLEC3A	Matrilysin (MMP-7) cleaves C-type lectin domain family 3 member A (CLEC3A) on tumor cell surface and modulates its cell adhesion activity. PMID: 19173304
C-type lectin domain family 3, member B	CLEC3B	Tetranectin knockout mice develop features of Parkinson disease. PMID: 25033953
C-type lectin domain family 4, member A	CLEC4A	Dendritic cell immunoreceptor: A novel receptor for intravenous immunoglobulin mediates induction of regulatory T cells. PMID: 24210883
C-type lectin domain family 4, member C	CLEC4C	Targeting antigens through blood dendritic cell antigen 2 on plasmacytoid dendritic cells promotes immunologic tolerance. PMID: 24829416
C-type lectin domain family 4, member D	CLEC4D	C-type lectin MCL is an FcRγ-coupled receptor that mediates the adjuvanticity of mycobacterial cord factor. PMID: 23602766
C-type lectin domain family 4, member E	CLEC4E	Macrophage-inducible C-type lectin underlies obesity-induced adipose tissue fibrosis. PMID: 25236782
C-type lectin domain family 4, member F	CLEC4F	CLEC4F is an inducible C-type lectin in F4/80-positive cells and is involved in alpha-galactosylceramide presentation in liver. PMID: 23762286
C-type lectin domain family 4, member G	CLEC4G	LSECtin expressed on melanoma cells promotes tumor progression by inhibiting antitumor T-cell responses. PMID: 24769443
C-type lectin domain family 4, member M	CLEC4M	The C-type lectin receptor CLEC4M binds, internalizes, and clears von Willebrand factor and contributes to the variation in plasma von Willebrand factor levels. PMID: 23529928

HGNC-approved name	Symbol	Target information
C-type lectin domain family 5, member A	CLEC5A	Myeloid DAP12-associating lectin (MDL)-1 regulates synovial inflammation and bone erosion associated with autoimmune arthritis. PMID: 20212065
C-type lectin domain family 6, member A	CLEC6A	Dectin-2 promotes house dust mite-induced T helper type 2 and type 17 cell differentiation and allergic airway inflammation in mice. PMID: 24588637
C-type lectin domain family 7, member A	CLEC7A	Disease: candidiasis, familial, 4 (CANDF4)
C-type lectin domain family 9, member A	CLEC9A	Targeting DNGR-1 (CLEC9A) with antibody/MUC1 peptide conjugates as a vaccine for carcinomas. PMID: 24648154
C-type lectin domain family 10, member A	CLEC10A	Tumor-associated Neu5Ac-Tn and Neu5Gc-Tn antigens bind to C-type lectin CLEC10A (CD301, MGL). PMID: 23507963
C-type lectin domain family 11, member A	CLEC11A	Proteomic detection of a large amount of SCGFα in the stroma of GISTs after imatinib therapy. PMID: 21943129
C-type lectin domain family 12, member A	CLEC12A	Clec12a is an inhibitory receptor for uric acid crystals that regulates inflammation in response to cell death. PMID: 24631154
C-type lectin domain family 12, member B	CLEC12B	Identification of CLEC12B, an inhibitory receptor on myeloid cells. PMID: 17562706
C-type lectin domain family 14, member A	CLEC14A	Human antibodies targeting the C-type lectin-like domain of the tumor endothelial cell marker clec14a regulate angiogenic properties *in vitro*. PMID: 23644659
C-type lectin domain family 16, member A	CLEC16A	Note: Insulin-dependent diabetes mellitus is significantly associated with variation within a 233-kb linkage disequilibrium block on chromosome 16p13 that includes KIAA0350.
C-type lectin domain family 17, member A	CLEC17A	Prolectin, a glycan-binding receptor on dividing B cells in germinal centers. PMID: 19419970
C-type lectin domain family 18, member A	CLEC18A	
C-type lectin domain family 18, member B	CLEC18B	
C-type lectin domain family 18, member C	CLEC18C	
C-type lectin domain family 19, member A	CLEC19A	
C-type lectin-like 1	CLECL1	Targeting C-type lectin-like molecule-1 for antibody-mediated immunotherapy in acute myeloid leukemia. PMID: 19648166
Dachsous cadherin-related 1	DCHS1	Disease: Van Maldergem syndrome 1 (VMLDS1)
Dachsous cadherin-related 2	DCHS2	
Desmocollin 1	DSC1	
Desmocollin 2	DSC2	Disease: arrhythmogenic right ventricular dysplasia, familial, 11 (ARVD11)
Desmocollin 3	DSC3	Disease: hypotrichosis and recurrent skin vesicles (HRSV)
Desmoglein 1	DSG1	The desmosomal protein desmoglein 1 aids recovery of epidermal differentiation after acute UV light exposure. PMID: 24594668. Disease: palmoplantar keratoderma 1, striate, focal, or diffuse (PPKS1)
Desmoglein 2	DSG2	Disease: arrhythmogenic right ventricular dysplasia, familial, 10 (ARVD10)
Desmoglein 3	DSG3	Desmosomal cadherins and signaling: Lessons from autoimmune disease. PMID: 24460203
Desmoglein 4	DSG4	Disease: hypotrichosis 6 (HYPT6)
FAT atypical cadherin 1	FAT1	Expression and function of the atypical cadherin FAT1 in chronic liver disease. PMID: 22959770
FAT atypical cadherin 2	FAT2	Sleeping giants: Emerging roles for the fat cadherins in health and disease. PMID: 23720094
FAT atypical cadherin 3	FAT3	

(Continued)

HGNC-approved name	Symbol	Target information
FAT atypical cadherin 4	FAT4	Disease: Van Maldergem syndrome 2 (VMLDS2)
Fibulin 1	FBLN1	Fibulin-1 is epigenetically down-regulated and related with bladder cancer recurrence. PMID: 25234557. Note: A chromosomal aberration involving FBLN1 is found in a complex type of synpolydactyly referred to as 3/3-prime/4 synpolydactyly associated with metacarpal and metatarsal synostoses. Reciprocal translocation t(1222)(p11.2q13.3) with RASSF8
Fibulin 2	FBLN2	Fibulin-2 is a driver of malignant progression in lung adenocarcinoma. PMID: 23785517
Fibulin 5	FBLN5	Effect of Fibulin-5 on cell proliferation and invasion in human gastric cancer patients. PMID: 25129461. Disease: cutis laxa, autosomal dominant, 2 (ADCL2)
Fibulin 7	FBLN7	A C-terminal fragment of fibulin-7 interacts with endothelial cells and inhibits their tube formation in culture. PMID: 24480309
Ficolin (collagen/fibrinogen domain containing) 1	FCN1	The pattern recognition molecule ficolin-1 exhibits differential binding to lymphocyte subsets, providing a novel link between innate and adaptive immunity. PMID: 24161415
Ficolin (collagen/fibrinogen domain containing lectin) 2	FCN2	Low Ficolin-2 levels in CVID Patients with bronchiectasis. PMID: 25251245
Ficolin (collagen/fibrinogen domain containing) 3	FCN3	Disease: Ficolin 3 deficiency (FCN3D)
Glypican 1	GPC1	Note: Associates (via the heparan sulfate side chains) with fibrillar APP-beta amyloid peptides in primitive and classic amyloid plaques and may be involved in the deposition of these senile plaques in the Alzheimer disease (AD) brain
Glypican 2	GPC2	
Glypican 3	GPC3	**RG7686** – antineoplastic agents. Disease: Simpson–Golabi–Behmel syndrome 1 (SGBS1)
Glypican 4	GPC4	Glypican-4 is increased in human subjects with impaired glucose tolerance and decreased in patients with newly diagnosed type 2 diabetes. PMID: 25240528
Glypican 5	GPC5	The role of GPC5 in lung metastasis of salivary adenoid cystic carcinoma. PMID: 25093697
Glypican 6	GPC6	Disease: omodysplasia 1 (OMOD1)
Intelectin 1 (galactofuranose binding)	ITLN1	Intelectin is required for IL-13-induced monocyte chemotactic protein-1 and -3 expression in lung epithelial cells and promotes allergic airway inflammation. PMID: 19965981
Intelectin 2	ITLN2	
Intercellular adhesion molecule 1	ICAM1	**Alicaforsen** – drugs for functional gastrointestinal disorders
Intercellular adhesion molecule 2	ICAM2	CXCL17 and ICAM2 are associated with a potential anti-tumor immune response in early intraepithelial stages of human pancreatic carcinogenesis. PMID: 20955708
Intercellular adhesion molecule 3	ICAM3	ICAM-3 endows anticancer drug resistance against microtubule-damaging agents via activation of the ICAM-3-AKT/ERK-CREB-2 pathway and blockage of apoptosis. PMID: 24177012
Intercellular adhesion molecule 4 (Landsteiner-Wiener blood group)	ICAM4	Red-cell ICAM-4 is a ligand for the monocyte/macrophage integrin CD11c/CD18: Characterization of the binding sites on ICAM-4. PMID: 16985175
Intercellular adhesion molecule 5 telencephalin	ICAM5	ICAM-5: A neuronal dendritic adhesion molecule involved in immune and neuronal functions. PMID: 25300135

HGNC-approved name	Symbol	Target information
Integrin, alpha 1	ITGA1	
Integrin, alpha 2 (CD49B, alpha 2 subunit of VLA-2 receptor)	ITGA2	**Vatelizumab** – anti-inflammatory and antirheumatic products
Integrin, alpha 2b (platelet glycoprotein IIb of IIb/IIIa complex, antigen CD41)	ITGA2B	**Abciximab** – antithrombotic agents. Disease: Glanzmann thrombasthenia (GT)
Integrin, alpha 3 (antigen CD49C, alpha 3 subunit of VLA-3 receptor)	ITGA3	Disease: interstitial lung disease, nephrotic syndrome, and epidermolysis bullosa, congenital (ILNEB)
Integrin, alpha 4 (antigen CD49D, alpha 4 subunit of VLA-4 receptor)	ITGA4	**Natalizumab** – anti-inflammatory and antirheumatic products
Integrin, alpha 5 (fibronectin receptor, alpha polypeptide)	ITGA5	**Cilengitide** – antineoplastic agents
Integrin, alpha 6	ITGA6	Disease: epidermolysis bullosa letalis, with pyloric atresia (EB-PA)
Integrin, alpha 7	ITGA7	Disease: muscular dystrophy congenital due to integrin alpha-7 deficiency (MDCI)
Integrin, alpha 8	ITGA8	Disease: renal hypodysplasia/aplasia 1 (RHDA1)
Integrin, alpha 9	ITGA9	Integrin alpha 9 (ITGA9) expression and epigenetic silencing in human breast tumors. PMID: 21975548
Integrin, alpha 10	ITGA10	
Integrin, alpha 11	ITGA11	Integrin alpha 11 regulates IGF2 expression in fibroblasts to enhance tumorigenicity of human non-small-cell lung cancer cells. PMID: 17600088
Integrin, alpha D	ITGAD	Treatment with an anti-CD11d integrin antibody reduces neuroinflammation and improves outcome in a rat model of repeated concussion. PMID: 23414334
Integrin, alpha E (antigen CD103, human mucosal lymphocyte antigen 1; alpha polypeptide)	ITGAE	**Anti-CD103** – anti-inflammatory and antirheumatic products
Integrin, alpha L (antigen CD11A (p180), lymphocyte function-associated antigen 1; alpha polypeptide)	ITGAL	**Efalizumab** – anti-inflammatory and antirheumatic products
Integrin, alpha M (complement component 3 receptor 3 subunit)	ITGAM	Disease: systemic lupus erythematosus 6 (SLEB6)
Integrin, alpha V	ITGAV	**Cilengitide** – antineoplastic agents
Integrin, alpha X (complement component 3 receptor 4 subunit)	ITGAX	A systems genetics approach identifies CXCL14, ITGAX, and LPCAT2 as novel aggressive prostate cancer susceptibility genes. PMID: 25411967
Integrin, beta 1 (fibronectin receptor, beta polypeptide, antigen CD29 includes MDF2, MSK12)	ITGB1	**Cilengitide** – antineoplastic agents
Integrin beta 1-binding protein 1	ITGB1BP1	Concepts and hypothesis: Integrin cytoplasmic domain-associated protein-1 (ICAP-1) as a potential player in cerebral cavernous malformation. PMID: 22711159
Integrin beta 1-binding protein (melusin) 2	ITGB1BP2	Altered melusin pathways involved in cardiac remodeling following acute myocardial infarction. PMID: 21546274
Integrin, beta 2 (complement component 3 receptor 3 and 4 subunit)	ITGB2	**Erlizumab** – antithrombotic agents. Disease: leukocyte adhesion deficiency 1 (LAD1)
Integrin, beta 3 (platelet glycoprotein IIIa, antigen CD61)	ITGB3	**Eptifibatide** – antithrombotic agents. Disease: Glanzmann thrombasthenia (GT)
Integrin, beta 4	ITGB4	Disease: epidermolysis bullosa letalis, with pyloric atresia (EB-PA)
Integrin, beta 5	ITGB5	**Cilengitide** – antineoplastic agents
Integrin, beta 6	ITGB6	**STX-100** – idiopathic pulmonary fibrosis

(Continued)

HGNC-approved name	Symbol	Target information
Integrin, beta 7	ITGB7	**Etrolizumab** – drugs for functional gastrointestinal disorders
Integrin, beta 8	ITGB8	Selective targeting of TGF-β activation to treat fibroinflammatory airway disease. PMID: 24944194
Junctional adhesion molecule 2	JAM2	The junctional adhesion molecule-B regulates JAM-C-dependent melanoma cell metastasis. PMID: 23068611
Junctional adhesion molecule 3	JAM3	A novel function of junctional adhesion molecule-C in mediating melanoma cell metastasis. PMID: 21593193. Disease: hemorrhagic destruction of the brain with subependymal calcification and cataracts (HDBSCC)
Laminin, alpha 1	LAMA1	
Laminin, alpha 2	LAMA2	Laminin alpha 2 enables glioblastoma stem cell growth. PMID: 23280793. Disease: merosin-deficient congenital muscular dystrophy 1A (MDC1A)
Laminin, alpha 3	LAMA3	Disease: epidermolysis bullosa, junctional, Herlitz type (H-JEB)
Laminin, alpha 4	LAMA4	Disease: cardiomyopathy, dilated 1JJ (CMD1JJ)
Laminin, alpha 5	LAMA5	A novel monoclonal antibody to human laminin α5 chain strongly inhibits integrin-mediated cell adhesion and migration on laminins 511 and 521. PMID: 23308268
Laminin, beta 1	LAMB1	Disease: lissencephaly 5 (LIS5)
Laminin, beta 2 (laminin S)	LAMB2	Disease: Pierson syndrome (PIERSS)
Laminin, beta 3	LAMB3	Disease: epidermolysis bullosa, junctional, Herlitz type (H-JEB)
Laminin, beta 4	LAMB4	Laminin gene LAMB4 is somatically mutated and expressionally altered in gastric and colorectal cancers. PMID: 25257191
Laminin, gamma 1 (formerly LAMB2)	LAMC1	Tumor-suppressive microRNA-29s inhibit cancer cell migration and invasion via targeting LAMC1 in prostate cancer. PMID: 24820027
Laminin, gamma 2	LAMC2	Laminin-5γ-2 (LAMC2) is highly expressed in anaplastic thyroid carcinoma and is associated with tumor progression, migration, and invasion by modulating signaling of EGFR. PMID: 24170107. Disease: epidermolysis bullosa, junctional, Herlitz type (H-JEB)
Laminin, gamma 3	LAMC3	Disease: cortical malformations occipital (OCCM)
Lectin, galactoside-binding, soluble, 1	LGALS1	Unraveling galectin-1 as a novel therapeutic target for cancer. PMID: 23953240
Lectin, galactoside-binding, soluble, 2	LGALS2	Lowered expression of galectin-2 is associated with lymph node metastasis in gastric cancer. PMID: 22015694
Lectin, galactoside-binding, soluble, 3	LGALS3	**Salirasib** – antineoplastic agents
Lectin, galactoside-binding, soluble, 3 binding protein	LGALS3BP	Functional screen for secreted proteins by monoclonal antibody library and identification of Mac-2 binding protein (Mac-2BP) as a potential therapeutic target and biomarker for lung cancer. PMID: 23184915
Lectin, galactoside-binding, soluble, 4	LGALS4	Galectin-4, a novel predictor for lymph node metastasis in lung adenocarcinoma. PMID: 24339976
Lectin, galactoside-binding, soluble, 7	LGALS7	Inhibition mechanism of human galectin-7 by a novel galactose-benzylphosphate inhibitor. PMID: 22059385
Lectin, galactoside-binding, soluble, 7B	LGALS7B	
Lectin, galactoside-binding, soluble, 8	LGALS8	Galectin 8 targets damaged vesicles for autophagy to defend cells against bacterial invasion. PMID: 22246324
Lectin, galactoside-binding, soluble, 9	LGALS9	Galectin-9 in tumor biology: A jack of multiple trades. PMID: 23648450
Charcot–Leyden crystal galectin	CLC	Galectin-10, a potential biomarker of eosinophilic airway inflammation. PMID: 22880030
Lectin, galactoside-binding, soluble, 9B	LGALS9B	
Lectin, galactoside-binding, soluble, 9C	LGALS9C	

HGNC-approved name	Symbol	Target information
Lectin, galactoside-binding, soluble, 12	LGALS12	Ablation of a galectin preferentially expressed in adipocytes increases lipolysis, reduces adiposity, and improves insulin sensitivity in mice. PMID: 21969596
Lectin, galactoside-binding, soluble, 13	LGALS13	Placental protein 13 (PP13): A new biological target shifting individualized risk assessment to personalized drug design combating pre-eclampsia. PMID: 23420029
Lectin, galactoside-binding, soluble, 14	LGALS14	
Lectin, galactoside-binding, soluble, 16	LGALS16	
Lectin, galactoside-binding-like	LGALSL	Identification of genes regulating alveolar morphogenesis by supression subtractive hybridization. PMID: 17348208
Lectin, mannose-binding, 1	LMAN1	Disease: Factor V and factor VIII combined deficiency 1 (F5F8D1)
Lectin, mannose-binding 2	LMAN2	Sugar-binding properties of VIP36, an intracellular animal lectin operating as a cargo receptor. PMID: 16129679
Lectin, mannose-binding, 1 like	LMAN1L	ERGL, a novel gene related to ERGIC-53 that is highly expressed in normal and neoplastic prostate and several other tissues. PMID: 11255007
Mannose receptor, C type 1	MRC1	Exploitation of the macrophage mannose receptor (CD206) in infectious disease diagnostics and therapeutics. PMID: 24672807
Mannose receptor, C type 2	MRC2	MRC2 expression correlates with TGFβ1 and survival in hepatocellular carcinoma. PMID: 25162823
Multiple EGF-like-domains 6	MEGF6	
Multiple EGF-like-domains 8	MEGF8	Disease: Carpenter syndrome 2 (CRPT2)
Multiple EGF-like-domains 9	MEGF9	
Multiple EGF-like-domains 10	MEGF10	Disease: myopathy, early-onset, areflexia, respiratory distress, and dysphagia (EMARDD)
Multiple EGF-like-domains 11	MEGF11	
Neural cell adhesion molecule 1	NCAM1	**Lorvotuzumab mertansine** – antineoplastic agents
Neural cell adhesion molecule 2	NCAM2	Neural cell adhesion molecule 2 as a target molecule for prostate and breast cancer gene therapy. PMID: 21214674
Ninjurin 1	NINJ1	Ninjurin1: A potential adhesion molecule and its role in inflammation and tissue remodeling. PMID: 20119872
Ninjurin 2	NINJ2	Association between NINJ2 gene polymorphisms and ischemic stroke: A family-based case-control study. PMID: 24664524
Plakophilin 1 (ectodermal dysplasia/skin fragility syndrome)	PKP1	Disease: ectodermal dysplasia–skin fragility syndrome (EDSFS)
Plakophilin 2	PKP2	Plakophilin-2 promotes activation of epidermal growth factor receptor. PMID: 25113561. Disease: arrhythmogenic right ventricular dysplasia, familial, 9 (ARVD9)
Plakophilin 3	PKP3	Plakophilins: Multifunctional proteins or just regulators of desmosomal adhesion? PMID: 16765467
Plakophilin 4	PKP4	FMRP regulates actin filament organization via the armadillo protein p0071. PMID: 24062571
Poliovirus receptor-related 1 (herpesvirus entry mediator C)	PVRL1	Disease: ectodermal dysplasia, Margarita Island type (EDMI)
Poliovirus receptor-related 2 (herpesvirus entry mediator B)	PVRL2	Expression of poliovirus receptor-related proteins PRR1 and PRR2 in acute myeloid leukemia: First report of surface marker analysis, contribution to diagnosis, prognosis and implications for future therapeutic strategies. PMID: 16313259

(Continued)

HGNC-approved name	Symbol	Target information
Poliovirus receptor-related 3	PVRL3	The cell adhesion gene PVRL3 is associated with congenital ocular defects. PMID: 21769484
Poliovirus receptor-related 4	PVRL4	**ASG-22ME** – antineoplastic agents. Disease: ectodermal dysplasia–syndactyly syndrome 1 (EDSS1)
Protocadherin 1	PCDH1	Protocadherin-1 – epithelial barrier dysfunction in asthma and eczema. PMID: 24585862
Protocadherin 7	PCDH7	Protocadherin-7 induces bone metastasis of breast cancer. PMID: 23751349
Protocadherin 8	PCDH8	Clinical significance of protocadherin 8 (PCDH8) promoter methylation in non-muscle invasive bladder cancer. PMID: 25145351
Protocadherin 9	PCDH9	Characterizing the role of PCDH9 in the regulation of glioma cell apoptosis and invasion. PMID: 24214103
Protocadherin 10	PCDH10	PCDH10 is required for the tumorigenicity of glioblastoma cells. PMID: 24406169
Protocadherin 11 X-linked	PCDH11X	Protocadherin 11 x regulates differentiation and proliferation of neural stem cell *in vitro* and in vivo. PMID: 24647733
Protocadherin 11 Y-linked	PCDH11Y	Note: A chromosomal aberration involving PCDH11Y is a cause of multiple congenital abnormalities, including severe bilateral vesicoureteral reflux (VUR) with ureterovesical junction defects. Translocation t(Y3)(p11p12) with ROBO2
Protocadherin 12	PCDH12	Endothelial permeability and VE-cadherin: A wacky comradeship. PMID: 25422846
Protocadherin-related 15	PCDH15	Disease: Usher syndrome 1F (USH1F)
Protocadherin 17	PCDH17	Protocadherin 17 acts as a tumour suppressor inducing tumour cell apoptosis and autophagy, and is frequently methylated in gastric and colorectal cancers. PMID: 22926751
Protocadherin 18	PCDH18	Protocadherin-18 is a novel differentiation marker and an inhibitory signaling receptor for CD8+ effector memory T cells. PMID: 22567129
Protocadherin 19	PCDH19	Disease: epileptic encephalopathy, early infantile, 9 (EIEE9)
Protocadherin 20	PCDH20	PCDH20 functions as a tumour-suppressor gene through antagonizing the Wnt/β-catenin signalling pathway in hepatocellular carcinoma. PMID: 24910204
Regenerating islet-derived 1 alpha	REG1A	REG Iα gene expression is linked with the poor prognosis of lung adenocarcinoma and squamous cell carcinoma patients via discrete mechanisms. PMID: 24065141
Regenerating islet-derived 1 beta	REG1B	Pancreatic stone protein/regenerating protein family in pancreatic and gastrointestinal diseases. PMID: 21804274
Regenerating islet-derived 3 alpha	REG3A	The antimicrobial protein REG3A regulates keratinocyte proliferation and differentiation after skin injury. PMID: 22727489
Regenerating islet-derived 3 gamma	REG3G	Innate Stat3-mediated induction of the antimicrobial protein Reg3γ is required for host defense against MRSA pneumonia. PMID: 23401489
Regenerating islet-derived family, member 4	REG4	The reg4 gene, amplified in the early stages of pancreatic cancer development, is a promising therapeutic target. PMID: 19834624
Sarcoglycan, alpha (50 kDa dystrophin-associated glycoprotein)	SGCA	Disease: limb-girdle muscular dystrophy 2D (LGMD2D)
Sarcoglycan, beta (43 kDa dystrophin-associated glycoprotein)	SGCB	Disease: limb-girdle muscular dystrophy 2E (LGMD2E)
Sarcoglycan, delta (35 kDa dystrophin-associated glycoprotein)	SGCD	Disease: limb-girdle muscular dystrophy 2F (LGMD2F)

HGNC-approved name	Symbol	Target information
Sarcoglycan, epsilon	SGCE	Disease: dystonia 11 (DYT11)
Sarcoglycan, gamma (35 kDa dystrophin-associated glycoprotein)	SGCG	Disease: limb-girdle muscular dystrophy 2C (LGMD2C)
Sarcoglycan, zeta	SGCZ	
Secreted protein, acidic, cysteine-rich (osteonectin)	SPARC	Molecular mechanisms underlying the divergent roles of SPARC in human carcinogenesis. PMID: 24675529
Sparc/osteonectin, cwcv and kazal-like domains proteoglycan (testican) 1	SPOCK1	SPOCK1 is a novel transforming growth factor-β target gene that regulates lung cancer cell epithelial-mesenchymal transition. PMID: 24134845
Sparc/osteonectin, cwcv and kazal-like domains proteoglycan (testican) 2	SPOCK2	Identification of SPOCK2 as a susceptibility gene for bronchopulmonary dysplasia. PMID: 21836138
Sparc/osteonectin, cwcv and kazal-like domains proteoglycan (testican) 3	SPOCK3	SPOCK3, a risk gene for adult ADHD and personality disorders. PMID: 24292267
Sialic acid binding Ig-like lectin 1, sialoadhesin	SIGLEC1	Increased expression of Siglec-1 on peripheral blood monocytes and its role in mononuclear cell reactivity to autoantigen in rheumatoid arthritis. PMID: 24196391
Sialic acid binding Ig-like lectin 5	SIGLEC5	Identification of the CD33-related Siglec receptor, Siglec-5 (CD170), as a useful marker in both normal myelopoiesis and acute myeloid leukaemias. PMID: 14617000
Sialic acid binding Ig-like lectin 6	SIGLEC6	Siglec-6 is expressed in gestational trophoblastic disease and affects proliferation, apoptosis and invasion. PMID: 23089140
Sialic acid binding Ig-like lectin 7	SIGLEC7	Role of Siglec-7 in apoptosis in human platelets. PMID: 25230315
Sialic acid binding Ig-like lectin 8	SIGLEC8	Siglec-8 and Siglec-F, the new therapeutic targets in asthma. PMID: 22324980
Sialic acid binding Ig-like lectin 9	SIGLEC9	Binding of the sialic acid-binding lectin, Siglec-9, to the membrane mucin, MUC1, induces recruitment of β-catenin and subsequent cell growth. PMID: 24045940
Sialic acid binding Ig-like lectin 10	SIGLEC10	Human Siglec-10 can bind to vascular adhesion protein-1 and serves as its substrate. PMID: 19861682
Sialic acid binding Ig-like lectin 11	SIGLEC11	Alleviation of neurotoxicity by microglial human Siglec-11. PMID: 20203208
Sialic acid binding Ig-like lectin 12 (gene/pseudogene)	SIGLEC12	SIGLEC12, a human-specific segregating (pseudo)gene, encodes a signaling molecule expressed in prostate carcinomas. PMID: 21555517
Sialic acid binding Ig-like lectin 14	SIGLEC14	Loss of Siglec-14 reduces the risk of chronic obstructive pulmonary disease exacerbation. PMID: 23519826
Sialic acid binding Ig-like lectin 15	SIGLEC15	Siglec-15 protein regulates formation of functional osteoclasts in concert with DNAX-activating protein of 12 kDa (DAP12). PMID: 22451653
Sialic acid binding Ig-like lectin 16 (gene/pseudogene)	SIGLEC16	SIGLEC16 encodes a DAP12-associated receptor expressed in macrophages that evolved from its inhibitory counterpart SIGLEC11 and has functional and non-functional alleles in humans. PMID: 18629938
SIGLEC family like 1	SIGLECL1	
Selectin E	SELE	Enhanced expression of the soluble form of E-selectin attenuates progression of lupus nephritis and vasculitis in MRL/lpr mice. PMID: 25400916
Selectin L	SELL	**Aselizumab** – selective immunosuppressants
Selectin P (granule membrane protein 140 kDa, antigen CD62)	SELP	**Inclacumab** – cardiac therapy. Disease: ischemic stroke (ISCHSTR)
Selectin P ligand	SELPLG	Targeting CD162 protects against streptococcal M1 protein-evoked neutrophil recruitment and lung injury. PMID: 24039252

(Continued)

HGNC-approved name	Symbol	Target information
Syndecan 1	SDC1	Syndecan-1 overexpression is associated with nonluminal subtypes and poor prognosis in advanced breast cancer. PMID: 24045542
Syndecan 2	SDC2	Shed syndecan-2 inhibits angiogenesis. PMID: 25179601
Syndecan 3	SDC3	Syndecan-3 is selectively pro-inflammatory in the joint and contributes to antigen-induced arthritis in mice. PMID: 25015005
Syndecan 4	SDC4	Syndecan 4 is required for endothelial alignment in flow and atheroprotective signaling. PMID: 25404299
Syndecan-binding protein (syntenin)	SDCBP	Mda-9/syntenin is expressed in uveal melanoma and correlates with metastatic progression. PMID: 22267972
Thrombospondin 1	THBS1	Thrombospondin 1 in tissue repair and fibrosis: TGF-β-dependent and independent mechanisms. PMID: 22266026
Thrombospondin 2	THBS2	Disease: intervertebral disc disease (IDD)
Thrombospondin 3	THBS3	Effects of THBS3, SPARC and SPP1 expression on biological behavior and survival in patients with osteosarcoma. PMID: 17022822
Thrombospondin 4	THBS4	Thrombospondin-4 expression is activated during the stromal response to invasive breast cancer. PMID: 23942617
Activated leukocyte cell adhesion molecule	ALCAM	Activated leukocyte cell-adhesion molecule (ALCAM) promotes malignant phenotypes of malignant mesothelioma. PMID: 22722789
Adhesion molecule with Ig-like domain 1	AMIGO1	AMIGO is expressed in multiple brain cell types and may regulate dendritic growth and neuronal survival. PMID: 21938721
Adhesion molecule, interacts with CXADR antigen 1	AMICA1	The junctional adhesion molecule JAML is a costimulatory receptor for epithelial gammadelta T cell activation. PMID: 20813954
Advanced glycosylation end product-specific receptor	AGER	**TTP448** – anti-dementia drugs
Amine oxidase, copper containing 3 (see also metabolic and related enzymes)	AOC3	**Vapaliximab** – anti-inflammatory and antirheumatic products
Basal cell adhesion molecule (Lutheran blood group)	BCAM	The FBI1/Akirin2 target gene, BCAM, acts as a suppressive oncogene. PMID: 24223164
Cerebral endothelial cell adhesion molecule	CERCAM	Cerebral cell adhesion molecule: A novel leukocyte adhesion determinant on blood-brain barrier capillary endothelium. PMID: 10608765
Chitinase domain containing 1	CHID1	Elevated YKL40 is associated with advanced prostate cancer (PCa) and positively regulates invasion and migration of PCa cells. PMID: 24981110
Chondrolectin	CHODL	Chondrolectin is a novel diagnostic biomarker and a therapeutic target for lung cancer. PMID: 22016508
Cytotoxic and regulatory T-cell molecule	CRTAM	CRTAM: A molecule involved in epithelial cell adhesion. PMID: 20556794
Dendrocyte expressed seven transmembrane protein	DCSTAMP	Lentivirus-mediated RNA interference of DC-STAMP expression inhibits the fusion and resorptive activity of human osteoclasts. PMID: 23525827
Endothelial cell adhesion molecule	ESAM	Endothelial cell-selective adhesion molecule in diabetic nephropathy. PMID: 22957687
Epithelial cell adhesion molecule	EPCAM	**Solitomab** – antineoplastic agents. Disease: diarrhea 5, with tufting enteropathy, congenital (DIAR5)
F11 receptor	F11R	JAM-A expression positively correlates with poor prognosis in breast cancer patients. PMID: 19533747
Hepatic and glial cell adhesion molecule	HEPACAM	Overexpression of HepaCAM inhibits cell viability and motility through suppressing nucleus translocation of androgen receptor and ERK signaling in prostate cancer. PMID: 24811146. Disease: leukoencephalopathy, megalencephalic, with subcortical cysts, 2A (MLC2A)

HGNC-approved name	Symbol	Target information
L1 cell adhesion molecule	L1CAM	A Fab fragment directed against the neural cell adhesion molecule L1 enhances functional recovery after injury of the adult mouse spinal cord. PMID: 24673421. Disease: hydrocephalus due to stenosis of the aqueduct of Sylvius (HSAS)
Mannose-binding lectin (protein C) 2, soluble	MBL2	MBL, a potential therapeutic target for type 1 diabetes? PMID: 25155995
Melanoma cell adhesion molecule	MCAM	CD146 is a coreceptor for VEGFR-2 in tumor angiogenesis. PMID: 22718841
Mucosal vascular addressin cell adhesion molecule 1	MADCAM1	**PF-00547,659** – drugs for functional gastrointestinal disorders
Nephronectin	NPNT	The functional properties of nephronectin: An adhesion molecule for cardiac tissue engineering. PMID: 22436799
Neuronal cell adhesion molecule	NRCAM	Association of the neuronal cell adhesion molecule (NRCAM) gene variants with autism. PMID: 18664314
Oxidized low-density lipoprotein (lectin-like) receptor 1	OLR1	The lectin-like oxidized low-density lipoprotein receptor-1 as therapeutic target for atherosclerosis, inflammatory conditions and longevity. PMID: 23738516. Note: Independent association genetic studies have implicated OLR1 gene variants in myocardial infarction
Paxillin	PXN	A small molecule that inhibits the interaction of paxillin and alpha 4 integrin inhibits accumulation of mononuclear leukocytes at a site of inflammation. PMID: 20097761
Platelet endothelial aggregation receptor 1	PEAR1	Genetic variation in PEAR1 is associated with platelet aggregation and cardiovascular outcomes. PMID: 23392654
Platelet/endothelial cell adhesion molecule 1	PECAM1	Platelet endothelial cell adhesion molecule targeted oxidant-resistant mutant thrombomodulin fusion protein with enhanced potency *in vitro* and *in vivo*. PMID: 23965383
Signal peptide, CUB domain, EGF-like 1	SCUBE1	Inhibition of the plasma SCUBE1, a novel platelet adhesive protein, protects mice against thrombosis. PMID: 24833801
Transmembrane and immunoglobulin domain containing 2	TMIGD2	Identification of IGPR-1 as a novel adhesion molecule involved in angiogenesis. PMID: 22419821
Vascular cell adhesion molecule 1	VCAM1	VCAM1 expression correlated with tumorigenesis and poor prognosis in high grade serous ovarian cancer. PMID: 23634244
Zona pellucida glycoprotein 3 (sperm receptor)	ZP3	Evidence for the inhibition of fertilization *in vitro* by anti-ZP3 antisera derived from DNA vaccine. PMID: 21596079

3.5   Host defence molecules

HGNC-approved name	Symbol	Target information
Bone marrow stromal cell antigen 1	BST1	Overexpression of CD157 contributes to epithelial ovarian cancer progression by promoting mesenchymal differentiation. PMID: 22916288
Bone marrow stromal cell antigen 2	BST2	Anti-inflammatory and anti-remodelling effects of ISU201, a modified form of the extracellular domain of human BST2, in experimental models of asthma: Association with inhibition of histone acetylation. PMID: 24594933
BPI fold-containing family A, member 1	BPIFA1	BPIFA1 (SPLUNC1) in airway host protection and respiratory disease. PMID: 25265466
BPI fold-containing family A, member 2	BPIFA2	Dual host-defence functions of SPLUNC2/PSP and synthetic peptides derived from the protein. PMID: 21787342
BPI fold-containing family A, member 3	BPIFA3	

(Continued)

HGNC-approved name	Symbol	Target information
BPI fold-containing family B, member 1	BPIFB1	BPIFB1 is a lung-specific autoantigen associated with interstitial lung disease. PMID: 24107778
BPI fold-containing family B, member 2	BPIFB2	
BPI fold-containing family B, member 3	BPIFB3	
BPI fold-containing family B, member 4	BPIFB4	Phylogenetic and evolutionary analysis of the PLUNC gene family. PMID: 14739326
BPI fold-containing family B, member 6	BPIFB6	
BPI fold-containing family C	BPIFC	
Butyrophilin, subfamily 1, member A1	BTN1A1	Immune modulation by butyrophilins. PMID: 25060581
Butyrophilin, subfamily 2, member A1	BTN2A1	The B7 homolog butyrophilin BTN2A1 is a novel ligand for DC-SIGN. PMID: 17785817
Butyrophilin, subfamily 2, member A2	BTN2A2	Butyrophilin Btn2a2 inhibits TCR activation and phosphatidylinositol 3-kinase/Akt pathway signaling and induces Foxp3 expression in I lymphocytes. PMID: 23589618
Butyrophilin, subfamily 3, member A1	BTN3A1	Butyrophilin 3A1 binds phosphorylated antigens and stimulates human $\gamma\delta$ T cells. PMID: 23872678
Butyrophilin, subfamily 3, member A2	BTN3A2	BTN3A2 expression in epithelial ovarian cancer is associated with higher tumor infiltrating T cells and a better prognosis. PMID: 22685580
Butyrophilin, subfamily 3, member A3	BTN3A3	
Butyrophilin-like 2 (MHC class II associated)	BTNL2	Disease: sarcoidosis 2 (SS2)
Butyrophilin-like 3	BTNL3	A butyrophilin family member critically inhibits T cell activation. PMID: 20944003
Butyrophilin-like 8	BTNL8	BTNL8, a butyrophilin-like molecule that costimulates the primary immune response. PMID: 24036152
Butyrophilin-like 9	BTNL9	
C1q and tumor necrosis factor-related protein 1	C1QTNF1	Tumor necrosis factor-alpha and interleukin-1beta increases CTRP1 expression in adipose tissue. PMID: 16806199
C1q and tumor necrosis factor-related protein 2	C1QTNF2	
C1q and tumor necrosis factor-related protein 3	C1QTNF3	The novel adipokine C1q/TNF-related protein-3 is expressed in human adipocytes and regulated by metabolic and infection-related parameters. PMID: 23174996
C1q and tumor necrosis factor-related protein 4	C1QTNF4	C1q/TNF-related protein 4 (CTRP4) is a unique secreted protein with two tandem C1q domains that functions in the hypothalamus to modulate food intake and body weight. PMID: 24366864
C1q and tumor necrosis factor-related protein 5	C1QTNF5	Disease: late-onset retinal degeneration (LORD)
C1q and tumor necrosis factor-related protein 6	C1QTNF6	C1qTNF-related protein-6 increases the expression of interleukin-10 in macrophages. PMID: 20652496
C1q and tumor necrosis factor-related protein 7	C1QTNF7	
C1q and tumor necrosis factor-related protein 8	C1QTNF8	C1q-tumour necrosis factor-related protein 8 (CTRP8) is a novel interaction partner of relaxin receptor RXFP1 in human brain cancer cells. PMID: 24014093
C1q and tumor necrosis factor-related protein 9	C1QTNF9	Identification and characterization of CTRP9, a novel secreted glycoprotein, from adipose tissue that reduces serum glucose in mice and forms heterotrimers with adiponectin. PMID: 18787108
C1q and tumor necrosis factor-related protein 9B	C1QTNF9B	CTRP8 and CTRP9B are novel proteins that hetero-oligomerize with C1q/TNF family members. PMID: 19666007

HGNC-approved name	Symbol	Target information
CD1a molecule	CD1A	**Antithymocyte globulin** – immunomodulating agents
CD1b molecule	CD1B	
CD1c molecule	CD1C	
CD1d molecule	CD1D	
CD1e molecule	CD1E	
CD2 molecule	CD2	**Alefacept** – selective immunosuppressants
CD2-associated protein	CD2AP	Disease: focal segmental glomerulosclerosis 3 (FSGS3)
CD3e molecule, epsilon (CD3-TCR complex)	CD3E	**Visilizumab** – selective immunosuppressants. Disease: immunodeficiency 18 (IMD18)
CD3d molecule, delta (CD3-TCR complex)	CD3D	Disease: severe combined immunodeficiency autosomal recessive T-cell-negative/B-cell-positive/NK-cell-positive (T(−)B(+)NK(+) SCID)
CD3g molecule, gamma (CD3-TCR complex)	CD3G	Disease: immunodeficiency 17 (IMD17)
CD4 molecule	CD4	**Priliximab** – selective immunosuppressants
CD5 molecule	CD5	
CD5 molecule-like	CD5L	Api6/AIM/Spα/CD5L overexpression in alveolar type II epithelial cells induces spontaneous lung adenocarcinoma. PMID: 21697282
CD6 molecule	CD6	**Itolizumab** – antipsoriatics
CD7 molecule	CD7	
CD8a molecule	CD8A	Disease: CD8 deficiency, familial (CD8 deficiency)
CD8b molecule	CD8B	
CD9 molecule	CD9	The nuclear pool of tetraspanin CD9 contributes to mitotic processes in human breast carcinoma. PMID: 25103498
CD14 molecule	CD14	Directed evolution of an LBP/CD14 inhibitory peptide and its anti-endotoxin activity. PMID: 25025695
CD19 molecule	CD19	**Binatumomab (bispecific with anti CD3)** – antineoplastic agents. Disease: immunodeficiency, common variable, 3 (CVID3)
Membrane-spanning 4-domains, subfamily A, member 1	MS4A1	**Rituximab** – antineoplastic agents. Disease: immunodeficiency, common variable, 5 (CVID5)
CD22 molecule	CD22	**Epratuzumab** – selective immunosuppressants
CD24 molecule	CD24	Amelioration of sepsis by inhibiting sialidase-mediated disruption of the CD24-SiglecG interaction. PMID: 21478876. Disease: multiple sclerosis (MS)
CD27 molecule	CD27	**CDX-1127** – antineoplastic agents. Disease: lymphoproliferative syndrome 2 (LPFS2)
CD28 molecule	CD28	Amplification of regulatory T Cells using a CD28 superagonist reduces brain damage after ischemic stroke in mice. PMID: 25378432
CD33 molecule	CD33	**Gemtuzumab ozogamicin** – antineoplastic agents
CD34 molecule	CD34	Circulating CD34+ progenitor cells and risk of mortality in a population with coronary artery disease. PMID: 25323857
CD36 molecule (thrombospondin receptor)	CD36	CD36 as a therapeutic target for endothelial dysfunction in stroke. PMID: 22574985. Disease: platelet glycoprotein IV deficiency (PG4D)
CD37 molecule	CD37	**Otlertuzumab** – antineoplastic agents
CD38 molecule	CD38	**Daratumumab** – multiple myeloma
Sialophorin	SPN	CD43 in the nucleus and cytoplasm of lung cancer is a potential therapeutic target. PMID: 23015282
CD44 molecule (Indian blood group)	CD44	**Bivatuzumab** – antineoplastic agents

(Continued)

HGNC-approved name	Symbol	Target information
CD46 molecule, complement regulatory protein	CD46	Disease: hemolytic uremic syndrome atypical 2 (AHUS2)
CD47 molecule	CD47	The CD47-SIRPα signalling system: Its physiological roles and therapeutic application. PMID: 24627525
CD48 molecule	CD48	The CD48 receptor mediates Staphylococcus aureus human and murine eosinophil activation. PMID: 25255823
CD52 molecule	CD52	**Alemtuzumab** – selective immunosuppressants
CD53 molecule	CD53	A genome-wide linkage scan reveals CD53 as an important regulator of innate TNF-alpha levels. PMID: 20407468
CD55 molecule, decay-accelerating factor for complement (Cromer blood group)	CD55	Decay-accelerating factor (CD55): A versatile acting molecule in human malignancies. PMID: 16784816
CD58 molecule	CD58	CD58, a novel surface marker, promotes self-renewal of tumor-initiating cells in colorectal cancer. PMID: 24727892
CD59 molecule, complement regulatory protein	CD59	Disease: hemolytic anemia, CD59-mediated, with or without polyneuropathy (HACD59)
CD63 molecule	CD63	Tetraspanin CD63 acts as a pro-metastatic factor via β-catenin stabilization. PMID: 25354204
CD68 molecule	CD68	Deletion of the murine scavenger receptor CD68. PMID: 21572087
CD72 molecule	CD72	CD72 negatively regulates mouse mast cell functions and down-regulates the expression of KIT and FcεRIα. PMID: 25239131
CD79a molecule, immunoglobulin-associated alpha	CD79A	Disease: agammaglobulinemia 3, autosomal recessive (AGM3)
CD79b molecule, immunoglobulin-associated beta	CD79B	Anti-CD79 antibody induces B cell anergy that protects against autoimmunity. PMID: 24442438. Disease: agammaglobulinemia 6, autosomal recessive (AGM6)
CD80 molecule	CD80	**Abatacept** – selective immunosuppressants
CD81 molecule	CD81	Disease: immunodeficiency, common variable, 6 (CVID6)
CD82 molecule	CD82	Identification of the tetraspanin CD82 as a new barrier to xenotransplantation. PMID: 23872050
CD83 molecule	CD83	Post-transcriptional regulation of CD83 expression by AUF1 proteins. PMID: 23161671
CD84 molecule	CD84	CD84 is a survival receptor for CLL cells. PMID: 23435417
CD86 molecule	CD86	**Abatacept** – selective immunosuppressants
Plasminogen activator, urokinase receptor	PLAUR	The urokinase receptor system as strategic therapeutic target: challenges for the 21st century. PMID: 21711231
Thy-1 cell surface antigen	THY1	Thy1 (CD90) controls adipogenesis by regulating activity of the Src family kinase, Fyn. PMID: 25416548
CD93 molecule	CD93	The characterization of a novel monoclonal antibody against CD93 unveils a new antiangiogenic target. PMID: 24809468
CD96 molecule	CD96	Disease: C syndrome (CSYN)
CD99 molecule	CD99	CD99 suppresses osteosarcoma cell migration through inhibition of ROCK2 activity. PMID: 23644663
CD99 molecule-like 2	CD99L2	Cutting edge: Endothelial-specific gene ablation of CD99L2 impairs leukocyte extravasation in vivo. PMID: 23293350
CD101 molecule	CD101	Evidence that Cd101 is an autoimmune diabetes gene in nonobese diabetic mice. PMID: 21613616
Thrombomodulin	THBD	**ART-123** – antithrombotic agents. Disease: thrombophilia due to thrombomodulin defect (THPH12)
Basigin (Ok blood group)	BSG	**Gavilimomab** – selective immunosuppressants

HGNC-approved name	Symbol	Target information
CD151 molecule (Raph blood group)	CD151	Integrin-associated CD151 drives ErbB2-evoked mammary tumor onset and metastasis. PMID: 22952421. Disease: nephropathy with pretibial epidermolysis bullosa and deafness (NPEBD)
Cytotoxic T-lymphocyte-associated protein 4	CTLA4	**Ipilimumab** – antineoplastic agents. Disease: systemic lupus erythematosus (SLE)
Poliovirus receptor	PVR	UPR decreases CD226 ligand CD155 expression and sensitivity to NK cell-mediated cytotoxicity in hepatoma cells. PMID: 25209846
CD160 molecule	CD160	CD160 isoforms and regulation of CD4 and CD8 T-cell responses. PMID: 25179432
CD163 molecule	CD163	Scavenger receptor CD163, a Jack-of-all-trades and potential target for cell-directed therapy. PMID: 20299103
CD163 molecule-like 1	CD163L1	
CD164 molecule, sialomucin	CD164	Inhibiting CD164 expression in colon cancer cell line HCT116 leads to reduced cancer cell proliferation, mobility, and metastasis *in vitro* and in vivo. PMID: 22409183
CD164 sialomucin-like 2	CD164L2	
CD177 molecule	CD177	Elevated neutrophil membrane expression of proteinase 3 is dependent upon CD177 expression. PMID: 20491791
CD180 molecule	CD180	The radioprotective 105/MD-1 complex contributes to diet-induced obesity and adipose tissue inflammation. PMID: 22396206
CD200 molecule	CD200	**Samalizumab** – antineoplastic agents
CD200 receptor 1	CD200R1	Aberrant CD200/CD200R1 expression and its potential role in Th17 cell differentiation, chemotaxis and osteoclastogenesis in rheumatoid arthritis. PMID: 25261692
CD200 receptor 1-like	CD200R1L	An interaction between CD200 and monoclonal antibody agonists to CD200R2 in development of dendritic cells that preferentially induce populations of CD4+CD25+ T regulatory cells. PMID: 18424714
Protein C receptor, endothelial	PROCR	Identification of multipotent mammary stem cells by protein C receptor expression. PMID: 25327250
CD226 molecule	CD226	Accelerated tumor growth in mice deficient in DNAM-1 receptor. PMID: 19029379
CD244 molecule, natural killer cell receptor 2B4	CD244	2B4 (CD244) induced by selective CD28 blockade functionally regulates allograft-specific CD8+ T cell responses. PMID: 24493803
CD247 molecule	CD247	Disease: immunodeficiency due to defect in CD3-zeta (CD3ZID)
CD248 molecule, endosialin	CD248	**Ontuxizumab** – antineoplastic agents
Programmed cell death 1 ligand 2	PDCD1LG2	**AMP-224** – antineoplastic agents
CD274 molecule	CD274	**MPDL3280A** – antineoplastic agents
CD276 molecule	CD276	B7-H3-mediated tumor immunology: Friend or foe? PMID: 24013874
Programmed cell death 1	PDCD1	**Nivolumab** – antineoplastic agents. Disease: systemic lupus erythematosus 2 (SLEB2)
CD300a molecule	CD300A	Mechanism of phosphatidylserine inhibition of IgE/FcεRI-dependent anaphylactic human basophil degranulation via CD300a. PMID: 24815424
CD300c molecule	CD300C	Human CD300C delivers an Fc receptor-γ-dependent activating signal in mast cells and monocytes and differs from CD300A in ligand recognition. PMID: 23372157
CD300e molecule	CD300E	Functional analysis of the CD300e receptor in human monocytes and myeloid dendritic cells. PMID: 20039296

(Continued)

HGNC-approved name	Symbol	Target information
CD300 molecule-like family member b	CD300LB	CD300b regulates the phagocytosis of apoptotic cells via phosphatidylserine recognition. PMID: 25034781
CD300 molecule-like family member d	CD300LD	Cloning and characterization of CD300d, a novel member of the human CD300 family of immune receptors. PMID: 22291008
CD300 molecule-like family member f	CD300LF	Sphingomyelin and ceramide are physiological ligands for human LMIR3/CD300f, inhibiting FcεRI-mediated mast cell activation. PMID: 24035150
CD300 molecule-like family member g	CD300LG	Dynamic changes in endothelial cell adhesion molecule nepmucin/CD300LG expression under physiological and pathological conditions. PMID: 24376728
CD320 molecule	CD320	Disease: methylmalonic aciduria type TCblR (MMATC)
Coagulation factor III (thromboplastin, tissue factor)	F3	**HuMax-TF-ADC** – antineoplastic agents
Coagulation factor VIII-associated 1	F8A1	
Coagulation factor VIII-associated 2	F8A2	
Coagulation factor VIII-associated 3	F8A3	
Coagulation factor XIII, A1 polypeptide	F13A1	Disease: factor XIII subunit A deficiency (FA13AD)
Coagulation factor XIII, B polypeptide	F13B	Disease: factor XIII subunit B deficiency (FA13BD)
Complement component 1, q subcomponent, A chain	C1QA	**Immune globulin** – immunostimulants. Disease: complement component C1q deficiency (C1QD)
Complement component 1, q subcomponent, B chain	C1QB	Disease: complement component C1q deficiency (C1QD)
Complement component 1, q subcomponent binding protein	C1QBP	Interactome analysis reveals that C1QBP (complement component 1, q subcomponent binding protein) is associated with cancer cell chemotaxis and metastasis. PMID: 23924515
Complement component 1, q subcomponent, C chain	C1QC	Disease: complement component C1q deficiency (C1QD)
Complement component 1, q subcomponent-like 1	C1QL1	
Complement component 1, q subcomponent-like 2	C1QL2	
Complement component 1, q subcomponent-like 3	C1QL3	
Complement component 1, q subcomponent-like 4	C1QL4	
Complement component 3	C3	Transient complement inhibition promotes a tumor-specific immune response through the implication of natural killer cells. PMID: 24778316. Disease: complement component 3 deficiency (C3D)
Complement component (3b/4b) receptor 1 (Knops blood group)	CR1	Soluble CR1 therapy improves complement regulation in C3 glomerulopathy. PMID: 23907509
Complement component (3b/4b) receptor 1-like	CR1L	
Complement component (3d/Epstein–Barr virus) receptor 2	CR2	Systemic human CR2-targeted complement alternative pathway inhibitor ameliorates mouse laser-induced choroidal neovascularization. PMID: 22309197. Disease: systemic lupus erythematosus 9 (SLEB9)
Complement component 4A (Rodgers blood group)	C4A	Complement anaphylatoxin C4a inhibits C5a-induced neointima formation following arterial injury. PMID: 24789665
Complement component 4B (Chido blood group), copy 2	C4B_2	Disease: systemic lupus erythematosus (SLE)
Complement component 4 binding protein, alpha	C4BPA	

HGNC-approved name	Symbol	Target information
Complement component 4 binding protein, beta	C4BPB	
Complement component 5	C5	**Eculizumab** – immunomodulating agents. Disease: complement component 5 deficiency (C5D)
Complement component 6	C6	Disease: complement component 6 deficiency (C6D)
Complement component 7	C7	Disease: complement component 7 deficiency (C7D)
Complement component 8, alpha polypeptide	C8A	Disease: complement component 8 deficiency, 1 (C8D1)
Complement component 8, beta polypeptide	C8B	Disease: complement component 8 deficiency, 2 (C8D2)
Complement component 8, gamma polypeptide	C8G	
Complement component 9	C9	Disease: complement component 9 deficiency (C9D)
Complement factor H	CFH	Complement factor H and age-related macular degeneration: the role of glycosaminoglycan recognition in disease pathology. PMID: 20863311. Disease: basal laminar drusen (BLD)
Complement factor H-related 1	CFHR1	
Complement factor H-related 2	CFHR2	Human factor H-related protein 2 (CFHR2) regulates complement activation. PMID: 24260121
Complement factor H-related 3	CFHR3	
Complement factor H-related 4	CFHR4	Factor H-related protein 4 activates complement by serving as a platform for the assembly of alternative pathway C3 convertase via its interaction with C3b protein. PMID: 22518841
Complement factor H-related 5	CFHR5	Note: Defects in CFHR5 have been found in patients with atypical hemolytic uremic syndrome and may contribute to the disease
Complement factor properdin	CFP	Genetic and therapeutic targeting of properdin in mice prevents complement-mediated tissue injury. PMID: 20941861. Disease: properdin deficiency (PFD)
Defensin, alpha 1	DEFA1	Defensins and sepsis. PMID: 25210703
Defensin, alpha 1B	DEFA1B	
Defensin, alpha 3, neutrophil-specific	DEFA3	
Defensin, alpha 4, corticostatin	DEFA4	
Defensin, alpha 5, Paneth cell-specific	DEFA5	Defensin-immunology in inflammatory bowel disease. PMID: 20117337
Defensin, alpha 6, Paneth cell-specific	DEFA6	Defensin-immunology in inflammatory bowel disease. PMID: 20117337
Defensin, beta 1	DEFB1	Human beta-defensin-1 suppresses tumor migration and invasion and is an independent predictor for survival of oral squamous cell carcinoma patients. PMID: 24658581
Defensin, beta 4A	DEFB4A	Overexpression of human β-defensin 2 promotes growth and invasion during esophageal carcinogenesis. PMID: 25226614
Defensin, beta 4B	DEFB4B	
Defensin, beta 103A	DEFB103A	
Defensin, beta 103B	DEFB103B	
Defensin, beta 104A	DEFB104A	
Defensin, beta 104B	DEFB104B	
Defensin, beta 105A	DEFB105A	
Defensin, beta 105B	DEFB105B	
Defensin, beta 106A	DEFB106A	
Defensin, beta 106B	DEFB106B	

(Continued)

HGNC-approved name	Symbol	Target information
Defensin, beta 107A	DEFB107A	
Defensin, beta 107B	DEFB107B	
Defensin, beta 108B	DEFB108B	
Defensin, beta 110 locus	DEFB110	
Defensin, beta 112	DEFB112	
Defensin, beta 113	DEFB113	
Defensin, beta 114	DEFB114	
Defensin, beta 115	DEFB115	
Defensin, beta 116	DEFB116	
Defensin, beta 118	DEFB118	
Defensin, beta 119	DEFB119	
Defensin, beta 121	DEFB121	
Defensin, beta 123	DEFB123	The novel beta-defensin DEFB123 prevents lipopolysaccharide-mediated effects *in vitro* and in vivo. PMID: 16790530
Defensin, beta 124	DEFB124	
Defensin, beta 125	DEFB125	
Defensin, beta 126	DEFB126	Human beta-defensin DEFB126 is capable of inhibiting LPS-mediated inflammation. PMID: 23229569
Defensin, beta 127	DEFB127	
Defensin, beta 128	DEFB128	
Defensin, beta 129	DEFB129	
Defensin, beta 130	DEFB130	
Defensin, beta 131	DEFB131	
Defensin, beta 132	DEFB132	
Defensin, beta 133	DEFB133	
Defensin, beta 134	DEFB134	
Defensin, beta 135	DEFB135	
Defensin, beta 136	DEFB136	
Fc fragment of IgG, high affinity Ia, receptor (CD64)	FCGR1A	Identification of cyclic peptides able to mimic the functional epitope of IgG1-Fc for human Fc gammaRI. PMID: 18957574
Fc fragment of IgG, high affinity Ib, receptor (CD64)	FCGR1B	
Fc fragment of IgG, low affinity IIa, receptor (CD32)	FCGR2A	Inhibition of destructive autoimmune arthritis in FcgammaRIIa transgenic mice by small chemical entities. PMID: 19030019
Fc fragment of IgG, low affinity IIb, receptor (CD32)	FCGR2B	Note: A chromosomal aberration involving FCGR2B is found in a follicular lymphoma. Translocation t(1;22)(q22;q11). The translocation leads to the hyperexpression of the receptor. This may play a role in the tumor progression. Disease: systemic lupus erythematosus (SLE)
Fc fragment of IgG, low affinity IIc, receptor for (CD32) (gene/pseudogene)	FCGR2C	
Fc fragment of IgG, low affinity IIIa, receptor (CD16a)	FCGR3A	A novel tetravalent bispecific TandAb (CD30/CD16A) efficiently recruits NK cells for the lysis of CD30+ tumor cells. PMID: 24670809
Fc fragment of IgG, low affinity IIIb, receptor (CD16b)	FCGR3B	Human basophils express the glycosylphosphatidylinositol-anchored low-affinity IgG receptor FcgammaRIIIB (CD16B). PMID: 19201911
Fc fragment of IgG, receptor, transporter, alpha	FCGRT	The immunologic functions of the neonatal Fc receptor for IgG. PMID: 22948741

HGNC-approved name	Symbol	Target information
Fc receptor, IgA, IgM, high affinity	FCAMR	Involvement of Fcα/μR (CD351) in autoantibody production. PMID: 24172225
Fc fragment of IgA, receptor for	FCAR	
Fc fragment of IgE, high affinity I, receptor for; alpha polypeptide	FCER1A	**Omalizumab** – drugs for obstructive airway diseases
Membrane-spanning 4-domains, subfamily A, member 2	MS4A2	The interaction between Lyn and FcεRIβ is indispensable for FcεRI-mediated human mast cell activation. PMID: 22845063
Fc fragment of IgE, high affinity I, receptor for; gamma polypeptide	FCER1G	
Fc fragment of IgE, low affinity II, receptor for (CD23)	FCER2	**Lumiliximab** – antineoplastic agents
Fc receptor-like 1	FCRL1	FCRL1 on chronic lymphocytic leukemia, hairy cell leukemia, and B-cell non-Hodgkin lymphoma as a target of immunotoxins. PMID: 17895404
Fc receptor-like 2	FCRL2	Ligation of human Fc receptor like-2 by monoclonal antibodies down-regulates B-cell receptor-mediated signalling. PMID: 24797767
Fc receptor-like 3	FCRL3	Disease: rheumatoid arthritis (RA)
Fc receptor-like 4	FCRL4	Note: A chromosomal aberration involving FCRL4 is found in non-Hodgkin lymphoma (NHG). Translocation t(11)(p36.3 q21.1-2). Note: A chromosomal aberration involving FCRL4 is found in multiple myeloma (MM)
Fc receptor-like 5	FCRL5	FcRL5 as a target of antibody-drug conjugates for the treatment of multiple myeloma. PMID: 22807577. Note: A chromosomal aberration involving FCRL5 has been found in cell lines with 1q21 abnormalities derived from Burkitt lymphoma. Duplication dup(1)(q21q3
Fc receptor-like 6	FCRL6	FCRL6 receptor: Expression and associated proteins. PMID: 20933011
Fc receptor-like A	FCRLA	Emerging roles for the FCRL family members in lymphocyte biology and disease. PMID: 25116094
Fc receptor-like B	FCRLB	FCGR2B and FCRLB gene polymorphisms associated with IgA nephropathy. PMID: 23593433
Glycoprotein Ib (platelet), alpha polypeptide	GP1BA	Disease: non-arteritic anterior ischemic optic neuropathy (NAION)
Glycoprotein Ib (platelet), beta polypeptide	GP1BB	Disease: Bernard–Soulier syndrome (BSS)
Glycoprotein V (platelet)	GP5	
Glycoprotein VI (platelet)	GP6	A novel antiplatelet antibody therapy that induces cAMP-dependent endocytosis of the GPVI/Fc receptor gamma-chain complex. PMID: 18382762. Disease: bleeding disorder, platelet-type 11 (BDPLT11)
Glycoprotein IX (platelet)	GP9	Disease: Bernard–Soulier syndrome (BSS)
Glycoprotein (transmembrane) nmb	GPNMB	**Glembatumumab vedotin** – antineoplastic agents. Note: Glioblastoma multiforme patients that exhibit increased mRNA and protein levels (>3-fold over normal brain) in biopsy samples have a significantly higher risk of death
Histatin 1	HTN1	Structure-activity analysis of histatin, a potent wound healing peptide from human saliva: Cyclization of histatin potentiates molar activity 1000-fold. PMID: 19652025
Histatin 3	HTN3	
Immunoglobulin lambda-like polypeptide 1	IGLL1	Disease: agammaglobulinemia 2, autosomal recessive (AGM2)
Immunoglobulin lambda-like polypeptide 5	IGLL5	

(Continued)

HGNC-approved name	Symbol	Target information
Immunoglobulin superfamily, member 6	IGSF6	
Immunoglobulin superfamily, member 8	IGSF8	Glioblastoma inhibition by cell surface immunoglobulin protein EWI-2, *in vitro* and in vivo. PMID: 19107234
Inducible T-cell co-stimulator	ICOS	Engagement of the ICOS pathway markedly enhances efficacy of CTLA-4 blockade in cancer immunotherapy. PMID: 24687957. Disease: immunodeficiency, common variable, 1 (CVID1)
Inducible T-cell co-stimulator ligand	ICOSLG	**MEDI5872** – selective immunosuppressants
Killer cell immunoglobulin-like receptor, two domains, long cytoplasmic tail, 1	KIR2DL1	**Lirilumab** – antineoplastic agents
Killer cell immunoglobulin-like receptor, two domains, long cytoplasmic tail, 2	KIR2DL2	**Lirilumab** – antineoplastic agents
Killer cell immunoglobulin-like receptor, two domains, long cytoplasmic tail, 3	KIR2DL3	**Lirilumab** – antineoplastic agents
Killer cell immunoglobulin-like receptor, two domains, long cytoplasmic tail, 4	KIR2DL4	KIR2DL4 (CD158d) – an activation receptor for HLA-G. PMID: 22934097
Killer cell immunoglobulin-like receptor, two domains, long cytoplasmic tail, 5A	KIR2DL5A	
Killer cell immunoglobulin-like receptor, two domains, long cytoplasmic tail, 5B	KIR2DL5B	Association of KIR2DL5B gene with celiac disease supports the susceptibility locus on 19q13.4. PMID: 17215859
Killer cell immunoglobulin-like receptor, two domains, short cytoplasmic tail, 1	KIR2DS1	Donor activating KIR2DS1 in leukemia. PMID: 25409391
Killer cell immunoglobulin-like receptor, two domains, short cytoplasmic tail, 2	KIR2DS2	NK cells with KIR2DS2 immunogenotype have a functional activation advantage to efficiently kill glioblastoma and prolong animal survival. PMID: 25381437
Killer cell immunoglobulin-like receptor, two domains, short cytoplasmic tail, 3	KIR2DS3	Association of KIR2DS1 and KIR2DS3 with fatal outcome in Ebola virus infection. PMID: 20878400
Killer cell immunoglobulin-like receptor, two domains, short cytoplasmic tail, 4	KIR2DS4	Presence of the full-length KIR2DS4 gene reduces the chance of rheumatoid arthritis patients to respond to methotrexate treatment. PMID: 25069714
Killer cell immunoglobulin-like receptor, two domains, short cytoplasmic tail, 5	KIR2DS5	The presence of KIR2DS5 confers protection against adult immune thrombocytopenia. PMID: 24571473
Killer cell immunoglobulin-like receptor, three domains, long cytoplasmic tail, 1	KIR3DL1	Upregulation of KIR3DL1 gene expression in intestinal mucosa in active celiac disease. PMID: 21616111
Killer cell immunoglobulin-like receptor, three domains, long cytoplasmic tail, 2	KIR3DL2	Killer-cell immunoglobulin-like receptors (KIR) and HLA-class I heavy chains in ankylosing spondylitis. PMID: 25381967
Killer cell immunoglobulin-like receptor, three domains, long cytoplasmic tail, 3	KIR3DL3	
Killer cell immunoglobulin-like receptor, three domains, short cytoplasmic tail, 1	KIR3DS1	Role of KIR3DS1 in human diseases. PMID: 23125843
Killer cell lectin-like receptor subfamily B, member 1	KLRB1	CD39 and CD161 modulate Th17 responses in Crohn's disease. PMID: 25172498
Killer cell lectin-like receptor subfamily C, member 1	KLRC1	**NN8765** – selective immunosuppressants
Killer cell lectin-like receptor subfamily C, member 2	KLRC2	Transplantation-induced cancers: Emerging evidence that clonal CMV-specific NK cells are causal immunogenic factors. PMID: 25050225
Killer cell lectin-like receptor subfamily C, member 3	KLRC3	
Killer cell lectin-like receptor subfamily C, member 4	KLRC4	Up-regulation of NKG2F receptor, a functionally unknown killer receptor, of human natural killer cells by interleukin-2 and interleukin-15. PMID: 20811687

HGNC-approved name	Symbol	Target information
Killer cell lectin-like receptor subfamily D, member 1	KLRD1	Low gene expression levels of activating receptors of natural killer cells (NKG2E and CD94) in patients with fulminant type 1 diabetes. PMID: 24177169
Killer cell lectin-like receptor subfamily F, member 1	KLRF1	Tumor cells of non-hematopoietic and hematopoietic origins express activation-induced C-type lectin, the ligand for killer cell lectin-like receptor F1. PMID: 20663776
Killer cell lectin-like receptor subfamily F, member 2	KLRF2	Structure of NKp65 bound to its keratinocyte ligand reveals basis for genetically linked recognition in natural killer gene complex. PMID: 23803857
Killer cell lectin-like receptor subfamily G, member 1	KLRG1	Transferrin' activation: Bonding with transferrin receptors tunes KLRG1 function. PMID: 24752778
Killer cell lectin-like receptor subfamily G, member 2	KLRG2	
Killer cell lectin-like receptor subfamily K, member 1	KLRK1	**NN8555** – selective immunosuppressants
Leukocyte-associated immunoglobulin-like receptor 1	LAIR1	The role of LAIR-1 (CD305) in T cells and monocytes/macrophages in patients with rheumatoid arthritis. PMID: 24380839
Leukocyte-associated immunoglobulin-like receptor 2	LAIR2	LAIR2 localizes specifically to sites of extravillous trophoblast invasion. PMID: 20692035
Leukocyte immunoglobulin-like receptor, subfamily A (with TM domain), member 1	LILRA1	Expression of activating and inhibitory leukocyte immunoglobulin-like receptors in rheumatoid synovium: Correlations to disease activity. PMID: 21388353
Leukocyte immunoglobulin-like receptor, subfamily A (with TM domain), member 2	LILRA2	
Leukocyte immunoglobulin-like receptor, subfamily A (without TM domain), member 3	LILRA3	Soluble LILRA3, a potential natural anti-inflammatory protein, is increased in patients with rheumatoid arthritis and is tightly regulated by interleukin 10, tumor necrosis factor-alpha, and interferon-gamma. PMID: 20595277
Leukocyte immunoglobulin-like receptor, subfamily A (with TM domain), member 4	LILRA4	Effect of immunoglobin-like transcript 7 cross-linking on plasmacytoid dendritic cells differentiation into antigen-presenting cells. PMID: 24586760
Leukocyte immunoglobulin-like receptor, subfamily A (with TM domain), member 5	LILRA5	LILRA5 is expressed by synovial tissue macrophages in rheumatoid arthritis, selectively induces pro-inflammatory cytokines and IL-10 and is regulated by TNF-alpha, IL-10 and IFN-gamma. PMID: 19009525
Leukocyte immunoglobulin-like receptor, subfamily A (with TM domain), member 6	LILRA6	
Leukocyte immunoglobulin-like receptor, subfamily B (with TM and ITIM domains), member 1	LILRB1	
Leukocyte immunoglobulin-like receptor, subfamily B (with TM and ITIM domains), member 2	LILRB2	Human LilrB2 is a β-amyloid receptor and its murine homolog PirB regulates synaptic plasticity in an Alzheimer's model. PMID: 24052308
Leukocyte immunoglobulin-like receptor, subfamily B (with TM and ITIM domains), member 3	LILRB3	
Leukocyte immunoglobulin-like receptor, subfamily B (with TM and ITIM domains), member 4	LILRB4	Crystal structure of leukocyte Ig-like receptor LILRB4 (ILT3/LIR-5/CD85k): A myeloid inhibitory receptor involved in immune tolerance. PMID: 21454581
Leukocyte immunoglobulin-like receptor, subfamily B (with TM and ITIM domains), member 5	LILRB5	CKM and LILRB5 are associated with serum levels of creatine kinase. PMID: 25214527

(Continued)

HGNC-approved name	Symbol	Target information
Linker for activation of T cells	LAT	Molecular editing of cellular responses by the high-affinity receptor for IgE. PMID: 24505132
Linker for activation of T cells family, member 2	LAT2	Molecular editing of cellular responses by the high-affinity receptor for IgE. PMID: 24505132. Note: LAT2 is located in the Williams-Beuren syndrome (WBS) critical region
MHC class I polypeptide-related sequence A	MICA	An six-amino acid motif in the alpha 3 domain of MICA is the cancer therapeutic target to inhibit shedding. PMID: 19615970. Note: Anti-MICA antibodies and ligand shedding are involved in the progression of monoclonal gammonathy of undetermined significance
MHC class I polypeptide-related sequence B	MICB	Disease: rheumatoid arthritis (RA)
Mucin 1, cell surface associated	MUC1	**SAR566658** – antineoplastic agents. Note: MUC1/CA 15-3 is used as a serological clinical marker of breast cancer to monitor response to breast cancer treatment and disease recurrence (PMID: 20816948). Decreased levels over time may be indicative of a positive response to treatment
Mucin 2, oligomeric mucus/gel-forming	MUC2	Suppression of mucin 2 promotes interleukin-6 secretion and tumor growth in an orthotopic immune-competent colon cancer animal model. PMID: 25322805
Mucin 3A, cell surface associated	MUC3A	Promoter hypomethylation contributes to the expression of MUC3A in cancer cells. PMID: 20510874
Mucin 3B, cell surface associated	MUC3B	
Mucin 4, cell surface associated	MUC4	The MUC4 membrane-bound mucin regulates esophageal cancer cell proliferation and migration properties: Implication for S100A4 protein. PMID: 21889495
Mucin 5AC, oligomeric mucus/gel-forming	MUC5AC	**Ensituximab** – antineoplastic agents
Mucin 5B, oligomeric mucus/gel-forming	MUC5B	Disease: pulmonary fibrosis, idiopathic (IPF)
Mucin 6, oligomeric mucus/gel-forming	MUC6	
Mucin 7, secreted	MUC7	Disease: asthma (ASTHMA)
Mucin 8	MUC8	Crosstalk between platelet-derived growth factor-induced Nox4 activation and MUC8 gene overexpression in human airway epithelial cells. PMID: 21255638
Mucin 12, cell surface associated	MUC12	
Mucin 13, cell surface associated	MUC13	Functions and regulation of MUC13 mucin in colon cancer cells. PMID: 24097071
Lymphocyte antigen 75	LY75	Induction of antigen-specific immunity with a vaccine targeting NY-ESO-1 to the dendritic cell receptor DEC-205. PMID: 24739759
Endomucin	EMCN	
Fibrinogen C domain containing 1	FIBCD1	Characterization of FIBCD1 as an acetyl group-binding receptor that binds chitin. PMID: 19710473
Mucin 15, cell surface associated	MUC15	Expression of the membrane mucins MUC4 and MUC15, potential markers of malignancy and prognosis, in papillary thyroid carcinoma. PMID: 21615302
Mucin 16, cell surface associated	MUC16	**Abagovomab** – antineoplastic agents
Mucin 17, cell surface associated	MUC17	Expression of MUC17 is regulated by HIF1α-mediated hypoxic responses and requires a methylation-free hypoxia responsible element in pancreatic cancer. PMID: 22970168
Mucin 19, oligomeric	MUC19	
Mucin 20, cell surface associated	MUC20	Role of MUC20 overexpression as a predictor of recurrence and poor outcome in colorectal cancer. PMID: 23787019
Mucin 21, cell surface associated	MUC21	Mucin 21 in esophageal squamous epithelia and carcinomas: analysis with glycoform-specific monoclonal antibodies. PMID: 22611128
Mucin 22	MUC22	

HGNC-approved name	Symbol	Target information
Natural cytotoxicity triggering receptor 1	NCR1	Expression of the activating receptor, NKp46 (CD335), in human natural killer and T-cell neoplasia. PMID: 24225754
Natural cytotoxicity triggering receptor 2	NCR2	NKp44L: A new tool for fighting cancer. PMID: 24800176
Natural cytotoxicity triggering receptor 3	NCR3	NCR3/NKp30 contributes to pathogenesis in primary Sjogren's syndrome. PMID: 23884468
Natural killer cell cytotoxicity receptor 3 ligand 1	NCR3LG1	Downregulation of the activating NKp30 ligand B7-H6 by HDAC inhibitors impairs tumor cell recognition by NK cells. PMID: 23801635
Natural killer cell triggering receptor	NKTR	
Peptidoglycan recognition protein 1	PGLYRP1	Peptidoglycan recognition protein 1 enhances experimental asthma by promoting Th2 and Th17 and limiting regulatory T cell and plasmacytoid dendritic cell responses. PMID: 23420883
Peptidoglycan recognition protein 2	PGLYRP2	Peptidoglycan recognition protein Pglyrp2 protects mice from psoriasis-like skin inflammation by promoting regulatory T cells and limiting Th17 responses. PMID: 22048773
Peptidoglycan recognition protein 3	PGLYRP3	Peptidoglycan recognition protein 3 (PglyRP3) has an anti-inflammatory role in intestinal epithelial cells. PMID: 22099350
Peptidoglycan recognition protein 4	PGLYRP4	
Pre-B lymphocyte 1	VPREB1	The potential role of VPREB1 gene copy number variation in susceptibility to rheumatoid arthritis. PMID: 21144590
Pre-B lymphocyte 3	VPREB3	
Serum amyloid A1	SAA1	Note: Reactive, secondary amyloidosis is characterized by the extracellular accumulation in various tissues of the SAA1 protein.
Serum amyloid A2	SAA2	Note: Reactive, secondary amyloidosis is characterized by the extracellular accumulation in various tissues of the SAA2 protein.
C-reactive protein, pentraxin-related	CRP	**ISIS-CRPRx** – cardiac therapy
Pentraxin 3, long	PTX3	Protective effect of the long pentraxin PTX3 against histone-mediated endothelial cell cytotoxicity in sepsis. PMID: 25227610
Neuronal pentraxin I	NPTX1	NPTX1 regulates neural lineage specification from human pluripotent stem cells. PMID: 24529709
Neuronal pentraxin II	NPTX2	Neuronal pentraxin 2 supports clear cell renal cell carcinoma by activating the AMPA-selective glutamate receptor-4. PMID: 24962026
Neuronal pentraxin receptor	NPTXR	Neuronal pentraxin receptor in cerebrospinal fluid as a potential biomarker for neurodegenerative diseases. PMID: 19368810
Signaling lymphocytic activation molecule family member 1	SLAMF1	Signaling lymphocyte activation molecule regulates development of colitis in mice. PMID: 22960654
SLAM family member 6	SLAMF6	Increased expression of SLAM receptors SLAMF3 and SLAMF6 in systemic lupus erythematosus T lymphocytes promotes Th17 differentiation. PMID: 22184727
SLAM family member 7	SLAMF7	**Elotuzumab** – antineoplastic agents
SLAM family member 8	SLAMF8	Cutting edge: Slamf8 is a negative regulator of Nox2 activity in macrophages. PMID: 22593622
SLAM family member 9	SLAMF9	Genetic approach to insight into the immunobiology of human dendritic cells and identification of CD84-H1, a novel CD84 homologue. PMID: 11300479
Stabilin 1	STAB1	Multifunctional receptor stabilin-1 in homeostasis and disease. PMID: 20953554
Stabilin 2	STAB2	Probing structural selectivity of synthetic heparin binding to Stabilin protein receptors. PMID: 22547069

(Continued)

HGNC-approved name	Symbol	Target information
Toll-like receptor 1	TLR1	Systemic injection of TLR1/2 agonist improves adoptive antigen-specific T cell therapy in glioma-bearing mice. PMID: 24928324
Toll-like receptor 2	TLR2	**OPN-305** – selective immunosuppressants
Toll-like receptor 3	TLR3	**IPH3102** – antineoplastic agents. Disease: herpes simplex encephalitis 2 (HSE2)
Toll-like receptor 4	TLR4	**Eritoran** – selective immunosuppressants. Disease: macular degeneration, age-related, 10 (ARMD10)
Toll-like receptor 5	TLR5	High expression of Toll-like receptor 5 correlates with better prognosis in non-small cell lung cancer. An anti-tumor effect of TLR5 signaling in non-small cell lung cancer. PMID: 24549739. Disease: systemic lupus erythematosus 1 (SLEB1)
Toll-like receptor 6	TLR6	Surface-expressed TLR6 participates in the recognition of diacylated lipopeptide and peptidoglycan in human cells. PMID: 15661917
Toll-like receptor 7	TLR7	**Isatoribine** – immunostimulants
Toll-like receptor 8	TLR8	**Resiquimod** – immunomodulating agents
Toll-like receptor 9	TLR9	**DV1179** – selective immunosuppressants
Toll-like receptor 10	TLR10	Human TLR10 is an anti-inflammatory pattern-recognition receptor. PMID: 25288745
Triggering receptor expressed on myeloid cells 1	TREM1	Triggering receptor expressed on myeloid cells type 1 as a potential therapeutic target in sepsis. PMID: 22156803
Triggering receptor expressed on myeloid cells 2	TREM2	Upregulation of TREM2 ameliorates neuropathology and rescues spatial cognitive impairment in a transgenic mouse model of Alzheimer's disease. PMID: 25047746. Disease: polycystic lipomembranous osteodysplasia with sclerosing leukoencephalopathy (PLOSL)
Triggering receptor expressed on myeloid cells-like 1	TREML1	Soluble TREM-like transcript-1 regulates leukocyte activation and controls microbial sepsis. PMID: 22551551
Triggering receptor expressed on myeloid cells-like 2	TREML2	Triggering receptor expressed on myeloid cell-like transcript 2 (TLT-2) is a counter-receptor for B7-H3 and enhances T-cell responses. PMID: 18650384
Triggering receptor expressed on myeloid cells-like 4	TREML4	Treml4, an Ig superfamily member, mediates presentation of several antigens to T cells in vivo, including protective immunity to HER2 protein. PMID: 22210914
UL16 binding protein 1	ULBP1	Immune evasion mediated by tumor-derived lactate dehydrogenase induction of NKG2D ligands on myeloid cells in glioblastoma patients. PMID: 25136121
UL16 binding protein 2	ULBP2	The NKG2D ligand ULBP2 is specifically regulated through an invariant chain-dependent endosomal pathway. PMID: 25024379
UL16 binding protein 3	ULBP3	The regulatory effect of UL-16 binding protein-3 expression on the cytotoxicity of NK cells in cancer patients. PMID: 25138242
Retinoic acid early transcript 1E	RAET1E	ULBP4 is a novel ligand for human NKG2D. PMID: 12732206
Retinoic acid early transcript 1G	RAET1G	Cellular expression, trafficking, and function of two isoforms of human ULBP5/RAET1G. PMID: 19223974
Retinoic acid early transcript 1L	RAET1L	ULBP6/RAET1L is an additional human NKG2D ligand. PMID: 19658097
B and T lymphocyte associated	BTLA	AAV-sBTLA facilitates HSP70 vaccine-triggered prophylactic antitumor immunity against a murine melanoma pulmonary metastasis model in vivo. PMID: 25153350
Bactericidal/permeability-increasing protein	BPI	
Beta-2-microglobulin	B2M	Identification of beta2-microglobulin as a potential target for ovarian cancer. PMID: 19829086. Disease: hypercatabolic hypoproteinemia (HYCATHYP)

HGNC-approved name	Symbol	Target information
Cathelicidin antimicrobial peptide	CAMP	LL-37 as a therapeutic target for late stage prostate cancer. PMID: 20957672
Clusterin	CLU	**Custirsen** – antineoplastic agents
Cysteine-rich secretory protein LCCL domain containing 2	CRISPLD2	CRISPLD2 is a target of progesterone receptor and its expression is decreased in women with endometriosis. PMID: 24955763
Dermcidin	DCD	Seriniquinone, a selective anticancer agent, induces cell death by autophagocytosis, targeting the cancer-protective protein dermcidin. PMID: 25271322
Granulysin	GNLY	Cytotoxic proteins and therapeutic targets in severe cutaneous adverse reactions. PMID: 24394640
HERV-H LTR-associating 2	HHLA2	HHLA2 is a member of the B7 family and inhibits human CD4 and CD8 T-cell function. PMID: 23716685
Integrin alpha FG-GAP repeat containing 1	ITFG1	TIP, a T-cell factor identified using high-throughput screening increases survival in a graft-versus-host disease model. PMID: 12598909
Lactotransferrin	LTF	Novel recombinant human lactoferrin: Differential activation of oxidative stress related gene expression. PMID: 24070904
Lck interacting transmembrane adaptor 1	LIME1	Deletion of the LIME adaptor protein minimally affects T and B cell development and function. PMID: 17918199
Lipopolysaccharide binding protein	LBP	Directed evolution of an LBP/CD14 inhibitory peptide and its anti-endotoxin activity. PMID: 25025695
Liver expressed antimicrobial peptide 2	LEAP2	Immune activation in HIV/HCV-infected patients is associated with low-level expression of liver expressed antimicrobial peptide-2 (LEAP-2). PMID: 23940131
Lymphocyte transmembrane adaptor 1	LAX1	Regulation of lymphocyte development and activation by the LAT family of adapter proteins. PMID: 19909357
Lymphocyte-activation gene 3	LAG3	**BMS-986016** – antineoplastic agents
Lymphocyte-specific protein 1	LSP1	Leukocyte specific protein-1: A novel regulator of hepatocellular proliferation and migration deleted in human HCC. PMID: 25234543
Macrophage receptor with collagenous structure	MARCO	HSV-1 exploits the innate immune scavenger receptor MARCO to enhance epithelial adsorption and infection. PMID: 23739639
Major histocompatibility complex, class I-related	MR1	Human T cells use CD1 and MR1 to recognize lipids and small molecules. PMID: 25271021
Mast cell immunoglobulin-like receptor 1	MILR1	Influence of MILR1 promoter polymorphism on expression levels and the phenotype of atopy. PMID: 25007884
Perforin 1 (pore forming protein)	PRF1	Disease: familial hemophagocytic lymphohistiocytosis 2 (FHL2)
Plasminogen receptor, C-terminal lysine transmembrane protein	PLGRKT	New insights into the role of Plg RKT in macrophage recruitment. PMID: 24529725
Polymeric immunoglobulin receptor	PIGR	Polymeric immunoglobulin receptor down-regulation in chronic obstructive pulmonary disease. Persistence in the cultured epithelium and role of transforming growth factor-β. PMID: 25078120
Protein S (alpha)	PROS1	Disease: thrombophilia due to protein S deficiency, autosomal dominant (THPH5)
Radical S-adenosyl methionine domain containing 2	RSAD2	The role of viperin in the innate antiviral response. PMID: 24157441
Scavenger receptor cysteine-rich domain containing (5 domains)	SSC5D	Molecular and functional characterization of mouse S5D-SRCRB: A new group B member of the scavenger receptor cysteine-rich superfamily. PMID: 21217009
Schlafen family member 5	SLFN5	Role of interferon {alpha} (IFN{alpha})-inducible Schlafen-5 in regulation of anchorage-independent growth and invasion of malignant melanoma cells. PMID: 20956525

(Continued)

HGNC-approved name	Symbol	Target information
SIVA1, apoptosis-inducing factor	SIVA1	Thromboxane A2 modulates cisplatin-induced apoptosis through a Siva1-dependent mechanism. PMID: 22343716
T cell immunoreceptor with Ig and ITIM domains	TIGIT	Agonistic anti-TIGIT treatment inhibits T cell responses in LDLr deficient mice without affecting atherosclerotic lesion development. PMID: 24376654
TAP binding protein (tapasin)	TAPBP	HLA-I antigen presentation and tapasin influence immune responses against malignant brain tumors – considerations for successful immunotherapy. PMID: 25175689
TLR4 interactor with leucine-rich repeats	TRIL	TRIL is involved in cytokine production in the brain following Escherichia coli infection. PMID: 25015823
TYRO protein tyrosine kinase binding protein	TYROBP	TYROBP in Alzheimer's disease. PMID: 25052481. Disease: polycystic lipomembranous osteodysplasia with sclerosing leukoencephalopathy (PLOSL)
von Willebrand factor	VWF	**Caplacizumab** – antithrombotic agents. Disease: von Willebrand disease 1 (VWD1)
V-set and transmembrane domain containing 1	VSTM1	VSTM1-v2, a novel soluble glycoprotein, promotes the differentiation and activation of Th17 cells. PMID: 22960280
V-set domain containing T cell activation inhibitor 1	VTCN1	**AMP-110** – selective immunosuppressants

3.6   Transporters and channels

3.6.1   ATP-binding cassette families and transporting ATPases

HGNC-approved name	Symbol	Target information
ATP-binding cassette, subfamily A (ABC1), member 1	ABCA1	**Probucol** – lipid-modifying agents. Disease: high density lipoprotein deficiency 1 (HDLD1)
ATP-binding cassette, subfamily A (ABC1), member 2	ABCA2	The ATP-binding cassette transporter-2 (ABCA2) regulates esterification of plasma membrane cholesterol by modulation of sphingolipid metabolism. PMID: 24201375
ATP-binding cassette, subfamily A (ABC1), member 3	ABCA3	Disease: pulmonary surfactant metabolism dysfunction 3 (SMDP3)
ATP-binding cassette, subfamily A (ABC1), member 4	ABCA4	Disease: Stargardt disease 1 (STGD1)
ATP-binding cassette, subfamily A (ABC1), member 5	ABCA5	ABCA5 regulates amyloid-β peptide production and is associated with Alzheimer's disease neuropathology. PMID: 25125465
ATP-binding cassette, subfamily A (ABC1), member 6	ABCA6	FoxO regulates expression of ABCA6, an intracellular ATP-binding-cassette transporter responsive to cholesterol. PMID: 24028821
ATP-binding cassette, subfamily A (ABC1), member 7	ABCA7	ABCA7 in Alzheimer's disease. PMID: 24878767
ATP-binding cassette, subfamily A (ABC1), member 8	ABCA8	ABCA8 stimulates sphingomyelin production in oligodendrocytes. PMID: 23560799
ATP-binding cassette, subfamily A (ABC1), member 9	ABCA9	Molecular structure of a novel cholesterol-responsive A subclass ABC transporter, ABCA9. PMID: 12150964
ATP-binding cassette, subfamily A (ABC1), member 10	ABCA10	Enhancement of aerosol cisplatin chemotherapy with gene therapy expressing ABC10 protein in respiratory system. PMID: 24723977
ATP-binding cassette, subfamily A (ABC1), member 12	ABCA12	Disease: ichthyosis, congenital, autosomal recessive 4A (ARCI4A)

HGNC-approved name	Symbol	Target information
ATP-binding cassette, subfamily A (ABC1), member 13	ABCA13	The effects of neurological disorder-related codon variations of ABCA13 on the function of the ABC protein. PMID: 23221702
ATP-binding cassette, subfamily B (MDR/TAP), member 1	ABCB1	**Zosuquidar** – antineoplastic agents. Disease: inflammatory bowel disease 13 (IBD13)
ATP-binding cassette, subfamily B (MDR/TAP), member 4	ABCB4	Disease: cholestasis, progressive familial intrahepatic, 3 (PFIC3)
ATP-binding cassette, subfamily B (MDR/TAP), member 5	ABCB5	Profiling of ABC transporters ABCB5, ABCF2 and nestin-positive stem cells in nevi, in situ and invasive melanoma. PMID: 22555176
ATP-binding cassette, subfamily B (MDR/TAP), member 6	ABCB6	Disease: microphthalmia, isolated, with coloboma, 7 (MCOPCB7)
ATP-binding cassette, subfamily B (MDR/TAP), member 7	ABCB7	Disease: anemia, sideroblastic, spinocerebellar ataxia (ASAT)
ATP-binding cassette, subfamily B (MDR/TAP), member 8	ABCB8	Mitochondrial ABC transporters function: The role of ABCB10 (ABC-me) as a novel player in cellular handling of reactive oxygen species. PMID: 22884976
ATP-binding cassette, subfamily B (MDR/TAP), member 9	ABCB9	The lysosomal polypeptide transporter TAPL: More than a housekeeping factor? PMID: 21194361
ATP-binding cassette, subfamily B (MDR/TAP), member 10	ABCB10	Structures of ABCB10, a human ATP-binding cassette transporter in apo- and nucleotide-bound states. PMID: 23716676
ATP-binding cassette, subfamily B (MDR/TAP), member 11	ABCB11	Disease: cholestasis, progressive familial intrahepatic, 2 (PFIC2)
Transporter 1, ATP-binding cassette, subfamily B (MDR/TAP)	TAP1	Disease: bare lymphocyte syndrome 1 (BLS1)
Transporter 2, ATP-binding cassette, subfamily B (MDR/TAP)	TAP2	Disease: bare lymphocyte syndrome 1 (BLS1)
ATP-binding cassette, subfamily C (CFTR/MRP), member 1	ABCC1	Multidrug resistance protein 1 (MRP1, ABCC1), a 'multitasking' ATP-binding cassette (ABC) transporter. PMID: 25281745
ATP-binding cassette, subfamily C (CFTR/MRP), member 2	ABCC2	Disease: Dubin–Johnson syndrome (DJS)
ATP-binding cassette, subfamily C (CFTR/MRP), member 3	ABCC3	MRP3: A molecular target for human glioblastoma multiforme immunotherapy. PMID: 20809959
ATP-binding cassette, subfamily C (CFTR/MRP), member 4	ABCC4	The ABCC4 gene is a promising target for pancreatic cancer therapy. PMID: 21989485
ATP-binding cassette, subfamily C (CFTR/MRP), member 5	ABCC5	ABCC5 supports osteoclast formation and promotes breast cancer metastasis to bone. PMID: 23174366
ATP-binding cassette, subfamily C (CFTR/MRP), member 6	ABCC6	Disease: pseudoxanthoma elasticum (PXE)
Cystic fibrosis transmembrane conductance regulator (ATP-binding cassette subfamily C, member 7)	CFTR	**Ivacaftor** – drugs for obstructive airway diseases. Disease: cystic fibrosis (CF)
ATP-binding cassette, subfamily C (CFTR/MRP), member 8	ABCC8	**Nateglinide** – drugs used in diabetes. Disease: leucine-induced hypoglycemia (LIH)
ATP-binding cassette, subfamily C (CFTR/MRP), member 9	ABCC9	Disease: cardiomyopathy, dilated 10 (CMD10)
ATP-binding cassette, subfamily C (CFTR/MRP), member 10	ABCC10	Tandutinib (MLN518) reverses multidrug resistance by inhibiting the efflux activity of the multidrug resistance protein 7 (ABCC10). PMID: 23525656
ATP-binding cassette, subfamily C (CFTR/MRP), member 11	ABCC11	Tomatoes cause under-arm odour. PMID: 24576684
ATP-binding cassette, subfamily C (CFTR/MRP), member 12	ABCC12	MRP9, an unusual truncated member of the ABC transporter superfamily, is highly expressed in breast cancer. PMID: 12011458

(Continued)

HGNC-approved name	Symbol	Target information
ATP-binding cassette, subfamily D (ALD), member 1	ABCD1	Disease: adrenoleukodystrophy (ALD)
ATP-binding cassette, subfamily D (ALD), member 2	ABCD2	Abcd2 is a strong modifier of the metabolic impairments in peritoneal macrophages of ABCD1-deficient mice. PMID: 25255441
ATP-binding cassette, subfamily D (ALD), member 3	ABCD3	A role for the human peroxisomal half-transporter ABCD3 in the oxidation of dicarboxylic acids. PMID: 24333844
ATP-binding cassette, subfamily D (ALD), member 4	ABCD4	Disease: methylmalonic aciduria and homocystinuria type cblJ (MAHCJ)
ATP-binding cassette, subfamily E (OABP), member 1	ABCE1	Knock-down of ABCE1 gene induces G1/S arrest in human oral cancer cells. PMID: 25337191
ATP-binding cassette, subfamily F (GCN20), member 1	ABCF1	ATP-binding cassette sub-family F member 1 (ABCF1) is identified as a putative therapeutic target of escitalopram in the inflammatory cytokine pathway. PMID: 23719290
ATP-binding cassette, subfamily F (GCN20), member 2	ABCF2	Clinical role of ABCF2 expression in breast cancer. PMID: 16827111
ATP-binding cassette, subfamily F (GCN20), member 3	ABCF3	hABCF3, a TPD52L2 interacting partner, enhances the proliferation of human liver cancer cell lines *in vitro*. PMID: 24052230
ATP-binding cassette, subfamily G (WHITE), member 1	ABCG1	Adipocyte ATP-binding cassette G1 promotes triglyceride storage, fat mass growth and human obesity. PMID: 25249572
ATP-binding cassette, subfamily G (WHITE), member 2	ABCG2	ABCG2: Recent discovery of potent and highly selective inhibitors. PMID: 23734686
ATP-binding cassette, subfamily G (WHITE), member 4	ABCG4	ABCG2- and ABCG4-mediated efflux of amyloid-β peptide 1–40 at the mouse blood-brain barrier. PMID: 22391220
ATP-binding cassette, subfamily G (WHITE), member 5	ABCG5	Disease: sitosterolemia (STSL)
ATP-binding cassette, subfamily G (WHITE), member 8	ABCG8	Disease: gallbladder disease 4 (GBD4)
ATPase, class V, type 10A	ATP10A	Disease: Angelman syndrome (AS)
ATPase, class V, type 10B	ATP10B	
ATPase, class V, type 10D	ATP10D	
ATPase, class VI, type 11A	ATP11A	Resistance to farnesyltransferase inhibitors in Bcr/Abl-positive lymphoblastic leukemia by increased expression of a novel ABC transporter homolog ATP11a. PMID: 15860663
ATPase, class VI, type 11B	ATP11B	ATP11B mediates platinum resistance in ovarian cancer. PMID: 23585472
ATPase, class VI, type 11C	ATP11C	ATP11C is critical for the internalization of phosphatidylserine and differentiation of B lymphocytes. PMID: 21423173
ATPase, H+/K+ transporting, nongastric, alpha polypeptide	ATP12A	Expression of the non-gastric H+/K+ ATPase ATP12A in normal and pathological human prostate tissue. PMID: 22179016
ATPase type 13A1	ATP13A1	
ATPase type 13A2	ATP13A2	Lysosomal dysfunction in Parkinson disease: ATP13A2 gets into the groove. PMID: 22885599. Disease: Kufor–Rakeb syndrome (KRS)
ATPase type 13A3	ATP13A3	
ATPase type 13A4	ATP13A4	Note: A chromosomal aberration involving ATP13A4 is found in 2 patients with specific language impairment (SLI) disorders. Paracentric inversion inv(3)(q25q29). The inversion produces a disruption of the protein
ATPase type 13A5	ATP13A5	

HGNC-approved name	Symbol	Target information
ATPase, Na+/K+ transporting, alpha 1 polypeptide	ATP1A1	**Digitoxin** – cardiac therapy
ATPase, Na+/K+ transporting, alpha 2 polypeptide	ATP1A2	Disease: migraine, familial hemiplegic, 2 (FHM2)
ATPase, Na+/K+ transporting, alpha 3 polypeptide	ATP1A3	Disease: dystonia 12 (DYT12)
ATPase, Na+/K+ transporting, alpha 4 polypeptide	ATP1A4	The Na,K-ATPase alpha 4 isoform from humans has distinct enzymatic properties and is important for sperm motility. PMID: 16861705
ATPase, Na+/K+ transporting, beta 1 polypeptide	ATP1B1	The beta1 subunit of the Na,K-ATPase pump interacts with megalencephalic leucoencephalopathy with subcortical cysts protein 1 (MLC1) in brain astrocytes: New insights into MLC pathogenesis. PMID: 20926452
ATPase, Na+/K+ transporting, beta 2 polypeptide	ATP1B2	Na$^+$/K$^+$-ATPase β2-subunit (AMOG) expression abrogates invasion of glioblastoma-derived brain tumor-initiating cells. PMID: 23887941
ATPase, Na+/K+ transporting, beta 3 polypeptide	ATP1B3	The beta3 subunit of the Na+,K+-ATPase mediates variable nociceptive sensitivity in the formalin test. PMID: 19464798
ATPase, Na+/K+ transporting, beta 4 polypeptide	ATP1B4	
ATPase, Ca++ transporting, cardiac muscle, fast twitch 1	ATP2A1	Disease: Brody myopathy (BRM)
ATPase, Ca++ transporting, cardiac muscle, slow twitch 2	ATP2A2	Sarcoplasmic reticulum Ca(2+) ATPase as a therapeutic target for heart failure. PMID: 20078230. Disease: acrokeratosis verruciformis (AKV)
ATPase, Ca++ transporting, ubiquitous	ATP2A3	
ATPase, Ca++ transporting, plasma membrane 1	ATP2B1	
ATPase, Ca++ transporting, plasma membrane 2	ATP2B2	
ATPase, Ca++ transporting, plasma membrane 3	ATP2B3	Disease: spinocerebellar ataxia, X-linked 1 (SCAX1)
ATPase, Ca++ transporting, plasma membrane 4	ATP2B4	Isolation and characterization of BetaM protein encoded by ATP1B4 – a unique member of the Na,K-ATPase β-subunit gene family. PMID: 21855530
ATPase, Ca++ transporting, type 2C, member 1	ATP2C1	Disease: Hailey–Hailey disease (HHD)
ATPase, Ca++ transporting, type 2C, member 2	ATP2C2	New insights into store-independent Ca(2+) entry: Secretory pathway calcium ATPase 2 in normal physiology and cancer. PMID: 23670239
ATPase, H+/K+ exchanging, alpha polypeptide	ATP4A	**Omeprazole** – drugs for acid-related disorders
ATPase, H+/K+ exchanging, beta polypeptide	ATP4B	**Omeprazole** – drugs for acid-related disorders
ATP synthase, H+ transporting, mitochondrial F1 complex, alpha subunit 1, cardiac muscle	ATP5A1	ATP synthase ecto-α subunit: A novel therapeutic target for breast cancer. PMID: 22152132. Disease: mitochondrial complex V deficiency, nuclear 4 (MC5DN4)
ATP synthase, H+ transporting, mitochondrial F1 complex, beta polypeptide	ATP5B	Deregulation of mitochondrial ATPsyn-β in acute myeloid leukemia cells and with increased drug resistance. PMID: 24391795
ATP synthase, H+ transporting, mitochondrial F1 complex, gamma polypeptide 1	ATP5C1	
ATP synthase, H+ transporting, mitochondrial F1 complex, delta subunit	ATP5D	Identification of a new Mpl-interacting protein, Atp5d. PMID: 24615392
ATP synthase, H+ transporting, mitochondrial F1 complex, epsilon subunit	ATP5E	Disease: mitochondrial complex V deficiency, nuclear 3 (MC5DN3)

(Continued)

HGNC-approved name	Symbol	Target information
ATP synthase, H+ transporting, mitochondrial Fo complex, subunit B1	ATP5F1	
ATP synthase, H+ transporting, mitochondrial Fo complex, subunit C1 (subunit 9)	ATP5G1	
ATP synthase, H+ transporting, mitochondrial Fo complex, subunit C2 (subunit 9)	ATP5G2	
ATP synthase, H+ transporting, mitochondrial Fo complex, subunit C3 (subunit 9)	ATP5G3	
ATP synthase, H+ transporting, mitochondrial Fo complex, subunit d	ATP5H	ATP5H/KCTD2 locus is associated with Alzheimer's disease risk. PMID: 23857120
ATP synthase, H+ transporting, mitochondrial Fo complex, subunit E	ATP5I	Antisense of ATP synthase subunit e inhibits the growth of human hepatocellular carcinoma cells. PMID: 11939412
ATP synthase, H+ transporting, mitochondrial Fo complex, subunit F6	ATP5J	Coupling factor 6-induced activation of ecto-F1F(o) complex induces insulin resistance, mild glucose intolerance and elevated blood pressure in mice. PMID: 22038518
ATP synthase, H+ transporting, mitochondrial Fo complex, subunit F2	ATP5J2	
ATP synthase, H+ transporting, mitochondrial Fo complex, subunit G	ATP5L	
ATP synthase, H+ transporting, mitochondrial Fo complex, subunit G2	ATP5L2	
ATP synthase, H+ transporting, mitochondrial F1 complex, O subunit	ATP5O	Genetic variation in ATP5O is associated with skeletal muscle ATP5O mRNA expression and glucose uptake in young twins. PMID: 19274082
ATP synthase, H+ transporting, mitochondrial Fo complex, subunit s (factor B)	ATP5S	
ATPase, H+ transporting, lysosomal accessory protein 1	ATP6AP1	V-ATPase subunit ATP6AP1 (Ac45) regulates osteoclast differentiation, extracellular acidification, lysosomal trafficking, and protease exocytosis in osteoclast-mediated bone resorption. PMID: 22467241
ATPase, H+ transporting, lysosomal accessory protein 1-like	ATP6AP1L	
ATPase, H+ transporting, lysosomal accessory protein 2	ATP6AP2	Signal transduction of the (pro)renin receptor as a novel therapeutic target for preventing end-organ damage. PMID: 20010781. Disease: mental retardation, X-linked, with epilepsy (MRXE)
ATPase, H+ transporting, lysosomal VO subunit a1	ATP6V0A1	POLR2F, ATP6V0A1 and PRNP expression in colorectal cancer: New molecules with prognostic significance? PMID: 18505059
ATPase, H+ transporting, lysosomal VO subunit a2	ATP6V0A2	Disease: cutis laxa, autosomal recessive, 2A (ARCL2A)
T-cell, immune regulator 1, ATPase, H+ transporting, lysosomal VO subunit a3	TCIRG1	Function of V-ATPase a subunit isoforms in invasiveness of MCF10a and MCF10CA1a human breast cancer cells. PMID: 24072707. Disease: osteopetrosis, autosomal recessive 1 (OPTB1)
ATPase, H+ transporting, lysosomal VO subunit a4	ATP6V0A4	Disease: renal tubular acidosis, distal, autosomal recessive (RTADR)
ATPase, H+ transporting, lysosomal 21 kDa, VO subunit b	ATP6V0B	
ATPase, H+ transporting, lysosomal 16 kDa, VO subunit c	ATP6V0C	**Bafilomycin A1** – drugs affecting bone structure and mineralization

HGNC-approved name	Symbol	Target information
ATPase, H+ transporting, lysosomal 38 kDa, V0 subunit d1	ATP6V0D1	
ATPase, H+ transporting, lysosomal 38 kDa, V0 subunit d2	ATP6V0D2	Luteolin inhibition of V-ATPase a3-d2 interaction decreases osteoclast resorptive activity. PMID: 23129004
ATPase, H+ transporting, lysosomal 9 kDa, V0 subunit e1	ATP6V0E1	
ATPase, H+ transporting V0 subunit e2	ATP6V0E2	Molecular cloning and characterization of a novel form of the human vacuolar H+-ATPase e-subunit: An essential proton pump component. PMID: 17350184
ATPase, H+ transporting, lysosomal 70 kDa, V1 subunit A	ATP6V1A	Vacuolar-type H+-ATPase V1A subunit is a molecular partner of Wolfram syndrome 1 (WFS1) protein, which regulates its expression and stability. PMID: 23035048
ATPase, H+ transporting, lysosomal 56/58 kDa, V1 subunit B1	ATP6V1B1	Disease: renal tubular acidosis, distal, with progressive nerve deafness (dRTA-D)
ATPase, H+ transporting, lysosomal 56/58 kDa, V1 subunit B2	ATP6V1B2	De novo mutation in ATP6V1B2 impairs lysosome acidification and causes dominant deafness-onychodystrophy syndrome. PMID: 24913193
ATPase, H+ transporting, lysosomal 42 kDa, V1 subunit C1	ATP6V1C1	Atp6v1c1 may regulate filament actin arrangement in breast cancer cells. PMID: 24454753
ATPase, H+ transporting, lysosomal 42 kDa, V1 subunit C2	ATP6V1C2	Differential expression of a V-type ATPase C subunit gene, Atp6v1c2, during culture of rat lung type II pneumocytes. PMID: 16283434
ATPase, H+ transporting, lysosomal 34 kDa, V1 subunit D	ATP6V1D	
ATPase, H+ transporting, lysosomal 31 kDa, V1 subunit E1	ATP6V1E1	
ATPase, H+ transporting, lysosomal 31 kDa, V1 subunit E2	ATP6V1E2	
ATPase, H+ transporting, lysosomal 14 kDa, V1 subunit F	ATP6V1F	
ATPase, H+ transporting, lysosomal 13 kDa, V1 subunit G1	ATP6V1G1	RILP regulates vacuolar ATPase through interaction with the V1G1 subunit. PMID: 24762812
ATPase, H+ transporting, lysosomal 13 kDa, V1 subunit G2	ATP6V1G2	
ATPase, H+ transporting, lysosomal 13 kDa, V1 subunit G3	ATP6V1G3	
ATPase, H+ transporting, lysosomal 50/57 kDa, V1 subunit H	ATP6V1H	Decreased expression of ATP6V1H in type 2 diabetes: A pilot report on the diabetes risk study in Mexican Americans. PMID: 21871445
Mitochondrially encoded ATP synthase 6	MT-ATP6	Disease: neuropathy, ataxia, and retinitis pigmentosa (NARP)
Mitochondrially encoded ATP synthase 8	MT-ATP8	Disease: mitochondrial complex V deficiency, mitochondrial 2 (MC5DM2)
ATPase, Cu++ transporting, alpha polypeptide	ATP7A	Unexpected role of the copper transporter ATP7A in PDGF-induced vascular smooth muscle cell migration. PMID: 20671235. Disease: Menkes disease (MNKD)
ATPase, Cu++ transporting, beta polypeptide	ATP7B	Disease: Wilson disease (WD)
ATPase, aminophospholipid transporter (APLT), class I, type 8A, member 1	ATP8A1	Atp8a1 deficiency is associated with phosphatidylserine externalization in hippocampus and delayed hippocampus-dependent learning. PMID: 22007859
ATPase, aminophospholipid transporter, class I, type 8A, member 2	ATP8A2	Disease: cerebellar ataxia, mental retardation, and disequilibrium syndrome 4 (CMARQ4)
ATPase, aminophospholipid transporter, class I, type 8B, member 1	ATP8B1	Disease: cholestasis, progressive familial intrahepatic, 1 (PFIC1)

(Continued)

HGNC-approved name	Symbol	Target information
ATPase, aminophospholipid transporter, class I, type 8B, member 2	ATP8B2	
ATPase, aminophospholipid transporter, class I, type 8B, member 3	ATP8B3	Expression of Atp8b3 in murine testis and its characterization as a testis specific P-type ATPase. PMID: 19017724
ATPase, class I, type 8B, member 4	ATP8B4	
ATPase, class II, type 9A	ATP9A	
ATPase, class II, type 9B	ATP9B	

3.6.2 Solute carrier (SLC) families

HGNC-approved name	Symbol	Target information
Solute carrier family 1 (neuronal/epithelial high-affinity glutamate transporter, system Xag), member 1	SLC1A1	Characterization of novel L-threo-beta-benzyloxyaspartate derivatives, potent blockers of the glutamate transporters. PMID: 15044631. Disease: schizophrenia 18 (SCZD18)
Solute carrier family 1 (glial high-affinity glutamate transporter), member 2	SLC1A2	Characterization of novel L-threo-beta-benzyloxyaspartate derivatives, potent blockers of the glutamate transporters. PMID: 15044631
Solute carrier family 1 (glial high-affinity glutamate transporter), member 3	SLC1A3	Characterization of novel L-threo-beta-benzyloxyaspartate derivatives, potent blockers of the glutamate transporters. PMID: 15044631. Disease: episodic ataxia 6 (EA6)
Solute carrier family 1 (glutamate/neutral amino acid transporter), member 4	SLC1A4	Arginine vasopressin regulated ASCT1 expression in astrocytes from stroke-prone spontaneously hypertensive rats and congenic SHRpch1_18 rats. PMID: 24613720
Solute carrier family 1 (neutral amino acid transporter), member 5	SLC1A5	Upregulated SLC1A5 promotes cell growth and survival in colorectal cancer. PMID: 25337245
Solute carrier family 1 (high-affinity aspartate/glutamate transporter), member 6	SLC1A6	Patterned neuroprotection in the Inpp4a(wbl) mutant mouse cerebellum correlates with the expression of Eaat4. PMID: 20011524
Solute carrier family 1 (glutamate transporter), member 7	SLC1A7	Possible roles of glutamate transporter EAAT5 in mouse cone depolarizing bipolar cell light responses. PMID: 24972005
Solute carrier family 2 (facilitated glucose transporter), member 1	SLC2A1	**Glufosfamide transporter** – antineoplastic agents. Disease: GLUT1 deficiency syndrome 1 (GLUT1DS1)
Solute carrier family 2 (facilitated glucose transporter), member 2	SLC2A2	**Dapagliflozin** – drugs used in diabetes. Disease: Fanconi–Bickel syndrome (FBS)
Solute carrier family 2 (facilitated glucose transporter), member 3	SLC2A3	GLUT3 is induced during epithelial-mesenchymal transition and promotes tumor cell proliferation in non-small cell lung cancer. PMID: 25097756
Solute carrier family 2 (facilitated glucose transporter), member 4	SLC2A4	Cellular glucose transport and glucotransporter 4 expression as a therapeutic target: clinical and experimental studies. PMID: 19885647. Disease: diabetes mellitus, non-insulin-dependent (NIDDM)
Solute carrier family 2 (facilitated glucose/fructose transporter), member 5	SLC2A5	Low glucose transporter SLC2A5-inhibited human normal adjacent lung adenocarcinoma cytoplasmic pro-B cell development mechanism network. PMID: 25326153
Solute carrier family 2 (facilitated glucose transporter), member 6	SLC2A6	Pathogenic GLUT9 mutations causing renal hypouricemia type 2 (RHUC2). PMID: 22132964
Solute carrier family 2 (facilitated glucose transporter), member 7	SLC2A7	GLUT7: A new intestinal facilitated hexose transporter. PMID: 18477702
Solute carrier family 2 (facilitated glucose transporter), member 8	SLC2A8	Glucose transporter 8 (GLUT8) mediates fructose-induced de novo lipogenesis and macrosteatosis. PMID: 24519932
Solute carrier family 2 (facilitated glucose transporter), member 9	SLC2A9	Disease: hypouricemia renal 2 (RHUC2)

HGNC-approved name	Symbol	Target information
Solute carrier family 2 (facilitated glucose transporter), member 10	SLC2A10	Disease: arterial tortuosity syndrome (ATS)
Solute carrier family 2 (facilitated glucose transporter), member 11	SLC2A11	Multiple myeloma exhibits novel dependence on GLUT4, GLUT8, and GLUT11: Implications for glucose transporter-directed therapy. PMID: 22452979
Solute carrier family 2 (facilitated glucose transporter), member 12	SLC2A12	GLUT12 functions as a basal and insulin-independent glucose transporter in the heart. PMID: 23041416
Solute carrier family 2 (facilitated glucose transporter), member 13	SLC2A13	Investigation of the H(+)-myo-inositol transporter (HMIT) as a neuronal regulator of phosphoinositide signalling. PMID: 19754467
Solute carrier family 2 (facilitated glucose transporter), member 14	SLC2A14	GLUT14, a duplicon of GLUT3, is specifically expressed in testis as alternative splice forms. PMID: 12504846
Solute carrier family 3 (amino acid transporter heavy chain), member 1	SLC3A1	Disease: cystinuria (CSNU)
Solute carrier family 3 (amino acid transporter heavy chain), member 2	SLC3A2	**IGN523** – antineoplastic agents
Solute carrier family 4 (anion exchanger), member 1	SLC4A1	Anti-tumour effects of small interfering RNA targeting anion exchanger 1 in experimental gastric cancer. PMID: 21649639.0 Disease: elliptocytosis 4 (EL4)
Solute carrier family 4 (anion exchanger), member 2	SLC4A2	Involvement of anion exchanger-2 in apoptosis of endothelial cells induced by high glucose through an mPTP-ROS-Caspase-3 dependent pathway. PMID: 20180022
Solute carrier family 4 (anion exchanger), member 3	SLC4A3	Resistance to cardiomyocyte hypertrophy in ae3-/- mice, deficient in the AE3 Cl-/HCO3- exchanger. PMID: 25047106
Solute carrier family 4 (sodium bicarbonate cotransporter), member 4	SLC4A4	Disease: Renal tubular acidosis, proximal, with ocular abnormalities and mental retardation (pRTA-OA)
Solute carrier family 4 (sodium bicarbonate cotransporter), member 5	SLC4A5	Targeted mutation of SLC4A5 induces arterial hypertension and renal metabolic acidosis. PMID: 22082831
Solute carrier family 4 (sodium bicarbonate cotransporter), member 7	SLC4A7	Contribution of Na+,HCO3(−)-cotransport to cellular pH control in human breast cancer: A role for the breast cancer susceptibility locus NBCn1 (SLC4A7). PMID: 22907202
Solute carrier family 4 (sodium bicarbonate cotransporter), member 8	SLC4A8	The sodium-driven chloride/bicarbonate exchanger NDCBE in rat brain is upregulated by chronic metabolic acidosis. PMID: 21195699
Solute carrier family 4 (sodium bicarbonate cotransporter), member 9	SLC4A9	
Solute carrier family 4 (sodium bicarbonate transporter), member 10	SLC4A10	Mice with targeted Slc4a10 gene disruption have small brain ventricles and show reduced neuronal excitability. PMID: 18165320
Solute carrier family 4 (sodium borate transporter), member 11	SLC4A11	Disease: corneal dystrophy and perceptive deafness (CDPD)
Solute carrier family 5 (sodium/glucose cotransporter), member 1	SLC5A1	**GSK 1614235** – drugs used in diabetes. Disease: congenital glucose/galactose malabsorption (GGM)
Solute carrier family 5 (sodium/glucose cotransporter), member 2	SLC5A2	**Canagliflozin** – Drugs used in diabetes. Disease: renal glucosuria (GLYS1)
Solute carrier family 5 (sodium/myo-inositol cotransporter), member 3	SLC5A3	Sodium/myo-inositol cotransporter 1 and myo-inositol are essential for osteogenesis and bone formation. PMID: 20818642
Solute carrier family 5 (glucose activated ion channel), member 4	SLC5A4	**Dapagliflozin** – drugs used in diabetes
Solute carrier family 5 (sodium/iodide cotransporter), member 5	SLC5A5	Disease: thyroid dyshormonogenesis 1 (TDH1)
Solute carrier family 5 (sodium/multivitamin and iodide cotransporter), member 6	SLC5A6	Sodium dependent multivitamin transporter (SMVT): A potential target for drug delivery. PMID: 22420308

(Continued)

HGNC-approved name	Symbol	Target information
Solute carrier family 5 (sodium/choline cotransporter), member 7	SLC5A7	Disease: neuronopathy, distal hereditary motor, 7A (IIMN7A)
Solute carrier family 5 (sodium/monocarboxylate cotransporter), member 8	SLC5A8	Molecular mechanism of SLC5A8 inactivation in breast cancer. PMID: 23918800
Solute carrier family 5 (sodium/sugar cotransporter), member 9	SLC5A9	SLC5A9/SGLT4, a new Na+-dependent glucose transporter, is an essential transporter for mannose, 1,5-anhydro-D-glucitol, and fructose. PMID: 15607332
Solute carrier family 5 (sodium/sugar cotransporter), member 10	SLC5A10	SGLT5 reabsorbs fructose in the kidney but its deficiency paradoxically exacerbates hepatic steatosis induced by fructose. PMID: 23451068
Solute carrier family 5 (sodium/inositol cotransporter), member 11	SLC5A11	The sodium-dependent glucose cotransporter SLC5A11 as an autoimmune modifier gene in SLE. PMID: 18069935
Solute carrier family 5 (sodium/monocarboxylate cotransporter), member 12	SLC5A12	Sodium-coupled monocarboxylate transporters in normal tissues and in cancer. PMID: 18446519
Solute carrier family 6 (neurotransmitter transporter), member 1	SLC6A1	**Tiagabine** – antiepileptics
Solute carrier family 6 (neurotransmitter transporter), member 2	SLC6A2	**Levomilnacipran** – antidepressants. Disease: orthostatic intolerance (OI)
Solute carrier family 6 (neurotransmitter transporter), member 3	SLC6A3	**Amphetamines** – Parasympathomimetics. Disease: parkinsonism–dystonia infantile (PKDYS)
Solute carrier family 6 (neurotransmitter transporter), member 4	SLC6A4	**Fluoxetine** – antidepressants
Solute carrier family 6 (neurotransmitter transporter), member 5	SLC6A5	Deletion of the mouse glycine transporter 2 results in a hyperekplexia phenotype and postnatal lethality. PMID: 14622583. Disease: hyperekplexia 3 (HKPX3)
Solute carrier family 6 (neurotransmitter transporter), member 6	SLC6A6	Role of SLC6A6 in promoting the survival and multidrug resistance of colorectal cancer. PMID: 24781822
Solute carrier family 6 (neurotransmitter transporter), member 7	SLC6A7	Novel inhibitors of the high-affinity L-proline transporter as potential therapeutic agents for the treatment of cognitive disorders. PMID: 25037917
Solute carrier family 6 (neurotransmitter transporter), member 8	SLC6A8	Disease: cerebral creatine deficiency syndrome 1 (CCDS1)
Solute carrier family 6 (neurotransmitter transporter, glycine), member 9	SLC6A9	**Bitopertin** – antipsychotics
Solute carrier family 6 (neurotransmitter transporter), member 11	SLC6A11	The antinociceptive effect of SNAP5114, a gamma-aminobutyric acid transporter-3 inhibitor, in rat experimental pain models. PMID: 23456665
Solute carrier family 6 (neurotransmitter transporter), member 12	SLC6A12	The betaine/GABA transporter and betaine: Roles in brain, kidney, and liver. PMID: 24795054
Solute carrier family 6 (neurotransmitter transporter), member 13	SLC6A13	High selectivity of the γ-aminobutyric acid transporter 2 (GAT-2, SLC6A13) revealed by structure-based approach. PMID: 22932902
Solute carrier family 6 (amino acid transporter), member 14	SLC6A14	Note: Genetic variations in SLC6A14 may be associated with obesity in some populations, as shown by significant differences in allele frequencies between obese and non-obese individuals
Solute carrier family 6 (neutral amino acid transporter), member 15	SLC6A15	Loratadine and analogues: Discovery and preliminary structure-activity relationship of inhibitors of the amino acid transporter B(0)AT2. PMID: 25318072
Solute carrier family 6, member 16	SLC6A16	
Solute carrier family 6 (neutral amino acid transporter), member 17	SLC6A17	Characterization of the transporterB0AT3 (Slc6a17) in the rodent central nervous system. PMID: 23672601
Solute carrier family 6 (neutral amino acid transporter), member 18	SLC6A18	Note: Genetic variations in SLC6A18 might contribute to the disease phentotype in some individuals with iminoglycinuria or hyperglycinuria, that carry variants in SLC36A2, SLC6A19 or SLC6A20

HGNC-approved name	Symbol	Target information
Solute carrier family 6 (neutral amino acid transporter), member 19	SLC6A19	Disease: Hartnup disorder (HND)
Solute carrier family 6 (proline IMINO transporter), member 20	SLC6A20	Disease: hyperglycinuria (HG)
Solute carrier family 7 (cationic amino acid transporter, y+ system), member 1	SLC7A1	Overexpression of cationic amino acid transporter-1 increases nitric oxide production in hypoxic human pulmonary microvascular endothelial cells. PMID: 21923750
Solute carrier family 7 (cationic amino acid transporter, y+ system), member 2	SLC7A2	Oncogenicity of L-type amino-acid transporter 1 (LAT1) revealed by targeted gene disruption in chicken DT40 cells: LAT1 is a promising molecular target for human cancer therapy. PMID: 21371427
Solute carrier family 7 (cationic amino acid transporter, y+ system), member 3	SLC7A3	
Solute carrier family 7, member 4	SLC7A4	
Solute carrier family 7 (amino acid transporter light chain, L system), member 5	SLC7A5	**IGN523** – antineoplastic agents
Solute carrier family 7 (amino acid transporter light chain, y+L system), member 6	SLC7A6	Targeting amino acid transport in metastatic castration-resistant prostate cancer: Effects on cell cycle, cell growth, and tumor development. PMID: 24052624
Solute carrier family 7 (amino acid transporter light chain, y+L system), member 7	SLC7A7	Disease: lysinuric protein intolerance (LPI)
Solute carrier family 7 (amino acid transporter light chain, L system), member 8	SLC7A8	Aminoaciduria, but normal thyroid hormone levels and signalling, in mice lacking the amino acid and thyroid hormone transporter Slc7a8. PMID: 21726201
Solute carrier family 7 (amino acid transporter light chain, bo,+ system), member 9	SLC7A9	Disease: cystinuria (CSNU)
Solute carrier family 7 (neutral amino acid transporter light chain, asc system), member 10	SLC7A10	*In vitro* characterization of a small molecule inhibitor of the alanine serine cysteine transporter -1 (SLC7A10). PMID: 24266811
Solute carrier family 7 (anionic amino acid transporter light chain, xc- system), member 11	SLC7A11	Glutamine sensitivity analysis identifies the xCT antiporter as a common triple-negative breast tumor therapeutic target. PMID: 24094812
Solute carrier family 7 (anionic amino acid transporter), member 13	SLC7A13	
Solute carrier family 7, member 14	SLC7A14	A chimera carrying the functional domain of the orphan protein SLC7A14 in the backbone of SLC7A2 mediates trans-stimulated arginine transport. PMID: 22787143
Solute carrier family 8 (sodium/calcium exchanger), member 1	SLC8A1	High level over-expression of different NCX isoforms in HEK293 cell lines and primary neuronal cultures is protective following oxygen glucose deprivation. PMID: 22561287
Solute carrier family 8 (sodium/calcium exchanger), member 2	SLC8A2	Neurounina-1, a novel compound that increases Na^+/Ca^{2+} exchanger activity, effectively protects against stroke damage. PMID: 23066092
Solute carrier family 8 (sodium/calcium exchanger), member 3	SLC8A3	NCX3 regulates mitochondrial Ca(2+) handling through the AKAP121-anchored signaling complex and prevents hypoxia-induced neuronal death. PMID: 24101730
Solute carrier family 8 (sodium/lithium/calcium exchanger), member B1	SLC8B1	

(Continued)

HGNC-approved name	Symbol	Target information
Solute carrier family 9, subfamily A (NHE1, cation proton antiporter 1), member 1	SLC9A1	**Rimeporide** – muscular dystrophy treatments
Solute carrier family 9, subfamily A (NHE2, cation proton antiporter 2), member 2	SLC9A2	Trefoil factor 2 requires Na/H exchanger 2 activity to enhance mouse gastric epithelial repair. PMID: 21900251
Solute carrier family 9, subfamily A (NHE3, cation proton antiporter 3), member 3	SLC9A3	**Tenapanor** – renal disease treatments
Solute carrier family 9, subfamily A (NHE4, cation proton antiporter 4), member 4	SLC9A4	Functional role of NHE4 as a pH regulator in rat and human colonic crypts. PMID: 22049213
Solute carrier family 9, subfamily A (NHE5, cation proton antiporter 5), member 5	SLC9A5	Endosomal acidification by Na+/H+ exchanger NHE5 regulates TrkA cell-surface targeting and NGF-induced PI3K signaling. PMID: 24006492
Solute carrier family 9, subfamily A (NHE6, cation proton antiporter 6), member 6	SLC9A6	Disease: mental retardation, X-linked, syndromic, Christianson type (MRXSCH)
Solute carrier family 9, subfamily A (NHE7, cation proton antiporter 7), member 7	SLC9A7	Organellar (Na+, K+)/H+ exchanger NHE7 regulates cell adhesion, invasion and anchorage-independent growth of breast cancer MDA-MB-231 cells. PMID: 22076128
Solute carrier family 9, subfamily A (NHE8, cation proton antiporter 8), member 8	SLC9A8	NHE8 plays an important role in mucosal protection via its effect on bacterial adhesion. PMID: 23657568
Solute carrier family 9, subfamily A (NHE9, cation proton antiporter 9), member 9	SLC9A9	Note: A chromosomal aberration involving SLC9A9 has been found in a family with early-onset behavioral/developmental disorder with features of attention deficit-hyperactivity disorder and intellectual disability. Inversion inv(3)(p14:q21)
Solute carrier family 9, subfamily B (NHA1, cation proton antiporter 1), member 1	SLC9B1	Cloning of a novel human NHEDC1 (Na+/H+ exchanger like domain containing 1) gene expressed specifically in testis. PMID: 16850186
Solute carrier family 9, subfamily B (NHA2, cation proton antiporter 2), member 2	SLC9B2	Model-guided mutagenesis drives functional studies of human NHA2, implicated in hypertension. PMID: 20053353
Solute carrier family 9, subfamily C (Na+-transporting carboxylic acid decarboxylase), member 1	SLC9C1	Proton channels and exchangers in cancer. PMID: 25449995
Solute carrier family 9, member C2 (putative)	SLC9C2	
Solute carrier family 10 (sodium/bile acid cotransporter), member 1	SLC10A1	The solute carrier family 10 (SLC10): Beyond bile acid transport. PMID: 23506869
Solute carrier family 10 (sodium/bile acid cotransporter), member 2	SLC10A2	**GSK 2330672** – drugs used in diabetes. Disease: primary bile acid malabsorption (PBAM)
Solute carrier family 10, member 3	SLC10A3	
Solute carrier family 10, member 4	SLC10A4	SLC10A4 is a vesicular amine-associated transporter modulating dopamine homeostasis. PMID: 25176177
Solute carrier family 10, member 5	SLC10A5	The novel putative bile acid transporter SLC10A5 is highly expressed in liver and kidney. PMID: 17632081
Solute carrier family 10 (sodium/bile acid cotransporter), member 6	SLC10A6	Cloning and functional characterization of human sodium-dependent organic anion transporter (SLC10A6). PMID: 17491011
Solute carrier family 10, member 7	SLC10A7	
Solute carrier family 11 (proton-coupled divalent metal ion transporter), member 1	SLC11A1	Slc11a1, formerly Nramp1, is expressed in dendritic cells and influences major histocompatibility complex class II expression and antigen-presenting cell function. PMID: 17620357
Solute carrier family 11 (proton-coupled divalent metal ion transporter), member 2	SLC11A2	H(+)-coupled divalent metal-ion transporter-1: Functional properties, physiological roles and therapeutics. PMID: 23177986. Disease: anemia, hypochromic microcytic, with iron overload 1 (AHMIO1)

HGNC-approved name	Symbol	Target information
Solute carrier family 12 (sodium/potassium/chloride transporter), member 1	SLC12A1	**Furosemide** – diuretics. Disease: Bartter syndrome 1 (BS1)
Solute carrier family 12 (sodium/potassium/chloride transporter), member 2	SLC12A2	**Bumetanide** – diuretics
Solute carrier family 12 (sodium/chloride transporter), member 3	SLC12A3	**Chlorothiazide** – diuretics. Disease: Gitelman syndrome (GS)
Solute carrier family 12 (potassium/chloride transporter), member 4	SLC12A4	**Bumetanide** – diuretics
Solute carrier family 12 (potassium/chloride transporter), member 5	SLC12A5	GABAergic disinhibition and impaired KCC2 cotransporter activity underlie tumor-associated epilepsy. PMID: 25066727
Solute carrier family 12 (potassium/chloride transporter), member 6	SLC12A6	Disease: agenesis of the corpus callosum, with peripheral neuropathy (ACCPN)
Solute carrier family 12 (potassium/chloride transporter), member 7	SLC12A7	IGF-1 upregulates electroneutral K-Cl cotransporter KCC3 and KCC4 which are differentially required for breast cancer cell proliferation and invasiveness. PMID: 17133354
Solute carrier family 12, member 8	SLC12A8	Note: SLC12A8 has been identified as a possible susceptibility gene for psoriasis mapped to chromosome 3q21 (PSORS5)
Solute carrier family 12, member 9	SLC12A9	
Solute carrier family 13 (sodium/sulfate symporter), member 1	SLC13A1	Increased lifespan in hyposulfatemic NaS1 null mice. PMID: 21651971
Solute carrier family 13 (sodium-dependent dicarboxylate transporter), member 2	SLC13A2	
Solute carrier family 13 (sodium-dependent dicarboxylate transporter), member 3	SLC13A3	PITX2 is involved in stress response in cultured human trabecular meshwork cells through regulation of SLC13A3. PMID: 21873665
Solute carrier family 13 (sodium/sulfate symporter), member 4	SLC13A4	
Solute carrier family 13 (sodium-dependent citrate transporter), member 5	SLC13A5	Deletion of the mammalian INDY homolog mimics aspects of dietary restriction and protects against adiposity and insulin resistance in mice. PMID: 21803289
Solute carrier family 14 (urea transporter), member 1	SLC14A1	Differential urinary specific gravity as a molecular phenotype of the bladder cancer genetic association in the urea transporter gene, SLC14A1. PMID: 23754249
Solute carrier family 14 (urea transporter), member 2	SLC14A2	Genetic variation in the human urea transporter-2 is associated with variation in blood pressure. PMID: 11590132
Solute carrier family 15 (oligopeptide transporter), member 1	SLC15A1	Proton-coupled oligopeptide transporter family SLC15: Physiological, pharmacological and pathological implications. PMID: 23506874
Solute carrier family 15 (oligopeptide transporter), member 2	SLC15A2	*S*-Nitrosothiol transport via PEPT2 mediates biological effects of nitric oxide gas exposure in macrophages. PMID: 23239496
Solute carrier family 15 (oligopeptide transporter), member 3	SLC15A3	Expression and regulation of the proton-coupled oligopeptide transporter PhT2 by LPS in macrophages and mouse spleen. PMID: 24754256
Solute carrier family 15 (oligopeptide transporter), member 4	SLC15A4	The histidine transporter SLC15A4 coordinates mTOR-dependent inflammatory responses and pathogenic antibody production. PMID: 25238095
Solute carrier family 15, member 5	SLC15A5	
Solute carrier family 16 (monocarboxylate transporter), member 1	SLC16A1	**AZD3965** – antineoplastic agents. Disease: symptomatic deficiency in lactate transport (SDLT)
Solute carrier family 16, member 2 (thyroid hormone transporter)	SLC16A2	Disease: monocarboxylate transporter 8 deficiency (MCT8 deficiency)

(Continued)

HGNC-approved name	Symbol	Target information
Solute carrier family 16 (monocarboxylate transporter), member 3	SLC16A3	In vivo pH in metabolic-defective Ras-transformed fibroblast tumors: Key role of the monocarboxylate transporter, MCT4, for inducing an alkaline intracellular pH. PMID: 21484790
Solute carrier family 16, member 4	SLC16A4	Cancer metabolism, stemness and tumor recurrence: MCT1 and MCT4 are functional biomarkers of metabolic symbiosis in head and neck cancer. PMID: 23574725
Solute carrier family 16 (monocarboxylate transporter), member 5	SLC16A5	
Solute carrier family 16, member 6	SLC16A6	Characterization of monocarboxylate transporter 6: Expression in human intestine and transport of the antidiabetic drug nateglinide. PMID: 23935065
Solute carrier family 16 (monocarboxylate transporter), member 7	SLC16A7	The inhibition of monocarboxylate transporter 2 (MCT2) by AR-C155858 is modulated by the associated ancillary protein. PMID: 20695846
Solute carrier family 16 (monocarboxylate transporter), member 8	SLC16A8	Altered visual function in monocarboxylate transporter 3 (Slc16a8) knockout mice. PMID: 18524945
Solute carrier family 16, member 9	SLC16A9	
Solute carrier family 16 (aromatic amino acid transporter), member 10	SLC16A10	T-type amino acid transporter TAT1 (Slc16a10) is essential for extracellular aromatic amino acid homeostasis control. PMID: 23045339
Solute carrier family 16, member 11	SLC16A11	Sequence variants in SLC16A11 are a common risk factor for type 2 diabetes in Mexico. PMID: 24390345. Disease: diabetes mellitus, non-insulin-dependent (NIDDM)
Solute carrier family 16, member 12	SLC16A12	Disease: cataract, juvenile, with microcornea and glucosuria (CJMG)
Solute carrier family 16, member 13	SLC16A13	Disease: diabetes mellitus, non-insulin-dependent (NIDDM)
Solute carrier family 16, member 14	SLC16A14	
Solute carrier family 17 (organic anion transporter), member 1	SLC17A1	NPT1/SLC17A1 is a renal urate exporter in humans and its common gain-of-function variant decreases the risk of renal underexcretion gout. PMID: 25252215
Solute carrier family 17, member 2	SLC17A2	
Solute carrier family 17 (organic anion transporter), member 3	SLC17A3	Human sodium phosphate transporter 4 (hNPT4/SLC17A3) as a common renal secretory pathway for drugs and urate. PMID: 20810651
Solute carrier family 17, member 4	SLC17A4	A Na+-phosphate cotransporter homologue (SLC17A4 protein) is an intestinal organic anion exporter. PMID: 22460716
Solute carrier family 17 (acidic sugar transporter), member 5	SLC17A5	Disease: Salla disease (SD)
Solute carrier family 17 (vesicular glutamate transporter), member 6	SLC17A6	VGLUT2-dependent sensory neurons in the TRPV1 population regulate pain and itch. PMID: 21040852
Solute carrier family 17 (vesicular glutamate transporter), member 7	SLC17A7	Normal distribution of VGLUT1 synapses on spinal motoneuron dendrites and their reorganization after nerve injury. PMID: 24599449
Solute carrier family 17 (vesicular glutamate transporter), member 8	SLC17A8	Disease: deafness, autosomal dominant, 25 (DFNA25)
Solute carrier family 17 (vesicular nucleotide transporter), member 9	SLC17A9	SLC17A9 protein functions as a lysosomal ATP transporter and regulates cell viability. PMID: 24962569
Solute carrier family 18 (vesicular monoamine transporter), member 1	SLC18A1	VMAT1 deletion causes neuronal loss in the hippocampus and neurocognitive deficits in spatial discrimination. PMID: 23201251
Solute carrier family 18 (vesicular monoamine transporter), member 2	SLC18A2	**Tetrabenazine** – anti-Parkinson drugs
Solute carrier family 18 (vesicular acetylcholine transporter), member 3	SLC18A3	Are vesicular neurotransmitter transporters potential treatment targets for temporal lobe epilepsy? PMID: 24009559
Solute carrier family 18, subfamily B, member 1	SLC18B1	Identification of a mammalian vesicular polyamine transporter. PMID: 25355561

HGNC-approved name	Symbol	Target information
Solute carrier family 19 (folate transporter), member 1	SLC19A1	Biology of the major facilitative folate transporters SLC19A1 and SLC46A1. PMID: 24745983
Solute carrier family 19 (thiamine transporter), member 2	SLC19A2	Disease: thiamine-responsive megaloblastic anemia syndrome (TRMA)
Solute carrier family 19 (thiamine transporter), member 3	SLC19A3	Disease: thiamine metabolism dysfunction syndrome 2, biotin- or thiamine-responsive type (THMD2)
Solute carrier family 20 (phosphate transporter), member 1	SLC20A1	Identification of a novel transport-independent function of PiT1/SLC20A1 in the regulation of TNF-induced apoptosis. PMID: 20817733
Solute carrier family 20 (phosphate transporter), member 2	SLC20A2	Disease: basal ganglia calcification, idiopathic, 1 (IBGC1)
Solute carrier organic anion transporter family, member 1A2	SLCO1A2	Organic anion-transporting polypeptides: A novel approach for cancer therapy. PMID: 23987090
Solute carrier organic anion transporter family, member 1B1	SLCO1B1	Disease: hyperbilirubinemia, Rotor type (HBLRR)
Solute carrier organic anion transporter family, member 1B3	SLCO1B3	Disease: hyperbilirubinemia, Rotor type (HBLRR)
Solute carrier organic anion transporter family, member 1B7 (non-functional)	SLCO1B7	Long evolutionary conservation and considerable tissue specificity of several atypical solute carrier transporters. PMID: 21044875
Solute carrier organic anion transporter family, member 1C1	SLCO1C1	
Solute carrier organic anion transporter family, member 2A1	SLCO2A1	Disease: hypertrophic osteoarthropathy, primary, autosomal recessive, 2 (PHOAR2)
Solute carrier organic anion transporter family, member 2B1	SLCO2B1	SLCO2B1 and SLCO1B3 may determine time to progression for patients receiving androgen deprivation therapy for prostate cancer. PMID: 21606417
Solute carrier organic anion transporter family, member 3A1	SLCO3A1	SLCO3A1, a novel Crohn's disease-associated gene, regulates NF-κB activity and associates with intestinal perforation. PMID: 24945726
Solute carrier organic anion transporter family, member 4A1	SLCO4A1	
Solute carrier organic anion transporter family, member 4C1	SLCO4C1	SLCO4C1 transporter eliminates uremic toxins and attenuates hypertension and renal inflammation. PMID: 19875811
Solute carrier organic anion transporter family, member 5A1	SLCO5A1	Characterization of SLCO5A1/OATP5A1, a solute carrier transport protein with non-classical function. PMID: 24376674
Solute carrier organic anion transporter family, member 6A1	SLCO6A1	Identification of the gonad-specific anion transporter SLCO6A1 as a cancer/testis (CT) antigen expressed in human lung cancer. PMID: 15546177
Solute carrier family 22 (organic cation transporter), member 1	SLC22A1	OCT1 is a high-capacity thiamine transporter that regulates hepatic steatosis and is a target of metformin. PMID: 24961373
Solute carrier family 22 (organic cation transporter), member 2	SLC22A2	Organic cation transporter 2 (SLC22A2), a low-affinity and high-capacity choline transporter, is preferentially enriched on synaptic vesicles in cholinergic neurons. PMID: 23958595
Solute carrier family 22 (organic cation transporter), member 3	SLC22A3	Organic cation transporter 3: Keeping the brake on extracellular serotonin in serotonin-transporter-deficient mice. PMID: 19033200
Solute carrier family 22 (organic cation/zwitterion transporter), member 4	SLC22A4	Disease: rheumatoid arthritis (RA)
Solute carrier family 22 (organic cation/carnitine transporter), member 5	SLC22A5	SLC22A5/OCTN2 expression in breast cancer is induced by estrogen via a novel intronic estrogen-response element (ERE). PMID: 22212555. Disease: systemic primary carnitine deficiency (CDSP)
Solute carrier family 22 (organic anion transporter), member 6	SLC22A6	**Probenecid** – antigout preparations

(Continued)

HGNC-approved name	Symbol	Target information
Solute carrier family 22 (organic anion transporter), member 7	SLC22A7	Organic cation transporter 2 and tumor budding as independent prognostic factors in metastatic colorectal cancer patients treated with oxaliplatin-based chemotherapy. PMID: 24427340
Solute carrier family 22 (organic anion transporter), member 8	SLC22A8	Multispecific drug transporter Slc22a8 (Oat3) regulates multiple metabolic and signaling pathways. PMID: 23920220
Solute carrier family 22 (organic anion transporter), member 9	SLC22A9	
Solute carrier family 22, member 10	SLC22A10	
Solute carrier family 22 (organic anion/urate transporter), member 11	SLC22A11	
Solute carrier family 22 (organic anion/urate transporter), member 12	SLC22A12	**Sulphinpyrazone** – antigout preparations. Disease: hypouricemia renal 1 (RHUC1)
Solute carrier family 22 (organic anion/urate transporter), member 13	SLC22A13	Isolation of ORCTL3 in a novel genetic screen for tumor-specific apoptosis inducers. PMID: 19282870
Solute carrier family 22, member 14	SLC22A14	
Solute carrier family 22, member 15	SLC22A15	
Solute carrier family 22 (organic cation/carnitine transporter), member 16	SLC22A16	The human carnitine transporter SLC22A16 mediates high affinity uptake of the anticancer polyamine analogue bleomycin-A5. PMID: 20037140
Solute carrier family 22, member 17	SLC22A17	Immunohistochemical detection of a specific receptor for lipocalin2 (solute carrier family 22 member 17, SLC22A17) and its prognostic significance in endometrial carcinoma. PMID: 21763306
Solute carrier family 22, member 18	SLC22A18	Disease: lung cancer (LNCR)
Solute carrier family 22, member 20	SLC22A20	Organic anion transporter 6 (Slc22a20) specificity and Sertoli cell-specific expression provide new insight on potential endogenous roles. PMID: 20519554
Solute carrier family 22, member 23	SLC22A23	Single-nucleotide polymorphisms in SLC22A23 are associated with ulcerative colitis in a Canadian white cohort. PMID: 24740203
Solute carrier family 22, member 24	SLC22A24	
Solute carrier family 22, member 25	SLC22A25	Identification of six putative human transporters with structural similarity to the drug transporter SLC22 family. PMID: 17714910
Solute carrier family 22, member 31	SLC22A31	
Solute carrier family 23 (ascorbic acid transporter), member 1	SLC23A1	SLC23A1 polymorphism rs6596473 in the vitamin C transporter SVCT1 is associated with aggressive periodontitis. PMID: 24708273
Solute carrier family 23 (ascorbic acid transporter), member 2	SLC23A2	Expression and/or activity of the SVCT2 ascorbate transporter may be decreased in many aggressive cancers, suggesting potential utility for sodium bicarbonate and dehydroascorbic acid in cancer therapy. PMID: 23916956
Solute carrier family 23, member 3	SLC23A3	
Solute carrier family 24 (sodium/potassium/calcium exchanger), member 1	SLC24A1	Disease: night blindness, congenital stationary, 1D (CSNB1D)
Solute carrier family 24 (sodium/potassium/calcium exchanger), member 2	SLC24A2	A critical role for the potassium-dependent sodium-calcium exchanger NCKX2 in protection against focal ischemic brain damage. PMID: 18305240
Solute carrier family 24 (sodium/potassium/calcium exchanger), member 3	SLC24A3	
Solute carrier family 24 (sodium/potassium/calcium exchanger), member 4	SLC24A4	Note: SLC24A4 mutations may be a cause of autosomal recessive hypomineralized amelogenesis imperfecta (AI), a defect of enamel formation characterized by variable degrees of incomplete mineralization of the enamel matrix
Solute carrier family 24 (sodium/potassium/calcium exchanger), member 5	SLC24A5	Disease: albinism, oculocutaneous, 6 (OCA6)

HGNC-approved name	Symbol	Target information
Solute carrier family 25 (mitochondrial carrier; citrate transporter), member 1	SLC25A1	Disease: combined D-2- and L-2-hydroxyglutaric aciduria (D2L2AD)
Solute carrier family 25 (mitochondrial carrier; ornithine transporter), member 2	SLC25A2	Cloning and characterization of human ORNT2: A second mitochondrial ornithine transporter that can rescue a defective ORNT1 in patients with the hyperornithinemia-hyperammonemia-homocitrullinuria syndrome, a urea cycle disorder. PMID: 12948741
Solute carrier family 25 (mitochondrial carrier; phosphate carrier), member 3	SLC25A3	Disease: mitochondrial phosphate carrier deficiency (MPCD)
Solute carrier family 25 (mitochondrial carrier; adenine nucleotide translocator), member 4	SLC25A4	**Clodronate** – drugs affecting bone structure and mineralization. Disease: progressive external ophthalmoplegia with mitochondrial DNA deletions, autosomal dominant, 2 (PEOA2)
Solute carrier family 25 (mitochondrial carrier; adenine nucleotide translocator), member 5	SLC25A5	The adenine nucleotide translocase 2, a mitochondrial target for anticancer biotherapy. PMID: 21269262
Solute carrier family 25 (mitochondrial carrier; adenine nucleotide translocator), member 6	SLC25A6	**Clodronate** – drugs affecting bone structure and mineralization
Uncoupling protein 1 (mitochondrial, proton carrier)	UCP1	ThermoMouse: An in vivo model to identify modulators of UCP1 expression in brown adipose tissue. PMID: 25466254
Uncoupling protein 2 (mitochondrial, proton carrier)	UCP2	Hydrogen peroxide regulates the mitochondrial content of uncoupling protein 5 in colon cancer cells. PMID: 19910678
Uncoupling protein 3 (mitochondrial, proton carrier)	UCP3	Disease: obesity (OBESITY)
Solute carrier family 25 (mitochondrial carrier; dicarboxylate transporter), member 10	SLC25A10	
Solute carrier family 25 (mitochondrial carrier; oxoglutarate carrier), member 11	SLC25A11	Stable over-expression of the 2-oxoglutarate carrier enhances neuronal cell resistance to oxidative stress via Bcl-2-dependent mitochondrial GSH transport. PMID: 24606213
Solute carrier family 25 (aspartate/glutamate carrier), member 12	SLC25A12	Disease: global cerebral hypomyelination (GCHM)
Solute carrier family 25 (aspartate/glutamate carrier), member 13	SLC25A13	Disease: citrullinemia 2 (CTLN2)
Solute carrier family 25 (mitochondrial carrier, brain), member 14	SLC25A14	Hydrogen peroxide regulates the mitochondrial content of uncoupling protein 5 in colon cancer cells. PMID: 19910678
Solute carrier family 25 (mitochondrial carrier; ornithine transporter), member 15	SLC25A15	Disease: hyperornithinemia–hyperammonemia–homocitrullinuria syndrome (HHH syndrome)
Solute carrier family 25 (mitochondrial carrier; Graves disease autoantigen), member 16	SLC25A16	Physiological and pathological roles of mitochondrial SLC25 carriers. PMID: 23988125
Solute carrier family 25 (mitochondrial carrier; peroxisomal membrane protein, 34 kDa), member 17	SLC25A17	
Solute carrier family 25 (glutamate carrier), member 18	SLC25A18	
Solute carrier family 25 (mitochondrial thiamine pyrophosphate carrier), member 19	SLC25A19	Disease: microcephaly, Amish type (MCPHA)
Solute carrier family 25 (carnitine/acylcarnitine translocase), member 20	SLC25A20	Disease: carnitine-acylcarnitine translocase deficiency (CACT deficiency)
Solute carrier family 25 (mitochondrial oxoadipate carrier), member 21	SLC25A21	Polyamine pathway inhibition as a novel therapeutic approach to treating neuroblastoma. PMID: 23181218

(Continued)

HGNC-approved name	Symbol	Target information
Solute carrier family 25 (mitochondrial carrier: glutamate), member 22	SLC25A22	Disease: epileptic encephalopathy, early infantile, 3 (EIEE3)
Solute carrier family 25 (mitochondrial carrier; phosphate carrier), member 23	SLC25A23	SLC25A23 augments mitochondrial Ca^{2+} uptake, interacts with MCU, and induces oxidative stress-mediated cell death. PMID: 24430870
Solute carrier family 25 (mitochondrial carrier; phosphate carrier), member 24	SLC25A24	
Solute carrier family 25 (mitochondrial carrier; phosphate carrier), member 25	SLC25A25	Inactivation of the mitochondrial carrier SLC25A25 (ATP-Mg^{2+}/Pi transporter) reduces physical endurance and metabolic efficiency in mice. PMID: 21296886
Solute carrier family 25 (S-adenosylmethionine carrier), member 26	SLC25A26	
Solute carrier family 25, member 27	SLC25A27	UCP4 is a target effector of the NF-κB c-Rel prosurvival pathway against oxidative stress. PMID: 22580300
Solute carrier family 25 (mitochondrial iron transporter), member 28	SLC25A28	Mitoferrin-2-dependent mitochondrial iron uptake sensitizes human head and neck squamous carcinoma cells to photodynamic therapy. PMID: 23135267
Solute carrier family 25 (mitochondrial carnitine/acylcarnitine carrier), member 29	SLC25A29	The human gene SLC25A29, of solute carrier family 25, encodes a mitochondrial transporter of basic amino acids. PMID: 24652292
Solute carrier family 25, member 30	SLC25A30	
Solute carrier family 25 (mitochondrial carrier; adenine nucleotide translocator), member 31	SLC25A31	Identification of adenine nucleotide translocase 4 inhibitors by molecular docking. PMID: 24056384
Solute carrier family 25 (mitochondrial folate carrier), member 32	SLC25A32	Effect of chronic alcohol exposure on folate uptake by liver mitochondria. PMID: 21956163
Solute carrier family 25 (pyrimidine nucleotide carrier), member 33	SLC25A33	The insulin-like growth factor-I-mTOR signaling pathway induces the mitochondrial pyrimidine nucleotide carrier to promote cell growth. PMID: 17596519
Solute carrier family 25, member 34	SLC25A34	
Solute carrier family 25, member 35	SLC25A35	
Solute carrier family 25 (pyrimidine nucleotide carrier), member 36	SLC25A36	
Solute carrier family 25 (mitochondrial iron transporter), member 37	SLC25A37	Mitoferrin is essential for erythroid iron assimilation. PMID: 16511496
Solute carrier family 25, member 38	SLC25A38	Disease: anemia, sideroblastic, pyridoxine-refractory, autosomal recessive (PRARSA)
Solute carrier family 25, member 39	SLC25A39	
Solute carrier family 25, member 40	SLC25A40	Joint linkage and association analysis with exome sequence data implicates SLC25A40 in hypertriglyceridemia. PMID: 24268658
Solute carrier family 25, member 41	SLC25A41	
Solute carrier family 25, member 42	SLC25A42	A novel member of solute carrier family 25 (SLC25A42) is a transporter of coenzyme A and adenosine 3′,5′-diphosphate in human mitochondria. PMID: 19429682
Solute carrier family 25, member 43	SLC25A43	The mitochondrial transport protein SLC25A43 affects drug efficacy and drug-induced cell cycle arrest in breast cancer cell lines. PMID: 23354756
Solute carrier family 25, member 44	SLC25A44	
Solute carrier family 25, member 45	SLC25A45	
Solute carrier family 25, member 46	SLC25A46	
Solute carrier family 25, member 47	SLC25A47	

HGNC-approved name	Symbol	Target information
Solute carrier family 25, member 48	SLC25A48	Cloning and identification of hepatocellular carcinoma down-regulated mitochondrial carrier protein, a novel liver-specific uncoupling protein. PMID: 15322095
Mitochondrial carrier 1	MTCH1	Mitochondrial carrier homolog 1 (Mtch1) antibodies in neuro-Behçet's disease. PMID: 24035008
Mitochondrial carrier 2	MTCH2	Molecular basis of the interaction between proapoptotic truncated BID (tBID) protein and mitochondrial carrier homologue 2 (MTCH2) protein: Key players in mitochondrial death pathway. PMID: 22416135
Solute carrier family 25, member 51	SLC25A51	
Solute carrier family 25, member 52	SLC25A52	
Solute carrier family 25, member 53	SLC25A53	
Solute carrier family 26 (anion exchanger), member 1	SLC26A1	Mis-trafficking of bicarbonate transporters: Implications to human diseases. PMID: 21455268
Solute carrier family 26 (anion exchanger), member 2	SLC26A2	Disease: diastrophic dysplasia (DTD)
Solute carrier family 26 (anion exchanger), member 3	SLC26A3	Disease: diarrhea 1, secretory chloride, congenital (DIAR1)
Solute carrier family 26 (anion exchanger), member 4	SLC26A4	Epithelial anion transporter pendrin contributes to inflammatory lung pathology in mouse models of Bordetella pertussis infection. PMID: 25069981. Disease: Pendred syndrome (PDS)
Solute carrier family 26 (anion exchanger), member 5	SLC26A5	Disease: deafness, autosomal recessive, 61 (DFNB61)
Solute carrier family 26 (anion exchanger), member 6	SLC26A6	Slc26a6 functions as an electrogenic Cl-/HCO3- exchanger in cardiac myocytes. PMID: 23933580
Solute carrier family 26 (anion exchanger), member 7	SLC26A7	Deletion of the chloride transporter slc26a7 causes distal renal tubular acidosis and impairs gastric acid secretion. PMID: 19723628
Solute carrier family 26 (anion exchanger), member 8	SLC26A8	Disease: spermatogenic failure 3 (SPGF3)
Solute carrier family 26 (anion exchanger), member 9	SLC26A9	SLC26A9-mediated chloride secretion prevents mucus obstruction in airway inflammation. PMID: 22945630
Solute carrier family 26, member 10	SLC26A10	
Solute carrier family 26 (anion exchanger), member 11	SLC26A11	Identification of a novel gene fusion RNF213-SLC26A11 in chronic myeloid leukemia by RNA-Seq. PMID: 23151010
Solute carrier family 27 (fatty acid transporter), member 1	SLC27A1	Fatty acid transport protein 1 (FATP1) localizes in mitochondria in mouse skeletal muscle and regulates lipid and ketone body disposal. PMID: 24858472
Solute carrier family 27 (fatty acid transporter), member 2	SLC27A2	Slc27a2 expression in peripheral blood mononuclear cells as a molecular marker for overweight development. PMID: 20142826
Solute carrier family 27 (fatty acid transporter), member 3	SLC27A3	Very long-chain acyl-CoA synthetase 3: Overexpression and growth dependence in lung cancer. PMID: 23936004
Solute carrier family 27 (fatty acid transporter), member 4	SLC27A4	Disease: ichthyosis prematurity syndrome (IPS)
Solute carrier family 27 (fatty acid transporter), member 5	SLC27A5	Targeted deletion of FATP5 reveals multiple functions in liver metabolism: Alterations in hepatic lipid homeostasis. PMID: 16618416
Solute carrier family 27 (fatty acid transporter), member 6	SLC27A6	A variant in the heart-specific fatty acid transport protein 6 is associated with lower fasting and postprandial TAG, blood pressure and left ventricular hypertrophy. PMID: 21920065
Solute carrier family 28 (concentrative nucleoside transporter), member 1	SLC28A1	CNT1 expression influences proliferation and chemosensitivity in drug-resistant pancreatic cancer cells. PMID: 21343396

(Continued)

HGNC-approved name	Symbol	Target information
Solute carrier family 28 (concentrative nucleoside transporter), member 2	SLC28A2	Ribavirin uptake into human hepatocyte HHL5 cells is enhanced by interferon-α via up-regulation of the human concentrative nucleoside transporter (hCNT2). PMID: 24957263
Solute carrier family 28 (concentrative nucleoside transporter), member 3	SLC28A3	All-trans-retinoic acid promotes trafficking of human concentrative nucleoside transporter-3 (hCNT3) to the plasma membrane by a TGF-beta1-mediated mechanism. PMID: 20172853
Solute carrier family 29 (equilibrative nucleoside transporter), member 1	SLC29A1	Prognostic value of human equilibrative nucleoside transporter1 in pancreatic cancer receiving gemcitabin-based chemotherapy: A meta-analysis. PMID: 24475233
Solute carrier family 29 (equilibrative nucleoside transporter), member 2	SLC29A2	Crosstalk between the equilibrative nucleoside transporter ENT2 and alveolar Adora2b adenosine receptors dampens acute lung injury. PMID: 23603835
Solute carrier family 29 (equilibrative nucleoside transporter), member 3	SLC29A3	Disease: histiocytosis–lymphadenopathy plus syndrome (HLAS)
Solute carrier family 29 (equilibrative nucleoside transporter), member 4	SLC29A4	Dipyridamole analogs as pharmacological inhibitors of equilibrative nucleoside transporters. Identification of novel potent and selective inhibitors of the adenosine transporter function of human equilibrative nucleoside transporter 4 (hENT4). PMID: 24021350
Solute carrier family 30 (zinc transporter), member 1	SLC30A1	Zinc transporter-1 concentrates at the postsynaptic density of hippocampal synapses. PMID: 24602382
Solute carrier family 30 (zinc transporter), member 2	SLC30A2	Disease: zinc deficiency, transient neonatal (TNZD)
Solute carrier family 30 (zinc transporter), member 3	SLC30A3	Zinc transporters ZnT3 and ZnT6 are downregulated in the spinal cords of patients with sporadic amyotrophic lateral sclerosis. PMID: 25284286
Solute carrier family 30 (zinc transporter), member 4	SLC30A4	Labile zinc and zinc transporter ZnT4 in mast cell granules: Role in regulation of caspase activation and NF-κB translocation. PMID: 15187159
Solute carrier family 30 (zinc transporter), member 5	SLC30A5	Zinc transporter Znt5/Slc30a5 is required for the mast cell-mediated delayed-type allergic reaction but not the immediate-type reaction. PMID: 19451265
Solute carrier family 30 (zinc transporter), member 6	SLC30A6	Zinc transporters ZnT3 and ZnT6 are downregulated in the spinal cords of patients with sporadic amyotrophic lateral sclerosis. PMID: 25284286
Solute carrier family 30 (zinc transporter), member 7	SLC30A7	Over-expression of ZnT7 increases insulin synthesis and secretion in pancreatic beta-cells by promoting insulin gene transcription. PMID: 20599947
Solute carrier family 30 (zinc transporter), member 8	SLC30A8	The diabetes-susceptible gene SLC30A8/ZnT8 regulates hepatic insulin clearance. PMID: 24051378
Solute carrier family 30 (zinc transporter), member 9	SLC30A9	
Solute carrier family 30, member 10	SLC30A10	Diabetes-linked zinc transporter ZnT8 is a homodimeric protein expressed by distinct rodent endocrine cell types in the pancreas and other glands. PMID: 19095428
Solute carrier family 31 (copper transporter), member 1	SLC31A1	Copper is required for oncogenic BRAF signalling and tumorigenesis. PMID: 24717435
Solute carrier family 31 (copper transporter), member 2	SLC31A2	The mammalian copper transporters CTR1 and CTR2 and their roles in development and disease. PMID: 23391749
Solute carrier family 32 (GABA vesicular transporter), member 1	SLC32A1	Glycine transporters: Essential regulators of synaptic transmission. PMID: 16417482

HGNC-approved name	Symbol	Target information
Solute carrier family 33 (acetyl-CoA transporter), member 1	SLC33A1	Disease: spastic paraplegia 42, autosomal dominant (SPG42)
Solute carrier family 34 (type II sodium/phosphate cotransporter), member 1	SLC34A1	Disease: nephrolithiasis/osteoporosis, hypophosphatemic, 1 (NPHLOP1)
Solute carrier family 34 (type II sodium/phosphate cotransporter), member 2	SLC34A2	**RG7599** – Antineoplastic agents. Disease: pulmonary alveolar microlithiasis (PALM)
Solute carrier family 34 (type II sodium/phosphate cotransporter), member 3	SLC34A3	Disease: hereditary hypophosphatemic rickets with hypercalciuria (HHRH)
Solute carrier family 35 (CMP-sialic acid transporter), member A1	SLC35A1	Disease: congenital disorder of glycosylation 2F (CDG2F)
Solute carrier family 35 (UDP-galactose transporter), member A2	SLC35A2	Disease: congenital disorder of glycosylation 2M (CDG2M)
Solute carrier family 35 (UDP-N-acetylglucosamine (UDP-GlcNAc) transporter), member A3	SLC35A3	Disease: arthrogryposis, mental retardation, and seizures (AMRS)
Solute carrier family 35, member A4	SLC35A4	
Solute carrier family 35, member A5	SLC35A5	
Solute carrier family 35, member B1	SLC35B1	
Solute carrier family 35 (adenosine 3′-phospho 5′-phosphosulfate transporter), member B2	SLC35B2	SLC35B2 expression is associated with a poor prognosis of invasive ductal breast carcinoma. PMID: 25124574
Solute carrier family 35 (adenosine 3′-phospho 5′-phosphosulfate transporter), member B3	SLC35B3	The 3′-phosphoadenosine 5′-phosphosulfate transporters, PAPST1 and 2, contribute to the maintenance and differentiation of mouse embryonic stem cells. PMID: 20011239
Solute carrier family 35 (UDP-xylose/UDP-N-acetylglucosamine transporter), member B4	SLC35B4	Deep congenic analysis identifies many strong, context-dependent QTLs, one of which, Slc35b4, regulates obesity and glucose homeostasis. PMID: 21507882
Solute carrier family 35 (GDP-fucose transporter), member C1	SLC35C1	Disease: congenital disorder of glycosylation 2C (CDG2C)
Solute carrier family 35 (GDP-fucose transporter), member C2	SLC35C2	Slc35c2 promotes Notch1 fucosylation and is required for optimal Notch signaling in mammalian cells. PMID: 20837470
Solute carrier family 35 (UDP-GlcA/UDP-GalNAc transporter), member D1	SLC35D1	Disease: Schneckenbecken dysplasia (SCHBCKD)
Solute carrier family 35 (UDP-GlcNAc/UDP-glucose transporter), member D2	SLC35D2	Identification and characterization of human Golgi nucleotide sugar transporter SLC35D2, a novel member of the SLC35 nucleotide sugar transporter family. PMID: 15607426
Solute carrier family 35, member D3	SLC35D3	Mutation of SLC35D3 causes metabolic syndrome by impairing dopamine signaling in striatal D1 neurons. PMID: 24550737
Solute carrier family 35, member E1	SLC35E1	
Solute carrier family 35, member E2	SLC35E2	
Solute carrier family 35, member E2B	SLC35E2B	
Solute carrier family 35, member E3	SLC35E3	
Solute carrier family 35, member E4	SLC35E4	
Solute carrier family 35, member F1	SLC35F1	
Solute carrier family 35, member F2	SLC35F2	Influence on the behavior of lung cancer H1299 cells by silencing SLC35F2 expression. PMID: 23879892
Solute carrier family 35, member F3	SLC35F3	Genetic implication of a novel thiamine transporter in human hypertension. PMID: 24509276

(Continued)

HGNC-approved name	Symbol	Target information
Solute carrier family 35, member F4	SLC35F4	
Solute carrier family 35, member F5	SLC35F5	
Solute carrier family 35, member F6	SLC35F6	Identification of C2orf18, termed ANT2BP (ANT2-binding protein), as one of the key molecules involved in pancreatic carcinogenesis. PMID: 19154410
Solute carrier family 35, member G1	SLC35G1	POST, partner of stromal interaction molecule 1 (STIM1), targets STIM1 to multiple transporters. PMID: 22084111
Solute carrier family 35, member G2	SLC35G2	Involvement of TMEM22 overexpression in the growth of renal cell carcinoma cells. PMID: 19148500
Solute carrier family 35, member G3	SLC35G3	
Solute carrier family 35, member G4	SLC35G4	
Solute carrier family 35, member G5	SLC35G5	
Solute carrier family 35, member G6	SLC35G6	
Solute carrier family 36 (proton/amino acid symporter), member 1	SLC36A1	Intracellular amino acid sensing and mTORC1-regulated growth: New ways to block an old target? PMID: 21154118
Solute carrier family 36 (proton/amino acid symporter), member 2	SLC36A2	Disease: hyperglycinuria (HG)
Solute carrier family 36, member 3	SLC36A3	
Solute carrier family 36 (proton/amino acid symporter), member 4	SLC36A4	Rab12 regulates mTORC1 activity and autophagy through controlling the degradation of amino-acid transporter PAT4. PMID: 23478338
Solute carrier family 37 (glucose-6-phosphate transporter), member 1	SLC37A1	SLC37A1 gene expression is up-regulated by epidermal growth factor in breast cancer cells. PMID: 19894109
Solute carrier family 37 (glucose-6-phosphate transporter), member 2	SLC37A2	The major facilitator superfamily member Slc37a2 is a novel macrophage-specific gene selectively expressed in obese white adipose tissue. PMID: 17356011
Solute carrier family 37, member 3	SLC37A3	
Solute carrier family 37 (glucose-6-phosphate transporter), member 4	SLC37A4	Disease: glycogen storage disease 1B (GSD1B)
Solute carrier family 38, member 1	SLC38A1	shRNA-mediated Slc38a1 silencing inhibits migration, but not invasiveness of human pancreatic cancer cells. PMID: 24255574
Solute carrier family 38, member 2	SLC38A2	Functional RNA interference (RNAi) screen identifies system A neutral amino acid transporter 2 (SNAT2) as a mediator of arsenic-induced endoplasmic reticulum stress. PMID: 22216663
Solute carrier family 38, member 3	SLC38A3	
Solute carrier family 38, member 4	SLC38A4	
Solute carrier family 38, member 5	SLC38A5	
Solute carrier family 38, member 6	SLC38A6	
Solute carrier family 38, member 7	SLC38A7	
Solute carrier family 38, member 8	SLC38A8	Disease: foveal hypoplasia 2 (FVH2)
Solute carrier family 38, member 9	SLC38A9	Metabolism. Lysosomal amino acid transporter SLC38A9 signals arginine sufficiency to mTORC1. PMID: 25567906
Solute carrier family 38, member 10	SLC38A10	
Solute carrier family 38, member 11	SLC38A11	
Solute carrier family 39 (zinc transporter), member 1	SLC39A1	Genetic inhibition of solute-linked carrier 39 family transporter 1 ameliorates aβ pathology in a Drosophila model of Alzheimer's disease. PMID: 22570624
Solute carrier family 39 (zinc transporter), member 2	SLC39A2	Up-regulation of Slc39A2(Zip2) mRNA in peripheral blood mononuclear cells from patients with pulmonary tuberculosis. PMID: 23686108

HGNC-approved name	Symbol	Target information
Solute carrier family 39 (zinc transporter), member 3	SLC39A3	The cytotoxic role of RREB1, ZIP3 zinc transporter, and zinc in human pancreatic adenocarcinoma. PMID: 25050557
Solute carrier family 39 (zinc transporter), member 4	SLC39A4	Disease: acrodermatitis enteropathica, zinc-deficiency type (AEZ)
Solute carrier family 39 (zinc transporter), member 5	SLC39A5	SLC39A5 mutations interfering with the BMP/TGF-β pathway in non-syndromic high myopia. PMID: 24891338
Solute carrier family 39 (zinc transporter), member 6	SLC39A6	**SGN-LIV1A** – antineoplastic agents
Solute carrier family 39 (zinc transporter), member 7	SLC39A7	The zinc transporter, Slc39a7 (Zip7) is implicated in glycaemic control in skeletal muscle cells. PMID: 24265765
Solute carrier family 39 (zinc transporter), member 8	SLC39A8	Regulation of the catabolic cascade in osteoarthritis by the zinc-ZIP8-MTF1 axis. PMID: 24529376
Solute carrier family 39, member 9	SLC39A9	Identification and characterization of membrane androgen receptors in the ZIP9 zinc transporter subfamily: II. Role of human ZIP9 in testosterone-induced prostate and breast cancer cell apoptosis. PMID: 25014355
Solute carrier family 39 (zinc transporter), member 10	SLC39A10	Zinc transporter SLC39A10/ZIP10 controls humoral immunity by modulating B-cell receptor signal strength. PMID: 25074919
Solute carrier family 39, member 11	SLC39A11	
Solute carrier family 39 (zinc transporter), member 12	SLC39A12	Neurulation and neurite extension require the zinc transporter ZIP12 (slc39a12). PMID: 23716681
Solute carrier family 39 (zinc transporter), member 13	SLC39A13	Disease: Ehlers–Danlos syndrome-like spondylocheirodysplasia (SCD-EDS)
Solute carrier family 39 (zinc transporter), member 14	SLC39A14	Zip14 expression induced by lipopolysaccharides in macrophages attenuates inflammatory response. PMID: 23052185
Solute carrier family 40 (iron-regulated transporter), member 1	SLC40A1	Ferroportin in the progression and prognosis of hepatocellular carcinoma. PMID: 24360312. Disease: hemochromatosis 4 (HFE4)
Solute carrier family 41 (magnesium transporter), member 1	SLC41A1	The SLC41 family of MgtE-like magnesium transporters. PMID: 23506895
Solute carrier family 41 (magnesium transporter), member 2	SLC41A2	
Solute carrier family 41, member 3	SLC41A3	
Solute carrier family 43 (amino acid system L transporter), member 1	SLC43A1	The small SLC43 family: Facilitator system l amino acid transporters and the orphan EEG1. PMID: 23268354
Solute carrier family 43 (amino acid system L transporter), member 2	SLC43A2	
Solute carrier family 43, member 3	SLC43A3	
Solute carrier family 44 (choline transporter), member 1	SLC44A1	Functional expression of choline transporter-like protein 1 (CTL1) in small cell lung carcinoma cells: A target molecule for lung cancer therapy. PMID: 23948665
Solute carrier family 44 (choline transporter), member 2	SLC44A2	
Solute carrier family 44, member 3	SLC44A3	
Solute carrier family 44, member 4	SLC44A4	**ASG-5ME** – antineoplastic agents. Note: An interstitial deletion causing the fusion of exon 10 of CTL4 with the 3′-UTR of NEU has been detected in two patients affected by sialidosis
Solute carrier family 44, member 5	SLC44A5	
Solute carrier family 45, member 1	SLC45A1	

(Continued)

HGNC-approved name	Symbol	Target information
Solute carrier family 45, member 2	SLC45A2	Disease: albinism, oculocutaneous, 4 (OCA4)
Solute carrier family 45, member 3	SLC45A3	Loss of SLC45A3 protein (prostein) expression in prostate cancer is associated with SLC45A3-ERG gene rearrangement and an unfavorable clinical course. PMID: 22821757
Solute carrier family 45, member 4	SLC45A4	
Solute carrier family 46 (folate transporter), member 1	SLC46A1	Disease: hereditary folate malabsorption (HFM)
Solute carrier family 46, member 2	SLC46A2	TSCOT+ thymic epithelial cell-mediated sensitive CD4 tolerance by direct presentation. PMID: 18684012
Solute carrier family 46, member 3	SLC46A3	
SLC47 Multidrug and Toxin Extrusion (MATE) family		
Solute carrier family 47 (multidrug and toxin extrusion), member 1	SLC47A1	Multidrug and toxin extrusion family SLC47: Physiological, pharmacokinetic and toxicokinetic importance of MATE1 and MATE2-K. PMID: 23506899
Solute carrier family 47 (multidrug and toxin extrusion), member 2	SLC47A2	
Solute carrier family 48 (heme transporter), member 1	SLC48A1	HRG1 is essential for heme transport from the phagolysosome of macrophages during erythrophagocytosis. PMID: 23395172
Solute carrier family 50 (sugar efflux transporter), member 1	SLC50A1	Glucose transport families SLC5 and SLC50. PMID: 23506865
Solute carrier family 51, alpha subunit	SLC51A	The heteromeric organic solute transporter, OSTα-OSTβ/SLC51: A transporter for steroid-derived molecules. PMID: 23506901
Solute carrier family 51, beta subunit	SLC51B	
Solute carrier family 52 (riboflavin transporter), member 1	SLC52A1	Disease: riboflavin deficiency (RBFVD)
Solute carrier family 52 (riboflavin transporter), member 2	SLC52A2	Disease: Brown–Vialetto–Van Laere syndrome 2 (BVVLS2)
Solute carrier family 52 (riboflavin transporter), member 3	SLC52A3	RFT2 is overexpressed in esophageal squamous cell carcinoma and promotes tumorigenesis by sustaining cell proliferation and protecting against cell death. PMID: 25045844. Disease: Brown–Vialetto–Van Laere syndrome 1 (BVVLS1)

3.6.3 Ligand-gated ion channels

HGNC-approved name	Symbol	Target information
5-Hydroxytryptamine (serotonin) receptor 3A, ionotropic	HTR3A	**Ondansetron** – antiemetics and antinauseants
5-Hydroxytryptamine (serotonin) receptor 3B, ionotropic	HTR3B	Variations in the 5-hydroxytryptamine type 3B receptor gene as predictors of the efficacy of antiemetic treatment in cancer patients. PMID: 12775740
5-Hydroxytryptamine (serotonin) receptor 3C, ionotropic	HTR3C	Serotonin receptor diversity in the human colon: Expression of serotonin type 3 receptor subunits 5-HT3C, 5-HT3D, and 5-HT3E. PMID: 21192076
5-Hydroxytryptamine (serotonin) receptor 3D, ionotropic	HTR3D	Serotonin receptor diversity in the human colon: Expression of serotonin type 3 receptor subunits 5-HT3C, 5-HT3D, and 5-HT3E. PMID: 21192076
5-Hydroxytryptamine (serotonin) receptor 3E, ionotropic	HTR3E	Serotonin receptor diversity in the human colon: Expression of serotonin type 3 receptor subunits 5-HT3C, 5-HT3D, and 5-HT3E. PMID: 21192076

HGNC-approved name	Symbol	Target information
Multi-subunit-binding drugs		**Nicotine** – parasympathomimetics
		Doxacurium – muscle relaxants
		Tubocurarine – muscle relaxants
Cholinergic receptor, nicotinic, alpha 1 (neuronal)	CHRNA1	Disease: multiple pterygium syndrome, lethal type (LMPS)
Cholinergic receptor, nicotinic, alpha 2 (neuronal)	CHRNA2	Disease: epilepsy, nocturnal frontal lobe, 4 (ENFL4)
Cholinergic receptor, nicotinic, alpha 3 (neuronal)	CHRNA3	AT-1001: A high affinity and selective α3β4 nicotinic acetylcholine receptor antagonist blocks nicotine self-administration in rats. PMID: 22278092
Cholinergic receptor, nicotinic, alpha 4 (neuronal)	CHRNA4	**Varenicline** – drugs used in addictive disorders. Disease: epilepsy, nocturnal frontal lobe, 1 (ENFL1)
Cholinergic receptor, nicotinic, alpha 5 (neuronal)	CHRNA5	CHRNA5 variants moderate the effect of nicotine deprivation on a neural index of cognitive control. PMID: 24934182
Cholinergic receptor, nicotinic, alpha 6 (neuronal)	CHRNA6	Nicotinic acetylcholine receptors containing the α6 subunit contribute to ethanol activation of ventral tegmental area dopaminergic neurons. PMID: 23811312
Cholinergic receptor, nicotinic, alpha 7 (neuronal)	CHRNA7	**ABT-126** – anti-dementia drugs
Cholinergic receptor, nicotinic, alpha 9 (neuronal)	CHRNA9	Novel small molecule α9α10 nicotinic receptor antagonist prevents and reverses chemotherapy-evoked neuropathic pain in rats. PMID: 22610850
Cholinergic receptor, nicotinic, alpha 10 (neuronal)	CHRNA10	**Pentolinium** – antihypertensives
Cholinergic receptor, nicotinic, beta 1 (neuronal)	CHRNB1	Disease: myasthenic syndrome, congenital, slow-channel (SCCMS)
Cholinergic receptor, nicotinic, beta 2 (neuronal)	CHRNB2	Disease: epilepsy, nocturnal frontal lobe, 3 (ENFL3)
Cholinergic receptor, nicotinic, beta 3 (neuronal)	CHRNB3	Significant association of CHRNB3 variants with nicotine dependence in multiple ethnic populations. PMID: 23319001
Cholinergic receptor, nicotinic, beta 4 (neuronal)	CHRNB4	Role of the nicotinic receptor β4 subunit in the antidepressant activity of novel N,6-dimethyltricyclo[5.2.1.0^{2,6}]decan-2-amine enantiomers. PMID: 23994392
Cholinergic receptor, nicotinic, gamma (neuronal)	CHRNG	Disease: multiple pterygium syndrome, lethal type (LMPS)
Cholinergic receptor, nicotinic, delta (neuronal)	CHRND	Disease: multiple pterygium syndrome, lethal type (LMPS)
Cholinergic receptor, nicotinic, epsilon (neuronal)	CHRNE	Note: The muscle AChR is the major target antigen in the autoimmune disease myasthenia gravis
Multi-subunit-binding drugs		**Diazepam** – hypnotics and sedatives
		Hexobarbital – hypnotics and sedatives
		Thiamylal – anaesthetics
Gamma-aminobutyric acid (GABA) A receptor, alpha 1	GABRA1	**Flumazenil** – antidotes. Disease: epilepsy, childhood absence 4 (ECA4)
Gamma-aminobutyric acid (GABA) A receptor, alpha 2	GABRA2	Deletion of the gabra2 gene results in hypersensitivity to the acute effects of ethanol but does not alter ethanol self administration. PMID: 23115637
Gamma-aminobutyric acid (GABA) A receptor, alpha 3	GABRA3	Tonic inhibition in principal cells of the amygdala: A central role for α3 subunit-containing GABAA receptors. PMID: 22723702
Gamma-aminobutyric acid (GABA) A receptor, alpha 4	GABRA4	Tonic inhibition of accumbal spiny neurons by extrasynaptic α4βδ GABAA receptors modulates the actions of psychostimulants. PMID: 24431441

(Continued)

HGNC-approved name	Symbol	Target information
Gamma-aminobutyric acid (GABA) A receptor, alpha 5	GABRA5	α5-GABAA receptors negatively regulate MYC-amplified medulloblastoma growth. PMID: 24196163
Gamma-aminobutyric acid (GABA) A receptor, alpha 6	GABRA6	Reduced GABA(A) receptor α6 expression in the trigeminal ganglion alters inflammatory TMJ hypersensitivity. PMID: 22521829
Gamma-aminobutyric acid (GABA) A receptor, beta 1	GABRB1	Mutations in the Gabrb1 gene promote alcohol consumption through increased tonic inhibition. PMID: 24281383
Gamma-aminobutyric acid (GABA) A receptor, beta 2	GABRB2	Social cognitive role of schizophrenia candidate gene GABRB2. PMID: 23638040
Gamma-aminobutyric acid (GABA) A receptor, beta 3	GABRB3	Disease: epilepsy, childhood absence 5 (ECA5)
Gamma-aminobutyric acid (GABA) A receptor, delta	GABRD	Disease: generalized epilepsy with febrile seizures plus 5 (GEFS+5)
Gamma-aminobutyric acid (GABA) A receptor, epsilon	GABRE	Trafficking and potential assembly patterns of epsilon-containing GABAA receptors. PMID: 17714454
Gamma-aminobutyric acid (GABA) A receptor, gamma 1	GABRG1	GABRG1 and GABRA2 as independent predictors for alcoholism in two populations. PMID: 18818659
Gamma-aminobutyric acid (GABA) A receptor, gamma 2	GABRG2	Disease: epilepsy, childhood absence 2 (ECA2)
Gamma-aminobutyric acid (GABA) A receptor, gamma 3	GABRG3	Association of GABRG3 with alcohol dependence. PMID: 14745296
Gamma-aminobutyric acid (GABA) A receptor, pi	GABRP	GABA(A) receptor pi (GABRP) stimulates basal-like breast cancer cell migration through activation of extracellular-regulated kinase 1/2 (ERK1/2). PMID: 25012653
Gamma-aminobutyric acid (GABA) A receptor, rho 1	GABRR1	Structural determinants for antagonist pharmacology that distinguish the rho1 GABAC receptor from GABAA receptors. PMID: 18599601
Gamma-aminobutyric acid (GABA) A receptor, rho 2	GABRR2	GABRR1 and GABRR2, encoding the GABA-A receptor subunits rho1 and rho2, are associated with alcohol dependence. PMID: 19536785
Gamma-aminobutyric acid (GABA) A receptor, rho 3	GABRR3	
Gamma-aminobutyric acid (GABA) A receptor, theta	GABRQ	mRNA and protein expression for novel GABAA receptors θ and ρ2 are altered in schizophrenia and mood disorders; relevance to FMRP-mGluR5 signaling pathway. PMID: 23778581
Multi-subunit-binding drugs		**Butabarbital** – hypnotics and sedatives
		Perampanel – antiepileptics
Glutamate receptor, ionotropic, AMPA 1	GRIA1	What causes aberrant salience in schizophrenia? A role for impaired short-term habituation and the GRIA1 (GluA1) AMPA receptor subunit. PMID: 25224260
Glutamate receptor, ionotropic, AMPA 2	GRIA2	Synthesis, pharmacological and structural characterization, and thermodynamic aspects of GluA2-positive allosteric modulators with a 3,4-dihydro-2H-1,2,4-benzothiadiazine 1,1-dioxide scaffold. PMID: 24131202
Glutamate receptor, ionotropic, AMPA 3	GRIA3	Disease: mental retardation, X-linked 94 (MRX94)
Glutamate receptor, ionotropic, AMPA 4	GRIA4	Unraveling genetic modifiers in the gria4 mouse model of absence epilepsy. PMID: 25010494
Glutamate receptor, ionotropic, delta 1	GRID1	The delta subfamily of glutamate receptors: Characterization of receptor chimeras and mutants. PMID: 23551821
Glutamate receptor, ionotropic, delta 2	GRID2	The delta subfamily of glutamate receptors: Characterization of receptor chimeras and mutants. PMID: 23551821
Glutamate receptor, ionotropic, kainate 1	GRIK1	mGluR5 negative allosteric modulators overview: A medicinal chemistry approach towards a series of novel therapeutic agents. PMID: 21261592
Glutamate receptor, ionotropic, kainate 2	GRIK2	Disease: mental retardation, autosomal recessive 6 (MRT6)

HGNC-approved name	Symbol	Target information
Glutamate receptor, ionotropic, kainate 3	GRIK3	GluR7 is an essential subunit of presynaptic kainate autoreceptors at hippocampal mossy fiber synapses. PMID: 17620617
Glutamate receptor, ionotropic, kainate 4	GRIK4	The GluK4 kainate receptor subunit regulates memory, mood, and excitotoxic neurodegeneration. PMID: 23357115
Glutamate receptor, ionotropic, kainate 5	GRIK5	Agonist binding to the GluK5 subunit is sufficient for functional surface expression of heteromeric GluK2/GluK5 kainate receptors. PMID: 23975096
Multi-subunit-binding drugs		**Memantine** – anti-dementia drugs
		Ketamine – anaesthetics
Glutamate receptor, ionotropic, N-methyl D-aspartate 1	GRIN1	Disease: mental retardation, autosomal dominant 8 (MRD8)
Glutamate receptor, ionotropic, N-methyl D-aspartate 2A	GRIN2A	Dextromethorphan – cough suppressants. Disease: epilepsy, focal, with speech disorder and with or without mental retardation (FESD)
Glutamate receptor, ionotropic, N-methyl D-aspartate 2B	GRIN2B	**Radiprodil** – analgesics. Disease: mental retardation, autosomal dominant 6 (MRD6)
Glutamate receptor, ionotropic, N-methyl D-aspartate 2C	GRIN2C	N-methyl-D-aspartate (NMDA) receptor NR2 subunit selectivity of a series of novel piperazine-2,3-dicarboxylate derivatives: Preferential blockade of extrasynaptic NMDA receptors in the rat hippocampal CA3-CA1 synapse. PMID: 19684252
Glutamate receptor, ionotropic, N-methyl D-aspartate 2D	GRIN2D	GluN2C/GluN2D subunit-selective NMDA receptor potentiator CIQ reverses MK-801-induced impairment in prepulse inhibition and working memory in Y-maze test in mice. PMID: 24236947
Glutamate receptor, ionotropic, N-methyl-D-aspartate 3A	GRIN3A	Influence of the NR3A subunit on NMDA receptor functions. PMID: 20097255
Glutamate receptor, ionotropic, N-methyl-D-aspartate 3B	GRIN3B	A recently discovered NMDA receptor gene, GRIN3B, is associated with duration mismatch negativity. PMID: 24814139
Glycine receptor, alpha 1	GLRA1	Presynaptic glycine receptors as a potential therapeutic target for hyperekplexia disease. PMID: 24390226. Disease: hyperekplexia 1 (HKPX1)
Glycine receptor, alpha 2	GLRA2	Subunit-specific inhibition of glycine receptors by curcumol. PMID: 22892339
Glycine receptor, alpha 3	GLRA3	Cannabinoids suppress inflammatory and neuropathic pain by targeting α3 glycine receptors. PMID: 22585736
Glycine receptor, beta	GLRB	Glycine receptor beta-subunit gene mutation in spastic mouse associated with LINE-1 element insertion. PMID: 7920630. Disease: hyperekplexia 2 (HKPX2)
Purinergic receptor P2X, ligand-gated ion channel, 1	P2RX1	The P2X1 ion channel in platelet function. PMID: 20201633
Purinergic receptor P2X, ligand-gated ion channel, 2	P2RX2	Disease: deafness, autosomal dominant, 41 (DFNA41)
Purinergic receptor P2X, ligand-gated ion channel, 3	P2RX3	**AF-219** – analgesics
Purinergic receptor P2X, ligand-gated ion channel, 4	P2RX4	P2X4 receptors induced in spinal microglia gate tactile allodynia after nerve injury. PMID: 12917686
Purinergic receptor P2X, ligand-gated ion channel, 5	P2RX5	A truncation variant of the cation channel P2RX5 is upregulated during T cell activation. PMID: 25181038
Purinergic receptor P2X, ligand-gated ion channel, 6	P2RX6	
Purinergic receptor P2X, ligand-gated ion channel, 7	P2RX7	The phenothiazine-class antipsychotic drugs prochlorperazine and trifluoperazine are potent allosteric modulators of the human P2X7 receptor. PMID: 23954492
Zinc-activated ligand-gated ion channel	ZACN	Cloning and expression of ligand-gated ion-channel receptor L2 in central nervous system. PMID: 16083862

3.6.4 Voltage-gated ion channels

HGNC-approved name	Symbol	Target information
Acid-sensing (proton-gated) ion channel 1	ASIC1	Acid-sensing ion channel-1a in the amygdala, a novel therapeutic target in depression-related behavior. PMID: 19403806
Acid-sensing (proton-gated) ion channel 2	ASIC2	Knockdown of ASIC2a subunit aggravates injury of rat C6 glioma cells in acidosis. PMID: 21061195
Acid-sensing (proton-gated) ion channel 3	ASIC3	Synthesis, structure-activity relationship, and pharmacological profile of analogs of the ASIC-3 inhibitor A-317567. PMID: 22778804
Acid-sensing (proton-gated) ion channel family member 4	ASIC4	Regulation of ASIC activity by ASIC4 – new insights into ASIC channel function revealed by a yeast two-hybrid assay. PMID: 18662336
Acid-sensing (proton-gated) ion channel family member 5	ASIC5	
Anoctamin 1, calcium-activated chloride channel	ANO1	**Crofelemer** – antidiarrheal agents
Anoctamin 2	ANO2	TMEM16 proteins: Unknown structure and confusing functions. PMID: 25451786
Anoctamin 3	ANO3	Disease: dystonia 24 (DYT24)
Anoctamin 4	ANO4	
Anoctamin 5	ANO5	Disease: gnathodiaphyseal dysplasia (GDD)
Anoctamin 6	ANO6	Disease: Scott syndrome (SCTS)
Anoctamin 7	ANO7	ANOs 3–7 in the anoctamin/Tmem16 Cl- channel family are intracellular proteins. PMID: 22075693
Anoctamin 8	ANO8	
Anoctamin 9	ANO9	
Anoctamin 10	ANO10	Disease: spinocerebellar ataxia, autosomal recessive, 10 (SCAR10)
Aquaporin 1 (Colton blood group)	AQP1	Aquaporin 1, a potential therapeutic target for migraine with aura. PMID: 20969805
Aquaporin 2 (collecting duct)	AQP2	Dynamic regulation and dysregulation of the water channel aquaporin-2: A common cause of and promising therapeutic target for water balance disorders. PMID: 24129558. Disease: diabetes insipidus, nephrogenic, autosomal (ANDI)
Aquaporin 3 (Gill blood group)	AQP3	Targeting aquaporin function: Potent inhibition of aquaglyceroporin-3 by a gold-based compound. PMID: 22624030
Aquaporin 4	AQP4	**Aquaporumab** – neuromyelitis optica treatments
Aquaporin 5	AQP5	Aquaporin 5 promotes the proliferation and migration of human gastric carcinoma cells. PMID: 23436048. Disease: keratoderma, palmoplantar, Bothnian type (PPKB)
Aquaporin 6, kidney specific	AQP6	Aquaporin-6 is expressed along the rat gastrointestinal tract and upregulated by feeding in the small intestine. PMID: 19811639
Aquaporin 7	AQP7	A gold coordination compound as a chemical probe to unravel aquaporin-7 function. PMID: 24891084
Aquaporin 8	AQP8	Expression of aquaporin8 in human astrocytomas: Correlation with pathologic grade. PMID: 24055034
Aquaporin 9	AQP9	Reduced hepatic aquaporin-9 and glycerol permeability are related to insulin resistance in non-alcoholic fatty liver disease. PMID: 24418844
Aquaporin 10	AQP10	Aquaporin-10 represents an alternative pathway for glycerol efflux from human adipocytes. PMID: 23382902

HGNC-approved name	Symbol	Target information
Aquaporin 11	AQP11	Human aquaporin-11 is a water and glycerol channel and localizes in the vicinity of lipid droplets in human adipocytes. PMID: 24845055
Aquaporin 12A	AQP12A	Identification of a novel aquaporin, AQP12, expressed in pancreatic acinar cells. PMID: 15809071
Aquaporin 12B	AQP12B	
Major intrinsic protein of lens fiber	MIP	Disease: cataract 15, multiple types (CTRCT15)
Bartter syndrome, infantile, with sensorineural deafness (Barttin)	BSND	Disease: Bartter syndrome 4A (BS4A)
Bestrophin 1	BEST1	Structure and selectivity in bestrophin ion channels. PMID: 25324390. Disease: vitelliform macular dystrophy 2 (VMD2)
Bestrophin 2	BEST2	Bestrophin-2 is involved in the generation of intraocular pressure. PMID: 18385076
Bestrophin 3	BEST3	C-terminal membrane association of Bestrophin 3 and its activation as a chloride channel. PMID: 23124946
Bestrophin 4	BEST4	Lineage-specific expression of bestrophin-2 and bestrophin-4 in human intestinal epithelial cells. PMID: 24223998
Calcium channel, voltage-dependent, alpha 2/delta subunit 1	CACNA2D1	**Pregabalin** – analgesics
Calcium channel, voltage-dependent, alpha 2/delta subunit 2	CACNA2D2	
Calcium channel, voltage-dependent, alpha 2/delta subunit 3	CACNA2D3	$\alpha2\delta3$ is essential for normal structure and function of auditory nerve synapses and is a novel candidate for auditory processing disorders. PMID: 24403143
Calcium channel, voltage-dependent, alpha 2/delta subunit 4	CACNA2D4	Disease: retinal cone dystrophy 4 (RCD4)
Calcium channel, voltage-dependent, beta 1 subunit	CACNB1	Domain cooperativity in the $\beta1a$ subunit is essential for dihydropyridine receptor voltage sensing in skeletal muscle. PMID: 23589859
Calcium channel, voltage-dependent, beta 2 subunit	CACNB2	Disease: Brugada syndrome 4 (BRGDA4)
Calcium channel, voltage-dependent, beta 3 subunit	CACNB3	Up-regulation of Cav$\beta3$ subunit in primary sensory neurons increases voltage-activated Ca^{2+} channel activity and nociceptive input in neuropathic pain. PMID: 22187436
Calcium channel, voltage-dependent, beta 4 subunit	CACNB4	Disease: epilepsy, idiopathic generalized 9 (EIG9)
Calcium channel, voltage-dependent, gamma subunit 1	CACNG1	Stargazin promotes closure of the AMPA receptor ligand-binding domain. PMID: 25422502
Multi-subunit-binding drugs		**Ibutilide** – cardiac therapy
		Verapamil – cardiac therapy
		Bepridil – cardiac therapy
		Ziconotide – analgesics
Calcium channel, voltage-dependent, P/Q type, alpha 1A subunit	CACNA1A	Disease: spinocerebellar ataxia 6 (SCA6)
Calcium channel, voltage-dependent, N type, alpha 1B subunit	CACNA1B	CACNA1B mutation is linked to unique myoclonus-dystonia syndrome. PMID: 25296916
Calcium channel, voltage-dependent, L type, alpha 1C subunit	CACNA1C	Antagonism of L-type Ca^{2+} channels CaV1.3 and CaV1.2 by 1,4-dihydropyrimidines and 4H-pyrans as dihydropyridine mimics. PMID: 23688558. Disease: Timothy syndrome (TS)

(Continued)

HGNC-approved name	Symbol	Target information
Calcium channel, voltage-dependent, L type, alpha 1D subunit	CACNA1D	Antagonism of L-type Ca²⁺ channels CaV1.3 and CaV1.2 by 1,4-dihydropyrimidines and 4*H*-pyrans as dihydropyridine mimics. PMID: 23688558. Disease: sinoatrial node dysfunction and deafness (SANDD)
Calcium channel, voltage-dependent, R type, alpha 1E subunit	CACNA1E	CACNA1E variants affect beta cell function in patients with newly diagnosed type 2 diabetes. The Verona newly diagnosed type 2 diabetes study (VNDS) 3. PMID: 22427875
Calcium channel, voltage-dependent, L type, alpha 1F subunit	CACNA1F	Disease: night blindness, congenital stationary, 2A (CSNB2A)
Calcium channel, voltage-dependent, T type, alpha 1G subunit	CACNA1G	**Ethosuximide** – antiepileptics
Calcium channel, voltage-dependent, T type, alpha 1H subunit	CACNA1H	**ABT-639** – analgesics. Disease: epilepsy, idiopathic generalized 6 (EIG6)
Calcium channel, voltage-dependent, T type, alpha 1I subunit	CACNA1I	Hydrogen sulfide inhibits Cav3.2 T-type Ca²⁺ channels. PMID: 25183670
Calcium channel, voltage-dependent, L type, alpha 1S subunit	CACNA1S	Disease: periodic paralysis hypokalemic 1 (HOKPP1)
Calcium channel, voltage-dependent, gamma subunit 2	CACNG2	Disease: mental retardation, autosomal dominant 10 (MRD10)
Calcium channel, voltage-dependent, gamma subunit 3	CACNG3	
Calcium channel, voltage-dependent, gamma subunit 4	CACNG4	A targeted mutation in Cacng4 exacerbates spike-wave seizures in stargazer (Cacng2) mice. PMID: 15677329
Calcium channel, voltage-dependent, gamma subunit 5	CACNG5	
Calcium channel, voltage-dependent, gamma subunit 6	CACNG6	
Calcium channel, voltage-dependent, gamma subunit 7	CACNG7	
Calcium channel, voltage-dependent, gamma subunit 8	CACNG8	
Cation channel, sperm associated 1	CATSPER1	All four CatSper ion channel proteins are required for male fertility and sperm cell hyperactivated motility. PMID: 17227845. Disease: spermatogenic failure 7 (SPGF7)
Cation channel, sperm associated 2	CATSPER2	Disease: Deafness–infertility syndrome (DIS)
Cation channel, sperm associated 3	CATSPER3	
Cation channel, sperm associated 4	CATSPER4	
Chloride channel, voltage-sensitive 1	CLCN1	Disease: myotonia congenita, autosomal dominant (MCD)
Chloride channel, voltage-sensitive 2	CLCN2	**Lubiprostone** – drugs for constipation. Disease: epilepsy, idiopathic generalized 11 (EIG11)
Chloride channel, voltage-sensitive 3	CLCN3	ClC-3 chloride channels are essential for cell proliferation and cell cycle progression in nasopharyngeal carcinoma cells. PMID: 20539936
Chloride channel, voltage-sensitive 4	CLCN4	Gene trapping identifies chloride channel 4 as a novel inducer of colon cancer cell migration, invasion and metastases. PMID: 20087350
Chloride channel, voltage-sensitive 5	CLCN5	Disease: hypophosphatemic rickets, X-linked recessive (XLRHR)
Chloride channel, voltage-sensitive 6	CLCN6	Distinct neuropathologic phenotypes after disrupting the chloride transport proteins ClC-6 or ClC-7/Ostm1. PMID: 21107136
Chloride channel, voltage-sensitive 7	CLCN7	CLC-7: A potential therapeutic target for the treatment of osteoporosis and neurodegeneration. PMID: 19393632. Disease: osteopetrosis, autosomal recessive 4 (OPTB4)

HGNC-approved name	Symbol	Target information
Chloride channel, voltage-sensitive Ka	CLCNKA	**Niflumic acid** – anti-inflammatory and antirheumatic products. Disease: Bartter syndrome 4B (BS4B)
Chloride channel, voltage-sensitive Kb	CLCNKB	Disease: Bartter syndrome 3 (BS3)
Chloride channel accessory 1	CLCA1	Secreted hCLCA1 is a signaling molecule that activates airway macrophages. PMID: 24349445
Chloride channel accessory 2	CLCA2	CLCA2, a target of the p53 family, negatively regulates cancer cell migration and invasion. PMID: 22990203
Chloride channel accessory 4	CLCA4	CLCA4 variants determine the manifestation of the cystic fibrosis basic defect in the intestine. PMID: 23073314
Chloride channel CLIC-like 1	CLCC1	
Chloride intracellular channel 1	CLIC1	Metformin repositioning as antitumoral agent: Selective antiproliferative effects in human glioblastoma stem cells, via inhibition of CLIC1-mediated ion current. PMID: 25361004
Chloride intracellular channel 2	CLIC2	Disease: mental retardation, X-linked, syndromic, 32 (MRXS32)
Chloride intracellular channel 3	CLIC3	CLIC3 controls recycling of late endosomal MT1-MMP and dictates invasion and metastasis in breast cancer. PMID: 25015290
Chloride intracellular channel 4	CLIC4	CLIC4 regulates cell adhesion and $\beta1$ integrin trafficking. PMID: 25344254
Chloride intracellular channel 5	CLIC5	CLIC5 mutant mice are resistant to diet-induced obesity and exhibit gastric hemorrhaging and increased susceptibility to torpor. PMID: 20357015
Chloride intracellular channel 6	CLIC6	
Cyclic nucleotide-gated channel alpha 1	CNGA1	Alveolar epithelial CNGA1 channels mediate cGMP-stimulated, amiloride-insensitive, lung liquid absorption. PMID: 21559843. Disease: retinitis pigmentosa 49 (RP49)
Cyclic nucleotide-gated channel alpha 2	CNGA2	Anxiety- and depressive-like behaviors in olfactory deficient Cnga2 knockout mice. PMID: 25192635
Cyclic nucleotide-gated channel alpha 3	CNGA3	Disease: achromatopsia 2 (ACHM2)
Cyclic nucleotide-gated channel alpha 4	CNGA4	Differential regulation by cyclic nucleotides of the CNGA4 and CNGB1b subunits in olfactory cyclic nucleotide-gated channels. PMID: 22786723
Cyclic nucleotide-gated channel beta 1	CNGB1	Disease: retinitis pigmentosa 45 (RP45)
Cyclic nucleotide-gated channel beta 3	CNGB3	Disease: Stargardt disease 1 (STGD1)
Hyperpolarization-activated cyclic nucleotide-gated potassium channel 1	HCN1	Novel blockers of hyperpolarization-activated current with isoform selectivity in recombinant cells and native tissue. PMID: 22091830
Hyperpolarization-activated cyclic nucleotide-gated potassium channel 2	HCN2	
Hyperpolarization-activated cyclic nucleotide-gated potassium channel 3	HCN3	HCN3 contributes to the ventricular action potential waveform in the murine heart. PMID: 21903939
Hyperpolarization-activated cyclic nucleotide-gated potassium channel 4	HCN4	**Ivabradine** – cardiac therapy. Disease: sick sinus syndrome 2 (SSS2)
Gap junction protein, alpha 1, 43 kDa	GJA1	The antiarrhythmic dipeptide ZP1609 (danegaptide) when given at reperfusion reduces myocardial infarct size in pigs. PMID: 23397587. Disease: oculodentodigital dysplasia (ODDD)
Gap junction protein, alpha 3, 46 kDa	GJA3	Disease: cataract 14, multiple types (CTRCT14)
Gap junction protein, alpha 4, 37 kDa	GJA4	Gap junction protein Cx37 interacts with endothelial nitric oxide synthase in endothelial cells. PMID: 20081116
Gap junction protein, alpha 5, 40 kDa	GJA5	Disease: atrial standstill (ATST)
Gap junction protein, alpha 8, 50 kDa	GJA8	Disease: cataract 1, multiple types (CTRCT1)
Gap junction protein, alpha 9, 59 kDa	GJA9	

(Continued)

HGNC-approved name	Symbol	Target information
Gap junction protein, alpha 10, 62 kDa	GJA10	
Gap junction protein, beta 1, 32 kDa	GJB1	Gap junction inhibition prevents drug-induced liver toxicity and fulminant hepatic failure. PMID: 22252509. Disease: Charcot–Marie–Tooth disease, X-linked dominant, 1 (CMTX1)
Gap junction protein, beta 2, 26 kDa	GJB2	Disease: deafness, autosomal recessive, 1A (DFNB1A)
Gap junction protein, beta 3, 31 kDa	GJB3	Disease: erythrokeratodermia variabilis (EKV)
Gap junction protein, beta 4, 30.3 kDa	GJB4	Disease: erythrokeratodermia variabilis (EKV)
Gap junction protein, beta 5, 31.1 kDa	GJB5	Connexin 31.1 degradation requires the Clathrin-mediated autophagy in NSCLC cell H1299. PMID: 25388970
Gap junction protein, beta 6, 30 kDa	GJB6	Disease: ectodermal dysplasia 2, Clouston type (ECTD2)
Gap junction protein, beta 7, 25 kDa	GJB7	
Gap junction protein, gamma 1, 45 kDa	GJC1	Connexin 43 potentiates osteoblast responsiveness to fibroblast growth factor 2 via a protein kinase C-delta/Runx2-dependent mechanism. PMID: 19339281
Gap junction protein, gamma 2, 47 kDa	GJC2	Disease: leukodystrophy, hypomyelinating, 2 (HLD2)
Gap junction protein, gamma 3, 30.2 kDa	GJC3	Functional properties of mouse connexin 30.2 expressed in the conduction system of the heart. PMID: 15879306
Gap junction protein, delta 2, 36 kDa	GJD2	Inhibition of connexin 36 hemichannels by glucose contributes to the stimulation of insulin secretion. PMID: 24735890
Gap junction protein, delta 3, 31.9 kDa	GJD3	
Gap junction protein, delta 4, 40.1 kDa	GJD4	
Gap junction protein, epsilon 1, 23 kDa	GJE1	
Hydrogen voltage-gated channel 1	HVCN1	The voltage-gated proton channel Hv1 enhances brain damage from ischemic stroke. PMID: 22388960
ORAI calcium release-activated calcium modulator 1	ORAI1	Antibody-mediated targeting of the orai1 calcium channel inhibits T cell function. PMID: 24376610. Disease: immunodeficiency 9 (IMD9)
ORAI calcium release-activated calcium modulator 2	ORAI2	CRACM1, CRACM2, and CRACM3 are store-operated Ca^{2+} channels with distinct functional properties. PMID: 17442569
ORAI calcium release-activated calcium modulator 3	ORAI3	Orai3 is an estrogen receptor α-regulated Ca^{2+} channel that promotes tumorigenesis. PMID: 22993197
Pannexin 1	PANX1	The role of pannexin 1 in the induction and resolution of inflammation. PMID: 24642372
Pannexin 2	PANX2	Pharmacological properties of homomeric and heteromeric pannexin hemichannels expressed in Xenopus oocytes. PMID: 15715654
Pannexin 3	PANX3	Pannexin 3 is a novel target for Runx2, expressed by osteoblasts and mature growth plate chondrocytes. PMID: 21915903
Multi-subunit-binding drugs		**Dalfampridine** – parasympathomimetics
		Dofetilide – cardiac therapy
Potassium voltage-gated channel, shaker-related subfamily, member 1 (episodic ataxia with myokymia)	KCNA1	Identification of selective inhibitors of the potassium channel Kv1.1-1.2((3)) by high-throughput virtual screening and automated patch clamp. PMID 22473914. Disease: episodic ataxia 1 (EA1)
Potassium voltage-gated channel, shaker-related subfamily, member 2	KCNA2	Identification of selective inhibitors of the potassium channel Kv1.1-1.2((3)) by high-throughput virtual screening and automated patch clamp. PMID 22473914
Potassium voltage-gated channel, shaker-related subfamily, member 3	KCNA3	Selective Kv1.3 channel blocker as therapeutic for obesity and insulin resistance. PMID: 23729813

HGNC-approved name	Symbol	Target information
Potassium voltage-gated channel, shaker-related subfamily, member 4	KCNA4	Reduced excitability of gp130-deficient nociceptors is associated with increased voltage-gated potassium currents and Kcna4 channel upregulation. PMID: 24463703
Potassium voltage-gated channel, shaker-related subfamily, member 5	KCNA5	**BMS-394136** – cardiac therapy. Disease: atrial fibrillation, familial, 7 (ATFB7)
Potassium voltage-gated channel, shaker-related subfamily, member 6	KCNA6	
Potassium voltage-gated channel, shaker-related subfamily, member 7	KCNA7	Block of Kv1.7 potassium currents increases glucose-stimulated insulin secretion. PMID: 22438204
Potassium voltage-gated channel, shaker-related subfamily, member 10	KCNA10	A null mutation of mouse Kcna10 causes significant vestibular and mild hearing dysfunction. PMID: 23528307
Potassium voltage-gated channel, Shab-related subfamily, member 1	KCNB1	Oxidation of KCNB1 K(+) channels in central nervous system and beyond. PMID: 24921000
Potassium voltage-gated channel, Shab-related subfamily, member 2	KCNB2	Kv2.2: A novel molecular target to study the role of basal forebrain GABAergic neurons in the sleep-wake cycle. PMID: 24293758
Potassium voltage-gated channel, Shaw-related subfamily, member 1	KCNC1	New positive Ca^{2+}-activated K^+ channel gating modulators with selectivity for KCa3.1. PMID: 24958817
Potassium voltage-gated channel, Shaw-related subfamily, member 2	KCNC2	Impaired long-range synchronization of gamma oscillations in the neocortex of a mouse lacking Kv3.2 potassium channels. PMID: 22539821
Potassium voltage-gated channel, Shaw-related subfamily, member 3	KCNC3	Disease: spinocerebellar ataxia 13 (SCA13)
Potassium voltage-gated channel, Shaw-related subfamily, member 4	KCNC4	Cell cycle-dependent expression of Kv3.4 channels modulates proliferation of human uterine artery smooth muscle cells. PMID: 20093253
Potassium voltage-gated channel, Shal-related subfamily, member 1	KCND1	Involvement of Kv4.1 K(+) channels in gastric cancer cell proliferation. PMID: 20930388
Potassium voltage-gated channel, Shal-related subfamily, member 2	KCND2	
Potassium voltage-gated channel, Shal-related subfamily, member 3	KCND3	Inhibition of A-type potassium current by the peptide toxin SNX-482. PMID: 25009251. Disease: spinocerebellar ataxia 19 (SCA19)
Potassium voltage-gated channel, subfamily F, member 1	KCNF1	
Potassium voltage-gated channel, subfamily G, member 1	KCNG1	
Potassium voltage-gated channel, subfamily G, member 2	KCNG2	
Potassium voltage-gated channel, subfamily G, member 3	KCNG3	Domain analysis of Kv6.3, an electrically silent channel. PMID: 16096342
Potassium voltage-gated channel, subfamily G, member 4	KCNG4	The electrically silent Kv6.4 subunit confers hyperpolarized gating charge movement in Kv2.1/Kv6.4 heterotetrameric channels. PMID: 22615922
Potassium voltage-gated channel, subfamily H (eag-related), member 1	KCNH1	KV10.1 K(+)-channel plasma membrane discrete domain partitioning and its functional correlation in neurons. PMID: 24269539
Potassium voltage-gated channel, subfamily H (eag-related), member 2	KCNH2	**Propafenone** – cardiac therapy. Disease: long QT syndrome 2 (LQT2)
Potassium voltage-gated channel, subfamily H (eag-related), member 3	KCNH3	Deletion of the potassium channel Kv12.2 causes hippocampal hyperexcitability and epilepsy. PMID: 20676103
Potassium voltage-gated channel, subfamily H (eag-related), member 4	KCNH4	

(Continued)

HGNC-approved name	Symbol	Target information
Potassium voltage-gated channel, subfamily H (eag-related), member 5	KCNH5	Multistate structural modeling and voltage-clamp analysis of epilepsy/ autism mutation Kv10.2-R327H demonstrate the role of this residue in stabilizing the channel closed state. PMID: 24133262
Potassium voltage-gated channel, subfamily H (eag-related), member 6	KCNH6	
Potassium voltage-gated channel, subfamily H (eag-related), member 7	KCNH7	Effects of the small molecule HERG activator NS1643 on Kv11.3 channels. PMID: 23226420
Potassium voltage-gated channel, subfamily H (eag-related), member 8	KCNH8	
Potassium voltage-gated channel, KQT-like subfamily, member 1	KCNQ1	**Indapamide** – antihypertensives. Disease: long QT syndrome 1 (LQT1)
Potassium voltage-gated channel, KQT-like subfamily, member 2	KCNQ2	**Retigabine** – antiepileptics. Disease: seizures, benign familial neonatal 1 (BFNS1)
Potassium voltage-gated channel, KQT-like subfamily, member 3	KCNQ3	**Retigabine** – antiepileptics. Disease: seizures, benign familial neonatal 2 (BFNS2)
Potassium voltage-gated channel, KQT-like subfamily, member 4	KCNQ4	Disease: deafness, autosomal dominant, 2A (DFNA2A)
Potassium voltage-gated channel, KQT-like subfamily, member 5	KCNQ5	Kv7.5 is the primary Kv7 subunit expressed in C-fibers. PMID: 22134895
Potassium voltage-gated channel, delayed-rectifier, subfamily S, member 1	KCNS1	Multiple chronic pain states are associated with a common amino acid-changing allele in KCNS1. PMID: 20724292
Potassium voltage-gated channel, delayed-rectifier, subfamily S, member 2	KCNS2	
Potassium voltage-gated channel, delayed-rectifier, subfamily S, member 3	KCNS3	Lower gene expression for KCNS3 potassium channel subunit in parvalbumin-containing neurons in the prefrontal cortex in schizophrenia. PMID: 24170294
Potassium channel, subfamily V, member 1	KCNV1	
Potassium channel, subfamily V, member 2	KCNV2	Disease: cone dystrophy retinal 3B (RCD3B)
Multi-subunit-binding drugs		**Dofetilide** – cardiac therapy
		Chlorpropamide – drugs used in diabetes
Potassium inwardly-rectifying channel, subfamily J, member 1	KCNJ1	Development of a selective small-molecule inhibitor of Kir1.1, the renal outer medullary potassium channel. PMID: 20926757. Disease: Bartter syndrome 2 (BS2)
Potassium inwardly-rectifying channel, subfamily J, member 2	KCNJ2	Selective inhibition of the K(ir)2 family of inward rectifier potassium channels by a small molecule probe: The discovery, SAR, and pharmacological characterization of ML133. PMID: 21615117. Disease: long QT syndrome 7 (LQT7)
Potassium inwardly-rectifying channel, subfamily J, member 3	KCNJ3	Characterization of in vivo and *in vitro* electrophysiological and antiarrhythmic effects of a novel IKACh blocker, NIP-151: A comparison with an IKr-blocker dofetilide. PMID: 18287884
Potassium inwardly-rectifying channel, subfamily J, member 4	KCNJ4	Tenidap, an agonist of the inwardly rectifying K+ channel Kir2.3, delays the onset of cortical epileptiform activity in a model of chronic temporal lobe epilepsy. PMID: 23561319
Potassium inwardly-rectifying channel, subfamily J, member 5	KCNJ5	Functional and biochemical evidence for G-protein-gated inwardly rectifying K+ (GIRK) channels composed of GIRK2 and GIRK3. PMID: 10956667. Disease: long QT syndrome 13 (LQT13)
Potassium inwardly-rectifying channel, subfamily J, member 6	KCNJ6	Functional and biochemical evidence for G-protein-gated inwardly rectifying K+ (GIRK) channels composed of GIRK2 and GIRK3. PMID: 10956667

HGNC-approved name	Symbol	Target information
Potassium inwardly-rectifying channel, subfamily J, member 8	KCNJ8	Note: Defects in KCNJ8 may be associated with susceptibility to J-wave syndromes, a group of heart disorders characterized by early repolarization events as indicated by abnormal J-wave manifestation on electrocardiogram (ECG)
Potassium inwardly-rectifying channel, subfamily J, member 9	KCNJ9	Functional and biochemical evidence for G-protein-gated inwardly rectifying K⁺ (GIRK) channels composed of GIRK2 and GIRK3. PMID: 10956667
Potassium inwardly-rectifying channel, subfamily J, member 10	KCNJ10	Inwardly rectifying potassium channel Kir4.1 is responsible for the native inward potassium conductance of satellite glial cells in sensory ganglia. PMID: 20074622. Disease: seizures, sensorineural deafness, ataxia, mental retardation, and electrolyte imbalance
Potassium inwardly-rectifying channel, subfamily J, member 11	KCNJ11	**Diazoxide** – antihypertensives. Disease: familial hyperinsulinemic hypoglycemia 2 (HHF2)
Potassium inwardly-rectifying channel, subfamily J, member 12	KCNJ12	Unconventional role of the inwardly rectifying potassium channel Kir2.2 as a constitutive activator of RelA in cancer. PMID: 23269273
Potassium inwardly-rectifying channel, subfamily J, member 13	KCNJ13	Mutations in KCNJ13 cause autosomal-dominant snowflake vitreoretinal degeneration. PMID: 18179896. Disease: snowflake vitreoretinal degeneration (SVD)
Potassium inwardly-rectifying channel, subfamily J, member 14	KCNJ14	Kir2.4 surface expression and basal current are affected by heterotrimeric G-proteins. PMID: 23339194
Potassium inwardly-rectifying channel, subfamily J, member 15	KCNJ15	Inhibition of glucose-stimulated insulin secretion by KCNJ15, a newly identified susceptibility gene for type 2 diabetes. PMID: 22566534
Potassium inwardly-rectifying channel, subfamily J, member 16	KCNJ16	S-Glutathionylation underscores the modulation of the heteromeric Kir4.1-Kir5.1 channel in oxidative stress. PMID: 22907060
Potassium inwardly-rectifying channel, subfamily J, member 18	KCNJ18	Disease: thyrotoxic periodic paralysis 2 (TTPP2)
Potassium channel, subfamily K, member 1	KCNK1	TWIK-1 contributes to the intrinsic excitability of dentate granule cells in mouse hippocampus. PMID: 25406588
Potassium channel, subfamily K, member 2	KCNK2	**Dofetilide** – cardiac therapy
Potassium channel, subfamily K, member 3	KCNK3	**Doxapram** – respiratory stimulants. Disease: pulmonary hypertension, primary, 4 (PPH4)
Potassium channel, subfamily K, member 4	KCNK4	Physical mechanism for gating and mechanosensitivity of the human TRAAK K⁺ channel. PMID: 25471887
Potassium channel, subfamily K, member 5	KCNK5	The two-pore domain potassium channel KCNK5 deteriorates outcome in ischemic neurodegeneration. PMID: 25315980
Potassium channel, subfamily K, member 6	KCNK6	TWIK-2 channel deficiency leads to pulmonary hypertension through a Rho-kinase-mediated process. PMID: 25245387
Potassium channel, subfamily K, member 7	KCNK7	
Potassium channel, subfamily K, member 9	KCNK9	**Doxapram** – respiratory stimulants. Disease: Birk-Barel mental retardation dysmorphism syndrome (BIBAS)
Potassium channel, subfamily K, member 10	KCNK10	Chemoselective regulation of TREK2 channel: Activation by sulfonate chalcones and inhibition by sulfonamide chalcones. PMID: 20570515
Potassium channel, subfamily K, member 12	KCNK12	Breaking the silence: Functional expression of the two-pore-domain potassium channel THIK-2. PMID: 24297522
Potassium channel, subfamily K, member 13	KCNK13	Tandem pore domain halothane-inhibited K⁺ channel subunits THIK1 and THIK2 assemble and form active channels. PMID: 25148687
Potassium channel, subfamily K, member 15	KCNK15	
Potassium channel, subfamily K, member 16	KCNK16	Pancreatic two P domain K⁺ channels TALK-1 and TALK-2 are activated by nitric oxide and reactive oxygen species. PMID: 15513946

(Continued)

HGNC-approved name	Symbol	Target information
Potassium channel, subfamily K, member 17	KCNK17	Gain-of-function mutation in TASK-4 channels and severe cardiac conduction disorder. PMID: 24972929
Potassium channel, subfamily K, member 18	KCNK18	TRESK channel as a potential target to treat T-cell mediated immune dysfunction. PMID: 19852929. Disease: migraine with or without aura 13 (MGR13)
Potassium large conductance calcium-activated channel, subfamily M, alpha member 1	KCNMA1	**Chlorzoxazone** – muscle relaxants. Disease: generalized epilepsy and paroxysmal dyskinesia (GEPD)
Potassium intermediate/small conductance calcium-activated channel, subfamily N, member 1	KCNN1	
Potassium intermediate/small conductance calcium-activated channel, subfamily N, member 2	KCNN2	
Potassium intermediate/small conductance calcium-activated channel, subfamily N, member 3	KCNN3	Pharmacological activation of KCa3.1/KCa2.3 channels produces endothelial hyperpolarization and lowers blood pressure in conscious dogs. PMID: 21699504
Potassium intermediate/small conductance calcium-activated channel, subfamily N, member 4	KCNN4	The K+ channel KCa3.1 as a novel target for idiopathic pulmonary fibrosis. PMID: 24392001
Potassium channel, subfamily T, member 1	KCNT1	Disease: epileptic encephalopathy, early infantile, 14 (EIEE14)
Potassium channel, subfamily T, member 2	KCNT2	Structure-activity relationship of fenamates as Slo2.1 channel activators. PMID: 22851714
Potassium channel, subfamily U, member 1	KCNU1	The Ca^{2+}-activated K+ current of human sperm is mediated by Slo3. PMID: 24670955
Kv channel-interacting protein 1	KCNIP1	Genome-wide copy number variation study reveals KCNIP1 as a modulator of insulin secretion. PMID: 24886904
Kv channel-interacting protein 2	KCNIP2	Stabilization of Kv4 protein by the accessory K(+) channel interacting protein 2 (KChIP2) subunit is required for the generation of native myocardial fast transient outward K(+) currents. PMID: 23713033
Kv channel-interacting protein 3, calsenilin	KCNIP3	The potassium channel interacting protein 3 (DREAM/KChIP3) heterodimerizes with and regulates calmodulin function. PMID: 23019329
Kv channel-interacting protein 4	KCNIP4	Integration of mouse and human genome-wide association data identifies KCNIP4 as an asthma gene. PMID: 23457522
Potassium channel regulator	KCNRG	Pro-apoptotic and antiproliferative activity of human KCNRG, a putative tumor suppressor in 13q14 region. PMID: 20237900
Potassium voltage-gated channel, Isk-related family, member 1	KCNE1	**Indapamide** – antihypertensives. Disease: Jervell and Lange-Nielsen syndrome 2 (JLNS2)
Potassium voltage-gated channel, Isk-related family, member 2	KCNE2	Disease: long QT syndrome 6 (LQT6)
Potassium voltage-gated channel, Isk-related family, member 3	KCNE3	Disease: Brugada syndrome 6 (BRGDA6)
Potassium voltage-gated channel, Isk-related family, member 4	KCNE4	KCNE4 is an inhibitory subunit to Kv1.1 and Kv1.3 potassium channels. PMID: 12944270
Potassium voltage-gated channel, Isk-related family, member 5	KCNE5	Disease: Alport syndrome with mental retardation, midface hypoplasia and elliptocytosis (ATS-MR)
Potassium large conductance calcium-activated channel, subfamily M, beta member 1	KCNMB1	Hypoxia-inducible factor-1α regulates KCNMB1 expression in human pulmonary artery smooth muscle cells. PMID: 22114151

HGNC-approved name	Symbol	Target information
Potassium large conductance calcium-activated channel, subfamily M, beta member 2	KCNMB2	
Potassium large conductance calcium-activated channel, subfamily M beta member 3	KCNMB3	Allelic association of a truncation mutation of the KCNMB3 gene with idiopathic generalized epilepsy. PMID: 16958040
Potassium large conductance calcium-activated channel, subfamily M, beta member 4	KCNMB4	Recurrent LRP1-SNRNP25 and KCNMB4-CCND3 fusion genes promote tumor cell motility in human osteosarcoma. PMID: 25300797
Potassium voltage-gated channel, shaker-related subfamily, beta member 1	KCNAB1	
Potassium voltage-gated channel, shaker-related subfamily, beta member 2	KCNAB2	Identification of voltage-gated K(+) channel beta 2 (Kvβ2) subunit as a novel interaction partner of the pain transducer Transient Receptor Potential Vanilloid 1 channel (TRPV1). PMID: 24036102
Potassium voltage-gated channel, shaker-related subfamily, beta member 3	KCNAB3	
Ryanodine receptor 1 (skeletal)	RYR1	**Dantrolene** – muscle relaxants. Disease: malignant hyperthermia 1 (MHS1)
Ryanodine receptor 2 (cardiac)	RYR2	**Tetracaine** – cardiac therapy. Disease: arrhythmogenic right ventricular dysplasia, familial, 2 (ARVD2)
Ryanodine receptor 3	RYR3	
Multi-subunit-binding drugs		**Amiloride** – antihypertensives
		Quinidine – cardiac therapy
		Prilocaine – anaesthetics
Sodium channel, voltage-gated, type I, alpha subunit	SCN1A	α-ENaC, a therapeutic target of dexamethasone on hydrogen sulphide induced acute pulmonary edema. PMID: 25195098. Disease: generalized epilepsy with febrile seizures plus 2 (GEFS+2)
Sodium channel, voltage-gated, type I, beta subunit	SCN1B	Disease: generalized epilepsy with febrile seizures plus 1 (GEFS+1)
Sodium channel, voltage-gated, type II, alpha subunit	SCN2A	Disease: seizures, benign familial infantile 3 (BFIS3)
Sodium channel, voltage-gated, type II, beta subunit	SCN2B	Disease: atrial fibrillation, familial, 14 (ATFB14)
Sodium channel, voltage-gated, type III, alpha subunit	SCN3A	Blockage of the upregulation of voltage-gated sodium channel nav1.3 improves outcomes after experimental traumatic brain injury. PMID: 24313291
Sodium channel, voltage-gated, type III, beta subunit	SCN3B	Disease: Brugada syndrome 7 (BRGDA7)
Sodium channel, voltage-gated, type IV, alpha subunit	SCN4A	**Flecainide** – cardiac therapy. Disease: paramyotonia congenita of von Eulenburg (PMC)
Sodium channel, voltage-gated, type IV, beta subunit	SCN4B	Disease: long QT syndrome 10 (LQT10)
Sodium channel, voltage-gated, type V, alpha subunit	SCN5A	**Benzonatate** – cough suppressants. Disease: progressive familial heart block 1A (PFHB1A)
Sodium channel, voltage-gated, type VII, alpha subunit	SCN7A	Enhanced SCN7A/Nax expression contributes to bone cancer pain by increasing excitability of neurons in dorsal root ganglion. PMID: 23026072
Sodium channel, voltage-gated, type VIII, alpha subunit	SCN8A	Disease: cognitive impairment with or without cerebellar ataxia (CIAT)
Sodium channel, voltage-gated, type IX, alpha subunit	SCN9A	**XEN402** – analgesics. Disease: primary erythermalgia (PERYTHM)

(Continued)

HGNC-approved name	Symbol	Target information
Sodium channel, voltage-gated, type X, alpha subunit	SCN10A	Disease: episodic pain syndrome, familial, 2 (FEPS2)
Sodium channel, voltage-gated, type XI, alpha subunit	SCN11A	Disease: neuropathy, hereditary sensory and autonomic, 7 (HSAN7)
Sodium channel, non-voltage-gated 1 alpha subunit	SCNN1A	Disease: pseudohypoaldosteronism 1, autosomal recessive (PHA1B)
Sodium channel, non-voltage-gated 1, beta subunit	SCNN1B	Disease: pseudohypoaldosteronism 1, autosomal recessive (PHA1B)
Sodium channel, non-voltage-gated 1, delta subunit	SCNN1D	δ ENaC: A novel divergent amiloride-inhibitable sodium channel. PMID: 22983350
Sodium channel, non-voltage-gated 1, gamma subunit	SCNN1G	Disease: Liddle syndrome (LIDDS)
Sodium leak channel, non-selective	NALCN	Disease: hypotonia, infantile, with psychomotor retardation and characteristic facies (IHPRF)
Transient receptor potential cation channel, subfamily A, member 1	TRPA1	**GRC 17536** – analgesics. Disease: episodic pain syndrome, familial, 1 (FEPS1)
Transient receptor potential cation channel, subfamily C, member 1	TRPC1	Silencing TRPC1 expression inhibits invasion of CNE2 nasopharyngeal tumor cells. PMID: 22367186
Transient receptor potential cation channel, subfamily C, member 3	TRPC3	Transient receptor potential canonical-3 channel-dependent fibroblast regulation in atrial fibrillation. PMID: 22992321
Transient receptor potential cation channel, subfamily C, member 4	TRPC4	Novel role for TRPC4 in regulation of macroautophagy by a small molecule in vascular endothelial cells. PMID: 25476892
Transient receptor potential cation channel, subfamily C, member 5	TRPC5	Inhibition of transient receptor potential channel 5 reverses 5-fluorouracil resistance in human colorectal cancer cells. PMID: 25404731
Transient receptor potential cation channel, subfamily C, member 6	TRPC6	Identification of TRPC6 as a possible candidate target gene within an amplicon at 11q21–q22.2 for migratory capacity in head and neck squamous cell carcinomas. PMID: 23497198. Disease: focal segmental glomerulosclerosis 2 (FSGS2)
Transient receptor potential cation channel, subfamily C, member 7	TRPC7	Critical role of canonical transient receptor potential channel 7 in initiation of seizures. PMID: 25049394
Transient receptor potential cation channel, subfamily M, member 1	TRPM1	Disease: night blindness, congenital stationary, 1C (CSNB1C)
Transient receptor potential cation channel, subfamily M, member 2	TRPM2	TRPM2: A multifunctional ion channel for calcium signalling. PMID: 21135052
Transient receptor potential cation channel, subfamily M, member 3	TRPM3	TRPM3. PMID: 24756716
Transient receptor potential cation channel, subfamily M, member 4	TRPM4	Disease: progressive familial heart block 1B (PFHB1B)
Transient receptor potential cation channel, subfamily M, member 5	TRPM5	Role of transient receptor potential melastatin-like subtype 5 channel in insulin secretion from rat β-cells. PMID: 24632551
Transient receptor potential cation channel, subfamily M, member 6	TRPM6	Disease: hypomagnesemia 1 (HOMG1)
Transient receptor potential cation channel, subfamily M, member 7	TRPM7	Current understanding of TRPM7 pharmacology and drug development for stroke. PMID: 22820907. Disease: amyotrophic lateral sclerosis–parkinsonism/dementia complex 1 (ALS-PDC1)
Transient receptor potential cation channel, subfamily M, member 8	TRPM8	**Menthol** – antipruritics

HGNC-approved name	Symbol	Target information
Mucolipin 1	MCOLN1	A small molecule restores function to TRPML1 mutant isoforms responsible for mucolipidosis type IV. PMID: 25119295. Disease: mucolipidosis type IV (MLIV)
Mucolipin 2	MCOLN2	
Mucolipin 3	MCOLN3	The TRPML3 channel: From gene to function. PMID: 21290299
Polycystic kidney disease 2 (autosomal dominant)	PKD2	Disease: polycystic kidney disease 2 (PKD2)
Polycystic kidney disease 2-like 1	PKD2L1	Inhibition of TRPP3 channel by amiloride and analogs. PMID: 17804601
Polycystic kidney disease 2-like 2	PKD2L2	Overexpression of Trpp5 contributes to cell proliferation and apoptosis probably through involving calcium homeostasis. PMID: 20043191
Transient receptor potential cation channel, subfamily V, member 1	TRPV1	**PHE 377** – analgesics
Transient receptor potential cation channel, subfamily V, member 2	TRPV2	TRPV2 activation induces apoptotic cell death in human T24 bladder cancer cells: A potential therapeutic target for bladder cancer. PMID: 20546877
Transient receptor potential cation channel, subfamily V, member 3	TRPV3	**Menthol** – antipruritics. Disease: Olmsted syndrome (OLMS)
Transient receptor potential cation channel, subfamily V, member 4	TRPV4	Analysis of responses to the TRPV4 agonist GSK1016790A in the pulmonary vascular bed of the intact-chest rat. PMID: 24186096. Disease: brachyolmia 3 (BRAC3)
Transient receptor potential cation channel, subfamily V, member 5	TRPV5	Inhibition of the TRPC5 ion channel protects the kidney filter. PMID: 24231357
Transient receptor potential cation channel, subfamily V, member 6	TRPV6	Calcium channel TRPV6 as a potential therapeutic target in estrogen receptor-negative breast cancer. PMID: 22807578
Tweety family member 1	TTYH1	Expression of Ttyh1, a member of the Tweety family in neurons *in vitro* and in vivo and its potential role in brain pathology. PMID: 20874767
Tweety family member 2	TTYH2	TTYH2, a human homologue of the Drosophila melanogaster gene tweety, is up-regulated in colon carcinoma and involved in cell proliferation and cell aggregation. PMID: 17569141
Tweety family member 3	TTYH3	
Two-pore segment channel 1	TPCN1	The voltage-gated sodium channel TPC1 confers endolysosomal excitability. PMID: 24776928
Two-pore segment channel 2	TPCN2	Lysosomal two-pore channel subtype 2 (TPC2) regulates skeletal muscle autophagic signaling. PMID: 25480788
Voltage-dependent anion channel 1	VDAC1	Is the mitochondrial outermembrane protein VDAC1 therapeutic target for Alzheimer's disease? PMID: 22995655
Voltage-dependent anion channel 2	VDAC2	The proteomic 2D-DIGE approach reveals the protein voltage-dependent anion channel 2 as a potential therapeutic target in epithelial thyroid tumours. PMID: 25617717
Voltage-dependent anion channel 3	VDAC3	Recombinant human voltage dependent anion selective channel isoform 3 (hVDAC3) forms pores with a very small conductance. PMID: 25171321

3.6.5   Assorted transport and carrier proteins

HGNC-approved name	Symbol	Target information
Apolipoprotein A-I	APOA1	Apolipoprotein A-I mimetic peptides: A potential new therapy for the prevention of atherosclerosis. PMID: 20395699. Disease: high-density lipoprotein deficiency 2 (HDLD2)
Apolipoprotein A-II	APOA2	Human apolipoprotein A-II protects against diet-induced atherosclerosis in transgenic rabbits. PMID: 23241412
Apolipoprotein A-IV	APOA4	Apolipoprotein A-IV improves glucose homeostasis by enhancing insulin secretion. PMID: 22619326
Apolipoprotein A-V	APOA5	The role of apolipoprotein A5 in obesity and the metabolic syndrome. PMID: 23279260. Disease: hypertriglyceridemia, familial (FHTR)
Apolipoprotein B	APOB	**Mipomersen** – lipid-modifying agents. Disease: hypobetalipoproteinemia, familial, 1 (FHBL1)
Apolipoprotein C-I	APOC1	Apolipoprotein C-I is an APOE genotype-dependent suppressor of glial activation. PMID: 22883744
Apolipoprotein C-II	APOC2	Disease: hyperlipoproteinemia 1B (HLPP1B)
Apolipoprotein C-III	APOC3	**ISIS-APOCIIIRx** – lipid-modifying agents. Disease: hyperalphalipoproteinemia 2 (HALP2)
Apolipoprotein C-IV	APOC4	Expression of apolipoprotein C-IV is regulated by Ku antigen/peroxisome proliferator-activated receptor gamma complex and correlates with liver steatosis. PMID: 18809223
Apolipoprotein D	APOD	Apolipoprotein D takes center stage in the stress response of the aging and degenerative brain. PMID: 24612673
Apolipoprotein E	APOE	Apolipoprotein E and Alzheimer disease: Risk, mechanisms and therapy. PMID: 23296339. Disease: hyperlipoproteinemia 3 (HLPP3)
Apolipoprotein F	APOF	Genetic manipulation of the ApoF/Stat2 locus supports an important role for type I interferon signaling in atherosclerosis. PMID: 24529150
Apolipoprotein H (beta-2-glycoprotein I)	APOH	APOH is increased in the plasma and liver of type 2 diabetic patients with metabolic syndrome. PMID: 19878946
Apolipoprotein L, 1	APOL1	Human apolipoprotein L1 (ApoL1) in cancer and chronic kidney disease. PMID: 22569246. Disease: focal segmental glomerulosclerosis 4 (FSGS4)
Apolipoprotein L, 2	APOL2	
Apolipoprotein L, 3	APOL3	
Apolipoprotein L, 4	APOL4	
Apolipoprotein L, 5	APOL5	
Apolipoprotein L, 6	APOL6	Apolipoprotein L6, induced in atherosclerotic lesions, promotes apoptosis and blocks Beclin 1-dependent autophagy in atherosclerotic cells. PMID: 21646352
Apolipoprotein M	APOM	Induction of insulin secretion by apolipoprotein M, a carrier for sphingosine 1-phosphate. PMID: 24814049
Apolipoprotein O	APOO	Apolipoprotein O is mitochondrial and promotes lipotoxicity in heart. PMID: 24743151
Apolipoprotein O-like	APOOL	APOOL is a cardiolipin-binding constituent of the Mitofilin/MINOS protein complex determining cristae morphology in mammalian mitochondria. PMID: 23704930
Apolipoprotein B receptor	APOBR	Pitavastatin inhibits remnant lipoprotein-induced macrophage foam cell formation through ApoB48 receptor-dependent mechanism. PMID: 15591219

HGNC-approved name	Symbol	Target information
Fatty acid binding protein 1, liver	FABP1	Hepatoprotective role of liver fatty acid binding protein in acetaminophen induced toxicity. PMID: 24606952
Fatty acid binding protein 2, intestinal	FABP2	Human intestinal fatty acid binding protein 2 expression is associated with fat intake and polymorphisms. PMID: 20534879
Fatty acid binding protein 3, muscle and heart (mammary-derived growth inhibitor)	FABP3	Inhibition of fatty acid binding proteins elevates brain anandamide levels and produces analgesia. PMID: 24705380
Fatty acid binding protein 4, adipocyte	FABP4	Treatment of diabetes and atherosclerosis by inhibiting fatty-acid-binding protein aP2. PMID: 17554340
Fatty acid binding protein 5 (psoriasis-associated)	FABP5	Inhibition of fatty acid binding proteins elevates brain anandamide levels and produces analgesia. PMID: 24705380
Fatty acid binding protein 6, ileal	FABP6	Evidence for the Thr79Met polymorphism of the ileal fatty acid binding protein (FABP6) to be associated with type 2 diabetes in obese individuals. PMID: 19744871
Fatty acid binding protein 7, brain	FABP7	Inhibition of fatty acid binding proteins elevates brain anandamide levels and produces analgesia. PMID: 24705380
Fatty acid binding protein 9, testis	FABP9	Mice lacking FABP9/PERF15 develop sperm head abnormalities but are fertile. PMID: 20920498
Fatty acid binding protein 12	FABP12	
Feline leukemia virus subgroup C cellular receptor 1	FLVCR1	Disease: posterior column ataxia with retinitis pigmentosa (PCARP)
Feline leukemia virus subgroup C cellular receptor family, member 2	FLVCR2	Disease: proliferative vasculopathy and hydranencephaly–hydrocephaly syndrome (PVHH)
Ferritin mitochondrial	FTMT	Mitochondrial ferritin, a new target for inhibiting neuronal tumor cell proliferation. PMID: 25213357
Ferritin, heavy polypeptide 1	FTH1	Disease: hemochromatosis 5 (HFE5)
Ferritin, light polypeptide	FTL	Disease: hereditary hyperferritinemia–cataract syndrome (HHCS)
Hemoglobin, alpha 1	HBA1	**Aes-103** – antisickling agents. Disease: Heinz body anemias (HEIBAN)
Hemoglobin, alpha 2	HBA2	Disease: Heinz body anemias (HEIBAN)
Hemoglobin, beta	HBB	Disease: Heinz body anemias (HEIBAN)
Hemoglobin, delta	HBD	
Hemoglobin, epsilon 1	HBE1	
Hemoglobin, gamma A	HBG1	
Hemoglobin, gamma G	HBG2	Disease: cyanosis transient neonatal (TNCY)
Hemoglobin, zeta	HBZ	
Low-density lipoprotein receptor	LDLR	Disease: familial hypercholesterolemia (FH)
Low-density lipoprotein receptor-related protein 1	LRP1	Low-density lipoprotein receptor-related protein-1: A serial clearance homeostatic mechanism controlling Alzheimer's amyloid β-peptide elimination from the brain. PMID: 20854368
Low-density lipoprotein receptor-related protein 2	LRP2	Disease: Donnai–Barrow syndrome (DBS)
Low-density lipoprotein receptor-related protein 4	LRP4	Disease: Cenani–Lenz syndactyly syndrome (CLSS)
Low-density lipoprotein receptor-related protein 8, apolipoprotein e receptor	LRP8	Disease: myocardial infarction 1 (MCI1)
Very-low-density lipoprotein receptor	VLDLR	Disease: cerebellar ataxia, mental retardation, and dysequilibrium syndrome 1 (CAMRQ1)

(Continued)

HGNC-approved name	Symbol	Target information
Magnesium transporter 1	MAGT1	Disease: immunodeficiency, X-linked, with magnesium defect, Epstein–Barr virus infection and neoplasia (XMEN)
Membrane magnesium transporter 1	MMGT1	Cellular magnesium homeostasis. PMID: 21640700
MRS2 magnesium transporter	MRS2	The role of Mg^{2+} in immune cells. PMID: 22990458
Mitochondrial calcium uniporter	MCU	A role for mitochondria in antigen processing and presentation. PMID: 25251370
Mitochondrial calcium uniporter regulator 1	MCUR1	MCUR1 is an essential component of mitochondrial Ca^{2+} uptake that regulates cellular metabolism. PMID: 23178883
Mitochondrial calcium uptake 1	MICU1	Note: Defects in MICU1 are the cause of brain and muscle disorder (BMDL): A syndrome characterized by proximal myopathy, learning difficulties and a progressive extrapyramidal movement disorder
Mitochondrial calcium uptake 2	MICU2	MICU2, a paralog of MICU1, resides within the mitochondrial uniporter complex to regulate calcium handling. PMID: 23409044
Mitochondrial pyruvate carrier 1	MPC1	Disease: mitochondrial pyruvate carrier deficiency (MPYCD)
Mitochondrial pyruvate carrier 1-like	MPC1L	
Mitochondrial pyruvate carrier 2	MPC2	A role for the mitochondrial pyruvate carrier as a repressor of the Warburg effect and colon cancer cell growth. PMID: 25458841
Niemann–Pick disease, type C1	NPC1	Disease: Niemann–Pick disease C1 (NPC1)
Niemann–Pick disease, type C2	NPC2	Disease: Niemann–Pick disease C2 (NPC2)
Non-imprinted in Prader–Willi/Angelman syndrome 1	NIPA1	Disease: spastic paraplegia 6, autosomal dominant (SPG6)
Non-imprinted in Prader–Willi/Angelman syndrome 2	NIPA2	Functional characterization of NIPA2, a selective Mg^{2+} transporter. PMID: 18667602
Orosomucoid 1	ORM1	A meta-analysis of genome-wide association studies identifies ORM1 as a novel gene controlling thrombin generation potential. PMID: 24357727
Orosomucoid 2	ORM2	The potential role of ORM2 in the development of colorectal cancer. PMID: 22363757
Oxysterol binding protein	OSBP	Oxysterol-binding protein-1 (OSBP1) modulates processing and trafficking of the amyloid precursor protein. PMID: 18348724
Oxysterol binding protein 2	OSBP2	Oxysterol-binding protein (OSBP)-related protein 4 (ORP4) is essential for cell proliferation and survival. PMID: 24742681
Oxysterol binding protein-like 1A	OSBPL1A	
Oxysterol binding protein-like 2	OSBPL2	OSBP-related protein 2 is a sterol receptor on lipid droplets that regulates the metabolism of neutral lipids. PMID: 19224871
Oxysterol binding protein-like 3	OSBPL3	OSBP-related protein 3 (ORP3) coupling with VAMP-associated protein A regulates R-Ras activity. PMID: 25447204
Oxysterol binding protein-like 5	OSBPL5	The role of oxysterol binding protein-related protein 5 in pancreatic cancer. PMID: 20128820
Oxysterol binding protein-like 6	OSBPL6	
Oxysterol binding protein-like 7	OSBPL7	OSBP-related protein 7 interacts with GATE-16 and negatively regulates GS28 protein stability. PMID: 21669198
Oxysterol binding protein-like 8	OSBPL8	Osbpl8 deficiency in mouse causes an elevation of high-density lipoproteins and gender-specific alterations of lipid metabolism. PMID: 23554939
Oxysterol binding protein-like 9	OSBPL9	
Oxysterol binding protein-like 10	OSBPL10	ORP10, a cholesterol binding protein associated with microtubules, regulates apolipoprotein B-100 secretion. PMID: 22906437

HGNC-approved name	Symbol	Target information
Oxysterol binding protein-like 11	OSBPL11	Association of OSBPL11 gene polymorphisms with cardiovascular disease risk factors in obesity. PMID: 19325544
Retinol binding protein 1, cellular	RBP1	CRBP-1 expression in ovarian cancer: A potential therapeutic target. PMID: 24982334
Retinol binding protein 2, cellular	RBP2	Retinoblastoma binding protein 2 (RBP2) promotes HIF-1α-VEGF-induced angiogenesis of non-small cell lung cancer via the Akt pathway. PMID: 25162518
Retinol binding protein 3, interstitial	RBP3	Disease: retinitis pigmentosa 66 (RP66)
Retinol binding protein 4, plasma	RBP4	Disease: retinal dystrophy, iris coloboma, and comedogenic acne syndrome (RDCCAS)
Retinol binding protein 5, cellular	RBP5	Down-regulation of retinol binding protein 5 is associated with aggressive tumor features in hepatocellular carcinoma. PMID: 17497168
Retinol binding protein 7, cellular	RBP7	
Rh-associated glycoprotein	RHAG	Disease: regulator type Rh-null hemolytic anemia (RHN)
Rh family, B glycoprotein (gene/pseudogene)	RHBG	Ammonia transport in the kidney by Rhesus glycoproteins. PMID: 24647713
Rh blood group, CcEe antigens	RHCE	
Rh family, C glycoprotein	RHCG	The role of the renal ammonia transporter Rhcg in metabolic responses to dietary protein. PMID: 24652796
Rh blood group, D antigen	RHD	**Roledumab** – immunosuppressants
Scavenger receptor class A, member 3	SCARA3	Scavenger receptor class A member 3 (SCARA3) in disease progression and therapy resistance in multiple myeloma. PMID: 23537707
Scavenger receptor class A, member 5 (putative)	SCARA5	L-ferritin binding to scara5: a new iron traffic pathway potentially implicated in retinopathy. PMID: 25259650
Scavenger receptor class B, member 1	SCARB1	Scavenger receptor class B, type I: A promising immunotherapy target. PMID: 21395381
Scavenger receptor class B, member 2	SCARB2	Disease: epilepsy, progressive myoclonic 4, with or without renal failure (EPM4)
Scavenger receptor class F, member 1	SCARF1	The scavenger receptor SCARF1 mediates the clearance of apoptotic cells and prevents autoimmunity. PMID: 23892722
Scavenger receptor class F, member 2	SCARF2	Disease: Van den Ende–Gupta syndrome (VDEGS)
Macrophage scavenger receptor 1	MSR1	Disease: prostate cancer (PC)
Sideroflexin 1	SFXN1	
Sideroflexin 2	SFXN2	
Sideroflexin 3	SFXN3	Serum autoantibody to sideroflexin 3 as a novel tumor marker for oral squamous cell carcinoma. PMID: 21136855
Sideroflexin 4	SFXN4	Disease: combined oxidative phosphorylation deficiency 18 (COXPD18)
Sideroflexin 5	SFXN5	Identification and characterization of a novel mitochondrial tricarboxylate carrier. PMID: 12150972
StAR-related lipid transfer (START) domain containing 3	STARD3	Mitochondrial proteases act on STARD3 to activate progesterone synthesis in human syncytiotrophoblast. PMID: 25459514
StAR-related lipid transfer (START) domain containing 4	STARD4	The StarD4 subfamily of steroidogenic acute regulatory-related lipid transfer (START) domain proteins: New players in cholesterol metabolism. PMID: 24440759
StAR-related lipid transfer (START) domain containing 5	STARD5	StAR-related lipid transfer domain protein 5 binds primary bile acids. PMID: 23018617

(Continued)

HGNC-approved name	Symbol	Target information
StAR-related lipid transfer (START) domain containing 6	STARD6	STARD6 is expressed in steroidogenic cells of the ovary and can enhance de novo steroidogenesis. PMID: 24595982
StAR-related lipid transfer (START) domain containing 7	STARD7	The lipid transfer protein StarD7: Structure, function, and regulation. PMID: 23507753
StAR-related lipid transfer (START) domain containing 8	STARD8	
StAR-related lipid transfer (START) domain containing 10	STARD10	Disruption of Stard10 gene alters the PPARα-mediated bile acid homeostasis. PMID: 23200860
STARD3 N-terminal like	STARD3NL	MLN64 and MENTHO, two mediators of endosomal cholesterol transport. PMID: 16709157
Transcobalamin I (vitamin B12 binding protein, R binder family)	TCN1	Vitamin B_{12} and its binding proteins in hepatocellular carcinoma and chronic liver diseases. PMID: 24958254
Transcobalamin II	TCN2	Transcobalamin 2 variant associated with poststroke homocysteine modifies recurrent stroke risk. PMID: 21975197. Disease: transcobalamin II deficiency (TCN2 deficiency)
Transferrin	TF	Disease: atransferrinemia (ATRAF)
Transferrin receptor	TFRC	Aberrant expression of TfR1/CD71 in thyroid carcinomas identifies a novel potential diagnostic marker and therapeutic target. PMID: 21323588
Transferrin receptor 2	TFR2	Disease: hemochromatosis 3 (HFE3)
Afamin	AFM	Afamin stimulates osteoclastogenesis and bone resorption via Gi-coupled receptor and Ca^{2+}/calmodulin-dependent protein kinase (CaMK) pathways. PMID: 23698732
Amnion-associated transmembrane protein	AMN	Disease: recessive hereditary megaloblastic anemia 1 (RH-MGA1)
ANKH inorganic pyrophosphate transport regulator	ANKH	Disease: chondrocalcinosis 2 (CCAL2)
Antigen p97 (melanoma associated) identified by monoclonal antibodies 133.2 and 96.5	MFI2	Melanotransferrin: Search for a function. PMID: 21933697
Cholesteryl ester transfer protein, plasma	CETP	**Anacetrapib** – lipid-modifying agents. Disease: hyperalphalipoproteinemia 1 (HALP1)
Copper chaperone for superoxide dismutase	CCS	Mitochondrial defects in transgenic mice expressing Cu,Zn superoxide dismutase mutations: The role of copper chaperone for SOD1. PMID: 24269091
cutC copper transporter	CUTC	
Cystinosin, lysosomal cystine transporter	CTNS	Disease: cystinosis, nephropathic type (CTNS)
Hemochromatosis	HFE	Disease: hemochromatosis 1 (HFE1)
Hemopexin	HPX	Acute-phase protein hemopexin is a negative regulator of Th17 response and experimental autoimmune encephalomyelitis development. PMID: 24154625
Microsomal triglyceride transfer protein	MTTP	**Lomitapide** – lipid-modifying agents. Disease: abetalipoproteinemia (ABL)
NIPA-like domain containing 4	NIPAL4	Disease: ichthyosis, congenital, autosomal recessive 6 (ARCI6)
NPC1-like 1	NPC1L1	**Ezetimibe** – lipid-modifying agents
Oculocutaneous albinism II	OCA2	Homology modelling and virtual screening of P-protein in a quest for novel antimelanogenic agent and in vitro assessments. PMID: 25236473. Disease: albinism, oculocutaneous, 2 (OCA2)
Organic solute carrier partner 1	OSCP1	Tumor suppressor gene Oxidored-nitro domain-containing protein 1 regulates nasopharyngeal cancer cell autophagy, metabolism, and apoptosis in vitro. PMID: 23831407

HGNC-approved name	Symbol	Target information
Phosphatidylcholine transfer protein	PCTP	Phosphatidylcholine transfer protein interacts with thioesterase superfamily member 2 to attenuate insulin signaling. PMID: 23901139
Phospholipid transfer protein	PLTP	Phospholipid transfer protein in diabetes, metabolic syndrome and obesity. PMID: 25107452
PQ loop repeat containing 2	PQLC2	Heptahelical protein PQLC2 is a lysosomal cationic amino acid exporter underlying the action of cysteamine in cystinosis therapy. PMID: 23169667
Retinaldehyde binding protein 1	RLBP1	Disease: retinitis pigmentosa autosomal recessive (ARRP)
Sarcalumenin	SRL	Sarcalumenin plays a critical role in age-related cardiac dysfunction due to decreases in SERCA2a expression and activity. PMID: 22119571
Sex hormone-binding globulin	SHBG	Sex hormone-binding globulin (SHBG) and estradiol cross-talk in breast cancer cells. PMID: 16700004
Tocopherol (alpha) transfer protein	TTPA	Disease: ataxia with isolated vitamin E deficiency (AVED)
Transthyretin	TTR	**ALN-TTR02** antiamyloidosis agents. Disease: amyloidosis, transthyretin-related (AMYL-TTR)

Chapter 4
Enzymes: Part 1

The next two chapters are devoted to enzymes, a protein class amenable to activation or inhibition by small-molecule drugs, due to their well-defined catalytic or allosteric sites. The human genome encodes several thousand enzymes, so to maximize clarity, each entry is grouped according to a particular biological function and the whole collection presented in two chapters. Many of the entries relate to proteins where catalytic activity has been demonstrated experimentally, but others have been assigned purely on the basis of electronic annotation, so these will not necessarily turn out to be *bona fide* enzymes. A number of entries have been shown to be catalytically inactive but are nevertheless listed together with related proteins known to have enzymatic activity. It should be noted that some of the enzyme references imply that their potential as a therapeutic target is unrelated to their catalytic activity and may involve protein–protein interactions. This chapter contains two main groups of enzymes, namely those involved in cell signalling and those that perform a number of protein modifications, such as glycosylation. As is the case elsewhere in this book, there is a degree of overlap between functions, so for example, some protein modification enzymes listed as peptidases may also participate in signalling pathways.

The signalling enzymes include protein kinases and phosphatases, cyclases and phosphodiesterases, also an extensive group of small GTP-binding proteins. Some lipid kinases and phosphatases relating to inositol signalling are listed with the protein kinases and protein phosphatases, respectively. Lastly, the 'other signalling enzymes' list entries (prenyltransferases and phospholipases for example), that do not readily fit into the other categories. The distribution of targets for current drugs is patchy throughout the above groups.

The protein kinases are relatively well served with clinical candidates through their importance in oncology and immunoinflammatory diseases and great progress has been made in producing specific inhibitors [1]. The phosphatases, on the other hand, have proven to be difficult targets which is reflected in the near absence of drug entries in the list. Varying degrees of success have been obtained with inhibitors of adenyl or guanyl cyclases or phosphodiesterases, but the potential for isoform-specific drugs has been realized, most notably with the development of PDE5 inhibitors. Small GTPases such as RAS have been considered as oncology targets for many years, but success in drug development has proved elusive (but maybe not impossible [2]). The extensive family of phospholipases contains a few enzymes that are targeted by current drugs; however, just browsing the references for the majority of the enzymes shows that there are plenty of opportunities to discover novel therapeutics, so long as compounds can be developed that can discriminate between closely related family members; this of course is also true for many of the other protein families listed in this book.

The protein modification enzymes are involved in protein synthesis and folding, as well as breakdown by peptidases and their inhibitors. There are also sections devoted to glycosylation, ubiquitylation (and related) modifications as well as specific (epigenetic) modifications to chromatin proteins. The section on 'other protein modifications' includes such diverse enzymes as the poly (ADP-ribose) polymerases, transglutaminases and tubulin tyrosine ligases.

Like the kinases in the signalling enzyme group, the peptidase family is diverse, extensive and the target of several well-established drugs. Peptidase inhibitors are included in this section because of their specific association with the peptidase enzymes. The inhibitors are important in controlling such vital areas as haemostasis and apoptosis; they also provide a number of targets for existing drugs and those with potential for treating conditions such as cancer and cardiovascular disease. The role of glycoproteins in disease has been well known for many years, so inhibitors of glycosylation enzymes should be useful drugs; the target opportunities exist, but significant medicinal chemistry challenges remain [3]. Cellular protein turnover is controlled by ubiquitylation or other post-translational modifications such as SUMOylation and neddylation. These small modifiers are added by a large family of specific ligases and removed by a family of peptidases. Both the ligases and peptidases are the subject of much pharmaceutical interest for oncology [4]. The chromatin-modifying enzymes represent another area of research because of their involvement in the epigenetic regulation of gene expression [5]. Inhibitors of histone methylation/demethylation or acetylation/deacetylation have potential utility in cancer and inflammatory diseases. The elongation factor enzymes involved in protein synthesis are therapeutic targets for anti-proliferative agents, as are the many proteins involved in protein folding. The section on protein synthesis and folding includes heat shock proteins, chaperonins and protein disulphide isomerases and targets for the immunosuppressant drugs cyclosporine A and FK506.

References

1. Fabbro D, Cowan-Jacob SW, Moebitz H. (2015) Ten things you should know about protein kinases. British Journal of Pharmacology. DOI: 10.1111/bph.13096
2. Cox AD, Fesik SW, Kimmelman AC, Luo J, Der CJ. (2014) Drugging the undruggable RAS: Mission possible? *Nature Reviews Drug Discovery* **13**, 828–51.
3. Büll C, Stoel MA, den Brok MH, Adema GJ. (2014) Sialic acids sweeten a tumor's life. *Cancer Research* **15**, 3199–204.
4. Liu J *et al.* (2014) Targeting the ubiquitin pathway for cancer treatment. *Biochimica et Biophysica Acta* **1855**, 50–60.
5. Zhang G, Pradhan S. (2014) Mammalian epigenetic mechanisms. *IUBMB Life* **66**, 240–56.

4.1　Signalling enzymes

4.1.1　Protein kinases

HGNC-approved name	Symbol	Target information
Tyrosine kinases		
3-Phosphoinositide-dependent protein kinase 1	PDPK1	Genetic and pharmacological inhibition of PDK1 in cancer cells: Characterization of a selective allosteric kinase inhibitor. PMID: 21118801
Anaplastic lymphoma receptor tyrosine kinase	ALK	**Crizotinib** – antineoplastic agents. Note: A chromosomal aberration involving ALK is found in a form of non-Hodgkin lymphoma. Translocation t(25)(p23q35) with NPM1

HGNC-approved name	Symbol	Target information
AXL receptor tyrosine kinase	AXL	The receptor tyrosine kinase Axl is an essential regulator of prostate cancer proliferation and tumor growth and represents a new therapeutic target. PMID: 22410775. Note: AXL and its ligand GAS6 are highly expressed in thyroid carcinoma tissues
Colony-stimulating factor 1 receptor	CSF1R	**AMG 820** – antineoplastic agents. Note: Aberrant expression of CSF1 or CSF1R can promote cancer cell proliferation, invasion and formation of metastases
Discoidin domain receptor tyrosine kinase 1	DDR1	Discoidin domain receptor 1 controls linear invadosome formation via a Cdc42-Tuba pathway. PMID: 25422375
Discoidin domain receptor tyrosine kinase 2	DDR2	**Regorafenib** (NB hits multiple kinases) – antineoplastic agents. Disease: spondyloepimetaphyseal dysplasia short limb-hand type (SEMD-SL)
EPH receptor A1	EPHA1	Lack of ephrin receptor A1 is a favorable independent prognostic factor in clear cell renal cell carcinoma. PMID: 25025847
EPH receptor A2	EPHA2	**MEDI-547** – antineoplastic agents. Disease: cataract 6, multiple types (CTRCT6)
EPH receptor A3	EPHA3	**KB004** – antineoplastic agents. Disease: colorectal cancer (CRC)
EPH receptor A4	EPHA4	EPHA4 is a disease modifier of amyotrophic lateral sclerosis in animal models and in humans. PMID: 22922411
EPH receptor A5	EPHA5	After repeated division, bone marrow stromal cells express inhibitory factors with osteogenic capabilities, and EphA5 is a primary candidate. PMID: 24029132
EPH receptor A6	EPHA6	Learning and memory impairment in Eph receptor A6 knockout mice. PMID: 18450376
EPH receptor A7	EPHA7	Mining the cancer genome uncovers therapeutic activity of EphA7 against lymphoma. PMID: 22356769
EPH receptor A8	EPHA8	Expression of EphA8-Fc in transgenic mouse embryos induces apoptosis of neural epithelial cells during brain development. PMID: 23696555
EPH receptor A10	EPHA10	Ephrin receptor A10 is a promising drug target potentially useful for breast cancers including triple negative breast cancers. PMID: 24946238
EPH receptor B1	EPHB1	Ligand-dependent EphB1 signaling suppresses glioma invasion and correlates with patient survival. PMID: 24121831
EPH receptor B2	EPHB2	Overexpression of EPH receptor B2 in malignant mesothelioma correlates with oncogenic behavior. PMID: 23887168. Disease: prostate cancer (PC)
EPH receptor B3	EPHB3	Silencing of the EPHB3 tumor-suppressor gene in human colorectal cancer through decommissioning of a transcriptional enhancer. PMID: 24707046
EPH receptor B4	EPHB4	**EphB4–HSA fusion protein** – antineoplastic agents
EPH receptor B6	EPHB6	Mutations of the EPHB6 receptor tyrosine kinase induce a pro-metastatic phenotype in non-small cell lung cancer. PMID: 23226491
Epidermal growth factor receptor	EGFR	**Lapatinib** – antineoplastic agents. Disease: lung cancer (LNCR)
v-erb-b2 avian erythroblastic leukemia viral oncogene homolog 2	ERBB2	**Lapatinib** – antineoplastic agents. Disease: hereditary diffuse gastric cancer (HDGC)
v-erb-b2 avian erythroblastic leukemia viral oncogene homolog 3	ERBB3	**Afatinib** – antineoplastic agents. Disease: lethal congenital contracture syndrome 2 (LCCS2)
v-erb-b2 avian erythroblastic leukemia viral oncogene homolog 4	ERBB4	**Afatinib** – antineoplastic agents. Disease: amyotrophic lateral sclerosis 19 (ALS19)

(Continued)

HGNC-approved name	Symbol	Target information
Fibroblast growth factor receptor 1	FGFR1	**Brivanib** – antineoplastic agents. Disease: Crouzon syndrome (CS)
Fibroblast growth factor receptor 2	FGFR2	**Brivanib** – antineoplastic agents. Disease: Crouzon syndrome (CS)
Fibroblast growth factor receptor 3	FGFR3	**Dovitinib** – antineoplastic agents. Disease: achondroplasia (ACH)
Fibroblast growth factor receptor 4	FGFR4	**ISIS-FGFR4Rx** – antiobesity preparations
fms-related tyrosine kinase 1	FLT1	**Axitinib** – antineoplastic agents. Note: Can contribute to cancer cell survival, proliferation, migration, and invasion, and tumor angiogenesis and metastasis
fms-related tyrosine kinase 3	FLT3	**Quizartinib** – antineoplastic agents. Disease: leukemia, acute myelogenous (AML)
fms-related tyrosine kinase 4	FLT4	**Axitinib** – antineoplastic agents. Disease: lymphedema, hereditary, 1A (LMPH1A)
Insulin-like growth factor 1 receptor	IGF1R	**Linsitinib** – antineoplastic agents. Disease: insulin-like growth factor 1 resistance (IGF1RES)
Insulin receptor	INSR	**Linsitinib** – antineoplastic agents. Disease: Rabson–Mendenhall syndrome (RMS)
Insulin receptor-related receptor	INSRR	Insulin receptor-related receptor as an extracellular pH sensor involved in the regulation of acid-base balance. PMID: 23220417
Kinase insert domain receptor (a type III receptor tyrosine kinase)	KDR	**Axitinib** – antineoplastic agents. Disease: hemangioma, capillary infantile (HCI)
v-kit Hardy–Zuckerman 4 feline sarcoma viral oncogene homolog	KIT	**Imatinib** – antineoplastic agents. Disease: Piebald trait (PBT)
Apoptosis-associated tyrosine kinase	AATK	LMTK1 regulates dendritic formation by regulating movement of Rab11A-positive endosomes. PMID: 24672056
Lemur tyrosine kinase 2	LMTK2	LMTK2-mediated phosphorylation regulates CFTR endocytosis in human airway epithelial cells. PMID: 24727471
Lemur tyrosine kinase 3	LMTK3	Kinome screening for regulators of the estrogen receptor identifies LMTK3 as a new therapeutic target in breast cancer. PMID: 21602804
Leukocyte receptor tyrosine kinase	LTK	Note: Genetic variations in LTK that cause up-regulation of the PI3K pathway may possibly contribute to susceptibility to abnormal proliferation of self-reactive B-cells and, therefore, to systemic lupus erythematosus (SLE)
c-mer proto-oncogene tyrosine kinase	MERTK	**UNC1062** – antineoplastic agents. Disease: retinitis pigmentosa 38 (RP38)
Macrophage-stimulating 1 receptor (c-met-related tyrosine kinase)	MST1R	The RON receptor tyrosine kinase is a potential therapeutic target in Burkitt lymphoma. PMID: 23360784
Muscle, skeletal, receptor tyrosine kinase	MUSK	The role of MuSK in synapse formation and neuromuscular disease. PMID: 23637281. Disease: myasthenic syndrome, congenital, associated with acetylcholine receptor deficiency (CMS-ACHRD)
Neurotrophic tyrosine kinase, receptor, type 1	NTRK1	**Milciclib maleate** – antineoplastic agents. Disease: congenital insensitivity to pain with anhidrosis (CIPA)
Neurotrophic tyrosine kinase, receptor, type 2	NTRK2	Identification of a low-molecular weight TrkB antagonist with anxiolytic and antidepressant activity in mice. PMID: 21505263. Disease: obesity, hyperphagia and developmental delay (OHPDD)
Neurotrophic tyrosine kinase, receptor, type 3	NTRK3	Small molecule ligands for active targeting of TrkC-expressing tumor cells. PMID: 23411915
Platelet-derived growth factor receptor, alpha polypeptide	PDGFRA	**Crenolanib** – antineoplastic agents. Note: A chromosomal aberration involving PDGFRA is found in some cases of hypereosinophilic syndrome. Interstitial chromosomal deletion del(4)(q12q12) causes the fusion of FIP1L1 and PDGFRA (FIP1L1-PDGFRA)

HGNC-approved name	Symbol	Target information
Platelet-derived growth factor receptor, beta polypeptide	PDGFRB	**Crenolanib** – antineoplastic agents. Note: A chromosomal aberration involving PDGFRB is found in a form of chronic myelomonocytic leukemia (CMML). Translocation t(512)(q33p13) with EVT6/TEL
Protein kinase domain containing, cytoplasmic	PKDCC	Short limbs, cleft palate, and delayed formation of flat proliferative chondrocytes in mice with targeted disruption of a putative protein kinase gene, Pkdcc (AW548124). PMID: 19097194
ret proto-oncogene	RET	**Vandetanib** – antineoplastic agents. Disease: colorectal cancer (CRC)
Receptor tyrosine kinase-like orphan receptor 1	ROR1	Ovarian cancer stem cells express ROR1, which can be targeted for anti-cancer-stem-cell therapy. PMID: 25411317
Receptor tyrosine kinase-like orphan receptor 2	ROR2	Orphan receptor tyrosine kinase ROR2 as a potential therapeutic target for osteosarcoma. PMID: 19486338. Disease: brachydactyly B1 (BDB1)
c-ros oncogene 1, receptor tyrosine kinase	ROS1	**Crizotinib** – antineoplastic agents. Note: A chromosomal aberration involving ROS1 is found in a glioblastoma multiforme sample
Receptor-like tyrosine kinase	RYK	Ryk is essential for Wnt-5a-dependent invasiveness in human glioma. PMID: 24621529
Serine/threonine/tyrosine kinase 1	STYK1	Overexpression of serine threonine tyrosine kinase 1/novel oncogene with kinase domain mRNA in patients with acute leukemia. PMID: 19409952
TEK tyrosine kinase, endothelial	TEK	**Rebastinib** – antineoplastic agents. Disease: dominantly inherited venous malformations (VMCM)
Tyrosine kinase with immunoglobulin-like and EGF-like domains 1	TIE1	Tie1 deletion inhibits tumor growth and improves angiopoietin antagonist therapy. PMID: 24430181
TYRO3 protein tyrosine kinase	TYRO3	The E3 ligase Cbl-b and TAM receptors regulate cancer metastasis via natural killer cells. PMID: 24553136
v-src avian sarcoma (Schmidt–Ruppin A-2) viral oncogene homolog	SRC	**Dasatinib** – antineoplastic agents. Note: SRC kinase activity has been shown to be increased in several tumor tissues and tumor cell lines such as colon carcinoma cells
c-abl oncogene 1, non-receptor tyrosine kinase	ABL1	**Imatinib** – antineoplastic agents. Disease: leukemia, chronic myeloid (CML)
c-abl oncogene 2, non-receptor tyrosine kinase	ABL2	Role of ABL family kinases in cancer: From leukaemia to solid tumours. PMID: 23842646
B lymphoid tyrosine kinase	BLK	Disease: maturity-onset diabetes of the young 11 (MODY11)
BMX non-receptor tyrosine kinase	BMX	Nonreceptor tyrosine kinase BMX maintains self-renewal and tumorigenic potential of glioblastoma stem cells by activating STAT3. PMID: 21481791
Bruton agammaglobulinemia tyrosine kinase	BTK	**Ibrutinib** – antineoplastic agents. Disease: X-linked agammaglobulinemia (XLA)
c-src tyrosine kinase	CSK	c-Src drives intestinal regeneration and transformation. PMID: 24788409
fer (fps/fes-related) tyrosine kinase	FER	Fer protein-tyrosine kinase promotes lung adenocarcinoma cell invasion and tumor metastasis. PMID: 23699534
Feline sarcoma oncogene	FES	Note: Has been shown to act as proto-oncogene in some types of cancer, possibly due to abnormal activation of the kinase. Has been shown to act as tumor suppressor in other types of cancer
Feline Gardner–Rasheed sarcoma viral oncogene homolog	FGR	Note: Mutations that cause aberrant kinase activation can confer oncogene activity and promote aberrant cell proliferation
fyn-related kinase	FRK	FRK suppresses the proliferation of human glioma cells by inhibiting cyclin D1 nuclear accumulation. PMID: 24792491
FYN oncogene related to SRC, FGR, YES	FYN	The aspartic acid of Fyn at 390 is critical for neuronal migration during corticogenesis. PMID: 25251774

(*Continued*)

HGNC-approved name	Symbol	Target information
Hemopoietic cell kinase	HCK	Identification of therapeutic targets for quiescent, chemotherapy-resistant human leukemia stem cells. PMID: 20371479. Note: Aberrant activation of HCK by HIV-1 protein Nef enhances HIV-1 replication and contributes to HIV-1 pathogenicity
IL-2-inducible T-cell kinase	ITK	**BMS-509722** – drugs for obstructive airway diseases. Disease: lymphoproliferative syndrome 1 (LPFS1)
Janus kinase 1	JAK1	**Ruxolitinib** – myelofibrosis treatments
Janus kinase 2	JAK2	A selective, orally bioavailable 1,2,4-triazolo[1,5-a]pyridine-based inhibitor of Janus kinase 2 for use in anticancer therapy: Discovery of CEP-33779. PMID: 22594690
Janus kinase 3	JAK3	**Tofacitinib** – anti-inflammatory and antirheumatic products. Disease: severe combined immunodeficiency autosomal recessive T-cell-negative/B-cell-positive/NK-cell-negative (T(−)B(+)NK(−) SCID)
Lymphocyte-specific protein tyrosine kinase	LCK	Effects of the novel and potent lymphocyte-specific protein tyrosine kinase inhibitor TKM0150 on mixed lymphocyte reaction and contact hypersensitivity in mice. PMID: 20533766. Note: A chromosomal aberration involving LCK is found in leukemias
v-yes-1 Yamaguchi sarcoma viral-related oncogene homolog	LYN	**Bafetinib** – antineoplastic agents. Note: Constitutively phosphorylated and activated in cells from a number of chronic myelogenous leukemia (CML) and acute myeloid leukemia (AML) patients. Mediates phosphorylation of the BCR-ABL fusion protein
Megakaryocyte-associated tyrosine kinase	MATK	Csk homologous kinase associates with RAFTK/Pyk2 in breast cancer cells and negatively regulates its activation and breast cancer cell migration. PMID: 12063569
NUAK family, SNF1-like kinase, 1	NUAK1	Characterization of WZ4003 and HTH-01-015 as selective inhibitors of the LKB1-tumour-suppressor-activated NUAK kinases. PMID: 24171924
NUAK family, SNF1-like kinase, 2	NUAK2	NUAK2: An emerging acral melanoma oncogene. PMID: 21911917
Protein tyrosine kinase 2	PTK2	**GSK2256098** – antineoplastic agents. Note: Aberrant PTK2/FAK1 expression may play a role in cancer cell proliferation, migration and invasion, in tumor formation and metastasis. PTK2/FAK1 overexpression is seen in many types of cancer
Protein tyrosine kinase 2 beta	PTK2B	Adrenergic signaling regulates mitochondrial Ca2+ uptake through Pyk2-dependent tyrosine phosphorylation of the mitochondrial Ca2+ uniporter. PMID: 24800979. Note: Aberrant PTK2B/PYK2 expression may play a role in cancer cell proliferation, migration and invasion, in tumor formation and metastasis
Protein tyrosine kinase 6	PTK6	Building a better understanding of the intracellular tyrosine kinase PTK6 – BRK by BRK. PMID: 20193745
Protein tyrosine kinase 7	PTK7	Protein tyrosine kinase 7 (PTK7) as a predictor of lymph node metastases and a novel prognostic biomarker in patients with prostate cancer. PMID: 24987951
Pseudopodium-enriched atypical kinase 1	PEAK1	Dynamic phosphorylation of tyrosine 665 in pseudopodium-enriched atypical kinase 1 (PEAK1) is essential for the regulation of cell migration and focal adhesion turnover. PMID: 23105102
src-related kinase lacking C-terminal regulatory tyrosine and N-terminal myristoylation sites	SRMS	The unique N-terminal region of SRMS regulates enzymatic activity and phosphorylation of its novel substrate docking protein 1. PMID: 23822091
Spleen tyrosine kinase	SYK	**Fostamatinib** – selective immunosuppressants
tec protein tyrosine kinase	TEC	The non-receptor tyrosine kinase tec controls assembly and activity of the noncanonical caspase-8 inflammasome. PMID: 25474208

HGNC-approved name	Symbol	Target information
Tyrosine kinase, non-receptor, 1	TNK1	Identification of activated Tnk1 kinase in Hodgkin's lymphoma. PMID: 20090780
Tyrosine kinase, non-receptor, 2	TNK2	ACK1 tyrosine kinase: Targeted inhibition to block cancer cell proliferation. PMID: 23597703
TXK tyrosine kinase	TXK	Selective expression rather than specific function of Txk and Itk regulate Th1 and Th2 responses. PMID: 18941202
Tyrosine kinase 2	TYK2	**PF-06263276** – antipsoriatics. Disease: protein tyrosine kinase 2 deficiency (TYK2 deficiency)
WEE1 G2 checkpoint kinase	WEE1	**MK-1775** – antineoplastic agents
WEE1 homolog 2 (*Schizosaccharomyces pombe*)	WEE2	Ser 15 of WEE1B is a potential PKA phosphorylation target in G2/M transition in one-cell stage mouse embryos. PMID: 23616086
v-yes-1 Yamaguchi sarcoma viral oncogene homolog 1	YES1	Identification of potent Yes1 kinase inhibitors using a library screening approach. PMID: 23787099
Zeta-chain (TCR) associated protein kinase 70 kDa	ZAP70	Disease: selective T-cell defect (STCD)
Serine/threonine kinases		
aarF domain containing kinase 1	ADCK1	
aarF domain containing kinase 2	ADCK2	Kinome-wide functional genomics screen reveals a novel mechanism of TNFα-induced nuclear accumulation of the HIF-1α transcription factor in cancer cells. PMID: 22355351
aarF domain containing kinase 3	ADCK3	Mitochondrial ADCK3 employs an atypical protein kinase-like fold to enable coenzyme Q biosynthesis. PMID: 25498144. Disease: coenzyme Q10 deficiency, primary, 4 (COQ10D4)
aarF domain containing kinase 4	ADCK4	ADCK4 'reenergizes' nephrotic syndrome. PMID: 24270414. Disease: nephrotic syndrome 9 (NPHS9)
aarF domain containing kinase 5	ADCK5	
Adrenergic, beta, receptor kinase 1	ADRBK1	Targeting cardiac β-adrenergic signaling via GRK2 inhibition for heart failure therapy. PMID: 24133451
Adrenergic, beta, receptor kinase 2	ADRBK2	GRK3 is essential for metastatic cells and promotes prostate tumor progression. PMID: 24434559
Activin A receptor, type I	ACVR1	**LDN 212854** – drugs affecting bone structure and mineralization. Disease: fibrodysplasia ossificans progressiva (FOP)
Activin A receptor, type IB	ACVR1B	Note: ACVRIB is abundantly expressed in systemic sclerosis patient fibroblasts and production of collagen is also induced by activin-A/INHBA. This suggests that the activin/ACRV1B signaling mechanism is involved in systemic sclerosis
Activin A receptor, type IC	ACVR1C	Roles of activin receptor-like kinase 7 signaling and its target, peroxisome proliferator-activated receptor γ, in lean and obese adipocytes. PMID: 24052900
Activin A receptor, type IIA	ACVR2A	**Sotatercept** – antianaemic preparations
Activin A receptor, type IIB	ACVR2B	**Bimagrumab** – sporadic inclusion body myositis treatments. Disease: heterotaxy, visceral, 4, autosomal (HTX4)
Activin A receptor type II-like 1	ACVRL1	ALK1 as an emerging target for antiangiogenic therapy of cancer. PMID: 21467543. Disease: telangiectasia, hereditary hemorrhagic, 2 (HHT2)
Alpha-kinase 1	ALPK1	Alpha-kinase 1, a new component in apical protein transport. PMID: 15883161

(Continued)

HGNC-approved name	Symbol	Target information
Alpha-kinase 2	ALPK2	ALPK2 is crucial for luminal apoptosis and DNA repair-related gene expression in a three-dimensional colonic-crypt model. PMID: 22641666
Alpha-kinase 3	ALPK3	Cardiomyopathy in α-kinase 3 (ALPK3)-deficient mice. PMID: 21441111
Aurora kinase A	AURKA	**Alisertib** – antineoplastic agents
Aurora kinase B	AURKB	**Barasertib** – antineoplastic agents. Note: Disruptive regulation of expression is a possible mechanism of the perturbation of chromosomal integrity in cancer cells through its dominant-negative effect on cytokinesis
Aurora kinase C	AURKC	Overexpression of active Aurora-C kinase results in cell transformation and tumour formation. PMID: 22046298. Disease: spermatogenic failure 5 (SPGF5)
Bone morphogenetic protein receptor, type IA	BMPR1A	DMH1, a small molecule inhibitor of BMP type i receptors, suppresses growth and invasion of lung cancer. PMID: 24603907. Disease: juvenile polyposis syndrome (JPS)
Bone morphogenetic protein receptor, type IB	BMPR1B	Endogenous BMPR-IB signaling is required for early osteoblast differentiation of human bone cells. PMID: 21136190. Disease: acromesomelic chondrodysplasia, with genital anomalies (AMDGA)
Bone morphogenetic protein receptor, type II (serine/threonine kinase)	BMPR2	Disease: pulmonary hypertension, primary, 1 (PPH1)
BR serine/threonine kinase 1	BRSK1	Decreased expression and prognostic role of cytoplasmic BRSK1 in human breast carcinoma: Correlation with Jab1 stability and PI3K/Akt pathway. PMID: 25036402
BR serine/threonine kinase 2	BRSK2	SAD-A kinase controls islet β-cell size and function as a mediator of mTORC1 signaling. PMID: 23922392
BUB1 mitotic checkpoint serine/threonine kinase	BUB1	Novel cycloalkenepyrazoles as inhibitors of bub1 kinase. PMID: 24900824
BUB1 mitotic checkpoint serine/threonine kinase B	BUB1B	Note: Defects in BUB1B are associated with tumor formation. Disease: premature chromatid separation trait (PCS)
Calcium/calmodulin-dependent protein kinase I	CAMK1	
Calcium/calmodulin-dependent protein kinase ID	CAMK1D	CAMK1D amplification implicated in epithelial-mesenchymal transition in basal-like breast cancer. PMID: 19383354
Calcium/calmodulin-dependent protein kinase IG	CAMK1G	CaMKII γ, a critical regulator of CML stem/progenitor cells, is a target of the natural product berbamine. PMID: 23074277
Calcium/calmodulin-dependent protein kinase II alpha	CAMK2A	αCaMKII controls the establishment of cocaine's reinforcing effects in mice and humans. PMID: 25290264
Calcium/calmodulin-dependent protein kinase II beta	CAMK2B	CaMKIIβ regulates oligodendrocyte maturation and CNS myelination. PMID: 23785157
Calcium/calmodulin-dependent protein kinase II gamma	CAMK2G	Ca(2+)/calmodulin-dependent protein kinase IIγ, a critical mediator of the NF-κB network, is a novel therapeutic target in non-small cell lung cancer. PMID: 24189456
Calcium/calmodulin-dependent protein kinase II delta	CAMK2D	The multifunctional Ca(2+)/calmodulin-dependent protein kinase II delta (CaMKIIδ) phosphorylates cardiac titin's spring elements. PMID: 23220127
Calcium/calmodulin-dependent protein kinase IV	CAMK4	KN-93, an inhibitor of calcium/calmodulin-dependent protein kinase IV, promotes generation and function of Foxp3(+) regulatory T cells in MRL/lpr mice. PMID: 24829059

HGNC-approved name	Symbol	Target information
Calcium/calmodulin-dependent protein kinase kinase 1, alpha	CAMKK1	Forebrain-specific constitutively active CaMKKα transgenic mice show deficits in hippocampus-dependent long-term memory. PMID: 21558011
Calcium/calmodulin-dependent protein kinase kinase 2, beta	CAMKK2	Calcium/calmodulin-dependent protein kinase kinase 2: Roles in signaling and pathophysiology. PMID: 22778263
Casein kinase 1, alpha 1	CSNK1A1	Role of casein kinase 1A1 in the biology and targeted therapy of del(5q) MDS. PMID: 25242043
Casein kinase 1, alpha 1-like	CSNK1A1L	
Casein kinase 1, gamma 1	CSNK1G1	Casein kinase 1γ1 inhibits the RIG-I/TLR signaling pathway through phosphorylating p65 and promoting its degradation. PMID: 24442433
Casein kinase 1, gamma 2	CSNK1G2	
Casein kinase 1, gamma 3	CSNK1G3	RNAi-based screening of the human kinome identifies Akt-cooperating kinases: A new approach to designing efficacious multitargeted kinase inhibitors. PMID: 16247451
Casein kinase 1, delta	CSNK1D	Disease: advanced sleep phase syndrome, familial, 2 (FASPS2)
Casein kinase 1, epsilon	CSNK1E	Acute inhibition of casein kinase 1δ/ε rapidly delays peripheral clock gene rhythms. PMID: 25245819
Casein kinase 2, alpha 1 polypeptide	CSNK2A1	**Silmitasertib** – antineoplastic agents
Casein kinase 2, alpha prime polypeptide	CSNK2A2	
Casein kinase 2, alpha 3 polypeptide	CSNK2A3	Functional polymorphism of the CK2alpha intronless gene plays oncogenic roles in lung cancer. PMID: 20625391
Casein kinase 2, beta polypeptide	CSNK2B	
CDC42 binding protein kinase alpha (DMPK-like)	CDC42BPA	Evidence for a role of MRCK in mediating HeLa cell elongation induced by the C1 domain ligand HMI-1a3. PMID: 24486483
CDC42 binding protein kinase beta (DMPK-like)	CDC42BPB	A novel small-molecule MRCK inhibitor blocks cancer cell invasion. PMID: 25288205
CDC42 binding protein kinase gamma (DMPK-like)	CDC42BPG	The actin-myosin regulatory MRCK kinases: Regulation, biological functions and associations with human cancer. PMID: 24553779
CDC-like kinase 1	CLK1	Human CDC2-like kinase 1 (CLK1): A novel target for Alzheimer's disease. PMID: 24568585
CDC-like kinase 2	CLK2	Cdc2-like kinase 2 is an insulin-regulated suppressor of hepatic gluconeogenesis. PMID: 20074525
CDC-like kinase 3	CLK3	
CDC-like kinase 4	CLK4	Specific CLK inhibitors from a novel chemotype for regulation of alternative splicing. PMID: 21276940
Checkpoint kinase 1	CHEK1	**RG7741** – antineoplastic agents
Checkpoint kinase 2	CHEK2	**Rabusertib** – antineoplastic agents. Disease: Li–Fraumeni syndrome 2 (LFS2)
Cyclin-dependent kinase 1	CDK1	**Dinaciclib** – antineoplastic agents
Cyclin-dependent kinase 2	CDK2	**Milciclib maleate** – antineoplastic agents
Cyclin-dependent kinase 3	CDK3	CDK3 expression and its clinical significance in human nasopharyngeal carcinoma. PMID: 24691537

(Continued)

HGNC-approved name	Symbol	Target information
Cyclin-dependent kinase 4	CDK4	**Palbociclib** – antineoplastic agents. Disease: melanoma, cutaneous malignant 3 (CMM3)
Cyclin-dependent kinase 5	CDK5	Cdk5 protein inhibition and Aβ42 increase BACE1 protein level in primary neurons by a post-transcriptional mechanism: Implications of CDK5 as a therapeutic target for Alzheimer disease. PMID: 22223639
Cyclin-dependent kinase 6	CDK6	**Palbociclib** – antineoplastic agents
Cyclin-dependent kinase 7	CDK7	Targeting transcription regulation in cancer with a covalent CDK7 inhibitor. PMID: 25043025
Cyclin-dependent kinase 8	CDK8	Revving the Throttle on an oncogene: CDK8 takes the driver seat. PMID: 19808961
Cyclin-dependent kinase 9	CDK9	**BAY 1143572** – antineoplastic agents. Note: Chronic activation of CDK9 causes cardiac myocyte enlargement leading to cardiac hypertrophy, and confers predisposition to heart failure
Cyclin-dependent kinase 10	CDK10	CDK10 functions as a tumor suppressor gene and regulates survivability of biliary tract cancer cells. PMID: 22209942
Cyclin-dependent kinase 11A	CDK11A	Cyclin-dependent kinase 11 (CDK11) is crucial in the growth of liposarcoma cells. PMID: 24007862
Cyclin-dependent kinase 11B	CDK11B	Analyses of CDC2L1 gene mutations in keloid tissue. PMID: 22188294
Cyclin-dependent kinase 12	CDK12	Note: Chromosomal aberrations involving CDK12 may be a cause of gastric cancer. Deletions within 17q12 region producing fusion transcripts with ERBB2, leading to CDK12-ERBB2 fusion leading to truncated CDK12 protein not in frame with ERBB2
Cyclin-dependent kinase 13	CDK13	Frequent amplification of CENPF, GMNN and CDK13 genes in hepatocellular carcinomas. PMID: 22912832
Cyclin-dependent kinase 14	CDK14	Overexpression of PFTK1 predicts resistance to chemotherapy in patients with oesophageal squamous cell carcinoma. PMID: 22333595
Cyclin-dependent kinase 15	CDK15	ALS2CR7 (CDK15) attenuates TRAIL induced apoptosis by inducing phosphorylation of survivin Thr34. PMID: 24866247
Cyclin-dependent kinase 16	CDK16	PCTAIRE1 phosphorylates p27 and regulates mitosis in cancer cells. PMID: 25203778
Cyclin-dependent kinase 17	CDK17	
Cyclin-dependent kinase 18	CDK18	PCTAIRE kinase 3/cyclin-dependent kinase 18 is activated through association with cyclin A and/or phosphorylation by protein kinase A. PMID: 24831015
Cyclin-dependent kinase 19	CDK19	
Cyclin-dependent kinase 20	CDK20	CCRK depletion inhibits glioblastoma cell proliferation in a cilium-dependent manner. PMID: 23743448
Cyclin-dependent kinase-like 1 (CDC2-related kinase)	CDKL1	A role for Cdkl1 in the development of gastric cancer. PMID: 22369697
Cyclin-dependent kinase-like 2 (CDC2-related kinase)	CDKL2	CDKL2 promotes epithelial-mesenchymal transition and breast cancer progression. PMID: 25333262
Cyclin-dependent kinase-like 3	CDKL3	Inactivition of CDKL3 mildly inhibits proliferation of cells at VZ/SVZ in brain. PMID: 25270654
Cyclin-dependent kinase-like 4	CDKL4	
Cyclin-dependent kinase-like 5	CDKL5	Note: Chromosomal aberrations involving CDKL5 are found in patients manifesting early-onset seizures and spams and psychomotor impairment. Translocation t(X;6)(p22.3;q14); translocation t(X;7)(p22.3;p15). Disease: epileptic encephalopathy, early infantile, 2 (EIEE2)

HGNC-approved name	Symbol	Target information
Death-associated protein kinase 1	DAPK1	Death-associated protein kinase 1 has a critical role in aberrant tau protein regulation and function. PMID: 24853415
Death-associated protein kinase 2	DAPK2	DAPK2 is a novel regulator of mTORC1 activity and autophagy. PMID: 25361081
Death-associated protein kinase 3	DAPK3	Fluorescence linked enzyme chemoproteomic strategy for discovery of a potent and selective DAPK1 and ZIPK inhibitor. PMID: 24070067
Doublecortin-like kinase 1	DCLK1	DCLK1 facilitates intestinal tumor growth via enhancing pluripotency and epithelial mesenchymal transition. PMID: 25211188
Doublecortin-like kinase 2	DCLK2	Mice lacking doublecortin and doublecortin-like kinase 2 display altered hippocampal neuronal maturation and spontaneous seizures. PMID: 19342486
Doublecortin-like kinase 3	DCLK3	
Endoplasmic reticulum to nucleus signaling 1	ERN1	**STF-083010** – antineoplastic agents
Endoplasmic reticulum to nucleus signaling 2	ERN2	The ER stress transducer IRE1β is required for airway epithelial mucin production. PMID: 23168839
Eukaryotic translation initiation factor 2-alpha kinase 1	EIF2AK1	Heme-regulated eIF2α kinase plays a crucial role in protecting erythroid cells against Pb-induced hemolytic stress. PMID: 25411909
Eukaryotic translation initiation factor 2-alpha kinase 2	EIF2AK2	Selective leukemia cell death by activation of the double-stranded RNA-dependent protein kinase PKR. PMID: 21468538
Eukaryotic translation initiation factor 2 alpha kinase 3	EIF2AK3	**GSK2656157** – antineoplastic agents Disease: Wolcott Rallison syndrome (WRS)
Eukaryotic translation initiation factor 2-alpha kinase 4	EIF2AK4	Suppression of eIF2α kinases alleviates Alzheimer's disease-related plasticity and memory deficits. PMID: 23933749. Disease: pulmonary venoocclusive disease 2, autosomal recessive (PVOD2)
Glycogen synthase kinase 3 alpha	GSK3A	Glycogen synthase kinase-3: A new therapeutic target in renal cell carcinoma. PMID: 19920820
Glycogen synthase kinase 3 beta	GSK3B	**LY2090314** – antineoplastic agents
G protein-coupled receptor kinase 1	GRK1	Disease: night blindness, congenital stationary, Oguchi type 2 (CSNBO2)
G protein-coupled receptor kinase 4	GRK4	Role of GRK4 in the regulation of arterial AT1 receptor in hypertension. PMID: 24218433
G protein-coupled receptor kinase 5	GRK5	Identification and characterization of amlexanox as a G protein-coupled receptor kinase 5 inhibitor. PMID: 25340299
G protein-coupled receptor kinase 6	GRK6	G protein-coupled receptor kinase 6 deficiency promotes angiogenesis, tumor progression, and metastasis. PMID: 23589623
G protein-coupled receptor kinase 7	GRK7	
Homeodomain-interacting protein kinase 1	HIPK1	HIPK1 drives p53 activation to limit colorectal cancer cell growth. PMID: 23676219
Homeodomain-interacting protein kinase 2	HIPK2	Synthesis and properties of a selective inhibitor of homeodomain-interacting protein kinase 2 (HIPK2). PMID: 24586573
Homeodomain-interacting protein kinase 3	HIPK3	Depletion of homeodomain-interacting protein kinase 3 impairs insulin secretion and glucose tolerance in mice. PMID: 22983607
Homeodomain-interacting protein kinase 4	HIPK4	Novel homeodomain-interacting protein kinase family member, HIPK4, phosphorylates human p53 at serine 9. PMID: 18022393

(Continued)

HGNC-approved name	Symbol	Target information
Inhibitor of kappa light polypeptide gene enhancer in B-cells, kinase beta	IKBKB	**Auranofin** – anti-inflammatory and antirheumatic products. Disease: immunodeficiency 15 (IMD15)
Inhibitor of kappa light polypeptide gene enhancer in B-cells, kinase epsilon	IKBKE	IκB kinase ε (IKKε): A therapeutic target in inflammation and cancer. PMID: 23333767
Interleukin-1 receptor-associated kinase 1	IRAK1	Targeting IRAK1 as a therapeutic approach for myelodysplastic syndrome. PMID: 23845443
Interleukin-1 receptor-associated kinase 2	IRAK2	A coding IRAK2 protein variant compromises Toll-like receptor (TLR) signaling and is associated with colorectal cancer survival. PMID: 24973222
Interleukin-1 receptor-associated kinase 3	IRAK3	Disease: asthma-related traits 5 (ASRT5)
Interleukin-1 receptor-associated kinase 4	IRAK4	Recent advances in the discovery of small molecule inhibitors of interleukin-1 receptor-associated kinase 4 (IRAK4) as a therapeutic target for inflammation and oncology disorders. PMID: 25479567. Disease: recurrent isolated invasive pneumococcal disease 1 (IPD1)
Large tumor suppressor kinase 1	LATS1	Expression of LATS1 contributes to good prognosis and can negatively regulate YAP oncoprotein in non-small-cell lung cancer. PMID: 24682895
Large tumor suppressor kinase 2	LATS2	Down-regulation of LATS2 in non-small cell lung cancer promoted the growth and motility of cancer cells. PMID: 25391426
Leucine-rich repeat kinase 1	LRRK1	Targeted disruption of leucine-rich repeat kinase 1 but not leucine-rich repeat kinase 2 in mice causes severe osteopetrosis. PMID: 23526378
Leucine-rich repeat kinase 2	LRRK2	Small molecule kinase inhibitors for LRRK2 and their application to Parkinson's disease models. PMID: 22860184. Disease: Parkinson disease 8 (PARK8)
LIM domain kinase 1	LIMK1	Damnacanthal, an effective inhibitor of LIM-kinase, inhibits cell migration and invasion. PMID: 24478456. Note: LIMK1 is located in the Williams–Beuren syndrome (WBS) critical region. WBS results from a hemizygous deletion of several genes on chromosome 7q
LIM domain kinase 2	LIMK2	LIM kinase-2 induces programmed necrotic neuronal death via dysfunction of DRP1-mediated mitochondrial fission. PMID: 24561342
MAP kinase-interacting serine/threonine kinase 1	MKNK1	First MNKs degrading agents block phosphorylation of eIF4E, induce apoptosis, inhibit cell growth, migration and invasion in triple negative and Her2-overexpressing breast cancer cell lines. PMID: 24504069
MAP kinase-interacting serine/threonine kinase 2	MKNK2	First MNKs degrading agents block phosphorylation of eIF4E, induce apoptosis, inhibit cell growth, migration and invasion in triple negative and Her2-overexpressing breast cancer cell lines. PMID: 24504069
MAP/microtubule affinity-regulating kinase 1	MARK1	Note: Genetic variations in MARK1 may be associated with susceptibility to autism. MARK1 is overexpressed in the prefrontal cortex of patients with autism and causes changes in the function of cortical dendrites
MAP/microtubule affinity-regulating kinase 2	MARK2	Elevated MARK2-dependent phosphorylation of Tau in Alzheimer's disease. PMID: 23001711
MAP/microtubule affinity-regulating kinase 3	MARK3	Loss of Par-1a/MARK3/C-TAK1 kinase leads to reduced adiposity, resistance to hepatic steatosis, and defective gluconeogenesis. PMID: 20733003
MAP/microtubule affinity-regulating kinase 4	MARK4	Inactivation of MARK4, an AMP-activated protein kinase (AMPK)-related kinase, leads to insulin hypersensitivity and resistance to diet-induced obesity. PMID: 22992738

HGNC-approved name	Symbol	Target information
Microtubule-associated serine/threonine kinase 1	MAST1	Functionally recurrent rearrangements of the MAST kinase and Notch gene families in breast cancer. PMID: 22101766
Microtubule-associated serine/threonine kinase 2	MAST2	Interference with the PTEN-MAST2 interaction by a viral protein leads to cellular relocalization of PTEN. PMID: 22894835
Microtubule-associated serine/threonine kinase 3	MAST3	MAST3: A novel IBD risk factor that modulates TLR4 signaling. PMID: 18650832
Microtubule-associated serine/threonine kinase family member 4	MAST4	
Microtubule-associated serine/threonine kinase-like	MASTL	Mastl kinase, a promising therapeutic target, promotes cancer recurrence. PMID: 25373736. Disease: thrombocytopenia 2 (THC2)
Mitogen-activated protein kinase 1	MAPK1	Slow inhibition and conformation selective properties of extracellular signal-regulated kinase 1 and 2 inhibitors. PMID: 25350931
Mitogen-activated protein kinase 3	MAPK3	ERK mutations confer resistance to mitogen-activated protein kinase pathway inhibitors. PMID: 25320010
Mitogen-activated protein kinase 4	MAPK4	
Mitogen-activated protein kinase 6	MAPK6	The non-classical MAP kinase ERK3 controls T cell activation. PMID: 24475167
Mitogen-activated protein kinase 7	MAPK7	Extracellular signal-regulated kinase 5: A potential therapeutic target for malignant mesotheliomas. PMID: 23446998
Mitogen-activated protein kinase 8	MAPK8	**Bentamapimod** – uterotonics
Mitogen-activated protein kinase 9	MAPK9	JNK2 regulates the functional plasticity of naturally occurring T regulatory cells and the enhancement of lung allergic responses. PMID: 25070841
Mitogen-activated protein kinase 10	MAPK10	JNK3 as a therapeutic target for neurodegenerative diseases. PMID: 21321401. Note: A chromosomal aberration involving MAPK10 has been found in a single patient with pharmacoresistant epileptic encephalopathy
Mitogen-activated protein kinase 11	MAPK11	**Ralimetinib** – antineoplastic agents
Mitogen-activated protein kinase 12	MAPK12	Note: MAPK is overexpressed in highly metastatic breast cancer cell lines and its expression is preferentially associated with basal-like and metastatic phenotypes of breast tumor samples
Mitogen-activated protein kinase 13	MAPK13	IL-13-induced airway mucus production is attenuated by MAPK13 inhibition. PMID: 23187130
Mitogen-activated protein kinase 14	MAPK14	**Ralimetinib** – antineoplastic agents
Mitogen-activated protein kinase 15	MAPK15	MAPK15/ERK8 stimulates autophagy by interacting with LC3 and GABARAP proteins. PMID: 22948227
Mitogen-activated protein kinase kinase 1	MAP2K1	**Pimasertib** – antineoplastic agents. Disease: cardiofaciocutaneous syndrome 3 (CFC3)
Mitogen-activated protein kinase kinase 2	MAP2K2	**Pimasertib** – antineoplastic agents. Disease: cardiofaciocutaneous syndrome 4 (CFC4)
Mitogen-activated protein kinase kinase 3	MAP2K3	Note: Defects in MAP2K3 may be involved in colon cancer
Mitogen-activated protein kinase kinase 4	MAP2K4	Mitogen-activated protein kinase kinase 4 (MAP2K4) promotes human prostate cancer metastasis. PMID: 25019290
Mitogen-activated protein kinase kinase 5	MAP2K5	Inhibition of MEK5 by BIX02188 induces apoptosis in cells expressing the oncogenic mutant FLT3-ITD. PMID: 21820407

(Continued)

HGNC-approved name	Symbol	Target information
Mitogen-activated protein kinase kinase 6	MAP2K6	Differential regulation of anti-inflammatory genes by p38 MAP kinase and MAP kinase kinase 6. PMID: 24855454
Mitogen-activated protein kinase kinase 7	MAP2K7	MKK7 mediates miR-493-dependent suppression of liver metastasis of colon cancer cells. PMID: 24533778
Mitogen-activated protein kinase kinase kinase 1, E3 ubiquitin protein ligase	MAP3K1	Disease: 46,XY sex reversal 6 (SRXY6)
Mitogen-activated protein kinase kinase kinase 2	MAP3K2	MEKK2 regulates focal adhesion stability and motility in invasive breast cancer cells. PMID: 24491810
Mitogen-activated protein kinase kinase kinase 3	MAP3K3	Mitogen activated protein kinase kinase kinase 3 (MAP3K3/MEKK3) overexpression is an early event in esophageal tumorigenesis and is a predictor of poor disease prognosis. PMID: 24383423
Mitogen-activated protein kinase kinase kinase 4	MAP3K4	Phosphorylation of Atg5 by the Gadd45β-MEKK4-p38 pathway inhibits autophagy. PMID: 23059785
Mitogen-activated protein kinase kinase kinase 5	MAP3K5	Apoptosis signal-regulating kinase 1 and cyclin D1 compose a positive feedback loop contributing to tumor growth in gastric cancer. PMID: 21187402
Mitogen-activated protein kinase kinase kinase 6	MAP3K6	Germline mutations in MAP3K6 are associated with familial gastric cancer. PMID: 25340522
Mitogen-activated protein kinase kinase kinase 7	MAP3K7	Targeting of TAK1 in inflammatory disorders and cancer. PMID: 22795313
Mitogen-activated protein kinase kinase kinase 8	MAP3K8	Cot/Tpl-2 protein kinase as a target for the treatment of inflammatory disease. PMID: 19689369
Mitogen-activated protein kinase kinase kinase 9	MAP3K9	**CEP-1347** – neuroprotective agents. Note: May play a role in esophageal cancer susceptibility and/or development
Mitogen-activated protein kinase kinase kinase 10	MAP3K10	**CEP-1347** – neuroprotective agents
Mitogen-activated protein kinase kinase kinase 11	MAP3K11	**CEP-1347** – neuroprotective agents
Mitogen-activated protein kinase kinase kinase 12	MAP3K12	Discovery of dual leucine zipper kinase (DLK, MAP3K12) inhibitors with activity in neurodegeneration models. PMID: 25341110
Mitogen-activated protein kinase kinase kinase 13	MAP3K13	
Mitogen-activated protein kinase kinase kinase 14	MAP3K14	The kinase NIK as a therapeutic target in multiple myeloma. PMID: 21204728
Mitogen-activated protein kinase kinase kinase 15	MAP3K15	
Mitogen-activated protein kinase kinase kinase 19	MAP3K19	
Mitogen-activated protein kinase kinase kinase kinase 1	MAP4K1	Hematopoietic progenitor kinase 1 (HPK1) is required for LFA-1-mediated neutrophil recruitment during the acute inflammatory response. PMID: 23460610
Mitogen-activated protein kinase kinase kinase kinase 2	MAP4K2	BAY61-3606 affects the viability of colon cancer cells in a genotype-directed manner. PMID: 22815993
Mitogen-activated protein kinase kinase kinase kinase 3	MAP4K3	The kinase GLK controls autoimmunity and NF-κB signaling by activating the kinase PKC-θ in T cells. PMID: 21983831
Mitogen-activated protein kinase kinase kinase kinase 4	MAP4K4	Silencing mitogen-activated protein 4 kinase 4 (MAP4K4) protects beta cells from tumor necrosis factor-alpha-induced decrease of IRS-2 and inhibition of glucose-stimulated insulin secretion. PMID: 19690174

HGNC-approved name	Symbol	Target information
Mitogen-activated protein kinase kinase kinase kinase 5	MAP4K5	
Mitogen-activated protein kinase-activated protein kinase 2	MAPKAPK2	**MMI-0100** – idiopathic pulmonary fibrosis treatments
Mitogen-activated protein kinase-activated protein kinase 3	MAPKAPK3	Organometallic titanocene-gold compounds as potential chemotherapeutics in renal cancer. Study of their protein kinase inhibitory properties. PMID: 25435644
Mitogen-activated protein kinase-activated protein kinase 5	MAPKAPK5	Structure and function of MK5/PRAK: The loner among the mitogen-activated protein kinase-activated protein kinases. PMID: 23729623
Myosin light chain kinase	MYLK	Nonmuscle myosin light chain kinase regulates murine asthmatic inflammation. PMID: 24428690. Disease: aortic aneurysm, familial thoracic 7 (AAT7)
Myosin light chain kinase 2	MYLK2	Disease: cardiomyopathy, familial hypertrophic (CMH)
Myosin light chain kinase 3	MYLK3	Regulation of calcium channels in smooth muscle: New insights into the role of myosin light chain kinase. PMID: 25483583
Myosin light chain kinase family, member 4	MYLK4	
p21 protein (Cdc42/Rac)-activated kinase 1	PAK1	P21-activated kinase 1 (PAK1) as a therapeutic target in BRAF wild-type melanoma. PMID: 23535073
p21 protein (Cdc42/Rac)-activated kinase 2	PAK2	Motility of select ovarian cancer cell lines: Effect of extra-cellular matrix proteins and the involvement of PAK2. PMID: 25050916
p21 protein (Cdc42/Rac)-activated kinase 3	PAK3	Disease: mental retardation, X-linked 30 (MRX30)
p21 protein (Cdc42/Rac)-activated kinase 4	PAK4	Small-molecule p21-activated kinase inhibitor PF-3758309 is a potent inhibitor of oncogenic signaling and tumor growth. PMID: 20439741
p21 protein (Cdc42/Rac)-activated kinase 6	PAK6	p21-activated kinase 6 (PAK6) inhibits prostate cancer growth via phosphorylation of androgen receptor and tumorigenic E3 ligase murine double minute-2 (Mdm2). PMID: 23132866
p21 protein (Cdc42/Rac)-activated kinase 7	PAK7	p21-activated kinase 7 is an oncogene in human osteosarcoma. PMID: 25052921
Phosphorylase kinase, alpha 1 (muscle)	PHKA1	Disease: glycogen storage disease 9D (GSD9D)
Phosphorylase kinase, alpha 2 (liver)	PHKA2	Disease: glycogen storage disease 9A (GSD9A)
Phosphorylase kinase, beta	PHKB	Disease: glycogen storage disease 9B (GSD9B)
Phosphorylase kinase, gamma 1 (muscle)	PHKG1	Identification of phosphorylase kinase as a novel therapeutic target through high-throughput screening for anti-angiogenesis compounds in zebrafish. PMID: 22179836
Phosphorylase kinase, gamma 2 (testis)	PHKG2	Disease: glycogen storage disease 9C (GSD9C)
Pim-1 oncogene	PIM1	A potential therapeutic target for FLT3-ITD AML: PIM1 kinase. PMID: 21802138
Pim-2 oncogene	PIM2	Antimyeloma activity of the sesquiterpene lactone cnicin: Impact on Pim-2 kinase as a novel therapeutic target. PMID: 22205266
Pim-3 oncogene	PIM3	Discovery of novel benzylidene-1,3-thiazolidine-2,4-diones as potent and selective inhibitors of the PIM-1, PIM-2, and PIM-3 protein kinases. PMID: 22727640
Polo-like kinase 1	PLK1	**Volasertib** – antineoplastic agents. Note: Defects in PLK1 are associated with some cancers, such as gastric, thyroid or B-cell lymphomas

(Continued)

HGNC-approved name	Symbol	Target information
Polo-like kinase 2	PLK2	Design and synthesis of highly selective, orally active polo-like kinase-2 (Plk-2) inhibitors. PMID: 23522834
Polo-like kinase 3	PLK3	Tenuigenin attenuates α-synuclein-induced cytotoxicity by down-regulating polo-like kinase 3. PMID: 23710708
Polo-like kinase 4	PLK4	Functional characterization of CFI-400945, a polo-like kinase 4 inhibitor, as a potential anticancer agent. PMID: 25043604
Polo-like kinase 5	PLK5	Plk5, a polo box domain-only protein with specific roles in neuron differentiation and glioblastoma suppression. PMID: 21245385
Protein kinase, AMP-activated, alpha 1 catalytic subunit	PRKAA1	**ETC-1002** – lipid-lowering agents
Protein kinase, AMP-activated, alpha 2 catalytic subunit	PRKAA2	AMPK in health and disease. PMID: 19584320
Pan-PKC inhibitor		**Sotrastaurin** – antipsoriatics
Protein kinase C, alpha	PRKCA	**Aprinocarsen** – antineoplastic agents
Protein kinase C, beta	PRKCB	Protein kinase Cβ is required for lupus development in Sle mice. PMID: 23280626
Protein kinase C, gamma	PRKCG	Disease: spinocerebellar ataxia 14 (SCA14)
Protein kinase C, delta	PRKCD	**Delcasertib** – reperfusion injury treatments. Disease: immunodeficiency, common variable, 9 (CVID9)
Protein kinase C, epsilon	PRKCE	Thienoquinolines as novel disruptors of the PKCε/RACK2 protein–protein interaction. PMID: 24712764
Protein kinase C, zeta	PRKCZ	Protein kinase Mζ-dependent maintenance of GluA2 at the synapse: A possible target for preventing or treating age-related memory decline? PMID: 23679685
Protein kinase C, eta	PRKCH	Disease: ischemic stroke (ISCHSTR)
Protein kinase C, theta	PRKCQ	Protein kinase C-theta inhibitors: A novel therapy for inflammatory disorders. PMID: 22830352
Protein kinase C, iota	PRKCI	Atypical protein kinase Cι as a human oncogene and therapeutic target. PMID: 24231509
Protein kinase, cAMP-dependent, catalytic, alpha	PRKACA	Detection of a recurrent DNAJB1-PRKACA chimeric transcript in fibrolamellar hepatocellular carcinoma. PMID: 24578576
Protein kinase, cAMP-dependent, catalytic, beta	PRKACB	PRKACB is downregulated in non-small cell lung cancer and exogenous PRKACB inhibits proliferation and invasion of LTEP-A2 cells. PMID: 23833645
Protein kinase, cAMP-dependent, catalytic, gamma	PRKACG	
Protein kinase, cGMP-dependent, type I	PRKG1	Protein kinase G type I alpha activity in human ovarian cancer cells significantly contributes to enhanced Src activation and DNA synthesis/cell proliferation. PMID: 20371672. Disease: aortic aneurysm, familial thoracic 8 (AAT8)
Protein kinase, cGMP-dependent, type II	PRKG2	Type II cGMP-dependent protein kinase inhibits ligand-induced activation of EGFR in gastric cancer cells. PMID: 24534906
Protein kinase D1	PRKD1	PKD1 negatively regulates cell invasion, migration and proliferation ability of human osteosarcoma. PMID: 22426824
Protein kinase D2	PRKD2	Protein kinase D2 is a novel regulator of glioblastoma growth and tumor formation. PMID: 21727210
Protein kinase D3	PRKD3	Elevated protein kinase D3 (PKD3) expression supports proliferation of triple-negative breast cancer cells and contributes to mTORC1-S6K1 pathway activation. PMID: 24337579

HGNC-approved name	Symbol	Target information
Protein kinase N1	PKN1	Protein kinase PKN1 represses Wnt/β-catenin signaling in human melanoma cells. PMID: 24114839
Protein kinase N2	PKN2	The Rho target PRK2 regulates apical junction formation in human bronchial epithelial cells. PMID: 20974804
Protein kinase N3	PKN3	**Atu027** – antineoplastic agents
Protein serine kinase H1	PSKH1	Identification of kinases regulating prostate cancer cell growth using an RNAi phenotypic screen. PMID: 22761715
Protein serine kinase H2	PSKH2	
Pyruvate dehydrogenase kinase, isozyme 1	PDK1	PDK1 inhibition is a novel therapeutic target in multiple myeloma. PMID: 23321518
Pyruvate dehydrogenase kinase, isozyme 2	PDK2	Dichloroacetate prevents restenosis in preclinical animal models of vessel injury. PMID: 24747400
Pyruvate dehydrogenase kinase, isozyme 3	PDK3	Disease: Charcot–Marie–Tooth disease, X-linked dominant, 6 (CMTX6)
Pyruvate dehydrogenase kinase, isozyme 4	PDK4	Metabolic and transcriptional profiling reveals pyruvate dehydrogenase kinase 4 as a mediator of epithelial-mesenchymal transition and drug resistance in tumor cells. PMID: 25379179
Receptor (TNFRSF)-interacting serine/threonine kinase 1	RIPK1	Intermediate domain of receptor-interacting protein kinase 1 (RIPK1) determines switch between necroptosis and RIPK1 kinase-dependent apoptosis. PMID: 22362767
Receptor-interacting serine/threonine kinase 2	RIPK2	In vivo inhibition of RIPK2 kinase alleviates inflammatory disease. PMID: 25213858
Receptor-interacting serine/threonine kinase 3	RIPK3	Activity of protein kinase RIPK3 determines whether cells die by necroptosis or apoptosis. PMID: 24557836
Receptor-interacting serine/threonine kinase 4	RIPK4	Phosphorylation of dishevelled by protein kinase RIPK4 regulates Wnt signaling. PMID: 23371553. Disease: popliteal pterygium syndrome, lethal type (PPS-L)
Rho-associated, coiled-coil-containing protein kinase 1	ROCK1	ROCK1 as a potential therapeutic target in osteosarcoma. PMID: 21387396
Rho-associated, coiled-coil-containing protein kinase 2	ROCK2	**Fasudil** – antihypertensives
Ribosomal protein S6 kinase, 90 kDa, polypeptide 1	RPS6KA1	RSK promotes prostate cancer progression in bone through ING3, CKAP2 and PTK6-mediated cell survival. PMID: 25189355
Ribosomal protein S6 kinase, 90 kDa, polypeptide 2	RPS6KA2	Synthetic lethality screen identifies RPS6KA2 as modifier of epidermal growth factor receptor activity in pancreatic cancer. PMID: 24403857
Ribosomal protein S6 kinase, 90 kDa, polypeptide 3	RPS6KA3	RSK2(Ser227) at N-terminal kinase domain is a potential therapeutic target for multiple myeloma. PMID: 23012246. Disease: Coffin–Lowry syndrome (CLS)
Ribosomal protein S6 kinase, 90 kDa, polypeptide 4	RPS6KA4	The versatile role of MSKs in transcriptional regulation. PMID: 19464896
Ribosomal protein S6 kinase, 90 kDa, polypeptide 5	RPS6KA5	Phosphorylation of mitogen- and stress-activated protein kinase-1 in astrocytic inflammation: A possible role in inhibiting production of inflammatory cytokines. PMID: 24349124
Ribosomal protein S6 kinase, 90 kDa, polypeptide 6	RPS6KA6	Ribosomal s6 protein kinase 4: A prognostic factor for renal cell carcinoma. PMID: 23942078

(Continued)

HGNC-approved name	Symbol	Target information
Ribosomal protein S6 kinase, 70 kDa, polypeptide 1	RPS6KB1	**AT13148** – antineoplastic agents
Ribosomal protein S6 kinase, 70 kDa, polypeptide 2	RPS6KB2	Inhibition of p70S6K2 down-regulates Hedgehog/GLI pathway in non-small cell lung cancer cell lines. PMID: 19575820
Ribosomal protein S6 kinase, 52 kDa, polypeptide 1	RPS6KC1	Growth inhibition by bupivacaine is associated with inactivation of ribosomal protein S6 kinase 1. PMID: 24605337
Ribosomal protein S6 kinase-like 1	RPS6KL1	Identification of candidate oncogenes in human colorectal cancers with microsatellite instability. PMID: 23684749
RIO kinase 1	RIOK1	The kinase activity of human Rio1 is required for final steps of cytoplasmic maturation of 40S subunits. PMID: 22072790
RIO kinase 2	RIOK2	A kinome-wide RNAi screen in Drosophila Glia reveals that the RIO kinases mediate cell proliferation and survival through TORC2-Akt signaling in glioblastoma. PMID: 23459592
RIO kinase 3	RIOK3	Hypoxic regulation of RIOK3 is a major mechanism for cancer cell invasion and metastasis. PMID: 25486436
Salt-inducible kinase 1	SIK1	Note: Defects in SIK1 may be associated with some cancers, such as breast cancers. Loss of SIK1 correlates with poor patient outcome in breast cancers. PMID: 19622832
Salt-inducible kinase 2	SIK2	SIK2 is critical in the regulation of lipid homeostasis and adipogenesis in vivo. PMID: 24898145
SIK family kinase 3	SIK3	Salt-inducible kinase 3 is a novel mitotic regulator and a target for enhancing antimitotic therapeutic-mediated cell death. PMID: 24743732
Serine/threonine kinase 3	STK3	The mammalian Ste20-like kinase 2 (Mst2) modulates stress-induced cardiac hypertrophy. PMID: 25035424
Serine/threonine kinase 4	STK4	Toward the development of a potent and selective organoruthenium mammalian sterile 20 kinase inhibitor. PMID: 19226137. Disease: T-cell immunodeficiency, recurrent infections, and autoimmunity with or without cardiac malformations (TIIAC)
Serine/threonine kinase 10	STK10	Disease: testicular germ cell tumor (TGCT)
Serine/threonine kinase 11	STK11	Disease: Peutz–Jeghers syndrome (PJS)
Serine/threonine kinase 16	STK16	Nucleocytoplasmic shuttling of STK16 (PKL12), a Golgi-resident serine/threonine kinase involved in VEGF expression regulation. PMID: 16310770
Serine/threonine kinase 17a	STK17A	Serine/threonine kinase 17A is a novel candidate for therapeutic targeting in glioblastoma. PMID: 24312360
Serine/threonine kinase 17b	STK17B	A novel DRAK inhibitor, SC82510, promotes axon branching of adult sensory neurons in vitro. PMID: 24407843
Serine/threonine kinase 19	STK19	
Serine/threonine kinase 24	STK24	MST3 kinase phosphorylates TAO1/2 to enable myosin va function in promoting spine synapse development. PMID: 25456499
Serine/threonine kinase 25	STK25	STK25 protein mediates TrkA and CCM2 protein-dependent death in pediatric tumor cells of neural origin. PMID: 22782892
Serine/threonine protein kinase 26	STK26	MST4 promotes hepatocellular carcinoma epithelial-mesenchymal transition and metastasis via activation of the p-ERK pathway. PMID: 24859810
Serine/threonine kinase 31	STK31	STK31 is a cell-cycle regulated protein that contributes to the tumorigenicity of epithelial cancer cells. PMID: 24667656
Serine/threonine kinase 32A	STK32A	

HGNC-approved name	Symbol	Target information
Serine/threonine kinase 32B	STK32B	
Serine/threonine kinase 32C	STK32C	
Serine/threonine kinase 33	STK33	STK33 promotes hepatocellular carcinoma through binding to c-Myc. PMID: 25398772
Serine/threonine kinase 35	STK35	Clik1: A novel kinase targeted to actin stress fibers by the CLP-36 PDZ-LIM protein. PMID: 11973348
Serine/threonine kinase 36	STK36	Fused has evolved divergent roles in vertebrate Hedgehog signalling and motile ciliogenesis. PMID: 19305393
Serine/threonine kinase 38	STK38	STK38 is a critical upstream regulator of MYC's oncogenic activity in human B-cell lymphoma. PMID: 23178486
Serine/threonine kinase 38-like	STK38L	The serine/threonine kinase Ndr2 controls integrin trafficking and integrin-dependent neurite growth. PMID: 24719112
Serine threonine kinase 39	STK39	Discovery of novel SPAK inhibitors that block WNK kinase signaling to cation chloride transporters. PMID: 25377078
Serine/threonine kinase 40	STK40	Deletion of STK40 protein in mice causes respiratory failure and death at birth. PMID: 23293024
Serum/glucocorticoid-regulated kinase 1	SGK1	**EMD638683** – antineoplastic agents
Serum/glucocorticoid-regulated kinase 2	SGK2	Serum- and glucocorticoid-regulated kinase 2 determines drug-activated pregnane X receptor to induce gluconeogenesis in human liver cells. PMID: 24204015
Serum/glucocorticoid-regulated kinase family, member 3	SGK3	Serum and glucocorticoid kinase 3 at 8q13.1 promotes cell proliferation and survival in hepatocellular carcinoma. PMID: 22262416
SH3 domain binding kinase 1	SBK1	Human SBK1 is dysregulated in multiple cancers and promotes survival of ovary cancer SK-OV-3 cells. PMID: 21104019
SH3 domain binding kinase family, member 2	SBK2	
SH3 domain binding kinase family, member 3	SBK3	
SRSF protein kinase 1	SRPK1	SRPK1 contributes to malignancy of hepatocellular carcinoma through a possible mechanism involving PI3K/Akt. PMID: 23644876
SRSF protein kinase 2	SRPK2	SRPK2 phosphorylates tau and mediates the cognitive defects in Alzheimer's disease. PMID: 23197718
SRSF protein kinase 3	SRPK3	
TAO kinase 1	TAOK1	A functional genomic screen identifies a role for TAO1 kinase in spindle-checkpoint signalling. PMID: 17417629
TAO kinase 2	TAOK2	Taok2 controls behavioral response to ethanol in mice. PMID: 22883308
TAO kinase 3	TAOK3	JNK pathway activation is controlled by Tao/TAOK3 to modulate ethanol sensitivity. PMID: 23227189
Tau tubulin kinase 1	TTBK1	The tau tubulin kinases TTBK1/2 promote accumulation of pathological TDP-43. PMID: 25473830
Tau tubulin kinase 2	TTBK2	Disease: spinocerebellar ataxia 11 (SCA11)
Testis-specific kinase 1	TESK1	Spatiotemporal expression of testicular protein kinase 1 after rat sciatic nerve injury. PMID: 22302232

(Continued)

HGNC-approved name	Symbol	Target information
Testis-specific kinase 2	TESK2	
Testis-specific serine kinase 1B	TSSK1B	Validation of a testis specific serine/threonine kinase (TSSK) family and the substrate of TSSK1 and 2, TSKS, as contraceptive targets. PMID: 17566264
Testis-specific serine kinase 2	TSSK2	Validation of a testis specific serine/threonine kinase (TSSK) family and the substrate of TSSK1 and 2, TSKS, as contraceptive targets. PMID: 17566264
Testis-specific serine kinase 3	TSSK3	
Testis-specific serine kinase 4	TSSK4	Tssk4 is essential for maintaining the structural integrity of sperm flagellum. PMID: 25361759
Testis-specific serine kinase 6	TSSK6	Tssk6 is required for Izumo relocalization and gamete fusion in the mouse. PMID: 19596796
Tousled-like kinase 1	TLK1	Silencing of Tousled-like kinase 1 sensitizes cholangiocarcinoma cells to cisplatin-induced apoptosis. PMID: 20381954
Tousled-like kinase 2	TLK2	Purification, crystallization and preliminary X-ray diffraction analysis of the kinase domain of human tousled-like kinase 2. PMID: 24598926
Transforming growth factor, beta receptor 1	TGFBR1	**LY2157299** – antineoplastic agents. Disease: Loeys–Dietz syndrome 1A (LDS1A)
Transforming growth factor, beta receptor II (70/80 kDa)	TGFBR2	**Avotermin** – dermatologicals. Disease: hereditary non-polyposis colorectal cancer 6 (HNPCC6)
unc-51-like autophagy-activating kinase 1	ULK1	UNC51-like kinase 1, autophagic regulator and cancer therapeutic target. PMID: 25327638
unc-51-like autophagy-activating kinase 2	ULK2	Distinct functions of Ulk1 and Ulk2 in the regulation of lipid metabolism in adipocytes. PMID: 24135897
unc-51-like kinase 3	ULK3	Protein kinase inhibitor SU6668 attenuates positive regulation of Gli proteins in cancer and multipotent progenitor cells. PMID: 24418624
unc-51-like kinase 4	ULK4	Recurrent deletions of ULK4 in schizophrenia: A gene crucial for neuritogenesis and neuronal motility. PMID: 24284070
Vaccinia-related kinase 1	VRK1	Disease: pontocerebellar hypoplasia 1A (PCH1A)
Vaccinia-related kinase 2	VRK2	Vaccinia-related kinase 2 mediates accumulation of polyglutamine aggregates via negative regulation of the chaperonin TRiC. PMID: 24298020
Vaccinia-related kinase 3	VRK3	Role of bivalent cations in structural stabilities of new drug targets Vaccinia-related kinases (VRK) from molecular dynamics simulations. PMID: 23082977
v-akt murine thymoma viral oncogene homolog 1	AKT1	**Afuresertib** – antineoplastic agents. Disease: breast cancer (BC)
v-akt murine thymoma viral oncogene homolog 2	AKT2	Note: Defects in AKT2 are a cause of susceptibility to breast cancer (BC). AKT2 promotes metastasis of tumor cells without affecting the latency of tumor development. With AKT3, plays also a pivotal role in the biology of glioblastoma. Disease: diabetes mellitus (NIDDM)
v-akt murine thymoma viral oncogene homolog 3	AKT3	Note: AKT3 is a key modulator of several tumors like melanoma, glioma and ovarian cancer. Active AKT3 increases progressively during melanoma tumor progression with highest levels present in advanced-stage metastatic melanomas
WNK lysine-deficient protein kinase 1	WNK1	Chloride sensing by WNK1 involves inhibition of autophosphorylation. PMID: 24803536. Disease: pseudohypoaldosteronism 2C (PHA2C)
WNK lysine-deficient protein kinase 2	WNK2	WNK2 kinase is a novel regulator of essential neuronal cation-chloride cotransporters. PMID: 21733846

HGNC-approved name	Symbol	Target information
WNK lysine-deficient protein kinase 3	WNK3	WNK3-SPAK interaction is required for the modulation of NCC and other members of the SLC12 family. PMID: 22415098
WNK lysine-deficient protein kinase 4	WNK4	Serine-threonine kinase with-no-lysine 4 (WNK4) controls blood pressure via transient receptor potential canonical 3 (TRPC3) in the vasculature. PMID: 21670282. Disease: pseudohypoaldosteronism 2B (PHA2B)
Ankyrin repeat and kinase domain containing 1	ANKK1	Updated findings of the association and functional studies of DRD2/ANKK1 variants with addictions. PMID: 25139281
Anti-Mullerian hormone receptor, type II	AMHR2	Disease: persistent Mullerian duct syndrome 2 (PMDS2)
AP2-associated kinase 1	AAK1	AAK1 identified as an inhibitor of neuregulin-1/ErbB4-dependent neurotrophic factor signaling using integrative chemical genomics and proteomics. PMID: 21802010
Ataxia telangiectasia and Rad3 related	ATR	Alternative lengthening of telomeres renders cancer cells hypersensitive to ATR inhibitors. PMID: 25593184. Disease: Seckel syndrome 1 (SCKL1)
Ataxia telangiectasia mutated	ATM	Cytoplasmic ATM protein kinase: An emerging therapeutic target for diabetes, cancer and neuronal degeneration. PMID: 21315178. Disease: ataxia telangiectasia (AT)
BMP2-inducible kinase	BMP2K	
Branched-chain ketoacid dehydrogenase kinase	BCKDK	Disease: branched-chain ketoacid dehydrogenase kinase deficiency (BCKDKD)
Calcium/calmodulin-dependent serine protein kinase (MAGUK family)	CASK	Disease: mental retardation and microcephaly with pontine and cerebellar hypoplasia (MICPCH)
Cell division cycle 7	CDC7	Targeting DNA replication before it starts: Cdc7 as a therapeutic target in p53-mutant breast cancers. PMID: 20724597
Citron (rho-interacting, serine/threonine kinase 21)	CIT	Negative control of keratinocyte differentiation by Rho/CRIK signaling coupled with up-regulation of KyoT1/2 (FHL1) expression. PMID: 16061799
Conserved helix–loop–helix ubiquitous kinase	CHUK	Disease: cocoon syndrome (COCOS)
Cyclin G-associated kinase	GAK	Cyclin-G-associated kinase modifies α-synuclein expression levels and toxicity in Parkinson's disease: Results from the GenePD Study. PMID: 21258085
Dystrophia myotonica-protein kinase	DMPK	Disease: dystrophia myotonica 1 (DM1)
Eukaryotic elongation factor 2 kinase	EEF2K	Eukaryotic elongation factor-2 kinase (eEF2K): A potential therapeutic target in cancer. PMID: 25023961
Fas-activated serine/threonine kinase	FASTK	Fas-activated serine/threonine phosphoprotein (FAST) is a regulator of alternative splicing. PMID: 17592127
Germ cell-associated 2 (haspin)	GSG2	Structure-activity relationship study of acridine analogs as haspin and DYRK2 kinase inhibitors. PMID: 20836251
Hormonally up-regulated Neu-associated kinase	HUNK	Hunk negatively regulates c-myc to promote Akt-mediated cell survival and mammary tumorigenesis induced by loss of Pten. PMID: 23520049
Integrin-linked kinase	ILK	Integrin-linked kinase silencing induces a S/G2/M phases cell cycle slowing and modulates metastasis-related genes in SGC7901 human gastric carcinoma cells. PMID: 23748822
Intestinal cell (MAK-like) kinase	ICK	Disease: endocrine-cerebroosteodysplasia (ECO)
Kalirin, RhoGEF kinase	KALRN	Kalirin signaling: Implications for synaptic pathology. PMID: 22194219. Disease: coronary heart disease 5 (CHDS5)

(Continued)

HGNC-approved name	Symbol	Target information
Male germ cell-associated kinase	MAK	Disease: retinitis pigmentosa 62 (RP62)
Maternal embryonic leucine zipper kinase	MELK	Maternal embryonic leucine zipper kinase: Key kinase for stem cell phenotype in glioma and other cancers. PMID: 24795222. Note: Defects in MELK are associated with some cancers, such as brain or breast cancers
Mechanistic target of rapamycin (serine/threonine kinase)	MTOR	**CC-223** – antineoplastic agents
Misshapen-like kinase 1	MINK1	Misshapen-like kinase 1 (MINK1) is a novel component of striatin-interacting phosphatase and kinase (STRIPAK) and is required for the completion of cytokinesis. PMID: 22665485
MOK protein kinase	MOK	Cdx2 homeodomain protein regulates the expression of MOK, a member of the mitogen-activated protein kinase superfamily, in the intestinal epithelial cells. PMID: 15327990
Nemo-like kinase	NLK	Clinical and biological significance of nemo-like kinase expression in glioma. PMID: 21177110
Nik-related kinase	NRK	Nrk, an X-linked protein kinase in the germinal center kinase family, is required for placental development and fetoplacental induction of labor. PMID: 21715335
NIM1 serine/threonine protein kinase	NIM1K	
Obscurin, cytoskeletal calmodulin and titin-interacting RhoGEF	OBSCN	Note: A chromosomal aberration involving OBSCN has been found in Wilms tumor. Translocation t(17)(q42p15) with PTHB1
Oxidative stress responsive 1	OXSR1	Regulation of OSR1 and the sodium, potassium, two chloride cotransporter by convergent signals. PMID: 24191005
PAS domain containing serine/threonine kinase	PASK	PAS kinase drives lipogenesis through SREBP-1 maturation. PMID: 25001282
PDLIM1-interacting kinase 1-like	PDIK1L	
PDZ-binding kinase	PBK	Novel TOPK inhibitor HI-TOPK-032 effectively suppresses colon cancer growth. PMID: 22523035
Pregnancy-up-regulated non-ubiquitous CaM kinase	PNCK	Increased expression of pregnancy up-regulated non-ubiquitous calmodulin kinase is associated with poor prognosis in clear cell renal cell carcinoma. PMID: 23634203
Pre-mRNA processing factor 4B	PRPF4B	Evaluation of cancer dependence and druggability of PRP4 kinase using cellular, biochemical, and structural approaches. PMID: 24003220
Protein kinase, DNA-activated, catalytic polypeptide	PRKDC	DNA-PK: A candidate driver of hepatocarcinogenesis and tissue biomarker that predicts response to treatment and survival. PMID: 25480831
Protein kinase, membrane-associated tyrosine/threonine 1	PKMYT1	A fluorescence anisotropy-based Myt1 kinase binding assay. PMID: 24229357
Protein kinase, X-linked	PRKX	Note: A chromosomal aberration involving PRKX is a cause of sex reversal disorder. Translocation t(XY)(p22p11) with PRKY
PTEN-induced putative kinase 1	PINK1	Parallel high-throughput RNA interference screens identify PINK1 as a potential therapeutic target for the treatment of DNA mismatch repair-deficient cancers. PMID: 21242281. Disease: Parkinson disease 6 (PARK6)
PX domain containing serine/threonine kinase	PXK	Characterization of PXK as a protein involved in epidermal growth factor receptor trafficking. PMID: 20086096
SMG1 phosphatidylinositol 3-kinase-related kinase	SMG1	SMG1 acts as a novel potential tumor suppressor with epigenetic inactivation in acute myeloid leukemia. PMID: 25257528
SNF-related kinase	SNRK	Identification of sucrose non-fermenting-related kinase (SNRK) as a suppressor of adipocyte inflammation. PMID: 23520131

HGNC-approved name	Symbol	Target information
SPEG complex locus	SPEG	SPEG interacts with myotubularin, and its deficiency causes centronuclear myopathy with dilated cardiomyopathy. PMID: 25087613
STE20-like kinase	SLK	Ste20-like kinase SLK, at the crossroads: A matter of life and death. PMID: 23154402
TANK-binding kinase 1	TBK1	The non-canonical IκB kinases IKKε and TBK1 as potential targets for the development of novel therapeutic drugs. PMID: 23157677. Disease: glaucoma 1, open angle, P (GLC1P)
TBC1 domain containing kinase	TBCK	TBCK influences cell proliferation, cell size and mTOR signaling pathway. PMID: 23977024
Titin	TTN	Disease: hereditary myopathy with early respiratory failure (HMERF)
TNNI3-interacting kinase	TNNI3K	Inhibition of the cardiomyocyte-specific kinase TNNI3K limits oxidative stress, injury, and adverse remodeling in the ischemic heart. PMID: 24132636
TP53-regulating kinase	TP53RK	A chemosensitization screen identifies TP53RK, a kinase that restrains apoptosis after mitotic stress. PMID: 20647325
TRAF2- and NCK-interacting kinase	TNIK	A novel aminothiazole KY-05009 with potential to inhibit Traf2- and Nck-interacting kinase (TNIK) attenuates TGF-β1-mediated epithelial-to-mesenchymal transition in human lung adenocarcinoma A549 cells. PMID: 25337707
TTK protein kinase	TTK	TTK/hMPS1 is an attractive therapeutic target for triple-negative breast cancer. PMID: 23700430
U2AF homology motif (UHM) kinase 1	UHMK1	The protein kinase KIS impacts gene expression during development and fear conditioning in adult mice. PMID: 22937132
v-mos Moloney murine sarcoma viral oncogene homolog	MOS	Mos in the oocyte: How to use MAPK independently of growth factors and transcription to control meiotic divisions. PMID: 21637374
v-raf murine sarcoma 3611 viral oncogene homolog	ARAF	Dimerization of the kinase ARAF promotes MAPK pathway activation and cell migration. PMID: 25097033
v-raf murine sarcoma viral oncogene homolog B	BRAF	**Vemurafenib** – antineoplastic agents. Note: Defects in BRAF are found in a wide range of cancers. Disease: colorectal cancer (CRC)
v-raf-1 murine leukemia viral oncogene homolog 1	RAF1	**Sorafenib** – antineoplastic agents. Disease: Noonan syndrome 5 (NS5)
NIMA-related kinase 1	NFK1	Increased Nek1 expression in renal cell carcinoma cells is associated with decreased sensitivity to DNA-damaging treatment. PMID: 24970796. Disease: short-rib thoracic dysplasia 6 with or without polydactyly (SRTD6)
NIMA-related kinase 2	NEK2	Nek2A contributes to tumorigenic growth and possibly functions as potential therapeutic target for human breast cancer. PMID: 22234886. Disease: retinitis pigmentosa 67 (RP67)
NIMA-related kinase 3	NEK3	The NIMA-family kinase Nek3 regulates microtubule acetylation in neurons. PMID: 19509051
NIMA-related kinase 4	NEK4	Nek4 regulates entry into replicative senescence and the response to DNA damage in human fibroblasts. PMID: 22851694
NIMA-related kinase 5	NEK5	Nek5, a novel substrate for caspase-3, promotes skeletal muscle differentiation by up-regulating caspase activity. PMID: 23727203
NIMA-related kinase 6	NEK6	Nek6 mediates human cancer cell transformation and is a potential cancer therapeutic target. PMID: 20407017
NIMA-related kinase 7	NEK7	Nek7 kinase accelerates microtubule dynamic instability. PMID: 23313050

(Continued)

HGNC-approved name	Symbol	Target information
NIMA-related kinase 8	NEK8	Disease: nephronophthisis 9 (NPHP9)
NIMA-related kinase 9	NEK9	NEK9 depletion induces catastrophic mitosis by impairment of mitotic checkpoint control and spindle dynamics. PMID: 23665325
NIMA-related kinase 10	NEK10	Nek10 mediates G2/M cell cycle arrest and MEK autoactivation in response to UV irradiation. PMID: 20956560
NIMA-related kinase 11	NEK11	Downregulation of NEK11 is associated with drug resistance in ovarian cancer. PMID: 24969318
Dual-specificity kinases		
Dual-specificity tyrosine-(Y)-phosphorylation-regulated kinase 1A	DYRK1A	Harmine is an ATP-competitive inhibitor for dual-specificity tyrosine phosphorylation-regulated kinase 1A (Dyrk1A). PMID: 21185805. Disease: mental retardation, autosomal dominant 7 (MRD7)
Dual-specificity tyrosine-(Y)-phosphorylation-regulated kinase 1B	DYRK1B	The kinase Mirk is a potential therapeutic target in osteosarcoma. PMID: 20042639
Dual-specificity tyrosine-(Y)-phosphorylation-regulated kinase 2	DYRK2	Structure-activity relationship study of acridine analogs as haspin and DYRK2 kinase inhibitors. PMID: 20836251
Dual-specificity tyrosine-(Y)-phosphorylation-regulated kinase 3	DYRK3	Dual specificity kinase DYRK3 couples stress granule condensation/dissolution to mTORC1 signaling. PMID: 23415227
Dual-specificity tyrosine-(Y)-phosphorylation-regulated kinase 4	DYRK4	The expression of the testis-specific Dyrk4 kinase is highly restricted to step 8 spermatids but is not required for male fertility in mice. PMID: 17292540
Dual serine/threonine and tyrosine protein kinase	DSTYK	Mutations in DSTYK and dominant urinary tract malformations. PMID: 23862974
Lipid kinases		
Acylglycerol kinase	AGK	Acylglycerol kinase augments JAK2/STAT3 signaling in esophageal squamous cells. PMID: 23676499. Disease: mitochondrial DNA depletion syndrome 10 (MTDPS10)
Diacylglycerol kinase, alpha 80 kDa	DGKA	Diacylglycerol kinase α is a critical signaling node and novel therapeutic target in glioblastoma and other cancers. PMID: 23558954
Diacylglycerol kinase, beta 90 kDa	DGKB	Diacylglycerol kinase β in neurons: Functional implications at the synapse and in disease. PMID: 22781745
Diacylglycerol kinase, gamma 90 kDa	DGKG	
Diacylglycerol kinase, delta 130 kDa	DGKD	Downregulation of diacylglycerol kinase delta contributes to hyperglycemia-induced insulin resistance. PMID: 18267070
Diacylglycerol kinase, epsilon 64 kDa	DGKE	Disease: nephrotic syndrome 7 (NPHS7)
Diacylglycerol kinase, zeta	DGKZ	Diacylglycerol kinase ζ limits B cell antigen receptor-dependent activation of ERK signaling to inhibit early antibody responses. PMID: 24129701
Diacylglycerol kinase, eta	DGKH	Overexpression of diacylglycerol kinase η enhances Gαq-coupled G protein-coupled receptor signaling. PMID: 24608858
Diacylglycerol kinase, theta 110 kDa	DGKQ	Diacylglycerol kinase θ: Regulation and stability. PMID: 23266086
Diacylglycerol kinase, iota	DGKI	Nuclear diacylglycerol kinases: Regulation and roles. PMID: 17981572
Diacylglycerol kinase, kappa	DGKK	Identification and characterization of a novel human type II diacylglycerol kinase, DGK kappa. PMID: 16210324
Diphosphoinositol pentakisphosphate kinase 1	PPIP5K1	PPIP5K1 modulates ligand competition between diphosphoinositol polyphosphates and PtdIns(3,4,5)P3 for polyphosphoinositide-binding domains. PMID: 23682967
Diphosphoinositol pentakisphosphate kinase 2	PPIP5K2	Synthetic inositol phosphate analogs reveal that PPIP5K2 has a surface-mounted substrate capture site that is a target for drug discovery. PMID: 24768307

HGNC-approved name	Symbol	Target information
Inositol-trisphosphate 3-kinase A	ITPKA	Identification of a new membrane-permeable inhibitor against inositol-1,4,5-trisphosphate-3-kinase A. PMID: 23981806
Inositol-trisphosphate 3-kinase B	ITPKB	Inositol trisphosphate 3-kinase B is increased in human Alzheimer brain and exacerbates mouse Alzheimer pathology. PMID: 24401760
Inositol-trisphosphate 3-kinase C	ITPKC	Disease: Kawasaki disease (KWD)
Inositol-tetrakisphosphate 1-kinase	ITPK1	Human ITPK1: A reversible inositol phosphate kinase/phosphatase that links receptor-dependent phospholipase C to Ca2+-activated chloride channels. PMID: 18272466
Inositol hexakisphosphate kinase 1	IP6K1	Inositol hexakisphosphate kinase-1 mediates assembly/disassembly of the CRL4-signalosome complex to regulate DNA repair and cell death. PMID: 25349427
Inositol hexakisphosphate kinase 2	IP6K2	p53-Mediated apoptosis requires inositol hexakisphosphate kinase-2. PMID: 21078964
Inositol hexakisphosphate kinase 3	IP6K3	
Pan-PI3kinase inhibitor		**GSK1059615** – antineoplastic agents
Phosphatidylinositol-4,5-bisphosphate 3-kinase, catalytic subunit alpha	PIK3CA	**GSK2636771** – antineoplastic agents. Disease: colorectal cancer (CRC)
Phosphatidylinositol-4,5-bisphosphate 3-kinase, catalytic subunit beta	PIK3CB	Discovery of new aminopyrimidine-based phosphoinositide 3-kinase beta (PI3Kβ) inhibitors with selectivity over PI3Kα. PMID: 22030027
Phosphatidylinositol-4,5-bisphosphate 3-kinase, catalytic subunit gamma	PIK3CG	PI3Kγ in hypertension: A novel therapeutic target controlling vascular myogenic tone and target organ damage. PMID: 22610309
Phosphatidylinositol-4,5-bisphosphate 3-kinase, catalytic subunit delta	PIK3CD	**Idelalisib** – antineoplastic agents. Disease: activated PI3K-delta syndrome (APDS)
Phosphatidylinositol-4-phosphate 3-kinase, catalytic subunit type 2 alpha	PIK3C2A	Endothelial PI3K-C2α, a class II PI3K, has an essential role in angiogenesis and vascular barrier function. PMID: 22983395
Phosphatidylinositol-4-phosphate 3-kinase, catalytic subunit type 2 beta	PIK3C2B	The class II phosphatidylinositol 3 kinase C2beta is required for the activation of the K+ channel KCa3.1 and CD4 T-cells. PMID: 19587117
Phosphatidylinositol-4-phosphate 3-kinase, catalytic subunit type 2 gamma	PIK3C2G	Identification of the macromolecular complex responsible for PI3Kgamma-dependent regulation of cAMP levels. PMID: 16856844
Phosphatidylinositol 3-kinase, catalytic subunit type 3	PIK3C3	ULK1 induces autophagy by phosphorylating Beclin-1 and activating VPS34 lipid kinase. PMID: 23685627
Phosphatidylinositol 4-kinase, catalytic, alpha	PI4KA	Pharmacological and genetic targeting of the PI4KA enzyme reveals its important role in maintaining plasma membrane phosphatidylinositol 4-phosphate and phosphatidylinositol 4,5-bisphosphate levels. PMID: 24415756
Phosphatidylinositol 4-kinase, catalytic, beta	PI4KB	Phosphatidylinositol 4-kinase III beta is essential for replication of human rhinovirus and its inhibition causes a lethal phenotype in vivo. PMID: 23650168
Phosphatidylinositol 4-kinase type 2 alpha	PI4K2A	Loss of phosphatidylinositol 4-kinase 2 alpha activity causes late onset degeneration of spinal cord axons. PMID: 19581584
Phosphatidylinositol 4-kinase type 2 beta	PI4K2B	Identification of phosphatidylinositol 4-kinase type II beta as HLA class II-restricted target in graft versus leukemia reactivity. PMID: 18316730
Phosphatidylinositol-5-phosphate 4-kinase, type II, alpha	PIP4K2A	A targeted knockdown screen of genes coding for phosphoinositide modulators identifies PIP4K2A as required for acute myeloid leukemia cell proliferation and survival. PMID: 24681948

(Continued)

HGNC-approved name	Symbol	Target information
Phosphatidylinositol-5-phosphate 4-kinase, type II, beta	PIP4K2B	Depletion of a putatively druggable class of phosphatidylinositol kinases inhibits growth of p53-null tumors. PMID: 24209622
Phosphatidylinositol-5-phosphate 4-kinase, type II, gamma	PIP4K2C	Distribution and neuronal expression of phosphatidylinositol phosphate kinase IIgamma in the mouse brain. PMID: 19757494
Phosphatidylinositol-4-phosphate 5-kinase, type I, alpha	PIP5K1A	The role of PI3K/AKT-related PIP5K1α and the discovery of its selective inhibitor for treatment of advanced prostate cancer. PMID: 25071204
Phosphatidylinositol-4-phosphate 5-kinase, type I, beta	PIP5K1B	Cis-silencing of PIP5K1B evidenced in Friedreich's ataxia patient cells results in cytoskeleton anomalies. PMID: 23552101
Phosphatidylinositol-4-phosphate 5-kinase, type I, gamma	PIP5K1C	The lipid kinase PIP5K1C regulates pain signaling and sensitization. PMID: 24853942. Disease: lethal congenital contracture syndrome 3 (LCCS3)
Phosphatidylinositol-4-phosphate 5-kinase-like 1	PIP5KL1	Overexpression of PIP5KL1 suppresses the growth of human cervical cancer cells *in vitro* and in vivo. PMID: 19947914
Phosphoinositide kinase, FYVE finger containing	PIKFYVE	**Apilimod** – immunomodulating agents. Disease: corneal dystrophy, fleck (CFD)

4.1.2   Protein phosphatases

HGNC-approved name	Symbol	Target information
Tyrosine phosphatases		
Protein tyrosine phosphatase, receptor type, A	PTPRA	Protein tyrosine phosphatase α mediates profibrotic signaling in lung fibroblasts through TGF-β responsiveness. PMID: 24650563
Protein tyrosine phosphatase, receptor type, B	PTPRB	Recurrent PTPRB and PLCG1 mutations in angiosarcoma. PMID: 24633157
Protein tyrosine phosphatase, receptor type, C	PTPRC	Disease: severe combined immunodeficiency autosomal recessive T-cell-negative/B-cell-positive/NK-cell-positive (T(−)B(+)NK(+) SCID)
Protein tyrosine phosphatase, receptor type, D	PTPRD	Loss of the tyrosine phosphatase PTPRD leads to aberrant STAT3 activation and promotes gliomagenesis. PMID: 24843164
Protein tyrosine phosphatase, receptor type, E	PTPRE	Tyrosine phosphatases epsilon and alpha perform specific and overlapping functions in regulation of voltage-gated potassium channels in Schwann cells. PMID: 16870705
Protein tyrosine phosphatase, receptor type, F	PTPRF	Functional genomics identified a novel protein tyrosine phosphatase receptor type F-mediated growth inhibition in hepatocarcinogenesis. PMID: 24470239
Protein tyrosine phosphatase, receptor type, G	PTPRG	Protein tyrosine phosphatase receptor type {gamma} is a functional tumor suppressor gene specifically downregulated in chronic myeloid leukemia. PMID: 20959494
Protein tyrosine phosphatase, receptor type, H	PTPRH	
Protein tyrosine phosphatase, receptor type, J	PTPRJ	Targeting density-enhanced phosphatase-1 (DEP-1) with antisense oligonucleotides improves the metabolic phenotype in high-fat diet-fed mice. PMID: 23889985
Protein tyrosine phosphatase, receptor type, K	PTPRK	Receptor-type protein tyrosine phosphatase κ directly dephosphorylates CD133 and regulates downstream AKT activation. PMID: 24882578
Protein tyrosine phosphatase, receptor type, M	PTPRM	Protein tyrosine phosphatase mu regulates glioblastoma cell growth and survival in vivo. PMID: 22505657

HGNC-approved name	Symbol	Target information
Protein tyrosine phosphatase, receptor type, N	PTPRN	Expression and function of IA-2 family proteins, unique neuroendocrine-specific protein-tyrosine phosphatases. PMID: 19550073
Protein tyrosine phosphatase, receptor type, N polypeptide 2	PTPRN2	Disturbances in the secretion of neurotransmitters in IA-2/IA-2beta null mice: Changes in behavior, learning and lifespan. PMID: 19361477
Protein tyrosine phosphatase, receptor type, O	PTPRO	Disease: nephrotic syndrome 6 (NPHS6)
Protein tyrosine phosphatase, receptor type, Q	PTPRQ	Identification of novel PTPRQ phosphatase inhibitors based on the virtual screening with docking simulations. PMID: 23981594. Disease: deafness, autosomal recessive, 84A (DFNB84A)
Protein tyrosine phosphatase, receptor type, R	PTPRR	Epigenetic silencing of PTPRR activates MAPK signaling, promotes metastasis and serves as a biomarker of invasive cervical cancer. PMID: 22330137
Protein tyrosine phosphatase, receptor type, S	PTPRS	A new role for RPTPsigma in spinal cord injury: Signaling chondroitin sulfate proteoglycan inhibition. PMID: 20179269
Protein tyrosine phosphatase, receptor type, T	PTPRT	PTPRT regulates high-fat diet-induced obesity and insulin resistance. PMID: 24949727
Protein tyrosine phosphatase, receptor type, U	PTPRU	Knockdown of protein tyrosine phosphatase receptor U inhibits growth and motility of gastric cancer cells. PMID: 25337216
Protein tyrosine phosphatase, receptor type, Z polypeptide 1	PTPRZ1	Increased expression of receptor phosphotyrosine phosphatase-β/ζ is associated with molecular, cellular, behavioral and cognitive schizophrenia phenotypes. PMID: 22832403
Protein tyrosine phosphatase, non-receptor type 1	PTPN1	**ISIS-PTP1BRx** – drugs used in diabetes
Protein tyrosine phosphatase, non-receptor type 2	PTPN2	PTPN2 is associated with Crohn's disease and its expression is regulated by NKX2-3. PMID: 22377701
Protein tyrosine phosphatase, non-receptor type 3	PTPN3	Protein tyrosine phosphatase PTPN3 inhibits lung cancer cell proliferation and migration by promoting EGFR endocytic degradation. PMID: 25263444
Protein tyrosine phosphatase, non-receptor type 4 (megakaryocyte)	PTPN4	Peptides targeting the PDZ domain of PTPN4 are efficient inducers of glioblastoma cell death. PMID: 22000519
Protein tyrosine phosphatase, non-receptor type 5 (striatum-enriched)	PTPN5	Inhibitor of the tyrosine phosphatase STEP reverses cognitive deficits in a mouse model of Alzheimer's disease. PMID: 25093460
Protein tyrosine phosphatase, non-receptor type 6	PTPN6	Novel SHP-1 inhibitors tyrosine phosphatase inhibitor-1 and analogs with preclinical anti-tumor activities as tolerated oral agents. PMID: 20421638
Protein tyrosine phosphatase, non-receptor type 7	PTPN7	Role of protein tyrosine phosphatase non-receptor type 7 in the regulation of TNF-α production in RAW 264.7 macrophages. PMID: 24265715
Protein tyrosine phosphatase, non-receptor type 9	PTPN9	A highly selective and potent PTP-MEG2 inhibitor with therapeutic potential for type 2 diabetes. PMID: 23075115
Protein tyrosine phosphatase, non-receptor type 11	PTPN11	Protein tyrosine phosphatase SHP2 regulates TGF-β1 production in airway epithelia and asthmatic airway remodeling in mice. PMID: 23057634. Disease: LEOPARD syndrome 1 (LEOPARD1)
Protein tyrosine phosphatase, non-receptor type 12	PTPN12	PTPN12 promotes resistance to oxidative stress and supports tumorigenesis by regulating FOXO signaling. PMID: 23435421
Protein tyrosine phosphatase, non-receptor type 13 (APO-1/CD95 (Fas)-associated phosphatase)	PTPN13	PTPN13/PTPL1: An important regulator of tumor aggressiveness. PMID: 21235435

(Continued)

HGNC-approved name	Symbol	Target Information
Protein tyrosine phosphatase, non-receptor type 14	PTPN14	Disease: choanal atresia and lymphedema (CHATLY)
Protein tyrosine phosphatase, non-receptor type 18 (brain-derived)	PTPN18	The catalytic region and PEST domain of PTPN18 distinctly regulate the HER2 phosphorylation and ubiquitination barcodes. PMID: 25081058
Protein tyrosine phosphatase, non-receptor type 20A	PTPN20A	Protein tyrosine phosphatase hPTPN20a is targeted to sites of actin polymerization. PMID: 15790311
Protein tyrosine phosphatase, non-receptor type 20B	PTPN20B	
Protein tyrosine phosphatase, non-receptor type 21	PTPN21	PTPD1 supports receptor stability and mitogenic signaling in bladder cancer cells. PMID: 20923765
Protein tyrosine phosphatase, non-receptor type 22 (lymphoid)	PTPN22	Lack of the phosphatase PTPN22 increases adhesion of murine regulatory T cells to improve their immunosuppressive function. PMID: 23193160. Disease: systemic lupus erythematosus (SLE)
Protein tyrosine phosphatase, non-receptor type 23	PTPN23	The catalytically inactive tyrosine phosphatase HD-PTP/PTPN23 is a novel regulator of SMN complex localization. PMID: 25392300
Magnesium-dependent phosphatase 1	MDP1	Magnesium-dependent phosphatase-1 is a protein-fructosamine-6-phosphatase potentially involved in glycation repair. PMID: 16670083
Protein tyrosine phosphatase domain containing 1	PTPDC1	
Protein tyrosine phosphatase, mitochondrial 1	PTPMT1	Structure-based virtual screening approach to the discovery of novel PTPMT1 phosphatase inhibitors. PMID: 22115589
Ubiquitin-associated and SH3 domain containing A	UBASH3A	Members of the novel UBASH3/STS/TULA family of cellular regulators suppress T-cell-driven inflammatory responses in vivo. PMID: 25047644
Ubiquitin-associated and SH3 domain containing B	UBASH3B	Protein tyrosine phosphatase UBASH3B is overexpressed in triple-negative breast cancer and promotes invasion and metastasis. PMID: 23784775

Serine/threonine phosphatases

HGNC-approved name	Symbol	Target Information
Calcineurin-like phosphoesterase domain containing 1	CPPED1	Downregulation of CPPED1 expression improves glucose metabolism *in vitro* in adipocytes. PMID: 23939394
CTD nuclear envelope phosphatase 1	CTDNEP1	A conserved phosphatase cascade that regulates nuclear membrane biogenesis. PMID: 17420445
CTD (carboxy-terminal domain, RNA polymerase II, polypeptide A) phosphatase, subunit 1	CTDP1	Disease: congenital cataracts, facial dysmorphism, and neuropathy (CCFDN)
CTD (carboxy-terminal domain, RNA polymerase II, polypeptide A) small phosphatase 1	CTDSP1	Selective inactivation of a human neuronal silencing phosphatase by a small molecule inhibitor. PMID: 21348431
CTD (carboxy-terminal domain, RNA polymerase II, polypeptide A) small phosphatase 2	CTDSP2	Intronic miR-26b controls neuronal differentiation by repressing its host transcript, ctdsp2. PMID: 22215807
CTD (carboxy-terminal domain, RNA polymerase II, polypeptide A) small phosphatase-like	CTDSPL	RBSP3 (HYA22) is a tumor suppressor gene implicated in major epithelial malignancies. PMID: 15051889
CTD (carboxy-terminal domain, RNA polymerase II, polypeptide A) small phosphatase-like 2	CTDSPL2	C-terminal domain (CTD) small phosphatase-like 2 modulates the canonical bone morphogenetic protein (BMP) signaling and mesenchymal differentiation via Smad dephosphorylation. PMID: 25100727

HGNC-approved name	Symbol	Target information
Ubiquitin-like domain containing CTD phosphatase 1	UBLCP1	UBLCP1 is a 26S proteasome phosphatase that regulates nuclear proteasome activity. PMID: 21949367
Integrin-linked kinase-associated serine/threonine phosphatase	ILKAP	Characterization of nuclear localization signal in the N terminus of integrin-linked kinase-associated phosphatase (ILKAP) and its essential role in the down-regulation of RSK2 protein signaling. PMID: 23329845
PH domain and leucine-rich repeat protein phosphatase 1	PHLPP1	Turning off AKT: PHLPP as a drug target. PMID: 24392697
PH domain and leucine rich repeat protein phosphatase 2	PHLPP2	Turning off AKT: PHLPP as a drug target. PMID: 24392697
Protein phosphatase 1, catalytic subunit, alpha isozyme	PPP1CA	Protein phosphatase 1 and its complexes in carcinogenesis. PMID: 24200083
Protein phosphatase 1, catalytic subunit, beta isozyme	PPP1CB	Penostatin derivatives, a novel kind of protein phosphatase 1b inhibitors isolated from solid cultures of the entomogenous fungus Isaria tenuipes. PMID: 24481115
Protein phosphatase 1, catalytic subunit, gamma isozyme	PPP1CC	Analysis of Ppp1cc-null mice suggests a role for PP1gamma2 in sperm morphogenesis. PMID: 17301292
Protein phosphatase 2, catalytic subunit, alpha isozyme	PPP2CA	Protein phosphatase 2A: A target for anticancer therapy. PMID: 23639323
Protein phosphatase 2, catalytic subunit, beta isozyme	PPP2CB	
Protein phosphatase 3, catalytic subunit, alpha isozyme	PPP3CA	Cellular and molecular consequences of calcineurin A alpha gene deletion. PMID: 17872390
Protein phosphatase 3, catalytic subunit, beta isozyme	PPP3CB	
Protein phosphatase 3, catalytic subunit, gamma isozyme	PPP3CC	PPP3CC gene: A putative modulator of antidepressant response through the B-cell receptor signaling pathway. PMID: 24709691
Protein phosphatase 4, catalytic subunit	PPP4C	Protein phosphatase 4 promotes hepatic lipogenesis through dephosphorylating acetyl-CoA carboxylase 1 on serine 79. PMID: 25050742
Protein phosphatase 5, catalytic subunit	PPP5C	Genetic disruption of protein phosphatase 5 in mice prevents high fat diet feeding-induced weight gain. PMID: 24220247
Protein phosphatase 6, catalytic subunit	PPP6C	Abrogation of protein phosphatase 6 promotes skin carcinogenesis induced by DMBA. PMID: 25486434
Protein phosphatase, EF-hand calcium-binding domain 1	PPEF1	Molecular cloning of a novel PPEF-1 gene variant from a T-cell lymphoblastic lymphoma cell line. PMID: 22292511
Protein phosphatase, EF-hand calcium-binding domain 2	PPEF2	Protein phosphatase with EF-hand domains 2 (PPEF2) is a potent negative regulator of apoptosis signal regulating kinase-1 (ASK1). PMID: 20674765
Protein phosphatase, Mg2+/Mn2+ dependent, 1A	PPM1A	Protein phosphatase magnesium dependent 1A governs the wound healing-inflammation-angiogenesis cross talk on injury. PMID: 25196308
Protein phosphatase, Mg2+/Mn2+ dependent, 1B	PPM1B	The serine/threonine phosphatase PPM1B (PP2Cβ) selectively modulates PPARγ activity. PMID: 23320500
Protein phosphatase, Mg2+/Mn2+ dependent, 1D	PPM1D	WIP1 phosphatase at the crossroads of cancer and aging. PMID: 19879149

(Continued)

HGNC-approved name	Symbol	Target information
Protein phosphatase, Mg2+/Mn2+ dependent, 1E	PPM1E	
Protein phosphatase, Mg2+/Mn2+ dependent, 1F	PPM1F	POPX2 phosphatase regulates the KIF3 kinesin motor complex. PMID: 24338362
Protein phosphatase, Mg2+/Mn2+ dependent, 1G	PPM1G	Protein phosphatase PPM1G regulates protein translation and cell growth by dephosphorylating 4E binding protein 1 (4E-BP1). PMID: 23814053
Protein phosphatase, Mg2+/Mn2+ dependent, 1H	PPM1H	PPM1H is a p27 phosphatase implicated in trastuzumab resistance. PMID: 22586611
Protein phosphatase, Mg2+/Mn2+ dependent, 1J	PPM1J	
Protein phosphatase, Mg2+/Mn2+ dependent, 1K	PPM1K	Disease: maple syrup urine disease, mild variant (MSUDMV)
Protein phosphatase, Mg2+/Mn2+ dependent, 1L	PPM1L	PPM1l encodes an inositol requiring-protein 1 (IRE1) specific phosphatase that regulates the functional outcome of the ER stress response. PMID: 24327956
Protein phosphatase, Mg2+/Mn2+ dependent, 1M	PPM1M	A mechanism for the suppression of interleukin-1-induced nuclear factor kappaB activation by protein phosphatase 2Ceta-2. PMID: 19594441
Protein phosphatase, Mg2+/Mn2+ dependent, 1N (putative)	PPM1N	
PTC7 protein phosphatase homolog (*Saccharomyces cerevisiae*)	PPTC7	
Pyruvate dehydrogenase phosphatase catalytic subunit 1	PDP1	Tyr-94 phosphorylation inhibits pyruvate dehydrogenase phosphatase 1 and promotes tumor growth. PMID: 24962578. Disease: pyruvate dehydrogenase phosphatase deficiency (PDP deficiency)
Pyruvate dehydrogenase phosphatase catalytic subunit 2	PDP2	
SSU72 RNA polymerase II CTD phosphatase homolog (*S. cerevisiae*)	SSU72	A gene-specific role for the Ssu72 RNAPII CTD phosphatase in HIV-1 Tat transactivation. PMID: 25319827
Dual-specificity phosphatases		
Acid phosphatase 1, soluble	ACP1	The role of low-molecular-weight protein tyrosine phosphatase (LMW-PTP ACP1) in oncogenesis. PMID: 23584899
Acid phosphatase, testicular	ACPT	Regulation of ErbB4 phosphorylation and cleavage by a novel histidine acid phosphatase. PMID: 15219672
CDC14 cell division cycle 14 homolog A (*S. cerevisiae*)	CDC14A	HCdc14A is involved in cell cycle regulation of human brain vascular endothelial cells following injury induced by high glucose, free fatty acids and hypoxia. PMID: 25463242
CDC14 cell division cycle 14 homolog B (*S. cerevisiae*)	CDC14B	The Cdc14B phosphatase displays oncogenic activity mediated by the Ras-Mek signaling pathway. PMID: 21502810
Cell division cycle 14C	CDC14C	Cdc14: A highly conserved family of phosphatases with non-conserved functions? PMID: 20720150
Cell division cycle 25 homolog A (*S. pombe*)	CDC25A	CDC25A and B dual-specificity phosphatase inhibitors: Potential agents for cancer therapy. PMID: 19442149
Cell division cycle 25 homolog B (*S. pombe*)	CDC25B	CDC25A and B dual-specificity phosphatase inhibitors: Potential agents for cancer therapy. PMID: 19442149
Cell division cycle 25 homolog C (*S. pombe*)	CDC25C	Recurrent CDC25C mutations drive malignant transformation in FPD/AML. PMID: 25159113
Cyclin-dependent kinase inhibitor 3	CDKN3	A critical role of CDKN3 in Bcr-Abl-mediated tumorigenesis. PMID: 25360622. Disease: hepatocellular carcinoma (HCC)

HGNC-approved name	Symbol	Target information
Dual-specificity phosphatase 1	DUSP1	Mitogen-activated protein kinase phosphatase-1: A potential therapeutic target in metabolic disease. PMID: 21058921
Dual-specificity phosphatase 2	DUSP2	Dual-specificity phosphatases 2: Surprising positive effect at the molecular level and a potential biomarker of diseases. PMID: 23190643
Dual-specificity phosphatase 3	DUSP3	DUSP3/VHR is a pro-angiogenic atypical dual-specificity phosphatase. PMID: 24886454
Dual-specificity phosphatase 4	DUSP4	MKP-2: Out of the DUSP-bin and back into the limelight. PMID: 22260697
Dual-specificity phosphatase 5	DUSP5	Dusp5 negatively regulates IL-33-mediated eosinophil survival and function. PMID: 25398911
Dual-specificity phosphatase 6	DUSP6	Disease: hypogonadotropic hypogonadism 19 with or without anosmia (HH19)
Dual-specificity phosphatase 7	DUSP7	Overexpression of the dual-specificity MAPK phosphatase PYST2 in acute leukemia. PMID: 12969791
Dual-specificity phosphatase 8	DUSP8	Differential regulation of M3/6 (DUSP8) signaling complexes in response to arsenite-induced oxidative stress. PMID: 23159405
Dual-specificity phosphatase 9	DUSP9	BMP4 signaling acts via dual-specificity phosphatase 9 to control ERK activity in mouse embryonic stem cells. PMID: 22305567
Dual-specificity phosphatase 10	DUSP10	Dual-specificity phosphatase 10 controls brown adipocyte differentiation by modulating the phosphorylation of p38 mitogen-activated protein kinase. PMID: 23977283
Dual-specificity phosphatase 11 (RNA/RNP complex 1-interacting)	DUSP11	Isolation and characterization of DUSP11, a novel p53 target gene. PMID: 19120688
Dual-specificity phosphatase 12	DUSP12	The dual-specificity phosphatase hYVH1 (DUSP12) is a novel modulator of cellular DNA content. PMID: 21521943
Dual-specificity phosphatase 13	DUSP13	DUSP13B/TMDP inhibits stress-activated MAPKs and suppresses AP-1-dependent gene expression. PMID: 21360282
Dual-specificity phosphatase 14	DUSP14	The dual-specificity phosphatase DUSP14 negatively regulates tumor necrosis factor- and interleukin-1-induced nuclear factor-κB activation by dephosphorylating the protein kinase TAK1. PMID: 23229544
Dual-specificity phosphatase 15	DUSP15	Identification of VHY/Dusp15 as a regulator of oligodendrocyte differentiation through a systematic genomics approach. PMID: 22792334
Dual-specificity phosphatase 16	DUSP16	Discovery of novel DUSP16 phosphatase inhibitors through virtual screening with homology modeled protein structure. PMID: 25245988
Dual-specificity phosphatase 18	DUSP18	Dual specificity phosphotase 18, interacting with SAPK, dephosphorylates SAPK and inhibits SAPK/JNK signal pathway in vivo. PMID: 16720344
Dual-specificity phosphatase 19	DUSP19	A novel dual specificity phosphatase SKRP1 interacts with the MAPK kinase MKK7 and inactivates the JNK MAPK pathway. Implication for the precise regulation of the particular MAPK pathway. PMID: 11959861
Dual-specificity phosphatase 21	DUSP21	RNA interference against cancer/testis genes identifies dual specificity phosphatase 21 as a potential therapeutic target in human hepatocellular carcinoma. PMID: 23929653
Dual-specificity phosphatase 22	DUSP22	The phosphatase JKAP/DUSP22 inhibits T-cell receptor signalling and autoimmunity by inactivating Lck. PMID: 24714587
Dual-specificity phosphatase 23	DUSP23	Dual-specificity phosphatase 23 mediates GCM1 dephosphorylation and activation. PMID: 20855292
Dual-specificity phosphatase 26 (putative)	DUSP26	Dual-specificity phosphatase 26 is a novel p53 phosphatase and inhibits p53 tumor suppressor functions in human neuroblastoma. PMID: 20562916

(Continued)

HGNC-approved name	Symbol	Target information
Dual-specificity phosphatase 27 (putative)	DUSP27	Impaired embryonic motility in dusp27 mutants reveals a developmental defect in myofibril structure. PMID: 24203884
Dual-specificity phosphatase 28	DUSP28	DUSP28 contributes to human hepatocellular carcinoma via regulation of the p38 MAPK signaling. PMID: 25230705
Dual-specificity phosphatase and pro-isomerase domain containing 1	DUPD1	
EYA transcriptional coactivator and phosphatase 1	EYA1	Disease: branchiootorenal syndrome 1 (BOR1)
EYA transcriptional coactivator and phosphatase 2	EYA2	Allosteric inhibitors of the Eya2 phosphatase are selective and inhibit Eya2-mediated cell migration. PMID: 24755226
EYA transcriptional coactivator and phosphatase 3	EYA3	The Eyes Absent proteins in development and disease. PMID: 22971774
EYA transcriptional coactivator and phosphatase 4	EYA4	Inhibition of Eyes Absent Homolog 4 expression induces malignant peripheral nerve sheath tumor necrosis. PMID: 19901965. Disease: deafness, autosomal dominant, 10 (DFNA10)
Myotubularin 1	MTM1	Inositol lipid phosphatases in membrane trafficking and human disease. PMID: 24966051. Disease: myopathy, centronuclear, X-linked (CNMX)
Myotubularin-related protein 1	MTMR1	Analysis of MTMR1 expression and correlation with muscle pathological features in juvenile/adult onset myotonic dystrophy type 1 (DM1) and in myotonic dystrophy type 2 (DM2). PMID: 20685272
Myotubularin-related protein 2	MTMR2	Disease: Charcot–Marie–Tooth disease 4B1 (CMT4B1)
Myotubularin-related protein 3	MTMR3	Myotubularin-related phosphatase 3 promotes growth of colorectal cancer cells. PMID: 25215329
Myotubularin-related protein 4	MTMR4	Myotubularin-related protein 4 (MTMR4) attenuates BMP/Dpp signaling by dephosphorylation of Smad proteins. PMID: 23150675
SET binding factor 1	SBF1	Disease: Charcot–Marie–Tooth disease 4B3 (CMT4B3)
Myotubularin-related protein 6	MTMR6	Phosphatidylinositol-3 phosphatase myotubularin-related protein 6 negatively regulates CD4 T cells. PMID: 16847315
Myotubularin-related protein 7	MTMR7	Genome-wide study links MTMR7 gene to variant Creutzfeldt-Jakob risk. PMID: 22137330
Myotubularin-related protein 8	MTMR8	Myotubularin-related protein (MTMR) 9 determines the enzymatic activity, substrate specificity, and role in autophagy of MTMR8. PMID: 22647598
Myotubularin-related protein 9	MTMR9	MTMR9 increases MTMR6 enzyme activity, stability, and role in apoptosis. PMID: 19038970
Myotubularin-related protein 10	MTMR10	
Myotubularin-related protein 11	MTMR11	
Myotubularin-related protein 12	MTMR12	
SET binding factor 2	SBF2	Disease: Charcot–Marie–Tooth disease 4B2 (CMT4B2)
Myotubularin-related protein 14	MTMR14	Muscle-specific inositide phosphatase (MIP/MTMR14) is reduced with age and its loss accelerates skeletal muscle aging process by altering calcium homeostasis. PMID: 20817957. Disease: myopathy, centronuclear, 1 (CNM1)
Phosphatase and tensin homolog	PTEN	Pills of PTEN? In and out for tumor suppression. PMID: 23917528. Disease: Cowden syndrome 1 (CWS1)
Tensin 1	TNS1	
Tensin 2	TNS2	C1-Ten is a protein tyrosine phosphatase of insulin receptor substrate 1 (IRS-1), regulating IRS-1 stability and muscle atrophy. PMID: 23401856

HGNC-approved name	Symbol	Target information
Transmembrane phosphatase with tensin homology	TPTE	
Transmembrane phosphoinositide 3-phosphatase and tensin homolog 2	TPTE2	Cell cycle arrest and apoptosis by expression of a novel TPIP (TPIP-C2) cDNA encoding a C2-domain in HEK-293 cells. PMID: 22311048
Protein tyrosine phosphatase type IVA, member 1	PTP4A1	Phosphatase of regenerating liver: A novel target for cancer therapy. PMID: 24579927
Protein tyrosine phosphatase type IVA, member 2	PTP4A2	PRL2/PTP4A2 phosphatase is important for haematopoietic stem cell self-renewal. PMID: 24753135
Protein tyrosine phosphatase type IVA, member 3	PTP4A3	Phosphatase of regenerating liver-3 as a convergent therapeutic target for lymph node metastasis in esophageal squamous cell carcinoma. PMID: 19960436
RNA polymerase II-associated protein 2	RPAP2	Rtr1 is a dual specificity phosphatase that dephosphorylates Tyr1 and Ser5 on the RNA polymerase II CTD. PMID: 24951832
Slingshot protein phosphatase 1	SSH1	Insulin receptor substrate-4 binds to Slingshot-1 phosphatase and promotes cofilin dephosphorylation. PMID: 25100728
Slingshot protein phosphatase 2	SSH2	Knockdown of Slingshot 2 (SSH2) serine phosphatase induces Caspase3 activation in human carcinoma cell lines with the loss of the Birt–Hogg–Dubé tumour suppressor gene (FLCN). PMID: 23416984
Slingshot protein phosphatase 3	SSH3	
Lipid phosphatases		
Inositol polyphosphate-1-phosphatase	INPP1	Inositol phosphate recycling regulates glycolytic and lipid metabolism that drives cancer aggressiveness. PMID: 24738946
Inositol polyphosphate-3-phosphatase	INPP3	
Inositol polyphosphate-4-phosphatase, type I, 107 kDa	INPP4A	The PtdIns(3,4)P(2) phosphatase INPP4A is a suppressor of excitotoxic neuronal death. PMID: 20463662
Inositol polyphosphate-4-phosphatase, type II, 105 kDa	INPP4B	INPP4B suppresses prostate cancer cell invasion. PMID: 25248616
Inositol polyphosphate-5-phosphatase, 40 kDa	INPP5A	Identification of inhibitors of inositol 5-phosphatases through multiple screening strategies. PMID: 24742366
Inositol polyphosphate-5-phosphatase, 75 kDa	INPP5B	Compensatory role of inositol 5-phosphatase INPP5B to OCRL in primary cilia formation in oculocerebrorenal syndrome of lowe. PMID: 23805271
Inositol polyphosphate-5-phosphatase, 145 kDa	INPP5D	Reduced proliferation of CD34(+) cells from patients with acute myeloid leukemia after gene transfer of INPP5D. PMID: 19148132
Inositol polyphosphate-5-phosphatase, 72 kDa	INPP5E	Inhibition of 72 kDa inositol polyphosphate 5-phosphatase E improves insulin signal transduction in diet-induced obesity. PMID: 23349329. Disease: Joubert syndrome 1 (JBTS1)
Inositol polyphosphate-5-phosphatase F	INPP5F	Inositol polyphosphate-5-phosphatase F (INPP5F) inhibits STAT3 activity and suppresses gliomas tumorigenicity. PMID: 25476455
Inositol polyphosphate-5-phosphatase J	INPP5J	
Inositol polyphosphate-5-phosphatase K	INPP5K	High SKIP expression is correlated with poor prognosis and cell proliferation of hepatocellular carcinoma. PMID: 23696020
Inositol polyphosphate phosphatase-like 1	INPPL1	Rational design and synthesis of 4-substituted 2-pyridin-2-ylamides with inhibitory effects on SH2 domain-containing inositol 5'-phosphatase 2 (SHIP2). PMID: 23434638. Disease: diabetes mellitus, non-insulin-dependent (NIDDM)

(Continued)

HGNC-approved name	Symbol	Target information
Synaptojanin 1	SYNJ1	Screening assay for small-molecule inhibitors of synaptojanin 1, a synaptic phosphoinositide phosphatase. PMID: 24186361. Disease: Parkinson disease 20, early-onset (PARK20)
Synaptojanin 2	SYNJ2	Genetic interactions within inositol-related pathways are associated with longitudinal changes in ventricle size. PMID: 24077433
Transmembrane protein 55A	TMEM55A	
Transmembrane protein 55B	TMEM55B	Transmembrane protein 55B is a novel regulator of cellular cholesterol metabolism. PMID: 25035345

4.1.3   Cyclic nucleotides and phosphodiesterases

HGNC-approved name	Symbol	Target information
Adenylate cyclase 1 (brain)	ADCY1	Extracellular calcium influx activates adenylate cyclase 1 and potentiates insulin secretion in MIN6 cells. PMID: 23282092
Adenylate cyclase 2 (brain)	ADCY2	
Adenylate cyclase 3	ADCY3	A gain-of-function mutation in adenylate cyclase 3 protects mice from diet-induced obesity. PMID: 25329148
Adenylate cyclase 4	ADCY4	
Adenylate cyclase 5	ADCY5	Inhibition of adenylyl cyclase type 5 prevents L-DOPA-induced dyskinesia in an animal model of Parkinson's disease. PMID: 25164669. Disease: dyskinesia, familial, with facial myokymia (FDFM)
Adenylate cyclase 6	ADCY6	Adenylyl cyclase type VI increases Akt activity and phospholamban phosphorylation in cardiac myocytes. PMID: 18838385
Adenylate cyclase 7	ADCY7	Type 7 adenylyl cyclase-mediated hypothalamic-pituitary-adrenal axis responsiveness: Influence of ethanol and sex. PMID: 20363852
Adenylate cyclase 8 (brain)	ADCY8	Multilevel control of glucose homeostasis by adenylyl cyclase 8. PMID: 25403481
Adenylate cyclase 9	ADCY9	
Adenylate cyclase 10 (soluble)	ADCY10	Crystal structures of human soluble adenylyl cyclase reveal mechanisms of catalysis and of its activation through bicarbonate. PMID: 24567411. Disease: hypercalciuria absorptive 2 (HCA2)
Mab-21 domain containing 1	MB21D1	Autophagy side of MB21D1/cGAS DNA sensor. PMID: 24879161
Guanylate cyclase 1, soluble, alpha 2	GUCY1A2	**Riociguat** – antihypertensives
Guanylate cyclase 1, soluble, alpha 3	GUCY1A3	**Riociguat** – antihypertensives
Guanylate cyclase 1, soluble, beta 3	GUCY1B3	**Riociguat** – antihypertensives
Guanylate cyclase 2C (heat-stable enterotoxin receptor)	GUCY2C	**Linaclotide** – drugs for functional gastrointestinal disorders. Disease: diarrhea 6 (DIAR6)
Guanylate cyclase 2D, membrane (retina-specific)	GUCY2D	Disease: Leber congenital amaurosis 1 (LCA1)
Guanylate cyclase 2F, retinal	GUCY2F	Gucy2f zebrafish knockdown – a model for Gucy2d-related leber congenital amaurosis. PMID: 22378290
Phosphodiesterase 1A, calmodulin-dependent	PDE1A	**ITI-214** – anti-dementia drugs

HGNC-approved name	Symbol	Target information
Phosphodiesterase 1B, calmodulin-dependent	PDE1B	Behavioral and neurochemical characterization of mice deficient in the phosphodiesterase-1B (PDE1B) enzyme. PMID: 17559891
Phosphodiesterase 1C, calmodulin-dependent 70 kDa	PDE1C	Phosphodiesterase 1C is dispensable for rapid response termination of olfactory sensory neurons. PMID: 19305400
Phosphodiesterase 2A, cGMP-stimulated	PDE2A	Anxiolytic effects of phosphodiesterase-2 inhibitors associated with increased cGMP signaling. PMID: 19684253
Phosphodiesterase 3A, cGMP-inhibited	PDE3A	**Cilostazol** – antithrombotic agents
Phosphodiesterase 3B, cGMP-inhibited	PDE3B	Targeting phosphodiesterase 3B enhances cisplatin sensitivity in human cancer cells. PMID: 24133626
Phosphodiesterase 4A, cAMP-specific	PDE4A	**Roflumilast** – drugs for obstructive airway diseases
Phosphodiesterase 4B, cAMP-specific	PDE4B	Discovery of triazines as selective PDE4B versus PDE4D inhibitors. PMID: 24998378
Phosphodiesterase 4C, cAMP-specific	PDE4C	Integrated analysis using methylation and gene expression microarrays reveals PDE4C as a prognostic biomarker in human glioma. PMID: 24842301
Phosphodiesterase 4D, cAMP-specific	PDE4D	The molecular basis for the inhibition of phosphodiesterase-4D by three natural resveratrol analogs. Isolation, molecular docking, molecular dynamics simulations, binding free energy, and bioassay. PMID: 23871879. Note: Genetic variations in PDE4D might be associated with susceptibility to stroke
Phosphodiesterase 5A, cGMP-specific	PDE5A	**Sildenafil** – urologicals
Phosphodiesterase 6A, cGMP-specific, rod, alpha	PDE6A	Disease: retinitis pigmentosa 43 (RP43)
Phosphodiesterase 6B, cGMP-specific, rod, beta	PDE6B	Disease: retinitis pigmentosa 40 (RP40)
Phosphodiesterase 6C, cGMP-specific, cone, alpha prime	PDE6C	Disease: cone dystrophy 4 (COD4)
Phosphodiesterase 6D, cGMP-specific, rod, delta	PDE6D	Small molecule inhibition of the KRAS-PDEδ interaction impairs oncogenic KRAS signalling. PMID: 23698361. Disease: Joubert syndrome 22 (JBTS22)
Phosphodiesterase 6G, cGMP-specific, rod, gamma	PDE6G	Disease: retinitis pigmentosa 57 (RP57)
Phosphodiesterase 6H, cGMP-specific, cone, gamma	PDE6H	Disease: cone dystrophy, retinal 3A (RCD3A)
Phosphodiesterase 7A	PDE7A	New methods for the discovery and synthesis of PDE7 inhibitors as new drugs for neurological and inflammatory disorders. PMID: 23570245
Phosphodiesterase 7B	PDE7B	**Dyphylline** – drugs for obstructive airway diseases
Phosphodiesterase 8A	PDE8A	cAMP-specific phosphodiesterases 8A and 8B, essential regulators of Leydig cell steroidogenesis. PMID: 22232524
Phosphodiesterase 8B	PDE8B	cAMP-specific phosphodiesterases 8A and 8B, essential regulators of Leydig cell steroidogenesis. PMID: 22232524. Disease: striatal degeneration, autosomal dominant (ADSD)
Phosphodiesterase 9A	PDE9A	**ASP4901** – urologicals
Phosphodiesterase 10A	PDE10A	**Dipyridamole** – urologicals
Phosphodiesterase 11A	PDE11A	Does phosphodiesterase 11A (PDE11A) hold promise as a future therapeutic target? PMID: 25159071. Disease: primary pigmented nodular adrenocortical disease 2 (PPNAD2)

4.1.4 GTPase signalling proteins

HGNC-approved name	Symbol	Target information
ADP-ribosylation factor 1	ARF1	The small GTPase Arf1 modulates mitochondrial morphology and function. PMID: 25190516
ADP-ribosylation factor 3	ARF3	
ADP-ribosylation factor 4	ARF4	Arf4 determines dentate gyrus-mediated pattern separation by regulating dendritic spine development. PMID: 23050017
ADP-ribosylation factor 5	ARF5	BRAG2/GEP100/IQSec1 interacts with clathrin and regulates α5β1 integrin endocytosis through activation of ADP ribosylation factor 5 (Arf5). PMID: 22815487
ADP-ribosylation factor 6	ARF6	ADP-ribosylated proteins as old and new drug targets for anticancer therapy: The example of ARF6. PMID: 23016858
Atlastin GTPase 1	ATL1	Homotypic fusion of ER membranes requires the dynamin-like GTPase atlastin. PMID: 19633650. Disease: spastic paraplegia 3, autosomal dominant (SPG3)
Atlastin GTPase 2	ATL2	Atlastin GTPases are required for Golgi apparatus and ER morphogenesis. PMID: 18270207
Atlastin GTPase 3	ATL3	Disease: neuropathy, hereditary sensory, 1F (HSN1F)
Dynamin 1-like	DNM1L	Note: May be associated with Alzheimer disease through beta-amyloid-induced increased S-nitrosylation of DNM1L, which triggers, directly or indirectly, excessive mitochondrial fission, synaptic loss and neuronal damage. Disease: encephalopathy, l
Optic atrophy 1 (autosomal dominant)	OPA1	OPA1 downregulation is involved in sorafenib-induced apoptosis in hepatocellular carcinoma. PMID: 23108376. Disease: optic atrophy 1 (OPA1)
GTPase, IMAP family member 1	GIMAP1	Putative GTPase GIMAP1 is critical for the development of mature B and T lymphocytes. PMID: 20194894
GTPase, IMAP family member 2	GIMAP2	
GTPase, IMAP family member 4	GIMAP4	Tubulin- and actin-associating GIMAP4 is required for IFN-γ secretion during Th cell differentiation. PMID: 25287446
GTPase, IMAP family member 5	GIMAP5	GTPase of the immune-associated nucleotide-binding protein 5 (GIMAP5) regulates calcium influx in T-lymphocytes by promoting mitochondrial calcium accumulation. PMID: 23098229
GTPase, IMAP family member 6	GIMAP6	
GTPase, IMAP family member 7	GIMAP7	Structural insights into the mechanism of GTPase activation in the GIMAP family. PMID: 23454188
GTPase, IMAP family member 8	GIMAP8	Generation and characterisation of mice deficient in the multi-GTPase domain containing protein, GIMAP8. PMID: 25329815
Guanylate binding protein 1, interferon-inducible	GBP1	Guanylate binding protein 1 is a novel effector of EGFR-driven invasion in glioblastoma. PMID: 22162832
Guanylate binding protein 2, interferon-inducible	GBP2	Interferon-inducible guanylate binding protein (GBP2) is associated with better prognosis in breast cancer and indicates an efficient T cell response. PMID: 23001506
Guanylate binding protein 3	GBP3	A new splice variant of the human guanylate-binding protein 3 mediates anti-influenza activity through inhibition of viral transcription and replication. PMID: 22106366
Guanylate binding protein 4	GBP4	Guanylate binding protein 4 negatively regulates virus-induced type I IFN and antiviral response by targeting IFN regulatory factor 7. PMID: 22095711

HGNC-approved name	Symbol	Target information
Guanylate binding protein 5	GBP5	GBP5 promotes NLRP3 inflammasome assembly and immunity in mammals. PMID: 22461501
Guanylate binding protein family, member 6	GBP6	
Guanylate binding protein 7	GBP7	
Guanine nucleotide binding protein (G protein), alpha 11 (Gq class)	GNA11	Disease: hypocalciuric hypercalcemia, familial 2 (HHC2)
Guanine nucleotide binding protein (G protein) alpha 12	GNA12	Breast cancer cell invasion mediated by Gα12 signaling involves expression of interleukins-6 and -8, and matrix metalloproteinase-2. PMID: 24976858
Guanine nucleotide binding protein (G protein), alpha 13	GNA13	The gep proto-oncogene Gα13 mediates lysophosphatidic acid-mediated migration of pancreatic cancer cells. PMID: 23508014
Guanine nucleotide binding protein (G protein), alpha 14	GNA14	TNF-α/TNFR1 signaling promotes gastric tumorigenesis through induction of Noxo1 and Gna14 in tumor cells. PMID: 23975421
Guanine nucleotide binding protein (G protein), alpha 15 (Gq class)	GNA15	
DIRAS family, GTP-binding RAS-like 1	DIRAS1	Downregulation of the novel tumor suppressor DIRAS1 predicts poor prognosis in esophageal squamous cell carcinoma. PMID: 23436800
DIRAS family, GTP-binding RAS-like 2	DIRAS2	
DIRAS family, GTP-binding RAS-like 3	DIRAS3	Over-expression of ARHI decreases tumor growth, migration, and invasion in human glioma. PMID: 24458808
ES cell-expressed Ras	ERAS	Role of the ERas gene in gastric cancer cells. PMID: 23612786
GTP binding protein overexpressed in skeletal muscle	GEM	Gem GTPase acts upstream Gmip/RhoA to regulate cortical actin remodeling and spindle positioning during early mitosis. PMID: 25173885
Harvey rat sarcoma viral oncogene homolog	HRAS	Disease: faciocutaneoskeletal syndrome (FCSS)
Kirsten rat sarcoma viral oncogene homolog	KRAS	K-Ras(G12C) inhibitors allosterically control GTP affinity and effector interactions. PMID: 24256730. Disease: leukemia, acute myelogenous (AML)
Muscle RAS oncogene homolog	MRAS	MRAS GTPase is a novel stemness marker that impacts mouse embryonic stem cell plasticity and Xenopus embryonic cell fate. PMID: 23863483
Neuroblastoma RAS viral (v-ras) oncogene homolog	NRAS	Disease: leukemia, juvenile myelomonocytic (JMML)
NFKB inhibitor-interacting Ras-like 1	NKIRAS1	Genetic and epigenetic changes of NKIRAS1 gene in human renal cell carcinomas. PMID: 20693965
NFKB inhibitor-interacting Ras-like 2	NKIRAS2	miR-125b controls apoptosis and temozolomide resistance by targeting TNFAIP3 and NKIRAS2 in glioblastomas. PMID: 24901050
Ras homolog enriched in brain	RHEB	In vivo AAV1 transduction with hRheb(S16H) protects hippocampal neurons by BDNF production. PMID: 25502903
Ras homolog enriched in brain-like 1	RHEBL1	Effects of RhebL1 silencing on the mTOR pathway. PMID: 21655954
RAS, dexamethasone-induced 1	RASD1	Dexras1 mediates glucocorticoid-associated adipogenesis and diet-induced obesity. PMID: 24297897
RASD family, member 2	RASD2	Rhes, a striatal-selective protein implicated in Huntington disease, binds beclin-1 and activates autophagy. PMID: 24324270
RAS-like, estrogen-regulated, growth inhibitor	RERG	ERβ- and prostaglandin E2-regulated pathways integrate cell proliferation via Ras-like and estrogen-regulated growth inhibitor in endometriosis. PMID: 24992181

(Continued)

HGNC-approved name	Symbol	Target information
RAS-like, family 10, member A	RASL10A	RRP22: A novel neural tumor suppressor for astrocytoma. PMID: 21264544
RAS-like, family 10, member B	RASL10B	Regulation of atrial natriuretic peptide secretion by a novel Ras-like protein. PMID: 17984325
RAS-like, family 11, member A	RASL11A	Chromatin association and regulation of rDNA transcription by the Ras-family protein RasL11a. PMID: 20168301
RAS-like, family 11, member B	RASL11B	
RAS-like, family 12	RASL12	
Related RAS viral (r-ras) oncogene homolog	RRAS	A genome-scale RNA-interference screen identifies RRAS signaling as a pathologic feature of Huntington's disease. PMID: 23209424
Related RAS viral (r-ras) oncogene homolog 2	RRAS2	R-RAS2 overexpression in tumors of the human central nervous system. PMID: 24148564. Disease: ovarian cancer (OC)
RERG/RAS-like	RERGL	
v-ral simian leukemia viral oncogene homolog A (ras related)	RALA	Discovery and characterization of small molecules that target the GTPase Ral. PMID: 25219851
v-ral simian leukemia viral oncogene homolog B	RALB	Discovery and characterization of small molecules that target the GTPase Ral. PMID: 25219851
RAB1A, member RAS oncogene family	RAB1A	Rab1A is an mTORC1 activator and a colorectal oncogene. PMID: 25446900
RAB1B, member RAS oncogene family	RAB1B	Rab1b overexpression modifies Golgi size and gene expression in HeLa cells and modulates the thyrotrophin response in thyroid cells in culture. PMID: 23325787
RAB2A, member RAS oncogene family	RAB2A	Rab2A is a pivotal switch protein that promotes either secretion or ER-associated degradation of (pro)insulin in insulin-secreting cells. PMID: 25377857
RAB2B, member RAS oncogene family	RAB2B	
RAB3A, member RAS oncogene family	RAB3A	Rab3a promotes brain tumor initiation and progression. PMID: 24965146
RAB3B, member RAS oncogene family	RAB3B	Rab3B protein is required for long-term depression of hippocampal inhibitory synapses and for normal reversal learning. PMID: 21844341
RAB3C, member RAS oncogene family	RAB3C	
RAB3D, member RAS oncogene family	RAB3D	Rab3D is critical for secretory granule maturation in PC12 cells. PMID: 23526941
RAB4A, member RAS oncogene family	RAB4A	Procathepsin L secretion, which triggers tumour progression, is regulated by Rab4a in human melanoma cells. PMID: 21501115
RAB4B, member RAS oncogene family	RAB4B	Rab4b is a small GTPase involved in the control of the glucose transporter GLUT4 localization in adipocyte. PMID: 19590752
RAB5A, member RAS oncogene family	RAB5A	Vacuolin-1 potently and reversibly inhibits autophagosome-lysosome fusion by activating RAB5A. PMID: 25483964
RAB5B, member RAS oncogene family	RAB5B	Rab-mediated endocytosis: Linking neurodegeneration, neuroprotection, and synaptic plasticity? PMID: 18077583
RAB5C, member RAS oncogene family	RAB5C	Rab5 isoforms orchestrate a 'division of labor' in the endocytic network; Rab5C modulates Rac-mediated cell motility. PMID: 24587345
RAB6A, member RAS oncogene family	RAB6A	Rab6a/a′ are important Golgi regulators of pro-inflammatory TNF secretion in macrophages. PMID: 23437303
RAB6B, member RAS oncogene family	RAB6B	
RAB6C, member RAS oncogene family	RAB6C	WTH3 is a direct target of the p53 protein. PMID: 17426708
RAB7A, member RAS oncogene family	RAB7A	Disease: Charcot–Marie–Tooth disease 2B (CMT2B)
RAB7B, member RAS oncogene family	RAB7B	A novel interaction between Rab7b and actomyosin reveals a dual role in intracellular transport and cell migration. PMID: 25217632

HGNC-approved name	Symbol	Target information
RAB8A, member RAS oncogene family	RAB8A	RAB8A a new biomarker for endometrial cancer? PMID: 25477298
RAB8B, member RAS oncogene family	RAB8B	RAB8B is required for activity and caveolar endocytosis of LRP6. PMID: 24035388
RAB9A, member RAS oncogene family	RAB9A	
RAB9B, member RAS oncogene family	RAB9B	
RAB10, member RAS oncogene family	RAB10	Rab10-mediated endocytosis of the hyaluronan synthase HAS3 regulates hyaluronan synthesis and cell adhesion to collagen. PMID: 24509846
RAB11A, member RAS oncogene family	RAB11A	Global ablation of the mouse rab11a gene impairs early embryogenesis and matrix metalloproteinase secretion. PMID: 25271168
RAB11B, member RAS oncogene family	RAB11B	Rab11 proteins in health and disease. PMID: 23176481
RAB12, member RAS oncogene family	RAB12	Rab12 regulates mTORC1 activity and autophagy through controlling the degradation of amino-acid transporter PAT4. PMID: 23478338
RAB13, member RAS oncogene family	RAB13	Rab13 is upregulated during osteoclast differentiation and associates with small vesicles revealing polarized distribution in resorbing cells. PMID: 22562557
RAB14, member RAS oncogene family	RAB14	A role for Rab14 in the endocytic trafficking of GLUT4 in 3T3-L1 adipocytes. PMID: 23444368
RAB15, member RAS oncogene family	RAB15	Rab15 expression correlates with retinoic acid-induced differentiation of neuroblastoma cells. PMID: 21491086
RAB17, member RAS oncogene family	RAB17	Small GTPase Rab17 regulates dendritic morphogenesis and postsynaptic development of hippocampal neurons. PMID: 22291024
RAB18, member RAS oncogene family	RAB18	Disease: Warburg micro syndrome 3 (WARBM3)
RAB19, member RAS oncogene family	RAB19	
RAB20, member RAS oncogene family	RAB20	The GTPase RAB20 is a HIF target with mitochondrial localization mediating apoptosis in hypoxia. PMID: 21056597
RAB21, member RAS oncogene family	RAB21	Rab21 attenuates EGF-mediated MAPK signaling through enhancing EGFR internalization and degradation. PMID: 22525675
RAB22A, member RAS oncogene family	RAB22A	Hypoxia-inducible factors and RAB22A mediate formation of microvesicles that stimulate breast cancer invasion and metastasis. PMID: 24938788
RAB23, member RAS oncogene family	RAB23	Disease: Carpenter syndrome 1 (CRPT1)
RAB24, member RAS oncogene family	RAB24	Rab24 is required for normal cell division. PMID: 23387408
RAB26, member RAS oncogene family	RAB26	RAB26 coordinates lysosome traffic and mitochondrial localization. PMID: 24413166
RAB27A, member RAS oncogene family	RAB27A	Disease: Griscelli syndrome 2 (GS2)
RAB27B, member RAS oncogene family	RAB27B	Vacuolar H+ ATPase expression and activity is required for Rab27B-dependent invasive growth and metastasis of breast cancer. PMID: 23390068
RAB28, member RAS oncogene family	RAB28	Disease: cone-rod dystrophy 18 (CORD18)
RAB29, member RAS oncogene family	RAB29	A role of Rab29 in the integrity of the trans-Golgi network and retrograde trafficking of mannose-6-phosphate receptor. PMID: 24788816
RAB30, member RAS oncogene family	RAB30	Rab30 is required for the morphological integrity of the Golgi apparatus. PMID: 22188167
RAB31, member RAS oncogene family	RAB31	Rab31 is expressed in neural progenitor cells and plays a role in their differentiation. PMID: 24999186
RAB32, member RAS oncogene family	RAB32	A novel anti-microbial function for a familiar Rab GTPase. PMID: 24321888
RAB33A, member RAS oncogene family	RAB33A	Rab33a mediates anterograde vesicular transport for membrane exocytosis and axon outgrowth. PMID: 22972995

(Continued)

HGNC-approved name	Symbol	Target information
RAB34, member RAS oncogene family	RAB34	RAB34 was a progression- and prognosis-associated biomarker in gliomas. PMID: 25501506
RAB35, member RAS oncogene family	RAB35	Rab35, acting through ACAP2 switching off Arf6, negatively regulates oligodendrocyte differentiation and myelination. PMID: 24600047
RAB36, member RAS oncogene family	RAB36	The Rab interacting lysosomal protein (RILP) homology domain functions as a novel effector domain for small GTPase Rab36: Rab36 regulates retrograde melanosome transport in melanocytes. PMID: 22740695
RAB37, member RAS oncogene family	RAB37	Release of TNF-α from macrophages is mediated by small GTPase Rab37. PMID: 21805469
RAB38, member RAS oncogene family	RAB38	RAB38 confers a poor prognosis, associated with malignant progression and subtype preference in glioma. PMID: 24026199
RAB39A, member RAS oncogene family	RAB39A	Rab39a interacts with phosphatidylinositol 3-kinase and negatively regulates autophagy induced by lipopolysaccharide stimulation in macrophages. PMID: 24349490
RAB40A, member RAS oncogene family	RAB40A	
RAB40B, member RAS oncogene family	RAB40B	Rab40b regulates trafficking of MMP2 and MMP9 during invadopodia formation and invasion of breast cancer cells. PMID: 23902685
RAB40C, member RAS oncogene family	RAB40C	Low-level expression of let-7a in gastric cancer and its involvement in tumorigenesis by targeting RAB40C. PMID: 21349817
RAB41, member RAS oncogene family	RAB41	Rab41 is a novel regulator of Golgi apparatus organization that is needed for ER-to-Golgi trafficking and cell growth. PMID: 23936529
RAB42, member RAS oncogene family	RAB42	
RAB43, member RAS oncogene family	RAB43	
RAB44, member RAS oncogene family	RAB44	
RAB, member of RAS oncogene family-like 2A	RABL2A	
RAB, member of RAS oncogene family-like 2B	RABL2B	
RAB, member of RAS oncogene family-like 3	RABL3	Evaluation of the novel gene Rabl3 in the regulation of proliferation and motility in human cancer cells. PMID: 20596630
RAB, member RAS oncogene family-like 6	RABL6	RBEL1 is a novel gene that encodes a nucleocytoplasmic Ras superfamily GTP-binding protein and is overexpressed in breast cancer. PMID: 17962191
RAN, member RAS oncogene family	RAN	Ran is a potential therapeutic target for cancer cells with molecular changes associated with activation of the PI3K/Akt/mTORC1 and Ras/MEK/ERK pathways. PMID: 22090358
RAP1A, member of RAS oncogene family	RAP1A	Rap1 stabilizes beta-catenin and enhances beta-catenin-dependent transcription and invasion in squamous cell carcinoma of the head and neck. PMID: 20028760
RAP1B, member of RAS oncogene family	RAP1B	Rap1 stabilizes beta-catenin and enhances beta-catenin-dependent transcription and invasion in squamous cell carcinoma of the head and neck. PMID: 20028760
RAP2A, member of RAS oncogene family	RAP2A	Over-expression of Rap2a inhibits glioma migration and invasion by down-regulating p-AKT. PMID: 24293123
RAP2B, member of RAS oncogene family	RAP2B	Rap2B promotes migration and invasion of human suprarenal epithelioma. PMID: 24951956
RAP2C, member of RAS oncogene family	RAP2C	
RAB3 GTPase activating protein subunit 1 (catalytic)	RAB3GAP1	RAB3GAP1 and RAB3GAP2 modulate basal and rapamycin-induced autophagy. PMID: 25495476

HGNC-approved name	Symbol	Target information
RAB3 GTPase activating protein subunit 2 (non-catalytic)	RAB3GAP2	RAB3GAP1 and RAB3GAP2 modulate basal and rapamycin-induced autophagy. PMID: 25495476
ras homolog family member A	RHOA	RhoA: A therapeutic target for chronic myeloid leukemia. PMID: 22443473
ras homolog family member B	RHOB	Critical functions of RhoB in support of glioblastoma tumorigenesis. PMID: 25216671
ras homolog family member C	RHOC	RhoC expression and head and neck cancer metastasis. PMID: 19861405
ras homolog family member D	RHOD	Atypical Rho GTPases RhoD and Rif integrate cytoskeletal dynamics and membrane trafficking. PMID: 24622787
ras homolog family member F	RHOF	The small Rho GTPase Rif and actin cytoskeletal remodelling. PMID: 22260703
ras homolog family member G	RHOG	**BA-210 (Cethrin)** – spinal cord injury treatments
ras homolog family member H	RHOH	Note: A chromosomal aberration involving RHOH is found in a non-Hodgkin lymphoma cell line. Translocation t(34)(q27p11) with BCL6
ras homolog family member J	RHOJ	Vascular RhoJ is an effective and selective target for tumor angiogenesis and vascular disruption. PMID: 24434213
ras homolog family member Q	RHOQ	RNA editing in RHOQ promotes invasion potential in colorectal cancer. PMID: 24663214
ras homolog family member T1	RHOT1	RhoT1 and Smad4 are correlated with lymph node metastasis and overall survival in pancreatic cancer. PMID: 22860091
ras homolog family member T2	RHOT2	Mitochondrial trafficking in neurons and the role of the Miro family of GTPase proteins. PMID: 24256248
ras homolog family member U	RHOU	The atypical Rho GTPase, RhoU, regulates cell-adhesion molecules during cardiac morphogenesis. PMID: 24607366
ras homolog family member V	RHOV	The RHOV gene is overexpressed in human non-small cell lung cancer. PMID: 24388711
Cell division cycle 42	CDC42	A crucial role for CDC42 in senescence-associated inflammation and atherosclerosis. PMID: 25057989
Rho-related BTB domain containing 1	RHOBTB1	The tumor suppressor gene RhoBTB1 is a novel target of miR-31 in human colon cancer. PMID: 23258531
Rho-related BTB domain containing 2	RHOBTB2	Ectopic expression of RhoBTB2 inhibits migration and invasion of human breast cancer cells. PMID: 20930524
Rho-related BTB domain containing 3	RHOBTB3	RhoBTB3 interacts with the 5-HT7a receptor and inhibits its proteasomal degradation. PMID: 22245496
ras-like without CAAX 1	RIT1	Rit subfamily small GTPases: Regulators in neuronal differentiation and survival. PMID: 23770287. Disease: Noonan syndrome 8 (NS8)
ras-like without CAAX 2	RIT2	The plasma membrane-associated GTPase Rin interacts with the dopamine transporter and is required for protein kinase C-regulated dopamine transporter trafficking. PMID: 21957239
RAS (RAD and GEM)-like GTP binding 1	REM1	Rem-GTPase regulates cardiac myocyte L-type calcium current. PMID: 22854599
RAS (RAD and GEM)-like GTP binding 2	REM2	A loss-of-function analysis reveals that endogenous Rem2 promotes functional glutamatergic synapse formation and restricts dendritic complexity. PMID: 23991227
ras-related C3 botulinum toxin substrate 1 (rho family, small GTP binding protein Rac1)	RAC1	Rac1 contributes to trastuzumab resistance of breast cancer cells: Rac1 as a potential therapeutic target for the treatment of trastuzumab-resistant breast cancer. PMID: 19509242

(Continued)

HGNC-approved name	Symbol	Target information
ras-related C3 botulinum toxin substrate 2 (rho family, small GTP binding protein Rac2)	RAC2	Disease: neutrophil immunodeficiency syndrome (NEUID)
ras-related C3 botulinum toxin substrate 3 (rho family, small GTP binding protein Rac3)	RAC3	A role for Rac3 GTPase in the regulation of autophagy. PMID: 21852230
Rho family GTPase 1	RND1	A novel testis-specific GTPase serves as a link to proteasome biogenesis: Functional characterization of RhoS/RSA-14-44 in spermatogenesis. PMID: 20980621
Rho family GTPase 2	RND2	
Rho family GTPase 3	RND3	Rnd3 regulates lung cancer cell proliferation through notch signaling. PMID: 25372032
Ras-related associated with diabetes	RRAD	RRAD promotes EGFR-mediated STAT3 activation and induces temozolomide resistance of malignant glioblastoma. PMID: 25313011
ras-related GTP binding A	RRAGA	Rag GTPases are cardioprotective by regulating lysosomal function. PMID: 24980141
ras-related GTP binding B	RRAGB	
ras-related GTP binding C	RRAGC	The folliculin tumor suppressor is a GAP for the RagC/D GTPases that signal amino acid levels to mTORC1. PMID: 24095279
ras-related GTP binding D	RRAGD	

4.1.5   Other signalling enzymes

HGNC-approved name	Symbol	Target information
2′-5′-Oligoadenylate synthetase 1, 40/46 kDa	OAS1	2′-5′-Oligoadenylate synthetase 1 polymorphism is associated with prostate cancer. PMID: 21638280
2′-5′-Oligoadenylate synthetase 2, 69/71 kDa	OAS2	
2′-5′-Oligoadenylate synthetase 3, 100 kDa	OAS3	The 2′-5′-oligoadenylate synthetase 3 enzyme potently synthesizes the 2′-5′-oligoadenylates required for RNase L activation. PMID: 25275129
2′-5′-Oligoadenylate synthetase-like	OASL	Antiviral activity of human OASL protein is mediated by enhancing signaling of the RIG-I RNA sensor. PMID: 24931123
Farnesyltransferase, CAAX box, alpha	FNTA	**Lonafarnib** – progeria treatments
Farnesyltransferase, CAAX box, beta	FNTB	**Lonafarnib** – progeria treatments
Protein geranylgeranyltransferase type I, beta subunit	PGGT1B	Dissecting the roles of DR4, DR5 and c-FLIP in the regulation of geranylgeranyltransferase I inhibition-mediated augmentation of TRAIL-induced apoptosis. PMID: 20113484
Rab geranylgeranyltransferase, alpha subunit	RABGGTA	Targeting protein prenylation for cancer therapy. PMID: 22020205
Rab geranylgeranyltransferase, beta subunit	RABGGTB	Genes associated with the tumour microenvironment are differentially expressed in cured versus primary chemotherapy-refractory diffuse large B-cell lymphoma. PMID: 18419622
Isoprenylcysteine carboxyl methyltransferase	ICMT	Targeting isoprenylcysteine methylation ameliorates disease in a mouse model of progeria. PMID: 23686339
Protein prenyltransferase alpha subunit repeat containing 1	PTAR1	

HGNC-approved name	Symbol	Target information
Hedgehog acyltransferase	HHAT	Inhibitors of Hedgehog acyltransferase block Sonic Hedgehog signaling. PMID: 23416332
Hedgehog acyltransferase-like	HHATL	Mammalian Gup1, a homolog of Saccharomyces cerevisiae glycerol uptake/transporter 1, acts as a negative regulator for N-terminal palmitoylation of Sonic hedgehog. PMID: 18081866
Nitric oxide synthase 1 (neuronal)	NOS1	**NXN-462** – analgesia
Nitric oxide synthase 2, inducible	NOS2	mNos2 deletion and human NOS2 replacement in Alzheimer disease models. PMID: 25003233
Nitric oxide synthase 3 (endothelial cell)	NOS3	Female resistance to pneumonia identifies lung macrophage nitric oxide synthase-3 as a therapeutic target. PMID: 25317947
Phosphatidic acid phosphatase type 2 domain containing 1A	PPAPDC1A	
Phosphatidic acid phosphatase type 2 domain containing 1B	PPAPDC1B	Characterization of the recurrent 8p11-12 amplicon identifies PPAPDC1B, a phosphatase protein, as a new therapeutic target in breast cancer. PMID: 18757432
Phosphatidic acid phosphatase type 2 domain containing 2	PPAPDC2	Design and synthesis of non-hydrolyzable homoisoprenoid α-monofluorophosphonate inhibitors of PPAPDC family integral membrane lipid phosphatases. PMID: 25150376
Phosphatidic acid phosphatase type 2 domain containing 3	PPAPDC3	Regulation of myoblast differentiation by the nuclear envelope protein NET39. PMID: 19704009
Phosphatidic acid phosphatase type 2A	PPAP2A	Lipid phosphate phosphohydrolase type 1 (LPP1) degrades extracellular lysophosphatidic acid in vivo. PMID: 19215222
Phosphatidic acid phosphatase type 2B	PPAP2B	Mice with targeted inactivation of ppap2b in endothelial and hematopoietic cells display enhanced vascular inflammation and permeability. PMID: 24504738
Phosphatidic acid phosphatase type 2C	PPAP2C	DNA methylation analysis of CD4+ T cells in patients with psoriasis. PMID: 24323136
Phospholipase A1 member A	PLA1A	
Phospholipase A2, group IB (pancreas)	PLA2G1B	**Niflumic acid** – anti-inflammatory and antirheumatic products
Phospholipase A2, group IIA (platelets, synovial fluid)	PLA2G2A	**Varespladib** – lipid-modifying agents
Phospholipase A2, group IID	PLA2G2D	Lymphoid tissue phospholipase A2 group IID resolves contact hypersensitivity by driving antiinflammatory lipid mediators. PMID: 23690440
Phospholipase A2, group IIE	PLA2G2E	The adipocyte-inducible secreted phospholipases PLA2G5 and PLA2G2E play distinct roles in obesity. PMID: 24910243
Phospholipase A2, group IIF	PLA2G2F	
Phospholipase A2, group III	PLA2G3	Mast cell maturation is driven via a group III phospholipase A2-prostaglandin D2-DP1 receptor paracrine axis. PMID: 23624557
Phospholipase A2, group IVA (cytosolic, calcium-dependent)	PLA2G4A	**Giripladib** – anti-inflammatory and antirheumatic products. Note: PLA2G4A mutations resulting in phospholipase A2 deficiency have been found in a patient affected by recurrent episodes of multiple complicated ulcers of the small intestine
Phospholipase A2, group IVB (cytosolic)	PLA2G4B	
Phospholipase A2, group IVC (cytosolic, calcium independent)	PLA2G4C	Cytosolic phospholipase A2 gamma is involved in hepatitis C virus replication and assembly. PMID: 23015700
Phospholipase A2, group IVD (cytosolic)	PLA2G4D	Cloning of a gene for a novel epithelium-specific cytosolic phospholipase A2, cPLA2delta, induced in psoriatic skin. PMID: 14709560

(Continued)

HGNC-approved name	Symbol	Target information
Phospholipase A2, group IVE	PLA2G4E	Cytosolic phospholipase $A_2\varepsilon$ drives recycling through the clathrin-independent endocytic route. PMID: 24413173
Phospholipase A2, group IVF	PLA2G4F	
Phospholipase A2, group V	PLA2G5	Disease: fleck retina, familial benign (FRFB)
Phospholipase A2, group VI (cytosolic, calcium-independent)	PLA2G6	Disease: neurodegeneration with brain iron accumulation 2B (NBIA2B)
Phospholipase A2, group VII (platelet-activating factor acetylhydrolase, plasma)	PLA2G7	**Darapladib** – lipid-modifying agents. Disease: platelet-activating factor acetylhydrolase deficiency (PAFAD)
Phospholipase A2, group X	PLA2G10	
Phospholipase A2, group XIIA	PLA2G12A	
Phospholipase A2, group XIIB	PLA2G12B	
Phospholipase A2, group XV	PLA2G15	Loss of lysophospholipase 3 increases atherosclerosis in apolipoprotein E-deficient mice. PMID: 15781238
Phospholipase A2, group XVI	PLA2G16	Pla2g16 phospholipase mediates gain-of-function activities of mutant p53. PMID: 25024203
Phospholipase B1	PLB1	
Phospholipase B domain containing 1	PLBD1	
Phospholipase B domain containing 2	PLBD2	
Phospholipase C, beta 1 (phosphoinositide-specific)	PLCB1	Nuclear phospholipase C β1 signaling, epigenetics and treatments in MDS. PMID: 23058275. Disease: epileptic encephalopathy, early infantile, 12 (EIEE12)
Phospholipase C, beta 2	PLCB2	WDR26 functions as a scaffolding protein to promote G$\beta\gamma$-mediated phospholipase C β2 (PLCβ2) activation in leukocytes. PMID: 23625927
Phospholipase C, beta 3 (phosphatidylinositol-specific)	PLCB3	Phospholipase C-β3 regulates FcεRI-mediated mast cell activation by recruiting the protein phosphatase SHP-1. PMID: 21683628
Phospholipase C, beta 4	PLCB4	Disease: auriculocondylar syndrome 2 (ARCND2)
Phospholipase C, gamma 1	PLCG1	Disruption of phosphoinositide-specific phospholipases Cγ1 contributes to extracellular matrix synthesis of human osteoarthritis chondrocytes. PMID: 25073093
Phospholipase C, delta 1	PLCD1	Disease: nail disorder, non-syndromic congenital, 3 (NDNC3)
Phospholipase C, delta 3	PLCD3	Simultaneous loss of phospholipase Cδ1 and phospholipase Cδ3 causes cardiomyocyte apoptosis and cardiomyopathy. PMID: 24810051
Phospholipase C, delta 4	PLCD4	
Phospholipase C, epsilon 1	PLCE1	Disease: nephrotic syndrome 3 (NPHS3)
Phospholipase C, zeta 1	PLCZ1	Sperm PLCζ: From structure to Ca2+ oscillations, egg activation and therapeutic potential. PMID: 24157362
Phospholipase C, eta 1	PLCH1	Phospholipase C-η activity may contribute to Alzheimer's disease-associated calciumopathy. PMID: 22475800
Phospholipase C, eta 2	PLCH2	Role of phosphoinositide-specific phospholipase C η2 in isolated and syndromic mental retardation. PMID: 21474938
Phospholipase D1, phosphatidylcholine-specific	PLD1	Further evaluation of novel structural modifications to scaffolds that engender PLD isoform selective inhibition. PMID: 25466173
Phospholipase D2	PLD2	Inhibition of phospholipase D2 induces autophagy in colorectal cancer cells. PMID: 25475140
Phospholipase D family, member 3	PLD3	PLD3 is accumulated on neuritic plaques in Alzheimer's disease brains. PMID: 25478031

HGNC-approved name	Symbol	Target information
Phospholipase D family, member 4	PLD4	PLD4 is involved in phagocytosis of microglia: Expression and localization changes of PLD4 are correlated with activation state of microglia. PMID: 22102906
Phospholipase D family, member 5	PLD5	
Phospholipase D family, member 6	PLD6	
Phospholipid scramblase 1	PLSCR1	Blockade of phospholipid scramblase 1 with its N-terminal domain antibody reduces tumorigenesis of colorectal carcinomas *in vitro* and in vivo. PMID: 23259795
Phospholipid scramblase 2	PLSCR2	
Phospholipid scramblase 3	PLSCR3	Tumor necrosis factor (TNF)-related apoptosis-inducing ligand (TRAIL) induced mitochondrial pathway to apoptosis and caspase activation is potentiated by phospholipid scramblase-3. PMID: 18491232
Phospholipid scramblase 4	PLSCR4	Biochemical and functional characterization of human phospholipid scramblase 4 (hPLSCR4). PMID: 23089641
Phospholipid scramblase family, member 5	PLSCR5	
Phosphatidylinositol-specific phospholipase C, X domain containing 1	PLCXD1	
Phosphatidylinositol-specific phospholipase C, X domain containing 2	PLCXD2	
Phosphatidylinositol specific phospholipase C, X domain containing 3	PLCXD3	Splice site SNPs of phospholipase PLCXD3 are significantly associated with variant and sporadic Creutzfeldt–Jakob disease. PMID: 24028506

4.2 Protein modification

4.2.1 Peptidases

HGNC-approved name	Symbol	Target information
Aspartyl peptidases		
Aspartic peptidase, retroviral-like 1	ASPRV1	SASPase regulates stratum corneum hydration through profilaggrin-to-filaggrin processing. PMID: 21542132
Beta-site APP-cleaving enzyme 1	BACE1	**MK-8931** – anti-dementia drugs
Beta-site APP-cleaving enzyme 2	BACE2	Inhibition of BACE2 counteracts hIAPP-induced insulin secretory defects in pancreatic β-cells. PMID: 25342134
Cathepsin D	CTSD	Disease: ceroid lipofuscinosis, neuronal, 10 (CLN10)
Cathepsin E	CTSE	Pancreatic cancer-associated Cathepsin E as a drug activator. PMID: 23422726
DNA-damage-inducible 1 homolog 1 (*S. cerevisiae*)	DDI1	The DNA damage-inducible UbL-UbA protein Ddi1 participates in Mec1-mediated degradation of Ho endonuclease. PMID: 15964793
DNA-damage-inducible 1 homolog 2 (*S. cerevisiae*)	DDI2	
Histocompatibility (minor) 13	HM13	

(Continued)

HGNC-approved name	Symbol	Target information
Napsin A aspartic peptidase	NAPSA	Frequent expression of napsin A in clear cell carcinoma of the endometrium: Potential diagnostic utility. PMID: 24145649
Nuclear receptor-interacting protein 2	NRIP2	
Nuclear receptor-interacting protein 3	NRIP3	
Pepsinogen 3, group I (pepsinogen A)	PGA3	**Sucralfate** – drugs for acid-related disorders
Pepsinogen 4, group I (pepsinogen A)	PGA4	
Pepsinogen 5, group I (pepsinogen A)	PGA5	
Presenilin 1	PSEN1	**PF-3084014** – antineoplastic agents. Disease: Alzheimer disease 3 (AD3)
Presenilin 2 (Alzheimer disease 4)	PSEN2	Disease: Alzheimer disease 4 (AD4)
Progastricsin (pepsinogen C)	PGC	
Prolactin-induced protein	PIP	Prolactin-induced protein in breast cancer. PMID: 25472539
Renin	REN	**Aliskiren** – antihypertensives. Disease: renal tubular dysgenesis (RTD)
Signal peptide peptidase-like 2A	SPPL2A	The intramembrane proteases signal Peptide peptidase-like 2a and 2b have distinct functions in vivo. PMID: 24492962
Signal peptide peptidase-like 2B	SPPL2B	The intramembrane proteases signal Peptide peptidase-like 2a and 2b have distinct functions in vivo. PMID: 24492962
Signal peptide peptidase-like 2C	SPPL2C	Mechanism, specificity, and physiology of signal peptide peptidase (SPP) and SPP-like proteases. PMID: 24099004
Signal peptide peptidase-like 3	SPPL3	A protease-independent function for SPPL3 in NFAT activation. PMID: 25384971
Cysteine peptidases		
Autophagy related 4A, cysteine peptidase	ATG4A	A mammosphere formation RNAi screen reveals that ATG4A promotes a breast cancer stem-like phenotype. PMID: 24229464
Autophagy related 4B, cysteine peptidase	ATG4B	A novel ATG4B antagonist inhibits autophagy and has a negative impact on osteosarcoma tumors. PMID: 25180520
Autophagy related 4C, cysteine peptidase	ATG4C	Tissue-specific autophagy alterations and increased tumorigenesis in mice deficient in Atg4C/autophagin-3. PMID: 17442669
Autophagy related 4D, cysteine peptidase	ATG4D	Atg4D at the interface between autophagy and apoptosis. PMID: 19713737
Calpain 1, (mu/I) large subunit	CAPN1	Calpain as a therapeutic target in traumatic brain injury. PMID: 20129495
Calpain 2, (m/II) large subunit	CAPN2	Preapoptotic protease calpain-2 is frequently suppressed in adult T-cell leukemia. PMID: 23538341
Calpain 3, (p94)	CAPN3	Disease: limb-girdle muscular dystrophy 2A (LGMD2A)
Calpain 5	CAPN5	Disease: vitreoretinopathy, neovascular inflammatory (VRNI)
Calpain 6	CAPN6	Calpain-6 is an endothelin-1 signaling dependent protective factor in chemoresistant osteosarcoma. PMID: 21681744
Calpain 7	CAPN7	Involvement of calpain-7 in epidermal growth factor receptor degradation via the endosomal sorting pathway. PMID: 24953135
Calpain 8	CAPN8	Calpain 8/nCL-2 and calpain 9/nCL-4 constitute an active protease complex, G-calpain, involved in gastric mucosal defense. PMID: 20686710
Calpain 9	CAPN9	Calpain 8/nCL-2 and calpain 9/nCL-4 constitute an active protease complex, G-calpain, involved in gastric mucosal defense. PMID: 20686710
Calpain 10	CAPN10	Disease: diabetes mellitus, non-insulin-dependent, 1 (NIDDM1)
Calpain 11	CAPN11	Calpain 11 is unique to mouse spermatogenic cells. PMID: 16541461
Calpain 12	CAPN12	
Calpain 13	CAPN13	

HGNC-approved name	Symbol	Target information
Calpain 14	CAPN14	
Calpain 15	CAPN15	
Pan-caspase inhibitor		**IDN-6556** – hepatic injury treatments
Caspase 1, apoptosis-related cysteine peptidase	CASP1	**VR-765** – antiepileptics
Caspase 2, apoptosis-related cysteine peptidase	CASP2	**QPI-1007** – dermatologicals
Caspase 3, apoptosis-related cysteine peptidase	CASP3	**M867** – antineoplastic agents
Caspase 4, apoptosis-related cysteine peptidase	CASP4	Caspase-12 ablation preserves muscle function in the mdx mouse. PMID: 24879640
Caspase 5, apoptosis-related cysteine peptidase	CASP5	The role of acid sphingomyelinase and caspase 5 in hypoxia-induced HuR cleavage and subsequent apoptosis in hepatocytes. PMID: 22906436
Caspase 6, apoptosis-related cysteine peptidase	CASP6	Extracellular caspase-6 drives murine inflammatory pain via microglial TNF-α secretion. PMID: 24531553
Caspase 7, apoptosis-related cysteine peptidase	CASP7	Non-apoptotic functions of caspase-7 during osteogenesis. PMID: 25118926
Caspase 8, apoptosis-related cysteine peptidase	CASP8	Caspase-8 as a therapeutic target in cancer. PMID: 20817393. Disease: caspase-8 deficiency (CASP8D)
Caspase 9, apoptosis-related cysteine peptidase	CASP9	Caspase-9 as a therapeutic target for treating cancer. PMID: 25255697
Caspase 10, apoptosis-related cysteine peptidase	CASP10	Disease: autoimmune lymphoproliferative syndrome 2A (ALPS2A)
Caspase 12 (gene/pseudogene)	CASP12	Modulation of plasmid DNA vaccine antigen clearance by caspase 12 RNA interference potentiates vaccination. PMID: 21325489
Caspase 14, apoptosis-related cysteine peptidase	CASP14	Caspase-14-deficient mice are more prone to the development of parakeratosis. PMID: 23014340
Caspase 16, apoptosis-related cysteine peptidase (putative)	CASP16	
Cathepsin B	CTSB	Cathepsin B as a cancer target. PMID: 23293836
Cathepsin C	CTSC	Therapeutic utility and medicinal chemistry of cathepsin C inhibitors. PMID: 20337582. Disease: Papillon–Lefevre syndrome (PLS)
Cathepsin F	CTSF	Disease: ceroid lipofuscinosis, neuronal, 13 (CLN13)
Cathepsin H	CTSH	CTSH regulates β-cell function and disease progression in newly diagnosed type 1 diabetes patients. PMID: 24982147
Cathepsin K	CTSK	**Odanacatib** – drugs affecting bone structure and mineralization. Disease: pycnodysostosis (PKND)
Cathepsin L	CTSL	Cathepsin L, target in cancer treatment? PMID: 19958782
Cathepsin O	CTSO	
Cathepsin S	CTSS	Inhibition of spinal microglial cathepsin S for the reversal of neuropathic pain. PMID: 17551020
Cathepsin V	CTSV	Decreased cathepsin V expression due to Fli1 deficiency contributes to the development of dermal fibrosis and proliferative vasculopathy in systemic sclerosis. PMID: 23287360
Cathepsin W	CTSW	
Cathepsin Z	CTSZ	Distinct functions of macrophage-derived and cancer cell-derived cathepsin Z combine to promote tumor malignancy via interactions with the extracellular matrix. PMID: 25274726

(Continued)

HGNC-approved name	Symbol	Target information
Desumoylating isopeptidase 1	DESI1	
Desumoylating isopeptidase 2	DESI2	High phosphorylation status of AKT/mTOR signal in DESI2-reduced pancreatic ductal adenocarcinoma. PMID: 25079376
Josephin domain containing 1	JOSD1	JosD1, a membrane-targeted deubiquitinating enzyme, is activated by ubiquitination and regulates membrane dynamics, cell motility, and endocytosis. PMID: 23625928
Josephin domain containing 2	JOSD2	
Secernin 1	SCRN1	Identification of secernin 1 as a novel immunotherapy target for gastric cancer using the expression profiles of cDNA microarray. PMID: 16630140
Secernin 2	SCRN2	
Secernin 3	SCRN3	
SUMO1/sentrin-specific peptidase 1	SENP1	SUMO-specific protease 1 promotes prostate cancer progression and metastasis. PMID: 22733136
SUMO1/sentrin/SMT3-specific peptidase 2	SENP2	Disruption of SUMO-specific protease 2 induces mitochondria mediated neurodegeneration. PMID: 25299344
SUMO1/sentrin/SMT3-specific peptidase 3	SENP3	De-SUMOylation of FOXC2 by SENP3 promotes the epithelial-mesenchymal transition in gastric cancer cells. PMID: 25216525
SUMO1/sentrin-specific peptidase 5	SENP5	Inhibition of SENP5 suppresses cell growth and promotes apoptosis in osteosarcoma cells. PMID: 24926368
SUMO1/sentrin-specific peptidase 6	SENP6	Negative regulation of TLR inflammatory signaling by the SUMO-deconjugating enzyme SENP6. PMID: 23825957
SUMO1/sentrin-specific peptidase 7	SENP7	The deSUMOylase SENP7 promotes chromatin relaxation for homologous recombination DNA repair. PMID: 24018422
SUMO/sentrin-specific peptidase family member 8	SENP8	Central role for endothelial human deneddylase-1/SENP8 in fine-tuning the vascular inflammatory response. PMID: 23209320
Ubiquitin-specific peptidase 1	USP1	Selective and cell-active inhibitors of the USP1/UAF1 deubiquitinase complex reverse cisplatin resistance in non-small cell lung cancer cells. PMID: 22118673
Ubiquitin-specific peptidase 2	USP2	Deubiquitylating enzyme USP2 counteracts Nedd4-2-mediated downregulation of KCNQ1 potassium channels. PMID: 22024150
Ubiquitin-specific peptidase 3	USP3	USP3 inhibits type I interferon signaling by deubiquitinating RIG-I-like receptors. PMID: 24366338
Ubiquitin-specific peptidase 4 (proto-oncogene)	USP4	Ubiquitin-specific protease 4 promotes TNF-α-induced apoptosis by deubiquitination of RIP1 in head and neck squamous cell carcinoma. PMID: 23313255
Ubiquitin-specific peptidase 5 (isopeptidase T)	USP5	The deubiquitinating enzyme USP5 modulates neuropathic and inflammatory pain by enhancing Cav3.2 channel activity. PMID: 25189210
Ubiquitin-specific peptidase 6 (Tre-2 oncogene)	USP6	Note: A chromosomal aberration involving USP6 is a common genetic feature of aneurysmal bone cyst, a benign osseous neoplasm. Translocation t(1617)(q22p13) with CDH11
Ubiquitin-specific peptidase 7 (herpes virus-associated)	USP7	Expression of HAUSP in gliomas correlates with disease progression and survival of patients. PMID: 23483195
Ubiquitin-specific peptidase 8	USP8	USP8 is a novel target for overcoming gefitinib resistance in lung cancer. PMID: 23748694
Ubiquitin-specific peptidase 9, X-linked	USP9X	Deubiquitinase USP9X stabilizes MCL1 and promotes tumour cell survival. PMID: 20023629

HGNC-approved name	Symbol	Target information
Ubiquitin-specific peptidase 9, Y-linked	USP9Y	Note: USP9Y is located in the 'azoospermia factor a' (AZFa) region on chromosome Y which is deleted in Sertoli cell-only syndrome. This is an infertility disorder in which no germ cells are visible in seminiferous tubules leading to azoospermia
Ubiquitin-specific peptidase 10	USP10	Beclin1 controls the levels of p53 by regulating the deubiquitination activity of USP10 and USP13. PMID: 21962518
Ubiquitin-specific peptidase 11	USP11	Mitoxantrone targets human ubiquitin-specific peptidase 11 (USP11) and is a potent inhibitor of pancreatic cancer cell survival. PMID: 23696131
Ubiquitin-specific peptidase 12	USP12	Deubiquitinating enzyme Usp12 is a novel co-activator of the androgen receptor. PMID: 24056413
Ubiquitin-specific peptidase 13 (isopeptidase T-3)	USP13	Beclin1 controls the levels of p53 by regulating the deubiquitination activity of USP10 and USP13. PMID: 21962518
Ubiquitin-specific peptidase 14 (tRNA-guanine transglycosylase)	USP14	Enhancement of proteasome activity by a small-molecule inhibitor of USP14. PMID: 20829789
Ubiquitin-specific peptidase 15	USP15	USP15 stabilizes TGF-β receptor I and promotes oncogenesis through the activation of TGF-β signaling in glioblastoma. PMID: 22344298
Ubiquitin-specific peptidase 16	USP16	Usp16 contributes to somatic stem-cell defects in Down's syndrome. PMID: 24025767. Note: A chromosomal aberration involving USP16 is a cause of Chronic myelomonocytic leukemia
Ubiquitin-specific peptidase 18	USP18	USP18 is crucial for IFN-γ-mediated inhibition of B16 melanoma tumorigenesis and antitumor immunity. PMID: 24884733
Ubiquitin-specific peptidase 19	USP19	Ubiquitin-specific protease 19 regulates the stability of the E3 ubiquitin ligase MARCH6. PMID: 25088257
Ubiquitin-specific peptidase 20	USP20	Ubiquitin-specific peptidase 20 targets TRAF6 and human T cell leukemia virus type 1 tax to negatively regulate NF-kappaB signaling. PMID: 21525354
Ubiquitin-specific peptidase 21	USP21	Deubiquitination and stabilization of IL-33 by USP21. PMID: 25197364
Ubiquitin-specific peptidase 22	USP22	Expression of cancer stem cell marker USP22 in laryngeal squamous cell carcinoma. PMID: 25241865
Ubiquitin-specific peptidase 24	USP24	Genetic evidence for ubiquitin-specific proteases USP24 and USP40 as candidate genes for late-onset Parkinson disease. PMID: 16917932
Ubiquitin-specific peptidase 25	USP25	Negative regulation of IL-17-mediated signaling and inflammation by the ubiquitin-specific protease USP25. PMID: 23042150
Ubiquitin-specific peptidase 26	USP26	Ubiquitin specific protease 26 (USP26) expression analysis in human testicular and extragonadal tissues indicates diverse action of USP26 in cell differentiation and tumorigenesis. PMID: 24922532
Ubiquitin-specific peptidase 27, X-linked	USP27X	
Ubiquitin-specific peptidase 28	USP28	The deubiquitinase USP28 controls intestinal homeostasis and promotes colorectal cancer. PMID: 24960159
Ubiquitin-specific peptidase 29	USP29	USP29 controls the stability of checkpoint adaptor Claspin by deubiquitination. PMID: 24632611
Ubiquitin-specific peptidase 30	USP30	The mitochondrial deubiquitinase USP30 opposes parkin-mediated mitophagy. PMID: 24896179
Ubiquitin-specific peptidase 31	USP31	Human ubiquitin specific protease 31 is a deubiquitinating enzyme implicated in activation of nuclear factor-kappaB. PMID: 16214042
Ubiquitin-specific peptidase 32	USP32	Deubiquitination and stabilization of T-bet by USP10. PMID: 24845384
Ubiquitin-specific peptidase 33	USP33	USP33 regulates centrosome biogenesis via deubiquitination of the centriolar protein CP110. PMID: 23486064

(Continued)

HGNC-approved name	Symbol	Target information
Ubiquitin-specific peptidase 34	USP34	The ubiquitin specific protease USP34 promotes ubiquitin signaling at DNA double-strand breaks. PMID: 23863847
Ubiquitin-specific peptidase 35	USP35	
Ubiquitin-specific peptidase 36	USP36	The deubiquitinating enzyme USP36 controls selective autophagy activation by ubiquitinated proteins. PMID: 22622177
Ubiquitin-specific peptidase 37	USP37	The deubiquitinating enzyme USP37 regulates the oncogenic fusion protein PLZF/RARA stability. PMID: 23208507
Ubiquitin-specific peptidase 38	USP38	
Ubiquitin-specific peptidase 39	USP39	Lentivirus-mediated inhibition of USP39 suppresses the growth of breast cancer cells *in vitro*. PMID: 24126978
Ubiquitin-specific peptidase 40	USP40	Genetic evidence for ubiquitin-specific proteases USP24 and USP40 as candidate genes for late-onset Parkinson disease. PMID: 16917932
Ubiquitin-specific peptidase 41	USP41	
Ubiquitin-specific peptidase 42	USP42	Ubiquitin specific peptidase 42 (USP42) functions to deubiquitylate histones and regulate transcriptional activity. PMID: 25336640
Ubiquitin-specific peptidase 43	USP43	
Ubiquitin-specific peptidase 44	USP44	The deubiquitinase USP44 is a tumor suppressor that protects against chromosome missegregation. PMID: 23187131
Ubiquitin-specific peptidase 45	USP45	USP45 deubiquitylase controls ERCC1-XPF endonuclease-mediated DNA damage responses. PMID: 25538220
Ubiquitin-specific peptidase 46	USP46	Ubiquitin-specific peptidase 46 (Usp46) regulates mouse immobile behavior in the tail suspension test through the GABAergic system. PMID: 22720038
Ubiquitin-specific peptidase 47	USP47	Selective dual inhibitors of the cancer-related deubiquitylating proteases USP7 and USP47. PMID: 24900381
Ubiquitin-specific peptidase 48	USP48	CSN-associated USP48 confers stability to nuclear NF-κB/RelA by trimming K48-linked Ub-chains. PMID: 25486460
Ubiquitin-specific peptidase 49	USP49	USP49 deubiquitinates histone H2B and regulates cotranscriptional pre-mRNA splicing. PMID: 23824326
Ubiquitin-specific peptidase 50	USP50	A screen for deubiquitinating enzymes involved in the G_2/M checkpoint identifies USP50 as a regulator of HSP90-dependent Wee1 stability. PMID: 20930503
Ubiquitin-specific peptidase 51	USP51	
Ubiquitin-specific peptidase 53	USP53	
Ubiquitin-specific peptidase 54	USP54	Ab initio protein modelling reveals novel human MIT domains. PMID: 19302785
Ubiquitin-specific peptidase 17-like family member 1	USP17L1	
Ubiquitin-specific peptidase 17-like family member 2	USP17L2	The deubiquitinating enzyme USP17 is highly expressed in tumor biopsies, is cell cycle regulated, and is required for G1-S progression. PMID: 20388806
Ubiquitin-specific peptidase 17-like family member 3	USP17L3	
Ubiquitin-specific peptidase 17-like family member 4	USP17L4	
Ubiquitin-specific peptidase 17-like family member 5	USP17L5	
Ubiquitin-specific peptidase 17-like family member 7	USP17L7	

HGNC-approved name	Symbol	Target information
Ubiquitin-specific peptidase 17-like family member 8	USP17L8	
Ubiquitin-specific peptidase 17-like family member 10	USP17L10	
Ubiquitin-specific peptidase 17-like family member 11	USP17L11	
Ubiquitin-specific peptidase 17-like family member 12	USP17L12	
Ubiquitin-specific peptidase 17-like family member 13	USP17L13	
Ubiquitin-specific peptidase 17-like family member 15	USP17L15	
Ubiquitin-specific peptidase 17-like family member 17	USP17L17	
Ubiquitin-specific peptidase 17-like family member 18	USP17L18	
Ubiquitin-specific peptidase 17-like family member 19	USP17L19	
Ubiquitin-specific peptidase 17-like family member 20	USP17L20	
Ubiquitin-specific peptidase 17-like family member 21	USP17L21	
Ubiquitin-specific peptidase 17-like family member 22	USP17L22	
Ubiquitin-specific peptidase 17-like family member 23	USP17L23	
Ubiquitin-specific peptidase 17-like family member 24	USP17L24	
Ubiquitin-specific peptidase 17-like family member 25	USP17L25	
Ubiquitin-specific peptidase 17-like family member 26	USP17L26	
Ubiquitin-specific peptidase 17-like family member 27	USP17L27	
Ubiquitin-specific peptidase 17-like family member 28	USP17L28	
Ubiquitin-specific peptidase 17-like family member 29	USP17L29	
Ubiquitin-specific peptidase 17-like family member 30	USP17L30	
UFM1-specific peptidase 1 (non-functional)	UFSP1	Two novel ubiquitin-fold modifier 1 (Ufm1)-specific proteases, UfSP1 and UfSP2. PMID: 17182609
UFM1-specific peptidase 2	UFSP2	Two novel ubiquitin-fold modifier 1 (Ufm1)-specific proteases, UfSP1 and UfSP2. PMID: 17182609
Ubiquitin-specific peptidase-like 1	USPL1	Ubiquitin-specific protease-like 1 (USPL1) is a SUMO isopeptidase with essential, non-catalytic functions. PMID: 22878415
Androglobin	ADGB	Androglobin knockdown inhibits growth of glioma cell lines. PMID: 24966926

(Continued)

HGNC-approved name	Symbol	Target information
Bleomycin hydrolase	BLMH	Cysteine proteases bleomycin hydrolase and cathepsin Z mediate N-terminal proteolysis and toxicity of mutant huntingtin. PMID: 21310951
Cylindromatosis (turban tumor syndrome)	CYLD	Disease: cylindromatosis, familial (FCYL)
Desert hedgehog	DHH	Disease: partial gonadal dysgenesis with minifascicular neuropathy 46,XY (PGD)
Extra spindle pole bodies homolog 1 (*S. cerevisiae*)	ESPL1	ESPL1 is a candidate oncogene of luminal B breast cancers. PMID: 25086634
Legumain	LGMN	Nuclear legumain activity in colorectal cancer. PMID: 23326369
Mucosa-associated lymphoid tissue lymphoma translocation gene 1	MALT1	Pharmacologic inhibition of MALT1 protease by phenothiazines as a therapeutic approach for the treatment of aggressive ABC-DLBCL. PMID: 23238017. Disease: immunodeficiency 12 (IMD12)
Penta-EF-hand domain containing 1	PEF1	
Valosin-containing protein (p97)/p47 complex interacting protein 1	VCPIP1	
Metallopeptidases		
ADAM-like, decysin 1	ADAMDEC1	ADAMDEC1 is a metzincin metalloprotease with dampened proteolytic activity. PMID: 23754285
ADAM metallopeptidase domain 2	ADAM2	Impaired sperm aggregation in Adam2 and Adam3 null mice. PMID: 20400072
ADAM metallopeptidase domain 7	ADAM7	Note: Has been found to be frequently mutated in melanoma. ADAM7 mutations may play a role in melanoma progression and metastasis
ADAM metallopeptidase domain 8	ADAM8	Protective effects of ADAM8 against cisplatin-mediated apoptosis in non-small-cell lung cancer. PMID: 23319321
ADAM metallopeptidase domain 9	ADAM9	RNAi-mediated ADAM9 gene silencing inhibits metastasis of adenoid cystic carcinoma cells. PMID: 20422344. Disease: cone-rod dystrophy 9 (CORD9)
ADAM metallopeptidase domain 10	ADAM10	ADAM-10 is overexpressed in rheumatoid arthritis synovial tissue and mediates angiogenesis. PMID: 23124962. Disease: reticulate acropigmentation of Kitamura (RAK)
ADAM metallopeptidase domain 11	ADAM11	Altered nociceptive response in ADAM11-deficient mice. PMID: 16729981
ADAM metallopeptidase domain 12	ADAM12	ADAM12 transmembrane and secreted isoforms promote breast tumor growth: A distinct role for ADAM12-S protein in tumor metastasis. PMID: 21493715
ADAM metallopeptidase domain 15	ADAM15	ADAM15 participates in fertilization through a physical interaction with acrogranin. PMID: 25392190
ADAM metallopeptidase domain 17	ADAM17	ADAM metallopeptidase domain 17 (ADAM17) is naturally processed through major histocompatibility complex (MHC) class I molecules and is a potential immunotherapeutic target in breast, ovarian and prostate cancers. PMID: 21175594. Disease: inflammatory skin and bowel disease, neonatal, 1 (NISBD1)
ADAM metallopeptidase domain 18	ADAM18	
ADAM metallopeptidase domain 19	ADAM19	High expression of the 'A Disintegrin And Metalloprotease' 19 (ADAM19), a sheddase for TNF-α in the mucosa of patients with inflammatory bowel diseases. PMID: 23429442
ADAM metallopeptidase domain 20	ADAM20	
ADAM metallopeptidase domain 21	ADAM21	A disintegrin and metalloprotease 21 (ADAM21) is associated with neurogenesis and axonal growth in developing and adult rodent CNS. PMID: 16052496

HGNC-approved name	Symbol	Target information
ADAM metallopeptidase domain 22	ADAM22	ADAM22 as a prognostic and therapeutic drug target in the treatment of endocrine-resistant breast cancer. PMID: 23810013
ADAM metallopeptidase domain 23	ADAM23	Intratumoral heterogeneity of ADAM23 promotes tumor growth and metastasis through LGI4 and nitric oxide signals. PMID: 24662834
ADAM metallopeptidase domain 28	ADAM28	Src plays a key role in ADAM28 expression in v-src-transformed epithelial cells and human carcinoma cells. PMID: 24007880
ADAM metallopeptidase domain 29	ADAM29	Note: Has been found to be frequently mutated in melanoma. ADAM7 mutations may play a role in melanoma progression and metastasis
ADAM metallopeptidase domain 30	ADAM30	
ADAM metallopeptidase domain 32	ADAM32	Identification and characterization of ADAM32 with testis-predominant gene expression. PMID: 12568724
ADAM metallopeptidase domain 33	ADAM33	Disease: asthma (ASTHMA)
ADAM metallopeptidase with thrombospondin type 1 motif, 1	ADAMTS1	Relevance of IGFBP2 proteolysis in glioma and contribution of the extracellular protease ADAMTS1. PMID: 24962328
ADAM metallopeptidase with thrombospondin type 1 motif, 2	ADAMTS2	Disease: Ehlers–Danlos syndrome 7C (EDS7C)
ADAM metallopeptidase with thrombospondin type 1 motif, 3	ADAMTS3	Increased serum ADAMTS-4 in knee osteoarthritis: A potential indicator for the diagnosis of osteoarthritis in early stages. PMID: 25501175
ADAM metallopeptidase with thrombospondin type 1 motif, 4	ADAMTS4	Drug insight: Aggrecanases as therapeutic targets for osteoarthritis. PMID: 18577998
ADAM metallopeptidase with thrombospondin type 1 motif, 5	ADAMTS5	Drug insight: Aggrecanases as therapeutic targets for osteoarthritis. PMID: 18577998
ADAM metallopeptidase with thrombospondin type 1 motif, 6	ADAMTS6	
ADAM metallopeptidase with thrombospondin type 1 motif, 7	ADAMTS7	ADAMTS7: A promising new therapeutic target in coronary heart disease. PMID: 23829786
ADAM metallopeptidase with thrombospondin type 1 motif, 8	ADAMTS8	The metalloprotease ADAMTS8 displays antitumor properties through antagonizing EGFR-MEK-ERK signaling and is silenced in carcinomas by CpG methylation. PMID: 24184540
ADAM metallopeptidase with thrombospondin type 1 motif, 9	ADAMTS9	Lentiviral shRNA knock-down of ADAMTS-5 and -9 restores matrix deposition in 3D chondrocyte culture. PMID: 20568084
ADAM metallopeptidase with thrombospondin type 1 motif, 10	ADAMTS10	Disease: Weill–Marchesani syndrome 1 (WMS1)
ADAM metallopeptidase with thrombospondin type 1 motif, 12	ADAMTS12	ADAMTS-12 metalloprotease is necessary for normal inflammatory response. PMID: 23019333
ADAM metallopeptidase with thrombospondin type 1 motif, 13	ADAMTS13	Disease: thrombotic thrombocytopenic purpura congenital (TTP)
ADAM metallopeptidase with thrombospondin type 1 motif, 14	ADAMTS14	Association of a nsSNP in ADAMTS14 to some osteoarthritis phenotypes. PMID: 18790654
ADAM metallopeptidase with thrombospondin type 1 motif, 15	ADAMTS15	Androgen regulates ADAMTS15 gene expression in prostate cancer cells. PMID: 20590445
ADAM metallopeptidase with thrombospondin type 1 motif, 16	ADAMTS16	Cryptorchidism and infertility in rats with targeted disruption of the Adamts16 locus. PMID: 24983376
ADAM metallopeptidase with thrombospondin type 1 motif, 17	ADAMTS17	Disease: Weill–Marchesani-like syndrome (WMLS)
ADAM metallopeptidase with thrombospondin type 1 motif, 18	ADAMTS18	ADAMTS-18: A metalloproteinase with multiple functions. PMID: 24896365. Disease: microcornea, myopic chorioretinal atrophy, and telecanthus (MMCAT)

(*Continued*)

HGNC-approved name	Symbol	Target information
ADAM metallopeptidase with thrombospondin type 1 motif, 19	ADAMTS19	
ADAM metallopeptidase with thrombospondin type 1 motif, 20	ADAMTS20	The secreted metalloprotease ADAMTS20 is required for melanoblast survival. PMID: 18454205
ADAMTS-like 1	ADAMTSL1	
ADAMTS-like 2	ADAMTSL2	Disease: geleophysic dysplasia 1 (GPHYSD1)
ADAMTS-like 3	ADAMTSL3	ADAMTSL3/punctin-2, a gene frequently mutated in colorectal tumors, is widely expressed in normal and malignant epithelial cells, vascular endothelial cells and other cell types, and its mRNA is reduced in colon cancer. PMID: 17597111
ADAMTS-like 4	ADAMTSL4	Disease: ectopia lentis 2, isolated, autosomal recessive (ECTOL2)
ADAMTS-like 5	ADAMTSL5	A disintegrin-like and metalloprotease domain containing thrombospondin type 1 motif-like 5 (ADAMTSL5) is a novel fibrillin-1-, fibrillin-2-, and heparin-binding member of the ADAMTS superfamily containing a netrin-like module. PMID: 23010571
Angiotensin I converting enzyme	ACE	**Lisinopril** – antihypertensives. Disease: ischemic stroke (ISCHSTR)
Angiotensin I converting enzyme 2	ACE2	ACE2, a promising therapeutic target for pulmonary hypertension. PMID: 21215698
Arginyl aminopeptidase (aminopeptidase B)	RNPEP	**Ubenimex** – antineoplastic agents
Arginyl aminopeptidase (aminopeptidase B)-like 1	RNPEPL1	Arginyl aminopeptidase-like 1 (RNPEPL1) is an alternatively processed aminopeptidase with specificity for methionine, glutamine, and citrulline residues. PMID: 19508204
Archaelysin family metallopeptidase 1	AMZ1	Identification and characterization of human archaemetzincin-1 and -2, two novel members of a family of metalloproteases widely distributed in Archaea. PMID: 15972818
Archaelysin family metallopeptidase 2	AMZ2	Identification and characterization of human archaemetzincin-1 and -2, two novel members of a family of metalloproteases widely distributed in Archaea. PMID: 15972818
ATP/GTP binding protein 1	AGTPBP1	Abnormal sperm development in pcd(3J)–/– mice: The importance of Agtpbp1 in spermatogenesis. PMID: 21110128
ATP/GTP binding protein-like 1	AGBL1	Disease: corneal dystrophy, Fuchs endothelial, 8 (FECD8)
ATP/GTP binding protein-like 2	AGBL2	Tumor suppressor RARRES1 interacts with cytoplasmic carboxypeptidase AGBL2 to regulate the α-tubulin tyrosination cycle. PMID: 21303978
ATP/GTP binding protein-like 3	AGBL3	The cytosolic carboxypeptidases CCP2 and CCP3 catalyze posttranslational removal of acidic amino acids. PMID: 25103237
ATP/GTP binding protein-like 4	AGBL4	Cytosolic carboxypeptidase CCP6 is required for megakaryopoiesis by modulating Mad2 polyglutamylation. PMID: 25332286
ATP/GTP binding protein-like 5	AGBL5	Cytosolic carboxypeptidase 5 removes α- and γ-linked glutamates from tubulin. PMID: 24022482
Carboxypeptidase A1 (pancreatic)	CPA1	Variants in CPA1 are strongly associated with early onset chronic pancreatitis. PMID: 23955596
Carboxypeptidase A2 (pancreatic)	CPA2	
Carboxypeptidase A3 (mast cell)	CPA3	Pivotal role of mast cell carboxypeptidase A in mediating protection against small intestinal ischemia-reperfusion injury in rats after ischemic preconditioning. PMID: 24953986

HGNC-approved name	Symbol	Target information
Carboxypeptidase A4	CPA4	Characterization of the substrate specificity of human carboxypeptidase A4 and implications for a role in extracellular peptide processing. PMID: 20385563
Carboxypeptidase A5	CPA5	
Carboxypeptidase A6	CPA6	Note: A chromosomal aberration involving CPA6 was found in a patient with Duane retraction syndrome. Translocation t(6;8) (q26;q13). Disease: epilepsy, familial temporal lobe, 5 (ETL5)
Carboxypeptidase B1 (tissue)	CPB1	
Carboxypeptidase B2 (plasma)	CPB2	**DS-1040** – vasoprotectives
Carboxypeptidase D	CPD	SiRNA-targeted carboxypeptidase D inhibits hepatocellular carcinoma growth. PMID: 23589395
Carboxypeptidase E	CPE	Downregulation of CPE regulates cell proliferation and chemosensitivity in pancreatic cancer. PMID: 25374060
Carboxypeptidase M	CPM	The potential of carboxypeptidase M as a therapeutic target in cancer. PMID: 23294303
Carboxypeptidase N, polypeptide 1	CPN1	Disease: carboxypeptidase N deficiency (CPND)
Carboxypeptidase N, polypeptide 2	CPN2	
Carboxypeptidase O	CPO	Carboxypeptidase O is a glycosylphosphatidylinositol-anchored intestinal peptidase with acidic amino acid specificity. PMID: 21921028
Carboxypeptidase Q	CPQ	
Carboxypeptidase X (M14 family), member 1	CPXM1	
Carboxypeptidase X (M14 family), member 2	CPXM2	
Carboxypeptidase Z	CPZ	Carboxypeptidase Z (CPZ) links thyroid hormone and Wnt signaling pathways in growth plate chondrocytes. PMID: 18847325
Carnosine dipeptidase 1 (metallopeptidase M20 family)	CNDP1	Carnosine as a protective factor in diabetic nephropathy: Association with a leucine repeat of the carnosinase gene CNDP1. PMID: 16046297
CNDP dipeptidase 2 (metallopeptidase M20 family)	CNDP2	Up-regulation of CNDP2 facilitates the proliferation of colon cancer. PMID: 24885395
COP9 signalosome subunit 5	COPS5	Targeting Jab1/CSN5 in nasopharyngeal carcinoma. PMID: 22867945
COP9 signalosome subunit 6	COPS6	CSN6 drives carcinogenesis by positively regulating Myc stability. PMID: 25395170
Dipeptidase 1 (renal)	DPEP1	**Cilastatin** – adjuvants
Dipeptidase 2	DPEP2	
Dipeptidase 3	DPEP3	Molecular characterization and expression of dipeptidase 3, a testis-specific membrane-bound dipeptidase: Complex formation with TEX101, a germ-cell-specific antigen in the mouse testis. PMID: 21724266
Endothelin-converting enzyme 1	ECE1	Disease: Hirschsprung disease cardiac defects and autonomic dysfunction (HSCRCDAD)
Endothelin-converting enzyme 2	ECE2	Endothelin-converting enzyme 2 differentially regulates opioid receptor activity. PMID: 24990314
Endothelin-converting enzyme-like 1	ECEL1	Disease: arthrogryposis, distal, 5D (DA5D)

(Continued)

HGNC-approved name	Symbol	Target information
Endoplasmic reticulum aminopeptidase 1	FRAP1	ERAP1 in the pathogenesis of ankylosing spondylitis. PMID: 25434650
Endoplasmic reticulum aminopeptidase 2	ERAP2	Rationally designed inhibitor targeting antigen-trimming aminopeptidases enhances antigen presentation and cytotoxic T-cell responses. PMID: 24248368
Folate hydrolase (prostate-specific membrane antigen) 1	FOLH1	**PSMA ADC** – antineoplastic agents
Folate hydrolase 1B	FOLH1B	
Glutaminyl-peptide cyclotransferase	QPCT	Glutaminyl cyclase inhibition attenuates pyroglutamate Abeta and Alzheimer's disease-like pathology. PMID: 18836460
Glutaminyl-peptide cyclotransferase-like	QPCTL	
Matrix metallopeptidase 1 (interstitial collagenase)	MMP1	Drug evaluation: Apratastat, a novel TACE/MMP inhibitor for rheumatoid arthritis. PMID: 17117591
Matrix metallopeptidase 2 (gelatinase A, 72 kDa gelatinase, 72 kDa type IV collagenase)	MMP2	O-phenyl carbamate and phenyl urea thiiranes as selective matrix metalloproteinase-2 inhibitors that cross the blood-brain barrier. PMID: 24028490. Disease: multicentric osteolysis, nodulosis, and arthropathy (MONA)
Matrix metallopeptidase 3 (stromelysin 1, progelatinase)	MMP3	Disease: coronary heart disease 6 (CHDS6)
Matrix metallopeptidase 7 (matrilysin, uterine)	MMP7	Matrix metalloproteinase-7 activates heparin-binding epidermal growth factor-like growth factor in cutaneous squamous cell carcinoma. PMID: 20586780
Matrix metallopeptidase 8 (neutrophil collagenase)	MMP8	An important role of matrix metalloproteinase-8 in angiogenesis *in vitro* and in vivo. PMID: 23512982
Matrix metallopeptidase 9 (gelatinase B, 92 kDa gelatinase, 92 kDa type IV collagenase)	MMP9	**GS-5745** – drugs for functional gastrointestinal disorders. Disease: intervertebral disc disease (IDD)
Matrix metallopeptidase 10 (stromelysin 2)	MMP10	Matrix metalloproteinase-10 is required for lung cancer stem cell maintenance, tumor initiation and metastatic potential. PMID: 22545096
Matrix metallopeptidase 11 (stromelysin 3)	MMP11	Knockdown of MMP11 inhibits proliferation and invasion of gastric cancer cells. PMID: 23755751
Matrix metallopeptidase 12 (macrophage elastase)	MMP12	**V85546** – drugs for obstructive airway diseases
Matrix metallopeptidase 13 (collagenase 3)	MMP13	Drug evaluation: Apratastat, a novel TACE/MMP inhibitor for rheumatoid arthritis. PMID: 17117591. Disease: spondyloepimetaphyseal dysplasia, Missouri type (SEMD-MO)
Matrix metallopeptidase 14 (membrane-inserted)	MMP14	Cancer cell-associated MT1-MMP promotes blood vessel invasion and distant metastasis in triple-negative mammary tumors. PMID: 21571860. Disease: Winchester syndrome (WNCHRS)
Matrix metallopeptidase 15 (membrane-inserted)	MMP15	MT2-MMP expression associates with tumor progression and angiogenesis in human lung cancer. PMID: 25031779
Matrix metallopeptidase 16 (membrane-inserted)	MMP16	Membrane-type-3 matrix metalloproteinase (MT3-MMP) functions as a matrix composition-dependent effector of melanoma cell invasion. PMID: 22164270
Matrix metallopeptidase 17 (membrane-inserted)	MMP17	Matrix metalloproteinase 17 is necessary for cartilage aggrecan degradation in an inflammatory environment. PMID: 21216815
Matrix metallopeptidase 19	MMP19	Note: May play a role in pathological processes participating in rheumatoid arthritis (RA)-associated joint tissue destruction. Autoantigen anti-MMP19 are frequent in RA patients
Matrix metallopeptidase 20	MMP20	Disease: amelogenesis imperfecta, hypomaturation type, 2A2 (AI2A2)

HGNC-approved name	Symbol	Target information
Matrix metallopeptidase 21	MMP21	Increased MMP-21 expression is associated with poor overall survival of patients with gastric cancer. PMID: 23275114
Matrix metallopeptidase 23B	MMP23B	Domain structure and function of matrix metalloprotease 23 (MMP23): Role in potassium channel trafficking. PMID: 23912897
Matrix metallopeptidase 24 (membrane-inserted)	MMP24	Metalloproteinase MT5-MMP is an essential modulator of neuro-immune interactions in thermal pain stimulation. PMID: 19805319
Matrix metallopeptidase 25	MMP25	Biochemical characterization and N-terminomics analysis of leukolysin, the membrane-type 6 matrix metalloprotease (MMP25): Chemokine and vimentin cleavages enhance cell migration and macrophage phagocytic activities. PMID: 22367194
Matrix metallopeptidase 26	MMP26	Regulation of chondrosarcoma invasion by MMP26. PMID: 25260492
Matrix metallopeptidase 27	MMP27	
Matrix metallopeptidase 28	MMP28	MMP28 (epilysin) as a novel promoter of invasion and metastasis in gastric cancer. PMID: 21615884
Membrane metallo-endopeptidase	MME	**Candoxatril** – antihypertensives
Membrane metallo-endopeptidase-like 1	MMEL1	A non-synonymous SNP within membrane metalloendopeptidase-like 1 (MMEL1) is associated with multiple sclerosis. PMID: 20574445
Meprin A, alpha (PABA peptide hydrolase)	MEP1A	Balance of meprin A and B in mice affects the progression of experimental inflammatory bowel disease. PMID: 21071511
Meprin A, beta	MEP1B	Metalloproteases meprin α and meprin β are C- and N-procollagen proteinases important for collagen assembly and tensile strength. PMID: 23940311
Methionyl aminopeptidase 1	METAP1	Pyridinylquinazolines selectively inhibit human methionine aminopeptidase-1 in cells. PMID: 23634668
Methionyl aminopeptidase type 1D (mitochondrial)	METAP1D	MAP1D, a novel methionine aminopeptidase family member is overexpressed in colon cancer. PMID: 16568094
Methionyl aminopeptidase 2	METAP2	**Beloranib** – antiobesity preparations
N-acetylated alpha-linked acidic dipeptidase 2	NAALAD2	
N-acetylated alpha-linked acidic dipeptidase-like 1	NAALADL1	
N-acetylated alpha-linked acidic dipeptidase-like 2	NAALADL2	N-acetyl-L-aspartyl-L-glutamate peptidase-like 2 is overexpressed in cancer and promotes a pro-migratory and pro-metastatic phenotype. PMID: 24240687
O-sialoglycoprotein endopeptidase	OSGEP	
O-sialoglycoprotein endopeptidase-like 1	OSGEPL1	Qri7/OSGEPL, the mitochondrial version of the universal Kae1/YgjD protein, is essential for mitochondrial genome maintenance. PMID: 19578062
Pregnancy-associated plasma protein A, pappalysin 1	PAPPA	Mice deficient in PAPP-A show resistance to the development of diabetic nephropathy. PMID: 23881937
Pappalysin 2	PAPPA2	Pregnancy associated plasma protein A2 (PAPP-A2) affects bone size and shape and contributes to natural variation in postnatal growth in mice. PMID: 23457539
Peptidase M20 domain containing 1	PM20D1	
Peptidase M20 domain containing 2	PM20D2	Metabolite proofreading in carnosine and homocarnosine synthesis: Molecular identification of PM20D2 as β-alanyl-lysine dipeptidase. PMID: 24891507

HGNC-approved name	Symbol	Target information
Peptidase (mitochondrial processing) alpha	PMPCA	Mitochondrial processing peptidase regulates PINK1 processing, import and Parkin recruitment. PMID: 22354088
Peptidase (mitochondrial processing) beta	PMPCB	
Pyroglutamyl-peptidase I	PGPEP1	
Pyroglutamyl-peptidase I-like	PGPEP1L	
STAM binding protein	STAMBP	Disease: microcephaly-capillary malformation syndrome (MICCAP)
STAM binding protein-like 1	STAMBPL1	An RNA interference screen identifies the deubiquitinase STAMBPL1 as a critical regulator of human T-cell leukemia virus type 1 tax nuclear export and NF-κB activation. PMID: 22258247
Tolloid-like 1	TLL1	Disease: atrial septal defect 6 (ASD6)
Tolloid-like 2	TLL2	Cross-species genetics converge to TLL2 for mouse avoidance behavior and human bipolar disorder. PMID: 23777486
TraB domain containing 2A	TRABD2A	Tiki1 is required for head formation via Wnt cleavage-oxidation and inactivation. PMID: 22726442
TraB domain containing 2B	TRABD2B	TIKI2 suppresses growth of osteosarcoma by targeting Wnt/β-catenin pathway. PMID: 24771064
X-prolyl aminopeptidase (aminopeptidase P) 1, soluble	XPNPEP1	
X-prolyl aminopeptidase (aminopeptidase P) 2, membrane-bound	XPNPEP2	A variant in XPNPEP2 is associated with angioedema induced by angiotensin I-converting enzyme inhibitors. PMID: 16175507
X-prolyl aminopeptidase (aminopeptidase P) 3, putative	XPNPEP3	Disease: nephronophthisis-like nephropathy 1 (NPHPL1)
AE binding protein 1	AEBP1	Aortic carboxypeptidase-like protein (ACLP) enhances lung myofibroblast differentiation through transforming growth factor β receptor-dependent and -independent pathways. PMID: 24344132
AFG3 ATPase family member 3-like 2 (S. cerevisiae)	AFG3L2	Disease: spinocerebellar ataxia 28 (SCA28)
Alanyl (membrane) aminopeptidase	ANPEP	**Ubenimex** – antineoplastic agents
Aminopeptidase puromycin sensitive	NPEPPS	Genetic associations and functional characterization of M1 aminopeptidases and immune-mediated diseases. PMID: 25142031
Aminopeptidase-like 1	NPEPL1	
Aspartyl aminopeptidase	DNPEP	Short forms of Ste20-related proline/alanine-rich kinase (SPAK) in the kidney are created by aspartyl aminopeptidase (Dnpep)-mediated proteolytic cleavage. PMID: 25164821
Astacin-like metallo-endopeptidase (M12 family)	ASTL	Ovastacin, a cortical granule protease, cleaves ZP2 in the zona pellucida to prevent polyspermy. PMID: 22472438
Bone morphogenetic protein 1	BMP1	Disease: osteogenesis imperfecta 13 (OI13)
BRCA1/BRCA2-containing complex, subunit 3	BRCC3	Deubiquitination of NLRP3 by BRCC3 critically regulates inflammasome activity. PMID: 23246432. Note: A chromosomal aberration involving BRCC3 is a cause of pro-lymphocytic T-cell leukemia (T-PLL). Translocation t(X14)(q28q11) with TCRA
Charged multivesicular body protein 1A	CHMP1A	Disease: pontocerebellar hypoplasia 8 (PCH8)
Chromosome 9 open reading frame 3	C9orf3	Identification of human aminopeptidase O, a novel metalloprotease with structural similarity to aminopeptidase B and leukotriene A4 hydrolase. PMID: 15687497
Dipeptidyl-peptidase 3	DPP3	Dipeptidyl peptidase III: A multifaceted oligopeptide N-end cutter. PMID: 21794094

HGNC-approved name	Symbol	Target information
Endoplasmic reticulum metallopeptidase 1	ERMP1	Fxna, a novel gene differentially expressed in the rat ovary at the time of folliculogenesis, is required for normal ovarian histogenesis. PMID: 17267443
Eukaryotic translation initiation factor 3, subunit F	EIF3F	eIF3f: A central regulator of the antagonism atrophy/hypertrophy in skeletal muscle. PMID: 23769948
Glutamyl aminopeptidase (aminopeptidase A)	ENPEP	A new strategy for treating hypertension by blocking the activity of the brain renin-angiotensin system with aminopeptidase A inhibitors. PMID: 24697296
Insulin-degrading enzyme	IDE	Anti-diabetic activity of insulin-degrading enzyme inhibitors mediated by multiple hormones. PMID: 24847884
Kell blood group, metallo-endopeptidase	KEL	
Leishmanolysin-like (metallopeptidase M8 family)	LMLN	Invadolysin, a conserved lipid-droplet-associated metalloproteinase, is required for mitochondrial function in Drosophila. PMID: 23943867
Leucine aminopeptidase 3	LAP3	LAP3 promotes glioma progression by regulating proliferation, migration and invasion of glioma cells. PMID: 25453285
Leucyl/cystinyl aminopeptidase	LNPEP	Disulfide cyclized tripeptide analogues of angiotensin IV as potent and selective inhibitors of insulin-regulated aminopeptidase (IRAP). PMID: 21047126
Leukotriene A4 hydrolase	LTA4H	**Ubenimex** – antineoplastic agents
Membrane-bound transcription factor peptidase, site 2	MBTPS2	Disease: IFAP syndrome with or without BRESHECK syndrome (IFAPS)
Mitochondrial intermediate peptidase	MIPEP	Gamma-secretase-regulated proteolysis of the Notch receptor by mitochondrial intermediate peptidase. PMID: 21685396
Myb-like, SWIRM and MPN domains 1	MYSM1	Epigenetic control of natural killer cell maturation by histone H2A deubiquitinase, MYSM1. PMID: 24062447
Nardilysin (N-arginine dibasic convertase)	NRD1	Nardilysin-dependent proteolysis of cell-associated VTCN1 (B7-H4) marks type 1 diabetes development. PMID: 24848066
Neurolysin (metallopeptidase M3 family)	NLN	Neurolysin knockout mice generation and initial phenotype characterization. PMID: 24719317
OMA1 zinc metallopeptidase	OMA1	Activation of mitochondrial protease OMA1 by Bax and Bak promotes cytochrome c release during apoptosis. PMID: 25275009
Peptidase D	PEPD	Disease: prolidase deficiency (PD)
Phosphate-regulating endopeptidase homolog, X-linked	PHEX	Disease: hypophosphatemic rickets, X-linked dominant (XLHR)
Pitrilysin metallopeptidase 1	PITRM1	The metalloendopeptidase gene Pitrm1 is regulated by hedgehog signaling in the developing mouse limb and is expressed in muscle progenitors. PMID: 19877269
RCE1 homolog, prenyl protein protease (S. cerevisiae)	RCE1	Mechanism of farnesylated CAAX protein processing by the intramembrane protease Rce1. PMID: 24291792
Spastic paraplegia 7 (pure and complicated autosomal recessive)	SPG7	Disease: spastic paraplegia 7, autosomal recessive (SPG7)
Thimet oligopeptidase 1	THOP1	Expression of THOP1 and its relationship to prognosis in non-small cell lung cancer. PMID: 25180910
Thrombospondin, type I, domain containing 4	THSD4	ADAMTSL-6 is a novel extracellular matrix protein that binds to fibrillin-1 and promotes fibrillin-1 fibril formation. PMID: 19940141
Thyrotropin-releasing hormone degrading enzyme	TRHDE	Discovery of a dual action first-in-class peptide that mimics and enhances CNS-mediated actions of thyrotropin-releasing hormone. PMID: 17418282
XRCC6 binding protein 1	XRCC6BP1	Glioma-amplified sequence KUB3 influences double-strand break repair after ionizing radiation. PMID: 23670597

(Continued)

HGNC-approved name	Symbol	Target information
YME1-like 1 ATPase	YME1L1	The i-AAA protease YME1L and OMA1 cleave OPA1 to balance mitochondrial fusion and fission. PMID: 24616225
Zinc metallopeptidase STE24	ZMPSTE24	ZMPSTE24, an integral membrane zinc metalloprotease with a connection to progeroid disorders. PMID: 19453269. Disease: mandibuloacral dysplasia with type B lipodystrophy (MADB)

Serine peptidases

Cathepsin A	CTSA	Disease: galactosialidosis (GSL)
Cathepsin G	CTSG	Cathepsin G-regulated release of formyl peptide receptor agonists modulate neutrophil effector functions. PMID: 22879591
Chymotrypsin C (caldecrin)	CTRC	Disease: pancreatitis, hereditary (PCTT)
Chymotrypsin-like	CTRL	
Chymotrypsin-like elastase family, member 1	CELA1	Effects of recombinant human type I pancreatic elastase on human atherosclerotic arteries. PMID: 25490419
Chymotrypsin-like elastase family, member 2A	CELA2A	
Chymotrypsin-like elastase family, member 2B	CELA2B	Inactivity of recombinant ELA2B provides a new example of evolutionary elastase silencing in humans. PMID: 16327289
Chymotrypsin-like elastase family, member 3A	CELA3A	
Chymotrypsin-like elastase family, member 3B	CELA3B	
Chymotrypsinogen B1	CTRB1	The CTRB1/2 locus affects diabetes susceptibility and treatment via the incretin pathway. PMID: 23674605
Chymotrypsinogen B2	CTRB2	Genome-wide association study identifies multiple susceptibility loci for pancreatic cancer. PMID: 25086665
Coagulation factor II (thrombin)	F2	**Ximelagatran** – antithrombotic agents. Disease: factor II deficiency (FA2D)
Coagulation factor V (proaccelerin, labile factor)	F5	**Xigris** – antithrombotic agents. Disease: factor V deficiency (FA5D)
Coagulation factor VII (serum prothrombin conversion accelerator)	F7	Disease: factor VII deficiency (FA7D)
Coagulation factor VIII, procoagulant component	F8	**Xigris** – antithrombotic agents. Disease: hemophilia A (HEMA)
Coagulation factor IX	F9	**Pegnivacogin** – antithrombotic agents. Disease: hemophilia B (HEMB)
Coagulation factor X	F10	**Dabigatran** – antithrombotic agents. Disease: factor X deficiency (FA10D)
Coagulation factor XI	F11	Disease: factor XI deficiency (FA11D)
Coagulation factor XII (Hageman factor)	F12	Disease: factor XII deficiency (FA12D)
Complement component 1, r subcomponent	C1R	Deciphering the fine details of c1 assembly and activation mechanisms: 'mission impossible'? PMID: 25414705
Complement component 1, r subcomponent-like	C1RL	A novel human complement-related protein, C1r-like protease (C1r-LP), specifically cleaves pro-C1s. PMID: 15527420
Complement component 1, s subcomponent	C1S	Disease: complement component C1s deficiency (C1SD)
Complement component 2	C2	Disease: complement component 2 deficiency (C2D)
Complement factor B	CFB	Disease: hemolytic uremic syndrome atypical 4 (AHUS4)
Complement factor D (adipsin)	CFD	**Lampalizumab** – antineovascularization agents. Disease: complement factor D deficiency (CFDD)

HGNC-approved name	Symbol	Target information
Complement factor I	CFI	Complement factor I promotes progression of cutaneous squamous cell carcinoma. PMID: 25184960. Disease: hemolytic uremic syndrome atypical 3 (AHUS3)
Dipeptidyl-peptidase 4	DPP4	**Sitagliptin** – drugs used in diabetes
Dipeptidyl-peptidase 6	DPP6	Disease: familial paroxysmal ventricular fibrillation 2 (VF2)
Dipeptidyl-peptidase 7	DPP7	Boro-norleucine as a P1 residue for the design of selective and potent DPP7 inhibitors. PMID: 16084722
Dipeptidyl-peptidase 8	DPP8	Grassypeptolides as natural inhibitors of dipeptidyl peptidase 8 and T cell activation. PMID: 24591193
Dipeptidyl-peptidase 9	DPP9	Dipeptidyl peptidase 9 subcellular localization and a role in cell adhesion involving focal adhesion kinase and paxillin. PMID: 25486458
Dipeptidyl-peptidase 10 (non-functional)	DPP10	Crystallization and preliminary X-ray diffraction analysis of human dipeptidyl peptidase 10 (DPPY), a component of voltage-gated potassium channels. PMID: 22298003. Disease: asthma (ASTHMA)
Granzyme A (granzyme 1, cytotoxic T-lymphocyte-associated serine esterase 3)	GZMA	Serine protease inhibition attenuates rIL-12-induced GZMA activity and proinflammatory events by modulating the Th2 profile from estrogen-treated mice. PMID: 24840346
Granzyme B (granzyme 2, cytotoxic T-lymphocyte-associated serine esterase 1)	GZMB	Granzyme B-activated p53 interacts with Bcl-2 to promote cytotoxic lymphocyte-mediated apoptosis. PMID: 25404359
Granzyme H (cathepsin G-like 2, protein h-CCPX)	GZMH	Natural killer cell-derived human granzyme H induces an alternative, caspase-independent cell-death program. PMID: 17409270
Granzyme K (granzyme 3; tryptase II)	GZMK	Unexpected role for granzyme K in CD56bright NK cell-mediated immunoregulation of multiple sclerosis. PMID: 21666061
Granzyme M (lymphocyte met-ase 1)	GZMM	NK cell intrinsic regulation of MIP-1α by granzyme M. PMID: 24625974
HtrA serine peptidase 1	HTRA1	HtrA serine proteases as potential therapeutic targets in cancer. PMID: 19519315. Disease: macular degeneration, age-related, 7 (ARMD7)
HtrA serine peptidase 2	HTRA2	Omi is a mammalian heat-shock protein that selectively binds and detoxifies oligomeric amyloid-beta. PMID: 19435805. Disease: Parkinson disease 13 (PARK13)
HtrA serine peptidase 3	HTRA3	Activity-modulating monoclonal antibodies to the human serine protease HtrA3 provide novel insights into regulating HtrA proteolytic activities. PMID: 25248123
HtrA serine peptidase 4	HTRA4	Upregulation of HtrA4 in the placentas of patients with severe pre-eclampsia. PMID: 22964307
IMP1 inner mitochondrial membrane peptidase-like (*S. cerevisiae*)	IMMP1L	
IMP2 inner mitochondrial membrane peptidase-like (*S. cerevisiae*)	IMMP2L	Interstitial 7q31.1 copy number variations disrupting IMMP2L gene are associated with a wide spectrum of neurodevelopmental disorders. PMID: 25478008. Disease: Gilles de la Tourette syndrome (GTS)
Kallikrein 1	KLK1	**Ecallantide** – treatment for hereditary angioedema
Kallikrein B, plasma (Fletcher factor) 1	KLKB1	**DX-2930** – drugs used in hereditary angioedema. Disease: prekallikrein deficiency (PKK deficiency)
Kallikrein-related peptidase 2	KLK2	Variants of the hK2 protein gene (KLK2) are associated with serum hK2 levels and predict the presence of prostate cancer at biopsy. PMID: 17085659

(Continued)

HGNC-approved name	Symbol	Target information
Kallikrein-related peptidase 3	KLK3	Development of peptides specifically modulating the activity of KLK2 and KLK3. PMID: 18627344
Kallikrein-related peptidase 4	KLK4	Disease: amelogenesis imperfecta, hypomaturation type, 2A1 (AI2A1)
Kallikrein-related peptidase 5	KLK5	Isomannide-based peptidomimetics as inhibitors for human tissue kallikreins 5 and 7. PMID: 24900785
Kallikrein-related peptidase 6	KLK6	Correlation of hK6 expression with tumor recurrence and prognosis in advanced gastric cancer. PMID: 23587030
Kallikrein-related peptidase 7	KLK7	Kallikrein-related peptidase 7 (KLK7) is a proliferative factor that is aberrantly expressed in human colon cancer. PMID: 25153388
Kallikrein-related peptidase 8	KLK8	Kallikrein-related peptidase-8 (KLK8) is an active serine protease in human epidermis and sweat and is involved in a skin barrier proteolytic cascade. PMID: 20940292
Kallikrein-related peptidase 9	KLK9	The prognostic value of the human kallikrein gene 9 (KLK9) in breast cancer. PMID: 12725415
Kallikrein-related peptidase 10	KLK10	Expression analysis and clinical evaluation of kallikrein-related peptidase 10 (KLK10) in colorectal cancer. PMID: 21487810
Kallikrein-related peptidase 11	KLK11	Identification and validation of Kallikrein-related peptidase 11 as a novel prognostic marker of gastric cancer based on immunohistochemistry. PMID: 21618246
Kallikrein-related peptidase 12	KLK12	Angiogenesis stimulated by human kallikrein-related peptidase 12 acting via a platelet-derived growth factor B-dependent paracrine pathway. PMID: 24225148
Kallikrein-related peptidase 13	KLK13	Decreased expression of kallikrein-related peptidase 13: Possible contribution to metastasis of human oral cancer. PMID: 23371469
Kallikrein-related peptidase 14	KLK14	Kallikrein-related peptidase 14 acts on proteinase-activated receptor 2 to induce signaling pathway in colon cancer cells. PMID: 21907696
Kallikrein-related peptidase 15	KLK15	Quantified KLK15 gene expression levels discriminate prostate cancer from benign tumors and constitute a novel independent predictor of disease progression. PMID: 23620432
Lon peptidase 1, mitochondrial	LONP1	ATP-dependent Lon protease controls tumor bioenergetics by reprogramming mitochondrial activity. PMID: 25017063
Lon peptidase 2, peroxisomal	LONP2	
Mannan-binding lectin serine peptidase 1 (C4/C2 activating component of Ra-reactive factor)	MASP1	MASP-1 induces a unique cytokine pattern in endothelial cells: A novel link between complement system and neutrophil granulocytes. PMID: 24489848. Disease: 3MC syndrome 1 (3MC1)
Mannan-binding lectin serine peptidase 2	MASP2	Disease: MASP2 deficiency (MASPD)
Ovochymase 1	OVCH1	
Ovochymase 2 (gene/pseudogene)	OVCH2	
Plasminogen activator, tissue	PLAT	**Tranexamic acid** – antithrombotic agents. Note: Increased activity of TPA results in increased fibrinolysis of fibrin blood clots that is associated with excessive bleeding. Defective release of TPA results in hypofibrinolysis that can lead to thrombosis or embolism
Plasminogen activator, urokinase	PLAU	Disease: Quebec platelet disorder (QPD)
Prolyl endopeptidase	PREP	**S-17092** – anti-dementia drugs
Prolyl endopeptidase-like	PREPL	Disease: hypotonia–cystinuria syndrome (HCS)
Proprotein convertase subtilisin/kexin type 1	PCSK1	Inhibition of prohormone convertases PC1/3 and PC2 by 2,5-dideoxystreptamine derivatives. PMID: 22169851. Disease: proprotein convertase 1 deficiency (PC1 deficiency)

HGNC-approved name	Symbol	Target information
Proprotein convertase subtilisin/kexin type 2	PCSK2	Inhibition of prohormone convertases PC1/3 and PC2 by 2,5-dideoxystreptamine derivatives. PMID: 22169851
Proprotein convertase subtilisin/kexin type 4	PCSK4	Enzymatic activity of sperm proprotein convertase is important for mammalian fertilization. PMID: 21302280
Proprotein convertase subtilisin/kexin type 5	PCSK5	Genetic variation at the proprotein convertase subtilisin/kexin type 5 gene modulates high-density lipoprotein cholesterol levels. PMID: 20031622
Proprotein convertase subtilisin/kexin type 6	PCSK6	Implications of proprotein convertases in ovarian cancer cell proliferation and tumor progression: Insights for PACE4 as a therapeutic target. PMID: 24818756
Proprotein convertase subtilisin/kexin type 7	PCSK7	Implication of the proprotein convertases in iron homeostasis: Proprotein convertase 7 sheds human transferrin receptor 1 and furin activates hepcidin. PMID: 23390091
Proprotein convertase subtilisin/kexin type 9	PCSK9	**Evolocumab** – lipid-modifying agents. Disease: hypercholesterolemia, autosomal dominant, 3 (HCHOLA3)
Protease, serine, 1 (trypsin 1)	PRSS1	Disease: pancreatitis, hereditary (PCTT)
Protease, serine, 2 (trypsin 2)	PRSS2	Trypsin-2 enhances carcinoma invasion by processing tight junctions and activating ProMT1-MMP. PMID: 22909050
Protease, serine, 3	PRSS3	PRSS3/mesotrypsin is a therapeutic target for metastatic prostate cancer. PMID: 23258495
Protease, serine, 8	PRSS8	The CAP1/Prss8 catalytic triad is not involved in PAR2 activation and protease nexin-1 (PN-1) inhibition. PMID: 25138159
Protease, serine, 12 (neurotrypsin, motopsin)	PRSS12	Disease: mental retardation, autosomal recessive 1 (MRT1)
Protease, serine, 16 (thymus)	PRSS16	Thymus-specific serine protease controls autoreactive CD4 T cell development and autoimmune diabetes in mice. PMID: 21505262
Protease, serine, 21 (testisin)	PRSS21	The glycosylphosphatidylinositol-anchored serine protease PRSS21 (testisin) imparts murine epididymal sperm cell maturation and fertilizing ability. PMID: 19571264
Protease, serine, 22	PRSS22	Urokinase-type plasminogen activator is a preferred substrate of the human epithelium serine protease tryptase epsilon/PRSS22. PMID: 15701722
Protease, serine, 23	PRSS23	Serine protease PRSS23 is upregulated by estrogen receptor α and associated with proliferation of breast cancer cells. PMID: 22291950
Protease, serine, 27	PRSS27	The serine protease marapsin is expressed in stratified squamous epithelia and is up-regulated in the hyperproliferative epidermis of psoriasis and regenerating wounds. PMID: 18948266
Protease, serine, 33	PRSS33	
Protease, serine, 35	PRSS35	
Protease, serine, 36	PRSS36	Expanding the complexity of the human degradome: Polyserases and their tandem serine protease domains. PMID: 17485402
Protease, serine, 37	PRSS37	Prss37 is required for male fertility in the mouse. PMID: 23553430
Protease, serine, 38	PRSS38	
Protease, serine, 41	PRSS41	A testis-specific serine protease, Prss41/Tessp-1, is necessary for the progression of meiosis during murine *in vitro* spermatogenesis. PMID: 24129193
Protease, serine, 42	PRSS42	Three testis-specific paralogous serine proteases play different roles in murine spermatogenesis and are involved in germ cell survival during meiosis. PMID: 23536369
Protease, serine, 43	PRSS43	

(Continued)

HGNC-approved name	Symbol	Target information
Protease, serine, 44	PRSS44	
Protease, serine, 45	PRSS45	
Protease, serine, 46	PRSS46	
Protease, serine, 48	PRSS48	
Protease, serine, 50[a]	PRSS50	Alantolactone induces cell apoptosis partially through down-regulation of testes-specific protease 50 expression. PMID: 24252419
Protease, serine, 53	PRSS53	Identification and characterization of human polyserase-3, a novel protein with tandem serine-protease domains in the same polypeptide chain. PMID: 16566820
Protease, serine, 54	PRSS54	
Protease, serine, 55	PRSS55	
Protease, serine, 56	PRSS56	Disease: microphthalmia, isolated, 6 (MCOP6)
Protease, serine, 57	PRSS57	
Protease, serine, 58	PRSS58	
Rhomboid 5 homolog 1 (*Drosophila*)	RHBDF1	Human rhomboid family-1 suppresses oxygen-independent degradation of hypoxia-inducible factor-1α in breast cancer. PMID: 24648344
Rhomboid 5 homolog 2 (*Drosophila*)	RHBDF2	iRHOM2 is a critical pathogenic mediator of inflammatory arthritis. PMID: 23348744. Disease: tylosis with esophageal cancer (TOC)
Rhomboid domain containing 1	RHBDD1	Lentiviral vector mediated delivery of RHBDD1 shRNA down regulated the proliferation of human glioblastoma cells. PMID: 23883433
Rhomboid domain containing 2	RHBDD2	RHBDD2: A 5-fluorouracil responsive gene overexpressed in the advanced stages of colorectal cancer. PMID: 22965880
Rhomboid domain containing 3	RHBDD3	Rhbdd3 controls autoimmunity by suppressing the production of IL-6 by dendritic cells via K27-linked ubiquitination of the regulator NEMO. PMID: 24859449
Rhomboid, veinlet-like 1 (*Drosophila*)	RHBDL1	Emerging role of rhomboid family proteins in mammalian biology and disease. PMID: 23562403
Rhomboid, veinlet-like 2 (*Drosophila*)	RHBDL2	RHBDL2 is a critical membrane protease for anoikis resistance in human malignant epithelial cells. PMID: 24977233
Rhomboid, veinlet-like 3 (*Drosophila*)	RHBDL3	
Transmembrane protease, serine 2	TMPRSS2	TMPRSS2:ERG blocks neuroendocrine and luminal cell differentiation to maintain prostate cancer proliferation. PMID: 25263440
Transmembrane protease, serine 3	TMPRSS3	Disease: deafness, autosomal recessive, 8 (DFNB8)
Transmembrane protease, serine 4	TMPRSS4	TMPRSS4: An emerging potential therapeutic target in cancer. PMID: 25203520
Transmembrane protease, serine 5	TMPRSS5	Note: Defects in TMPRSS5 may be a cause of deafness
Transmembrane protease, serine 6	TMPRSS6	Disease: iron-refractory iron deficiency anemia (IRIDA)
Transmembrane protease, serine 7	TMPRSS7	
Transmembrane protease, serine 9	TMPRSS9	Polyserase-1/TMPRSS9 induces pro-tumor effects in pancreatic cancer cells by activation of pro-uPA. PMID: 24756697
Transmembrane protease, serine 11A	TMPRSS11A	ECRG1 and its relationship with esophageal cancer: A brief review. PMID: 23548972
Transmembrane protease, serine 11B	TMPRSS11B	
Transmembrane protease, serine 11D	TMPRSS11D	Human airway trypsin-like protease increases mucin gene expression in airway epithelial cells. PMID: 14500256
Transmembrane protease, serine 11E	TMPRSS11E	DESC1 and MSPL activate influenza A viruses and emerging coronaviruses for host cell entry. PMID: 25122802
Transmembrane protease, serine 11F	TMPRSS11F	

HGNC-approved name	Symbol	Target information
Transmembrane (C-terminal) protease, serine 12	TMPRSS12	
Transmembrane protease, serine 13	TMPRSS13	TMPRSS13 deficiency impairs stratum corneum formation and epidermal barrier acquisition. PMID: 24832573
Transmembrane protease, serine 15	TMPRSS15	Disease: enterokinase deficiency (ENTKD)
Tripeptidyl peptidase I	TPP1	Recombinant human tripeptidyl peptidase-1 infusion to the monkey CNS: Safety, pharmacokinetics, and distribution. PMID: 24642058. Disease: ceroid lipofuscinosis, neuronal, 2 (CLN2)
Tripeptidyl peptidase II	TPP2	Tripeptidyl peptidase II in human oral squamous cell carcinoma. PMID: 22986808
Tryptase alpha/beta 1	TPSAB1	Mast cell tryptase beta as a target in allergic inflammation: An evolving story. PMID: 17313363
Tryptase beta 2 (gene/pseudogene)	TPSB2	
Tryptase gamma 1	TPSG1	Importance of mast cell Prss31/transmembrane tryptase/tryptase-γ in lung function and experimental chronic obstructive pulmonary disease and colitis. PMID: 24821729
Tryptase delta 1	TPSD1	
Acrosin	ACR	Synthesis and acrosin inhibitory activities of 5-phenyl-1H-pyrazole-3-carboxylic acid amide derivatives. PMID: 23746472
Acylaminoacyl-peptide hydrolase	APEH	Click-generated triazole ureas as ultrapotent in vivo-active serine hydrolase inhibitors. PMID: 21572424
Azurocidin 1	AZU1	Leukotriene B4-induced changes in vascular permeability are mediated by neutrophil release of heparin-binding protein (HBP/CAP37/azurocidin). PMID: 19151333
Carboxypeptidase, vitellogenic-like	CPVL	Serine carboxypeptidases in regulation of vasoconstriction and elastogenesis. PMID: 19467448
Caseinolytic mitochondrial matrix peptidase proteolytic subunit	CLPP	Disease: Perrault syndrome 3 (PRLTS3)
Chymase 1, mast cell	CMA1	**ASB17061** – agents for dermatitis
Corin, serine peptidase	CORIN	Disease: pre-eclampsia/eclampsia 5 (PEE5)
Elastase, neutrophil expressed	ELANE	**Alvelestat** – drugs for obstructive airway diseases. Disease: cyclic haematopoiesis (CH)
Fibroblast activation protein, alpha	FAP	**Sibrotuzumab** – antineoplastic agents
Furin (paired basic amino acid cleaving enzyme)	FURIN	X-ray structures of human furin in complex with competitive inhibitors. PMID: 24666235
Hepsin	HPN	**Bentiromide** – diagnostic agents
HGF activator	HGFAC	Identification of hepatocyte growth factor activator (Hgfac) gene as a target of HNF1α in mouse β-cells. PMID: 22877752
Hyaluronan binding protein 2	HABP2	Hyaluronic acid binding protein 2 is a novel regulator of vascular integrity. PMID: 20042707
Lipoprotein, Lp(a)	LPA	Lipoprotein(a) as a therapeutic target in cardiovascular disease. PMID: 24848373
Membrane-bound transcription factor peptidase, site 1	MBTPS1	Pharmacologic inhibition of site 1 protease activity inhibits sterol regulatory element-binding protein processing and reduces lipogenic enzyme gene expression and lipid synthesis in cultured cells and experimental animals. PMID: 18577702
Plasminogen	PLG	**Alteplase** – antithrombotic agents. Disease: plasminogen deficiency (PLGD)

(Continued)

HGNC-approved name	Symbol	Target information
Presenilin associated, rhomboid-like	PARL	The mitochondrial rhomboid protease PSARL is a new candidate gene for type 2 diabetes. PMID: 15729572
Prolylcarboxypeptidase (angiotensinase C)	PRCP	Design and synthesis of prolylcarboxypeptidase (PrCP) inhibitors to validate PrCP as a potential target for obesity. PMID: 20857914
Protein C (inactivator of coagulation factors Va and VIIIa)	PROC	Disease: thrombophilia due to protein C deficiency, autosomal dominant (THPH3)
Proteinase 3	PRTN3	Elevated neutrophil membrane expression of proteinase 3 is dependent upon CD177 expression. PMID: 20491791
Reelin	RELN	Disease: lissencephaly 2 (LIS2)
Serine carboxypeptidase 1	SCPEP1	Serine carboxypeptidases in regulation of vasoconstriction and elastogenesis. PMID: 19467448
Suppression of tumorigenicity 14 (colon carcinoma)	ST14	Overexpression of matriptase correlates with poor prognosis in esophageal squamous cell carcinoma. PMID: 24248283. Disease: ichthyosis, autosomal recessive, with hypotrichosis (ARIH)
Trypsin domain containing 1	TYSND1	Novel peroxisomal protease Tysnd1 processes PTS1- and PTS2-containing enzymes involved in beta-oxidation of fatty acids. PMID: 17255948

Threonine peptidases

HGNC-approved name	Symbol	Target information
Proteasome (prosome, macropain) subunit, alpha type, 1	PSMA1	
Proteasome (prosome, macropain) subunit, alpha type, 2	PSMA2	
Proteasome (prosome, macropain) subunit, alpha type, 3	PSMA3	
Proteasome (prosome, macropain) subunit, alpha type, 4	PSMA4	
Proteasome (prosome, macropain) subunit, alpha type, 5	PSMA5	
Proteasome (prosome, macropain) subunit, alpha type, 6	PSMA6	
Proteasome (prosome, macropain) subunit, alpha type, 7	PSMA7	PSMA7 inhibits the tumorigenicity of A549 human lung adenocarcinoma cells. PMID: 22584585
Proteasome (prosome, macropain) subunit, alpha type, 8	PSMA8	
Proteasome (prosome, macropain) subunit, beta type, 1	PSMB1	**Bortezomib** – antineoplastic agents
Proteasome (prosome, macropain) subunit, beta type, 2	PSMB2	The over-expression of the β2 catalytic subunit of the proteasome decreases homologous recombination and impairs DNA double-strand break repair in human cells. PMID: 21660142
Proteasome (prosome, macropain) subunit, beta type, 3	PSMB3	
Proteasome (prosome, macropain) subunit, beta type, 4	PSMB4	Gene expression analysis of the 26S proteasome subunit PSMB4 reveals significant upregulation, different expression and association with proliferation in human pulmonary neuroendocrine tumours. PMID: 25157275
Proteasome (prosome, macropain) subunit, beta type, 5	PSMB5	**Bortezomib** – antineoplastic agents
Proteasome (prosome, macropain) subunit, beta type, 6	PSMB6	Proteomic analysis reveals that proteasome subunit beta 6 is involved in hypoxia-induced pulmonary vascular remodeling in rats. PMID: 23844134
Proteasome (prosome, macropain) subunit, beta type, 7	PSMB7	Knockdown of PSMB7 induces autophagy in cardiomyocyte cultures: Possible role in endoplasmic reticulum stress. PMID: 23969338

HGNC-approved name	Symbol	Target information
Proteasome (prosome, macropain) subunit, beta type, 8	PSMB8	**ONX-0914** – anti-inflammatory and antirheumatic products. Disease: Nakajo syndrome (NKJO)
Proteasome (prosome, macropain) subunit, beta type, 9	PSMB9	**NCT00672542** – antineoplastic agents
Proteasome (prosome, macropain) subunit, beta type, 10	PSMB10	**NCT00672542** – antineoplastic agents
Proteasome (prosome, macropain) subunit, beta type, 11	PSMB11	Expression of thymoproteasome subunit β5t in type AB thymoma. PMID: 24293611
Proteasome (prosome, macropain) 26S subunit, non-ATPase, 1	PSMD1	**Bortezomib** – antineoplastic agents
Proteasome (prosome, macropain) 26S subunit, non-ATPase, 2	PSMD2	**Bortezomib** – antineoplastic agents
Proteasome (prosome, macropain) 26S subunit, non-ATPase, 3	PSMD3	Genome-wide association study identifies a PSMD3 variant associated with neutropenia in interferon-based therapy for chronic hepatitis C. PMID: 25515861
Proteasome (prosome, macropain) 26S subunit, non-ATPase, 4	PSMD4	Ube3a, the E3 ubiquitin ligase causing Angelman syndrome and linked to autism, regulates protein homeostasis through the proteasomal shuttle Rpn10. PMID: 24292889
Proteasome (prosome, macropain) 26S subunit, non-ATPase, 5	PSMD5	Role of S5b/PSMD5 in proteasome inhibition caused by TNF-α/NFκB in higher eukaryotes. PMID: 22921402
Proteasome (prosome, macropain) 26S subunit, non-ATPase, 6	PSMD6	The 19S proteasome subunit Rpn7 stabilizes DNA damage foci upon genotoxic insult. PMID: 22473755
Proteasome (prosome, macropain) 26S subunit, non-ATPase, 7	PSMD7	Crystal structure of the proteasomal deubiquitylation module Rpn8-Rpn11. PMID: 24516147
Proteasome (prosome, macropain) 26S subunit, non-ATPase, 8	PSMD8	Incorporation of the Rpn12 subunit couples completion of proteasome regulatory particle lid assembly to lid-base joining. PMID: 22195964
Proteasome (prosome, macropain) 26S subunit, non-ATPase, 9	PSMD9	PSMD9 expression predicts radiotherapy response in breast cancer. PMID: 24673853
Proteasome (prosome, macropain) 26S subunit, non-ATPase, 10	PSMD10	Up-regulated oncoprotein P28GANK correlates with proliferation and poor prognosis of human glioma. PMID: 22913315
Proteasome (prosome, macropain) 26S subunit, non-ATPase, 11	PSMD11	Increased proteasome activity in human embryonic stem cells is regulated by PSMD11. PMID: 22972301
Proteasome (prosome, macropain) 26S subunit, non-ATPase, 12	PSMD12	
Proteasome (prosome, macropain) 26S subunit, non-ATPase, 13	PSMD13	Investigations into the role of 26S proteasome non-ATPase regulatory subunit 13 in neuroinflammation. PMID: 24642793
Proteasome (prosome, macropain) 26S subunit, non-ATPase, 14	PSMD14	Disassembly of Lys11- and mixed-linkage polyubiquitin conjugates provide insights into function of proteasomal deubiquitinases Rpn11 and Ubp6. PMID: 25389291
Asparaginase like 1	ASRGL1	The human asparaginase-like protein 1 hASRGL1 is an Ntn hydrolase with beta-aspartyl peptidase activity. PMID: 19839645
Taspase, threonine aspartase, 1	TASP1	Taspase1 cleaves MLL1 to activate cyclin E for HER2/neu breast tumorigenesis. PMID: 25266805
Miscellaneous peptidases		
Kyphoscoliosis peptidase	KY	
LON peptidase N-terminal domain and ring finger 1	LONRF1	

(Continued)

HGNC-approved name	Symbol	Target information
LON peptidase N-terminal domain and ring finger 2	LONRF2	
LON peptidase N-terminal domain and ring finger 3	LONRF3	
MPN domain containing	MPND	
Transmembrane protein 59	TMEM59	TMEM59 defines a novel ATG16L1-binding motif that promotes local activation of LC3. PMID: 23376921

4.2.1.1 Peptidase inhibitors

HGNC-approved name	Symbol	Target information
Pan-peptidase inhibitors		
Alpha-2-macroglobulin	A2M	Alteration of prolyl oligopeptidase and activated α-2-macroglobulin in multiple sclerosis subtypes and in the clinically isolated syndrome. PMID: 23643808
Alpha-2-macroglobulin-like 1	A2ML1	The protease inhibitor alpha-2-macroglobulin-like-1 is the p170 antigen recognized by paraneoplastic pemphigus autoantibodies in human. PMID: 20805888
Pregnancy zone protein	PZP	Serum levels of pregnancy zone protein are elevated in presymptomatic Alzheimer's disease. PMID: 21879768
Secreted phosphoprotein 2, 24 kDa	SPP2	The bone matrix protein secreted phosphoprotein 24 kDa (Spp24): Bone metabolism regulator and starting material for biotherapeutic materials. PMID: 25339413
Cysteine peptidase inhibitors		
Baculoviral IAP repeat containing 5	BIRC5	**Gataparsen** – antineoplastic agents
Baculoviral IAP repeat containing 7	BIRC7	Knockdown of Livin inhibits growth and invasion of gastric cancer cells through blockade of the MAPK pathway *in vitro* and in vivo. PMID: 24220265
Cystatin SN	CST1	Upregulation of the cysteine protease inhibitor, cystatin SN, contributes to cell proliferation and cathepsin inhibition in gastric cancer. PMID: 19463800
Cystatin SA	CST2	Global secretome analysis identifies novel mediators of bone metastasis. PMID: 22688892
Cystatin C	CST3	Prognostic value of cystatin C in acute coronary syndromes: Enhancer of atherosclerosis and promising therapeutic target. PMID: 21605013. Disease: amyloidosis 6 (AMYL6)
Cystatin S	CST4	
Cystatin D	CST5	Cystatin D is a candidate tumor suppressor gene induced by vitamin D in human colon cancer cells. PMID: 19662683
Cystatin E/M	CST6	TBX2 represses CST6 resulting in uncontrolled legumain activity to sustain breast cancer proliferation: A novel cancer-selective target pathway with therapeutic opportunities. PMID: 24742492
Cystatin F (leukocystatin)	CST7	Microglial cystatin F expression is a sensitive indicator for ongoing demyelination with concurrent remyelination. PMID: 21344476
Cystatin 8 (cystatin-related epididymal specific)	CST8	*In vitro* regulatory effect of epididymal serpin CRES on protease activity of proprotein convertase PC4/PCSK4. PMID: 22827436

HGNC-approved name	Symbol	Target information
Cystatin 9 (testatin)	CST9	Testatin transgenic and knockout mice exhibit normal sex-differentiation. PMID: 16427609
Cystatin 9-like	CST9L	
Cystatin 11	CST11	Cystatin 11: A new member of the cystatin type 2 family. PMID: 12072414
Cystatin A (stefin A)	CSTA	Disease: ichthyosis, exfoliative, autosomal recessive, ichthyosis bullosa of Siemens-like (AREI)
Cystatin B (stefin B)	CSTB	Disease: epilepsy, progressive myoclonic 1 (EPM1)
Cystatin-like 1	CSTL1	
Alpha-2-HS-glycoprotein	AHSG	Fetuin-A: A novel link between obesity and related complications. PMID: 25468829
Calpastatin	CAST	Calpastatin-mediated inhibition of calpains in the mouse brain prevents mutant ataxin 3 proteolysis, nuclear localization and aggregation, relieving Machado–Joseph disease. PMID: 22843411
Histidine-rich glycoprotein	HRG	Disease: thrombophilia due to histidine-rich glycoprotein deficiency (THPH11)
Kininogen 1	KNG1	Disease: high molecular weight kininogen deficiency (HMWK deficiency)
NLR family, apoptosis inhibitory protein	NAIP	NAIPs: Building an innate immune barrier against bacterial pathogens. NAIPs function as sensors that initiate innate immunity by detection of bacterial proteins in the host cell cytosol. PMID: 22513803
Tumor necrosis factor, alpha-induced protein 8	TNFAIP8	SCC-S2 is overexpressed in colon cancers and regulates cell proliferation. PMID: 22886548
X-linked inhibitor of apoptosis	XIAP	Small molecule XIAP inhibitors enhance TRAIL-induced apoptosis and antitumor activity in preclinical models of pancreatic carcinoma. PMID: 19258513. Disease: lymphoproliferative syndrome, X-linked, 2 (XLP2)
Serine peptidase inhibitors		
Inter-alpha-trypsin inhibitor heavy chain 1	ITIH1	
Inter-alpha-trypsin inhibitor heavy chain 2	ITIH2	
Inter-alpha-trypsin inhibitor heavy chain 3	ITIH3	ITIH3 polymorphism may confer susceptibility to psychiatric disorders by altering the expression levels of GLT8D1. PMID: 24373612
Inter-alpha-trypsin inhibitor heavy chain family, member 4	ITIH4	Inter-alpha-trypsin inhibitor heavy chain H4 as a diagnostic and prognostic indicator in patients with hepatitis B virus-associated hepatocellular carcinoma. PMID: 24836184
Inter-alpha-trypsin inhibitor heavy chain family, member 5	ITIH5	Decreased ITIH5 expression is associated with poor prognosis in primary gastric cancer. PMID: 24913813
Inter-alpha-trypsin inhibitor heavy chain family, member 6	ITIH6	
Peptidase inhibitor 3, skin-derived	PI3	Protease inhibitors derived from elafin and SLPI and engineered to have enhanced specificity towards neutrophil serine proteases. PMID: 19241385
Peptidase inhibitor 15	PI15	Protease inhibitor 15, a candidate gene for abdominal aortic internal elastic lamina ruptures in the rat. PMID: 24790086

(Continued)

HGNC-approved name	Symbol	Target information
Peptidase inhibitor 16	PI16	PI16 is expressed by a subset of human memory Treg with enhanced migration to CCL17 and CCL20. PMID: 22533972
Serpin peptidase inhibitor, clade A (alpha-1 antiproteinase, antitrypsin), member 1	SERPINA1	Disease: alpha-1-antitrypsin deficiency (A1ATD)
Serpin peptidase inhibitor, clade A (alpha-1 antiproteinase, antitrypsin), member 2 (gene/pseudogene)	SERPINA2	SERPINA2 is a novel gene with a divergent function from SERPINA1. PMID: 23826168
Serpin peptidase inhibitor, clade A (alpha-1 antiproteinase, antitrypsin), member 3	SERPINA3	Serine protease inhibitor A3 in atherosclerosis and aneurysm disease. PMID: 22580763
Serpin peptidase inhibitor, clade A (alpha-1 antiproteinase, antitrypsin), member 4	SERPINA4	Novel role of kallistatin in vascular repair by promoting mobility, viability, and function of endothelial progenitor cells. PMID: 25237049
Serpin peptidase inhibitor, clade A (alpha-1 antiproteinase, antitrypsin), member 5	SERPINA5	SERPINA5 inhibits tumor cell migration by modulating the fibronectin-integrin β1 signaling pathway in hepatocellular carcinoma. PMID: 24388360
Serpin peptidase inhibitor, clade A (alpha-1 antiproteinase, antitrypsin), member 6	SERPINA6	Disease: corticosteroid-binding globulin deficiency (CBG deficiency)
Serpin peptidase inhibitor, clade A (alpha-1 antiproteinase, antitrypsin), member 7	SERPINA7	Disease: thyroxine-binding globulin deficiency (TBG deficiency)
Angiotensinogen (serpin peptidase inhibitor, clade A, member 8)	AGT	Disease: essential hypertension (EHT)
Serpin peptidase inhibitor, clade A (alpha-1 antiproteinase, antitrypsin), member 9	SERPINA9	Expression of the serpin centerin defines a germinal center phenotype in B-cell lymphomas. PMID: 18550480
Serpin peptidase inhibitor, clade A (alpha-1 antiproteinase, antitrypsin), member 10	SERPINA10	Heparin is a major activator of the anticoagulant serpin, protein Z-dependent protease inhibitor. PMID: 21220417
Serpin peptidase inhibitor, clade A (alpha-1 antiproteinase, antitrypsin), member 11	SERPINA11	
Serpin peptidase inhibitor, clade A (alpha-1 antiproteinase, antitrypsin), member 12	SERPINA12	Vaspin (serpinA12) in obesity, insulin resistance, and inflammation. PMID: 24596079
Serpin peptidase inhibitor, clade B (ovalbumin), member 1	SERPINB1	Serine protease inhibitor (SERPIN) B1 suppresses cell migration and invasion in glioma cells. PMID: 24968089
Serpin peptidase inhibitor, clade B (ovalbumin), member 2	SERPINB2	Serpins promote cancer cell survival and vascular co-option in brain metastasis. PMID: 24581498
Serpin peptidase inhibitor, clade B (ovalbumin), member 3	SERPINB3	SerpinB3/B4: Mediators of Ras-driven inflammation and oncogenesis. PMID: 25485489
Serpin peptidase inhibitor, clade B (ovalbumin), member 4	SERPINB4	Intracellular serine protease inhibitor SERPINB4 inhibits granzyme M-induced cell death. PMID: 21857942
Serpin peptidase inhibitor, clade B (ovalbumin), member 5	SERPINB5	Supramolecular assembly of multifunctional maspin-mimetic nanostructures as a potent peptide-based angiogenesis inhibitor. PMID: 25462852
Serpin peptidase inhibitor, clade B (ovalbumin), member 6	SERPINB6	Disease: deafness, autosomal recessive, 91 (DFNB91)
Serpin peptidase inhibitor, clade B (ovalbumin), member 7	SERPINB7	Disease: keratoderma, palmoplantar, Nagashima type (PPKN)

HGNC-approved name	Symbol	Target information
Serpin peptidase inhibitor, clade B (ovalbumin), member 8	SERPINB8	
Serpin peptidase inhibitor, clade B (ovalbumin), member 9	SERPINB9	Decreased expression of protease inhibitor 9, a granzyme B inhibitor, in celiac disease: A potential mechanism in enterocyte destruction and villous atrophy. PMID: 24355225
Serpin peptidase inhibitor, clade B (ovalbumin), member 10	SERPINB10	Bomapin is a redox-sensitive nuclear serpin that affects responsiveness of myeloid progenitor cells to growth environment. PMID: 20433722
Serpin peptidase inhibitor, clade B (ovalbumin), member 11 (gene/pseudogene)	SERPINB11	The spatiotemporal expression and localization implicates a potential role for SerpinB11 in the process of mouse spermatogenesis and apoptosis. PMID: 24785531
Serpin peptidase inhibitor, clade B (ovalbumin), member 12	SERPINB12	Serpin regulation of fibrinolytic system: Implications for therapeutic applications in cardiovascular diseases. PMID: 25374013
Serpin peptidase inhibitor, clade B (ovalbumin), member 13	SERPINB13	Headpin: A serpin with endogenous and exogenous suppression of angiogenesis. PMID: 16357159
Serpin peptidase inhibitor, clade C (antithrombin), member 1	SERPINC1	**Enoxaparin** – antithrombotic agents. Disease: antithrombin III deficiency (AT3D)
Serpin peptidase inhibitor, clade D (heparin cofactor), member 1	SERPIND1	**Ardeparin** – antithrombotic agents. Disease: thrombophilia due to heparin cofactor 2 deficiency (THPH10)
Serpin peptidase inhibitor, clade E (nexin, plasminogen activator inhibitor type 1), member 1	SERPINE1	Intra-airway administration of small interfering RNA targeting plasminogen activator inhibitor-1 attenuates allergic asthma in mice. PMID: 21926267. Disease: plasminogen activator inhibitor-1 deficiency (PAI-1D)
Serpin peptidase inhibitor, clade E (nexin, plasminogen activator inhibitor type 1), member 2	SERPINE2	The serine protease inhibitor serpinE2 is a novel target of ERK signaling involved in human colorectal tumorigenesis. PMID: 20942929
Serpin peptidase inhibitor, clade E (nexin, plasminogen activator inhibitor type 1), member 3	SERPINE3	
Serpin peptidase inhibitor, clade F (alpha-2 antiplasmin, pigment epithelium-derived factor), member 1	SERPINF1	Pigment epithelium-derived factor (PEDF) as a therapeutic target in cardiovascular disease. PMID: 19694500. Disease: osteogenesis imperfecta 6 (OI6)
Serpin peptidase inhibitor, clade F (alpha-2 antiplasmin, pigment epithelium-derived factor), member 2	SERPINF2	Disease: alpha-2-plasmin inhibitor deficiency (APLID)
Serpin peptidase inhibitor, clade G (C1 inhibitor), member 1	SERPING1	Disease: hereditary angioedema (HAE)
Serpin peptidase inhibitor, clade H (heat shock protein 47), member 1, (collagen binding protein 1)	SERPINH1	Disease: osteogenesis imperfecta 10 (OI10)
Serpin peptidase inhibitor, clade I (neuroserpin), member 1	SERPINI1	Serpins promote cancer cell survival and vascular co-option in brain metastasis. PMID: 24581498. Disease: encephalopathy, familial, with neuroserpin inclusion bodies (FENIB)
Serpin peptidase inhibitor, clade I (pancpin), member 2	SERPINI2	
Serine peptidase inhibitor, Kazal type 1	SPINK1	Therapeutic targeting of SPINK1-positive prostate cancer. PMID: 21368222. Disease: pancreatitis, hereditary (PCTT)
Serine peptidase inhibitor, Kazal type 2 (acrosin–trypsin inhibitor)	SPINK2	Identification of trypsin-inhibitory site and structure determination of human SPINK2 serine proteinase inhibitor. PMID: 19422058
Serine peptidase inhibitor, Kazal type 4	SPINK4	

(*Continued*)

HGNC-approved name	Symbol	Target information
Serine peptidase inhibitor, Kazal type 5	SPINK5	Disease: Netherton syndrome (NETH)
Serine peptidase inhibitor, Kazal type 6	SPINK6	Isolation of SPINK6 in human skin: Selective inhibitor of kallikrein-related peptidases. PMID: 20667819
Serine peptidase inhibitor, Kazal type 7 (putative)	SPINK7	ECRG2 regulates ECM degradation and uPAR/FPRL1 pathway contributing cell invasion/migration. PMID: 19796867
Serine peptidase inhibitor, Kazal type 8 (putative)	SPINK8	Novel epididymal protease inhibitors with Kazal or WAP family domain. PMID: 16930550
Serine peptidase inhibitor, Kazal type 9	SPINK9	SPINK9 stimulates metalloprotease/EGFR-dependent keratinocyte migration via purinergic receptor activation. PMID: 24441102
Serine peptidase inhibitor, Kazal type 13 (putative)	SPINK13	Spink13, an epididymis-specific gene of the Kazal-type serine protease inhibitor (SPINK) family, is essential for the acrosomal integrity and male fertility. PMID: 23430248
Serine peptidase inhibitor, Kazal type 14 (putative)	SPINK14	
Serine peptidase inhibitor, Kunitz type 1	SPINT1	The role of hepatocyte growth factor activator inhibitor-1 (HAI-1) as a prognostic indicator in cervical cancer. PMID: 19578736
Serine peptidase inhibitor, Kunitz type 2	SPINT2	Disease: diarrhea 3, secretory sodium, congenital (DIAR3)
Serine peptidase inhibitor, Kunitz type 3	SPINT3	Three genes expressing Kunitz domains in the epididymis are related to genes of WFDC-type protease inhibitors and semen coagulum proteins in spite of lacking similarity between their protein products. PMID: 21988899
Serine peptidase inhibitor, Kunitz type 4	SPINT4	
Tissue factor pathway inhibitor (lipoprotein-associated coagulation inhibitor)	TFPI	**Concizumab** – haemophilia treatments
Tissue factor pathway inhibitor 2	TFPI2	Secretome-based identification of TFPI2, a novel serum biomarker for detection of ovarian clear cell adenocarcinoma. PMID: 23805888
WAP, follistatin/kazal, immunoglobulin, kunitz and netrin domain containing 1	WFIKKN1	Latent myostatin has significant activity and this activity is controlled more efficiently by WFIKKN1 than by WFIKKN2. PMID: 23829672
WAP, follistatin/kazal, immunoglobulin, kunitz and netrin domain containing 2	WFIKKN2	
WAP four-disulfide core domain 1	WFDC1	WFDC1 is a key modulator of inflammatory and wound repair responses. PMID: 25219356
WAP four-disulfide core domain 2	WFDC2	Identification of human epididymis protein-4 as a fibroblast-derived mediator of fibrosis. PMID: 23353556
WAP four-disulfide core domain 3	WFDC3	
WAP four-disulfide core domain 5	WFDC5	
WAP four-disulfide core domain 6	WFDC6	
WAP four-disulfide core domain 8	WFDC8	
WAP four-disulfide core domain 9	WFDC9	
WAP four-disulfide core domain 10A	WFDC10A	Differential expression and antibacterial activity of WFDC10A in the monkey epididymis. PMID: 16996203
WAP four-disulfide core domain 10B	WFDC10B	
WAP four-disulfide core domain 11	WFDC11	Identification and characterization of Wfdc gene expression in the male reproductive tract of the rat. PMID: 21796715
WAP four-disulfide core domain 12	WFDC12	In silico study of whey-acidic-protein domain containing oral protease inhibitors. PMID: 18360692
WAP four-disulfide core domain 13	WFDC13	

HGNC-approved name	Symbol	Target information
Alpha-1-microglobulin/bikunin precursor	AMBP	The proteoglycan bikunin has a defined sequence. PMID: 21983600
Amyloid beta (A4) precursor-like protein 2	APLP2	Amyloid precursor protein (APP)/APP-like protein 2 (APLP2) expression is required to initiate endosome-nucleus-autophagosome trafficking of glypican-1-derived heparan sulfate. PMID: 24898256
C3- and PZP-like, alpha-2-macroglobulin domain containing 8	CPAMD8	
CD109 molecule	CD109	CD109 plays a role in osteoclastogenesis. PMID: 23593435
Epididymal peptidase inhibitor	EPPIN	Epididymal protein targets: A brief history of the development of epididymal protease inhibitor as a contraceptive. PMID: 21441428
Histocompatibility (minor) serpin domain containing	HMSD	
Kallmann syndrome 1 sequence	KAL1	FGF-2 and Anosmin-1 are selectively expressed in different types of multiple sclerosis lesions. PMID: 22016523. Disease: hypogonadotropic hypogonadism 1 with or without anosmia (HH1)
Kazal-type serine peptidase inhibitor domain 1	KAZALD1	Epigenetic silencing of KAZALD1 confers a better prognosis and is associated with malignant transformation/progression in glioma. PMID: 24002581
Papilin, proteoglycan-like sulfated glycoprotein	PAPLN	
Proprotein convertase subtilisin/kexin type 1 inhibitor	PCSK1N	A novel function for proSAAS as an amyloid anti-aggregant in Alzheimer's disease. PMID: 24102330
R3H domain containing-like	R3HDML	
Secretory leukocyte peptidase inhibitor	SLPI	Protease inhibitors derived from elafin and SLPI and engineered to have enhanced specificity towards neutrophil serine proteases. PMID: 19241385

Metallopeptidase inhibitors

SEC11 homolog A (*S. cerevisiae*)	SEC11A	Signal peptidase complex 18, encoded by SEC11A, contributes to progression via TGF-α secretion in gastric cancer. PMID: 23995782
SEC11 homolog C (*S. cerevisiae*)	SEC11C	
TIMP metallopeptidase inhibitor 1	TIMP1	TIMP1 in conditioned media of human adipose stromal cells protects neurons against oxygen-glucose deprivation injury. PMID: 25281791
TIMP metallopeptidase inhibitor 2	TIMP2	TIMP-2-derived 18-mer peptide inhibits endothelial cell proliferation and migration through cAMP/PKA-dependent mechanism. PMID: 24252252
TIMP metallopeptidase inhibitor 3	TIMP3	TIMP3 is the primary TIMP to regulate agonist-induced vascular remodelling and hypertension. PMID: 23524300. Disease: Sorsby fundus dystrophy (SFD)
TIMP metallopeptidase inhibitor 4	TIMP4	Targeted overexpression of tissue inhibitor of matrix metalloproteinase-4 modifies post-myocardial infarction remodeling in mice. PMID: 24637197
Fetuin-B	FETUB	Fetuin-B, a liver-derived plasma protein is essential for fertilization. PMID: 23562279
Latexin	LXN	Latexin sensitizes leukemogenic cells to gamma-irradiation-induced cell-cycle arrest and cell death through Rps3 pathway. PMID: 25341047
Lumican	LUM	Lumican: A new inhibitor of matrix metalloproteinase-14 activity. PMID: 25304424

(Continued)

HGNC-approved name	Symbol	Target information
Proline rich, lacrimal 1	PROL1	The opiorphin gene (ProL1) and its homologues function in erectile physiology. PMID: 18410445
Retinoic acid receptor responder (tazarotene induced) 1	RARRES1	Tumor suppressor RARRES1 interacts with cytoplasmic carboxypeptidase AGBL2 to regulate the α-tubulin tyrosination cycle. PMID: 21303978
Reversion-inducing-cysteine-rich protein with kazal motifs	RECK	The potential of RECK inducers as antitumor agents for glioma. PMID: 22753763

4.2.2 Glycosylation

HGNC-approved name	Symbol	Target information
ALG1, chitobiosyldiphosphodolichol beta-mannosyltransferase	ALG1	Disease: congenital disorder of glycosylation 1K (CDG1K)
ALG1, chitobiosyldiphosphodolichol beta-mannosyltransferase-like	ALG1L	
ALG1, chitobiosyldiphosphodolichol beta-mannosyltransferase-like 2	ALG1L2	
ALG2, alpha-1,3/1,6-mannosyltransferase	ALG2	Disease: congenital disorder of glycosylation 1I (CDG1I)
ALG3, alpha-1,3-mannosyltransferase	ALG3	Disease: congenital disorder of glycosylation 1D (CDG1D)
ALG5, dolichylphosphate beta-glucosyltransferase	ALG5	
ALG6, alpha-1,3-glucosyltransferase	ALG6	Disease: congenital disorder of glycosylation 1C (CDG1C)
ALG8, alpha-1,3-glucosyltransferase	ALG8	Disease: congenital disorder of glycosylation 1H (CDG1H)
ALG9, alpha-1,2-mannosyltransferase	ALG9	Note: A chromosomal aberration involving ALG9 is found in a family with bipolar affective disorder. Translocation t(911)(p24q23). However, common variations in ALG9 do not play a major role in predisposition to bipolar affective disorder
ALG10, alpha-1,2-glucosyltransferase	ALG10	HERG is protected from pharmacological block by alpha-1,2-glucosyltransferase function. PMID: 17189275
ALG10B, alpha-1,2-glucosyltransferase	ALG10B	A point mutation in the gene for asparagine-linked glycosylation 10B (Alg10b) causes nonsyndromic hearing impairment in mice (Mus musculus). PMID: 24303013
ALG11, alpha-1,2-mannosyltransferase	ALG11	Disease: congenital disorder of glycosylation 1P (CDG1P)
ALG12, alpha-1,6-mannosyltransferase	ALG12	Disease: congenital disorder of glycosylation 1G (CDG1G)
ALG13, UDP-N-acetylglucosaminyltransferase subunit	ALG13	Disease: congenital disorder of glycosylation 1S (CDG1S)
ALG14, UDP-N-acetylglucosaminyltransferase subunit	ALG14	Congenital myasthenic syndromes due to mutations in ALG2 and ALG14. PMID: 23404334
ABO blood group (transferase A, alpha 1-3-N-acetylgalactosaminyltransferase; transferase B, alpha 1-3-galactosyltransferase)	ABO	Possible role of ABO system in age-related diseases and longevity: A narrative review. PMID: 25512760
Alpha 1,3-galactosyltransferase 2	A3GALT2	Humans lack iGb3 due to the absence of functional iGb3-synthase: Implications for NKT cell development and transplantation. PMID: 18630988
Alpha 1,4-galactosyltransferase	A4GALT	
Alpha-1,4-N-acetylglucosaminyltransferase	A4GNT	Dual roles of gastric gland mucin-specific O-glycans in prevention of gastric cancer. PMID: 24761044

HGNC-approved name	Symbol	Target information
Beta-1,3-glucuronyltransferase 1 (glucuronosyltransferase P)	B3GAT1	Beta-1,3-Glucuronyltransferase-1 gene implicated as a candidate for a schizophrenia-like psychosis through molecular analysis of a balanced translocation. PMID: 12874601
Beta-1,3-glucuronyltransferase 2 (glucuronosyltransferase S)	B3GAT2	Candidate gene analysis of the human natural killer-1 carbohydrate pathway and perineuronal nets in schizophrenia: B3GAT2 is associated with disease risk and cortical surface area. PMID: 20950796
Beta-1,3-glucuronyltransferase 3 (glucuronosyltransferase I)	B3GAT3	Disease: multiple joint dislocations, short stature, craniofacial dysmorphism, and congenital heart defects (JDSSDHD)
Beta-1,3-galactosyltransferase-like	B3GALTL	Disease: Peters-plus syndrome (PpS)
Beta-1,3-N-acetylgalactosaminyltransferase 1 (globoside blood group)	B3GALNT1	TINAGL1 and B3GALNT1 are potential therapy target genes to suppress metastasis in non-small cell lung cancer. PMID: 25521548
Beta-1,3-N-acetylgalactosaminyltransferase 2	B3GALNT2	Disease: muscular dystrophy–dystroglycanopathy congenital with brain and eye anomalies A11 (MDDGA11)
Beta-1,4-N-acetylgalactosaminyltransferase 1	B4GALNT1	Disease: spastic paraplegia 26, autosomal recessive (SPG26)
Beta-1,4-N-acetylgalactosaminyltransferase 2	B4GALNT2	Therapeutic adenoviral gene transfer of a glycosyltransferase for prevention of peritoneal dissemination and metastasis of gastric cancer. PMID: 25213663
Beta-1,4-N-acetylgalactosaminyltransferase 3	B4GALNT3	β1, 4-N-acetylgalactosaminyltransferase III modulates cancer stemness through EGFR signaling pathway in colon cancer cells. PMID: 25003232
Beta-1,4-N-acetylgalactosaminyltransferase 4	B4GALNT4	
Carbohydrate (keratan sulfate Gal-6) sulfotransferase 1	CHST1	Upregulation of chondroitin 6-sulphotransferase 1 facilitates Schwann cell migration during axonal growth. PMID: 16495484
Carbohydrate (N-acetylglucosamine 6-O) sulfotransferase 2	CHST2	
Carbohydrate (chondroitin 6) sulfotransferase 3	CHST3	Disease: spondyloepiphyseal dysplasia with congenital joint dislocations (SEDC-JD)
Carbohydrate (N-acetylglucosamine 6-O) sulfotransferase 4	CHST4	A HEV-restricted sulfotransferase is expressed in rheumatoid arthritis synovium and is induced by lymphotoxin-alpha/beta and TNF-alpha in cultured endothelial cells. PMID: 15752429
Carbohydrate (N-acetylglucosamine 6-O) sulfotransferase 5	CHST5	Matrix morphogenesis in cornea is mediated by the modification of keratan sulfate by GlcNAc 6-O-sulfotransferase. PMID: 16938051
Carbohydrate (N-acetylglucosamine 6-O) sulfotransferase 6	CHST6	Disease: macular dystrophy, corneal 1 (MCDC1)
Carbohydrate (N-acetylglucosamine 6-O) sulfotransferase 7	CHST7	6-Sulphated chondroitins have a positive influence on axonal regeneration. PMID: 21747937
Carbohydrate (N-acetylgalactosamine 4-0) sulfotransferase 8	CHST8	Note: CHST8 mutations may be a cause of generalized non-inflammatory peeling skin syndrome type A. PMID: 22289416. Peeling skin syndrome (PSS) is a genodermatosis characterized by continuous shedding of the outer layers of the epidermis
Carbohydrate (N-acetylgalactosamine 4-0) sulfotransferase 9	CHST9	Examination of copy number variations of CHST9 in multiple types of hematologic malignancies. PMID: 21156230
Carbohydrate sulfotransferase 10	CHST10	Mechanism of regulation and suppression of melanoma invasiveness by novel retinoic acid receptor-gamma target gene carbohydrate sulfotransferase 10. PMID: 19470764
Carbohydrate (chondroitin 4) sulfotransferase 11	CHST11	The roles of chondroitin-4-sulfotransferase-1 in development and disease. PMID: 20807643. Note: A chromosomal aberration involving CHST11 is found in B-cell chronic lymphocytic leukemias. Translocation t(1214)(q23q32) with IgH

(Continued)

HGNC-approved name	Symbol	Target information
Carbohydrate (chondroitin 4) sulfotransferase 12	CHST12	
Carbohydrate (chondroitin 4) sulfotransferase 13	CHST13	
Carbohydrate (N-acetylgalactosamine 4-0) sulfotransferase 14	CHST14	Disease: Ehlers–Danlos syndrome, musculocontractural type 1 (EDSMC1)
Carbohydrate (N-acetylgalactosamine 4-sulfate 6-O) sulfotransferase 15	CHST15	
Chondroitin sulfate N-acetylgalactosaminyltransferase 1	CSGALNACT1	Alterations in the chondroitin sulfate chain in human osteoarthritic cartilage of the knee. PMID: 24280246
Chondroitin sulfate N-acetylgalactosaminyltransferase 2	CSGALNACT2	
Collagen beta(1-O)galactosyltransferase 1	COLGALT1	Core glycosylation of collagen is initiated by two beta(1-O)galactosyltransferases. PMID: 19075007
Collagen beta(1-O)galactosyltransferase 2	COLGALT2	
Dolichyl-phosphate mannosyltransferase polypeptide 1, catalytic subunit	DPM1	Disease: congenital disorder of glycosylation 1E (CDG1E)
Dolichyl-phosphate mannosyltransferase polypeptide 2, regulatory subunit	DPM2	Disease: congenital disorder of glycosylation 1U (CDG1U)
Dolichyl-phosphate mannosyltransferase polypeptide 3	DPM3	Disease: congenital disorder of glycosylation 1O (CDG1O)
dpy-19-like 1 (*Caenorhabditis elegans*)	DPY19L1	Dpy19l1, a multi-transmembrane protein, regulates the radial migration of glutamatergic neurons in the developing cerebral cortex. PMID: 22028030
dpy-19-like 2 (*C. elegans*)	DPY19L2	Disease: spermatogenic failure 9 (SPGF9)
dpy-19-like 3 (*C. elegans*)	DPY19L3	
dpy-19-like 4 (*C. elegans*)	DPY19L4	
Exostosin glycosyltransferase 1	EXT1	Disease: hereditary multiple exostoses 1 (EXT1)
Exostosin glycosyltransferase 2	EXT2	Disease: hereditary multiple exostoses 2 (EXT2)
Exostosin-like glycosyltransferase 1	EXTL1	Glycotranscriptome study reveals an enzymatic switch modulating glycosaminoglycan synthesis during B-cell development and activation. PMID: 22076801
Exostosin-like glycosyltransferase 2	EXTL2	EXTL2 controls liver regeneration and aortic calcification through xylose kinase-dependent regulation of glycosaminoglycan biosynthesis. PMID: 24176719
Exostosin-like glycosyltransferase 3	EXTL3	Regenerating islet-derived 1α (Reg-1α) protein is new neuronal secreted factor that stimulates neurite outgrowth via exostosin tumor-like 3 (EXTL3) receptor. PMID: 22158612
Fucosyltransferase 1 (galactoside 2-alpha-L-fucosyltransferase, H blood group)	FUT1	RNA-mediated gene silencing of FUT1 and FUT2 influences expression and activities of bovine and human fucosylated nucleolin and inhibits cell adhesion and proliferation. PMID: 20506485
Fucosyltransferase 2 (secretor status included)	FUT2	RNA-mediated gene silencing of FUT1 and FUT2 influences expression and activities of bovine and human fucosylated nucleolin and inhibits cell adhesion and proliferation. PMID: 20506485
Fucosyltransferase 3 (galactoside 3(4)-L-fucosyltransferase, Lewis blood group)	FUT3	Down-regulation of FUT3 and FUT5 by shRNA alters Lewis antigens expression and reduces the adhesion capacities of gastric cancer cells. PMID: 21978830
Fucosyltransferase 4 (alpha (1,3) fucosyltransferase, myeloid-specific)	FUT4	HSF1 and Sp1 regulate FUT4 gene expression and cell proliferation in breast cancer cells. PMID: 23959823
Fucosyltransferase 5 (alpha (1,3) fucosyltransferase)	FUT5	

HGNC-approved name	Symbol	Target information
Fucosyltransferase 6 (alpha (1,3) fucosyltransferase)	FUT6	The biosynthesis of the selectin-ligand sialyl Lewis x in colorectal cancer tissues is regulated by fucosyltransferase VI and can be inhibited by an RNA interference-based approach. PMID: 20965272
Fucosyltransferase 7 (alpha (1,3) fucosyltransferase)	FUT7	Fucosylation with fucosyltransferase VI or fucosyltransferase VII improves cord blood engraftment. PMID: 24094497
Fucosyltransferase 8 (alpha (1,6) fucosyltransferase)	FUT8	Overexpression of α (1,6) fucosyltransferase associated with aggressive prostate cancer. PMID: 24906821
Fucosyltransferase 9 (alpha (1,3) fucosyltransferase)	FUT9	Silencing α1,3-fucosyltransferases in human leukocytes reveals a role for FUT9 enzyme during E-selectin-mediated cell adhesion. PMID: 23192350
Fucosyltransferase 10 (alpha (1,3) fucosyltransferase)	FUT10	The Lewis X-related α1,3-fucosyltransferase, Fut10, is required for the maintenance of stem cell populations. PMID: 23906452
Fucosyltransferase 11 (alpha (1,3) fucosyltransferase)	FUT11	FUT11 as a potential biomarker of clear cell renal cell carcinoma progression based on meta-analysis of gene expression data. PMID: 24318988
Galactose-3-*O*-sulfotransferase 1	GAL3ST1	Expression of a testis-specific form of Gal3st1 (CST), a gene essential for spermatogenesis, is regulated by the CTCF paralogous gene BORIS. PMID: 20231363
Galactose-3-*O*-sulfotransferase 2	GAL3ST2	Gal3ST-2 involved in tumor metastasis process by regulation of adhesion ability to selectins and expression of integrins. PMID: 15921657
Galactose-3-*O*-sulfotransferase 3	GAL3ST3	
Galactose-3-*O*-sulfotransferase 4	GAL3ST4	Evidence for GAL3ST4 mutation as the potential cause of pectus excavatum. PMID: 23147795
Glucosaminyl (*N*-acetyl) transferase 1, core 2	GCNT1	Increased expression of GCNT1 is associated with altered O-glycosylation of PSA, PAP, and MUC1 in human prostate cancers. PMID: 24854630
Glucosaminyl (*N*-acetyl) transferase 2, I-branching enzyme (I blood group)	GCNT2	Engagement of I-branching {beta}-1, 6-*N*-acetylglucosaminyltransferase 2 in breast cancer metastasis and TGF-{beta} signaling. PMID: 21750175
Glucosaminyl (*N*-acetyl) transferase 3, mucin type	GCNT3	Clinical relevance of the differential expression of the glycosyltransferase gene GCNT3 in colon cancer. PMID: 25466507
Glucosaminyl (*N*-acetyl) transferase 4, core 2	GCNT4	
Glucosaminyl (*N*-acetyl) transferase 6	GCNT6	
Glucosaminyl (*N*-acetyl) transferase family member 7	GCNT7	
Glucoside xylosyltransferase 1	GXYLT1	*In vitro* assays of orphan glycosyltransferases and their application to identify Notch xylosyltransferases. PMID: 23765671
Glucoside xylosyltransferase 2	GXYLT2	
Glycosyltransferase 1 domain containing 1	GLT1D1	
Glycosyltransferase 6 domain containing 1	GLT6D1	A genome-wide association study identifies GLT6D1 as a susceptibility locus for periodontitis. PMID: 19897590
Glycosyltransferase 8 domain containing 1	GLT8D1	ITIH3 polymorphism may confer susceptibility to psychiatric disorders by altering the expression levels of GLT8D1. PMID: 24373612
Glycosyltransferase 8 domain containing 2	GLT8D2	Glycosyltransferase GLT8D2 positively regulates ApoB100 protein expression in hepatocytes. PMID: 24173238
Heparan sulfate (glucosamine) 3-*O*-sulfotransferase 1	HS3ST1	Mice deficient in heparan sulfate 3-*O*-sulfotransferase-1: Normal hemostasis with unexpected perinatal phenotypes. PMID: 12975616

(Continued)

HGNC-approved name	Symbol	Target information
Heparan sulfate (glucosamine) 3-*O*-sulfotransferase 2	HS3ST2	HS3ST2 modulates breast cancer cell invasiveness via MAP kinase- and Tcf4 (Tcf7l2)-dependent regulation of protease and cadherin expression. PMID: 24752740
Heparan sulfate (glucosamine) 3-*O*-sulfotransferase 3A1	HS3ST3A1	
Heparan sulfate (glucosamine) 3-*O*-sulfotransferase 3B1	HS3ST3B1	Heparan sulfate D-glucosaminyl 3-*O*-sulfotransferase-3B1 (HS3ST3B1) promotes angiogenesis and proliferation by induction of VEGF in acute myeloid leukemia cells. PMID: 25536282
Heparan sulfate (glucosamine) 3-*O*-sulfotransferase 4	HS3ST4	
Heparan sulfate (glucosamine) 3-*O*-sulfotransferase 5	HS3ST5	The biosynthesis of anticoagulant heparan sulfate by the heparan sulfate 3-*O*-sulfotransferase isoform 5. PMID: 15026143
Heparan sulfate (glucosamine) 3-*O*-sulfotransferase 6	HS3ST6	Characterization of heparan sulphate 3-*O*-sulphotransferase isoform 6 and its role in assisting the entry of herpes simplex virus type 1. PMID: 15303968
Heparan sulfate 2-*O*-sulfotransferase 1	HS2ST1	
Heparan sulfate 6-*O*-sulfotransferase 1	HS6ST1	Disease: hypogonadotropic hypogonadism 15 with or without anosmia (HH15)
Heparan sulfate 6-*O*-sulfotransferase 2	HS6ST2	Overexpression of heparan sulfate 6-*O*-sulfotransferase-2 in colorectal cancer. PMID: 24649258
Heparan sulfate 6-*O*-sulfotransferase 3	HS6ST3	
Mannosidase, alpha, class 1A, member 1	MAN1A1	
Mannosidase, alpha, class 1C, member 1	MAN1C1	
Mannosidase, alpha, class 2A, member 1	MAN2A1	
Mannosidase, alpha, class 2A, member 2	MAN2A2	The in vivo role of alpha-mannosidase IIx and its role in processing of *N*-glycans in spermatogenesis. PMID: 12943224
Mannosidase, alpha, class 2B, member 1	MAN2B1	Disease: mannosidosis, alpha B, lysosomal (MANSA)
Mannosidase, alpha, class 2B, member 2	MAN2B2	
Mannosidase, alpha, class 2C, member 1	MAN2C1	Calystegine B3 as a specific inhibitor for cytoplasmic alpha-mannosidase, Man2C1. PMID: 21217149
Mannosyl (alpha-1,3-)-glycoprotein beta-1,2-*N*-acetylglucosaminyltransferase	MGAT1	Golgi *N*-glycan branching *N*-acetylglucosaminyltransferases I, V and VI promote nutrient uptake and metabolism. PMID: 25395405
Mannosyl (alpha-1,6-)-glycoprotein beta-1,2-*N*-acetylglucosaminyltransferase	MGAT2	Disease: congenital disorder of glycosylation 2A (CDG2A)
Mannosyl (beta-1,4-)-glycoprotein beta-1,4-*N* acetylglucosaminyltransferase	MGAT3	All-trans-retinoic acid modulates ICAM-1 *N*-glycan composition by influencing GnT-III levels and inhibits cell adhesion and trans-endothelial migration. PMID: 23300837
Mannosyl (alpha-1,3-)-glycoprotein beta-1,4-*N*-acetylglucosaminyltransferase, isozyme A	MGAT4A	The transcription of MGAT4A glycosyl transferase is increased in white cells of peripheral blood of type 2 diabetes patients. PMID: 17953760
Mannosyl (alpha-1,3-)-glycoprotein beta-1,4-*N*-acetylglucosaminyltransferase, isozyme B	MGAT4B	Aberrant expression of *N*-acetylglucosaminyltransferase-IVa and IVb (GnT-IVa and b) in pancreatic cancer. PMID: 16434023
Mannosyl (alpha-1,3-)-glycoprotein beta-1,4-*N*-acetylglucosaminyltransferase, isozyme C (putative)	MGAT4C	
Mannosyl (alpha-1,6-)-glycoprotein beta-1,6-*N*-acetylglucosaminyltransferase	MGAT5	Knockdown of Mgat5 inhibits CD133+ human pulmonary adenocarcinoma cell growth *in vitro* and in vivo. PMID: 21631992
Mannosyl (alpha-1,6-)-glycoprotein beta-1,6-*N*-acetylglucosaminyltransferase, isozyme B	MGAT5B	MGAT5 and disease severity in progressive multiple sclerosis. PMID: 21115203

HGNC-approved name	Symbol	Target information
N-deacetylase/N-sulfotransferase (heparan glucosaminyl) 1	NDST1	NDST1 missense mutations in autosomal recessive intellectual disability. PMID: 25125150
N-deacetylase/N-sulfotransferase (heparan glucosaminyl) 2	NDST2	
N-deacetylase/N-sulfotransferase (heparan glucosaminyl) 3	NDST3	Altered heparan sulfate structure in mice with deleted NDST3 gene function. PMID: 18385133
N-deacetylase/N-sulfotransferase (heparan glucosaminyl) 4	NDST4	NDST4 is a novel candidate tumor suppressor gene at chromosome 4q26 and its genetic loss predicts adverse prognosis in colorectal cancer. PMID: 23825612
Polypeptide N-acetylgalactosaminyltransferase 1	GALNT1	BCMab1, a monoclonal antibody against aberrantly glycosylated integrin α3β1, has potent antitumor activity of bladder cancer in vivo. PMID: 25002124
Polypeptide N-acetylgalactosaminyltransferase 2	GALNT2	Polypeptide N-acetylgalactosaminyltransferase 2 regulates cellular metastasis-associated behavior in gastric cancer. PMID: 22992780
Polypeptide N-acetylgalactosaminyltransferase 3	GALNT3	Polypeptide N-acetylgalactosaminyl transferase 3 independently predicts high-grade tumours and poor prognosis in patients with renal cell carcinomas. PMID: 23799843. Disease: tumoral calcinosis, hyperphosphatemic, familial (HFTC)
Polypeptide N-acetylgalactosaminyltransferase 4	GALNT4	GALNT4 predicts clinical outcome in patients with clear cell renal cell carcinoma. PMID: 24769034
Polypeptide N-acetylgalactosaminyltransferase 5	GALNT5	Clinical significance of polypeptide N-acetylgalactosaminyl transferase-5 (GalNAc-T5) expression in patients with gastric cancer. PMID: 24619076
Polypeptide N-acetylgalactosaminyltransferase 6	GALNT6	Polypeptide N-acetylgalactosaminyltransferase 6 disrupts mammary acinar morphogenesis through O-glycosylation of fibronectin. PMID: 21472136
Polypeptide N-acetylgalactosaminyltransferase 7	GALNT7	MicroRNA-214 suppresses growth and invasiveness of cervical cancer cells by targeting UDP-N-acetyl-α-D-galactosamine:polypeptide N-acetylgalactosaminyltransferase 7. PMID: 22399294
Polypeptide N-acetylgalactosaminyltransferase 8	GALNT8	Single-nucleotide polymorphisms in GALNT8 are associated with the response to interferon therapy for chronic hepatitis C. PMID: 23034592
Polypeptide N-acetylgalactosaminyltransferase 9	GALNT9	GALNT9 gene expression is a prognostic marker in neuroblastoma patients. PMID: 23136245
Polypeptide N-acetylgalactosaminyltransferase 10	GALNT10	Elevated expression of N-acetylgalactosaminyltransferase 10 predicts poor survival and early recurrence of patients with clear-cell renal cell carcinoma. PMID: 18301266
Polypeptide N-acetylgalactosaminyltransferase 11	GALNT11	Note: Defects in GALNT11 may be a cause of heterotaxy, a congenital heart disease resulting from abnormalities in left-right (LR) body patterning. PMID: 21282601
Polypeptide N-acetylgalactosaminyltransferase 12	GALNT12	Disease: colorectal cancer 1 (CRCS1)
Polypeptide N-acetylgalactosaminyltransferase 13	GALNT13	ppGalNAc-T13: A new molecular marker of bone marrow involvement in neuroblastoma. PMID: 16873292
Polypeptide N-acetylgalactosaminyltransferase 14	GALNT14	GALNT14 mediates tumor invasion and migration in breast cancer cell MCF-7. PMID: 24962947
Polypeptide N-acetylgalactosaminyltransferase 15	GALNT15	MicroRNA-34a/c function as tumor suppressors in Hep-2 laryngeal carcinoma cells and may reduce GALNT7 expression. PMID: 24482044
Polypeptide N-acetylgalactosaminyltransferase 16	GALNT16	

(Continued)

HGNC-approved name	Symbol	Target information
Polypeptide N-acetylgalactosaminyltransferase 18	GALNT18	
Polypeptide N-acetylgalactosaminyltransferase-like 5	GALNTL5	Note: Defects in GALNTL5 have been found in a patient with primary infertility due to asthenozoospermia. PMID: 24398516
Polypeptide N-acetylgalactosaminyltransferase-like 6	GALNTL6	
Protein O-fucosyltransferase 1	POFUT1	Disease: Dowling–Degos disease 2 (DDD2)
Protein O-fucosyltransferase 2	POFUT2	Structure of human POFUT2: Insights into thrombospondin type 1 repeat fold and O-fucosylation. PMID: 22588082
Protein O-linked mannose N-acetylglucosaminyltransferase 1 (beta 1,2-)	POMGNT1	Disease: muscular dystrophy–dystroglycanopathy congenital with brain and eye anomalies A3 (MDDGA3)
Protein O-linked mannose N-acetylglucosaminyltransferase 2 (beta 1,4-)	POMGNT2	Disease: muscular dystrophy–dystroglycanopathy congenital with brain and eye anomalies A8 (MDDGA8)
Protein-O-mannosyltransferase 1	POMT1	Disease: muscular dystrophy–dystroglycanopathy congenital with mental retardation B1 (MDDGB1)
Protein-O-mannosyltransferase 2	POMT2	Disease: muscular dystrophy–dystroglycanopathy congenital with brain and eye anomalies A2 (MDDGA2)
ST3 beta-galactoside alpha-2, 3-sialyltransferase 1	ST3GAL1	
ST3 beta-galactoside alpha-2, 3-sialyltransferase 2	ST3GAL2	
ST3 beta-galactoside alpha-2, 3-sialyltransferase 3	ST3GAL3	Disease: mental retardation, autosomal recessive 12 (MRT12)
ST3 beta-galactoside alpha-2, 3-sialyltransferase 4	ST3GAL4	TNF regulates sialyl-Lewis^x and 6-sulfo-sialyl-Lewis^x expression in human lung through up-regulation of ST3GAL4 transcript isoform BX. PMID: 22691873
ST3 beta-galactoside alpha-2, 3-sialyltransferase 5	ST3GAL5	Disease: Amish infantile epilepsy syndrome (AIES)
ST3 beta-galactoside alpha-2, 3-sialyltransferase 6	ST3GAL6	The sialyltransferase ST3GAL6 influences homing and survival in multiple myeloma. PMID: 25061176
ST6 beta-galactosamide alpha-2, 6-sialyltranferase 1	ST6GAL1	Gene silencing of β-galactosamide α-2,6-sialyltransferase 1 inhibits human influenza virus infection of airway epithelial cells. PMID: 24670114
ST6 beta-galactosamide alpha-2, 6-sialyltranferase 2	ST6GAL2	Transcriptional regulation of the human ST6GAL2 gene in cerebral cortex and neuronal cells. PMID: 19768537
ST6 (alpha-N-acetyl-neuraminyl-2, 3-beta-galactosyl-1,3)-N-acetylgalactosaminide alpha-2,6-sialyltransferase 1	ST6GALNAC1	Enhancement of metastatic ability by ectopic expression of ST6GalNAcI on a gastric cancer cell line in a mouse model. PMID: 22228572
ST6 (alpha-N-acetyl-neuraminyl-2, 3-beta-galactosyl-1,3)-N-acetylgalactosaminide alpha-2,6-sialyltransferase 2	ST6GALNAC2	Sticking to sugars at the metastatic site: Sialyltransferase ST6GalNAc2 acts as a breast cancer metastasis suppressor. PMID: 24596201
ST6 (alpha-N-acetyl-neuraminyl-2, 3-beta-galactosyl-1,3)-N-acetylgalactosaminide alpha-2,6-sialyltransferase 3	ST6GALNAC3	
ST6 (alpha-N-acetyl-neuraminyl-2, 3-beta-galactosyl-1,3)-N-acetylgalactosaminide alpha-2,6-sialyltransferase 4	ST6GALNAC4	Aberrant glycosylation promotes lung cancer metastasis through adhesion to galectins in the metastatic niche. PMID: 25421439
ST6 (alpha-N-acetyl-neuraminyl-2, 3-beta-galactosyl-1,3)-N-acetylgalactosaminide alpha-2,6-sialyltransferase 5	ST6GALNAC5	Overexpression of ST6GalNAcV, a ganglioside-specific alpha2,6-sialyltransferase, inhibits glioma growth in vivo. PMID: 20616019

HGNC-approved name	Symbol	Target information
ST6 (alpha-*N*-acetyl-neuraminyl-2, 3-beta-galactosyl-1,3)-*N*-acetylgalactosaminide alpha-2,6-sialyltransferase 6	ST6GALNAC6	
ST8 alpha-*N*-acetyl-neuraminide alpha-2, 8-sialyltransferase 1	ST8SIA1	Elimination of GD3 synthase improves memory and reduces amyloid-beta plaque load in transgenic mice. PMID: 18258340
ST8 alpha-*N*-acetyl-neuraminide alpha-2, 8-sialyltransferase 2	ST8SIA2	Characterisation of genetic variation in ST8SIA2 and its interaction region in NCAM1 in patients with bipolar disorder. PMID: 24651862
ST8 alpha-*N*-acetyl-neuraminide alpha-2, 8-sialyltransferase 3	ST8SIA3	
ST8 alpha-*N*-acetyl-neuraminide alpha-2, 8-sialyltransferase 4	ST8SIA4	
ST8 alpha-*N*-acetyl-neuraminide alpha-2, 8-sialyltransferase 5	ST8SIA5	
ST8 alpha-*N*-acetyl-neuraminide alpha-2, 8-sialyltransferase 6	ST8SIA6	
STT3A, subunit of the oligosaccharyltransferase complex (catalytic)	STT3A	Disease: congenital disorder of glycosylation 1W (CDG1W)
STT3B, subunit of the oligosaccharyltransferase complex (catalytic)	STT3B	Disease: congenital disorder of glycosylation 1X (CDG1X)
Tyrosylprotein sulfotransferase 1	TPST1	Reduced body weight and increased postimplantation fetal death in tyrosylprotein sulfotransferase-1-deficient mice. PMID: 11904405
Tyrosylprotein sulfotransferase 2	TPST2	Targeted disruption of tyrosylprotein sulfotransferase-2, an enzyme that catalyzes post-translational protein tyrosine O-sulfation, causes male infertility. PMID: 16469738
UDP-Gal:betaGlcNAc beta-1, 3-galactosyltransferase, polypeptide 1	B3GALT1	B4GALT family mediates the multidrug resistance of human leukemia cells by regulating the hedgehog pathway and the expression of p-glycoprotein and multidrug resistance-associated protein 1. PMID: 23744354
UDP-Gal:betaGlcNAc beta-1, 3-galactosyltransferase, polypeptide 2	B3GALT2	
UDP-Gal:betaGlcNAc beta-1, 3-galactosyltransferase, polypeptide 4	B3GALT4	Beta1,3-galactosyltransferases-4/5 are novel tumor markers for gynecological cancers. PMID: 19225246
UDP-Gal:betaGlcNAc beta-1, 3-galactosyltransferase, polypeptide 5	B3GALT5	Enhanced expression of beta 3-galactosyltransferase 5 activity is sufficient to induce in vivo synthesis of extended type 1 chains on lactosylceramides of selected human colonic carcinoma cell lines. PMID: 19136585
UDP-Gal:betaGal beta-1, 3-galactosyltransferase polypeptide 6	B3GALT6	Disease: Ehlers–Danlos syndrome, progeroid type, 2 (EDSP2)
UDP-Gal:betaGlcNAc beta-1, 4-galactosyltransferase, polypeptide 1	B4GALT1	Disease: congenital disorder of glycosylation 2D (CDG2D)
UDP-Gal:betaGlcNAc beta-1, 4-galactosyltransferase, polypeptide 2	B4GALT2	Beta4GalT-II increases cisplatin-induced apoptosis in HeLa cells depending on its Golgi localization. PMID: 17470362
UDP-Gal:betaGlcNAc beta-1, 4-galactosyltransferase, polypeptide 3	B4GALT3	β-1,4-Galactosyltransferase III suppresses β1 integrin-mediated invasive phenotypes and negatively correlates with metastasis in colorectal cancer. PMID: 24403309
UDP-Gal:betaGlcNAc beta-1, 4-galactosyltransferase, polypeptide 4	B4GALT4	
UDP-Gal:betaGlcNAc beta-1, 4-galactosyltransferase, polypeptide 5	B4GALT5	Early lethality of beta-1,4-galactosyltransferase V-mutant mice by growth retardation. PMID: 19114028

(Continued)

HGNC-approved name	Symbol	Target information
UDP-Gal:betaGlcNAc beta-1, 4-galactosyltransferase, polypeptide 6	B4GALT6	Regulation of astrocyte activation by glycolipids drives chronic CNS inflammation. PMID: 25216636
Xylosylprotein beta-1,4-galactosyltransferase, polypeptide 7	B4GALT7	Disease: Ehlers–Danlos syndrome, progeroid type, 1 (EDSP1)
UDP-GlcNAc:betaGal beta-1, 3-N-acetylglucosaminyltransferase 2	B3GNT2	Beta3GnT2 (B3GNT2), a major polylactosamine synthase: Analysis of B3GNT2-deficient mice. PMID: 20816167
UDP-GlcNAc:betaGal beta-1, 3-N-acetylglucosaminyltransferase 3	B3GNT3	B3GNT3 expression suppresses cell migration and invasion and predicts favorable outcomes in neuroblastoma. PMID: 24118321
UDP-GlcNAc:betaGal beta-1, 3-N-acetylglucosaminyltransferase 4	B3GNT4	
UDP-GlcNAc:betaGal beta-1, 3-N-acetylglucosaminyltransferase 5	B3GNT5	Multiple phenotypic changes in mice after knockout of the B3gnt5 gene, encoding Lc3 synthase – A key enzyme in lacto-neolacto ganglioside synthesis. PMID: 21087515
UDP-GlcNAc:betaGal beta-1, 3-N-acetylglucosaminyltransferase 6 (core 3 synthase)	B3GNT6	
UDP-GlcNAc:betaGal beta-1, 3-N-acetylglucosaminyltransferase 7	B3GNT7	A novel beta1,3-N-acetylglucosaminyltransferase involved in invasion of cancer cells as assayed in vitro. PMID: 12061784
UDP-GlcNAc:betaGal beta-1, 3-N-acetylglucosaminyltransferase 8	B3GNT8	Regulation of MMP-2 expression and activity by β-1, 3-N-acetylglucosaminyltransferase-8 in AGS gastric cancer cells. PMID: 20963502
UDP-GlcNAc:betaGal beta-1, 3-N-acetylglucosaminyltransferase 9	B3GNT9	
UDP-GlcNAc:betaGal beta-1, 3-N-acetylglucosaminyltransferase-like 1	B3GNTL1	High expression of β3GnT8 is associated with the metastatic potential of human glioma. PMID: 24715095
UDP-glucose glycoprotein glucosyltransferase 1	UGGT1	UDP-glucose:glycoprotein glucosyltransferase (UGGT1) promotes substrate solubility in the endoplasmic reticulum. PMID: 23864712
UDP-glucose glycoprotein glucosyltransferase 2	UGGT2	Both isoforms of human UDP-glucose:glycoprotein glucosyltransferase are enzymatically active. PMID: 24415556
Xylosyltransferase I	XYLT1	XYLT1 mutations in Desbuquois dysplasia type 2. PMID: 24581741
Xylosyltransferase II	XYLT2	Polycystic disease caused by deficiency in xylosyltransferase 2, an initiating enzyme of glycosaminoglycan biosynthesis. PMID: 17517600
Aspartylglucosaminidase	AGA	Disease: aspartylglucosaminuria (AGU)
C1GALT1-specific chaperone 1	C1GALT1C1	Disease: Tn polyagglutination syndrome (TNPS)
coiled-coil domain containing 126	CCDC126	
Core 1 synthase, glycoprotein-N-acetylgalactosamine 3-beta-galactosyltransferase, 1	C1GALT1	C1GALT1 promotes invasive phenotypes of hepatocellular carcinoma cells by modulating integrin β1 glycosylation and activity. PMID: 25089569
Defender against cell death 1	DAD1	Oligosaccharyltransferase isoforms that contain different catalytic STT3 subunits have distinct enzymatic properties. PMID: 12887896
Dolichyldiphosphatase 1	DOLPP1	
Dolichyl-phosphate (UDP-N-acetylglucosamine) N-acetylglucosaminephosphotransferase 1 (GlcNAc-1-P transferase)	DPAGT1	Disease: congenital disorder of glycosylation 1J (CDG1J)
EGF domain-specific O-linked N-acetylglucosamine (GlcNAc) transferase	EOGT	Disease: Adams–Oliver syndrome 4 (AOS4)

HGNC-approved name	Symbol	Target information
Endo-beta-N-acetylglucosaminidase	ENGASE	
ER degradation enhancer, mannosidase alpha-like 3	EDEM3	EDEM3, a soluble EDEM homolog, enhances glycoprotein endoplasmic reticulum-associated degradation and mannose trimming. PMID: 16431915
Fucose-1-phosphate guanylyltransferase	FPGT	
Fukutin-related protein	FKRP	Disease: muscular dystrophy–dystroglycanopathy congenital with brain and eye anomalies A5 (MDDGA5)
Globoside alpha-1, 3-N-acetylgalactosaminyltransferase 1	GBGT1	Expression of GBGT1 is epigenetically regulated by DNA methylation in ovarian cancer cells. PMID: 25294702
Glycosyltransferase-like 1B	GYLTL1B	The glycosyltransferase LARGE2 is repressed by snail and ZEB1 in prostate cancer. PMID: 25455932
Glycosyltransferase-like domain containing 1	GTDC1	
Heparan-alpha-glucosaminide N-acetyltransferase	HGSNAT	Disease: mucopolysaccharidosis 3C (MPS3C)
KIAA2018	KIAA2018	
Like-glycosyltransferase	LARGE	Disease: muscular dystrophy–dystroglycanopathy congenital with mental retardation B6 (MDDGB6)
Mannosidase, beta A, lysosomal	MANBA	Note: Defects in MANBA are the cause of a mild disorder that affects peripheral and central nervous system myelin. Disease: mannosidosis, beta A, lysosomal (MANSB)
Mannosidase, endo-alpha	MANEA	The α-endomannosidase gene (MANEA) is associated with panic disorder and social anxiety disorder. PMID: 24473444
Mannosidase, endo-alpha-like	MANEAL	
mannosyl-oligosaccharide glucosidase	MOGS	Disease: type IIb congenital disorder of glycosylation (CDGIIb)
LFNG O-fucosylpeptide 3-beta-N-acetylglucosaminyltransferase	LFNG	Disease: spondylocostal dysostosis 3, autosomal recessive (SCDO3)
MFNG O-fucosylpeptide 3-beta-N-acetylglucosaminyltransferase	MFNG	Fringe glycosyltransferases differentially modulate Notch1 proteolysis induced by Delta1 and Jagged1. PMID: 15574878
RFNG O-fucosylpeptide 3-beta-N-acetylglucosaminyltransferase	RFNG	Fringe glycosyltransferases differentially modulate Notch1 proteolysis induced by Delta1 and Jagged1. PMID: 15574878
N-glycanase 1	NGLY1	Disease: congenital disorder of glycosylation 1V (CDG1V)
O-linked N-acetylglucosamine (GlcNAc) transferase	OGT	Note: Regulation of OGT activity and altered O-GlcNAcylations are implicated in diabetes and Alzheimer disease. O-GlcNAcylation of AKT1 affects insulin signaling and, possibly diabetes
Parkinson protein 7	PARK7	Parkinsonism-associated protein DJ-1/park7 is a major protein deglycase that repairs methylglyoxal- and glyoxal-glycated cysteine, arginine, and lysine residues. PMID: 25416785
Protein O-glucosyltransferase 1	POGLUT1	Disease: Dowling–Degos disease 4 (DDD4)
Protein-O-mannose kinase	POMK	Disease: muscular dystrophy–dystroglycanopathy congenital with brain and eye anomalies A12 (MDDGA12)
Renin binding protein	RENBP	Physiological relevance of renin/prorenin binding and uptake. PMID: 16025735
Ribophorin I	RPN1	Malectin forms a complex with ribophorin I for enhanced association with misfolded glycoproteins. PMID: 22988243
Ribophorin II	RPN2	RPN2 gene confers osteosarcoma cell malignant phenotypes and determines clinical prognosis. PMID: 25181275
Uronyl-2-sulfotransferase	UST	Role of heparan sulfate 2-o-sulfotransferase in prostate cancer cell proliferation, invasion, and growth factor signaling. PMID: 22135748

(Continued)

HGNC-approved name	Symbol	Target information
WD repeat and FYVE domain containing 3	WDFY3	Loss of Wdfy3 in mice alters cerebral cortical neurogenesis reflecting aspects of the autism pathology. PMID: 25198012
Williams–Beuren syndrome chromosome region 17	WBSCR17	Note: WBSCR17 is located in the Williams–Beuren syndrome (WBS) critical region
Xyloside xylosyltransferase 1	XXYLT1	Identification of glycosyltransferase 8 family members as xylosyltransferases acting on O-glucosylated notch epidermal growth factor repeats. PMID: 19940119

4.2.3   Ubiquitylation and related modifications

HGNC-approved name	Symbol	Target information
Ariadne RBR E3 ubiquitin protein ligase 1	ARIH1	Structure of HHARI, a RING-IBR-RING ubiquitin ligase: Autoinhibition of an Ariadne-family E3 and insights into ligation mechanism. PMID: 23707686
Ariadne RBR E3 ubiquitin protein ligase 2	ARIH2	ARIH2 is essential for embryogenesis, and its hematopoietic deficiency causes lethal activation of the immune system. PMID: 23179078
Ataxin 3	ATXN3	The deubiquitylase ataxin-3 restricts PTEN transcription in lung cancer cells. PMID: 24292675. Disease: spinocerebellar ataxia 3 (SCA3)
Ataxin 3-like	ATXN3L	
Cbl proto-oncogene, E3 ubiquitin protein ligase	CBL	An E3 ubiquitin ligase: c-Cbl: A new therapeutic target of lung cancer. PMID: 21607942. Disease: Noonan syndrome-like disorder with or without juvenile myelomonocytic leukemia (NSLL)
Cbl proto-oncogene B, E3 ubiquitin protein ligase	CBLB	E3 ubiquitin ligase Cbl-b suppresses proallergic T cell development and allergic airway inflammation. PMID: 24508458
Cbl proto-oncogene C, E3 ubiquitin protein ligase	CBLC	Enigma prevents Cbl-c-mediated ubiquitination and degradation of RETMEN2A. PMID: 24466333
Cbl proto-oncogene-like 1, E3 ubiquitin protein ligase	CBLL1	Biological influence of Hakai in cancer: A 10-year review. PMID: 22349934
Deltex 1, E3 ubiquitin ligase	DTX1	Deltex1 promotes protein kinase Cθ degradation and sustains Casitas B-lineage lymphoma expression. PMID: 25000980
Deltex 2, E3 ubiquitin ligase	DTX2	T cells develop normally in the absence of both Deltex1 and Deltex2. PMID: 16923970
Deltex 3, E3 ubiquitin ligase	DTX3	
Deltex 3-like, E3 ubiquitin ligase	DTX3L	DTX3L and ARTD9 inhibit IRF1 expression and mediate in cooperation with ARTD8 survival and proliferation of metastatic prostate cancer cells. PMID: 24886089
Deltex 4, E3 ubiquitin ligase	DTX4	NLRP4 negatively regulates type I interferon signaling by targeting the kinase TBK1 for degradation via the ubiquitin ligase DTX4. PMID: 22388039
Fanconi anemia, complementation group L	FANCL	Disease: Fanconi anemia complementation group L (FANCL)
Fanconi anemia, complementation group M	FANCM	Disease: Fanconi anemia complementation group M (FANCM)
HECT and RLD domain containing E3 ubiquitin protein ligase family member 1	HERC1	Resistance to UV-induced apoptosis by β-HPV5 E6 involves targeting of activated BAK for proteolysis by recruitment of the HERC1 ubiquitin ligase. PMID: 25408501
HECT and RLD domain containing E3 ubiquitin protein ligase 2	HERC2	Disease: mental retardation, autosomal recessive 38 (MRT38)

HGNC-approved name	Symbol	Target information
HECT and RLD domain containing E3 ubiquitin protein ligase 3	HERC3	Progressive Purkinje cell degeneration in tambaleante mutant mice is a consequence of a missense mutation in HERC1 E3 ubiquitin ligase. PMID: 20041218
HECT and RLD domain containing E3 ubiquitin protein ligase 4	HERC4	Expression of E3 ligase HERC4 in breast cancer and its clinical implications. PMID: 25176076
HECT and RLD domain containing E3 ubiquitin protein ligase 5	HERC5	Identification of HERC5 and its potential role in NSCLC progression. PMID: 25353388
HECT and RLD domain containing E3 ubiquitin protein ligase family member 6	HERC6	Murine Herc6 plays a critical role in protein ISGylation in vivo and has an ISGylation-independent function in seminal vesicles. PMID: 25406959
HECT, C2 and WW domain containing E3 ubiquitin protein ligase 1	HECW1	Muscle atrophy and motor neuron degeneration in human NEDL1 transgenic mice. PMID: 20976258
HECT, C2 and WW domain containing E3 ubiquitin protein ligase 2	HECW2	The HECT type ubiquitin ligase NEDL2 is degraded by anaphase-promoting complex/cyclosome (APC/C)-Cdh1, and its tight regulation maintains the metaphase to anaphase transition. PMID: 24163370
HECT domain containing E3 ubiquitin protein ligase 1	HECTD1	HectD1 E3 ligase modifies adenomatous polyposis coli (APC) with polyubiquitin to promote the APC-axin interaction. PMID: 23277359
HECT domain containing E3 ubiquitin protein ligase 2	HECTD2	HECTD2 is associated with susceptibility to mouse and human prion disease. PMID: 19214206
HECT domain containing E3 ubiquitin protein ligase 3	HECTD3	The HECTD3 E3 ubiquitin ligase facilitates cancer cell survival by promoting K63-linked polyubiquitination of caspase-8. PMID: 24287696
HECT domain containing E3 ubiquitin protein ligase 4	HECTD4	
Makorin ring finger protein 1	MKRN1	Acceleration of gastric tumorigenesis through MKRN1-mediated posttranslational regulation of p14ARF. PMID: 23104211
Makorin ring finger protein 2	MKRN2	Ubiquitous expression of MAKORIN-2 in normal and malignant hematopoietic cells and its growth promoting activity. PMID: 24675897
Makorin ring finger protein 3	MKRN3	Disease: precocious puberty, central 2 (CPPB2)
Membrane-associated ring finger (C3HC4) 1, E3 ubiquitin protein ligase	MARCH1	MARCH1-mediated MHCII ubiquitination promotes dendritic cell selection of natural regulatory T cells. PMID: 23712430
Membrane-associated ring finger (C3HC4) 2, E3 ubiquitin protein ligase	MARCH2	MARCH2 promotes endocytosis and lysosomal sorting of carvedilol-bound β(2)-adrenergic receptors. PMID: 23166351
Membrane-associated ring finger (C3HC4) 3, E3 ubiquitin protein ligase	MARCH3	MARCH-III is a novel component of endosomes with properties similar to those of MARCH-II. PMID: 16428329
Membrane-associated ring finger (C3HC4) 4, E3 ubiquitin protein ligase	MARCH4	Membrane-Associated RING-CH proteins associate with Bap31 and target CD81 and CD44 to lysosomes. PMID: 21151997
Membrane-associated ring finger (C3HC4) 5	MARCH5	Roles of mitochondrial ubiquitin ligase MITOL/MARCH5 in mitochondrial dynamics and diseases. PMID: 24616159
Membrane-associated ring finger (C3HC4) 6, E3 ubiquitin protein ligase	MARCH6	The E3 ubiquitin ligase MARCH6 degrades squalene monooxygenase and affects 3-hydroxy-3-methyl-glutaryl coenzyme A reductase and the cholesterol synthesis pathway. PMID: 24449766
Membrane-associated ring finger (C3HC4) 7, E3 ubiquitin protein ligase	MARCH7	Axotrophin/MARCH7 acts as an E3 ubiquitin ligase and ubiquitinates tau protein in vitro impairing microtubule binding. PMID: 24905733
Membrane-associated ring finger (C3HC4) 8, E3 ubiquitin protein ligase	MARCH8	Ubiquitination by the membrane-associated RING-CH-8 (MARCH-8) ligase controls steady-state cell surface expression of tumor necrosis factor-related apoptosis inducing ligand (TRAIL) receptor 1. PMID: 23300075
Membrane-associated ring finger (C3HC4) 9	MARCH9	MARCH-IX mediates ubiquitination and downregulation of ICAM-1. PMID: 17174307

(Continued)

HGNC-approved name	Symbol	Target information
Membrane-associated ring finger (C3HC4) 10, E3 ubiquitin protein ligase	MARCH10	Membrane-associated RING-CH 10 (MARCH10 protein) is a microtubule-associated E3 ubiquitin ligase of the spermatid flagella. PMID: 21937444
Membrane-associated ring finger (C3HC4) 11	MARCH11	Identification of SAMT family proteins as substrates of MARCH11 in mouse spermatids. PMID: 22075566
Midline 1	MID1	MID1 catalyzes the ubiquitination of protein phosphatase 2A and mutations within its Bbox1 domain disrupt polyubiquitination of alpha4 but not of PP2Ac. PMID: 25207814. Disease: Opitz GBBB syndrome 1 (OGS1)
Midline 2	MID2	Disease: mental retardation, X-linked 101 (MRX101)
Mindbomb E3 ubiquitin protein ligase 1	MIB1	The E3 ubiquitin ligase mind bomb 1 ubiquitinates and promotes the degradation of survival of motor neuron protein. PMID: 23615451. Disease: left ventricular non-compaction 7 (LVNC7)
Mindbomb E3 ubiquitin protein ligase 2	MIB2	The E3 ubiquitin ligase mind bomb-2 (MIB2) protein controls B-cell CLL/lymphoma 10 (BCL10)-dependent NF-κB activation. PMID: 21896478
Neural precursor cell expressed, developmentally down-regulated 4, E3 ubiquitin protein ligase	NEDD4	Identification and rescue of α-synuclein toxicity in Parkinson patient-derived neurons. PMID: 24158904
Neural precursor cell expressed, developmentally down-regulated 4-like, E3 ubiquitin protein ligase	NEDD4L	NEDD4-2 (NEDD4L): The ubiquitin ligase for multiple membrane proteins. PMID: 25433090
Neural precursor cell expressed, developmentally down-regulated 8	NEDD8	Biochemical and cellular effects of inhibiting Nedd8 conjugation. PMID: 20603103
Neuralized E3 ubiquitin protein ligase 1	NEURL1	Neuralized1 causes apoptosis and downregulates Notch target genes in medulloblastoma. PMID: 20847082
Neuralized E3 ubiquitin protein ligase 1B	NEURL1B	
Neuralized E3 ubiquitin protein ligase 2	NEURL2	Neuralized-2: Expression in human and rodents and interaction with Delta-like ligands. PMID: 19723503
Neuralized E3 ubiquitin protein ligase 3	NEURL3	A novel inflammation-induced ubiquitin E3 ligase in alveolar type II cells. PMID: 15936721
Neuralized E3 ubiquitin protein ligase 4	NEURL4	Interaction proteomics identify NEURL4 and the HECT E3 ligase HERC2 as novel modulators of centrosome architecture. PMID: 22261722
OTU deubiquitinase 1	OTUD1	A putative OTU domain-containing protein 1 deubiquitinating enzyme is differentially expressed in thyroid cancer and identifies less-aggressive tumours. PMID: 24937664
OTU deubiquitinase 3	OTUD3	
OTU deubiquitinase 4	OTUD4	Ataxia, dementia, and hypogonadotropism caused by disordered ubiquitination. PMID: 23656588
OTU deubiquitinase 5	OTUD5	Deubiquitinase OTUD5 mediates the sequential activation of PDCD5 and p53 in response to genotoxic stress. PMID: 25499082
OTU deubiquitinase 6A	OTUD6A	
OTU deubiquitinase 7A	OTUD7A	Snail1-dependent transcriptional repression of Cezanne2 in hepatocellular carcinoma. PMID: 23792447
OTU deubiquitinase 7B	OTUD7B	OTUD7B controls non-canonical NF-κB activation through deubiquitination of TRAF3. PMID: 23334419
OTU deubiquitinase, ubiquitin aldehyde binding 1	OTUB1	OTUB1 promotes metastasis and serves as a marker of poor prognosis in colorectal cancer. PMID: 25431208
OTU deubiquitinase, ubiquitin aldehyde binding 2	OTUB2	Fine-tuning of DNA damage-dependent ubiquitination by OTUB2 supports the DNA repair pathway choice. PMID: 24560272

HGNC-approved name	Symbol	Target information
PDZ domain containing ring finger 3	PDZRN3	PDZRN3/LNX3 is a novel target of human papillomavirus type 16 (HPV-16) and HPV-18 E6. PMID: 25355882
PDZ domain containing ring finger 4	PDZRN4	
Pellino E3 ubiquitin protein ligase 1	PELI1	Silencing of Pellino1 improves post-infarct cardiac dysfunction and attenuates left ventricular remodelling in mice. PMID: 24442869
Pellino E3 ubiquitin protein ligase family member 2	PELI2	Pellino 2 is critical for Toll-like receptor/interleukin-1 receptor (TLR/IL-1R)-mediated post-transcriptional control. PMID: 22669975
Pellino E3 ubiquitin protein ligase family member 3	PELI3	Autophagy-dependent PELI3 degradation inhibits proinflammatory IL1B expression. PMID: 25483963
Praja ring finger 1, E3 ubiquitin protein ligase	PJA1	The ubiquitin ligase Praja1 reduces NRAGE expression and inhibits neuronal differentiation of PC12 cells. PMID: 23717400
Praja ring finger 2, E3 ubiquitin protein ligase	PJA2	Control of PKA stability and signalling by the RING ligase praja2. PMID: 21423175
Ring finger protein 2	RNF2	RNF2/Ring1b negatively regulates p53 expression in selective cancer cell types to promote tumor development. PMID: 23319651
Ring finger protein 4	RNF4	RING finger protein 4 (RNF4) derepresses gene expression from DNA methylation. PMID: 25355316
Ring finger protein 5, E3 ubiquitin protein ligase	RNF5	The E3 ubiquitin ligase RNF5 targets virus-induced signaling adaptor for ubiquitination and degradation. PMID: 20483786
Ring finger protein (C3H2C3 type) 6	RNF6	Disease: esophageal cancer (ESCR)
Ring finger protein 8, E3 ubiquitin protein ligase	RNF8	The ubiquitin ligases RNF8 and RNF168 display rapid but distinct dynamics at DNA repair foci in living cells. PMID: 25304081
Ring finger protein 13	RNF13	RNF13: An emerging RING finger ubiquitin ligase important in cell proliferation. PMID: 21078127
Ring finger protein 14	RNF14	Ring finger protein 14 is a new regulator of TCF/β-catenin-mediated transcription and colon cancer cell survival. PMID: 23449499
Ring finger protein 19A, RBR E3 ubiquitin protein ligase	RNF19A	Dorfin ameliorates phenotypes in a transgenic mouse model of amyotrophic lateral sclerosis. PMID: 19610091
Ring finger protein 19B	RNF19B	
Ring finger protein 25	RNF25	RING finger protein AO7 supports NF-kappaB-mediated transcription by interacting with the transactivation domain of the p65 subunit. PMID: 12748188
Ring finger protein 31	RNF31	E3 ubiquitin ligase HOIP attenuates apoptotic cell death induced by cisplatin. PMID: 24686174
Ring finger protein 34, E3 ubiquitin protein ligase	RNF34	Ring finger protein 34 (RNF34) interacts with and promotes γ-aminobutyric acid type-A receptor degradation via ubiquitination of the γ2 subunit. PMID: 25193658
Ring finger protein 38	RNF38	RNF38 encodes a nuclear ubiquitin protein ligase that modifies p53. PMID: 23973461
Ring finger protein 41, E3 ubiquitin protein ligase	RNF41	Nrdp1 inhibits growth of colorectal cancer cells by nuclear retention of p27. PMID: 24867101
Ring finger protein 43	RNF43	RNF43 is frequently mutated in colorectal and endometrial cancers. PMID: 25344691
Ring finger protein 111	RNF111	RNF111/Arkadia is a SUMO-targeted ubiquitin ligase that facilitates the DNA damage response. PMID: 23751493
Ring finger protein 114	RNF114	The RING ubiquitin E3 RNF114 interacts with A20 and modulates NF-κB activity and T-cell activation. PMID: 25165885

(Continued)

HGNC-approved name	Symbol	Target information
Ring finger protein 115	RNF115	RNF115/BCA2 E3 ubiquitin ligase promotes breast cancer cell proliferation through targeting p21Waf1/Cip1 for ubiquitin-mediated degradation. PMID: 24027428
Ring finger protein 123	RNF123	Ubiquitin ligase RNF123 mediates degradation of heterochromatin protein 1α and β in lamin A/C knock-down cells. PMID: 23077635
Ring finger protein 125, E3 ubiquitin protein ligase	RNF125	T-cell regulator RNF125/TRAC-1 belongs to a novel family of ubiquitin ligases with zinc fingers and a ubiquitin-binding domain. PMID: 17990982
Ring finger protein 126	RNF126	E3 ubiquitin ligase RNF126 promotes cancer cell proliferation by targeting the tumor suppressor p21 for ubiquitin-mediated degradation. PMID: 23026136
Ring finger protein 128, E3 ubiquitin protein ligase	RNF128	The E3 ubiquitin ligase GRAIL regulates T cell tolerance and regulatory T cell function by mediating T cell receptor-CD3 degradation. PMID: 20493730
Ring finger protein 130	RNF130	Goliath family E3 ligases regulate the recycling endosome pathway via VAMP3 ubiquitylation. PMID: 23353890
Ring finger protein 133	RNF133	Mouse RING finger protein Rnf133 is a testis-specific endoplasmic reticulum-associated E3 ubiquitin ligase. PMID: 18574499
Ring finger protein 135	RNF135	Disease: macrocephaly, macrosomia, facial dysmorphism syndrome (MMFD)
Ring finger protein 138, E3 ubiquitin protein ligase	RNF138	A novel gene RNF138 expressed in human gliomas and its function in the glioma cell line U251. PMID: 22155992
Ring finger protein 139	RNF139	Disease: renal cell carcinoma (RCC)
Ring finger protein 144A	RNF144A	RNF144A, an E3 ubiquitin ligase for DNA-PKcs, promotes apoptosis during DNA damage. PMID: 24979766
Ring finger protein 144B	RNF144B	IBRDC2, an IBR-type E3 ubiquitin ligase, is a regulatory factor for Bax and apoptosis activation. PMID: 20300062
Ring finger protein 146	RNF146	Note: Defects in RNF146 are a cause of susceptibility to breast cancer
Ring finger protein 149	RNF149	Ring finger protein 149 is an E3 ubiquitin ligase active on wild-type v-Raf murine sarcoma viral oncogene homolog B1 (BRAF). PMID: 22628551
Ring finger protein 151	RNF151	RNF151, a testis-specific RING finger protein, interacts with dysbindin. PMID: 17577571
Ring finger protein 152	RNF152	RNF152, a novel lysosome localized E3 ligase with pro-apoptotic activities. PMID: 21203937
Ring finger protein 167	RNF167	Ubiquitin ligase RNF167 regulates AMPA receptor-mediated synaptic transmission. PMID: 23129617
Ring finger protein 168, E3 ubiquitin protein ligase	RNF168	Disease: Riddle syndrome (RIDDLES)
Ring finger protein 169	RNF169	Ring finger protein RNF169 antagonizes the ubiquitin-dependent signalling cascade at sites of DNA damage. PMID: 22733822
Ring finger protein 170	RNF170	Disease: ataxia, sensory, 1, autosomal dominant (SNAX1)
Ring finger protein 180	RNF180	Rines E3 ubiquitin ligase regulates MAO-A levels and emotional responses. PMID: 23926250
Ring finger protein 181	RNF181	RN181, a novel ubiquitin E3 ligase that interacts with the KVGFFKR motif of platelet integrin alpha(IIb)beta3. PMID: 18331836
Ring finger protein 182	RNF182	A novel brain-enriched E3 ubiquitin ligase RNF182 is up regulated in the brains of Alzheimer's patients and targets ATP6V0C for degradation. PMID: 18298843
Ring finger protein 185	RNF185	The E3 ligase RNF185 negatively regulates osteogenic differentiation by targeting Dvl2 for degradation. PMID: 24727453
Ring finger protein 187	RNF187	Identification of a co-activator that links growth factor signalling to c-Jun/AP-1 activation. PMID: 20852630

HGNC-approved name	Symbol	Target information
Ring finger protein 208	RNF208	
Ring finger protein 212	RNF212	Antagonistic roles of ubiquitin ligase HEI10 and SUMO ligase RNF212 regulate meiotic recombination. PMID: 24390283
Ring finger protein 213	RNF213	Disease: Moyamoya disease 2 (MYMY2)
Ring finger protein 216	RNF216	Disease: Gordon Holmes syndrome (GDHS)
Ring finger protein 217	RNF217	Identification and characterization of OSTL (RNF217) encoding a RING-IBR-RING protein adjacent to a translocation breakpoint involving ETV6 in childhood ALL. PMID: 25298122
Ring finger protein 220	RNF220	The ubiquitin ligase RNF220 enhances canonical Wnt signaling through USP7-mediated deubiquitination of β-catenin. PMID: 25266658
Siah E3 ubiquitin protein ligase 1	SIAH1	SIAH1 induced apoptosis by activation of the JNK pathway and inhibited invasion by inactivation of the ERK pathway in breast cancer cells. PMID: 19775288
Siah E3 ubiquitin protein ligase 2	SIAH2	Possible role of death receptor-mediated apoptosis by the E3 ubiquitin ligases Siah2 and POSH. PMID: 21586138
Siah E3 ubiquitin protein ligase family member 3	SIAH3	High-content genome-wide RNAi screens identify regulators of parkin upstream of mitophagy. PMID: 24270810
SMAD-specific E3 ubiquitin protein ligase 1	SMURF1	SMURF1 silencing diminishes a CD44-high cancer stem cell-like population in head and neck squamous cell carcinoma. PMID: 25471937
SMAD-specific E3 ubiquitin protein ligase 2	SMURF2	Smurf2 E3 ubiquitin ligase modulates proliferation and invasiveness of breast cancer cells in a CNKSR2 dependent manner. PMID: 25191523
Small ubiquitin-like modifier 1	SUMO1	Dicarbol non syndromic orofacial cleft 10 (OFC10)
Small ubiquitin-like modifier 2	SUMO2	Modification of DBC1 by SUMO2/3 is crucial for p53-mediated apoptosis in response to DNA damage. PMID: 25406032
Small ubiquitin-like modifier 3	SUMO3	SUMO3 modification accelerates the aggregation of ALS-linked SOD1 mutants. PMID: 24971881
Small ubiquitin-like modifier 4	SUMO4	Inflammatory factor-specific sumoylation regulates NF-κB signalling in glomerular cells from diabetic rats. PMID: 24173240
TNF receptor-associated factor 6, E3 ubiquitin protein ligase	TRAF6	TRAF6 is upregulated in colon cancer and promotes proliferation of colon cancer cells. PMID: 24755241
TNF receptor-associated factor 7, E3 ubiquitin protein ligase	TRAF7	Downregulation of ubiquitin E3 ligase TNF receptor-associated factor 7 leads to stabilization of p53 in breast cancer. PMID: 23128672
Tripartite motif containing 2	TRIM2	Disease: Charcot–Marie–Tooth disease 2R (CMT2R)
Tripartite motif containing 8	TRIM8	TRIM8 anti-proliferative action against chemo-resistant renal cell carcinoma. PMID: 25277184
Tripartite motif containing 9	TRIM9	Negative regulation of NF-κB activity by brain-specific TRIpartite Motif protein 9. PMID: 25190485
Tripartite motif containing 11	TRIM11	TRIM11 is overexpressed in high-grade gliomas and promotes proliferation, invasion, migration and glial tumor growth. PMID: 23178488
Tripartite motif containing 13	TRIM13	TRIM13 regulates ubiquitination and turnover of NEMO to suppress TNF induced NF-κB activation. PMID: 25152375
Tripartite motif containing 17	TRIM17	Trim17-mediated ubiquitination and degradation of Mcl-1 initiate apoptosis in neurons. PMID: 22976837
Tripartite motif containing 22	TRIM22	TRIM22 can activate the noncanonical NF-κB pathway by affecting IKKα. PMID: 25510414
Tripartite motif containing 23	TRIM23	Polyubiquitin conjugation to NEMO by tripartite motif protein 23 (TRIM23) is critical in antiviral defense. PMID: 20724660

(Continued)

HGNC-approved name	Symbol	Target information
Tripartite motif containing 25	TRIM25	Trim25 is an RNA-specific activator of Lin28a/TuT4-mediated uridylation. PMID: 25457611
Tripartite motif containing 27	TRIM27	Disease: thyroid papillary carcinoma (TPC)
Tripartite motif containing 31	TRIM31	TRIM31 is downregulated in non-small cell lung cancer and serves as a potential tumor suppressor. PMID: 24566900
Tripartite motif containing 32	TRIM32	TRIM32 promotes neural differentiation through retinoic acid receptor-mediated transcription. PMID: 21984809. Disease: limb-girdle muscular dystrophy 2H (LGMD2H)
Tripartite motif containing 33	TRIM33	Disease: thyroid papillary carcinoma (TPC)
Tripartite motif containing 36	TRIM36	Haprin, a novel haploid germ cell-specific RING finger protein involved in the acrosome reaction. PMID: 12917430
Tripartite motif containing 37	TRIM37	TRIM37 is a new histone H2A ubiquitin ligase and breast cancer oncoprotein. PMID: 25470042. Disease: Mulibrey nanism (MUL)
Tripartite motif containing 39	TRIM39	Ubiquitylation of p53 by the APC/C inhibitor Trim39. PMID: 23213260
Tripartite motif containing 40	TRIM40	TRIM40 promotes neddylation of IKKγ and is downregulated in gastrointestinal cancers. PMID: 21474709
Tripartite motif containing 41	TRIM41	Amplitude control of protein kinase C by RINCK, a novel E3 ubiquitin ligase. PMID: 17893151
Tripartite motif containing 50	TRIM50	TRIM50 protein regulates vesicular trafficking for acid secretion in gastric parietal cells. PMID: 22872646
Tripartite motif containing 56	TRIM56	TRIM56 is an essential component of the TLR3 antiviral signaling pathway. PMID: 22948160
Tripartite motif containing 59	TRIM59	TRIM59 is up-regulated in gastric tumors, promoting ubiquitination and degradation of p53. PMID: 25046164
Tripartite motif containing 62	TRIM62	Loss of the novel tumour suppressor and polarity gene Trim62 (Dear1) synergizes with oncogenic Ras in invasive lung cancer. PMID: 24890125
Tripartite motif containing 63, E3 ubiquitin protein ligase	TRIM63	Cooperative control of striated muscle mass and metabolism by MuRF1 and MuRF2. PMID: 18157088
Tripartite motif containing 68	TRIM68	TRIM68 negatively regulates IFN-β production by degrading TRK fused gene, a novel driver of IFN-β downstream of anti-viral detection systems. PMID: 24999993
Tripartite motif containing 69	TRIM69	Characterisation of human RING finger protein TRIM69, a novel testis E3 ubiquitin ligase and its subcellular localisation. PMID: 23131556
Tripartite motif containing 71, E3 ubiquitin protein ligase	TRIM71	The ubiquitin ligase human TRIM71 regulates let-7 microRNA biogenesis via modulation of Lin28B protein. PMID: 24602972
Tripartite motif family-like 1	TRIML1	Characterization and potential function of a novel pre-implantation embryo-specific RING finger protein: TRIML1. PMID: 19156909
Tripartite motif family-like 2	TRIML2	Identification of TRIML2, a novel p53 target, that enhances p53-SUMOylation and regulates the transactivation of pro-apoptotic genes. PMID: 25256710
Ubiquitin B	UBB	Restoration of cellular ubiquitin reverses impairments in neuronal development caused by disruption of the polyubiquitin gene Ubb. PMID: 25280998
Ubiquitin C	UBC	
Ubiquitin D	UBD	Increased FAT10 expression is related to poor prognosis in pancreatic ductal adenocarcinoma. PMID: 24492942
Ubiquitin carboxyl-terminal esterase L1 (ubiquitin thiolesterase)	UCHL1	Ubiquitin C-terminal hydrolase-L1 potentiates cancer chemosensitivity by stabilizing NOXA. PMID: 23499448. Disease: Parkinson disease 5 (PARK5)

HGNC-approved name	Symbol	Target information
Ubiquitin carboxyl-terminal esterase L3 (ubiquitin thiolesterase)	UCHL3	Ubiquitin C-terminal hydrolase-L3 regulates EMT process and cancer metastasis in prostate cell lines. PMID: 25194810
Ubiquitin carboxyl-terminal hydrolase L5	UCHL5	Deubiquitinase inhibition as a cancer therapeutic strategy. PMID: 25444757
Ubiquitin-conjugating enzyme E2A	UBE2A	Homology modeling and virtual screening of ubiquitin conjugation enzyme E2A for designing a novel selective antagonist against cancer. PMID: 25316404. Disease: mental retardation, X-linked, syndromic, Nascimento type (MRXSN)
Ubiquitin-conjugating enzyme E2B	UBE2B	Design, synthesis and *in vitro* anticancer evaluation of 4,6-diamino-1,3,5-triazine-2-carbohydrazides and -carboxamides. PMID: 24153206
Ubiquitin-conjugating enzyme E2C	UBE2C	UbcH10 expression in hepatocellular carcinoma and its clinicopathological significance. PMID: 21354912
Ubiquitin-conjugating enzyme E2D 1	UBE2D1	Cadmium toxicity is caused by accumulation of p53 through the down-regulation of Ube2d family genes *in vitro* and in vivo. PMID: 21467746
Ubiquitin-conjugating enzyme E2D 2	UBE2D2	The ubiquitin-conjugating enzymes UBE2N, UBE2L3 and UBE2D2/3 are essential for Parkin-dependent mitophagy. PMID: 24906799
Ubiquitin-conjugating enzyme E2D 3	UBE2D3	Crystal structure of the PRC1 ubiquitylation module bound to the nucleosome. PMID: 25355358
Ubiquitin-conjugating enzyme E2D 4 (putative)	UBE2D4	
Ubiquitin-conjugating enzyme E2E 1	UBE2E1	
Ubiquitin-conjugating enzyme E2E 2	UBE2E2	Anti-Ro52 autoantibodies from patients with Sjögren's syndrome inhibit the Ro52 E3 ligase activity by blocking the E3/E2 interface PMID: 21062500
Ubiquitin-conjugating enzyme E2E 3	UBE2E3	UBE2E ubiquitin-conjugating enzymes and ubiquitin isopeptidase Y regulate TDP-43 protein ubiquitination. PMID: 24825905
Ubiquitin-conjugating enzyme E2F (putative)	UBE2F	Inhibition of a NEDD8 cascade restores restriction of HIV by APOBEC3G. PMID: 23300442
Ubiquitin-conjugating enzyme E2G 1	UBE2G1	
Ubiquitin-conjugating enzyme E2G 2	UBE2G2	The human ubiquitin conjugating enzyme UBE2J2 (Ubc6) is a substrate for proteasomal degradation. PMID: 25083800
Ubiquitin-conjugating enzyme E2H	UBE2H	Mutation screening and association study of the UBE2H gene on chromosome 7q32 in autistic disorder. PMID: 14639049
Ubiquitin-conjugating enzyme E2I	UBE2I	SUMO-conjugating enzyme UBC9 promotes proliferation and migration of fibroblast like synoviocytes in rheumatoid arthritis. PMID: 24634869
Ubiquitin conjugating enzyme E2, J1	UBE2J1	The E2 ubiquitin-conjugating enzyme UBE2J1 is required for spermiogenesis in mice. PMID: 25320092
Ubiquitin-conjugating enzyme E2, J2	UBE2J2	Ube2j2 ubiquitinates hydroxylated amino acids on ER-associated degradation substrates. PMID: 19951915
Ubiquitin-conjugating enzyme E2K	UBE2K	Hip2 ubiquitin-conjugating enzyme overcomes radiation-induced G2/M arrest. PMID: 23933584
Ubiquitin-conjugating enzyme E2L 3	UBE2L3	UbcH7 regulates 53BP1 stability and DSB repair. PMID: 25422456
Ubiquitin-conjugating enzyme E2L 6	UBE2L6	Identification of a loss-of-function mutation in Ube2l6 associated with obesity resistance. PMID: 23557705
Ubiquitin-conjugating enzyme E2M	UBE2M	Inactivating UBE2M impacts the DNA damage response and genome integrity involving multiple cullin ligases. PMID: 25025768
Ubiquitin-conjugating enzyme E2N	UBE2N	Aged monkey brains reveal the role of ubiquitin-conjugating enzyme UBE2N in the synaptosomal accumulation of mutant huntingtin. PMID: 25343992

(Continued)

HGNC-approved name	Symbol	Target information
Ubiquitin-conjugating enzyme E2N-like (gene/pseudogene)	UBE2NL	
Ubiquitin-conjugating enzyme E2O	UBE2O	Fine-tuning BMP7 signalling in adipogenesis by UBE2O/E2-230K-mediated monoubiquitination of SMAD6. PMID: 23455153
Ubiquitin-conjugating enzyme E2Q family member 1	UBE2Q1	Expression of the novel human gene, UBE2Q1, in breast tumors. PMID: 22167327
Ubiquitin-conjugating enzyme E2Q family member 2	UBE2Q2	Ubiquitin-conjugating enzyme UBE2Q2 suppresses cell proliferation and is down-regulated in recurrent head and neck cancer. PMID: 19723876
Ubiquitin-conjugating enzyme E2Q family member 2-like	UBE2Q2L	
Ubiquitin-conjugating enzyme E2Q family-like 1	UBE2QL1	UBE2QL1 is disrupted by a constitutional translocation associated with renal tumor predisposition and is a novel candidate renal tumor suppressor gene. PMID: 24000165
Ubiquitin-conjugating enzyme E2R 2	UBE2R2	
Ubiquitin-conjugating enzyme E2S	UBE2S	The ubiquitin-conjugating enzyme E2-EPF is overexpressed in cervical cancer and associates with tumor growth. PMID: 22895574
Ubiquitin-conjugating enzyme E2T (putative)	UBE2T	Hypoxia disrupts the Fanconi anemia pathway and sensitizes cells to chemotherapy through regulation of UBE2T. PMID: 21722982
Ubiquitin-conjugating enzyme E2U (putative)	UBE2U	
Ubiquitin-conjugating enzyme E2 variant 1	UBE2V1	Inhibition of proliferation and survival of diffuse large B-cell lymphoma cells by a small-molecule inhibitor of the ubiquitin-conjugating enzyme Ubc13-Uev1A. PMID: 22791293
Ubiquitin-conjugating enzyme E2 variant 2	UBE2V2	hMMS2 serves a redundant role in human PCNA polyubiquitination. PMID: 18284681
Ubiquitin-conjugating enzyme E2W (putative)	UBE2W	Intrinsic disorder drives N-terminal ubiquitination by Ube2w. PMID: 25436519
Ubiquitin-conjugating enzyme E2Z	UBE2Z	Altered social behavior and neuronal development in mice lacking the Uba6-Use1 ubiquitin transfer system. PMID: 23499007
Ubiquitin protein ligase E3 component n-recognin 1	UBR1	Disease: Johanson–Blizzard syndrome (JBS)
Ubiquitin protein ligase E3 component n-recognin 2	UBR2	Signaling mechanism of tumor cell-induced up-regulation of E3 ubiquitin ligase UBR2. PMID: 23568773
Ubiquitin protein ligase E3 component n-recognin 3 (putative)	UBR3	Ubiquitin ligase UBR3 regulates cellular levels of the essential DNA repair protein APE1 and is required for genome stability. PMID: 21933813
Ubiquitin protein ligase E3 component n-recognin 4	UBR4	Time-of-day- and light-dependent expression of ubiquitin protein ligase E3 component n-recognin 4 (UBR4) in the suprachiasmatic nucleus circadian clock. PMID: 25084275
Ubiquitin protein ligase E3 component n-recognin 5	UBR5	UBR5-mediated ubiquitination of ATMIN is required for ionizing radiation-induced ATM signaling and function. PMID: 25092319
Ubiquitin protein ligase E3 component n-recognin 7 (putative)	UBR7	Identification and characterization of RING-finger ubiquitin ligase UBR7 in mammalian spermatozoa. PMID: 24664117
Ubiquitin-like modifier-activating enzyme 1	UBA1	The ubiquitin-activating enzyme E1 as a therapeutic target for the treatment of leukemia and multiple myeloma. PMID: 20075161. Disease: spinal muscular atrophy X-linked 2 (SMAX2)
Ubiquitin-like modifier-activating enzyme 2	UBA2	Human factors and pathways essential for mediating epigenetic gene silencing. PMID: 25147916
Ubiquitin-like modifier-activating enzyme 3	UBA3	Mutations in UBA3 confer resistance to the NEDD8-activating enzyme inhibitor MLN4924 in human leukemic cells. PMID: 24691136

HGNC-approved name	Symbol	Target information
Ubiquitin-like modifier-activating enzyme 5	UBA5	The Ufm1-activating enzyme Uba5 is indispensable for erythroid differentiation in mice. PMID: 21304510
Ubiquitin-like modifier-activating enzyme 6	UBA6	Impairment of social behavior and communication in mice lacking the Uba6-dependent ubiquitin activation system. PMID: 25523030
Ubiquitin-like modifier-activating enzyme 7	UBA7	Down-regulation of epidermal growth factor receptor by curcumin-induced UBE1L in human bronchial epithelial cells. PMID: 24445050
Ubiquitin-like with PHD and ring finger domains 2, E3 ubiquitin protein ligase	UHRF2	Note: Associated with various cancers. DNA copy number loss is found in multiple kinds of malignancies originating from the brain, breast, stomach, kidney, hematopoietic tissue and lung
Ubiquitin protein ligase E3A	UBE3A	Proteomic identification of E6AP as a molecular target of tamoxifen in MCF7 cells. PMID: 22589186. Disease: Angelman syndrome (AS)
Ubiquitin protein ligase E3B	UBE3B	Disease: blepharophimosis–ptosis–intellectual disability syndrome (BPIDS)
Ubiquitin protein ligase E3C	UBE3C	Clinical significance of the ubiquitin ligase UBE3C in hepatocellular carcinoma revealed by exome sequencing. PMID: 24425307
Ubiquitin protein ligase E3D	UBE3D	A novel UbcH10-binding protein facilitates the ubiquitinylation of cyclin B *in vitro*. PMID: 15749827
Ubiquitination factor E4A	UBE4A	Ubiquitin ligase Ufd2 is required for efficient degradation of Mps1 kinase. PMID: 22045814
Ubiquitination factor E4B	UBE4B	Regulation of p53 level by UBE4B in breast cancer. PMID: 24587254
WW domain containing E3 ubiquitin protein ligase 1	WWP1	Overexpression of WWP1 is associated with the estrogen receptor and insulin-like growth factor receptor 1 in breast carcinoma. PMID: 19267401
WW domain containing E3 ubiquitin protein ligase 2	WWP2	Notch3 interactome analysis identified WWP2 as a negative regulator of Notch3 signaling in ovarian cancer. PMID: 25356737
Zinc and ring finger 1, E3 ubiquitin protein ligase	ZNRF1	ZNRF1 promotes Wallerian degeneration by degrading AKT to induce GSK3B-dependent CRMP2 phosphorylation. PMID: 22057101
Zinc and ring finger 2	ZNRF2	ZNRF2 is released from membranes by growth factors and, together with ZNRF1, regulates the Na+/K+ATPase. PMID: 22797923
Apoptosis-resistant E3 ubiquitin protein ligase 1	AREL1	Identification of a novel anti-apoptotic E3 ubiquitin ligase that ubiquitinates antagonists of inhibitor of apoptosis proteins SMAC, HtrA2, and ARTS. PMID: 23479728
Arginyltransferase 1	ATE1	Small molecule inhibitors of arginyltransferase regulate arginylation-dependent protein degradation, cell motility, and angiogenesis. PMID: 22200815
Autocrine motility factor receptor, E3 ubiquitin protein ligase	AMFR	The E3 ubiquitin ligase AMFR and INSIG1 bridge the activation of TBK1 kinase by modifying the adaptor STING. PMID: 25526307
Baculoviral IAP repeat containing 6	BIRC6	Comparative proteomics of colon cancer stem cells and differentiated tumor cells identifies BIRC6 as a potential therapeutic target. PMID: 21788403
Beta-transducin repeat containing E3 ubiquitin protein ligase	BTRC	β-Transducin repeat-containing protein 1 (β-TrCP1)-mediated silencing mediator of retinoic acid and thyroid hormone receptor (SMRT) protein degradation promotes tumor necrosis factor α (TNFα)-induced inflammatory gene expression. PMID: 23861398
Breast cancer 1, early onset	BRCA1	Disease: breast cancer (BC)
BRCA1-associated protein-1 (ubiquitin carboxy-terminal hydrolase)	BAP1	Frequent mutation of BAP1 in metastasizing uveal melanomas. PMID: 21051595. Disease: mesothelioma, malignant (MESOM)
Checkpoint with forkhead and ring finger domains, E3 ubiquitin protein ligase	CHFR	CHFR protein expression predicts outcomes to taxane-based first line therapy in metastatic NSCLC. PMID: 23386692

(Continued)

HGNC-approved name	Symbol	Target information
Chromobox homolog 4	CBX4	Cbx4 governs HIF-1α to potentiate angiogenesis of hepatocellular carcinoma by its SUMO E3 ligase activity. PMID: 24434214
Cyclin B1-interacting protein 1, E3 ubiquitin protein ligase	CCNB1IP1	Antagonistic roles of ubiquitin ligase HEI10 and SUMO ligase RNF212 regulate meiotic recombination. PMID: 24390283
Denticleless E3 ubiquitin protein ligase homolog (*Drosophila*)	DTL	Ubiquitin E3 ligase CRL4(CDT2/DCAF2) as a potential chemotherapeutic target for ovarian surface epithelial cancer. PMID: 23995842
G2/M-phase-specific E3 ubiquitin protein ligase	G2E3	G2E3 is a dual function ubiquitin ligase required for early embryonic development. PMID: 18511420
HECT domain and ankyrin repeat containing E3 ubiquitin protein ligase 1	HACE1	Note: Defects in HACE1 are a cause of Wilms tumor (WT). WT is a pediatric malignancy of kidney and one of the most common solid cancers in childhood. HACE1 is epigenetically down-regulated in sporadic Wilms tumor
HECT, UBA and WWE domain containing 1, E3 ubiquitin protein ligase	HUWE1	Huwe1 as a therapeutic target for neural injury. PMID: 25036176. Disease: mental retardation, X-linked, syndromic, Turner type (MRXST)
ISG15 ubiquitin-like modifier	ISG15	ISG15 is a critical microenvironmental factor for pancreatic cancer stem cells. PMID: 25368022
Itchy E3 ubiquitin protein ligase	ITCH	The E3 ubiquitin ligase Itch regulates tumor suppressor protein RASSF5/NORE1 stability in an acetylation-dependent manner. PMID: 23538446. Disease: syndromic multisystem autoimmune disease (SMAD)
Ligand of numb-protein X 1, E3 ubiquitin protein ligase	LNX1	Ligand of Numb proteins LNX1p80 and LNX2 interact with the human glycoprotein CD8α and promote its ubiquitylation and endocytosis. PMID: 22045731
LIM domain 7	LMO7	Decreased expression of LMO7 and its clinicopathological significance in human lung adenocarcinoma. PMID: 22977619
Listerin E3 ubiquitin protein ligase 1	LTN1	Ubiquitylation by the Ltn1 E3 ligase protects 60S ribosomes from starvation-induced selective autophagy. PMID: 24616224
Mahogunin ring finger 1, E3 ubiquitin protein ligase	MGRN1	Mahogunin ring finger 1 suppresses misfolded polyglutamine aggregation and cytotoxicity. PMID: 24769000
Male-specific lethal 2 homolog (*Drosophila*)	MSL2	Msl2 is a novel component of the vertebrate DNA damage response. PMID: 23874665
MDM2 proto-oncogene, E3 ubiquitin protein ligase	MDM2	**RG7388** – antineoplastic agents. Note: Seems to be amplified in certain tumors (including soft tissue sarcomas, osteosarcomas and gliomas)
Mex-3 RNA binding family member C	MEX3C	Note: Genetic variations in MEX3C may be associated with susceptibility to essential hypertension
Mitochondrial E3 ubiquitin protein ligase 1	MUL1	MUL1 acts in parallel to the PINK1/parkin pathway in regulating mitofusin and compensates for loss of PINK1/parkin. PMID: 24898855
MYC binding protein 2, E3 ubiquitin protein ligase	MYCBP2	The ubiquitin ligase MYCBP2 regulates transient receptor potential vanilloid receptor 1 (TRPV1) internalization through inhibition of p38 MAPK signaling. PMID: 21098484
NHL repeat containing E3 ubiquitin protein ligase 1	NHLRC1	Disease: epilepsy, progressive myoclonic 2 (EPM2)
Non-SMC element 2, MMS21 homolog (*S. cerevisiae*)	NSMCE2	Hypomorphism in human NSMCE2 linked to primordial dwarfism and insulin resistance. PMID: 25105364
OTU deubiquitinase with linear linkage specificity	OTULIN	OTULIN restricts Met1-linked ubiquitination to control innate immune signaling. PMID: 23806334
Parkin RBR E3 ubiquitin protein ligase	PARK2	Can parkin be a target for future treatment of Parkinson's disease? PMID: 23930597. Disease: Parkinson disease (PARK)
Potassium channel modulatory factor 1	KCMF1	The zinc-finger protein KCMF1 is overexpressed during pancreatic cancer development and downregulation of KCMF1 inhibits pancreatic cancer development in mice. PMID: 20473331
RAD18 homolog (*S. cerevisiae*)	RAD18	DNA damage-specific deubiquitination regulates Rad18 functions to suppress mutagenesis. PMID: 25023518

HGNC-approved name	Symbol	Target information
Retinoblastoma binding protein 6	RBBP6	Silencing RBBP6 (Retinoblastoma Binding Protein 6) sensitises breast cancer cells MCF7 to staurosporine and camptothecin-induced cell death. PMID: 24703106
Ring finger and CHY zinc finger domain containing 1, E3 ubiquitin protein ligase	RCHY1	A novel oncoprotein Pirh2: Rising from the shadow of MDM2. PMID: 21284766
Ring finger and FYVE-like domain containing E3 ubiquitin protein ligase	RFFL	Different Raf protein kinases mediate different signaling pathways to stimulate E3 ligase RFFL gene expression in cell migration regulation. PMID: 24114843
Ring finger and WD repeat domain 2, E3 ubiquitin protein ligase	RFWD2	COP1 and GSK3β cooperate to promote c-Jun degradation and inhibit breast cancer cell tumorigenesis. PMID: 24027432
Ring finger and WD repeat domain 3	RFWD3	E3 ligase RFWD3 participates in replication checkpoint control. PMID: 21504906
Ring-box 1, E3 ubiquitin protein ligase	RBX1	Knockdown of regulator of cullins-1 (ROC1) expression induces bladder cancer cell cycle arrest at the G2 phase and senescence. PMID: 23667514
SH3 domain containing ring finger 1	SH3RF1	Possible role of death receptor-mediated apoptosis by the E3 ubiquitin ligases Siah2 and POSH. PMID: 21586138
SH3 domain containing ring finger 2	SH3RF2	SH3RF2 functions as an oncogene by mediating PAK4 protein stability. PMID: 24130170
SNF2 histone linker PHD RING helicase, E3 ubiquitin protein ligase	SHPRH	Human SHPRH is a ubiquitin ligase for Mms2-Ubc13-dependent polyubiquitylation of proliferating cell nuclear antigen. PMID: 17108083
STIP1 homology and U-box-containing protein 1, E3 ubiquitin protein ligase	STUB1	The ubiquitin ligase Stub1 negatively modulates regulatory T cell suppressive activity by promoting degradation of the transcription factor Foxp3. PMID: 23973223
SUMO1-activating enzyme subunit 1	SAE1	Identification of sumoylation activating enzyme 1 inhibitors by structure-based virtual screening. PMID: 23544417
Synovial apoptosis inhibitor 1, synoviolin	SYVN1	RING-finger type E3 ubiquitin ligase inhibitors as novel candidates for the treatment of rheumatoid arthritis. PMID: 22992760
Thyroid hormone receptor interactor 12	TRIP12	RNA-Seq analysis identifies aberrant RNA splicing of TRIP12 in acute myeloid leukemia patients at remission. PMID: 24961348
Topoisomerase I binding, arginine/serine-rich, E3 ubiquitin protein ligase	TOPORS	Disease: retinitis pigmentosa 31 (RP31)
TRAF-interacting protein	TRAIP	TRAIP is a regulator of the spindle assembly checkpoint. PMID: 25335891
Transmembrane protein 129, E3 ubiquitin protein ligase	TMEM129	TMEM129 is a Derlin-1 associated ERAD E3 ligase essential for virus-induced degradation of MHC-I. PMID: 25030448
Tumor necrosis factor, alpha-induced protein 3	TNFAIP3	A20 is overexpressed in glioma cells and may serve as a potential therapeutic target. PMID: 19492975
Ubiquitin A-52 residue ribosomal protein fusion product 1	UBA52	Altered dynamics of ubiquitin hybrid proteins during tumor cell apoptosis. PMID: 22258406
Ubiquitin-fold modifier-conjugating enzyme 1	UFC1	Crystal structure of Ufc1, the Ufm1-conjugating enzyme. PMID: 17825256
Ubiquitin-fold modifier 1	UFM1	The ufm1 cascade. PMID: 24921187
Ubiquitin-related modifier 1	URM1	The dual role of ubiquitin-like protein Urm1 as a protein modifier and sulfur carrier. PMID: 21904977
U-box domain containing 5	UBOX5	
Unkempt family zinc finger-like	UNKL	
von Hippel–Lindau tumor suppressor, E3 ubiquitin protein ligase	VHL	Disease: pheochromocytoma (PCC)
WD repeat, sterile alpha motif and U-box domain containing 1	WDSUB1	

(Continued)

HGNC-approved name	Symbol	Target information
YOD1 deubiquitinase	YOD1	
ZFP91 zinc finger protein	ZFP91	ZFP91-A newly described gene potentially involved in prostate pathology. PMID: 24272675
Zinc finger protein 645	ZNF645	Structure of a novel phosphotyrosine-binding domain in Hakai that targets E-cadherin. PMID: 22252131
Zinc finger, RAN-binding domain containing 1	ZRANB1	Identification of small molecule TRABID deubiquitinase inhibitors by computation-based virtual screen. PMID: 22584113
Zinc finger, SWIM-type containing 2	ZSWIM2	MEX is a testis-specific E3 ubiquitin ligase that promotes death receptor-induced apoptosis. PMID: 16522193

4.2.4   Chromatin modification

HGNC-approved name	Symbol	Target information
DNA (cytosine-5-)-methyltransferase 1	DNMT1	**Decitabine** – myelodysplastic syndrome. Disease: neuropathy, hereditary sensory, 1E (HSN1E)
DNA (cytosine-5-)-methyltransferase 3 alpha	DNMT3A	**Azacitidine** – myelodysplastic syndrome
DNA (cytosine-5-)-methyltransferase 3 beta	DNMT3B	**Azacitidine** – myelodysplastic syndrome. Disease: immunodeficiency-centromeric instability-facial anomalies syndrome 1 (ICF1)
DNA (cytosine-5-)-methyltransferase 3-like	DNMT3L	DNMT3L interacts with transcription factors to target DNMT3L/DNMT3B to specific DNA sequences: Role of the DNMT3L/DNMT3B/p65-NFκB complex in the (de-)methylation of TRAF1. PMID: 24952347
Establishment of sister chromatid cohesion N-acetyltransferase 1	ESCO1	
Establishment of sister chromatid cohesion N-acetyltransferase 2	ESCO2	Disease: Roberts syndrome (RBS)
Euchromatic histone-lysine N-methyltransferase 1	EHMT1	Disease: Kleefstra syndrome (KLESTS)
Euchromatic histone-lysine N-methyltransferase 2	EHMT2	Reversal of H3K9me2 by a small-molecule inhibitor for the G9a histone methyltransferase. PMID: 17289593
Enhancer of zeste homolog 1 (*Drosophila*)	EZH1	Selective inhibition of EZH2 and EZH1 enzymatic activity by a small molecule suppresses MLL-rearranged leukemia. PMID: 25395428
Enhancer of zeste homolog 2 (*Drosophila*)	EZH2	EZH2 inhibition as a therapeutic strategy for lymphoma with EZH2-activating mutations. PMID: 23051747. Disease: Weaver syndrome (WVS)
Histone deacetylase 1	HDAC1	**Mocetinostat** – antineoplastic agents
Histone deacetylase 2	HDAC2	Computational design of a time-dependent histone deacetylase 2 selective inhibitor. PMID: 25546141
Histone deacetylase 3	HDAC3	Histone deacetylase 3 as a novel therapeutic target in multiple myeloma. PMID: 23913134
Histone deacetylase 4	HDAC4	Histone deacetylase 4 alters cartilage homeostasis in human osteoarthritis. PMID: 25515592. Disease: brachydactyly–mental retardation syndrome (BDMR)
Histone deacetylase 5	HDAC5	HDAC5 is a repressor of angiogenesis and determines the angiogenic gene expression pattern of endothelial cells. PMID: 19351956
Histone deacetylase 6	HDAC6	**ACY-1215** – antineoplastic agents. Disease: chondrodysplasia with platyspondyly, distinctive brachydactyly, hydrocephaly, and microphthalmia (CDP-PBHM)

HGNC-approved name	Symbol	Target information
Histone deacetylase 7	HDAC7	Histone deacetylase 7 (Hdac7) suppresses chondrocyte proliferation and β-catenin activity during endochondral ossification. PMID: 25389289
Histone deacetylase 8	HDAC8	**PCI-34051** – antineoplastic agents. Disease: Cornelia de Lange syndrome 5 (CDLS5)
Histone deacetylase 9	HDAC9	Histone deacetylase 9 represses cholesterol efflux and alternatively activated macrophages in atherosclerosis development. PMID: 25035344. Note: A chromosomal aberration involving HDAC9 is found in a family with Peters anomaly. Translocation t(17)(q41p21) with TGFB2 resulting in lack of HDAC9 protein
Histone deacetylase 10	HDAC10	Histone deacetylase 10-promoted autophagy as a druggable point of interference to improve the treatment response of advanced neuroblastomas. PMID: 24145760
Histone deacetylase 11	HDAC11	Histone deacetylase 11: A novel epigenetic regulator of myeloid derived suppressor cell expansion and function. PMID: 25155994
Jumonji domain containing 1C	JMJD1C	shRNA screening identifies JMJD1C as being required for leukemia maintenance. PMID: 24501218
Jumonji domain containing 6	JMJD6	JMJD6 promotes colon carcinogenesis through negative regulation of p53 by hydroxylation. PMID: 24667498
K(lysine) acetyltransferase 2A	KAT2A	K-Lysine acetyltransferase 2a regulates a hippocampal gene expression network linked to memory formation. PMID: 25024434
K(lysine) acetyltransferase 2B	KAT2B	Identification of structural features of 2-alkylidene-1,3-dicarbonyl derivatives that induce inhibition and/or activation of histone acetyltransferases KAT3B/p300 and KAT2B/PCAF. PMID: 25333655
K(lysine) acetyltransferase 5	KAT5	Role of Tip60 in human melanoma cell migration, metastasis, and patient survival. PMID: 22673729
K(lysine) acetyltransferase 6A	KAT6A	Note: Chromosomal aberrations involving KAT6A may be a cause of acute myeloid leukemias. Translocation t(816)(p11p13) with CREBBP translocation t(822)(p11q13) with EP300
K(lysine) acetyltransferase 6B	KAT6B	Note: A chromosomal aberration involving KAT6B may be a cause of acute myeloid leukemias. Translocation t(10;16)(q22;p13) with CREBBP. Disease: Ohdo syndrome, SBBYS variant (SBBYSS)
K(lysine) acetyltransferase 7	KAT7	Cell cycle-dependent chromatin shuttling of HBO1-JADE1 histone acetyl transferase (HAT) complex. PMID: 24739512
K(lysine) acetyltransferase 8	KAT8	The histone acetyltransferase hMOF acetylates Nrf2 and regulates anti-drug responses in human non-small cell lung cancer. PMID: 24571402
Lysine (K)-specific demethylase 1A	KDM1A	**ORY-1001** – antineoplastic agents
Lysine (K)-specific demethylase 1B	KDM1B	Inhibition of histone demethylase, LSD2 (KDM1B), attenuates DNA methylation and increases sensitivity to DNMT inhibitor-induced apoptosis in breast cancer cells. PMID: 24924415
Lysine (K)-specific demethylase 2A	KDM2A	KDM2A promotes lung tumorigenesis by epigenetically enhancing ERK1/2 signaling. PMID: 24200691
Lysine (K)-specific demethylase 2B	KDM2B	NDY1/KDM2B functions as a master regulator of polycomb complexes and controls self-renewal of breast cancer stem cells. PMID: 24853546
Lysine (K)-specific demethylase 3A	KDM3A	The histone demethylase enzyme KDM3A is a key estrogen receptor regulator in breast cancer. PMID: 25488809
Lysine (K)-specific demethylase 3B	KDM3B	KDM3B is the H3K9 demethylase involved in transcriptional activation of lmo2 in leukemia. PMID: 22615488
Lysine (K)-specific demethylase 4A	KDM4A	Expression of JMJD2A in infiltrating duct carcinoma was markedly higher than fibroadenoma, and associated with expression of ARHI, p53 and ER in infiltrating duct carcinoma. PMID: 23678541

(Continued)

HGNC-approved name	Symbol	Target information
Lysine (K)-specific demethylase 4B	KDM4B	KDM4B as a target for prostate cancer: Structural analysis and selective inhibition by a novel inhibitor. PMID: 24971742
Lysine (K)-specific demethylase 4C	KDM4C	Substrate- and cofactor-independent inhibition of histone demethylase KDM4C. PMID: 25014588
Lysine (K)-specific demethylase 4D	KDM4D	Regulation of tumor suppressor p53 and HCT116 cell physiology by histone demethylase JMJD2D/KDM4D. PMID: 22514644
Lysine (K)-specific demethylase 4E	KDM4E	Investigations on the oxygen dependence of a 2-oxoglutarate histone demethylase. PMID: 23092293
Lysine (K)-specific demethylase 5A	KDM5A	Histone lysine demethylase JARID1a activates CLOCK-BMAL1 and influences the circadian clock. PMID: 21960634
Lysine (K)-specific demethylase 5B	KDM5B	Identification of small molecule inhibitors of Jumonji AT-rich interactive domain 1B (JARID1B) histone demethylase by a sensitive high throughput screen. PMID: 23408432
Lysine (K)-specific demethylase 5C	KDM5C	Disease: mental retardation, X-linked, syndromic, Claes-Jensen type (MRXSCJ)
Lysine (K)-specific demethylase 5D	KDM5D	HDAC inhibition imparts beneficial transgenerational effects in Huntington's disease mice via altered DNA and histone methylation. PMID: 25535382
Lysine (K)-specific demethylase 6A	KDM6A	A selective jumonji H3K27 demethylase inhibitor modulates the proinflammatory macrophage response. PMID: 22842901. Disease: Kabuki syndrome 2 (KABUK2)
Lysine (K)-specific demethylase 6B	KDM6B	Histone demethylase Jumonji D3 (JMJD3/KDM6B) at the nexus of epigenetic regulation of inflammation and the aging process. PMID: 24925089
Lysine (K)-specific demethylase 7A	KDM7A	Identification of the KDM2/7 histone lysine demethylase subfamily inhibitor and its antiproliferative activity. PMID: 23964788
Lysine (K)-specific demethylase 8	KDM8	JMJD5 regulates PKM2 nuclear translocation and reprograms HIF-1α-mediated glucose metabolism. PMID: 24344305
Lysine (K)-specific methyltransferase 2A	KMT2A	Menin-MLL inhibitors reverse oncogenic activity of MLL fusion proteins in leukemia. PMID: 22286128. Disease: Wiedemann–Steiner syndrome (WDSTS)
Lysine (K)-specific methyltransferase 2B	KMT2B	Histone-methyltransferase MLL2 (KMT2B) is required for memory formation in mice. PMID: 23426673
Lysine (K)-specific methyltransferase 2C	KMT2C	Crucial roles of MLL3 and MLL4 as epigenetic switches of the hepatic circadian clock controlling bile acid homeostasis. PMID: 25346535
Lysine (K)-specific methyltransferase 2D	KMT2D	UTX and MLL4 coordinately regulate transcriptional programs for cell proliferation and invasiveness in breast cancer cells. PMID: 24491801. Disease: Kabuki syndrome 1 (KABUK1)
Lysine (K)-specific methyltransferase 2E	KMT2E	Targeted silencing of MLL5β inhibits tumor growth and promotes gamma-irradiation sensitization in HPV16/18 associated cervical cancers. PMID: 25172963
Methyltransferase-like 8	METTL8	TIPs are tension-responsive proteins involved in myogenic versus adipogenic differentiation. PMID: 15992539
Methyltransferase-like 9	METTL9	Novel radiation response genes identified in gene-trapped MCF10A mammary epithelial cells. PMID: 17390725
Methyltransferase-like 10	METTL10	Selenium-based S-adenosylmethionine analog reveals the mammalian seven-beta-strand methyltransferase METTL10 to be an EF1A1 lysine methyltransferase. PMID: 25144183
N(alpha)-acetyltransferase 10, NatA catalytic subunit	NAA10	Arrest defective-1 controls tumor cell behavior by acetylating myosin light chain kinase. PMID: 19826488. Disease: N-terminal acetyltransferase deficiency (NATD)
N(alpha)-acetyltransferase 11, NatA catalytic subunit	NAA11	Characterization of hARD2, a processed hARD1 gene duplicate, encoding a human protein N-alpha-acetyltransferase. PMID: 16638120

HGNC-approved name	Symbol	Target information
N(alpha)-acetyltransferase 15, NatA auxiliary subunit	NAA15	Induction of apoptosis in human cells by RNAi-mediated knockdown of hARD1 and NATH, components of the protein *N*-alpha-acetyltransferase complex. PMID: 16518407
N(alpha)-acetyltransferase 20, NatB catalytic subunit	NAA20	
N(alpha)-acetyltransferase 30, NatC catalytic subunit	NAA30	
N(alpha)-acetyltransferase 40, NatD catalytic subunit	NAA40	The human *N*-alpha-acetyltransferase 40 (hNaa40p/hNatD) is conserved from yeast and N-terminally acetylates histones H2A and H4. PMID: 21935442
N(alpha)-acetyltransferase 50, NatE catalytic subunit	NAA50	Structure of a ternary Naa50p (NAT5/SAN) N-terminal acetyltransferase complex reveals the molecular basis for substrate-specific acetylation. PMID: 21900231
N(alpha)-acetyltransferase 60, NatF catalytic subunit	NAA60	NatF contributes to an evolutionary shift in protein N-terminal acetylation and is important for normal chromosome segregation. PMID: 21750686
N-6 adenine-specific DNA methyltransferase 1 (putative)	N6AMT1	Involvement of *N*-6 adenine-specific DNA methyltransferase 1 (N6AMT1) in arsenic biomethylation and its role in arsenic-induced toxicity. PMID: 21193388
N-6 adenine-specific DNA methyltransferase 2 (putative)	N6AMT2	
Nuclear receptor coactivator 1	NCOA1	Steroid receptor coactivator-1: A versatile regulator and promising therapeutic target for breast cancer. PMID: 23474438. Note: A chromosomal aberration involving NCOA1 is a cause of rhabdomyosarcoma
Nuclear receptor coactivator 2	NCOA2	Note: Chromosomal aberrations involving NCOA2 may be a cause of acute myeloid leukemias. Inversion inv(8)(p11q13) generates the KAT6A-NCOA2 oncogene, which consists of the N-terminal part of KAT6A and the C-terminal part of NCOA2/TIF2
Nuclear receptor coactivator 3	NCOA3	Steroid receptor coactivator-3 as a potential molecular target for cancer therapy. PMID: 22924430
PHD finger protein 2	PHF2	The histone demethylase PHF2 promotes fat cell differentiation as an epigenetic activator of both C/EBPα and C/EBPδ. PMID: 25266703
PHD finger protein 8	PHF8	The histone demethylase PHF8 is an oncogenic protein in human non-small cell lung cancer. PMID: 25065740. Disease: mental retardation, X-linked, syndromic, Siderius type (MRXSSD)
PR domain containing 2, with ZNF domain	PRDM2	Retinoblastoma protein-interacting zinc-finger gene 1 (RIZ1) dysregulation in human malignant meningiomas. PMID: 22614009
PR domain containing 6	PRDM6	Prdm6 is essential for cardiovascular development in vivo. PMID: 24278461
PR domain containing 7	PRDM7	
PR domain containing 8	PRDM8	Prdm8 regulates the morphological transition at multipolar phase during neocortical development. PMID: 24489718
PR domain containing 9	PRDM9	Characterization of the histone methyltransferase PRDM9 using biochemical, biophysical and chemical biology techniques. PMID: 24785241
PR domain containing 10	PRDM10	Recurrent PRDM10 gene fusions in undifferentiated pleomorphic sarcoma. PMID: 25516889
PR domain containing 11	PRDM11	Loss of PRDM11 promotes MYC-driven lymphomagenesis. PMID: 25499759
Protein arginine methyltransferase 1	PRMT1	Virtual screening and biological evaluation of novel small molecular inhibitors against protein arginine methyltransferase 1 (PRMT1). PMID: 25348815

(Continued)

HGNC-approved name	Symbol	Target information
Protein arginine methyltransferase 2	PRMT2	PRMT2 and RORγ expression are associated with breast cancer survival outcomes. PMID: 24911119
Protein arginine methyltransferase 3	PRMT3	Exploiting an allosteric binding site of PRMT3 yields potent and selective inhibitors. PMID: 23445220
Protein arginine methyltransferase 5	PRMT5	Protein arginine methyltransferase 5 is a key regulator of the MYCN oncoprotein in neuroblastoma cells. PMID: 25475372
Protein arginine methyltransferase 6	PRMT6	PRMT6 mediates CSE induced inflammation and apoptosis. PMID: 25481537
Protein arginine methyltransferase 7	PRMT7	PRMT7 induces epithelial-to-mesenchymal transition and promotes metastasis in breast cancer. PMID: 25136067
Protein arginine methyltransferase 8	PRMT8	PRMT1 and PRMT8 regulate retinoic acid dependent neuronal differentiation with implications to neuropathology. PMID: 25388207
Protein arginine methyltransferase 9	PRMT9	
SET and MYND domain containing 1	SMYD1	The myosin-interacting protein SMYD1 is essential for sarcomere organization. PMID: 21852424
SET and MYND domain containing 2	SMYD2	RB1 methylation by SMYD2 enhances cell cycle progression through an increase of RB1 phosphorylation. PMID: 22787429
SET and MYND domain containing 3	SMYD3	SMYD3 links lysine methylation of MAP3K2 to Ras-driven cancer. PMID: 24847881
SET and MYND domain containing 4	SMYD4	Identification of Smyd4 as a potential tumor suppressor gene involved in breast cancer development. PMID: 19383909
SMYD family member 5	SMYD5	
SET domain containing 1A	SETD1A	Histone methyltransferase hSETD1A is a novel regulator of metastasis in breast cancer. PMID: 25373480
SET domain containing 1B	SETD1B	Ott1(Rbm15) regulates thrombopoietin response in hematopoietic stem cells through alternative splicing of c-Mpl. PMID: 25468569
SET domain containing 2	SETD2	Disease: renal cell carcinoma (RCC)
SET domain containing 3	SETD3	The role of a newly identified SET domain-containing protein, SETD3, in oncogenesis. PMID: 23065515
SET domain containing 4	SETD4	SET domain-containing protein 4 (SETD4) is a newly identified cytosolic and nuclear lysine methyltransferase involved in breast cancer cell proliferation. PMID: 24738023
SET domain containing 5	SETD5	Disease: mental retardation, autosomal dominant 23 (MRD23)
SET domain containing 6	SETD6	SETD6 controls the expression of estrogen-responsive genes and proliferation of breast carcinoma cells. PMID: 24751716
SET domain containing (lysine methyltransferase) 7	SETD7	(R)-PFI-2 is a potent and selective inhibitor of SETD7 methyltransferase activity in cells. PMID: 25136132
SET domain containing (lysine methyltransferase) 8	SETD8	Structure-activity relationship studies of SETD8 inhibitors. PMID: 25554733
SET domain containing 9	SETD9	
SET domain, bifurcated 1	SETDB1	H3K9 histone methyltransferase, KMT1E/SETDB1, cooperates with the SMAD2/3 pathway to suppress lung cancer metastasis. PMID: 25477335
SET domain, bifurcated 2	SETDB2	The methyltransferase Setdb2 mediates virus-induced susceptibility to bacterial superinfection. PMID: 25419628
Sirtuin 1	SIRT1	**GSK2245840** – antipsoriatics
Sirtuin 2	SIRT2	SIRT2 interacts with β-catenin to inhibit Wnt signaling output in response to radiation-induced stress. PMID: 24866770
Sirtuin 3	SIRT3	Sirtuin-3 (SIRT3), a novel potential therapeutic target for oral cancer. PMID: 21472714

HGNC-approved name	Symbol	Target information
Sirtuin 4	SIRT4	Sirtuin 4 is a lipoamidase regulating pyruvate dehydrogenase complex activity. PMID: 25525879
Sirtuin 5	SIRT5	Protective role of SIRT5 against motor deficit and dopaminergic degeneration in MPTP-induced mice model of Parkinson's disease. PMID: 25541039
Sirtuin 6	SIRT6	SIRT6 regulates TNF-α secretion through hydrolysis of long-chain fatty acyl lysine. PMID: 23552949
Sirtuin 7	SIRT7	Sirtuin 7 in cell proliferation, stress and disease: Rise of the Seventh Sirtuin! PMID: 25435428
Suppressor of variegation 3-9 homolog 1 (*Drosophila*)	SUV39H1	Histone lysine methyltransferase SUV39H1 is a potent target for epigenetic therapy of hepatocellular carcinoma. PMID: 24844570
Suppressor of variegation 3-9 homolog 2 (*Drosophila*)	SUV39H2	Activity and specificity of the human SUV39H2 protein lysine methyltransferase. PMID: 25459750
Suppressor of variegation 4-20 homolog 1 (*Drosophila*)	SUV420H1	Genetic alterations of histone lysine methyltransferases and their significance in breast cancer. PMID: 25537518
Suppressor of variegation 4-20 homolog 2 (*Drosophila*)	SUV420H2	SUV420H2-mediated H4K20 trimethylation enforces RNA polymerase II promoter-proximal pausing by blocking hMOF-dependent H4K16 acetylation. PMID: 21321083
SWI/SNF-related, matrix-associated, actin-dependent regulator of chromatin, subfamily a, member 1	SMARCA1	Singular v dual inhibition of SNF2L and its isoform, SNF2LT, have similar effects on DNA damage but opposite effects on the DNA damage response, cancer cell growth arrest and apoptosis. PMID: 22577152
SWI/SNF-related, matrix-associated actin-dependent regulator of chromatin, subfamily a, member 2	SMARCA2	A synthetic lethality-based strategy to treat cancers harboring a genetic deficiency in the chromatin remodeling factor BRG1. PMID: 23872584. Disease: Nicolaides–Baraitser syndrome (NCBRS)
SWI/SNF-related, matrix-associated, actin-dependent regulator of chromatin, subfamily a, member 4	SMARCA4	BRG1 is a prognostic marker and potential therapeutic target in human breast cancer. PMID: 23533649. Disease: rhabdoid tumor predisposition syndrome 2 (RTPS2)
SWI/SNF-related, matrix-associated, actin-dependent regulator of chromatin, subfamily a, member 5	SMARCA5	
SWI/SNF-related, matrix-associated actin-dependent regulator of chromatin, subfamily a, containing DEAD/H box 1	SMARCAD1	Disease: adermatoglyphia (ADERM)
SWI/SNF-related, matrix-associated, actin-dependent regulator of chromatin, subfamily a like 1	SMARCAL1	Disease: Schimke immuno-osseous dysplasia (SIOD)
Wolf–Hirschhorn syndrome candidate 1	WHSC1	MMSET is highly expressed and associated with aggressiveness in neuroblastoma. PMID: 21527557. Note: A chromosomal aberration involving WHSC1 is a cause of multiple myeloma tumors. Translocation t(414)(p16.3q32.3) with IgH
Wolf–Hirschhorn syndrome candidate 1-like 1	WHSC1L1	Note: Defects in WHSC1L1 may be involved in non small cell lung carcinomas (NSCLC). Amplified or overexpressed in NSCLC. Note: A chromosomal aberration involving WHSC1L1 is found in childhood acute myeloid leukemia
Ash1 (absent, small, or homeotic)-like (*Drosophila*)	ASH1L	Human ASH-1 promotes neuroendocrine differentiation in androgen deprivation conditions and interferes with androgen responsiveness in prostate cancer cells. PMID: 23657976
Chromodomain protein, Y-like	CDYL	Cdyl, a new partner of the inactive X chromosome and potential reader of H3K27me3 and H3K9me2. PMID: 24144980

(Continued)

HGNC-approved name	Symbol	Target information
Coactivator-associated arginine methyltransferase 1	CARM1	Identification of a novel inhibitor of coactivator-associated arginine methyltransferase 1 (CARM1)-mediated methylation of histone H3 Arg-17. PMID: 20022955
CSRP2 binding protein	CSRP2BP	The double-histone-acetyltransferase complex ATAC is essential for mammalian development. PMID: 19103755
DOT1-like histone H3K79 methyltransferase	DOT1L	**EPZ-5676** – antineoplastic agents
Elongator acetyltransferase complex subunit 3	ELP3	Regulation of G6PD acetylation by SIRT2 and KAT9 modulates NADPH homeostasis and cell survival during oxidative stress. PMID: 24769394
Embryonic ectoderm development	EED	Astemizole arrests the proliferation of cancer cells by disrupting the EZH2–EED interaction of polycomb repressive complex 2. PMID: 25369470
General transcription factor IIIC, polypeptide 4, 90 kDa	GTF3C4	Influenza A virus matrix protein 1 interacts with hTFIIIC102-s, a short isoform of the polypeptide 3 subunit of human general transcription factor IIIC. PMID: 19521658
Hair growth associated	HR	Disease: alopecia universalis congenita (ALUNC)
Histone acetyltransferase 1	HAT1	RNAi screening identifies HAT1 as a potential drug target in esophageal squamous cell carcinoma. PMID: 25120766
O-6-methylguanine-DNA methyltransferase	MGMT	S-alkylthiolation of O-6-methylguanine-DNA-methyltransferase (MGMT) to sensitize cancer cells to anticancer therapy. PMID: 17298293
MYC-induced nuclear antigen	MINA	Potential effects of Mina53 on tumor growth in human pancreatic cancer. PMID: 24522517
Nuclear receptor binding SET domain protein 1	NSD1	Disease: Sotos syndrome 1 (SOTOS1)
SET domain and mariner transposase fusion gene	SETMAR	Metnase mediates chromosome decatenation in acute leukemia cells. PMID: 19458360
Snf2-related CREBBP activator protein	SRCAP	The human SRCAP chromatin remodeling complex promotes DNA-end resection. PMID: 25176633. Disease: Floating-Harbor syndrome (FLHS)
SPO11 meiotic protein covalently bound to DSB	SPO11	
Ubiquitously transcribed tetratricopeptide repeat containing, Y-linked	UTY	Human UTY(KDM6C) is a male-specific Nε-methyl lysyl demethylase. PMID: 24798337

4.2.5 Protein synthesis and folding

HGNC-approved name	Symbol	Target information
t-complex 1	TCP1	Human TRiC complex purified from HeLa cells contains all eight CCT subunits and is active in vitro. PMID: 23011926
Chaperonin containing TCP1, subunit 2 (beta)	CCT2	Targeting β-tubulin:CCT-β complexes incurs Hsp90- and VCP-related protein degradation and induces ER stress-associated apoptosis by triggering capacitative Ca2+ entry, mitochondrial perturbation and caspase overactivation. PMID: 23190606
Chaperonin containing TCP1, subunit 3 (gamma)	CCT3	
Chaperonin containing TCP1, subunit 4 (delta)	CCT4	Biochemical characterization of mutants in chaperonin proteins CCT4 and CCT5 associated with hereditary sensory neuropathy. PMID: 25124038
Chaperonin containing TCP1, subunit 5 (epsilon)	CCT5	Disease: neuropathy, hereditary sensory, with spastic paraplegia, autosomal recessive (HSNSP)
Chaperonin containing TCP1, subunit 6A (zeta 1)	CCT6A	
Chaperonin containing TCP1, subunit 6B (zeta 2)	CCT6B	

HGNC-approved name	Symbol	Target information
Chaperonin containing TCP1, subunit 7 (eta)	CCT7	Increased CCT-eta expression is a marker of latent and active disease and a modulator of fibroblast contractility in Dupuytren's contracture. PMID: 23292503
Chaperonin containing TCP1, subunit 8 (theta)	CCT8	Chaperonin containing TCP1, subunit 8 (CCT8) is upregulated in hepatocellular carcinoma and promotes HCC proliferation. PMID: 24862099
Chaperonin containing TCP1, subunit 8 (theta)-like 2	CCT8L2	
ERO1-like (*S. cerevisiae*)	ERO1L	Inactivation of mammalian Ero1α is catalysed by specific protein disulfide-isomerases. PMID: 24758166
ERO1-like beta (*S. cerevisiae*)	ERO1LB	Deregulation of pancreas-specific oxidoreductin ERO1β in the pathogenesis of diabetes mellitus. PMID: 24469402
Eukaryotic translation elongation factor 1 alpha 1	EEF1A1	A pentapeptide monocyte locomotion inhibitory factor protects brain ischemia injury by targeting the eEF1A1/endothelial nitric oxide synthase pathway. PMID: 22829547
Eukaryotic translation elongation factor 1 alpha 2	EEF1A2	eEF1A2 promotes cell migration, invasion and metastasis in pancreatic cancer by upregulating MMP-9 expression through Akt activation. PMID: 23739844
Eukaryotic translation elongation factor 2	EEF2	Disease: spinocerebellar ataxia 26 (SCA26)
FK506 binding protein 1A, 12 kDa	FKBP1A	**Sirolimus** – immunomodulating agents
FK506 binding protein 1B, 12.6 kDa	FKBP1B	FK506-binding protein 1b/12.6: A key to aging-related hippocampal Ca2+ dysregulation? PMID: 24291098
FK506 binding protein 2, 13 kDa	FKBP2	The FK506 binding protein 13 kDa (FKBP13) interacts with the C-chain of complement C1q. PMID: 15533007
FK506 binding protein 3, 25 kDa	FKBP3	FKBP25, a novel regulator of the p53 pathway, induces the degradation of MDM2 and activation of p53. PMID: 19166840
FK506 binding protein 4, 59 kDa	FKBP4	Immunophilin FKBP52 induces Tau-P301L filamentous assembly *in vitro* and modulates its activity in a model of tauopathy. PMID: 24623856
FK506 binding protein 5	FKBP5	The role of FKBP5, a co-chaperone of the glucocorticoid receptor in the pathogenesis and therapy of affective and anxiety disorders. PMID: 19560279
FK506 binding protein 6, 36 kDa	FKBP6	Note: FKBP6 is located in the Williams–Beuren syndrome (WBS) critical region
FK506 binding protein 7	FKBP7	Mouse FKBP23 mediates conformer-specific functions of BiP by catalyzing Pro117 cis/trans isomerization. PMID: 21637344
FK506 binding protein 8, 38 kDa	FKBP8	FK506 binding protein 8 peptidylprolyl isomerase activity manages a late stage of cystic fibrosis transmembrane conductance regulator (CFTR) folding and stability. PMID: 22474283
FK506 binding protein 9, 63 kDa	FKBP9	
FK506 binding protein 10, 65 kDa	FKBP10	Disease: osteogenesis imperfecta 11 (OI11)
FK506 binding protein 11, 19 kDa	FKBP11	Identification of FKBP11 as a biomarker for hepatocellular carcinoma. PMID: 23749938
FK506 binding protein 14, 22 kDa	FKBP14	Disease: Ehlers–Danlos syndrome, with progressive kyphoscoliosis, myopathy, and hearing loss (EDSKMH)
FK506 binding protein 15, 133 kDa	FKBP15	The ulcerative colitis marker protein WAFL interacts with accessory proteins in endocytosis. PMID: 20376207
FK506 binding protein-like	FKBPL	Targeting treatment-resistant breast cancer stem cells with FKBPL and its peptide derivative, AD-01, via the CD44 pathway. PMID: 23741069
G elongation factor, mitochondrial 1	GFM1	The effect of small molecules on nuclear-encoded translation diseases. PMID: 24012549. Disease: combined oxidative phosphorylation deficiency 1 (COXPD1)

(*Continued*)

HGNC-approved name	Symbol	Target information
G elongation factor, mitochondrial 2	GFM2	
G1 to S phase transition 1	GSPT1	Nicotine-mediated invasion and migration of non-small cell lung carcinoma cells by modulating STMN3 and GSPT1 genes in an ID1-dependent manner. PMID: 25028095
G1 to S phase transition 2	GSPT2	eRF3b, a biomarker for hepatocellular carcinoma, influences cell cycle and phosphoralation status of 4E-BP1. PMID: 24466059
Heat shock protein 90 kDa alpha (cytosolic), class A member 1	HSP90AA1	**Retaspimycin** – antineoplastic agents
Heat shock protein 90 kDa alpha (cytosolic), class B member 1	HSP90AB1	Identification of novel HSP90α/β isoform selective inhibitors using structure-based drug design. demonstration of potential utility in treating CNS disorders such as Huntington's disease. PMID: 24673104
Heat shock protein 90 kDa beta (Grp94), member 1	HSP90B1	Glycoprotein 96 perpetuates the persistent inflammation of rheumatoid arthritis. PMID: 22777994
Heat shock 70 kDa protein 1A	HSPA1A	Proteomic study identified HSP 70 kDa protein 1A as a possible therapeutic target, in combination with histone deacetylase inhibitors, for lymphoid neoplasms. PMID: 22123078
Heat shock 70 kDa protein 1B	HSPA1B	Targeting the testis-specific heat-shock protein 70-2 (HSP70-2) reduces cellular growth, migration, and invasion in renal cell carcinoma cells. PMID: 25213699
Heat shock 70 kDa protein 1-like	HSPA1L	Synergistic effect and VEGF/HSP70-hom haplotype analysis: Relationship to prostate cancer risk and clinical outcome. PMID: 20096741
Heat shock 70 kDa protein 2	HSPA2	Expression of HSPA2 in human hepatocellular carcinoma and its clinical significance. PMID: 25117073
Heat shock 70 kDa protein 4	HSPA4	Targeted disruption of Hspa4 gene leads to cardiac hypertrophy and fibrosis. PMID: 22884543
Heat shock 70 kDa protein 4-like	HSPA4L	Respiratory distress and early neonatal lethality in Hspa4l/Hspa4 double-mutant mice. PMID: 23980576
Heat shock 70 kDa protein 5 (glucose-regulated protein, 78 kDa)	HSPA5	Glucose-regulated proteins in cancer: Molecular mechanisms and therapeutic potential. PMID: 24658275. Note: Autoantigen in rheumatoid arthritis
Heat shock 70 kDa protein 6 (HSP70B')	HSPA6	Localization of heat shock protein HSPA6 (HSP70B') to sites of transcription in cultured differentiated human neuronal cells following thermal stress. PMID: 25319762
Heat shock 70 kDa protein 8	HSPA8	Heat shock cognate 70 protein secretion as a new growth arrest signal for cancer cells. PMID: 19802014
Heat shock 70 kDa protein 9 (mortalin)	HSPA9	Inhibition of mortalin expression reverses cisplatin resistance and attenuates growth of ovarian cancer cells. PMID: 23665506
Heat shock 70 kDa protein 12A	HSPA12A	
Heat shock 70 kDa protein 12B	HSPA12B	HSPA12B inhibits lipopolysaccharide-induced inflammatory response in human umbilical vein endothelial cells. PMID: 25545050
Heat shock protein 70 kDa family, member 13	HSPA13	Overexpression of the Hspa13 (Stch) gene reduces prion disease incubation time in mice. PMID: 22869728
Heat shock 70 kDa protein 14	HSPA14	Efficient induction of a Her2-specific anti-tumor response by dendritic cells pulsed with a Hsp70L1-Her2(341-456) fusion protein. PMID: 21785448
Heat shock 27 kDa protein 1	HSPB1	**OGX-427** – antineoplastic agents. Disease: Charcot–Marie–Tooth disease 2F (CMT2F)
Heat shock 27 kDa protein 2	HSPB2	The small heat shock protein HspB2 is a novel anti-apoptotic protein that inhibits apical caspase activation in the extrinsic apoptotic pathway. PMID: 20087649
Heat shock 27 kDa protein 3	HSPB3	Disease: neuronopathy, distal hereditary motor, 2C (HMN2C)

HGNC-approved name	Symbol	Target information
Heat shock protein, alpha-crystallin-related, B6	HSPB6	Heat shock protein 20 (HSPB6) regulates TNF-α-induced intracellular signaling pathway in human hepatocellular carcinoma cells. PMID: 25447820
Heat shock 27 kDa protein family, member 7 (cardiovascular)	HSPB7	Genetic association study identifies HSPB7 as a risk gene for idiopathic dilated cardiomyopathy. PMID: 20975947
Heat shock 22 kDa protein 8	HSPB8	Disease: neuronopathy, distal hereditary motor, 2A (HMN2A)
Heat shock protein, alpha-crystallin-related, B9	HSPB9	Testis-specific human small heat shock protein HSPB9 is a cancer/testis antigen, and potentially interacts with the dynein subunit TCTEL1. PMID: 15503857
Heat shock protein family B (small), member 11	HSPB11	IFT25 links the signal-dependent movement of Hedgehog components to intraflagellar transport. PMID: 22595669
Heat shock 60 kDa protein 1 (chaperonin)	HSPD1	Chaperonopathies and chaperonotherapy. Hsp60 as therapeutic target in cancer: Potential benefits and risks. PMID: 22920896. Disease: spastic paraplegia 13, autosomal dominant (SPG13)
Heat shock 10 kDa protein 1	HSPE1	Suppression of Cpn10 increases mitochondrial fission and dysfunction in neuroblastoma cells. PMID: 25390895
Heat shock 105/110 kDa protein 1	HSPH1	Serological identification of HSP105 as a novel non-Hodgkin lymphoma therapeutic target. PMID: 21860023
Mitochondrial ribosome-associated GTPase 1	MTG1	Human G-proteins, ObgH1 and Mtg1, associate with the large mitochondrial ribosome subunit and are involved in translation and assembly of respiratory complexes. PMID: 23396448
Mitochondrial ribosome-associated GTPase 2	MTG2	Human G-proteins, ObgH1 and Mtg1, associate with the large mitochondrial ribosome subunit and are involved in translation and assembly of respiratory complexes. PMID: 23396448
Multiple cyclophilin inhibitor		**Cyclosporine A** – immunomodulating agents
Peptidylprolyl isomerase A (cyclophilin A)	PPIA	Extracellular cyclophilin A activates platelets via EMMPRIN (CD147) and PI3K/Akt signaling, which promotes platelet adhesion and thrombus formation *in vitro* and in vivo. PMID: 25550208
Peptidylprolyl isomerase B (cyclophilin B)	PPIB	Disease: osteogenesis imperfecta 9 (OI9)
Peptidylprolyl isomerase C (cyclophilin C)	PPIC	Roles of cyclophilins in cancers and other organ systems. PMID: 15706440
Peptidylprolyl isomerase D	PPID	Structural mechanisms of cyclophilin D-dependent control of the mitochondrial permeability transition pore. PMID: 25445707
Peptidylprolyl isomerase E (cyclophilin E)	PPIE	Cyclophilin E functions as a negative regulator to influenza virus replication by impairing the formation of the viral ribonucleoprotein complex. PMID: 21887220
Peptidylprolyl isomerase F	PPIF	Molecular mechanisms of cell death: Central implication of ATP synthase in mitochondrial permeability transition. PMID: 24727893
Peptidylprolyl isomerase G (cyclophilin G)	PPIG	The human nuclear SRcyp is a cell cycle-regulated cyclophilin. PMID: 15016823
Peptidylprolyl isomerase H (cyclophilin H)	PPIH	
Peptidylprolyl isomerase (cyclophilin)-like 1	PPIL1	The crystal structure of PPIL1 bound to cyclosporine A suggests a binding mode for a linear epitope of the SKIP protein. PMID: 20368803
Peptidylprolyl isomerase (cyclophilin)-like 2	PPIL2	A genome wide analysis of ubiquitin ligases in APP processing identifies a novel regulator of BACE1 mRNA levels. PMID: 16978875
Peptidylprolyl isomerase (cyclophilin)-like 3	PPIL3	Interaction with Ppil3 leads to the cytoplasmic localization of apoptin in tumor cells. PMID: 18474220
Peptidylprolyl isomerase (cyclophilin)-like 4	PPIL4	
Peptidylprolyl isomerase A (cyclophilin A)-like 4A	PPIAL4A	

(Continued)

HGNC-approved name	Symbol	Target information
Peptidylprolyl isomerase A (cyclophilin A)-like 4B	PPIAL4B	
Peptidylprolyl isomerase A (cyclophilin A)-like 4C	PPIAL4C	
Peptidylprolyl isomerase A (cyclophilin A)-like 4D	PPIAL4D	
Peptidylprolyl isomerase A (cyclophilin A)-like 4E	PPIAL4E	
Peptidylprolyl isomerase A (cyclophilin A)-like 4F	PPIAL4F	
Peptidylprolyl isomerase A (cyclophilin A)-like 4G	PPIAL4G	
Peptidylprolyl isomerase (cyclophilin)-like 6	PPIL6	
Peptidylprolyl isomerase domain and WD repeat containing 1	PPWD1	The crystal structure of human WD40 repeat-containing peptidylprolyl isomerase (PPWD1). PMID: 18397323
Peptidylprolyl cis/trans isomerase, NIMA-interacting 1	PIN1	Peptidyl-prolyl cis-trans isomerase Pin1 in ageing, cancer and Alzheimer disease. PMID: 21682951
Protein (peptidylprolyl cis/trans isomerase) NIMA-interacting, 4 (parvulin)	PIN4	Par14 protein associates with insulin receptor substrate 1 (IRS-1), thereby enhancing insulin-induced IRS-1 phosphorylation and metabolic actions. PMID: 23720771
Protein disulfide isomerase family A, member 2	PDIA2	Inhibitors of protein disulfide isomerase suppress apoptosis induced by misfolded proteins. PMID: 21079601
Protein disulfide isomerase family A, member 3	PDIA3	Impaired bone formation in Pdia3 deficient mice. PMID: 25405762
Protein disulfide isomerase family A, member 4	PDIA4	The protein disulfide isomerases PDIA4 and PDIA6 mediate resistance to cisplatin-induced cell death in lung adenocarcinoma. PMID: 24464223
Protein disulfide isomerase family A, member 5	PDIA5	Endoplasmic reticulum stress-activated transcription factor ATF6α requires the disulfide isomerase PDIA5 to modulate chemoresistance. PMID: 24636989
Protein disulfide isomerase family A, member 6	PDIA6	The protein disulfide isomerases PDIA4 and PDIA6 mediate resistance to cisplatin-induced cell death in lung adenocarcinoma. PMID: 24464223
Protein disulfide isomerase-like, testis expressed	PDILT	Structure of the substrate-binding b′ domain of the protein disulfide isomerase-like protein of the testis. PMID: 24662985
Ribosomal modification protein rimK-like family member A	RIMKLA	N-acetylaspartylglutamate synthetase II synthesizes N-acetylaspartylglutamylglutamate. PMID: 21454531
Ribosomal modification protein rimK-like family member B	RIMKLB	Transplantation of N-acetyl aspartyl-glutamate synthetase-activated neural stem cells after experimental traumatic brain injury significantly improves neurological recovery. PMID: 24356585
Torsin family 1, member A (torsin A)	TOR1A	Chemical enhancement of torsinA function in cell and animal models of torsion dystonia. PMID: 20223934. Disease: dystonia 1, torsion, autosomal dominant (DYT1)
Torsin family 1, member B (torsin B)	TOR1B	Arresting a Torsin ATPase reshapes the endoplasmic reticulum. PMID: 24275647
Torsin family 2, member A	TOR2A	The roles of salusins in atherosclerosis and related cardiovascular diseases. PMID: 21925457
Torsin family 3, member A	TOR3A	
Torsin family 4, member A	TOR4A	
Coiled-coil-helix-coiled-coil-helix domain containing 4	CHCHD4	Human CHCHD4 mitochondrial proteins regulate cellular oxygen consumption rate and metabolism and provide a critical role in hypoxia signaling and tumor progression. PMID: 22214851

HGNC-approved name	Symbol	Target information
CWC27 spliceosome-associated protein homolog (*S. cerevisiae*)	CWC27	
Growth factor, augmenter of liver regeneration	GFER	Disease: myopathy, mitochondrial progressive, with congenital cataract, hearing loss and developmental delay (MPMCHD)
GUF1 GTPase homolog (*S. cerevisiae*)	GUF1	
HBS1-like translational GTPase	HBS1L	Dom34:hbs1 plays a general role in quality-control systems by dissociation of a stalled ribosome at the 3′ end of aberrant mRNA. PMID: 22503425
Immature colon carcinoma transcript 1	ICT1	Solution structure of the catalytic domain of the mitochondrial protein ICT1 that is essential for cell vitality. PMID: 20869366
Interferon, gamma-inducible protein 30	IFI30	MHC class II-restricted presentation of the major house dust mite allergen Der p 1 Is GILT-dependent: Implications for allergic asthma. PMID: 23326313
Large 60S subunit nuclear export GTPase 1	LSG1	Human Lsg1 defines a family of essential GTPases that correlates with the evolution of compartmentalization. PMID: 16209721
Methyltransferase-like 17	METTL17	
Mitochondrial translational initiation factor 2	MTIF2	
Peptide deformylase (mitochondrial)	PDF	Inhibition of human mitochondrial peptide deformylase causes apoptosis in c-myc-overexpressing hematopoietic cancers. PMID: 24675470
Tetratricopeptide repeat domain 9B	TTC9B	
Tu translation elongation factor, mitochondrial	TUFM	Disease: combined oxidative phosphorylation deficiency 4 (COXPD4)

4.2.6 Other protein modifications

HGNC-approved name	Symbol	Target information
ADP-ribosyltransferase 1	ART1	ADP-ribosyltransferase-specific modification of human neutrophil peptide-1. PMID: 16627471
ADP-ribosyltransferase 3	ART3	Genome-wide expression of azoospermia testes demonstrates a specific profile and implicates ART3 in genetic susceptibility. PMID: 18266473
ADP-ribosyltransferase 4 (Dombrock blood group)	ART4	The Dombrock blood group system: A review. PMID: 20932078
ADP-ribosyltransferase 5	ART5	Ecto ADP ribosyltransferases (ARTS): Emerging actors in cell communication and signaling. PMID: 15078170
alkB, alkylation repair homolog 4 (*Escherichia coli*)	ALKBH4	ALKBH4 depletion in mice leads to spermatogenic defects. PMID: 25153837
alkB, alkylation repair homolog 6 (*E. coli*)	ALKBH6	
alkB, alkylation repair homolog 7 (*E. coli*)	ALKBH7	The atomic resolution structure of human AlkB homolog 7 (ALKBH7), a key protein for programmed necrosis and fat metabolism. PMID: 25122757
Katanin p60 (ATPase containing) subunit A 1	KATNA1	Katanin p60 contributes to microtubule instability around the midbody and facilitates cytokinesis in rat cells. PMID: 24303010
Katanin p60 subunit A-like 1	KATNAL1	KATNAL1 regulation of sertoli cell microtubule dynamics is essential for spermiogenesis and male fertility. PMID: 22654668
Katanin p60 subunit A-like 2	KATNAL2	Meta-analysis of genome-wide association studies for personality. PMID: 21173776
Lipoyltransferase 1	LIPT1	Mutations in the lipoyltransferase LIPT1 gene cause a fatal disease associated with a specific lipoylation defect of the 2-ketoacid dehydrogenase complexes. PMID: 24256811

(Continued)

HGNC-approved name	Symbol	Target information
Lipoyl(octanoyl) transferase 2 (putative)	LIPT2	Lipoic acid biosynthesis defects. PMID: 24777537
MACRO domain containing 1	MACROD1	Note: A chromosomal aberration involving MACROD1 is found in acute leukemia. Translocation t(1121)(q13q22) that forms a RUNX1-MACROD1 fusion protein
MACRO domain containing 2	MACROD2	MACROD2 overexpression mediates estrogen independent growth and tamoxifen resistance in breast cancers. PMID: 25422431
Methyltransferase-like 11B	METTL11B	NRMT2 is an N-terminal monomethylase that primes for its homologue NRMT1. PMID: 24090352
Methyltransferase-like 18	METTL18	
Methyltransferase-like 20	METTL20	Human METTL20 is a mitochondrial lysine methyltransferase that targets the β subunit of electron transfer flavoprotein (ETFβ) and modulates its activity. PMID: 25416781
Methyltransferase-like 21A	METTL21A	Identification and characterization of a novel human methyltransferase modulating Hsp70 protein function through lysine methylation. PMID: 23921388
Methyltransferase-like 21B	METTL21B	The multiple sclerosis whole blood mRNA transcriptome and genetic associations indicate dysregulation of specific T cell pathways in pathogenesis. PMID: 20190274
Methyltransferase-like 21C	METTL21C	METTL21C is a potential pleiotropic gene for osteoporosis and sarcopenia acting through the modulation of the NF-κB signaling pathway. PMID: 24677265
Methyltransferase-like 22	METTL22	Methylation of the DNA/RNA-binding protein Kin17 by METTL22 affects its association with chromatin. PMID: 24140279
Methyltransferase-like 23	METTL23	Disruption of the methyltransferase-like 23 gene METTL23 causes mild autosomal recessive intellectual disability. PMID: 24626631
N-acetyltransferase 6 (GCN5-related)	NAT6	
N-acetyltransferase 8 (GCN5-related, putative)	NAT8	The role of N-acetyltransferase 8 in mesenchymal stem cell-based therapy for liver ischemia/reperfusion injury in rats. PMID: 25057902
N-acetyltransferase 8B (GCN5-related, putative, gene/pseudogene)	NAT8B	Post-translational regulation of CD133 by ATase1/ATase2-mediated lysine acetylation. PMID: 24556617
N-acetyltransferase 9 (GCN5-related, putative)	NAT9	
N-acetyltransferase 10 (GCN5-related)	NAT10	Chemical inhibition of NAT10 corrects defects of laminopathic cells. PMID: 24786082
N-acetyltransferase 16 (GCN5-related, putative)	NAT16	
N-myristoyltransferase 1	NMT1	Global profiling of co- and post-translationally N-myristoylated proteomes in human cells. PMID: 25255805
N-myristoyltransferase 2	NMT2	N-myristoyltransferase 2 expression in human colon cancer: Cross-talk between the calpain and caspase system. PMID: 16530191
Palmitoyl-protein thioesterase 1	PPT1	An anti-neuroinflammatory that targets dysregulated glia enhances the efficacy of CNS-directed gene therapy in murine infantile neuronal ceroid lipofuscinosis. PMID: 25253854. Disease: ceroid lipofuscinosis, neuronal, 1 (CLN1)
Palmitoyl-protein thioesterase 2	PPT2	
Peptidyl arginine deiminase, type I	PADI1	Inhibition of peptidyl-arginine deiminases reverses protein-hypercitrullination and disease in mouse models of multiple sclerosis. PMID: 23118341
Peptidyl arginine deiminase, type II	PADI2	PADI2 is significantly associated with rheumatoid arthritis. PMID: 24339914

HGNC-approved name	Symbol	Target information
Peptidyl arginine deiminase, type III	PADI3	NF-Y and Sp1/Sp3 are involved in the transcriptional regulation of the peptidylarginine deiminase type III gene (PADI3) in human keratinocytes. PMID: 16671893
Peptidyl arginine deiminase, type IV	PADI4	Discovery of a new class of inhibitors for the protein arginine deiminase type 4 (PAD4) by structure-based virtual screening. PMID: 23282142. Disease: rheumatoid arthritis (RA)
Peptidyl arginine deiminase, type VI	PADI6	Identification and structural characterization of two 14-3-3 binding sites in the human peptidylarginine deiminase type VI. PMID: 22634725
Phosphatidylinositol glycan anchor biosynthesis, class K	PIGK	Recent developments in the molecular, biochemical and functional characterization of GPI8 and the GPI-anchoring mechanism [review]. PMID: 16785205
Phosphatidylinositol glycan anchor biosynthesis, class S	PIGS	
Phosphatidylinositol glycan anchor biosynthesis, class T	PIGT	Disease: multiple congenital anomalies–hypotonia–seizures syndrome 3 (MCAHS3)
Phosphatidylinositol glycan anchor biosynthesis, class U	PIGU	
Poly (ADP-ribose) polymerase 1	PARP1	**Olaparib** – antineoplastic agents
Poly (ADP-ribose) polymerase 2	PARP2	**Veliparib** – antineoplastic agents
Poly (ADP-ribose) polymerase family, member 3	PARP3	Poly (ADP-ribose) polymerase 3 (PARP3), a potential repressor of telomerase activity. PMID: 24528514
Poly (ADP-ribose) polymerase family, member 4	PARP4	Expression profiles of vault components MVP, TEP1 and vPARP and their correlation to other multidrug resistance proteins in ovarian cancer. PMID: 23739867
Poly (ADP-ribose) polymerase family, member 6	PARP6	PARP6, a mono(ADP-ribosyl) transferase and a negative regulator of cell proliferation, is involved in colorectal cancer development. PMID: 23042038
Poly (ADP-ribose) polymerase family, member 8	PARP8	
Poly (ADP-ribose) polymerase family, member 9	PARP9	Note: Overexpressed at significantly higher levels in fatal high-risk diffuse large B-cell lymphomas (DLB-CL) compared to cured low-risk tumors
Poly (ADP-ribose) polymerase family, member 10	PARP10	Function and regulation of the mono-ADP-ribosyltransferase ARTD10. PMID: 24878761
Poly (ADP-ribose) polymerase family, member 11	PARP11	Structural insight into the interaction of ADP-ribose with the PARP WWE domains. PMID: 23010590
Poly (ADP-ribose) polymerase family, member 12	PARP12	PARP12, an interferon-stimulated gene involved in the control of protein translation and inflammation. PMID: 25086041
Poly (ADP-ribose) polymerase family, member 14	PARP14	Poly(ADP-ribose) polymerase family member 14 (PARP14) is a novel effector of the JNK2-dependent pro-survival signal in multiple myeloma. PMID: 23045269
Poly (ADP-ribose) polymerase family, member 15	PARP15	Activity-based assay for human mono-ADP-ribosyltransferases ARTD7/PARP15 and ARTD10/PARP10 aimed at screening and profiling inhibitors. PMID: 23485441
Poly (ADP-ribose) polymerase family, member 16	PARP16	PARP16 is a tail-anchored endoplasmic reticulum protein required for the PERK- and IRE1α-mediated unfolded protein response. PMID: 23103912
Prenylcysteine oxidase 1	PCYOX1	Proteomic analysis of human low-density lipoprotein reveals the presence of prenylcysteine lyase, a hydrogen peroxide-generating enzyme. PMID: 19253276
Prenylcysteine oxidase 1-like	PCYOX1L	
Procollagen-lysine, 2-oxoglutarate 5-dioxygenase 1	PLOD1	Disease: Ehlers–Danlos syndrome 6 (EDS6)

(Continued)

HGNC-approved name	Symbol	Target information
Procollagen-lysine, 2-oxoglutarate 5-dioxygenase 2	PLOD2	Procollagen lysyl hydroxylase 2 is essential for hypoxia-induced breast cancer metastasis. PMID: 23378577. Disease: Bruck syndrome 2 (BRKS2)
Procollagen-lysine, 2-oxoglutarate 5-dioxygenase 3	PLOD3	Disease: lysyl hydroxylase 3 deficiency (LH3 deficiency)
Protein-L-isoaspartate (D-aspartate) O-methyltransferase	PCMT1	Protein L-isoaspartyl methyltransferase regulates p53 activity. PMID: 22735455
Protein-L-isoaspartate (D-aspartate) O-methyltransferase domain containing 1	PCMTD1	
Protein-L-isoaspartate (D-aspartate) O-methyltransferase domain containing 2	PCMTD2	
Tankyrase, TRF1-interacting ankyrin-related ADP-ribose polymerase	TNKS	**JW55** – antineoplastic agents
Tankyrase, TRF1-interacting ankyrin-related ADP-ribose polymerase 2	TNKS2	**JW55** – antineoplastic agents
Transglutaminase 1	TGM1	Disease: ichthyosis, congenital, autosomal recessive 1 (ARCI1)
Transglutaminase 2	TGM2	Transglutaminase 2 inhibitors and their therapeutic role in disease states. PMID: 17582505
Transglutaminase 3	TGM3	Reduced inflammatory threshold indicates skin barrier defect in transglutaminase 3 knockout mice. PMID: 23884312
Transglutaminase 4	TGM4	Prostate transglutaminase: A unique transglutaminase and its role in prostate cancer. PMID: 21657838
Transglutaminase 5	TGM5	Disease: peeling skin syndrome, acral type (APSS)
Transglutaminase 6	TGM6	Disease: spinocerebellar ataxia 35 (SCA35)
Transglutaminase 7	TGM7	
Tubulin polyglutamylase complex subunit 1	TPGS1	Post-translational modifications of microtubules. PMID: 20930140
Tubulin polyglutamylase complex subunit 2	TPGS2	
Tubulin tyrosine ligase	TTL	Sesterterpenes as tubulin tyrosine ligase inhibitors. First insight of structure-activity relationships and discovery of new lead. PMID: 19459643
Tubulin tyrosine ligase-like family, member 1	TTLL1	Tubulin tyrosine ligase-like 1 deficiency results in chronic rhinosinusitis and abnormal development of spermatid flagella in mice. PMID: 20442420
Tubulin tyrosine ligase-like family, member 2	TTLL2	
Tubulin tyrosine ligase-like family, member 3	TTLL3	Tubulin glycylases are required for primary cilia, control of cell proliferation and tumor development in colon. PMID: 25180231
Tubulin tyrosine ligase-like family, member 4	TTLL4	Involvement of the tubulin tyrosine ligase-like family member 4 polyglutamylase in PELP1 polyglutamylation and chromatin remodeling in pancreatic cancer cells. PMID: 20442285
Tubulin tyrosine ligase-like family, member 5	TTLL5	Biallelic variants in TTLL5, encoding a tubulin glutamylase, cause retinal dystrophy. PMID: 24791901
Tubulin tyrosine ligase-like family, member 6	TTLL6	Amyloid-β oligomers induce synaptic damage via Tau-dependent microtubule severing by TTLL6 and spastin. PMID: 24065130
Tubulin tyrosine ligase-like family, member 7	TTLL7	TTLL7 is a mammalian beta-tubulin polyglutamylase required for growth of MAP2-positive neurites. PMID: 16901895
Tubulin tyrosine ligase-like family, member 8	TTLL8	TTLL10 can perform tubulin glycylation when co-expressed with TTLL8. PMID: 19427864
Tubulin tyrosine ligase-like family, member 9	TTLL9	
Tubulin tyrosine ligase-like family, member 10	TTLL10	TTLL10 is a protein polyglycylase that can modify nucleosome assembly protein 1. PMID: 18331838

HGNC-approved name	Symbol	Target information
Tubulin tyrosine ligase-like family, member 11	TTLL11	
Tubulin tyrosine ligase-like family, member 12	TTLL12	Tubulin tyrosine ligase like 12, a TTLL family member with SET- and TTL-like domains and roles in histone and tubulin modifications and mitosis. PMID: 23251473
Zinc finger, DHHC-type containing 1	ZDHHC1	The ER-associated protein ZDHHC1 is a positive regulator of DNA virus-triggered, MITA/STING-dependent innate immune signaling. PMID: 25299331
Zinc finger, DHHC-type containing 2	ZDHHC2	Note: Mutations in ZDHHC2 are found in hepatocellular carcinoma and colorectal cancer
Zinc finger, DHHC-type containing 3	ZDHHC3	Palmitoylation by DHHC3 is critical for the function, expression, and stability of integrin α6β4. PMID: 22314500
Zinc finger, DHHC-type containing 4	ZDHHC4	
Zinc finger, DHHC-type containing 5	ZDHHC5	Somatostatin receptor 5 is palmitoylated by the interacting ZDHHC5 palmitoyltransferase. PMID: 21820437
Zinc finger, DHHC-type containing 6	ZDHHC6	Palmitoylated calnexin is a key component of the ribosome-translocon complex. PMID: 22314232
Zinc finger, DHHC-type containing 7	ZDHHC7	DHHC-7 and -21 are palmitoylacyltransferases for sex steroid receptors. PMID: 22031296
Zinc finger, DHHC-type containing 8	ZDHHC8	DHHC8-dependent PICK1 palmitoylation is required for induction of cerebellar long-term synaptic depression. PMID: 24068808
Zinc finger, DHHC-type containing 9	ZDHHC9	Disease: mental retardation, X-linked, syndromic, ZDHHC9-related (MRXSZ)
Zinc finger, DHHC-type containing 11	ZDHHC11	
Zinc finger, DHHC-type containing 11B	ZDHHC11B	
Zinc finger, DHHC-type containing 12	ZDHHC12	Putting proteins in their place: Palmitoylation in Huntington disease and other neuropsychiatric diseases. PMID: 22155432
Zinc finger, DHHC-type containing 13	ZDHHC13	Palmitoyl acyltransferase, Zdhhc13, facilitates bone mass acquisition by regulating postnatal epiphyseal development and endochondral ossification: A mouse model. PMID: 24637783
Zinc finger, DHHC-type containing 14	ZDHHC14	Overexpression of ZDHHC14 promotes migration and invasion of scirrhous type gastric cancer. PMID: 24807047
Zinc finger, DHHC-type containing 15	ZDHHC15	Disease: mental retardation, X-linked 91 (MRX91)
Zinc finger, DHHC-type containing 16	ZDHHC16	Aph2, a protein with a zf-DHHC motif, interacts with c-Abl and has pro-apoptotic activity. PMID: 12021275
Zinc finger, DHHC-type containing 17	ZDHHC17	Altered palmitoylation and neuropathological deficits in mice lacking HIP14. PMID: 21775500
Zinc finger, DHHC-type containing 18	ZDHHC18	
Zinc finger, DHHC-type containing 19	ZDHHC19	Palmitoylation of R-Ras by human DHHC19, a palmitoyl transferase with a CaaX box. PMID: 20074548
Zinc finger, DHHC-type containing 20	ZDHHC20	DHHC20: A human palmitoyl acyltransferase that causes cellular transformation. PMID: 20334580
Zinc finger, DHHC-type containing 21	ZDHHC21	DHHC-7 and -21 are palmitoylacyltransferases for sex steroid receptors. PMID: 22031296
Zinc finger, DHHC-type containing 22	ZDHHC22	Distinct acyl protein transferases and thioesterases control surface expression of calcium-activated potassium channels. PMID: 22399288
Zinc finger, DHHC-type containing 23	ZDHHC23	Distinct acyl protein transferases and thioesterases control surface expression of calcium-activated potassium channels. PMID: 22399288
Zinc finger, DHHC-type containing 24	ZDHHC24	

(Continued)

HGNC-approved name	Symbol	Target information
Alpha tubulin acetyltransferase 1	ATAT1	ATAT1/MEC-17 acetyltransferase and HDAC6 deacetylase control a balance of acetylation of alpha-tubulin and cortactin and regulate MT1-MMP trafficking and breast tumor cell invasion. PMID: 22902175
Calmodulin-lysine N-methyltransferase	CAMKMT	Human calmodulin methyltransferase: Expression, activity on calmodulin, and Hsp90 dependence. PMID: 23285036
Eukaryotic elongation factor 2 lysine methyltransferase	EEF2KMT	Identification and characterization of a novel evolutionarily conserved lysine-specific methyltransferase targeting eukaryotic translation elongation factor 2 (eEF2). PMID: 25231979
Erythrocyte membrane protein band 4.2	EPB42	Disease: spherocytosis 5 (SPH5)
FIC domain containing	FICD	
Gamma-glutamyl carboxylase	GGCX	**Anisindione** – antithrombotic agents. Disease: combined deficiency of vitamin K-dependent clotting factors 1 (VKCFD1)
HemK methyltransferase family member 1	HEMK1	Cloning and primarily function study of two novel putative N5-glutamine methyltransferase (Hemk) splice variants from mouse stem cells. PMID: 19116772
Leucine carboxyl methyltransferase 1	LCMT1	The structure of human leucine carboxyl methyltransferase 1 that regulates protein phosphatase PP2A. PMID: 21206058
Lipoic acid synthetase	LIAS	Disease: pyruvate dehydrogenase lipoic acid synthetase deficiency (PDHLD)
NADH dehydrogenase (ubiquinone) complex I, assembly factor 7	NDUFAF7	NDUFAF7 methylates arginine 85 in the NDUFS2 subunit of human complex I. PMID: 24089531
N-terminal Xaa-Pro-Lys N-methyltransferase 1	NTMT1	Identification of protein N-terminal methyltransferases in yeast and humans. PMID: 20481588
O-acyl-ADP-ribose deacylase 1	OARD1	Deficiency of terminal ADP-ribose protein glycohydrolase TARG1/ C6orf130 in neurodegenerative disease. PMID: 23481255
Protein phosphatase methylesterase 1	PPME1	PME-1 protects extracellular signal-regulated kinase pathway activity from protein phosphatase 2A-mediated inactivation in human malignant glioma. PMID: 19293187
Valosin-containing protein lysine (K) methyltransferase	VCPKMT	Discovery of a novel tumour metastasis-promoting gene, NVM-1. PMID: 21744341
WDYHV motif containing 1	WDYHV1	
Zinc finger, CCHC domain containing 4	ZCCHC4	

Chapter 5
Enzymes: Part 2

The introductory comments about the organization of enzyme entries presented in Chapter 4 are equally applicable to those listed in this chapter, which covers the enzymes that catalyse cellular metabolism. The entries are broadly divided into enzymes affecting lipids, amino acids, nucleotides, carbohydrates, vitamins and cofactors and nucleic acids. Other enzymes related to stress responses and homeostasis are listed separately, as are a small number of miscellaneous proteins which are not readily categorized. Many of the entries were assigned to their groups on the basis of their GO annotation as listed in UniProt, although some enzymes perform functions across boundaries, for example, between lipid and carbohydrate syntheses. Some areas of metabolism are strongly associated with particular therapeutic areas; disordered lipid and carbohydrate metabolism leads to metabolic and cardiovascular diseases, while the enzymes of nucleotide and nucleic acid metabolism include key targets for anti-proliferative drugs and modulators of the DNA damage response. There has been a recent upsurge of interest in the 'Warburg effect' in cancer because key enzymes for oxidative metabolism, such as isocitrate dehydrogenase, are mutated in some tumours and could be useful therapeutic targets [1].

The 'stress response and homeostasis' section contains enzymes involved in drug metabolism, including the cytochrome P450s. Although some of these may be primary drug targets for a particular disease, the majority modify therapeutic responses to existing drugs and therefore fall within the remit of pharmacogenetics [2].

References

1. Kee IIJ, Cheong JH. (2014) Tumor bioenergetics: an emerging avenue for cancer metabolism targeted therapy. *BMB Reports* **47**, 158–66.
2. Samer CF, Lorenzini K, Rollason V, Daali Y, Desmeules JA. (2013) Applications of CYP450 testing in the clinical setting. *Molecular Diagnosis and Therapy* **17**, 165–84.

5.1 Lipids and related

HGNC-approved name	Symbol	Target information
1-Acylglycerol-3-phosphate *O*-acyltransferase 1	AGPAT1	A role for 1-acylglycerol-3-phosphate-*O*-acyltransferase-1 in myoblast differentiation. PMID: 20561744
1-Acylglycerol-3-phosphate *O*-acyltransferase 2	AGPAT2	**CT-32615** – antineoplastic agents. Disease: congenital generalized lipodystrophy 1 (CGL1)
1-Acylglycerol-3-phosphate *O*-acyltransferase 3	AGPAT3	Lysophospholipid acyltransferases: 1-Acylglycerol-3-phosphate *O*-acyltransferases. From discovery to disease. PMID: 22777291
1-Acylglycerol-3-phosphate *O*-acyltransferase 4	AGPAT4	A novel lysophosphatidic acid acyltransferase enzyme (LPAAT4) with a possible role for incorporating docosahexaenoic acid into brain glycerophospholipids. PMID: 24333445
1-Acylglycerol-3-phosphate *O*-acyltransferase 5	AGPAT5	Enzymatic activities of the human AGPAT isoform 3 and isoform 5: Localization of AGPAT5 to mitochondria. PMID: 21173190
1-Acylglycerol-3-phosphate *O*-acyltransferase 6	AGPAT6	AGPAT6 is a novel microsomal glycerol-3-phosphate acyltransferase. PMID: 18238778
1-Acylglycerol-3-phosphate *O*-acyltransferase 9	AGPAT9	The epigenetic drug 5-azacytidine interferes with cholesterol and lipid metabolism. PMID: 24855646
7-Dehydrocholesterol reductase	DHCR7	Rapid suppression of 7-dehydrocholesterol reductase activity in keratinocytes by vitamin D. PMID: 25500071. Disease: Smith–Lemli–Opitz syndrome (SLOS)
24-Dehydrocholesterol reductase	DHCR24	Hippocampal DHCR24 down regulation in a rat model of streptozotocin-induced cognitive decline. PMID: 25541351. Disease: desmosterolosis (DESMOS)
2,4-Dienoyl-CoA reductase 1, mitochondrial	DECR1	Elevated expression of DecR1 impairs ErbB2/Neu-induced mammary tumor development. PMID: 17636013
2,4-Dienoyl-CoA reductase 2, peroxisomal	DECR2	Studies of human 2,4-dienoyl CoA reductase shed new light on peroxisomal β-oxidation of unsaturated fatty acids. PMID: 22745130
3-Hydroxyacyl-CoA dehydratase 1	HACD1	Congenital myopathy is caused by mutation of HACD1. PMID: 2393373
3-Hydroxyacyl-CoA dehydratase 2	HACD2	Characterization of four mammalian 3-hydroxyacyl-CoA dehydratases involved in very long-chain fatty acid synthesis. PMID: 18554506
3-Hydroxyacyl-CoA dehydratase 3	HACD3	Human B-ind1 gene promoter: Cloning and regulation by histone deacetylase inhibitors. PMID: 16516406
3-Hydroxyacyl-CoA dehydratase 4	HACD4	PTPLAD2 is a tumor suppressor in esophageal squamous cell carcinogenesis. PMID: 24530685
3-Hydroxy-3-methylglutaryl-CoA synthase 1 (soluble)	HMGCS1	Controlled sumoylation of the mevalonate pathway enzyme HMGS-1 regulates metabolism during aging. PMID: 25187565
3-Hydroxy-3-methylglutaryl-CoA synthase 2 (mitochondrial)	HMGCS2	Disease: HMG-CoA synthase deficiency (HMGCS deficiency)
3-Hydroxybutyrate dehydrogenase, type 1	BDH1	
3-Hydroxybutyrate dehydrogenase, type 2	BDH2	Human BDH2, an anti-apoptosis factor, is a novel poor prognostic factor for de novo cytogenetically normal acute myeloid leukemia. PMID: 23941109
3-Hydroxymethyl-3-methylglutaryl-CoA lyase	HMGCL	Disease: 3-hydroxy-3-methylglutaryl-CoA lyase deficiency (HMGCLD)
3-Hydroxymethyl-3-methylglutaryl-CoA lyase-like 1	HMGCLL1	Identification and characterization of an extramitochondrial human 3-hydroxy-3-methylglutaryl-CoA lyase. PMID: 22865860
3-Oxoacid CoA transferase 1	OXCT1	Disease: succinyl-CoA-3-ketoacid-CoA transferase deficiency (SCOTD)
3-Oxoacid CoA transferase 2	OXCT2	Ketolytic and glycolytic enzymatic expression profiles in malignant gliomas: Implication for ketogenic diet therapy. PMID: 23829383

HGNC-approved name	Symbol	Target information
Abhydrolase domain containing 1	ABHD1	Alpha/beta hydrolase 1 is upregulated in D5 dopamine receptor knockout mice and reduces O2- production of NADPH oxidase. PMID: 19073140
Abhydrolase domain containing 2	ABHD2	Antisense oligonucleotides targeting abhydrolase domain containing 2 block human hepatitis B virus propagation. PMID: 21466387
Abhydrolase domain containing 3	ABHD3	Metabolomics annotates ABHD3 as a physiologic regulator of medium-chain phospholipids. PMID: 21926997
Abhydrolase domain containing 4	ABHD4	A genome wide shRNA screen identifies α/β hydrolase domain containing 4 (ABHD4) as a novel regulator of anoikis resistance. PMID: 22488300
Abhydrolase domain containing 5	ABHD5	Disease: Chanarin–Dorfman syndrome (CDS)
Abhydrolase domain containing 6	ABHD6	Optimization and characterization of triazole urea inhibitors for abhydrolase domain containing protein 6 (ABHD6). In: Probe Reports from the NIH Molecular Libraries Program. PMID: 23762934
Abhydrolase domain containing 8	ABHD8	
Abhydrolase domain containing 10	ABHD10	Probe development efforts to identify novel inhibitors of ABHD10. PMID: 23762952
Abhydrolase domain containing 11	ABHD11	Click-generated triazole ureas as ultrapotent in vivo-active serine hydrolase inhibitors. PMID: 21572424. Note: ABHD11 is located in the Williams-Beuren syndrome (WBS) critical region. WBS results from a hemizygous deletion of several genes on chromosome 7
Abhydrolase domain containing 12	ABHD12	Disease: polyneuropathy, hearing loss, ataxia, retinitis pigmentosa, and cataract (PHARC)
Abhydrolase domain containing 12B	ABHD12B	
Abhydrolase domain containing 13	ABHD13	
Abhydrolase domain containing 14A	ABHD14A	Dorz1, a novel gene expressed in differentiating cerebellar granule neurons, is down-regulated in Zic1-deficient mouse. PMID: 14667578
Abhydrolase domain containing 14B	ABHD14B	Localization of sporadic neuroendocrine tumors by gene expression analysis of their metastases. PMID: 21681495
Abhydrolase domain containing 15	ABHD15	α/β-Hydrolase domain containing protein 15 (ABHD15) – an adipogenic protein protecting from apoptosis. PMID: 24236098
Abhydrolase domain containing 16A	ABHD16A	Biochemical and pharmacological characterization of the human lymphocyte antigen B-associated transcript 5 (BAT5/ABHD16A) PMID: 25290914
Abhydrolase domain containing 16B	ABHD16B	
Abhydrolase domain containing 17A	ABHD17A	
Abhydrolase domain containing 17B	ABHD17B	
Abhydrolase domain containing 17C	ABHD17C	
Acetyl-CoA acetyltransferase 1	ACAT1	**Avasimibe** – lipid-modifying agents. Disease: 3-ketothiolase deficiency (3KTD)
Acetyl-CoA acetyltransferase 2	ACAT2	Intestine-specific MTP and global ACAT2 deficiency lowers acute cholesterol absorption with chylomicrons and HDLs. PMID: 25030663
Acetyl-CoA acyltransferase 1	ACAA1	Effects of endotoxin exposure on childhood asthma risk are modified by a genetic polymorphism in ACAA1. PMID: 22151743
Acetyl-CoA acyltransferase 2	ACAA2	**Trimetazidine** – cardiac therapy
Acetyl-CoA carboxylase alpha	ACACA	**PF-05175157** – drugs used in diabetes. Disease: acetyl-CoA carboxylase 1 deficiency (ACACAD)

(Continued)

HGNC-approved name	Symbol	Target information
Acetyl-CoA carboxylase beta	ACACB	Targeting acetyl-CoA carboxylases: Small molecular inhibitors and their therapeutic potential. PMID: 22339356
Acyl-CoA dehydrogenase family, member 8	ACAD8	Disease: isobutyryl-CoA dehydrogenase deficiency (IBDD)
Acyl-CoA dehydrogenase family, member 9	ACAD9	Disease: acyl-CoA dehydrogenase family, member 9, deficiency (ACAD9 deficiency)
Acyl-CoA dehydrogenase family, member 10	ACAD10	Identification and characterization of new long chain acyl-CoA dehydrogenases. PMID: 21237683
Acyl-CoA dehydrogenase family, member 11	ACAD11	Identification and characterization of new long chain acyl-CoA dehydrogenases. PMID: 21237683
Acyl-CoA dehydrogenase, C-2 to C-3 short chain	ACADS	Metabolic heritability at birth: Implications for chronic disease research. PMID: 24850141. Disease: acyl-CoA dehydrogenase short-chain deficiency (ACADSD)
Acyl-CoA dehydrogenase, C-4 to C-12 straight chain	ACADM	Disease: acyl-CoA dehydrogenase medium-chain deficiency (ACADMD)
Acyl-CoA dehydrogenase, long chain	ACADL	Disease: acyl-CoA dehydrogenase very long-chain deficiency (ACADVLD)
Acyl-CoA dehydrogenase, short/branched chain	ACADSB	Disease: short/branched-chain acyl-CoA dehydrogenase deficiency (SBCADD)
Acyl-CoA dehydrogenase, very long chain	ACADVL	Disease: acyl-CoA dehydrogenase very long-chain deficiency (ACADVLD)
Acyl-CoA oxidase 1, palmitoyl	ACOX1	Disease: adrenoleukodystrophy, pseudoneonatal (Pseudo-NALD)
Acyl-CoA oxidase 2, branched chain	ACOX2	Identification of ACOX2 as a shared genetic risk factor for preeclampsia and cardiovascular disease. PMID: 21343950
Acyl-CoA oxidase 3, pristanoyl	ACOX3	
Acyl-CoA oxidase-like	ACOXL	
Acyl-CoA synthetase bubblegum family member 1	ACSBG1	Acyl-CoA metabolism and partitioning. PMID: 24819326
Acyl-CoA synthetase bubblegum family member 2	ACSBG2	The second member of the human and murine bubblegum family is a testis- and brainstem-specific acyl-CoA synthetase. PMID: 16371355
Acyl-CoA synthetase family member 2	ACSF2	
Acyl-CoA synthetase family member 3	ACSF3	Disease: combined malonic and methylmalonic aciduria (CMAMMA)
Acyl-CoA synthetase long-chain family member 1	ACSL1	Deficiency of cardiac Acyl CoA synthetase-1 induces diastolic dysfunction, but pathologic hypertrophy is reversed by rapamycin. PMID: 24631848
Acyl-CoA synthetase long-chain family member 3	ACSL3	Increased long chain acyl-Coa synthetase activity and fatty acid import is linked to membrane synthesis for development of picornavirus replication organelles. PMID: 23762027
Acyl-CoA synthetase long-chain family member 4	ACSL4	Disease: mental retardation, X-linked 63 (MRX63)
Acyl-CoA synthetase long-chain family member 5	ACSL5	High ACSL5 transcript levels associate with systemic lupus erythematosus and apoptosis in Jurkat T lymphocytes and peripheral blood cells. PMID: 22163040
Acyl-CoA synthetase long-chain family member 6	ACSL6	Note: A chromosomal aberration involving ACSL6 may be a cause of myelodysplastic syndrome with basophilia. Translocation t(5;12)(q31;p13) with ETV6

HGNC-approved name	Symbol	Target information
Acyl-CoA synthetase medium-chain family member 1	ACSM1	15-prostaglandin dehydrogenase expression alone or in combination with ACSM1 defines a subgroup of the apocrine molecular subtype of breast carcinoma. PMID: 18632593
Acyl-CoA synthetase medium-chain family member 2A	ACSM2A	Structural snapshots for the conformation-dependent catalysis by human medium-chain acyl-coenzyme A synthetase ACSM2A. PMID: 19345228
Acyl-CoA synthetase medium-chain family member 2B	ACSM2B	Comparative analyses of disease risk genes belonging to the acyl-CoA synthetase medium-chain (ACSM) family in human liver and cell lines. PMID: 19634011
Acyl-CoA synthetase medium-chain family member 3	ACSM3	Impaired butyrate oxidation in ulcerative colitis is due to decreased butyrate uptake and a defect in the oxidation pathway. PMID: 21987487
Acyl-CoA synthetase medium-chain family member 4	ACSM4	ACSM4 polymorphisms are associated with rapid AIDS progression in HIV infected patients. PMID: 23982661
Acyl-CoA synthetase medium-chain family member 5	ACSM5	
Acyl-CoA synthetase medium-chain family member 6	ACSM6	
Acyl-CoA synthetase short-chain family member 1	ACSS1	
Acyl-CoA synthetase short-chain family member 2	ACSS2	Acetate is a bioenergetic substrate for human glioblastoma and brain metastases. PMID: 25525878
Acyl-CoA synthetase short-chain family member 3	ACSS3	
Acyl-CoA thioesterase 1	ACOT1	Protective effects of Acyl-coA thioesterase 1 on diabetic heart via PPARα/PGC1α signaling. PMID: 23226270
Acyl-CoA thioesterase 2	ACOT2	Acyl-CoA thioesterase-2 facilitates mitochondrial fatty acid oxidation in the liver. PMID: 25114170
Acyl-CoA thioesterase 4	ACOT4	The emerging role of acyl-CoA thioesterases and acyltransferases in regulating peroxisomal lipid metabolism. PMID: 22465940
Acyl-CoA thioesterase 6	ACOT6	
Acyl-CoA thioesterase 7	ACOT7	Acyl coenzyme A thioesterase 7 regulates neuronal fatty acid metabolism to prevent neurotoxicity. PMID: 23459938
Acyl-CoA thioesterase 8	ACOT8	Acyl-CoA thioesterase 8 is a specific protein related to nodal metastasis and prognosis of lung adenocarcinoma. PMID: 23540296
Acyl-CoA thioesterase 9	ACOT9	Acyl-CoA thioesterase 9 (ACOT9) in mouse may provide a novel link between fatty acid and amino acid metabolism in mitochondria. PMID: 23864032
Acyl-CoA thioesterase 11	ACOT11	Targeted deletion of thioesterase superfamily member 1 promotes energy expenditure and protects against obesity and insulin resistance. PMID: 22427358
Acyl-CoA thioesterase 12	ACOT12	Structural basis for regulation of the human acetyl-CoA thioesterase 12 and interactions with the steroidogenic acute regulatory protein-related lipid transfer (START) domain. PMID: 25002576
Acyl-CoA thioesterase 13	ACOT13	Phosphatidylcholine transfer protein interacts with thioesterase superfamily member 2 to attenuate insulin signaling. PMID: 23901139

(Continued)

HGNC-approved name	Symbol	Target information
Acyl-CoA wax alcohol acyltransferase 1	AWAT1	Identification of two novel human acyl-CoA wax alcohol acyltransferases: Members of the diacylglycerol acyltransferase 2 (DGAT2) gene superfamily. PMID: 15671038
Acyl-CoA wax alcohol acyltransferase 2	AWAT2	
Aldehyde dehydrogenase 3 family, member A2	ALDH3A2	Disease: Sjögren–Larsson syndrome (SLS)
Aldehyde dehydrogenase 3 family, member B1	ALDH3B1	Substrate specificity, plasma membrane localization, and lipid modification of the aldehyde dehydrogenase ALDH3B1. PMID: 23721920
Aldo-keto reductase family 1, member B15	AKR1B15	Aldo-keto reductase 1B15 (AKR1B15): A mitochondrial human aldo-keto reductase with activity towards steroids and 3-keto-acyl-CoA conjugates. PMID: 25577493
Aldo-keto reductase family 1, member C1	AKR1C1	Inhibitors of human 20α-hydroxysteroid dehydrogenase (AKR1C1). PMID: 21050889
Aldo-keto reductase family 1, member C2	AKR1C2	Disease: 46,XY sex reversal 8 (SRXY8)
Aldo-keto reductase family 1, member C3	AKR1C3	Development of potent and selective indomethacin analogues for the inhibition of AKR1C3 (Type 5 17β-hydroxysteroid dehydrogenase/prostaglandin F synthase) in castrate-resistant prostate cancer. PMID: 23432095
Aldo-keto reductase family 1, member C4	AKR1C4	Disease: 46,XY sex reversal 8 (SRXY8)
Aldo-keto reductase family 1, member D1	AKR1D1	Disease: congenital bile acid synthesis defect 2 (CBAS2)
Alkaline ceramidase 1	ACER1	
Alkaline ceramidase 2	ACER2	Alkaline ceramidase 2 (ACER2) and its product dihydrosphingosine mediate the cytotoxicity of N-(4-hydroxyphenyl)retinamide in tumor cells. PMID: 20628055
Alkaline ceramidase 3	ACER3	Alkaline ceramidase 3 (ACER3) hydrolyzes unsaturated long-chain ceramides, and its down-regulation inhibits both cell proliferation and apoptosis. PMID: 20068046
Arachidonate lipoxygenase 3	ALOXE3	Disease: ichthyosis, congenital, autosomal recessive 3 (ARCI3)
Arachidonate 5-lipoxygenase	ALOX5	**Zileuton** – drugs for obstructive airway diseases
Arachidonate 5-lipoxygenase-activating protein	ALOX5AP	**Fiboflapon** – drugs for obstructive airway diseases. Disease: ischemic stroke (ISCHSTR)
Arachidonate 12-lipoxygenase	ALOX12	Arachidonate 12-lipoxygenase may serve as a potential marker and therapeutic target for prostate cancer stem cells. PMID: 21225230. Disease: esophageal cancer (ESCR)
Arachidonate 12-lipoxygenase, 12R type	ALOX12B	Disease: ichthyosis, congenital, autosomal recessive 2 (ARCI2)
Arachidonate 15-lipoxygenase	ALOX15	Note: Disease susceptibility may be associated with variations affecting the gene represented in this entry. Met at position 560 may confer interindividual susceptibility to coronary artery disease (CAD) (PMID: 17959182)
Arachidonate 15-lipoxygenase, type B	ALOX15B	Arachidonate 15-lipoxygenase type B knockdown leads to reduced lipid accumulation and inflammation in atherosclerosis. PMID: 22912809
Beta-carotene oxygenase 1	BCO1	Loss of β-carotene 15,15′-oxygenase in developing mouse tissues alters esterification of retinol, cholesterol and diacylglycerols. PMID: 23988655. Disease: hypercarotenemia and vitamin A deficiency, autosomal dominant (ADHVAD)
Beta-carotene oxygenase 2	BCO2	Evidence for compartmentalization of mammalian carotenoid metabolism. PMID: 25002123

HGNC-approved name	Symbol	Target information
Carbonyl reductase 1	CBR1	Carbonyl reductase 1 offers a novel therapeutic target to enhance leukemia treatment by arsenic trioxide. PMID: 22719067
Carbonyl reductase 3	CBR3	Metabolism of doxorubicin to the cardiotoxic metabolite doxorubicinol is increased in a mouse model of chronic glutathione deficiency: A potential role for carbonyl reductase 3. PMID: 25446851
Carbonyl reductase 4	CBR4	Human carbonyl reductases. PMID: 20942781
Carnitine palmitoyltransferase 1A (liver)	CPT1A	**Perhexiline** – cardiac therapy. Disease: carnitine palmitoyltransferase 1A deficiency (CPT1AD)
Carnitine palmitoyltransferase 1B (muscle)	CPT1B	Inhibition of carnitine palymitoyltransferase1b induces cardiac hypertrophy and mortality in mice. PMID: 24330405
Carnitine palmitoyltransferase 1C	CPT1C	Depletion of the novel p53-target gene carnitine palmitoyltransferase 1C delays tumor growth in the neurofibromatosis type I tumor model. PMID: 23412344
Carnitine palmitoyltransferase 2	CPT2	**Perhexiline** – cardiac therapy. Disease: carnitine palmitoyltransferase 2 deficiency late-onset (CPT2D)
CDP-diacylglycerol synthase (phosphatidate cytidylyltransferase) 1	CDS1	Distinct properties of the two isoforms of CDP-diacylglycerol synthase. PMID: 25375833
CDP-diacylglycerol synthase (phosphatidate cytidylyltransferase) 2	CDS2	Distinct properties of the two isoforms of CDP-diacylglycerol synthase. PMID: 25375833
Ceramide kinase	CERK	**NVP-231** – antineoplastic agents
Ceramide kinase-like	CERKL	Disease: retinitis pigmentosa 26 (RP26)
Ceramide synthase 1	CERS1	Impairment of ceramide synthesis causes a novel progressive myoclonus epilepsy. PMID: 24782409
Ceramide synthase 2	CERS2	Reduced ceramide synthase 2 activity causes progressive myoclonic epilepsy. PMID: 25356388
Ceramide synthase 3	CERS3	Disease: ichthyosis, congenital, autosomal recessive 9 (ARCI9)
Ceramide synthase 4	CERS4	Ceramide synthase 4 deficiency in mice causes lipid alterations in sebum and results in alopecia. PMID: 24738593
Ceramide synthase 5	CERS5	LASS5 is the predominant ceramide synthase isoform involved in de novo sphingolipid synthesis in lung epithelia. PMID: 15772421
Ceramide synthase 6	CERS6	CerS2 haploinsufficiency inhibits β-oxidation and confers susceptibility to diet-induced steatohepatitis and insulin resistance. PMID: 25295789
Choline kinase alpha	CHKA	**TCD-717** – antineoplastic agents
Choline kinase beta	CHKB	Choline kinase beta mutant mice exhibit reduced phosphocholine, elevated osteoclast activity and low bone mass. PMID: 25451916
Coenzyme Q2 4-hydroxybenzoate polyprenyltransferase	COQ2	Disease: coenzyme Q10 deficiency, primary, 1 (COQ10D1)
Coenzyme Q3 methyltransferase	COQ3	
Coenzyme Q5 homolog, methyltransferase (Saccharomyces cerevisiae)	COQ5	Molecular characterization of the human COQ5 C-methyltransferase in coenzyme Q10 biosynthesis. PMID: 25152161
Coenzyme Q6 monooxygenase	COQ6	Disease: coenzyme Q10 deficiency, primary, 6 (COQ10D6)
DDHD domain containing 1	DDHD1	Disease: spastic paraplegia 28, autosomal recessive (SPG28)
DDHD domain containing 2	DDHD2	The hereditary spastic paraplegia-related enzyme DDHD2 is a principal brain triglyceride lipase. PMID: 25267624. Disease: spastic paraplegia 54, autosomal recessive (SPG54)

(Continued)

HGNC-approved name	Symbol	Target information
Delta(4)-desaturase, sphingolipid 1	DEGS1	Dihydroceramide desaturase 1, the gatekeeper of ceramide induced lipotoxicity. PMID: 25283058
Delta(4)-desaturase, sphingolipid 2	DEGS2	Dihydroceramide-based response to hypoxia. PMID: 21914808
Diacylglycerol lipase, alpha	DAGLA	A novel fluorophosphonate inhibitor of the biosynthesis of the endocannabinoid 2-arachidonoylglycerol with potential anti-obesity effects. PMID: 23072382. Disease: spinocerebellar ataxia 20 (SCA20)
Diacylglycerol lipase, beta	DAGLB	DAGLβ inhibition perturbs a lipid network involved in macrophage inflammatory responses. PMID: 23103940
Diacylglycerol O-acyltransferase 1	DGAT1	**Pradigastat** – familial chylomicronemia syndrome
Diacylglycerol O-acyltransferase 2	DGAT2	**ISIS-DGAT2Rx** – nonalcoholic steatohepatitis
Diacylglycerol O-acyltransferase 2-like 6	DGAT2L6	
Ectonucleotide pyrophosphatase/phosphodiesterase 2	ENPP2	Serum autotaxin is increased in pruritus of cholestasis, but not of other origin, and responds to therapeutic interventions. PMID: 22473838
Ectonucleotide pyrophosphatase/phosphodiesterase 4 (putative)	ENPP4	NPP4 is a procoagulant enzyme on the surface of vascular endothelium. PMID: 22995898
Ectonucleotide pyrophosphatase/phosphodiesterase 5 (putative)	ENPP5	Cellular function and molecular structure of ecto-nucleotidases. PMID: 22555564
Ectonucleotide pyrophosphatase/phosphodiesterase 6	ENPP6	Fluorescence probe for lysophospholipase C/NPP6 activity and a potent NPP6 inhibitor. PMID: 21721554
Ectonucleotide pyrophosphatase/phosphodiesterase 7	ENPP7	Enhanced colonic tumorigenesis in alkaline sphingomyelinase (NPP7) knockout mice. PMID: 25381265
ELOVL fatty acid elongase 1	ELOVL1	Lorenzo's oil inhibits ELOVL1 and lowers the level of sphingomyelin with a saturated very long-chain fatty acid. PMID: 24489110
ELOVL fatty acid elongase 2	ELOVL2	Elovl2 ablation demonstrates that systemic DHA is endogenously produced and is essential for lipid homeostasis in mice. PMID: 24489111
ELOVL fatty acid elongase 3	ELOVL3	Ablation of the very-long-chain fatty acid elongase ELOVL3 in mice leads to constrained lipid storage and resistance to diet-induced obesity. PMID: 20605947
ELOVL fatty acid elongase 4	ELOVL4	Disease: Stargardt disease 3 (STGD3)
ELOVL fatty acid elongase 5	ELOVL5	Elovl5 regulates the mTORC2-Akt-FOXO1 pathway by controlling hepatic cis-vaccenic acid synthesis in diet-induced obese mice. PMID: 23099444
ELOVL fatty acid elongase 6	ELOVL6	Elovl6 promotes nonalcoholic steatohepatitis. PMID: 22753171
ELOVL fatty acid elongase 7	ELOVL7	Biochemical characterization of the very long-chain fatty acid elongase ELOVL7. PMID: 21959040
Emopamil binding protein (sterol isomerase)	EBP	Novel 4-(4-aryl)cyclohexyl-1-(2-pyridyl)piperazines as Delta(8)-Delta(7) sterol isomerase (emopamil binding protein) selective ligands with antiproliferative activity. PMID: 19053780. Disease: chondrodysplasia punctata 2, X-linked dominant (CDPX2)
Emopamil binding protein-like	EBPL	
Enoyl-CoA hydratase 1, peroxisomal	ECH1	Enoyl coenzyme A hydratase 1 is an important factor in the lymphatic metastasis of tumors. PMID: 21616630
Enoyl-CoA hydratase domain containing 1	ECHDC1	Ethylmalonyl-CoA decarboxylase, a new enzyme involved in metabolite proofreading. PMID: 22016388

HGNC-approved name	Symbol	Target information
Enoyl-CoA hydratase domain containing 2	ECHDC2	Enoyl coenzyme a hydratase domain-containing 2, a potential novel regulator of myocardial ischemia injury. PMID: 24108764
Enoyl-CoA hydratase domain containing 3	ECHDC3	
Enoyl-CoA hydratase, short chain, 1, mitochondrial	ECHS1	ECHS1 mutations cause combined respiratory chain deficiency resulting in Leigh syndrome. PMID: 25393721
Enoyl-CoA delta-isomerase 1	ECI1	Systems virology identifies a mitochondrial fatty acid oxidation enzyme, dodecenoyl coenzyme A delta isomerase, required for hepatitis C virus replication and likely pathogenesis. PMID: 21917952
Enoyl-CoA delta-isomerase 2	ECI2	
Ethanolamine kinase 1	ETNK1	
Ethanolamine kinase 2	ETNK2	Testis development, fertility, and survival in Ethanolamine kinase 2-deficient mice. PMID: 18755794
Fatty acid amide hydrolase	FAAH	**PF-04457845** – analgesics
Fatty acid amide hydrolase 2	FAAH2	Lipid droplets are novel sites of N-acylethanolamine inactivation by fatty acid amide hydrolase-2. PMID: 19926788
Fatty acid desaturase 1	FADS1	Delta-5 and delta-6 desaturases: Crucial enzymes in polyunsaturated fatty acid-related pathways with pleiotropic influences in health and disease. PMID: 25038994
Fatty acid desaturase 2	FADS2	Loss of FADS2 function severely impairs the use of HeLa cells as an *in vitro* model for host response studies involving fatty acid effects. PMID: 25549244
Fatty acid desaturase 3	FADS3	The 51 kDa FADS3 is secreted in the ECM of hepatocytes and blood in rat. PMID: 23966218
Fatty acid desaturase 6	FADS6	
Fatty acyl-CoA reductase 1	FAR1	Topogenesis and homeostasis of fatty acyl-CoA reductase 1. PMID: 24108123
Fatty acyl-CoA reductase 2	FAR2	Genetic analysis of mesangial matrix expansion in aging mice and identification of Far2 as a candidate gene. PMID: 24009241
Glycerol kinase	GK	Disease: glycerol kinase deficiency (GKD)
Glycerol kinase 2	GK2	
Glycerol kinase 5 (putative)	GK5	
Glycerol-3-phosphate acyltransferase, mitochondrial	GPAM	Integration of metabolomics and expression of glycerol-3-phosphate acyltransferase (GPAM) in breast cancer-link to patient survival, hormone receptor status, and metabolic profiling. PMID: 22070544
Glycerol-3-phosphate acyltransferase 2, mitochondrial	GPAT2	Glycerol-3-phosphate acyltranferase-2 behaves as a cancer testis gene and promotes growth and tumorigenicity of the breast cancer MDA-MB-231 cell line. PMID: 24967918
Glycerol-3-phosphate dehydrogenase 1 (soluble)	GPD1	Disease: hypertriglyceridemia, transient infantile (HTGTI)
Glycerol-3-phosphate dehydrogenase 1-like	GPD1L	Disease: Brugada syndrome 2 (BRGDA2)
Glycerol-3-phosphate dehydrogenase 2 (mitochondrial)	GPD2	Metformin suppresses gluconeogenesis by inhibiting mitochondrial glycerophosphate dehydrogenase. PMID: 24847880
Glycerophosphodiester phosphodiesterase 1	GDE1	Anandamide biosynthesis catalyzed by the phosphodiesterase GDE1 and detection of glycerophospho-N-acyl ethanolamine precursors in mouse brain. PMID: 18227059

(Continued)

HGNC-approved name	Symbol	Target information
Glycerophosphodiester phosphodiesterase domain containing 1	GDPD1	New members of the mammalian glycerophosphodiester phosphodiesterase family: GDE4 and GDE7 produce lysophosphatidic acid by lysophospholipase D activity. PMID: 25528375
Glycerophosphodiester phosphodiesterase domain containing 2	GDPD2	The developmentally regulated osteoblast phosphodiesterase GDE3 is glycerophosphoinositol-specific and modulates cell growth. PMID: 19596859
Glycerophosphodiester phosphodiesterase domain containing 3	GDPD3	
Glycerophosphodiester phosphodiesterase domain containing 4	GDPD4	Isolation and characterization of two serpentine membrane proteins containing glycerophosphodiester phosphodiesterase, GDE2 and GDE6. PMID: 15276213
Glycerophosphodiester phosphodiesterase domain containing 5	GDPD5	GDE2 promotes neurogenesis by glycosylphosphatidylinositol-anchor cleavage of RECK. PMID: 23329048
HRAS-like suppressor	HRASLS	New players in the fatty acyl ethanolamide metabolism. PMID: 24747663
HRAS-like suppressor 2	HRASLS2	Characterization of the human tumor suppressors TIG3 and HRASLS2 as phospholipid-metabolizing enzymes. PMID: 19615464
HRAS-like suppressor family, member 5	HRASLS5	Expression of the Ha-ras suppressor family member 5 gene in the maturing rat testis. PMID: 18460797
Hydroxyacid oxidase (glycolate oxidase) 1	HAO1	Exploring glycolate oxidase (GOX) as an antiurolithic drug target: Molecular modeling and in vitro inhibitor study. PMID: 21458484
Hydroxyacid oxidase 2 (long chain)	HAO2	High resolution crystal structure of rat long chain hydroxy acid oxidase in complex with the inhibitor 4-carboxy-5-[(4-chlorophenyl)sulfanyl]-1, 2, 3-thiadiazole. Implications for inhibitor specificity and drug design. PMID: 22342614
Hydroxyacyl-CoA dehydrogenase	HADH	Disease: 3-alpha-hydroxyacyl-CoA dehydrogenase deficiency (HADH deficiency)
Hydroxyacyl-CoA dehydrogenase/3-ketoacyl-CoA thiolase/enoyl-CoA hydratase (trifunctional protein), alpha subunit	HADHA	Disease: trifunctional protein deficiency (TFP deficiency)
Hydroxyacyl-CoA dehydrogenase/3-ketoacyl-CoA thiolase/enoyl-CoA hydratase (trifunctional protein), beta subunit	HADHB	Disease: trifunctional protein deficiency (TFP deficiency)
Hydroxy-delta-5-steroid dehydrogenase, 3 beta- and steroid delta-isomerase 1	HSD3B1	**Trilostane** – antiadrenal preparations
Hydroxy-delta-5-steroid dehydrogenase, 3 beta- and steroid delta-isomerase 2	HSD3B2	**Trilostane** – antiadrenal preparations. Disease: adrenal hyperplasia 2 (AH2)
Hydroxy-delta-5-steroid dehydrogenase, 3 beta- and steroid delta-isomerase 7	HSD3B7	Disease: congenital bile acid synthesis defect 1 (CBAS1)
Hydroxysteroid (11-beta) dehydrogenase 1	HSD11B1	11β-Hydroxysteroid dehydrogenase type 1: Potential therapeutic target for metabolic syndrome. PMID: 23238463. Disease: cortisone reductase deficiency (CRD)
Hydroxysteroid (11-beta) dehydrogenase 1-like	HSD11B1L	
Hydroxysteroid (11-beta) dehydrogenase 2	HSD11B2	Disease: apparent mineralocorticoid excess (AME)
Hydroxysteroid (17-beta) dehydrogenase 1	HSD17B1	Increased expression of 17-beta-hydroxysteroid dehydrogenase type 1 in non-small cell lung cancer. PMID: 25564397
Hydroxysteroid (17-beta) dehydrogenase 2	HSD17B2	Reductive 17beta-hydroxysteroid dehydrogenases which synthesize estradiol and inactivate dihydrotestosterone constitute major and concerted players in ER+ breast cancer cells. PMID: 25257817

HGNC-approved name	Symbol	Target information
Hydroxysteroid (17-beta) dehydrogenase 3	HSD17B3	Disease: male pseudohermaphrodism with gynecomastia (MPH)
Hydroxysteroid (17-beta) dehydrogenase 4	HSD17B4	Disease: D-bifunctional protein deficiency (DBPD)
Hydroxysteroid (17-beta) dehydrogenase 6	HSD17B6	Estrogen receptor β and 17β-hydroxysteroid dehydrogenase type 6, a growth regulatory pathway that is lost in prostate cancer. PMID: 22114194
Hydroxysteroid (17-beta) dehydrogenase 7	HSD17B7	The stimulation of HSD17B7 expression by estradiol provides a powerful feed-forward mechanism for estradiol biosynthesis in breast cancer cells. PMID: 21372145
Hydroxysteroid (17-beta) dehydrogenase 8	HSD17B8	Insights into mitochondrial fatty acid synthesis from the structure of heterotetrameric 3-ketoacyl-ACP reductase/3R-hydroxyacyl-CoA dehydrogenase. PMID: 25203508
Hydroxysteroid (17-beta) dehydrogenase 10	HSD17B10	ABAD: A potential therapeutic target for Abeta-induced mitochondrial dysfunction in Alzheimer's disease. PMID: 19601895. Disease: 2 methyl-3-hydroxybutyryl-CoA dehydrogenase deficiency (MHBD deficiency)
Hydroxysteroid (17-beta) dehydrogenase 11	HSD17B11	17beta-hydroxysteroid dehydrogenase type 11 (Pan1b) expression in human prostate cancer. PMID: 19469652
Hydroxysteroid (17-beta) dehydrogenase 12	HSD17B12	17β Hydroxysteroid dehydrogenase type 12 (HSD17B12) is a marker of poor prognosis in ovarian carcinoma. PMID: 22903146
Hydroxysteroid (17-beta) dehydrogenase 13	HSD17B13	
Hydroxysteroid (17-beta) dehydrogenase 14	HSD17B14	17β-Hydroxysteroid dehydrogenase type 14 is a predictive marker for tamoxifen response in oestrogen receptor positive breast cancer. PMID: 22792371
Isopentenyl-diphosphate delta-isomerase 1	IDI1	
Isopentenyl-diphosphate delta-isomerase 2	IDI2	Covalent modification of reduced flavin mononucleotide in type-2 isopentenyl diphosphate isomerase by active-site-directed inhibitors. PMID: 22158896
Lipase A, lysosomal acid, cholesterol esterase	LIPA	Lysosomal acid lipase deficiency – An under-recognized cause of dyslipidaemia and liver dysfunction. PMID: 24792990. Disease: Wolman disease (WOD)
Lipase B, lysosomal acid	LIPB	Extended use of a selective inhibitor of acid lipase for the diagnosis of Wolman disease and cholesteryl ester storage disease. PMID: 24508470
Lipase, hepatic	LIPC	Disease: hepatic lipase deficiency (HL deficiency)
Lipase, hormone-sensitive	LIPE	Phospholipase C-related catalytically inactive protein (PRIP) regulates lipolysis in adipose tissue by modulating the phosphorylation of hormone-sensitive lipase. PMID: 24945349
Lipase, gastric	LIPF	**Orlistat** – antiobesity preparations
Lipase, endothelial	LIPG	Design and synthesis of boronic acid inhibitors of endothelial lipase. PMID: 22225633
Lipase, member H	LIPH	Disease: hypotrichosis 7 (HYPT7)
Lipase, member I	LIPI	Disease: hypertriglyceridemia, familial (FHTR)
Lipase, family member J	LIPJ	
Lipase, family member K	LIPK	
Lipase, family member M	LIPM	
Lipase, family member N	LIPN	Disease: ichthyosis, congenital, autosomal recessive 8 (ARCI8)

(Continued)

HGNC-approved name	Symbol	Target information
Lipin 1	LPIN1	Disease: myoglobinuria, acute recurrent, autosomal recessive (ARARM)
Lipin 2	LPIN2	Disease: Majeed syndrome (MAJEEDS)
Lipin 3	LPIN3	Lipin-1 and lipin-3 together determine adiposity in vivo. PMID: 24634820
Lysophosphatidylcholine acyltransferase 1	LPCAT1	AAV-mediated lysophosphatidylcholine acyltransferase 1 (Lpcat1) gene replacement therapy rescues retinal degeneration in rd11 mice. PMID: 24557352
Lysophosphatidylcholine acyltransferase 2	LPCAT2	Selective inhibitors of a PAF biosynthetic enzyme lysophosphatidylcholine acyltransferase 2. PMID: 24850807
Lysophosphatidylcholine acyltransferase 3	LPCAT3	Lysophosphatidylcholine acyltransferase 3 knockdown-mediated liver lysophosphatidylcholine accumulation promotes very low density lipoprotein production by enhancing microsomal triglyceride transfer protein expression. PMID: 22511767
Lysophosphatidylcholine acyltransferase 4	LPCAT4	Accumulated phosphatidylcholine (16:0/16:1) in human colorectal cancer; possible involvement of LPCAT4. PMID: 23815430
Lysophospholipase I	LYPLA1	An in vivo active carbamate-based dual inhibitor of lysophospholipase 1 (LYPLA1) and lysophospholipase 2 (LYPLA2). PMID: 25506974
Lysophospholipase II	LYPLA2	Identification of the major prostaglandin glycerol ester hydrolase in human cancer cells. PMID: 25301951
Lysophospholipase-like 1	LYPLAL1	
Membrane-bound O-acyltransferase domain containing 1	MBOAT1	
Membrane-bound O-acyltransferase domain containing 2	MBOAT2	Combined gene expression analysis of whole-tissue and microdissected pancreatic ductal adenocarcinoma identifies genes specifically overexpressed in tumor epithelia. PMID: 19260470
Membrane-bound O-acyltransferase domain containing 4	MBOAT4	A small molecule antagonist of ghrelin O-acyltransferase (GOAT). PMID: 21594284
Membrane-bound O-acyltransferase domain containing 7	MBOAT7	LPIAT1 regulates arachidonic acid content in phosphatidylinositol and is required for cortical lamination in mice. PMID: 23097495
Monoacylglycerol O-acyltransferase 1	MOGAT1	Hepatic monoacylglycerol O-acyltransferase 1 as a promising therapeutic target for steatosis, obesity, and type 2 diabetes. PMID: 24643205
Monoacylglycerol O-acyltransferase 2	MOGAT2	Intestine-specific deletion of acyl-CoA:monoacylglycerol acyltransferase (MGAT) 2 protects mice from diet-induced obesity and glucose intolerance. PMID: 24784138
Monoacylglycerol O-acyltransferase 3	MOGAT3	MGAT3 mRNA: A biomarker for prognosis and therapy of Alzheimer's disease by vitamin D and curcuminoids. PMID: 21368380
N-Acylsphingosine amidohydrolase (acid ceramidase) 1	ASAH1	Acid ceramidase as a therapeutic target in metastatic prostate cancer. PMID: 23423838. Disease: Farber lipogranulomatosis (FL)
N-Acylsphingosine amidohydrolase (non-lysosomal ceramidase) 2	ASAH2	Involvement of neutral sphingomyelinase in the angiotensin II signaling pathway. PMID: 25354938
N-Acylsphingosine amidohydrolase (non-lysosomal ceramidase) 2B	ASAH2B	A novel gene derived from a segmental duplication shows perturbed expression in Alzheimer's disease. PMID: 17334805
N-Acylsphingosine amidohydrolase (non-lysosomal ceramidase) 2C	ASAH2C	

HGNC-approved name	Symbol	Target information
Pantothenate kinase 1	PANK1	Passenger deletions generate therapeutic vulnerabilities in cancer. PMID: 22895339
Pantothenate kinase 2	PANK2	Disease: neurodegeneration with brain iron accumulation 1 (NBIA1)
Pantothenate kinase 3	PANK3	Modulation of pantothenate kinase 3 activity by small molecules that interact with the substrate/allosteric regulatory domain. PMID: 20797618
Pantothenate kinase 4	PANK4	PanK4 inhibits pancreatic beta-cell apoptosis by decreasing the transcriptional level of pro-caspase-9. PMID: 17971806
Patatin-like phospholipase domain containing 1	PNPLA1	Disease: ichthyosis, congenital, autosomal recessive 10 (ARCI10)
Patatin-like phospholipase domain containing 2	PNPLA2	Adipose triglyceride lipase contributes to cancer-associated cachexia. PMID: 21680814. Note: Genetic variations in PNPLA2 may be associated with risk of diabetes mellitus type 2. Disease: neutral lipid storage disease with myopathy (NLSDM)
Patatin-like phospholipase domain containing 3	PNPLA3	Disease: non-alcoholic fatty liver disease 1 (NAFLD1)
Patatin-like phospholipase domain containing 4	PNPLA4	
Patatin-like phospholipase domain containing 5	PNPLA5	Neutral lipid stores and lipase PNPLA5 contribute to autophagosome biogenesis. PMID: 24613307
Patatin-like phospholipase domain containing 6	PNPLA6	Disease: spastic paraplegia 39, autosomal recessive (SPG39)
Patatin-like phospholipase domain containing 7	PNPLA7	Neuropathy target esterase (NTE): Overview and future. PMID: 23220002
Patatin-like phospholipase domain containing 8	PNPLA8	Loss of function variants in human PNPLA8 encoding calcium independent phospholipase A2 γ recapitulate the mitochondriopathy of the homologous null mouse. PMID: 25512002
Phosphatase, orphan 1	PHOSPHO1	Design, synthesis and evaluation of benzoisothiazolones as selective inhibitors of PHOSPHO1. PMID: 25124115
Phosphatase, orphan 2	PHOSPHO2	Probing the substrate specificities of human PHOSPHO1 and PHOSPHO2. PMID: 16054448
Phosphate cytidylyltransferase 1, choline, alpha	PCYT1A	Mutations in PCYT1A cause spondylometaphyseal dysplasia with cone-rod dystrophy. PMID: 24387991
Phosphate cytidylyltransferase 1, choline, beta	PCYT1B	Mass spectrometric identification of human phosphate cytidylyltransferase 1 as a novel calcium oxalate crystal growth inhibitor purified from human renal stone matrix. PMID: 19595683
Phosphate cytidylyltransferase 2, ethanolamine	PCYT2	Meclizine inhibits mitochondrial respiration through direct targeting of cytosolic phosphoethanolamine metabolism. PMID: 24142790
Phosphatidylinositol glycan anchor biosynthesis, class A	PIGA	Disease: paroxysmal nocturnal hemoglobinuria 1 (PNH1)
Phosphatidylinositol glycan anchor biosynthesis, class B	PIGB	Discovery of genetic biomarkers contributing to variation in drug response of cytidine analogues using human lymphoblastoid cell lines. PMID: 24483146
Phosphatidylinositol glycan anchor biosynthesis, class C	PIGC	
Phosphatidylinositol glycan anchor biosynthesis, class F	PIGF	

(Continued)

HGNC-approved name	Symbol	Target information
Phosphatidylinositol glycan anchor biosynthesis, class G	PIGG	
Phosphatidylinositol glycan anchor biosynthesis, class H	PIGH	
Phosphatidylinositol glycan anchor biosynthesis, class L	PIGL	Disease: coloboma, congenital heart disease, ichthyosiform dermatosis, mental retardation and ear anomalies syndrome (CHIME)
Phosphatidylinositol glycan anchor biosynthesis, class M	PIGM	Disease: glycosylphosphatidylinositol deficiency (GPID)
Phosphatidylinositol glycan anchor biosynthesis, class N	PIGN	Disease: multiple congenital anomalies–hypotonia–seizures syndrome 1 (MCAHS1)
Phosphatidylinositol glycan anchor biosynthesis, class O	PIGO	Disease: hyperphosphatasia with mental retardation syndrome 2 (HPMRS2)
Phosphatidylinositol glycan anchor biosynthesis, class P	PIGP	The histone methyltransferase Wbp7 controls macrophage function through GPI glycolipid anchor synthesis. PMID: 22483804
Phosphatidylinositol glycan anchor biosynthesis, class Q	PIGQ	Clinical whole-genome sequencing in severe early-onset epilepsy reveals new genes and improves molecular diagnosis. PMID: 24463883
Phosphatidylinositol glycan anchor biosynthesis, class V	PIGV	Disease: hyperphosphatasia with mental retardation syndrome 1 (HPMRS1)
Phosphatidylinositol glycan anchor biosynthesis, class W	PIGW	Glycosylphosphatidylinositol (GPI) anchor deficiency caused by mutations in PIGW is associated with West syndrome and hyperphosphatasia with mental retardation syndrome. PMID: 24367057
Phosphatidylinositol glycan anchor biosynthesis, class X	PIGX	
Phosphatidylinositol glycan anchor biosynthesis, class Y	PIGY	Silencing of genes required for glycosylphosphatidylinositol anchor biosynthesis in Burkitt lymphoma. PMID: 19302917
Phosphatidylinositol glycan anchor biosynthesis, class Z	PIGZ	Human Smp3p adds a fourth mannose to yeast and human glycosylphosphatidylinositol precursors in vivo. PMID: 15208306
Phosphatidylserine synthase 1	PTDSS1	Defining the importance of phosphatidylserine synthase-1 (PSS1): Unexpected viability of PSS1-deficient mice. PMID: 18343815. Disease: Lenz–Majewski hyperostotic dwarfism (LMHD)
Phosphatidylserine synthase 2	PTDSS2	Phosphatidylserine synthase 2: High efficiency for synthesizing phosphatidylserine containing docosahexaenoic acid. PMID: 23071296
Phytanoyl-CoA 2-hydroxylase	PHYH	Disease: Refsum disease (RD)
Phytanoyl-CoA dioxygenase domain containing 1	PHYHD1	
Platelet-activating factor acetylhydrolase 1b, catalytic subunit 2 (30 kDa)	PAFAH1B2	Loss of PAFAH1B2 reduces amyloid-β generation by promoting the degradation of amyloid precursor protein C-terminal fragments. PMID: 23238734
Platelet-activating factor acetylhydrolase 1b, catalytic subunit 3 (29 kDa)	PAFAH1B3	Metabolic profiling reveals PAFAH1B3 as a critical driver of breast cancer pathogenicity. PMID: 24954006
Platelet-activating factor acetylhydrolase 2 (40 kDa)	PAFAH2	Click-generated triazole ureas as ultrapotent in vivo-active serine hydrolase inhibitors. PMID: 21572424
Prenyl (decaprenyl) diphosphate synthase, subunit 1	PDSS1	Disease: coenzyme Q10 deficiency, primary, 2 (COQ10D2)
Prenyl (decaprenyl) diphosphate synthase, subunit 2	PDSS2	Clinical utility of PDSS2 expression to stratify patients at risk for recurrence of hepatocellular carcinoma. PMID: 25189544. Disease: coenzyme Q10 deficiency, primary, 3 (COQ10D3)

HGNC-approved name	Symbol	Target information
Propionyl CoA carboxylase, alpha polypeptide	PCCA	Disease: propionic acidemia type I (PA-1)
Propionyl CoA carboxylase, beta polypeptide	PCCB	Feasibility of nonsense mutation readthrough as a novel therapeutical approach in propionic acidemia. PMID: 22334403. Disease: propionic acidemia type II (PA-2)
Prostaglandin D2 synthase 21 kDa (brain)	PTGDS	Does prostaglandin D2 hold the cure to male pattern baldness? PMID: 24521203
Prostaglandin E synthase	PTGES	**LY3023703** – analgesics
Prostaglandin E synthase 2	PTGES2	
Prostaglandin E synthase 3 (cytosolic)	PTGES3	
Prostaglandin E synthase 3 (cytosolic) like	PTGES3L	
Prostaglandin I2 (prostacyclin) synthase	PTGIS	**Phenylbutazone** – anti-inflammatory and antirheumatic products
Prostaglandin reductase 1	PTGR1	An enzyme that inactivates the inflammatory mediator leukotriene b4 restricts mycobacterial infection. PMID: 23874453
Prostaglandin reductase 2	PTGR2	Prostaglandin reductase 2 modulates ROS-mediated cell death and tumor transformation of gastric cancer cells and is associated with higher mortality in gastric cancer patients. PMID: 22998775
Prostaglandin-endoperoxide synthase 1 (prostaglandin G/H synthase and cyclooxygenase)	PTGS1	**Diclofenac** – anti-inflammatory and antirheumatic products
Prostaglandin-endoperoxide synthase 2 (prostaglandin G/H synthase and cyclooxygenase)	PTGS2	**Etoricoxib** – anti-inflammatory and antirheumatic products
Retinol dehydrogenase 5 (11-cis/9-cis)	RDH5	Disease: retinitis punctata albescens (RPA)
Retinol dehydrogenase 8 (all-trans)	RDH8	Retinol dehydrogenases (RDHs) in the visual cycle. PMID: 20801113
Retinol dehydrogenase 10 (all-trans)	RDH10	The retinaldehyde reductase activity of DHRS3 is reciprocally activated by retinol dehydrogenase 10 to control retinoid homeostasis. PMID: 24733397
Retinol dehydrogenase 11 (all-trans/9-cis/11-cis)	RDH11	New syndrome with retinitis pigmentosa is caused by nonsense mutations in retinol dehydrogenase RDH11. PMID: 24916380
Retinol dehydrogenase 12 (all-trans/9-cis/11-cis)	RDH12	Disease: Leber congenital amaurosis 13 (LCA13)
Retinol dehydrogenase 13 (all-trans/9-cis)	RDH13	Retinol dehydrogenase 13 protects the mouse retina from acute light damage. PMID: 22605914
Retinol dehydrogenase 14 (all-trans/9-cis/11-cis)	RDH14	Human pancreas protein 2 (PAN2) has retinal reductase activity and is ubiquitously expressed in human tissues. PMID: 12435598
Retinol dehydrogenase 16 (all-trans)	RDH16	13-cis-Retinoic acid competitively inhibits 3 alpha-hydroxysteroid oxidation by retinol dehydrogenase RoDH-4: A mechanism for its anti-androgenic effects in sebaceous glands? PMID: 12646198
Serine palmitoyltransferase, long-chain base subunit 1	SPTLC1	Disease: neuropathy, hereditary sensory and autonomic, 1A (HSAN1A)
Serine palmitoyltransferase, long-chain base subunit 2	SPTLC2	Disease: neuropathy, hereditary sensory and autonomic, 1C (HSAN1C)
Serine palmitoyltransferase, long-chain base subunit 3	SPTLC3	The SPTLC3 subunit of serine palmitoyltransferase generates short chain sphingoid bases. PMID: 19648650
Serine palmitoyltransferase, small subunit A	SPTSSA	Identification of small subunit of serine palmitoyltransferase a as a lysophosphatidylinositol acyltransferase 1-interacting protein. PMID: 23510452

(Continued)

HGNC-approved name	Symbol	Target information
Serine palmitoyltransferase, small subunit B	SPTSSB	Molecular cloning and characterization of a novel androgen repressible gene expressed in the prostate epithelium. PMID: 15777716
Short-chain dehydrogenase/reductase family 9C, member 7	SDR9C7	SDR9C7 promotes lymph node metastases in patients with esophageal squamous cell carcinoma. PMID: 23341893
Short-chain dehydrogenase/reductase family 42E, member 1	SDR42E1	
Short-chain dehydrogenase/reductase family 42E, member 2	SDR42E2	
Sphingomyelin phosphodiesterase 1, acid lysosomal	SMPD1	Transformation-associated changes in sphingolipid metabolism sensitize cells to lysosomal cell death induced by inhibitors of acid sphingomyelinase. PMID: 24029234. Disease: Niemann–Pick disease A (NPDA)
Sphingomyelin phosphodiesterase 2, neutral membrane (neutral sphingomyelinase)	SMPD2	Stress-induced ceramide generation and apoptosis via the phosphorylation and activation of nSMase1 by JNK signaling. PMID: 25168245
Sphingomyelin phosphodiesterase 3, neutral membrane (neutral sphingomyelinase II)	SMPD3	Roles and regulation of neutral sphingomyelinase-2 in cellular and pathological processes. PMID: 25465297
Sphingomyelin phosphodiesterase 4, neutral membrane (neutral sphingomyelinase-3)	SMPD4	Neutral sphingomyelinase-3 mediates TNF-stimulated oxidant activity in skeletal muscle. PMID: 25180167
Sphingomyelin phosphodiesterase, acid-like 3A	SMPDL3A	Sphingomyelin phosphodiesterase acid-like 3A (SMPDL3A) is a novel nucleotide phosphodiesterase regulated by cholesterol in human macrophages. PMID: 25288789
Sphingomyelin phosphodiesterase, acid-like 3B	SMPDL3B	Sphingomyelinase-like phosphodiesterase 3b expression levels determine podocyte injury phenotypes in glomerular disease. PMID: 24925721
Sphingomyelin synthase 1	SGMS1	Identification of small molecule sphingomyelin synthase inhibitors. PMID: 24374347
Sphingomyelin synthase 2	SGMS2	Sphingomyelin synthase 2, but not sphingomyelin synthase 1, is involved in HIV-1 envelope-mediated membrane fusion. PMID: 25231990
Sphingosine kinase 1	SPHK1	When the sphingosine kinase 1/sphingosine 1-phosphate pathway meets hypoxia signaling: New targets for cancer therapy. PMID: 19383898
Sphingosine kinase 2	SPHK2	Sphingosine kinase 2 promotes acute lymphoblastic leukemia by enhancing MYC expression. PMID: 24686171
Sphingosine-1-phosphate phosphatase 1	SGPP1	miRNA-95 mediates radioresistance in tumors by targeting the sphingolipid phosphatase SGPP1. PMID: 24145350
Sphingosine-1-phosphate phosphatase 2	SGPP2	Sphingosine 1-phosphate phosphatase 2 is induced during inflammatory responses. PMID: 17113265
Stearoyl-CoA desaturase (delta-9-desaturase)	SCD	Stearoyl-CoA desaturase 1 is a novel molecular therapeutic target for clear cell renal cell carcinoma. PMID: 23633458
Stearoyl-CoA desaturase 5	SCD5	StearoylCoA desaturase-5: A novel regulator of neuronal cell proliferation and differentiation. PMID: 22745828
Steroid-5-alpha-reductase, alpha polypeptide 1 (3-oxo-5 alpha-steroid delta 4-dehydrogenase alpha 1)	SRD5A1	**Finasteride** – anti-alopecia agents
Steroid-5-alpha-reductase, alpha polypeptide 2 (3-oxo-5 alpha-steroid delta 4-dehydrogenase alpha 2)	SRD5A2	Disease: pseudovaginal perineoscrotal hypospadias (PPSH)
Steroid 5 alpha-reductase 3	SRD5A3	Disease: congenital disorder of glycosylation 1Q (CDG1Q)

HGNC-approved name	Symbol	Target information
Sterol *O*-acyltransferase 1	SOAT1	**Ezetimibe** – lipid-modifying agents
Sterol *O*-acyltransferase 2	SOAT2	Deletion of sterol *O*-acyltransferase 2 (SOAT2) function in mice deficient in lysosomal acid lipase (LAL) dramatically reduces esterified cholesterol sequestration in the small intestine and liver. PMID: 25450374
Thioesterase superfamily member 4	THEM4	Small interfering RNA directed against CTMP reduces acute traumatic brain injury in a mouse model by activating Akt. PMID: 24670215
Thioesterase superfamily member 5	THEM5	Acyl coenzyme A thioesterase Them5/Acot15 is involved in cardiolipin remodeling and fatty liver development. PMID: 22586271
Thioesterase superfamily member 6	THEM6	
Trans-2,3-enoyl-CoA reductase	TECR	Dual functions of the trans-2-enoyl-CoA reductase TER in the sphingosine 1 phosphate metabolic pathway and in fatty acid elongation. PMID: 25049234. Disease: mental retardation, autosomal recessive 14 (MRT14)
Trans-2,3-enoyl-CoA reductase-like	TECRL	
2-Hydroxyacyl-CoA lyase 1	HACL1	Thiamine and magnesium deficiencies: Keys to disease. PMID: 25542071
3-Hydroxy-3-methylglutaryl-CoA reductase	HMGCR	**Atorvastatin** – lipid-modifying agents
3-Ketodihydrosphingosine reductase	KDSR	Note: A chromosomal aberration involving KDSR is a cause of follicular lymphoma also known as type II chronic lymphatic leukemia. Translocation t(2;18)(p11;q21) with a Ig 1 kappa chain region
3-Oxoacyl-ACP synthase, mitochondrial	OXSM	
Acetoacetyl-CoA synthetase	AACS	Acetoacetyl-CoA synthetase, a ketone body-utilizing enzyme, is controlled by SREBP-2 and affects serum cholesterol levels. PMID: 22985732
Acid phosphatase 6, lysophosphatidic	ACP6	Expression of ACP6 is an independent prognostic factor for poor survival in patients with esophageal squamous cell carcinoma. PMID: 16685394
Acyloxyacyl hydrolase (neutrophil)	AOAH	Altered inactivation of commensal LPS due to acyloxyacyl hydrolase deficiency in colonic dendritic cells impairs mucosal Th17 immunity. PMID: 24344308
Alkylglycerol monooxygenase	AGMO	Alkylglycerol monooxygenase as a potential modulator for PAF synthesis in macrophages. PMID: 23743196
Alkylglycerone phosphate synthase	AGPS	Role and mechanism of the alkylglycerone phosphate synthase in suppressing the invasion potential of human glioma and hepatic carcinoma cells *in vitro*. PMID: 24841318. Disease: rhizomelic chondrodysplasia punctata 3 (RCDP3)
Alpha-methylacyl-CoA racemase	AMACR	AMACR overexpression as a poor prognostic factor in patients with nasopharyngeal carcinoma. PMID: 24833092. Disease: alpha-methylacyl-CoA racemase deficiency (AMACRD)
Aminoadipate-semialdehyde dehydrogenase-phosphopantetheinyl transferase	AASDHPPT	Identifying novel autoantibody signatures in ovarian cancer using high-density protein microarrays. PMID: 19094976
Bile acid CoA: amino acid *N*-acyltransferase (glycine-*N*-choloyltransferase)	BAAT	Disease: familial hypercholanemia (FHCA)
Carboxyl ester lipase	CEL	Bile salt-stimulated lipase plays an unexpected role in arthritis development in rodents. PMID: 23071697. Disease: maturity-onset diabetes of the young 8 with exocrine dysfunction (MODY8)

(Continued)

HGNC-approved name	Symbol	Target information
Cardiolipin synthase 1	CRLS1	E3 ligase subunit Fbxo15 and PINK1 kinase regulate cardiolipin synthase 1 stability and mitochondrial function in pneumonia. PMID: 24703837
Carnitine *O*-acetyltransferase	CRAT	Obesity and lipid stress inhibit carnitine acetyltransferase activity. PMID: 24395925
Carnitine *O*-octanoyltransferase	CROT	Changes in carnitine octanoyltransferase activity induce alteration in fatty acid metabolism. PMID: 21619872
CDP-diacylglycerol–inositol 3-phosphatidyltransferase	CDIPT	Clinical significance of phosphatidyl inositol synthase overexpression in oral cancer. PMID: 20426864
Cholesterol 25-hydroxylase	CH25H	Inflammation. 25-Hydroxycholesterol suppresses interleukin-1-driven inflammation downstream of type I interferon. PMID: 25104388
Choline/ethanolamine phosphotransferase 1	CEPT1	PC and PE synthesis: Mixed micellar analysis of the cholinephosphotransferase and ethanolaminephosphotransferase activities of human choline/ethanolamine phosphotransferase 1 (CEPT1). PMID: 12216837
CoA synthase	COASY	Disease: neurodegeneration with brain iron accumulation 6 (NBIA6)
Choline *O*-acetyltransferase	CHAT	Disease: myasthenic syndrome, congenital, associated with episodic apnea (CMSEA)
Choline phosphotransferase 1	CHPT1	Lipid catabolism via CPT1 as a therapeutic target for prostate cancer. PMID: 25122071
Crystallin, lambda 1	CRYL1	Reduced CRYL1 expression in hepatocellular carcinoma confers cell growth advantages and correlates with adverse patient prognosis. PMID: 19927314
Dephospho-CoA kinase domain containing	DCAKD	
Enoyl-CoA, hydratase/3-hydroxyacyl-CoA dehydrogenase	EHHADH	Disease: Fanconi renotubular syndrome 3 (FRTS3)
Ethanolaminephosphotransferase 1 (CDP-ethanolamine-specific)	EPT1	Identification and characterization of human ethanolaminephosphotransferase1. PMID: 17132865
Family with sequence similarity 213, member B	FAM213B	Preferential localization of prostamide/prostaglandin F synthase in myelin sheaths of the central nervous system. PMID: 20950588
Farnesyl-diphosphate synthase	FDPS	**Alendronate** – drugs affecting bone structure and mineralization
Farnesyl-diphosphate farnesyltransferase 1	FDFT1	Squalene synthase inhibitors: Clinical pharmacology and cholesterol-lowering potential. PMID: 17209661
Fatty acid 2-hydroxylase	FA2H	Disease: spastic paraplegia 35, autosomal recessive (SPG35)
Fatty acid hydroxylase domain containing 2	FAXDC2	
Fatty acid synthase	FASN	**Orlistat** – antiobesity preparations
Ferredoxin reductase	FDXR	Both human ferredoxins 1 and 2 and ferredoxin reductase are important for iron-sulfur cluster biogenesis. PMID: 22101253
Geranylgeranyl diphosphate synthase 1	GGPS1	**Zoledronate** – drugs affecting bone structure and mineralization
Glutaryl-CoA dehydrogenase	GCDH	Disease: glutaric aciduria 1 (GA1)
Glyceronephosphate *O*-acyltransferase	GNPAT	Disease: rhizomelic chondrodysplasia punctata 2 (RCDP2)
Glycerophosphocholine phosphodiesterase GDE1 homolog (*S. cerevisiae*)	GPCPD1	
Hematopoietic prostaglandin D synthase	HPGDS	Discovery of an oral potent selective inhibitor of hematopoietic prostaglandin D synthase (HPGDS). PMID: 24900177
Histidine triad nucleotide binding protein 2	HINT2	Disruption of the histidine triad nucleotide-binding hint2 gene in mice affects glycemic control and mitochondrial function. PMID: 22961760

HGNC-approved name	Symbol	Target information
Hydroxyprostaglandin dehydrogenase 15-(NAD)	HPGD	Structure-activity relationship studies and biological characterization of human NAD(+)-dependent 15-hydroxyprostaglandin dehydrogenase inhibitors. PMID: 24360556. Disease: hypertrophic osteoarthropathy, primary, autosomal recessive, 1 (PHOAR1)
Isoamyl acetate-hydrolyzing esterase 1 homolog (*S. cerevisiae*)	IAH1	
Isoprenoid synthase domain containing	ISPD	Disease: muscular dystrophy–dystroglycanopathy congenital with brain and eye anomalies A7 (MDDGA7)
Lanosterol synthase (2,3-oxidosqualene–lanosterol cyclase)	LSS	Sustained and selective suppression of intestinal cholesterol synthesis by Ro 48-8071, an inhibitor of 2,3-oxidosqualene:lanosterol cyclase, in the BALB/c mouse. PMID: 24486573
Lecithin retinol acyltransferase (phosphatidylcholine–retinol *O*-acyltransferase)	LRAT	Disease: Leber congenital amaurosis 14 (LCA14)
Lecithin–cholesterol acyltransferase	LCAT	**ACP-501** – lipid-modifying agents. Disease: lecithin–cholesterol acyltransferase deficiency (LCATD)
Leukotriene C4 synthase	LTC4S	Crystal structures of leukotriene C4 synthase in complex with product analogs: Implications for the enzyme mechanism. PMID: 24366866. Note: LTC4 synthase deficiency is associated with a neurometabolic developmental disorder
Lipoprotein lipase	LPL	Design and synthesis of boronic acid inhibitors of endothelial lipase. PMID: 22225633. Disease: lipoprotein lipase deficiency (LPL deficiency)
Lysocardiolipin acyltransferase 1	LCLAT1	The mitochondrial cardiolipin remodeling enzyme lysocardiolipin acyltransferase is a novel target in pulmonary fibrosis. PMID: 24779708
Malonyl CoA:ACP acyltransferase (mitochondrial)	MCAT	
Malonyl-CoA decarboxylase	MLYCD	Malonyl-CoA decarboxylase inhibition is selectively cytotoxic to human breast cancer cells. PMID: 19543323. Disease: malonyl-CoA decarboxylase deficiency (MLYCD deficiency)
Metallophosphoesterase 1	MPPE1	Association between polymorphisms in the metallophosphoesterase (MPPE1) gene and bipolar disorder. PMID: 19859903
Methylmalonyl CoA epimerase	MCEE	Disease: methylmalonyl-CoA epimerase deficiency (MCEED)
Methylmalonyl CoA mutase	MUT	Disease: methylmalonic aciduria type mut (MMAM)
Methylsterol monooxygenase 1	MSMO1	Modulation of ERG25 expression by LDL in vascular cells. PMID: 12667060
Methyltransferase like 7B	METTL7B	Identification and characterization of associated with lipid droplet protein 1: A novel membrane-associated protein that resides on hepatic lipid droplets. PMID: 17004324
Mevalonate (diphospho) decarboxylase	MVD	
Mevalonate kinase	MVK	Disease: mevalonic aciduria (MEVA)
Mitochondrial trans-2-enoyl-CoA reductase	MECR	Myocardial overexpression of Mecr, a gene of mitochondrial FAS II leads to cardiac dysfunction in mouse. PMID: 19440339
Monoglyceride lipase	MGLL	Therapeutic potential of monoacylglycerol lipase inhibitors. PMID: 23142242
N-Acetyltransferase 8-like (GCN5-related, putative)	NAT8L	NAT8L (*N*-acetyltransferase 8-like) accelerates lipid turnover and increases energy expenditure in brown adipocytes. PMID: 24155240. Disease: *N*-acetylaspartate deficiency (NACED)
N-Acylethanolamine acid amidase	NAAA	Synthesis, biological evaluation, and 3D QSAR study of 2-methyl-4-oxo-3-oxetanylcarbamic acid esters as *N*-acylethanolamine acid amidase (NAAA) inhibitors. PMID: 25380517

(Continued)

HGNC-approved name	Symbol	Target information
NAD(P-dependent steroid dehydrogenase-like	NSDHL	Disease: congenital hemidysplasia with ichthyosiform erythroderma and limb defects (CHILD)
Neutral cholesterol ester hydrolase 1	NCEH1	Synthesis and structure-activity relationship of (1-halo-2-naphthyl) carbamate-based inhibitors of KIAA1363 (NCEH1/ AADACL1). PMID: 22877630
Nuclear undecaprenyl pyrophosphate synthase 1 homolog (*S. cerevisiae*)	NUS1	Expression of NgBR is highly associated with estrogen receptor alpha and survivin in breast cancer. PMID: 24223763
Oleoyl-ACP hydrolase	OLAH	
Pancreatic lipase	PNLIP	**Orlistat** – antiobesity preparations
Peroxisomal trans-2-enoyl-CoA reductase	PECR	
Phosphatidylethanolamine *N*-methyltransferase	PEMT	Vagus nerve contributes to the development of steatohepatitis and obesity in phosphatidylethanolamine *N*-methyltransferase deficient mice. PMID: 25433161
Phosphatidylglycerophosphate synthase 1	PGS1	
Phosphatidylserine decarboxylase	PISD	
Phosphomevalonate kinase	PMVK	Molecular docking and NMR binding studies to identify novel inhibitors of human phosphomevalonate kinase. PMID: 23146631
Protein tyrosine phosphatase-like (proline instead of catalytic arginine), member A	PTPLA	Human recombinant cementum attachment protein (hrPTPLa/CAP) promotes hydroxyapatite crystal formation *in vitro* and bone healing in vivo. PMID: 25263524
Retinol saturase (all-trans-retinol 13,14-reductase)	RETSAT	Retinol saturase promotes adipogenesis and is downregulated in obesity. PMID: 19139408
Serine active site containing 1	SERAC1	Disease: 3-methylglutaconic aciduria with deafness, encephalopathy, and Leigh-like syndrome (MEGDEL)
Sphingosine-1-phosphate lyase 1	SGPL1	**LX3305** – anti-inflammatory and antirheumatic products
Squalene epoxidase	SQLE	Squalene epoxidase, located on chromosome 8q24.1, is upregulated in 8q+ breast cancer and indicates poor clinical outcome in stage I and II disease. PMID: 18728668
Sterile alpha motif domain containing 8	SAMD8	Sphingomyelin synthase-related protein SMSr is a suppressor of ceramide-induced mitochondrial apoptosis. PMID: 24259670
Steroid sulfatase (microsomal), isozyme S	STS	Steroid sulfatase inhibitors for estrogen- and androgen-dependent cancers. PMID: 21859802. Disease: ichthyosis, X-linked (IXL)
Sterol carrier protein 2	SCP2	Disease: leukoencephalopathy, with dystonia and motor neuropathy (LDMN)
Sterol-C5-desaturase	SC5D	Disease: lathosterolosis (LATHST)
Tafazzin	TAZ	Disease: Barth syndrome (BTHS)
Thromboxane A synthase 1 (platelet)	TBXAS1	**Ridogrel** – antithrombotic agents. Disease: Ghosal hematodiaphyseal dysplasia (GHDD)
Transmembrane 7 superfamily member 2	TM7SF2	A novel role for Tm7sf2 gene in regulating TNFα expression. PMID: 23935851
Transmembrane protein 86B	TMEM86B	Purification, identification, and cloning of lysoplasmalogenase, the enzyme that catalyzes hydrolysis of the vinyl ether bond of lysoplasmalogen. PMID: 21515882
UDP-glucose ceramide glucosyltransferase	UGCG	**Miglustat** – Gaucher's disease

5.2 Amino acids and related

HGNC-approved name	Symbol	Target information
4-Hydroxyphenylpyruvate dioxygenase	HPD	**Nitisinone** – antityrosinemia agents. Disease: tyrosinemia 3 (TYRSN3)
4-Hydroxyphenylpyruvate dioxygenase-like	HPDL	
5-Methyltetrahydrofolate-homocysteine methyltransferase	MTR	Disease: homocystinuria–megaloblastic anemia, cblG complementation type (HMAG)
5-Methyltetrahydrofolate-homocysteine methyltransferase reductase	MTRR	Disease: homocystinuria–megaloblastic anemia, cblE complementation type (HMAE)
Acetylserotonin O-methyltransferase	ASMT	Molecular docking studies for the identification of novel melatoninergic inhibitors for acetylserotonin-O-methyltransferase using different docking routines. PMID: 24156411
Acetylserotonin O-methyltransferase-like	ASMTL	
Adenosylhomocysteinase	AHCY	Disease: hypermethioninemia with S-adenosylhomocysteine hydrolase deficiency (HMAHCHD)
Adenosylhomocysteinase-like 1	AHCYL1	Enzyme regulation. IRBIT is a novel regulator of ribonucleotide reductase in higher eukaryotes. PMID: 25237103
Adenosylhomocysteinase-like 2	AHCYL2	AHCYL2 (long-IRBIT) as a potential regulator of the electrogenic Na(+)-HCO3(−) cotransporter NBCe1-B. PMID: 24472682
Alanine–glyoxylate aminotransferase	AGXT	Disease: hyperoxaluria primary 1 (HP1)
Alanine–glyoxylate aminotransferase 2	AGXT2	AGXT2: A promiscuous aminotransferase. PMID: 25294000
Aldehyde dehydrogenase 4 family, member A1	ALDH4A1	Disease: hyperprolinemia 2 (HP-2)
Aldehyde dehydrogenase 5 family, member A1	ALDH5A1	RNA-Seq of human breast ductal carcinoma in situ models reveals aldehyde dehydrogenase isoform 5A1 as a novel potential target. PMID: 23236365. Disease: succinic semialdehyde dehydrogenase deficiency (SSADHD)
Aldehyde dehydrogenase 6 family, member A1	ALDH6A1	Disease: methylmalonate semialdehyde dehydrogenase deficiency (MMSDHD)
Aldehyde dehydrogenase 9 family, member A1	ALDH9A1	Identification of ALDH4 as a p53-inducible gene and its protective role in cellular stresses. PMID: 14986171
Aldehyde dehydrogenase 18 family, member A1	ALDH18A1	Disease: cutis laxa, autosomal recessive, 3A (ARCL3A)
Arginase 1	ARG1	Increased arginase levels in heart failure represent a therapeutic target to rescue microvascular perfusion. PMID: 23075998. Disease: argininemia (ARGIN)
Arginase 2	ARG2	Arginase: A key enzyme in the pathophysiology of allergic asthma opening novel therapeutic perspectives. PMID: 19703164
Asparagine synthetase (glutamine-hydrolyzing)	ASNS	Asparagine synthetase is an independent predictor of surgical survival and a potential therapeutic target in hepatocellular carcinoma. PMID: 23764751. Disease: asparagine synthetase deficiency (ASNSD)
Asparagine synthetase domain containing 1	ASNSD1	
Aspartate beta-hydroxylase	ASPH	A cell-surface β-hydroxylase is a biomarker and therapeutic target for hepatocellular carcinoma. PMID: 24954865
Aspartate beta-hydroxylase domain containing 1	ASPHD1	
Aspartate beta-hydroxylase domain containing 2	ASPHD2	

(Continued)

HGNC-approved name	Symbol	Target information
Aspartoacylase	ASPA	Aspartoacylase catalytic deficiency as the cause of Canavan disease: A structural perspective. PMID: 25003821. Disease: Canavan disease (CAND)
Aminoacylase 1	ACY1	Disease: aminoacylase-1 deficiency (ACY1D)
Aspartoacylase (aminocyclase) 3	ACY3	Differential aminoacylase expression in neuroblastoma. PMID: 21128244
Betaine–homocysteine *S*-methyltransferase	BHMT	The development of a new class of inhibitors for betaine-homocysteine *S*-methyltransferase. PMID: 23727536
Betaine–homocysteine *S*-methyltransferase 2	BHMT2	Double-headed sulfur-linked amino acids as first inhibitors for betaine-homocysteine *S*-methyltransferase 2. PMID: 22775318
Branched-chain amino-acid transaminase 1, cytosolic	BCAT1	BCAT1 promotes cell proliferation through amino acid catabolism in gliomas carrying wild-type IDH1. PMID: 23793099
Branched-chain amino-acid transaminase 2, mitochondrial	BCAT2	Systematic identification of single amino acid variants in glioma stem-cell-derived chromosome 19 proteins. PMID: 25399873
Branched-chain keto acid dehydrogenase E1, alpha polypeptide	BCKDHA	Branched-chain amino acids in metabolic signalling and insulin resistance. PMID: 25287287. Disease: maple syrup urine disease 1A (MSUD1A)
Branched-chain keto acid dehydrogenase E1, beta polypeptide	BCKDHB	Disease: maple syrup urine disease 1B (MSUD1B)
Catechol-*O*-methyltransferase	COMT	**Entacapone** – anti-Parkinson drugs
Catechol-*O*-methyltransferase domain containing 1	COMTD1	
Creatine kinase, mitochondrial 1A	CKMT1A	Ubiquitous mitochondrial creatine kinase downregulated in oral squamous cell carcinoma. PMID: 16479256
Creatine kinase, mitochondrial 1B	CKMT1B	High ubiquitous mitochondrial creatine kinase expression in hepatocellular carcinoma denotes a poor prognosis with highly malignant potential. PMID: 24174293
Creatine kinase, mitochondrial 2 (sarcomeric)	CKMT2	
Creatine kinase, brain	CKB	Neuroprotective activity of creatylglycine ethyl ester fumarate. PMID: 25561316
Creatine kinase, muscle	CKM	Association of the muscle-specific creatine kinase (CKMM) gene polymorphism with physical performance of athletes PMID: 22567844
Cysteine conjugate-beta lyase, cytoplasmic	CCBL1	
Cysteine conjugate-beta lyase 2	CCBL2	Homology modeling of human kynurenine aminotransferase III and observations on inhibitor binding using molecular docking. PMID: 24739074
D-Dopachrome tautomerase	DDT	Targeting distinct tautomerase sites of D-DT and MIF with a single molecule for inhibition of neutrophil lung recruitment. PMID: 25016026
D-Dopachrome tautomerase-like	DDTL	
Deiodinase, iodothyronine, type I	DIO1	Strong induction of iodothyronine deiodinases by chemotherapeutic selenocompounds. PMID: 25579002
Deiodinase, iodothyronine, type II	DIO2	Adult mice lacking the type 2 iodothyronine deiodinase have increased subchondral bone but normal articular cartilage. PMID: 25549200
Deiodinase, iodothyronine, type III	DIO3	Type 3 deiodinase and consumptive hypothyroidism: A common mechanism for a rare disease. PMID: 24027558

HGNC-approved name	Symbol	Target information
Dihydrofolate reductase	DHFR	**Methotrexate** – antineoplastic agents. Disease: megaloblastic anemia due to dihydrofolate reductase deficiency (DHFRD)
Dihydrofolate reductase-like 1	DHFRL1	The former annotated human pseudogene dihydrofolate reductase-like 1 (DHFRL1) is expressed and functional. PMID: 21876184
Dimethylarginine dimethylaminohydrolase 1	DDAH1	Disruption of methylarginine metabolism impairs vascular homeostasis. PMID: 17273169
Dimethylarginine dimethylaminohydrolase 2	DDAH2	The hydrolase DDAH2 enhances pancreatic insulin secretion by transcriptional regulation of secretagogin through a Sirt1-dependent mechanism in mice. PMID: 23430976
Diphthamide biosynthesis 5	DPH5	The biosynthesis and biological function of diphthamide. PMID: 23971743
Diphthamine biosynthesis 6	DPH6	
Formimidoyltransferase cyclodeaminase	FTCD	Using a yeast two-hybrid system to identify FTCD as a new regulator for HIF-1α in HepG2 cells. PMID: 24686083. Disease: glutamate formiminotransferase deficiency (FIGLU-URIA)
Formiminotransferase cyclodeaminase N-terminal-like	FTCDNL1	
Glutamate decarboxylase 1 (brain, 67 kDa)	GAD1	Glutamate acid decarboxylase 1 promotes metastasis of human oral cancer by β-catenin translocation and MMP7 activation. PMID: 24261884. Disease: cerebral palsy, spastic quadriplegic 1 (CPSQ1)
Glutamate decarboxylase 2 (pancreatic islets and brain, 65 kDa)	GAD2	Deficiency of the 65 kDa isoform of glutamic acid decarboxylase impairs extinction of cued but not contextual fear memory. PMID: 20016086
Glutamate decarboxylase-like 1	GADL1	Role of glutamate decarboxylase-like protein 1 (GADL1) in taurine biosynthesis. PMID: 23038267
Glutamate dehydrogenase 1	GLUD1	Disease: familial hyperinsulinemic hypoglycemia 6 (HHF6)
Glutamate dehydrogenase 2	GLUD2	Hominoid-specific enzyme GLUD2 promotes growth of IDH1R132H glioma. PMID: 25225364
Glutamic-oxaloacetic transaminase 1, soluble	GOT1	Inducible glutamate oxaloacetate transaminase as a therapeutic target against ischemic stroke. PMID: 25343301
Glutamic-oxaloacetic transaminase 1-like 1	GOT1L1	Is D-aspartate produced by glutamic-oxaloacetic transaminase-1 like 1 (Got1l1). A putative aspartate racemase? PMID: 25287256
Glutamic-oxaloacetic transaminase 2, mitochondrial	GOT2	Glutamine supports pancreatic cancer growth through a KRAS-regulated metabolic pathway. PMID: 23535601
Glutaminase	GLS	Dibenzophenanthridines as inhibitors of glutaminase C and cancer cell proliferation. PMID: 22496480
Glutaminase 2 (liver, mitochondrial)	GLS2	Dibenzophenanthridines as inhibitors of glutaminase C and cancer cell proliferation. PMID: 22496480
Glycine-N-acyltransferase	GLYAT	Glycine conjugation: Importance in metabolism, the role of glycine N-acyltransferase, and factors that influence interindividual variation. PMID: 23650932
Glycine-N-acyltransferase-like 1	GLYATL1	
Glycine-N-acyltransferase-like 2	GLYATL2	Reversible lysine acetylation regulates activity of human glycine N-acyltransferase-like 2 (hGLYATL2): Implications for production of glycine-conjugated signaling molecules. PMID: 22408254
Glycine-N-acyltransferase-like 3	GLYATL3	

(Continued)

HGNC-approved name	Symbol	Target information
Glutamic-pyruvate transaminase (alanine aminotransferase)	GPT	Isoform-specific alanine aminotransferase measurement can distinguish hepatic from extrahepatic injury in humans. PMID: 22922605
Glutamic-pyruvate transaminase (alanine aminotransferase) 2	GPT2	Activating transcription factor 4 mediates up-regulation of alanine aminotransferase 2 gene expression under metabolic stress. PMID: 24418603
Indoleamine 2,3-dioxygenase 1	IDO1	**INCB24360** – antineoplastic agents
Indoleamine 2,3-dioxygenase 2	IDO2	IDO2 in immunomodulation and autoimmune disease. PMID: 25477879
Lysyl oxidase	LOX	Lysyl oxidases play a causal role in vascular remodeling in clinical and experimental pulmonary arterial hypertension. PMID: 24833797
Lysyl oxidase-like 1	LOXL1	Disease: exfoliation syndrome (XFS)
Lysyl oxidase-like 2	LOXL2	**Simtuzumab** – hepatic fibrosis treatments
Lysyl oxidase-like 3	LOXL3	The potential for targeting extracellular LOX proteins in human malignancy. PMID: 24348049
Lysyl oxidase-like 4	LOXL4	Lysyl oxidase-like 4 (LOXL4) promotes proliferation and metastasis of gastric cancer via FAK/Src pathway. PMID: 25215901
Methionine adenosyltransferase I, alpha	MAT1A	Disease: methionine adenosyltransferase deficiency (MATD)
Methionine adenosyltransferase II, alpha	MAT2A	2′,6′-Dihalostyrylanilines, pyridines, and pyrimidines for the inhibition of the catalytic subunit of methionine S-adenosyltransferase-2. PMID: 24950374
Methionine adenosyltransferase II, beta	MAT2B	Lentivirus mediated shRNA interference targeting MAT2B induces growth-inhibition and apoptosis in hepatocellular carcinoma. PMID: 18698677
Methionine sulfoxide reductase A	MSRA	Methionine sulfoxide reductase: Chemistry, substrate binding, recycling process and oxidase activity. PMID: 25108804
Methionine sulfoxide reductase B1	MSRB1	Unraveling the specificities of the different human methionine sulfoxide reductases. PMID: 24737740
Methionine sulfoxide reductase B2	MSRB2	Identification of the mitochondrial MSRB2 as a binding partner of LG72. PMID: 25078755
Methionine sulfoxide reductase B3	MSRB3	Disease: deafness, autosomal recessive, 74 (DFNB74)
Methylcrotonoyl-CoA carboxylase 1 (alpha)	MCCC1	Disease: methylcrotonoyl-CoA carboxylase 1 deficiency (MCC1D)
Methylcrotonoyl-CoA carboxylase 2 (beta)	MCCC2	Disease: methylcrotonoyl-CoA carboxylase 2 deficiency (MCC2D)
Methylenetetrahydrofolate reductase (NAD(P)H)	MTHFR	Disease: methylenetetrahydrofolate reductase deficiency (MTHFRD)
Methylenetetrahydrofolate dehydrogenase (NADP+ dependent) 1, Methenyltetrahydrofolate cyclohydrolase, formyltetrahydrofolate synthetase	MTHFD1	Disease: folate-sensitive neural tube defects (FS-NTD)
Methylenetetrahydrofolate dehydrogenase (NADP+ dependent) 1-like	MTHFD1L	Cancer. Systems biology, metabolomics, and cancer metabolism. PMID: 22628644
Methylenetetrahydrofolate dehydrogenase (NADP+ dependent) 2, methenyltetrahydrofolate cyclohydrolase	MTHFD2	Increased MTHFD2 expression is associated with poor prognosis in breast cancer. PMID: 24870594
Methylenetetrahydrofolate dehydrogenase (NADP+ dependent) 2-like	MTHFD2L	Mitochondrial MTHFD2L is a dual redox cofactor-specific methylenetetrahydrofolate dehydrogenase/methenyltetrahydrofolate cyclohydrolase expressed in both adult and embryonic tissues. PMID: 24733394
Monoamine oxidase A	MAOA	**Phenelzine** – antidepressants. Disease: Brunner syndrome (BRUNS)

HGNC-approved name	Symbol	Target information
Monoamine oxidase B	MAOB	**Rasagiline** – anti-Parkinson drugs
Nitrilase 1	NIT1	Mammalian nitrilase 1 homologue Nit1 is a negative regulator in T cells. PMID: 19395373
Nitrilase family, member 2	NIT2	Structural insights into the catalytic active site and activity of human Nit2/ω-amidase: Kinetic assay and molecular dynamics simulation. PMID: 22674578
Ornithine decarboxylase 1	ODC1	**Eflornithine** – antihirsuitism agents
Proline dehydrogenase (oxidase) 1	PRODH	P53 family members modulate the expression of PRODH, but not PRODH2, via intronic p53 response elements. PMID: 23861960. Disease: hyperprolinemia 1 (HP-1)
Proline dehydrogenase (oxidase) 2	PRODH2	
Prolyl 3-hydroxylase 1	P3H1	Collagen prolyl 3-hydroxylation: A major role for a minor post-translational modification? PMID: 23772978. Disease: osteogenesis imperfecta 8 (OI8)
Prolyl 3-hydroxylase 2	P3H2	The prolyl 3-hydroxylases P3H2 and P3H3 are novel targets for epigenetic silencing in breast cancer. PMID: 19436308. Disease: myopia, high, with cataract and vitreoretinal degeneration (MCVD)
Prolyl 3-hydroxylase 3	P3H3	The collagen prolyl hydroxylases are novel transcriptionally silenced genes in lymphoma. PMID: 22955849
Prolyl 3-hydroxylase 4	P3H4	
Pterin-4 alpha-carbinolamine dehydratase/dimerization cofactor of hepatocyte nuclear factor 1 alpha	PCDD1	Disease: hyperphenylalaninemia, BH4-deficient, D (HPABH4D)
Pterin-4 alpha-carbinolamine dehydratase/ dimerization cofactor of hepatocyte nuclear factor 1 alpha (TCF1) 2	PCBD2	Biochemical and structural basis for partially redundant enzymatic and transcriptional functions of DCoH and DCoH2. PMID: 15182178
Pyrroline-5-carboxylate reductase 1	PYCR1	Disease: cutis laxa, autosomal recessive, 2B (ARCL2B)
Pyrroline-5-carboxylate reductase family, member 2	PYCR2	Functional specialization in proline biosynthesis of melanoma. PMID: 23024808
Pyrroline-5-carboxylate reductase-like	PYCRL	
Selenophosphate synthetase 1	SEPHS1	
Selenophosphate synthetase 2	SEPHS2	Attenuation of hepatic expression and secretion of selenoprotein P by metformin. PMID: 19370170
Serine dehydratase	SDS	
Serine dehydratase-like	SDSL	A catalytic mechanism that explains a low catalytic activity of serine dehydratase like-1 from human cancer cells: Crystal structure and site-directed mutagenesis studies. PMID: 18342636
Serine hydroxymethyltransferase 1 (soluble)	SHMT1	SHMT1 knockdown induces apoptosis in lung cancer cells by causing uracil misincorporation. PMID: 25412303
Serine hydroxymethyltransferase 2 (mitochondrial)	SHMT2	Cancer. Systems biology, metabolomics, and cancer metabolism. PMID: 22628644
Spermidine/spermine N1-acetyltransferase 1	SAT1	The polyamine catabolic enzyme SAT1 modulates tumorigenesis and radiation response in GBM. PMID: 25277523. Disease: keratosis follicularis spinulosa decalvans X-linked (KFSDX)
Spermidine/spermine N1-acetyl transferase-like 1	SATL1	

(Continued)

HGNC-approved name	Symbol	Target information
Spermidine/spermine N1-acetyltransferase family member 2	SAT2	Antitumor agent PX-12 inhibits HIF-1α protein levels through an Nrf2/PMF-1-mediated increase in spermidine/spermine acetyl transferase. PMID: 21069338
Threonine synthase-like 2 (S. cerevisiae)	THNSL2	Secreted osteoclastogenic factor of activated T cells (SOFAT), a novel osteoclast activator, in chronic periodontitis. PMID: 23619471
Tryptophan hydroxylase 1	TPH1	Pharmacological inhibition of gut-derived serotonin synthesis is a potential bone anabolic treatment for osteoporosis. PMID: 20139991
Tryptophan hydroxylase 2	TPH2	Tryptophan hydroxylase-2: An emerging therapeutic target for stress disorders. PMID: 23435356. Disease: major depressive disorder (MDD)
Tyrosinase	TYR	**MHY884** – dermatologicals. Disease: albinism, oculocutaneous, 1A (OCA1A)
Tyrosinase-related protein 1	TYRP1	**Flanvotumab** – antineoplastic agents. Disease: albinism, oculocutaneous, 3 (OCA3)
2-Aminoethanethiol (cysteamine) dioxygenase	ADO	Discovery and characterization of a second mammalian thiol dioxygenase, cysteamine dioxygenase. PMID: 17581819
3-Hydroxyanthranilate 3,4-dioxygenase	HAAO	Anticonvulsant effects of the 3-hydroxyanthranilic acid dioxygenase inhibitor NCR-631. PMID: 11026503
3-Hydroxyisobutyrate dehydrogenase	HIBADH	Characterization of 3-hydroxyisobutyrate dehydrogenase, HIBADH, as a sperm-motility marker. PMID: 23423614
3-Hydroxyisobutyryl-CoA hydrolase	HIBCH	Disease: HIBCH deficiency (HIBCHD)
4-Aminobutyrate aminotransferase	ABAT	**Vigabatrin** – antiepileptics. Disease: GABA transaminase deficiency (GABATD)
4-Hydroxy-2-oxoglutarate aldolase 1	HOGA1	Disease: hyperoxaluria primary 3 (HP3)
5-Phosphohydroxy-L-lysine phospho-lyase	PHYKPL	Disease: phosphohydroxylysinuria (PHLU)
6-Pyruvoyltetrahydropterin synthase	PTS	Disease: hyperphenylalaninemia, BH4-deficient, A (HPABH4A)
Acetylcholinesterase (Yt blood group)	ACHE	**Galantamine** – parasympathomimetics
Acireductone dioxygenase 1	ADI1	Expression and function of the human androgen-responsive gene ADI1 in prostate cancer. PMID: 17786183
Adenosylmethionine decarboxylase 1	AMD1	**SAM486A** – antineoplastic agents
Agmatine ureohydrolase (agmatinase)	AGMAT	Arginine decarboxylase and agmatinase: An alternative pathway for de novo biosynthesis of polyamines for development of mammalian conceptuses. PMID: 24648395
Allantoicase	ALLC	Genome-wide association study identifies ALLC polymorphisms correlated with FEV$_1$ change by corticosteroid. PMID: 24792382
Amidohydrolase domain containing 1	AMDHD1	
Aminoadipate aminotransferase	AADAT	**PF-04859989** – antipsychotics
Aminoadipate-semialdehyde dehydrogenase	AASDH	Vertebrate Acyl CoA synthetase family member 4 (ACSF4-U26) is a β-alanine-activating enzyme homologous to bacterial non-ribosomal peptide synthetase. PMID: 24467666
Aminoadipate-semialdehyde synthase	AASS	Disease: hyperlysinemia, 1 (HYPLYS1)
Aminocarboxymuconate semialdehyde decarboxylase	ACMSD	Involvement of the kynurenine pathway in human glioma pathophysiology. PMID: 25415278
Aminomethyltransferase	AMT	Disease: non-ketotic hyperglycinemia (NKH)
Antizyme inhibitor 2	AZIN2	Restorative benefits of transplanting human mesenchymal stromal cells overexpressing arginine decarboxylase genes after spinal cord injury. PMID: 25442787
Aralkylamine N-acetyltransferase	AANAT	Disease: delayed sleep phase syndrome (DSPS)

HGNC-approved name	Symbol	Target information
Argininosuccinate lyase	ASL	Disease: argininosuccinic aciduria (ARGINSA)
Argininosuccinate synthase 1	ASS1	Prognostic and therapeutic impact of argininosuccinate synthetase 1 control in bladder cancer as monitored longitudinally by PET imaging. PMID: 24285724. Disease: citrullinemia 1 (CTLN1)
Arylformamidase	AFMID	
Asparaginase	ASPG	**Asparaginase** – antineoplastic agents
Aspartate dehydrogenase domain containing	ASPDH	
AU RNA binding protein/enoyl-CoA hydratase	AUH	Disease: 3-methylglutaconic aciduria 1 (MGA1)
Butyrobetaine (gamma), 2-oxoglutarate dioxygenase (gamma-butyrobetaine hydroxylase) 1	BBOX1	Targeting carnitine biosynthesis: Discovery of new inhibitors against γ-butyrobetaine hydroxylase. PMID: 24571165
Butyrylcholinesterase	BCHE	**Hexafluronium bromide** – muscle relaxants. Disease: butyrylcholinesterase deficiency (BChE deficiency)
Carbamoyl-phosphate synthase 1, mitochondrial	CPS1	**Carglumic acid** – urea cycle disorder treatments. Disease: carbamoyl phosphate synthetase 1 deficiency (CPS1D)
Carnosine synthase 1	CARNS1	Molecular identification of carnosine synthase as ATP-grasp domain-containing protein 1 (ATPGD1). PMID: 20097752
Choline dehydrogenase	CHDH	Choline dehydrogenase interacts with SQSTM1/p62 to recruit LC3 and stimulate mitophagy. PMID: 25483962
Cystathionase (cystathionine gamma-lyase)	CTH	Dysregulation of cystathionine γ-lyase (CSE)/hydrogen sulfide pathway contributes to ox-LDL-induced inflammation in macrophage. PMID: 23872072. Disease: cystathioninuria (CSTNU)
Cystathionine-beta-synthase	CBS	The CBS/CSE system: A potential therapeutic target in NAFLD? PMID: 25493326. Disease: cystathionine beta-synthase deficiency (CBSD)
Cysteine dioxygenase type 1	CDO1	Frequent inactivation of cysteine dioxygenase type 1 contributes to survival of breast cancer cells and resistance to anthracyclines. PMID: 23630167
Cysteine sulfinic acid decarboxylase	CSAD	Development of a novel cysteine sulfinic Acid decarboxylase knockout mouse: Dietary taurine reduces neonatal mortality. PMID: 24639894
D-Amino-acid oxidase	DAO	D-Amino acid oxidase controls motoneuron degeneration through D-serine. PMID: 22203986
D-Aspartate oxidase	DDO	Free D-aspartate regulates neuronal dendritic morphology, synaptic plasticity, gray matter volume and brain activity in mammals. PMID: 25072322
Deoxyhypusine hydroxylase/ monooxygenase	DOHH	A hypusine-eIF5A-PEAK1 switch regulates the pathogenesis of pancreatic cancer. PMID: 25261239
Deoxyhypusine synthase	DHPS	Deoxyhypusine synthase (DHPS) inhibitor GC7 induces p21/Rb-mediated inhibition of tumor cell growth and DHPS expression correlates with poor prognosis in neuroblastoma patients. PMID: 25315710
Dihydrolipoamide branched-chain transacylase E2	DBT	Disease: maple syrup urine disease 2 (MSUD2)
Dihydrolipoamide dehydrogenase	DLD	Disease: dihydrolipoamide dehydrogenase deficiency (DLDD)
Dihydrolipoamide S-succinyltransferase (E2 component of 2-oxoglutarate complex)	DLST	The alpha-ketoglutarate-dehydrogenase complex: A mediator between mitochondria and oxidative stress in neurodegeneration. PMID: 15953811
Dimethylglycine dehydrogenase	DMGDH	Disease: DMGDH deficiency (DMGDHD)

(Continued)

HGNC-approved name	Symbol	Target information
Dopa decarboxylase (aromatic L-amino acid decarboxylase)	DDC	**Carbidopa** – anti-Parkinson drugs. Disease: aromatic L-amino-acid decarboxylase deficiency (AADCD)
Dopamine beta-hydroxylase (dopamine beta-monooxygenase)	DBH	**Etamicastat** – antihypertensives. Disease: dopamine beta-hydroxylase deficiency (DBH deficiency)
Enolase superfamily member 1	ENOSF1	A candidate gene study of capecitabine-related toxicity in colorectal cancer identifies new toxicity variants at DPYD and a putative role for ENOSF1 rather than TYMS. PMID: 24647007
Enolase-phosphatase 1	ENOPH1	Analysis of quantitative trait loci in mice suggests a role of Enoph1 in stress reactivity. PMID: 24236849
Ethanolamine-phosphate phospho-lyase	ETNPPL	Strict reaction and substrate specificity of AGXT2L1, the human O-phosphoethanolamine phospho-lyase. PMID: 23761375
Ethylmalonic encephalopathy 1	ETHE1	Disease: ethylmalonic encephalopathy (EE)
Fumarylacetoacetate hydrolase (fumarylacetoacetase)	FAH	Disease: tyrosinemia 1 (TYRSN1)
Gamma-glutamyl hydrolase (conjugase, folylpolygammaglutamyl hydrolase)	GGH	γ-Glutamyl hydrolase modulation and folate influence chemosensitivity of cancer cells to 5-fluorouracil and methotrexate. PMID: 24045662
Gamma-glutamylamine cyclotransferase	GGACT	γ-Glutamylamines and neurodegenerative diseases. PMID: 22407484
Glutamate-ammonia ligase	GLUL	Disease: congenital systemic glutamine deficiency (CSGD)
Glutamate-cysteine ligase, catalytic subunit	GCLC	Disease: hemolytic anemia due to gamma glutamylcysteine synthetase deficiency (HAGGSD)
Glutamate-cysteine ligase, modifier subunit	GCLM	Cardiovascular and renal manifestations of glutathione depletion induced by buthionine sulfoximine. PMID: 22223042
Glycine amidinotransferase (L-arginine:glycine amidinotransferase)	GATM	Disease: cerebral creatine deficiency syndrome 3 (CCDS3)
Glycine C-acetyltransferase	GCAT	
Glycine cleavage system protein H (aminomethyl carrier)	GCSH	Disease: non-ketotic hyperglycinemia (NKH)
Glycine dehydrogenase (decarboxylating)	GLDC	Disease: non-ketotic hyperglycinemia (NKH)
Glycine-N-methyltransferase	GNMT	Disease: glycine N-methyltransferase deficiency (GNMT deficiency)
Guanidinoacetate N-methyltransferase	GAMT	Disease: cerebral creatine deficiency syndrome 2 (CCDS2)
Histamine N-methyltransferase	HNMT	Histamine pharmacogenomics. PMID: 19450133
Histidine ammonia-lyase	HAL	Disease: histidinemia (HISTID)
Histidine decarboxylase	HDC	L-Histidine decarboxylase and Tourette's syndrome. PMID: 20445167
Homogentisate 1,2-dioxygenase	HGD	Disease: alkaptonuria (AKU)
Hydroxylysine kinase	HYKK	Molecular identification of hydroxylysine kinase and of ammoniophospholyases acting on 5-phosphohydroxy-L-lysine and phosphoethanolamine. PMID: 22241472
IBA57, iron–sulfur cluster assembly homolog (S. cerevisiae)	IBA57	Disease: multiple mitochondrial dysfunctions syndrome 3 (MMDS3)
Interleukin 4 induced 1	IL4I1	Differential expression of the immunosuppressive enzyme IL4I1 induce Aiolos+, but not natural Helios+, FOXP3+ Treg cells. PMID: 25446972
Iodotyrosine deiodinase	IYD	Disease: thyroid dyshormonogenesis 4 (TDH4)
Isovaleryl-CoA dehydrogenase	IVD	Disease: isovaleric acidemia (IVA)
Kynureninase	KYNU	Crystal structure of the Homo sapiens kynureninase-3-hydroxyhippuric acid inhibitor complex: Insights into the molecular basis of kynureninase substrate specificity. PMID: 19143568
Kynurenine 3-monooxygenase (kynurenine 3-hydroxylase)	KMO	**UPF648** – neuroprotective agents

HGNC-approved name	Symbol	Target information
L-2-Hydroxyglutarate dehydrogenase	L2HGDH	Disease: L-2-hydroxyglutaric aciduria (L2HGA)
L-3-Hydroxyproline dehydratase (trans-)	L3HYPDH	Identification of a human trans-3-hydroxy-L-proline dehydratase, the first characterized member of a novel family of proline racemase-like enzymes. PMID: 22528483
Lengsin, lens protein with glutamine synthetase domain	LGSN	Novel spliced form of a lens protein as a novel lung cancer antigen, Lengsin splicing variant 4. PMID: 19459848
Leucine rich transmembrane and O-methyltransferase domain containing	LRTOMT	Disease: deafness, autosomal recessive, 63 (DFNB63)
Methylthioribose-1-phosphate isomerase 1	MRI1	Structure of mediator of RhoA-dependent invasion (MRDI) explains its dual function as a metabolic enzyme and a mediator of cell invasion. PMID: 23859498
N-Acetylglutamate synthase	NAGS	Disease: N-acetylglutamate synthase deficiency (NAGSD)
NADP-dependent oxidoreductase domain containing 1	NOXRED1	
NFS1 cysteine desulfurase	NFS1	
N-Terminal asparagine amidase	NTAN1	Stimulation of ubiquitin-proteasome pathway through the expression of amidohydrolase for N-terminal asparagine (Ntan1) in cultured rat hippocampal neurons exposed to static magnetism. PMID: 16539681
Ornithine aminotransferase	OAT	Diazonamide toxins reveal an unexpected function for ornithine delta-amino transferase in mitotic cell division. PMID: 17287350. Disease: hyperornithinemia with gyrate atrophy of choroid and retina (HOGA)
Ornithine carbamoyltransferase	OTC	Disease: ornithine carbamoyltransferase deficiency (OTCD)
Oxoglutarate (alpha-ketoglutarate) dehydrogenase (lipoamide)	OGDH	
Phenylalanine hydroxylase	PAH	Disease: phenylketonuria (PKU)
Phenylethanolamine N-methyltransferase	PNMT	
Phosphoglycerate dehydrogenase	PHGDH	Phosphoglycerate dehydrogenase: Potential therapeutic target and putative metabolic oncogene. PMID: 25574168. Disease: phosphoglycerate dehydrogenase deficiency (PHGDH deficiency)
Phosphoserine aminotransferase 1	PSAT1	PSAT1 regulates cyclin D1 degradation and sustains proliferation of non-small cell lung cancer cells. PMID: 25142862. Disease: phosphoserine aminotransferase deficiency (PSATD)
Phosphoserine phosphatase	PSPH	Disease: phosphoserine phosphatase deficiency (PSPHD)
Pipecolic acid oxidase	PIPOX	The role of sarcosine metabolism in prostate cancer progression. PMID: 23633921
Pyruvate dehydrogenase phosphatase regulatory subunit	PDPR	
Quinoid dihydropteridine reductase	QDPR	Disease: hyperphenylalaninemia, BH4-deficient, C (HPABH4C)
Sarcosine dehydrogenase	SARDH	Disease: sarcosinemia (SARCOS)
Selenocysteine lyase	SCLY	Mice lacking selenoprotein P and selenocysteine lyase exhibit severe neurological dysfunction, neurodegeneration, and audiogenic seizures. PMID: 24519931
Serine racemase	SRR	Malonate-based inhibitors of mammalian serine racemase: Kinetic characterization and structure-based computational study. PMID: 25462239
Spermidine synthase	SRM	Chemoprevention of B-cell lymphomas by inhibition of the Myc target spermidine synthase. PMID: 20103729

(Continued)

HGNC-approved name	Symbol	Target information
Spermine oxidase	SMOX	Reactive oxygen species spermine metabolites generated from amine oxidases and radiation represent a therapeutic gain in cancer treatments. PMID: 23857253
Spermine synthase	SMS	Disease: X-linked syndromic mental retardation Snyder–Robinson type (MRXSSR)
Sulfide quinone reductase-like (yeast)	SQRDL	The vertebrate homolog of sulfide-quinone reductase is expressed in mitochondria of neuronal tissues. PMID: 22067608
Sulfite oxidase	SUOX	Disease: isolated sulfite oxidase deficiency (ISOD)
Trimethyllysine hydroxylase, epsilon	TMLHE	Disease: epsilon-trimethyllysine hydroxylase deficiency (TMLHED)
Tryptophan 2,3-dioxygenase	TDO2	Targeting key dioxygenases in tryptophan-kynurenine metabolism for immunomodulation and cancer chemotherapy. PMID: 25478733
Tyrosine aminotransferase	TAT	Disease: tyrosinemia 2 (TYRSN2)
Tyrosine hydroxylase	TH	**Metyrosine** – antihypertensives. Disease: Segawa syndrome autosomal recessive (ARSEGS)
Ureidopropionase, beta	UPB1	Disease: beta-ureidopropionase deficiency (BUPD)
Urocanate hydratase 1	UROC1	Disease: urocanase deficiency (UROD)

5.3   Nucleotides and related

HGNC-approved name	Symbol	Target information
5′-Nucleotidase, cytosolic IA	NT5C1A	The role of soluble 5′-nucleotidases in the conversion of nucleotide analogs: Metabolic and therapeutic aspects. PMID: 23992311
5′-Nucleotidase, cytosolic IB	NT5C1B	
5′-Nucleotidase, cytosolic II	NT5C2	Identification and characterization of inhibitors of cytoplasmic 5′-nucleotidase cN-II issued from virtual screening. PMID: 23220537. Disease: spastic paraplegia 45, autosomal recessive (SPG45)
5′-Nucleotidase, cytosolic IIIA	NT5C3A	Disease: P5N deficiency (P5ND)
5′-Nucleotidase, cytosolic IIIB	NT5C3B	Novel genes for airway wall thickness identified with combined genome-wide association and expression analyses. PMID: 25517131
5′-Nucleotidase, ecto (CD73)	NT5E	Anti-CD73 antibody therapy inhibits breast tumor growth and metastasis. PMID: 20080644. Disease: calcification of joints and arteries (CALJA)
5′,3′-Nucleotidase, cytosolic	NT5C	Cytosolic and mitochondrial deoxyribonucleotidases: Activity with substrate analogs, inhibitors and implications for therapy. PMID: 12907246
5′,3′-Nucleotidase, mitochondrial	NT5M	Cytosolic and mitochondrial deoxyribonucleotidases: Activity with substrate analogs, inhibitors and implications for therapy. PMID: 12907246
Adenosine deaminase	ADA	**Pentostatin** – antineoplastic agents. Disease: severe combined immunodeficiency autosomal recessive T-cell-negative/B-cell-negative/NK-cell-negative due to adenosine deaminase deficiency (ADASCID)
Adenosine deaminase-like	ADAL	Mutations in adenosine deaminase-like (ADAL) protein confer resistance to the antiproliferative agents N6-cyclopropyl-PMEDAP and GS-9219. PMID: 23645737
Adenosine monophosphate deaminase 1	AMPD1	AMPD1: A novel therapeutic target for reversing insulin resistance. PMID: 25511531. Disease: myopathy due to myoadenylate deaminase deficiency (MMDD)
Adenosine monophosphate deaminase 2	AMPD2	Proteinuria in AMPD2-deficient mice. PMID: 22212473

HGNC-approved name	Symbol	Target information
Adenosine monophosphate deaminase 3	AMPD3	AMP deaminase 3 plays a critical role in remote reperfusion lung injury. PMID: 23542464. Disease: adenosine monophosphate deaminase deficiency erythrocyte type (AMPDDE)
Adenylate kinase 1	AK1	Neuropathogenic role of adenylate kinase-1 in Aβ-mediated tau phosphorylation via AMPK and GSK3β. PMID: 22419736. Disease: hemolytic anemia due to adenylate kinase deficiency (HAAKD)
Adenylate kinase 2	AK2	The DUSP26 phosphatase activator adenylate kinase 2 regulates FADD phosphorylation and cell growth. PMID: 24548998. Disease: reticular dysgenesis (RDYS)
Adenylate kinase 3	AK3	Adenylate kinase 3 sensitizes cells to cigarette smoke condensate vapor induced cisplatin resistance. PMID: 21698293
Adenylate kinase 4	AK4	Adenylate kinase-4 is a marker of poor clinical outcomes that promotes metastasis of lung cancer by downregulating the transcription factor ATF3. PMID: 23002211
Adenylate kinase 5	AK5	Inhibition of adenylyl cyclase type 5 prevents L-DOPA-induced dyskinesia in an animal model of Parkinson's disease. PMID: 25164669
Adenylate kinase 6	AK6	The crystal structure of human adenylate kinase 6: An adenylate kinase localized to the cell nucleus. PMID: 15630091
Adenylate kinase 7	AK7	New adenylate kinase 7 (AK7) mutation in primary ciliary dyskinesia. PMID: 22801010
Adenylate kinase 8	AK8	The characterization of human adenylate kinases 7 and 8 demonstrates differences in kinetic parameters and structural organization among the family of adenylate kinase isoenzymes. PMID: 21080915
Adenylate kinase 9	AK9	The human adenylate kinase 9 is a nucleoside mono- and diphosphate kinase. PMID: 23416111
Adenylosuccinate synthase	ADSS	
Adenylosuccinate synthase-like 1	ADSSL1	Molecular cloning and characterization of a novel muscle adenylosuccinate synthetase, AdSSL1, from human bone marrow stromal cells. PMID: 15786719
CTP synthase 1	CTPS1	CTP synthase 1 deficiency in humans reveals its central role in lymphocyte proliferation. PMID: 24870241
CTP synthase 2	CTPS2	Low cytosine triphosphate synthase 2 expression renders resistance to 5-fluorouracil in colorectal cancer. PMID: 21378502
Cytidine monophosphate (UMP–CMP) kinase 1, cytosolic	CMPK1	Effect of genetic polymorphisms on therapeutic response and clinical outcomes in pancreatic cancer patients treated with gemcitabine. PMID: 22020950
Cytidine monophosphate (UMP–CMP) kinase 2, mitochondrial	CMPK2	Human UMP-CMP kinase 2, a novel nucleoside monophosphate kinase localized in mitochondria. PMID: 17999954
Dihydropyrimidinase	DPYS	Disease: dihydropyrimidinase deficiency (DHPD)
Dihydropyrimidinase-like 2	DPYSL2	**Lacosamide** – antiepileptics
Dihydropyrimidinase-like 3	DPYSL3	Dihydropyrimidinase-like 3 facilitates malignant behavior of gastric cancer. PMID: 25096402
Dihydropyrimidinase-like 4	DPYSL4	Identification of dihydropyrimidinase-related protein 4 as a novel target of the p53 tumor suppressor in the apoptotic response to DNA damage. PMID: 20499313
Dihydropyrimidinase-like 5	DPYSL5	Collapsin response mediator protein 5 (CRMP5) induces mitophagy, thereby regulating mitochondrion numbers in dendrites. PMID: 24324268
Dihydrouridine synthase 1-like (S. cerevisiae)	DUS1L	

(Continued)

HGNC-approved name	Symbol	Target information
Dihydrouridine synthase 2	DUS2	A novel human tRNA-dihydrouridine synthase involved in pulmonary carcinogenesis. PMID: 15994936
Dihydrouridine synthase 3-like (*S. cerevisiae*)	DUS3L	
Dihydrouridine synthase 4-like (*S. cerevisiae*)	DUS4L	Novel fusion transcripts in human gastric cancer revealed by transcriptome analysis. PMID: 24240688
Ectonucleoside triphosphate diphosphohydrolase 1	ENTPD1	ENTPD1/CD39 is a promising therapeutic target in oncology. PMID: 22751118
Ectonucleoside triphosphate diphosphohydrolase 2	ENTPD2	Role of the ectonucleotidase NTPDase2 in taste bud function. PMID: 23959882
Ectonucleoside triphosphate diphosphohydrolase 3	ENTPD3	Ectonucleotidase NTPDase3 is abundant in pancreatic β-cells and regulates glucose-induced insulin secretion. PMID: 24085034
Ectonucleoside triphosphate diphosphohydrolase 4	ENTPD4	YND1 interacts with CDC55 and is a novel mediator of E4orf4-induced toxicity. PMID: 16227198
Ectonucleoside triphosphate diphosphohydrolase 5	ENTPD5	ENTPD5-mediated modulation of ATP results in altered metabolism and decreased survival in glioblastoma multiforme. PMID: 22992974
Ectonucleoside triphosphate diphosphohydrolase 6 (putative)	ENTPD6	Ectonucleoside triphosphate diphosphohydrolase 6 expression in testis and testicular cancer and its implication in cisplatin resistance. PMID: 21519793
Ectonucleoside triphosphate diphosphohydrolase 7	ENTPD7	Ecto-nucleoside triphosphate diphosphohydrolase 7 controls Th17 cell responses through regulation of luminal ATP in the small intestine. PMID: 23241884
Ectonucleoside triphosphate diphosphohydrolase 8	ENTPD8	
Guanosine monophosphate reductase	GMPR	A purine nucleotide biosynthesis enzyme guanosine monophosphate reductase is a suppressor of melanoma invasion. PMID: 24139804
Guanosine monophosphate reductase 2	GMPR2	Crystal structure of human guanosine monophosphate reductase 2 (GMPR2) in complex with GMP. PMID: 16359702
Histidine triad nucleotide binding protein 1	HINT1	HINT1 protein: A new therapeutic target to enhance opioid antinociception and block mechanical allodynia. PMID: 25445489. Disease: neuromyotonia and axonal neuropathy, autosomal recessive (NMAN)
Histidine triad nucleotide binding protein 3	HINT3	Evidence that human histidine triad nucleotide binding protein 3 (Hint3) is a distinct branch of the histidine triad (HIT) superfamily. PMID: 17870088
IMP (inosine 5'-monophosphate) dehydrogenase 1	IMPDH1	**Mycophenolic acid** – immunosuppressants. Disease: retinitis pigmentosa 10 (RP10)
IMP (inosine 5'-monophosphate) dehydrogenase 2	IMPDH2	Identification of IMPDH2 as a tumor-associated antigen in colorectal cancer using immunoproteomics analysis. PMID: 19597826
NME/NM23 nucleoside diphosphate kinase 1	NME1	Nm23-h1 binds to gelsolin and inactivates its actin-severing capacity to promote tumor cell motility and metastasis. PMID: 23940300
NME/NM23 nucleoside diphosphate kinase 2	NME2	Non-metastatic 2 (NME2)-mediated suppression of lung cancer metastasis involves transcriptional regulation of key cell adhesion factor vinculin. PMID: 25249619
NME/NM23 nucleoside diphosphate kinase 3	NME3	Inhibitory effect of upregulated DR-nm23 expression on invasion and metastasis in colorectal cancer. PMID: 23765094
NME/NM23 nucleoside diphosphate kinase 4	NME4	Dual function of mitochondrial Nm23-H4 protein in phosphotransfer and intermembrane lipid transfer: A cardiolipin-dependent switch. PMID: 23150663
NME/NM23 nucleoside diphosphate kinase 6	NME6	A shRNA functional screen reveals Nme6 and Nme7 are crucial for embryonic stem cell renewal. PMID: 22899353
NME/NM23 family member 7	NME7	NME7 is a functional component of the γ-tubulin ring complex. PMID: 24807905
NME/NM23 family member 8	NME8	Disease: ciliary dyskinesia, primary, 6 (CILD6)

HGNC-approved name	Symbol	Target information
NME/NM23 family member 9	NME9	Thioredoxin-like protein 2 is overexpressed in colon cancer and promotes cancer cell metastasis by interaction with ran. PMID: 23311631
Nudix (nucleoside diphosphate linked moiety X)-type motif 1	NUDT1	Stereospecific targeting of MTH1 by (S)-crizotinib as an anticancer strategy. PMID: 24695225
Nudix (nucleoside diphosphate linked moiety X)-type motif 2	NUDT2	Nudix-type motif 2 in human breast carcinoma: A potent prognostic factor associated with cell proliferation. PMID: 20533549
Nudix (nucleoside diphosphate linked moiety X)-type motif 3	NUDT3	What model organisms and interactomics can reveal about the genetics of human obesity. PMID: 22618246
Nudix (nucleoside diphosphate linked moiety X)-type motif 4	NUDT4	
Nudix (nucleoside diphosphate linked moiety X)-type motif 5	NUDT5	Lowered nudix type 5 (NUDT5) expression leads to cell cycle retardation in HeLa cells. PMID: 22200976
Nudix (nucleoside diphosphate linked moiety X)-type motif 8	NUDT8	
Nudix (nucleoside diphosphate linked moiety X)-type motif 9	NUDT9	The crystal structure and mutational analysis of human NUDT9. PMID: 12948489
Nudix (nucleoside diphosphate linked moiety X)-type motif 10	NUDT10	The enzymes of human diphosphoinositol polyphosphate metabolism. PMID: 24152294
Nudix (nucleoside diphosphate linked moiety X)-type motif 11	NUDT11	Genetic and functional analyses implicate the NUDT11, HNF1B, and SLC22A3 genes in prostate cancer pathogenesis. PMID: 22730461
Nudix (nucleoside diphosphate linked moiety X)-type motif 14	NUDT14	
Nudix (nucleoside diphosphate linked moiety X)-type motif 15	NUDT15	A common missense variant in NUDT15 confers susceptibility to thiopurine-induced leukopenia. PMID: 25108385
Nudix (nucleoside diphosphate linked moiety X)-type motif 17	NUDT17	
Nudix (nucleoside diphosphate linked moiety X)-type motif 18	NUDT18	Human MTH3 (NUDT18) protein hydrolyzes oxidized forms of guanosine and deoxyguanosine diphosphates: Comparison with MTH1 and MTH2. PMID: 22556419
Phosphoribosyl pyrophosphate synthetase 1	PRPS1	Hearing loss and PRPS1 mutations: Wide spectrum of phenotypes and potential therapy. PMID: 23190330. Disease: phosphoribosylpyrophosphate synthetase superactivity (PRPS1 superactivity)
Phosphoribosyl pyrophosphate synthetase 1-like 1	PRPS1L1	
Phosphoribosyl pyrophosphate synthetase 2	PRPS2	Purine and nucleotide biosynthesis are coupled by a single rate-limiting enzyme, PRPS2, to drive cancer. PMID: 24855946
Ribonucleotide reductase M1	RRM1	**Fludarabine** – antineoplastic agents
Ribonucleotide reductase M2	RRM2	AKT-induced tamoxifen resistance is overturned by RRM2 inhibition. PMID: 24362250
Ribonucleotide reductase M2 B (TP53 inducible)	RRM2B	Regulation of p53R2 and its role as potential target for cancer therapy. PMID: 18760875. Disease: mitochondrial DNA depletion syndrome 8A (MTDPS8A)
Thymidine kinase 1, soluble	TK1	N3-substituted thymidine bioconjugates for cancer therapy and imaging. PMID: 23617430
Thymidine kinase 2, mitochondrial	TK2	Disease: mitochondrial DNA depletion syndrome 2 (MTDPS2)
Uridine phosphorylase 1	UPP1	Modulation of 5-fluorouracil host-toxicity and chemotherapeutic efficacy against human colon tumors by 5-(Phenylthio)acyclouridine, a uridine phosphorylase inhibitor. PMID: 16528530

(Continued)

HGNC-approved name	Symbol	Target information
Uridine phosphorylase 2	UPP2	A novel structural mechanism for redox regulation of uridine phosphorylase 2 activity. PMID: 21855639
Uridine–cytidine kinase 1	UCK1	Expression of nucleoside-metabolizing enzymes in myelodysplastic syndromes and modulation of response to azacitidine. PMID: 24192812
Uridine–cytidine kinase 1-like 1	UCKL1	
Uridine–cytidine kinase 2	UCK2	Mechanisms of resistance to azacitidine in human leukemia cell lines. PMID: 24368162
2′,3′-Cyclic nucleotide 3′ phosphodiesterase	CNP	Expression of 2′,3′-cyclic nucleotide 3′-phosphodiesterase (CNPase) and its roles in activated microglia in vivo and *in vitro*. PMID: 25148928
2′-Deoxynucleoside 5′-phosphate N-hydrolase 1	DNPH1	6-(Hetero)Arylpurine nucleotides as inhibitors of the oncogenic target DNPH1: Synthesis, structural studies and cytotoxic activities. PMID: 25108359
3′(2′), 5′-Bisphosphate nucleotidase 1	BPNT1	Alteration of lithium pharmacology through manipulation of phosphoadenosine phosphate metabolism. PMID: 15583009
5-Aminoimidazole-4-carboxamide ribonucleotide formyltransferase/ IMP cyclohydrolase	ATIC	**Pemetrexed** – antineoplastic agents. Disease: ATIC transformylase/IMP cyclohydrolase deficiency (AICAR)
Acid phosphatase, prostate	ACPP	**Sipuleucel-T (DC vaccine)** – antineoplastic agents
Adenine phosphoribosyltransferase	APRT	Disease: adenine phosphoribosyltransferase deficiency (APRTD)
Adenosine kinase	ADK	ABT-702, an adenosine kinase inhibitor, attenuates inflammation in diabetic retinopathy. PMID: 23770229. Disease: hypermethioninemia due to adenosine kinase deficiency (HMAKD)
Adenylosuccinate lyase	ADSL	Disease: adenylosuccinase deficiency (ADSL deficiency)
Carbamoyl-phosphate synthetase 2, aspartate transcarbamylase, and dihydroorotase	CAD	The nucleotide synthesis enzyme CAD inhibits NOD2 antibacterial function in human intestinal epithelial cells. PMID: 22387394
Cat eye syndrome chromosome region, candidate 1	CECR1	Human adenosine deaminase 2 induces differentiation of monocytes into macrophages and stimulates proliferation of T helper cells and macrophages. PMID: 20453107
Collapsin response mediator protein 1	CRMP1	CRMP1 interacted with Spy1 during the collapse of growth cones induced by Sema3A and acted on regeneration after sciatic nerve crush. PMID: 25526860
Cytidine deaminase	CDA	**Tetrahydrouridine** – antineoplastic agents
dCMP deaminase	DCTD	DNA mismatch repair (MMR)-dependent 5-fluorouracil cytotoxicity and the potential for new therapeutic targets. PMID: 19775280
dCTP pyrophosphatase 1	DCTPP1	Triptolide directly inhibits dCTP pyrophosphatase. PMID: 21671327
Deoxycytidine kinase	DCK	Hypoxia-induced deoxycytidine kinase contributes to epithelial proliferation in pulmonary fibrosis. PMID: 25358054
Deoxyguanosine kinase	DGUOK	Disease: mitochondrial DNA depletion syndrome 3 (MTDPS3)
Deoxyribose-phosphate aldolase (putative)	DERA	DERA is the human deoxyribose phosphate aldolase and is involved in stress response. PMID: 25229427
Deoxythymidylate kinase (thymidylate kinase)	DTYMK	Metabolic and functional genomic studies identify deoxythymidylate kinase as a target in LKB1-mutant lung cancer. PMID: 23715154
Deoxyuridine triphosphatase	DUT	**TAS-114** – antineoplastic agents
Dihydroorotate dehydrogenase (quinone)	DHODH	**Leflunomide** – anti-inflammatory and antirheumatic products. Disease: postaxial acrofacial dysostosis (POADS)
Dihydropyrimidine dehydrogenase	DPYD	**Gimeracil** – antineoplastic agents. Disease: dihydropyrimidine dehydrogenase deficiency (DPYDD)
Ectonucleotide pyrophosphatase/ phosphodiesterase 1	ENPP1	Disease: ossification of the posterior longitudinal ligament of the spine (OPLL)

HGNC-approved name	Symbol	Target information
Fragile histidine triad	FHIT	Note: A chromosomal aberration involving FHIT has been found in a lymphoblastoid cell line established from a family with renal cell carcinoma and thyroid carcinoma. Translocation t(38)(p14.2q24.1) with RNF139. Although the 3p14.2 breakpoint has been shown to interrupt FHIT in its 5-prime non-coding region, it is unlikely that FHIT is causally related to renal or other malignancies
Guanine deaminase	GDA	Structural basis of the substrate specificity of cytidine deaminase superfamily Guanine deaminase. PMID: 24083949
Guanine monophosphate synthase	GMPS	Note: A chromosomal aberration involving GMPS is found in acute myeloid leukemias. Translocation t(3,11)(q25,q23) with KMT2A/MLL1
Guanylate kinase 1	GUK1	A guanylate kinase/HSV-1 thymidine kinase fusion protein enhances prodrug-mediated cell killing. PMID: 16810197
Haloacid dehalogenase-like hydrolase domain containing 1	HDHD1	HDHD1, which is often deleted in X-linked ichthyosis, encodes a pseudouridine-5′-phosphatase. PMID: 20722631
HD domain containing 3	HDDC3	
Hypoxanthine-guanine phosphoribosyltransferase	HPRT1	**Azathioprine** – antineoplastic agents. Disease: Lesch–Nyhan syndrome (LNS)
Inosine triphosphatase (nucleoside triphosphate pyrophosphatase)	ITPA	Disease: inosine triphosphate pyrophosphohydrolase deficiency (ITPAD)
Leucine carboxyl methyltransferase 2	LCMT2	
Methylthioadenosine phosphorylase	MTAP	Disease: diaphyseal medullary stenosis with malignant fibrous histiocytoma (DM3MFH)
Nucleoside-triphosphatase, cancer-related	NTPCR	On the cytotoxicity of HCR-NTPase in the neuroblastoma cell line SH-SY5Y. PMID: 19519914
Phosphoribosyl pyrophosphate amidotransferase	PPAT	**Mercaptopurine** – antineoplastic agents
Phosphoribosylaminoimidazole carboxylase, phosphoribosylaminoimidazole succinocarboxamide synthetase	PAICS	Octameric structure of the human bifunctional enzyme PAICS in purine biosynthesis. PMID: 17224163
Phosphoribosylformylglycinamidine synthase	PFAS	Reversible compartmentalization of de novo purine biosynthetic complexes in living cells. PMID: 18388293
Phosphoribosylglycinamide formyltransferase, phosphoribosylglycinamide synthetase, phosphoribosylaminoimidazole synthetase	GART	
Purine nucleoside phosphorylase	PNP	**Forodesine** – antineoplastic agents. Disease: purine nucleoside phosphorylase deficiency (PNPD)
SAM domain and HD domain 1	SAMHD1	Disease: Aicardi–Goutieres syndrome 5 (AGS5)
TDP-glucose 4,6-dehydratase	TGDS	Homozygous and compound-heterozygous mutations in TGDS cause Catel–Manzke syndrome. PMID: 25480037
Thiopurine S-methyltransferase	TPMT	Disease: thiopurine S-methyltransferase deficiency (TPMT deficiency)
Thymidylate synthetase	TYMS	**Leucovorin** – antineoplastic adjuncts
Thymidine phosphorylase	TYMP	Thymidine phosphorylase inhibitors: Recent developments and potential therapeutic applications. PMID: 16375757. Disease: mitochondrial DNA depletion syndrome 1, MNGIE type (MTDPS1)

(Continued)

HGNC-approved name	Symbol	Target information
Uracil phosphoribosyltransferase (FUR1) homolog (*S. cerevisiae*)	UPRT	Complete regression of glioblastoma by mesenchymal stem cells mediated prodrug gene therapy simulating clinical therapeutic scenario. PMID: 24038033
Ureidoimidazoline (2-oxo-4-hydroxy-4-carboxy-5-) decarboxylase	URAD	Completing the uric acid degradation pathway through phylogenetic comparison of whole genomes. PMID: 16462750
Uridine monophosphate synthetase	UMPS	Disease: orotic aciduria 1 (ORAC1)
Xanthine dehydrogenase	XDH	**Allopurinol** – antigout preparations. Disease: xanthinuria 1 (XU1)

5.4   Carbohydrates and related

HGNC-approved name	Symbol	Target information
3′-Phosphoadenosine 5′-phosphosulfate synthase 1	PAPSS1	Human 3′-phosphoadenosine 5′-phosphosulfate (PAPS) synthase: Biochemistry, molecular biology and genetic deficiency. PMID: 12716056
3′-Phosphoadenosine 5′-phosphosulfate synthase 2	PAPSS2	Disease: spondyloepimetaphyseal dysplasia Pakistani type (SEMD-PA)
6-Phosphofructo-2-kinase/fructose-2,6-biphosphatase 1	PFKFB1	Hypoxia, glucose metabolism and the Warburg's effect. PMID: 17661163
6-Phosphofructo-2-kinase/fructose-2,6-biphosphatase 2	PFKFB2	A fundamental trade-off in covalent switching and its circumvention by enzyme bifunctionality in glucose homeostasis. PMID: 24634222
6-Phosphofructo-2-kinase/fructose-2,6-biphosphatase 3	PFKFB3	Partial and transient reduction of glycolysis by PFKFB3 blockade reduces pathological angiogenesis. PMID: 24332967
6-Phosphofructo-2-kinase/fructose-2,6-biphosphatase 4	PFKFB4	RNAi screening in glioma stem-like cells identifies PFKFB4 as a key molecule important for cancer cell survival. PMID: 22056879
Aldolase A, fructose-bisphosphate	ALDOA	Role of aldolase A in osteosarcoma progression and metastasis: *in vitro* and in vivo evidence. PMID: 25213699. Disease: glycogen storage disease 12 (GSD12)
Aldolase B, fructose-bisphosphate	ALDOB	Disease: hereditary fructose intolerance (HFI)
Aldolase C, fructose-bisphosphate	ALDOC	Detailed expression pattern of aldolase C (Aldoc) in the cerebellum, retina and other areas of the CNS studied in Aldoc-Venus knock-in mice. PMID: 24475166
Amylase, alpha 1A (salivary)	AMY1A	Amylase α-1A (AMY1A): A novel immunohistochemical marker to differentiate chromophobe renal cell carcinoma from benign oncocytoma. PMID: 24225843
Amylase, alpha 1B (salivary)	AMY1B	Low copy number of the salivary amylase gene predisposes to obesity. PMID: 24686848
Amylase, alpha 1C (salivary)	AMY1C	
Amylase, alpha 2A (pancreatic)	AMY2A	**Acarbose** – drugs used in diabetes
Amylase, alpha 2B (pancreatic)	AMY2B	
Chitinase 1 (chitotriosidase)	CHIT1	Chitinase 1 is a biomarker for and therapeutic target in scleroderma-associated interstitial lung disease that augments TGF-β1 signaling. PMID: 22826322
Chitinase 3-like 1 (cartilage glycoprotein-39)	CHI3L1	Chitinase 3 like 1 is associated with tumor angiogenesis in cervical cancer. PMID: 24691276. Disease: asthma-related traits 7 (ASRT7)

HGNC-approved name	Symbol	Target information
Chitinase 3-like 2	CHI3L2	Human cartilage chitinase 3-like protein 2: Cloning, expression, and production of polyclonal and monoclonal antibodies for osteoarthritis detection and identification of potential binding partners. PMID: 24111862
Chitinase, acidic	CHIA	Targeting AMCase reduces esophageal eosinophilic inflammation and remodeling in a mouse model of egg induced eosinophilic esophagitis. PMID: 24239745
Chondroitin polymerizing factor	CHPF	Inhibiting glycosaminoglycan chain polymerization decreases the inhibitory activity of astrocyte-derived chondroitin sulfate proteoglycans. PMID: 18160657
Chondroitin polymerizing factor 2	CHPF2	Chondroitin synthases I, II, III and chondroitin sulfate glucuronyltransferase expression in colorectal cancer. PMID: 21468578
Chondroitin sulfate synthase 1	CHSY1	Loss of CHSY1, a secreted FRINGE enzyme, causes syndromic brachydactyly in humans via increased NOTCH signaling. PMID: 21129727. Disease: temtamy preaxial brachydactyly syndrome (TPBS)
Chondroitin sulfate synthase 3	CHSY3	Identification of chondroitin sulfate glucuronyltransferase as chondroitin synthase-3 involved in chondroitin polymerization: Chondroitin polymerization is achieved by multiple enzyme complexes consisting of chondroitin synthase family members. PMID: 18316376
Dermatan sulfate epimerase	DSE	Dermatan sulfate epimerase 1 deficient mice as a model for human abdominal wall defects. PMID: 25186462. Disease: Ehlers–Danlos syndrome, musculocontractural type 2 (EDSMC2)
Dermatan sulfate epimerase-like	DSEL	Biosynthesis of dermatan sulfate. Chondroitin-glucuronate C5-epimerase is identical to SART2. PMID: 16505484
Enolase 1 (alpha)	ENO1	Surface α-enolase promotes extracellular matrix degradation and tumor metastasis and represents a new therapeutic target. PMID: 23894455
Enolase 2 (gamma, neuronal)	ENO2	Passenger deletions generate therapeutic vulnerabilities in cancer. PMID: 22895339
Enolase 3 (beta, muscle)	ENO3	Disease: glycogen storage disease 13 (GSD13)
Enolase family member 4	ENO4	Disruption of a spermatogenic cell-specific mouse enolase 4 (eno4) gene causes sperm structural defects and male infertility. PMID: 23446454
Fructosamine 3 kinase	FN3K	Fructosamine 3-kinase and glyoxalase I polymorphisms and their association with soluble RAGE and adhesion molecules in diabetes. PMID: 24908254
Fructosamine 3 kinase-related protein	FN3KRP	The physiological substrates of fructosamine-3-kinase-related-protein (FN3KRP) are intermediates of nonenzymatic reactions between biological amines and ketose sugars (fructation products). PMID: 21924559
Fructose-1,6-bisphosphatase 1	FBP1	Fructose-1,6-bisphosphatase opposes renal carcinoma progression. PMID: 25043030. Disease: fructose-1,6-bisphosphatase deficiency (FBPD)
Fructose-1,6-bisphosphatase 2	FBP2	Decreased fructose-1,6-bisphosphatase-2 expression promotes glycolysis and growth in gastric cancer cells. PMID: 24063558
Fucosidase, alpha-L- 1, tissue	FUCA1	Decreased expression of alpha-L-fucosidase gene FUCA1 in human colorectal tumors. PMID: 23965968. Disease: fucosidosis (FUCA1D)

(*Continued*)

HGNC-approved name	Symbol	Target information
Fucosidase, alpha-L- 2, plasma	FUCA2	Identification of predictive biomarkers for response to trastuzumab using plasma FUCA activity and *N*-glycan identified by MALDI-TOF-MS. PMID: 19140672
Galactokinase 1	GALK1	Disease: galactosemia II (GALCT2)
Galactokinase 2	GALK2	Evaluation of analogues of GalNAc as substrates for enzymes of the mammalian GalNAc salvage pathway. PMID: 22276930
Galactosidase, alpha	GLA	Disease: Fabry disease (FD)
Galactosidase, beta 1	GLB1	Disease: GM1-gangliosidosis 1 (GM1G1)
Galactosidase, beta 1-like	GLB1L	
Galactosidase, beta 1-like 2	GLB1L2	
Galactosidase, beta 1-like 3	GLB1L3	Altered expression of β-galactosidase-1-like protein 3 (Glb1l3) in the retinal pigment epithelium (RPE)-specific 65-kDa protein knock-out mouse model of Leber's congenital amaurosis. PMID: 21633714
GDP-mannose pyrophosphorylase A	GMPPA	Disease: alacrima, achalasia, and mental retardation syndrome (AAMR)
GDP-mannose pyrophosphorylase B	GMPPB	Disease: muscular dystrophy–dystroglycanopathy congenital with brain and eye anomalies A14 (MDDGA14)
Glucosamine-6-phosphate deaminase 1	GNPDA1	Allosteric kinetics of the isoform 1 of human glucosamine-6-phosphate deaminase. PMID: 21807125
Glucosamine-6-phosphate deaminase 2	GNPDA2	
Glucose-6-phosphatase, catalytic, 3	G6PC3	Disease: neutropenia, severe congenital 4, autosomal recessive (SCN4)
Glucose-6-phosphatase, catalytic subunit	G6PC	Glucose-6-phosphatase is a key metabolic regulator of glioblastoma invasion. PMID: 25001192. Disease: glycogen storage disease 1A (GSD1A)
Glucose-6-phosphatase, catalytic, 2	G6PC2	Moving on from GWAS: Functional studies on the G6PC2 gene implicated in the regulation of fasting blood glucose. PMID: 24142592
Glucosidase, alpha; acid	GAA	**Acarbose** – drugs used in diabetes. Disease: glycogen storage disease 2 (GSD2)
Glucosidase, alpha; neutral AB	GANAB	
Glucosidase, alpha; neutral C	GANC	
Glucosidase, beta (bile acid) 2	GBA2	Disease: spastic paraplegia 46, autosomal recessive (SPG46)
Glucosidase, beta, acid	GBA	Augmenting CNS glucocerebrosidase activity as a therapeutic strategy for parkinsonism and other Gaucher-related synucleinopathies. PMID: 23297226. Disease: Gaucher disease (GD)
Glucosidase, beta, acid 3 (gene/pseudogene)	GBA3	Exploring functional cyclophellitol analogues as human retaining beta-glucosidase inhibitors. PMID: 25156485
Glutamine–fructose-6-phosphate transaminase 1	GFPT1	Disease: myasthenic syndrome, congenital, with tubular aggregates, 1 (CMSTA1)
Glutamine–fructose-6-phosphate transaminase 2	GFPT2	
Glycogen synthase 1 (muscle)	GYS1	Disease: muscle glycogen storage disease 0 (GSD0b)
Glycogen synthase 2 (liver)	GYS2	Disease: glycogen storage disease 0 (GSD0)
Glycogenin 1	GYG1	Disease: glycogen storage disease 15 (GSD15)
Glycogenin 2	GYG2	A hemizygous GYG2 mutation and Leigh syndrome: A possible link? PMID: 24100632
Heparanase	HPSE	**PG545** – antineoplastic agents
Heparanase 2	HPSE2	Disease: urofacial syndrome 1 (UFS1)

HGNC-approved name	Symbol	Target information
Hexokinase 1	HK1	Disease: hexokinase deficiency (HK deficiency)
Hexokinase 2	HK2	Cerulenin-induced apoptosis is mediated by disrupting the interaction between AIF and hexokinase II. PMID: 22426850
Hexokinase 3 (white cell)	HK3	CEBPA-dependent HK3 and KLF5 expression in primary AML and during AML differentiation. PMID: 24584857
Hexosaminidase (glycosyl hydrolase family 20, catalytic domain) containing	HEXDC	The recently identified hexosaminidase D enzyme substantially contributes to the elevated hexosaminidase activity in rheumatoid arthritis. PMID: 23099419
Hexosaminidase A (alpha polypeptide)	HEXA	Disease: GM2-gangliosidosis 1 (GM2G1)
Hexosaminidase B (beta polypeptide)	HEXB	Disease: GM2-gangliosidosis 2 (GM2G2)
Hyaluronan synthase 1	HAS1	Aberrant posttranscriptional processing of hyaluronan synthase 1 in malignant transformation and tumor progression. PMID: 25081526
Hyaluronan synthase 2	HAS2	Note: A chromosomal aberration involving HAS2 may be a cause of lipoblastomas, which are benign tumors resulting from transformation of adipocytes, usually diagnosed in children. 8q12.1 to 8q24.1 intrachromosomal rearrangement with PLAG1
Hyaluronan synthase 3	HAS3	Accumulation of extracellular hyaluronan by hyaluronan synthase 3 promotes tumor growth and modulates the pancreatic cancer microenvironment. PMID: 25147816
Hyaluronoglucosaminidase 1	HYAL1	Disease: mucopolysaccharidosis 9 (MPS9)
Hyaluronoglucosaminidase 2	HYAL2	Platelet hyaluronidase-2: An enzyme that translocates to the surface upon activation to function in extracellular matrix degradation PMID: 25411425
Hyaluronoglucosaminidase 3	HYAL3	Hyaluronidase 3 (HYAL3) knockout mice do not display evidence of hyaluronan accumulation. PMID: 18762256
Hyaluronoglucosaminidase 4	HYAL4	Identification of amino acid residues required for the substrate specificity of human and mouse chondroitin sulfate hydrolase (conventional hyaluronidase-4). PMID: 23086929
Inositol(myo)-1(or 4)-monophosphatase 1	IMPA1	Inositol-related gene knockouts mimic lithium's effect on mitochondrial function. PMID: 23924600
Inositol(myo)-1(or 4)-monophosphatase 2	IMPA2	In silico study on the substrate binding manner in human myo-inositol monophosphatase 2. PMID: 21213002
Inositol monophosphatase domain containing 1	IMPAD1	Disease: chondrodysplasia with joint dislocations, GPAPP type (CDP-GPAPP)
Isocitrate dehydrogenase 1 (NADP+), soluble	IDH1	**AGH-120** – antineoplastic agents. Disease: glioma (GLM)
Isocitrate dehydrogenase 2 (NADP+), mitochondrial	IDH2	**AG-221** – antineoplastic agents. Disease: D-2-hydroxyglutaric aciduria 2 (D2HGA2)
Isocitrate dehydrogenase 3 (NAD+) alpha	IDH3A	
Isocitrate dehydrogenase 3 (NAD+) beta	IDH3B	Disease: retinitis pigmentosa 46 (RP46)
Isocitrate dehydrogenase 3 (NAD+) gamma	IDH3G	Genome-wide shRNA screening identifies host factors involved in early endocytic events for HIV-1-induced CD4 down-regulation. PMID: 25496667
Lactase	LCT	Disease: congenital lactase deficiency (COLACD)
Lactase-like	LCTL	Short hairpin RNA screen indicates that Klotho beta/FGF19 protein overcomes stasis in human colonic epithelial cells. PMID: 22020932

(Continued)

HGNC-approved name	Symbol	Target information
Lactate dehydrogenase A	LDHA	Optimization of 5-(2,6-dichlorophenyl)-3-hydroxy-2-mercaptocyclohex-2-enones as potent inhibitors of human lactate dehydrogenase. PMID: 25466195. Disease: glycogen storage disease 11 (GSD11)
Lactate dehydrogenase A-like 6A	LDHAL6A	Identification of a novel human lactate dehydrogenase gene LDHAL6A, which activates transcriptional activities of AP1(PMA). PMID: 18351441
Lactate dehydrogenase A-like 6B	LDHAL6B	
Lactate dehydrogenase B	LDHB	
Lactate dehydrogenase C	LDHC	Lactate dehydrogenase C and energy metabolism in mouse sperm. PMID: 21565994
Lactate dehydrogenase D	LDHD	
Lysozyme	LYZ	Disease: amyloidosis 8 (AMYL8)
Lysozyme-like 1	LYZL1	
Lysozyme-like 2	LYZL2	Molecular cloning and characterization of three novel lysozyme-like genes, predominantly expressed in the male reproductive system of humans, belonging to the c-type lysozyme/alpha-lactalbumin family. PMID: 16014814
Lysozyme-like 4	LYZL4	
Lysozyme-like 6	LYZL6	Characterisation of Lyzls in mice and antibacterial properties of human LYZL6. PMID: 24013621
Malate dehydrogenase 1, NAD (soluble)	MDH1	Cytosolic malate dehydrogenase regulates senescence in human fibroblasts. PMID: 22971926
Malate dehydrogenase 1B, NAD (soluble)	MDH1B	
Malate dehydrogenase 2, NAD (mitochondrial)	MDH2	Visnagin protects against doxorubicin-induced cardiomyopathy through modulation of mitochondrial malate dehydrogenase. PMID: 25504881
Malic enzyme 1, NADP(+)-dependent, cytosolic	ME1	Enhanced gastrointestinal expression of cytosolic malic enzyme (ME1) induces intestinal and liver lipogenic gene expression and intestinal cell proliferation in mice. PMID: 25402228
Malic enzyme 2, NAD(+)-dependent, mitochondrial	ME2	Knockdown of malic enzyme 2 suppresses lung tumor growth, induces differentiation and impacts PI3K/AKT signaling. PMID: 24957098
Malic enzyme 3, NADP(+)-dependent, mitochondrial	ME3	Mitochondrial malic enzyme 3 is important for insulin secretion in pancreatic beta cells. PMID: 25594249
N-Acetylglucosamine-1-phosphate transferase, alpha and beta subunits	GNPTAB	Disease: mucolipidosis type II (MLII)
N-Acetylglucosamine-1-phosphate transferase, gamma subunit	GNPTG	Disease: mucolipidosis type III complementation group C (MLIIIC)
NADH dehydrogenase (ubiquinone) 1 alpha subcomplex, 4, 9 kDa	NDUFA4	NDUFA4 mutations underlie dysfunction of a cytochrome c oxidase subunit linked to human neurological disease. PMID: 23746447
NADH dehydrogenase (ubiquinone) Fe–S protein 1, 75 kDa (NADH-coenzyme Q reductase)	NDUFS1	Disease: mitochondrial complex I deficiency (MT-C1D)
NADH dehydrogenase (ubiquinone) Fe–S protein 2, 49 kDa (NADH-coenzyme Q reductase)	NDUFS2	Disease: mitochondrial complex I deficiency (MT-C1D)
NADH dehydrogenase (ubiquinone) Fe–S protein 3, 30 kDa (NADH-coenzyme Q reductase)	NDUFS3	Human mitochondrial NDUFS3 protein bearing Leigh syndrome mutation is more prone to aggregation than its wild-type. PMID: 24028823

HGNC-approved name	Symbol	Target information
NADH dehydrogenase (ubiquinone) Fe–S protein 7, 20 kDa (NADH-coenzyme Q reductase)	NDUFS7	Disease: Leigh syndrome (LS)
NADH dehydrogenase (ubiquinone) Fe–S protein 8, 23 kDa (NADH-coenzyme Q reductase)	NDUFS8	Disease: Leigh syndrome (LS)
NADH dehydrogenase (ubiquinone) flavoprotein 1, 51 kDa	NDUFV1	Disease: Leigh syndrome (LS)
NADH dehydrogenase (ubiquinone) flavoprotein 2, 24 kDa	NDUFV2	Mitochondrial targeting of human NADH dehydrogenase (ubiquinone) flavoprotein 2 (NDUFV2) and its association with early-onset hypertrophic cardiomyopathy and encephalopathy. PMID: 21548921
Otogelin	OTOG	Disease: deafness, autosomal recessive, 18B (DFNB18B)
Otogelin-like	OTOGL	Disease: deafness, autosomal recessive, 84B (DFNB84B)
Phosphoenolpyruvate carboxykinase 1 (soluble)	PCK1	Disease: cytosolic phosphoenolpyruvate carboxykinase deficiency (C-PEPCKD)
Phosphoenolpyruvate carboxykinase 2 (mitochondrial)	PCK2	PCK2 activation mediates an adaptive response to glucose depletion in lung cancer. PMID: 24632615. Disease: mitochondrial phosphoenolpyruvate carboxykinase deficiency (M-PEPCKD)
Phosphofructokinase, liver	PFKL	A fresh view of glycolysis and glucokinase regulation: History and current status. PMID: 24637025
Phosphofructokinase, muscle	PFKM	Phosphofructokinase 1 glycosylation regulates cell growth and metabolism. PMID: 22923583. Disease: glycogen storage disease 7 (GSD7)
Phosphofructokinase, platelet	PFKP	Anti-human protein S antibody induces tissue factor expression through a direct interaction with platelet phosphofructokinase. PMID: 24331211
Phosphoglucomutase 1	PGM1	Disease: glycogen storage disease 14 (GSD14)
Phosphoglucomutase 2	PGM2	
Phosphoglucomutase 2-like 1	PGM2L1	
Phosphoglucomutase 3	PGM3	Hyper-IgE syndromes: Reviewing PGM3 deficiency. PMID: 25365149
Phosphoglucomutase 5	PGM5	
Phosphoglycerate mutase 1 (brain)	PGAM1	Quantitative proteomics identification of phosphoglycerate mutase 1 at a novel therapeutic target in hepatocellular carcinoma. PMID: 20403181
Phosphoglycerate mutase 2 (muscle)	PGAM2	Disease: glycogen storage disease 10 (GSD10)
Phosphoglycerate mutase family member 4	PGAM4	A single nucleotide polymorphism within the novel sex-linked testis-specific retrotransposed PGAM4 gene influences human male fertility. PMID: 22590500
Phosphoglycerate mutase family member 5	PGAM5	Genetic deficiency of the mitochondrial protein PGAM5 causes a Parkinson's-like movement disorder. PMID: 25222142
Phosphomannomutase 1	PMM1	Mammalian phosphomannomutase PMM1 is the brain IMP-sensitive glucose-1,6-bisphosphatase. PMID: 18927083
Phosphomannomutase 2	PMM2	Disease: congenital disorder of glycosylation 1A (CDG1A)
Phosphorylase, glycogen, liver	PYGL	Glucose utilization via glycogen phosphorylase sustains proliferation and prevents premature senescence in cancer cells. PMID: 23177934. Disease: glycogen storage disease 6 (GSD6)

(Continued)

HGNC-approved name	Symbol	Target information
Phosphorylase, glycogen, muscle	PYGM	Disease: glycogen storage disease 5 (GSD5)
Phosphorylase, glycogen; brain	PYGB	Increased sensitivity to glucose starvation correlates with down-regulation of glycogen phosphorylase isoform PYGB in tumor cell lines resistant to 2-deoxy-ᴅ-glucose. PMID: 24292700
Pyruvate dehydrogenase (lipoamide) alpha 1	PDHA1	Disease: pyruvate dehydrogenase E1-alpha deficiency (PDHAD)
Pyruvate dehydrogenase (lipoamide) alpha 2	PDHA2	Demethylation of the coding region triggers the activation of the human testis-specific PDHA2 gene in somatic tissues. PMID: 22675509
Pyruvate dehydrogenase (lipoamide) beta	PDHB	Disease: pyruvate dehydrogenase E1-beta deficiency (PDHBD)
Pyruvate dehydrogenase complex, component X	PDHX	Disease: pyruvate dehydrogenase E3-binding protein deficiency (PDHXD)
Pyruvate kinase, liver and RBC	PKLR	Disease: pyruvate kinase hyperactivity (PKHYP)
Pyruvate kinase, muscle	PKM	**AG-348** – antianaemic preparations, antineoplastic agents
Ribulose-5-phosphate-3-epimerase	RPE	Conversion of ᴅ-ribulose 5-phosphate to ᴅ-xylulose 5-phosphate: New insights from structural and biochemical studies on human RPE. PMID: 20923965
Ribulose-5-phosphate-3-epimerase-like 1	RPEL1	
Sialidase 1 (lysosomal sialidase)	NEU1	Therapeutic targeting of Neu1 sialidase with oseltamivir phosphate (Tamiflu®) disables cancer cell survival in human pancreatic cancer with acquired chemoresistance. PMID: 24470763. Disease: sialidosis (SIALIDOSIS)
Sialidase 2 (cytosolic sialidase)	NEU2	Inhibitory effects and specificity of synthetic sialyldendrimers toward recombinant human cytosolic sialidase 2 (NEU2). PMID: 23363739
Sialidase 3 (membrane sialidase)	NEU3	Human airway epithelia express catalytically active NEU3 sialidase. PMID: 24658138
Sialidase 4	NEU4	Identification of selective nanomolar inhibitors of the human neuraminidase, NEU4. PMID: 24900705
Succinate-CoA ligase, ADP-forming, beta subunit	SUCLA2	Disease: mitochondrial DNA depletion syndrome 5 (MTDPS5)
Succinate-CoA ligase, alpha subunit	SUCLG1	Disease: mitochondrial DNA depletion syndrome 9 (MTDPS9)
Succinate-CoA ligase, GDP-forming, beta subunit	SUCLG2	SUCLG2 identified as both a determinator of CSF Aβ1-42 levels and an attenuator of cognitive decline in Alzheimer's disease. PMID: 25027320
Sulfatase 1	SULF1	Activation of the TGFβ/SMAD transcriptional pathway underlies a novel tumor promoting role of sulfatase1 in hepatocellular carcinoma. PMID: 25503294
Sulfatase 2	SULF2	Sulf-2: An extracellular modulator of cell signaling and a cancer target candidate. PMID: 20629619
Transketolase	TKT	Altered proteolysis in fibroblasts of Alzheimer patients with predictive implications for subjects at risk of disease. PMID: 24949214
Transketolase-like 1	TKTL1	The TKTL1 gene influences total transketolase activity and cell proliferation in human colon cancer LoVo cells. PMID: 17351395
Transketolase-like 2	TKTL2	
UDP-N-acetylglucosamine pyrophosphorylase 1	UAP1	UAP1 is overexpressed in prostate cancer and is protective against inhibitors of N-linked glycosylation. PMID: 25241896
UDP-N-acetylglucosamine pyrophosphorylase 1-like 1	UAP1L1	

HGNC-approved name	Symbol	Target information
2,3-Bisphosphoglycerate mutase	BPGM	Disease: bisphosphoglycerate mutase deficiency (BPGMD)
2-Phosphoxylose phosphatase 1	PXYLP1	
6-Phosphogluconolactonase	PGLS	
Aconitase 1, soluble	ACO1	Passenger deletions generate therapeutic vulnerabilities in cancer. PMID: 22895339
Aconitase 2, mitochondrial	ACO2	Disease: infantile cerebellar–retinal degeneration (ICRD)
ADP-dependent glucokinase	ADPGK	Zinc finger nuclease mediated knockout of ADP-dependent glucokinase in cancer cell lines: Effects on cell survival and mitochondrial oxidative metabolism. PMID: 23799003
ADP-ribose/CDP-alcohol diphosphatase, manganese-dependent	ADPRM	
Amidohydrolase domain containing 2	AMDHD2	
Amylo-alpha-1, 6-glucosidase, 4 alpha glucanotransferase	AGL	Disease: glycogen storage disease 3 (GSD3)
ATH1, acid trehalase-like 1 (yeast)	ATHL1	
ATP citrate lyase	ACLY	**ETC-1002** – lipid-lowering agents
Beta-1,4-glucuronyltransferase 1	B4GAT1	The glucuronyltransferase B4GAT1 is required for initiation of LARGE-mediated α-dystroglycan functional glycosylation. PMID: 25279699
Chitobiase, di-N-acetyl-	CTBS	Lysosomal di-N-acetylchitobiase-deficient mouse tissues accumulate Man2GlcNAc2 and Man3GlcNAc2. PMID: 22465033
Citrate lyase beta-like	CLYBL	CLYBL is a polymorphic human enzyme with malate synthase and β-methylmalate synthase activity. PMID: 24334609
Citrate synthase	CS	Citrate synthase expression affects tumor phenotype and drug resistance in human ovarian carcinoma. PMID: 25545012
Cytidine monophosphate N-acetylneuraminic acid synthetase	CMAS	Nuclear localization signal of murine CMP-Neu5Ac synthetase includes residues required for both nuclear targeting and enzymatic activity. PMID: 11893746
D-2-Hydroxyglutarate dehydrogenase	D2HGDH	Disease: D-2-hydroxyglutaric aciduria 1 (D2HGA1)
Dehydrodolichyl diphosphate synthase	DHDDS	Disease: retinitis pigmentosa 59 (RP59)
Dehydrogenase E1 and transketolase domain containing 1	DHTKD1	Disease: Charcot–Marie–Tooth disease 2Q (CMT2Q)
Dicarbonyl/L-xylulose reductase	DCXR	Note: The enzyme defect in pentosuria has been shown to involve L-xylulose reductase. Essential pentosuria is an inborn error of metabolism characterized by the excessive urinary excretion of the pentose L-xylulose
Dihydrodiol dehydrogenase (dimeric)	DHDH	
Dihydrolipoamide S-acetyltransferase	DLAT	Note: Primary biliary cirrhosis is a chronic, progressive cholestatic liver disease characterized by the presence of antimitochondrial autoantibodies in patients' serum
Dihydroxyacetone kinase 2 homolog (S. cerevisiae)	DAK	Negative regulation of MDA5- but not RIG-I-mediated innate antiviral signaling by the dihydroxyacetone kinase. PMID: 17600090
Dolichol kinase	DOLK	Disease: congenital disorder of glycosylation 1M (CDG1M)
FGGY carbohydrate kinase domain containing	FGGY	Disease: amyotrophic lateral sclerosis (ALS)
Fucokinase	FUK	
Fucose mutarotase	FUOM	Male-like sexual behavior of female mouse lacking fucose mutarotase. PMID: 20609214

(Continued)

HGNC-approved name	Symbol	Target information
Fumarate hydratase	FH	Disease: fumarase deficiency (FHD)
Galactosamine (N-acetyl)-6-sulfate sulfatase	GALNS	**Elosulfase alfa** – anti-mucopolysaccharidosis treatments. Disease: mucopolysaccharidosis 4A (MPS4A)
Galactose mutarotase (aldose 1-epimerase)	GALM	Genome-wide copy number variation analysis in extended families and unrelated individuals characterized for musical aptitude and creativity in music. PMID: 23460800
Galactose-1-phosphate uridylyltransferase	GALT	Disease: galactosemia (GALCT)
Galactosylceramidase	GALC	Disease: leukodystrophy, globoid cell (GLD)
GDP-D-glucose phosphorylase 1	GDPGP1	A novel GDP-D-glucose phosphorylase involved in quality control of the nucleoside diphosphate sugar pool in Caenorhabditis elegans and mammals. PMID: 21507950
GDP-mannose 4,6-dehydratase	GMDS	Mutation of GDP-mannose-4,6-dehydratase in colorectal cancer metastasis. PMID: 23922970
Glucan (1,4-alpha-), branching enzyme 1	GBE1	Disease: glycogen storage disease 4 (GSD4)
Glucokinase (hexokinase 4)	GCK	**Piragliatin** – drugs used in diabetes. Disease: maturity-onset diabetes of the young 2 (MODY2)
Glucokinase (hexokinase 4) regulator	GCKR	**AMG-1694** – drugs used in diabetes
Glucosamine (N-acetyl)-6-sulfatase	GNS	Elosulfase alfa – anti-mucopolysaccharidosis treatments. Disease: mucopolysaccharidosis 3D (MPS3D)
Glucosamine (UDP-N-acetyl)-2-epimerase/N-acetylmannosamine kinase	GNE	Disease: sialuria (SIALURIA)
Glucosamine-phosphate N-acetyltransferase 1	GNPNAT1	Cell-free expression of human glucosamine 6-phosphate N-acetyltransferase (HsGNA1) for inhibitor screening. PMID: 23036358
Glucose-6-phosphate dehydrogenase	G6PD	**6-Anicotinamide** – antineoplastic agents. Disease: anemia, non-spherocytic hemolytic, due to G6PD deficiency (NSHA)
Glucose-6-phosphate isomerase	GPI	Disease: hemolytic anemia, non-spherocytic, due to glucose phosphate isomerase deficiency (HA-GPID)
Glucuronic acid epimerase	GLCE	GLCE regulates PC12 cell neuritogenesis induced by nerve growth factor through activating SMAD/ID3 signalling. PMID: 24499487
Glucuronidase, beta	GUSB	Disease: mucopolysaccharidosis 7 (MPS7)
Glyceraldehyde-3-phosphate dehydrogenase	GAPDH	**Omigapil** – muscular dystrophy treatments
Glyceraldehyde-3-phosphate dehydrogenase, spermatogenic	GAPDHS	Development and implementation of a high throughput screen for the human sperm-specific isoform of glyceraldehyde 3-phosphate dehydrogenase (GAPDHS). PMID: 21760877
Glycerate kinase	GLYCTK	Disease: D-glyceric aciduria (D-GA)
Glyoxalase I	GLO1	Structural investigation into the inhibitory mechanisms of indomethacin and its analogues towards human glyoxalase I. PMID: 21689932
Glyoxylate reductase 1 homolog (Arabidopsis)	GLYR1	LSD2/KDM1B and its cofactor NPAC/GLYR1 endow a structural and molecular model for regulation of H3K4 demethylation. PMID: 23260659
Glyoxylate reductase/hydroxypyruvate reductase	GRHPR	Disease: hyperoxaluria primary 2 (HP2)
Hexose-6-phosphate dehydrogenase (glucose 1-dehydrogenase)	H6PD	Disease: cortisone reductase deficiency (CRD)

HGNC-approved name	Symbol	Target information
Hydroxypyruvate isomerase (putative)	HYI	
Iduronate 2-sulfatase	IDS	Disease: mucopolysaccharidosis 2 (MPS2)
Iduronidase, alpha-L-	IDUA	Disease: mucopolysaccharidosis 1H (MPS1H)
Immunoresponsive 1 homolog (mouse)	IRG1	Immune responsive gene 1, a novel oncogene, increases the growth and tumorigenicity of glioma. PMID: 25216059
Inositol 1,3,4,5,6-pentakisphosphate 2-kinase	IPPK	Human genome-wide RNAi screen identifies an essential role for inositol pyrophosphates in Type-I interferon response. PMID: 24586175
Inositol polyphosphate multikinase	IPMK	Convergence of IPMK and LKB1-AMPK signaling pathways on metformin action. PMID: 24877601
Inositol-3-phosphate synthase 1	ISYNA1	Inositol synthesis regulates the activation of GSK-3α in neuronal cells. PMID: 25345501
Ketohexokinase (fructokinase)	KHK	Pyrimidinopyrimidine inhibitors of ketohexokinase: Exploring the ring C2 group that interacts with Asp-27B in the ligand binding pocket. PMID: 22795331. Disease: fructosuria (FRUCT)
KIAA1161	KIAA1161	NET37, a nuclear envelope transmembrane protein with glycosidase homology, is involved in myoblast differentiation. PMID: 19706595
Lactalbumin, alpha-	LALBA	
Maltase-glucoamylase (alpha-glucosidase)	MGAM	**Acarbose** – drugs used in diabetes
Mannose phosphate isomerase	MPI	Disease: congenital disorder of glycosylation 1B (CDG1B)
Meningioma expressed antigen 5 (hyaluronidase)	MGEA5	Increasing O-GlcNAc slows neurodegeneration and stabilizes tau against aggregation. PMID: 22366723
Mitochondrially encoded cytochrome c oxidase I	MT-CO1	Disease: Leber hereditary optic neuropathy (LHON)
Myo-inositol oxygenase	MIOX	Disruption of renal tubular mitochondrial quality control by myo-inositol oxygenase in diabetic kidney disease. PMID: 25270067
N-Acetylgalactosaminidase, alpha-	NAGA	Disease: Schindler disease (SCHIND)
N-Acetylglucosamine kinase	NAGK	Upregulation of dendritic arborization by N-acetyl-D-glucosamine kinase is not dependent on its kinase activity. PMID: 24722415
N-Acetylglucosamine-1-phosphodiester alpha-N-acetylglucosaminidase	NAGPA	A role for inherited metabolic deficits in persistent developmental stuttering. PMID: 22884963
N-Acetylglucosaminidase, alpha	NAGLU	Disease: mucopolysaccharidosis 3B (MPS3B)
N-Acetylneuraminate pyruvate lyase (dihydrodipicolinate synthase)	NPL	
N-Acetylneuraminic acid phosphatase	NANP	Design, synthesis, functional and structural characterization of an inhibitor of N-acetylneuraminate-9-phosphate phosphatase: Observation of extensive dynamics in an enzyme/inhibitor complex. PMID: 23747226
N-Acetylneuraminic acid synthase	NANS	Manifold decrease of sialic acid synthase in fetal Down syndrome brain. PMID: 16729195
Notum pectinacetylesterase homolog (Drosophila)	NOTUM	Human homolog of NOTUM, overexpressed in hepatocellular carcinoma, is regulated transcriptionally by beta-catenin/TCF. PMID: 18429952
N-Sulfoglucosamine sulfohydrolase	SGSH	Disease: mucopolysaccharidosis 3A (MPS3A)
Oxoglutarate dehydrogenase-like	OGDHL	OGDHL is a modifier of AKT-dependent signaling and NF-κB function. PMID: 23152800

(Continued)

HGNC-approved name	Symbol	Target information
Phosphogluconate dehydrogenase	PGD	Lysine acetylation activates 6-phosphogluconate dehydrogenase to promote tumor growth. PMID: 25042803
Phosphoglycerate kinase 1	PGK1	Disease: phosphoglycerate kinase 1 deficiency (PGK1D)
Phosphoglycolate phosphatase	PGP	
Poly (ADP-ribose) glycohydrolase	PARG	Structures of the human poly (ADP-ribose) glycohydrolase catalytic domain confirm catalytic mechanism and explain inhibition by ADP-HPD derivatives. PMID: 23251397
Pyruvate carboxylase	PC	Pyruvate carboxylase is required for glutamine-independent growth of tumor cells. PMID: 21555572. Disease: pyruvate carboxylase deficiency (PC deficiency)
Ribokinase	RBKS	Identification and characterization of human ribokinase and comparison of its properties with E. coli ribokinase and human adenosine kinase. PMID: 17585908
Ribose 5-phosphate isomerase A	RPIA	Disease: ribose 5-phosphate isomerase deficiency (RPID)
Sedoheptulokinase	SHPK	The sedoheptulose kinase CARKL directs macrophage polarization through control of glucose metabolism. PMID: 22682222
Sialic acid acetylesterase	SIAE	Disease: autoimmune disease 6 (AIS6)
Sorbitol dehydrogenase	SORD	Sorbitol dehydrogenase inhibitor protects the liver from ischemia/reperfusion-induced injury via elevated glycolytic flux and enhanced sirtuin 1 activity. PMID: 25333577
Sperm adhesion molecule 1 (PH-20 hyaluronidase, zona pellucida binding)	SPAM1	Functional characterization of double-knockout mouse sperm lacking SPAM1 and ACR or SPAM1 and PRSS21 in fertilization. PMID: 22362218
Succinyl-CoA:glutarate-CoA transferase	SUGCT	Disease: glutaric aciduria 3 (GA3)
Sucrase–isomaltase (alpha-glucosidase)	SI	**Acarbose** – drugs used in diabetes. Disease: congenital sucrase–isomaltase deficiency (CSID)
Tissue-specific transplantation antigen P35B	TSTA3	The crystal structure of human GDP-L-fucose synthase. PMID: 23774504
Transaldolase 1	TALDO1	Disease: transaldolase 1 deficiency (TALDO1 deficiency)
Trehalase (brush-border membrane glycoprotein)	TREH	Note: Deficiency of TREH results in isolated trehalose intolerance that causes gastrointestinal symptoms after ingestion of edible mushrooms
Triosephosphate isomerase 1	TPI1	Disease: triosephosphate isomerase deficiency (TPID)
UDP-galactose-4-epimerase	GALE	Disease: epimerase-deficiency galactosemia (EDG)
UDP-glucose 6-dehydrogenase	UGDH	UDP-glucose dehydrogenase modulates proteoglycan synthesis in articular chondrocytes: Its possible involvement and regulation in osteoarthritis. PMID: 25465897
UDP-glucose pyrophosphorylase 2	UGP2	Gene regulation of UDP-galactose synthesis and transport: potential rate-limiting processes in initiation of milk production in humans. PMID: 22649065
UDP-glucuronate decarboxylase 1	UXS1	UDP xylose synthase 1 is required for morphogenesis and histogenesis of the craniofacial skeleton. PMID: 20226781
UEV and lactate/malate dehyrogenase domains	UEVLD	Identification and characterization of UEV3, a human cDNA with similarities to inactive E2 ubiquitin-conjugating enzymes. PMID: 12427560
Xylulokinase homolog (*Haemophilus influenzae*)	XYLB	Structure and function of human xylulokinase, an enzyme with important roles in carbohydrate metabolism. PMID: 23179721

5.5 Vitamins, cofactors and related

HGNC-approved name	Symbol	Target information
Aldehyde dehydrogenase 1 family, member A1	ALDH1A1	Increased expression of ALDH1A1 protein is associated with poor prognosis in clear cell renal cell carcinoma. PMID: 23585015
Aldehyde dehydrogenase 1 family, member A2	ALDH1A2	The enzymatic activity of human aldehyde dehydrogenases 1A2 and 2 (ALDH1A2 and ALDH2) is detected by Aldefluor, inhibited by diethylaminobenzaldehyde and has significant effects on cell proliferation and drug resistance. PMID: 22079344
Aldehyde dehydrogenase 1 family, member A3	ALDH1A3	Disease: microphthalmia, isolated, 8 (MCOP8)
Aldehyde dehydrogenase 1 family, member L1	ALDH1L1	Epigenetic silencing of ALDH1L1, a metabolic regulator of cellular proliferation, in cancers. PMID: 21779486
Aldehyde dehydrogenase 1 family, member L2	ALDH1L2	ALDH1L2 is the mitochondrial homolog of 10-formyltetrahydrofolate dehydrogenase. PMID: 20498374
Aldehyde dehydrogenase 8 family, member A1	ALDH8A1	Acetaldehyde and retinaldehyde-metabolizing enzymes in colon and pancreatic cancers. PMID: 25427913
Aminolevulinate, delta, synthase 1	ALAS1	ALAS1 gene expression is down-regulated by Akt-mediated phosphorylation and nuclear exclusion of FOXO1 by vanadate in diabetic mice. PMID: 22070747
Aminolevulinate, delta, synthase 2	ALAS2	Disease: anemia, sideroblastic, X-linked (XLSA)
1-Aminocyclopropane-1-carboxylate synthase homolog (*Arabidopsis*) (non-functional)	ACCS	
1-Aminocyclopropane-1-carboxylate synthase homolog (*Arabidopsis*)(non functional)-like	ACCSL	
Cytochrome c oxidase assembly homolog 10 (yeast)	COX10	Disease: mitochondrial complex IV deficiency (MT-C4D)
Cytochrome c oxidase assembly homolog 15 (yeast)	COX15	In vivo correction of COX deficiency by activation of the AMPK/PGC-1α axis. PMID: 21723506. Disease: cardioencephalomyopathy, fatal infantile, due to cytochrome c oxidase deficiency 2 (CEMCOX2)
Gamma-glutamyltransferase 1	GGT1	**NOV-002** – antineoplastic agents. Disease: glutathionuria (GLUTH)
Gamma-glutamyltransferase 2	GGT2	Human GGT2 does not autocleave into a functional enzyme: A cautionary tale for interpretation of microarray data on redox signaling. PMID: 23682772
Gamma-glutamyltransferase 5	GGT5	Immunolabelling of gamma-glutamyl transferase 5 in normal human tissue reveals that expression and localization differ from gamma-glutamyl transferase 1. PMID: 25377544
Gamma-glutamyltransferase 6	GGT6	
Gamma-glutamyltransferase 7	GGT7	
Gamma-glutamyltransferase light chain 1	GGTLC1	The human gamma-glutamyltransferase gene family. PMID: 18357469
Gamma-glutamyltransferase light chain 2	GGTLC2	
Gamma-glutamyltransferase light chain 3	GGTLC3	
Methylmalonic aciduria (cobalamin deficiency) cblA type	MMAA	Disease: methylmalonic aciduria type cblA (MMAA)

(Continued)

HGNC-approved name	Symbol	Target information
Methylmalonic aciduria (cobalamin deficiency) cblB type	MMAB	Disease: methylmalonic aciduria type cblB (MMAB)
Molybdenum cofactor synthesis 1	MOCS1	Long-term rescue of a lethal inherited disease by adeno-associated virus-mediated gene transfer in a mouse model of molybdenum-cofactor deficiency. PMID: 17236133. Disease: molybdenum cofactor deficiency, complementation group A (MOCODA)
Molybdenum cofactor synthesis 2	MOCS2	Disease: molybdenum cofactor deficiency, complementation group B (MOCODB)
Molybdenum cofactor synthesis 3	MOCS3	Dual role of the molybdenum cofactor biosynthesis protein MOCS3 in tRNA thiolation and molybdenum cofactor biosynthesis in humans. PMID: 22453920
NAD kinase	NADK	A second target of benzamide riboside: Dihydrofolate reductase. PMID: 22954684
NAD kinase 2, mitochondrial	NADK2	Mitochondrial NADP(H) deficiency due to a mutation in NADK2 causes dienoyl-CoA reductase deficiency with hyperlysinemia. PMID: 24847004
Nicotinamide nucleotide adenylyltransferase 1	NMNAT1	Disease: Leber congenital amaurosis 9 (LCA9)
Nicotinamide nucleotide adenylyltransferase 2	NMNAT2	Endogenous Nmnat2 is an essential survival factor for maintenance of healthy axons. PMID: 20126265
Nicotinamide nucleotide adenylyltransferase 3	NMNAT3	Insight into molecular and functional properties of NMNAT3 reveals new hints of NAD homeostasis within human mitochondria. PMID: 24155910
Nicotinamide riboside kinase 1	NMRK1	Nicotinamide riboside promotes Sir2 silencing and extends lifespan via Nrk and Urh1/Pnp1/Meu1 pathways to NAD+. PMID: 17482543
Nicotinamide riboside kinase 2	NMRK2	Nrk2b-mediated NAD+ production regulates cell adhesion and is required for muscle morphogenesis in vivo: Nrk2b and NAD+ in muscle morphogenesis. PMID: 20566368
Nudix (nucleoside diphosphate linked moiety X)-type motif 7	NUDT7	The nudix hydrolase 7 is an Acyl-CoA diphosphatase involved in regulating peroxisomal coenzyme A homeostasis. PMID: 18799520
Nudix (nucleoside diphosphate linked moiety X)-type motif 12	NUDT12	Mammalian NADH diphosphatases of the Nudix family: Cloning and characterization of the human peroxisomal NUDT12 protein. PMID: 12790796
Nudix (nucleoside diphosphate linked moiety X)-type motif 19	NUDT19	
Vanin 1	VNN1	Vanin-1 is a key activator for hepatic gluconeogenesis. PMID: 24550194
Vanin 2	VNN2	Linkage between coenzyme a metabolism and inflammation: Roles of pantetheinase. PMID: 23978960
Vanin 3	VNN3	Expression of the vanin gene family in normal and inflamed human skin: Induction by proinflammatory cytokines. PMID: 19322213
5,10-Methenyltetrahydrofolate synthetase (5-formyltetrahydrofolate cyclo-ligase)	MTHFS	Mthfs is an essential gene in mice and a component of the purinosome. PMID: 22303332
Aminolevulinate dehydratase	ALAD	**Aminolevulinic acid** – photosensitizing agents. Disease: acute hepatic porphyria (AHEPP)
Biotinidase	BTD	Disease: biotinidase deficiency (BTD deficiency)
Carbohydrate kinase domain containing	CARKD	Occurrence and subcellular distribution of the NADPHX repair system in mammals. PMID: 24611804
Coproporphyrinogen oxidase	CPOX	Disease: hereditary coproporphyria (HCP)
Flavin adenine dinucleotide synthetase 1	FLAD1	Use of whole-exome sequencing to determine the genetic basis of multiple mitochondrial respiratory chain complex deficiencies. PMID: 25058219
Folylpolyglutamate synthase	FPGS	Binding of a Smad4/Ets-1 complex to a novel intragenic regulatory element in exon12 of FPGS underlies decreased gene expression and antifolate resistance in leukemia. PMID: 25229333

HGNC-approved name	Symbol	Target information
GTP cyclohydrolase 1	GCH1	Disease: GTP cyclohydrolase 1 deficiency (GCH1D)
Holocarboxylase synthetase (biotin-(propionyl-CoA-carboxylase (ATP-hydrolysing)) ligase)	HLCS	Disease: holocarboxylase synthetase deficiency (HLCS deficiency)
Hydroxymethylbilane synthase	HMBS	Disease: acute intermittent porphyria (AIP)
Methenyltetrahydrofolate synthetase domain containing	MTHFSD	Axonal guidance signaling pathway interacting with smoking in modifying the risk of pancreatic cancer: A gene- and pathway-based interaction analysis of GWAS data. PMID: 24419231
Molybdenum cofactor sulfurase	MOCOS	Disease: xanthinuria 2 (XU2)
NAD synthetase 1	NADSYN1	Polymorphisms in GC and NADSYN1 genes are associated with vitamin D status and metabolic profile in non-diabetic adults. PMID: 24073860
Nicotinamide nucleotide transhydrogenase	NNT	Nicotinamide nucleotide transhydrogenase (Nnt) links the substrate requirement in brain mitochondria for hydrogen peroxide removal to the thioredoxin/peroxiredoxin (Trx/Prx) system. PMID: 24722990. Disease: glucocorticoid deficiency 4 (GCCD4)
Nicotinamide phosphoribosyltransferase	NAMPT	Pharmacological inhibition of nicotinamide phosphoribosyltransferase (NAMPT), an enzyme essential for NAD+ biosynthesis, in human cancer cells: Metabolic basis and potential clinical implications. PMID: 3239881
Nicotinate phosphoribosyltransferase	NAPRT	FK866, a highly specific noncompetitive inhibitor of nicotinamide phosphoribosyltransferase, represents a novel mechanism for induction of tumor cell apoptosis. PMID: 14612543
Phosphopantothenoylcysteine decarboxylase	PPCDC	
Phosphopantothenoylcysteine synthetase	PPCS	Kinetic characterization of human phosphopantothenoylcysteine synthetase. PMID: 19683078
Polyamine oxidase (exo-N4-amino)	PAOX	Suppression of acetylpolyamine oxidase by selected AP-1 members regulates DNp73 abundance: Mechanistic insights for overcoming DNp73-mediated resistance to chemotherapeutic drugs. PMID: 24722210
Protoporphyrinogen oxidase	PPOX	Disease: variegate porphyria (VP)
Pyridoxal (pyridoxine, vitamin B6) kinase	PDXK	Effects of vitamin B6 metabolism on oncogenesis, tumor progression and therapeutic responses. PMID: 23334322
Pyridoxal (pyridoxine, vitamin B6) phosphatase	PDXP	
Pyridoxamine 5′-phosphate oxidase	PNPO	Disease: pyridoxine-5′-phosphate oxidase deficiency (PNPO deficiency)
Quinolinate phosphoribosyltransferase	QPRT	The endogenous tryptophan metabolite and NAD+ precursor quinolinic acid confers resistance of gliomas to oxidative stress. PMID: 23548271
Regucalcin	RGN	Involvement of regucalcin as a suppressor protein in human carcinogenesis: Insight into the gene therapy. PMID: 25230901
Riboflavin kinase	RFK	Importance of riboflavin kinase in the pathogenesis of stroke. PMID: 22925047
Short-chain dehydrogenase/reductase family 16C, member 5	SDR16C5	Biochemical characterization of human epidermal retinol dehydrogenase 2. PMID: 18926804
Thiamin pyrophosphokinase 1	TPK1	Disease: thiamine metabolism dysfunction syndrome 5, episodic encephalopathy type (THMD5)
Thiamine triphosphatase	THTPA	The iron-regulated metastasis suppressor, Ndrg-1: Identification of novel molecular targets. PMID: 18582504
Uroporphyrinogen decarboxylase	UROD	Disease: familial porphyria cutanea tarda (FPCT)
Uroporphyrinogen III synthase	UROS	Disease: congenital erythropoietic porphyria (CEP)

5.6 DNA-processing enzymes

HGNC-approved name	Symbol	Target information
AlkB, alkylation repair homolog 1 (*Escherichia coli*)	ALKBH1	ALKBH1 is a histone H2A dioxygenase involved in neural differentiation. PMID: 22961808
AlkB, alkylation repair homolog 2 (*E. coli*)	ALKBH2	The DNA repair protein ALKBH2 mediates temozolomide resistance in human glioblastoma cells. PMID: 23258843
AlkB, alkylation repair homolog 3 (*E. coli*)	ALKBH3	PCA-1/ALKBH3 contributes to pancreatic cancer by supporting apoptotic resistance and angiogenesis. PMID: 22826605
APEX nuclease (multifunctional DNA repair enzyme) 1	APEX1	The DNA base excision repair protein Ape1/Ref-1 as a therapeutic and chemopreventive target. PMID: 17560642
APEX nuclease (apurinic/apyrimidinic endonuclease) 2	APEX2	Apurinic/apyrimidinic endonuclease 2 regulates the expansion of germinal centers by protecting against activation-induced cytidine deaminase-independent DNA damage in B cells. PMID: 24935922
Apolipoprotein B mRNA-editing enzyme, catalytic polypeptide-like 3A	APOBEC3A	APOBEC3A functions as a restriction factor of human papillomavirus. PMID: 25355878
Apolipoprotein B mRNA-editing enzyme, catalytic polypeptide-like 3B	APOBEC3B	APOBEC3B is an enzymatic source of mutation in breast cancer. PMID: 23389445
Apolipoprotein B mRNA-editing enzyme, catalytic polypeptide-like 3C	APOBEC3C	APOBEC3 multimerization correlates with HIV-1 packaging and restriction activity in living cells. PMID: 24361275
Apolipoprotein B mRNA-editing enzyme, catalytic polypeptide-like 3D	APOBEC3D	Expression of DNA cytosine deaminase APOBEC3 proteins, a potential source for producing mutations, in gastric, colorectal and prostate cancers. PMID: 25296601
Apolipoprotein B mRNA-editing enzyme, catalytic polypeptide-like 3F	APOBEC3F	APOBEC3D and APOBEC3F potently promote HIV-1 diversification and evolution in humanized mouse model. PMID: 25330146
Apolipoprotein B mRNA-editing enzyme, catalytic polypeptide-like 3G	APOBEC3G	HIV-1 Vif inhibits G to A hypermutations catalyzed by virus-encapsidated APOBEC3G to maintain HIV-1 infectivity. PMID: 25304135
Apolipoprotein B mRNA-editing enzyme, catalytic polypeptide-like 3H	APOBEC3H	Sequence and structural determinants of human APOBEC3H deaminase and anti-HIV-1 activities. PMID: 25614027
Chromodomain helicase DNA binding protein 1	CHD1	**Epirubicin** – antineoplastic agents
Chromodomain helicase DNA binding protein 1-like	CHD1L	CHD1L: A novel oncogene. PMID: 24359616
Chromodomain helicase DNA binding protein 2	CHD2	Disease: epileptic encephalopathy, childhood-onset (EEOC)
Chromodomain helicase DNA binding protein 3	CHD3	Opposing ISWI- and CHD-class chromatin remodeling activities orchestrate heterochromatic DNA repair. PMID: 25533843
Chromodomain helicase DNA binding protein 4	CHD4	CHD4 in the DNA-damage response and cell cycle progression: Not so NuRDy now. PMID: 23697937
Chromodomain helicase DNA binding protein 5	CHD5	Note: Defects in CHD5 may be a cause of the development of cancers from epithelial, neural and hematopoietic origin. CHD5 is one of the missing genes in the del(1p36), a deletion which is extremely common in this type of cancers. A decrease in its expression results in increased susceptibility of cells to Ras-mediated transformation *in vitro* and in vivo
Chromodomain helicase DNA binding protein 6	CHD6	Note: A chromosomal aberration disrupting CHD6 has been found in a patient with mild to moderate mental retardation and minor facial anomalies. Translocation t(1820)(q21.1q11.2) with TCF4 producing a CHD6-TCF4 fusion transcript (PMID: 18627065)
Chromodomain helicase DNA binding protein 7	CHD7	Disease: CHARGE syndrome (CHARGES)

HGNC-approved name	Symbol	Target information
Chromodomain helicase DNA binding protein 8	CHD8	Disease: autism 18 (AUTS18)
Chromodomain helicase DNA binding protein 9	CHD9	Expression and regulation of CReMM, a chromodomain helicase-DNA-binding (CHD), in marrow stroma derived osteoprogenitors. PMID: 16523501
Deoxyribonuclease I	DNASE1	Disease: systemic lupus erythematosus (SLE)
Deoxyribonuclease I-like 1	DNASE1L1	DNase X is a glycosylphosphatidylinositol-anchored membrane enzyme that provides a barrier to endocytosis-mediated transfer of a foreign gene. PMID: 17416904
Deoxyribonuclease I-like 2	DNASE1L2	Essential role of the keratinocyte-specific endonuclease DNase1L2 in the removal of nuclear DNA from hair and nails. PMID: 21307874
Deoxyribonuclease I-like 3	DNASE1L3	Disease: systemic lupus erythematosus 16 (SLEB16)
Deoxyribonuclease II, lysosomal	DNASE2	Eradication of human ovarian cancer cells by transgenic expression of recombinant DNASF1, DNASE1L3, DNASE2, and DFFD controlled by EGFR promoter: Novel strategy for targeted therapy of cancer. PMID: 24587967
Deoxyribonuclease II beta	DNASE2B	HSF4 regulates DLAD expression and promotes lens de-nucleation. PMID: 23507146
DNA cross-link repair 1A	DCLRE1A	Snm1-deficient mice exhibit accelerated tumorigenesis and susceptibility to infection. PMID: 16260620
DNA cross-link repair 1B	DCLRE1B	Disease: Hoyeraal–Hreidarsson syndrome (HHS)
DNA cross-link repair 1C	DCLRE1C	Disease: severe combined immunodeficiency autosomal recessive T-cell-negative/B-cell-negative/NK-cell-positive with sensitivity to ionizing radiation (RSSCID)
Endonuclease G	ENDOG	
Endonuclease domain containing 1	ENDOD1	
Essential meiotic structure-specific endonuclease 1	EME1	
Essential meiotic structure-specific endonuclease subunit 2	EME2	
Excision repair cross-complementation group 1	ERCC1	DNA repair endonuclease ERCC1-XPF as a novel therapeutic target to overcome chemoresistance in cancer therapy. PMID: 22941649. Disease: cerebro-oculo-facio-skeletal syndrome 4 (COFS4)
Excision repair cross-complementation group 2	ERCC2	Disease: xeroderma pigmentosum complementation group D (XP-D)
Excision repair cross-complementation group 3	ERCC3	Disease: xeroderma pigmentosum complementation group B (XP-B)
Excision repair cross-complementation group 4	ERCC4	DNA repair endonuclease ERCC1-XPF as a novel therapeutic target to overcome chemoresistance in cancer therapy. PMID: 22941649. Disease: xeroderma pigmentosum complementation group F (XP-F)
Excision repair cross-complementation group 5	ERCC5	Disease: xeroderma pigmentosum complementation group G (XP-G)
Excision repair cross-complementation group 6	ERCC6	Disease: Cockayne syndrome B (CSB)
Excision repair cross-complementation group 6-like	ERCC6L	
Excision repair cross-complementation group 6-like 2	ERCC6L2	Note: ERCC6L2 mutations may be involved in a bone marrow failure syndrome characterized by bone marrow hypocellularity, learning difficulties, and developmental delay

(Continued)

HGNC-approved name	Symbol	Target information
Exonuclease 3′-5′ domain containing 1	EXD1	
Exonuclease 3′-5′ domain containing 2	EXD2	
Exonuclease 3′-5′ domain containing 3	EXD3	
Fidgetin-like 1	FIGNL1	FIGNL1-containing protein complex is required for efficient homologous recombination repair. PMID: 23754376
Fidgetin-like 2	FIGNL2	
Helicase (DNA) B	HELB	Identifying candidate genes for discrimination of ulcerative colitis and Crohn's disease. PMID: 25182475
Helicase with zinc finger	HELZ	The putative RNA helicase HELZ promotes cell proliferation, translation initiation and ribosomal protein S6 phosphorylation. PMID: 21765940
Helicase with zinc finger 2, transcriptional coactivator	HELZ2	Protection against high-fat diet-induced obesity in Helz2-deficient male mice due to enhanced expression of hepatic leptin receptor. PMID: 25004093
Helicase, lymphoid-specific	HELLS	Chromatin remodelers HELLS and UHRF1 mediate the epigenetic deregulation of genes that drive retinoblastoma tumor progression. PMID: 25338120
Helicase, POLQ-like	HELQ	Human DNA helicase HELQ participates in DNA interstrand crosslink tolerance with ATR and RAD51 paralogs. PMID: 24005565
High-mobility group AT-hook 1	HMGA1	High mobility group A1 and cancer: Potential biomarker and therapeutic target. PMID: 22419021. Note: A chromosomal aberration involving HMGA1 is found in pulmonary chondroid hamartoma. Translocation t(614) (p21q23-24) with RAD51B
High-mobility group AT-hook 2	HMGA2	HMGA2 protein expression in ovarian serous carcinoma effusions, primary tumors, and solid metastases. PMID: 22476403. Note: A chromosomal aberration involving HMGA2 is associated with a subclass of benign mesenchymal tumors known as lipomas
Ligase I, DNA, ATP-dependent	LIG1	DNA ligase I is not essential for mammalian cell viability. PMID: 24726358
Ligase III, DNA, ATP-dependent	LIG3	Structure and function of the DNA ligases encoded by the mammalian LIG3 gene. PMID: 24013086
Ligase IV, DNA, ATP-dependent	LIG4	Disease: LIG4 syndrome (LIG4S)
Minichromosome maintenance complex component 2	MCM2	MCM-2 is a therapeutic target of Trichostatin A in colon cancer cells. PMID: 23770000
Minichromosome maintenance complex component 3	MCM3	A p53 drug response signature identifies prognostic genes in high-risk neuroblastoma. PMID: 24348903
Minichromosome maintenance complex component 4	MCM4	Disease: natural killer cell and glucocorticoid deficiency with DNA repair defect (NKGCD)
Minichromosome maintenance complex component 5	MCM5	MicroRNA-10b and minichromosome maintenance complex component 5 gene as prognostic biomarkers in breast cancer. PMID: 25596707
Minichromosome maintenance complex component 6	MCM6	Plasma minichromosome maintenance complex component 6 is a novel biomarker for hepatocellular carcinoma patients. PMID: 24451028
Minichromosome maintenance complex component 7	MCM7	Minichromosome maintenance protein 7 is a potential therapeutic target in human cancer and a novel prognostic marker of non-small cell lung cancer. PMID: 21619671
Minichromosome maintenance complex component 8	MCM8	Exome sequencing reveals MCM8 mutation underlies ovarian failure and chromosomal instability. PMID: 25437880
Minichromosome maintenance complex component 9	MCM9	Mcm8 and Mcm9 form a complex that functions in homologous recombination repair induced by DNA interstrand crosslinks. PMID: 22771115
Minichromosome maintenance complex component 3 associated protein	MCM3AP	Selective cell death of p53-insufficient cancer cells is induced by knockdown of the mRNA export molecule GANP. PMID: 22395445

HGNC-approved name	Symbol	Target information
Nei endonuclease VIII-like 1 (*E. coli*)	NEIL1	Loss of NEIL1 causes defects in olfactory function in mice. PMID: 25448603
Nei endonuclease VIII-like 2 (*E. coli*)	NEIL2	NEIL2 protects against oxidative DNA damage induced by sidestream smoke in human cells. PMID: 24595271
Nei endonuclease VIII-like 3 (*E. coli*)	NEIL3	Human NEIL3 is mainly a monofunctional DNA glycosylase removing spiroimindiohydantoin and guanidinohydantoin. PMID: 23755964
PMS1 postmeiotic segregation increased 1 (*S. cerevisiae*)	PMS1	DNA mismatch repair enzymes: Genetic defects and autoimmunity. PMID: 25619773
PMS2 postmeiotic segregation increased 2 (*S. cerevisiae*)	PMS2	Disease: hereditary non-polyposis colorectal cancer 4 (HNPCC4)
Polymerase (DNA directed), alpha 1, catalytic subunit	POLA1	**Fludarabine** – antineoplastic agents
Polymerase (DNA directed), beta	POLB	**Cytarabine** – antineoplastic agents
Polymerase (DNA directed), gamma	POLG	Disease: progressive external ophthalmoplegia with mitochondrial DNA deletions, autosomal dominant, 1 (PEOA1)
Polymerase (DNA directed), delta 1, catalytic subunit	POLD1	Disease: colorectal cancer 10 (CRCS10)
Polymerase (DNA directed), epsilon, catalytic subunit	POLE	Disease: colorectal cancer 12 (CRCS12)
Polymerase (DNA directed), epsilon 2, accessory subunit	POLE2	
Polymerase (DNA directed), eta	POLH	Disease: xeroderma pigmentosum variant type (XPV)
Polymerase (DNA directed), theta	POLQ	DNA polymerase POLQ and cellular defense against DNA damage. PMID: 23219161
Polymerase (DNA directed), iota	POLI	Overexpression of DNA polymerase iota (Polι) in esophageal squamous cell carcinoma. PMID: 22509890
Polymerase (DNA directed), kappa	POLK	DNA polymerase κ-dependent DNA synthesis at stalled replication forks is important for CHK1 activation. PMID: 23799366
Polymerase (DNA directed), lambda	POLL	Structural basis for the binding and incorporation of nucleotide analogs with ʟ-stereochemistry by human DNA polymerase λ. PMID: 25015085
Polymerase (DNA directed), mu	POLM	Increased learning and brain long-term potentiation in aged mice lacking DNA polymerase μ. PMID: 23301049
Polymerase (DNA directed), nu	POLN	DNA polymerase POLN participates in cross link repair and homologous recombination. PMID: 19995904
Primase, DNA, polypeptide 1 (49 kDa)	PRIM1	Inhibition of DNA primase and induction of apoptosis by 3,3′-diethyl-9-methylthia-carbocyanine iodide in hepatocellular carcinoma BEL-7402 cells. PMID: 14966908
Primase, DNA, polypeptide 2 (58 kDa)	PRIM2	
RAD1 homolog (*Schizosaccharomyces pombe*)	RAD1	Sumoylation of the Rad1 nuclease promotes DNA repair and regulates its DNA association. PMID: 24753409
RAD9 homolog A (*S. pombe*)	RAD9A	Rad9A is required for G2 decatenation checkpoint and to prevent endoreduplication in response to topoisomerase II inhibition. PMID: 20305300
RAD17 homolog (*S. pombe*)	RAD17	v-Src inhibits the interaction between Rad17 and Rad9 and induces replication fork collapse. PMID: 24971543
RAD51 recombinase	RAD51	RAD51 as a potential biomarker and therapeutic target for pancreatic cancer. PMID: 21807066. Disease: breast cancer (BC)

(*Continued*)

HGNC-approved name	Symbol	Target information
RAD51 paralog B	RAD51B	Note: A chromosomal aberration involving RAD51B is found in pulmonary chondroid hamartoma. Translocation t(614)(p21q23-24) with HMGA1. Note: A chromosomal aberration involving RAD51B is found in uterine leiomyoma
RAD51 paralog C	RAD51C	Disease: Fanconi anemia complementation group O (FANCO)
RAD51 paralog D	RAD51D	Disease: breast–ovarian cancer, familial, 4 (BROVCA4)
RAD54 homolog B (*S. cerevisiae*)	RAD54B	Rad54B serves as a scaffold in the DNA damage response that limits checkpoint strength. PMID: 25384516
RAD54-like (*S. cerevisiae*)	RAD54L	From strand exchange to branch migration; bypassing of non-homologous sequences by human Rad51 and Rad54. PMID: 21056573
RAD54-like 2 (*S. cerevisiae*)	RAD54L2	Transcriptional suppression by transient recruitment of ARIP4 to sumoylated nuclear receptor Ad4BP/SF-1. PMID: 19692572
RecQ helicase-like	RECQL	RECQL1 DNA repair helicase: A potential therapeutic target and a proliferative marker against ovarian cancer. PMID: 23951333
RecQ protein-like 4	RECQL4	Disease: Rothmund–Thomson syndrome (RTS)
RecQ protein-like 5	RECQL5	RECQL5 controls transcript elongation and suppresses genome instability associated with transcription stress. PMID: 24836610
REV1, polymerase (DNA directed)	REV1	Starvation promotes REV1 SUMOylation and p53-dependent sensitization of melanoma and breast cancer cells. PMID: 25614517
REV3-like, polymerase (DNA directed), zeta, catalytic subunit	REV3L	REV7 is essential for DNA damage tolerance via two REV3L binding sites in mammalian DNA polymerase ζ. PMID: 25567983
RuvB-like AAA ATPase 1	RUVBL1	RUVBL1 directly binds actin filaments and induces formation of cell protrusions to promote pancreatic cancer cell invasion. PMID: 24728183
RuvB-like AAA ATPase 2	RUVBL2	Reptin is required for the transcription of telomerase reverse transcriptase and over-expressed in gastric cancer. PMID: 20509972
TatD DNase domain containing 1	TATDN1	
TatD DNase domain containing 2	TATDN2	
TatD DNase domain containing 3	TATDN3	
Tet methylcytosine dioxygenase 1	TET1	TET1 plays an essential oncogenic role in MLL-rearranged leukemia. PMID: 23818607. Note: A chromosomal aberration involving TET1 may be a cause of acute leukemias. Translocation t(1011)(q22q23) with KMT2A/MLL1
Tet methylcytosine dioxygenase 2	TET2	Note: TET2 is frequently mutated in myeloproliferative disorders (MPD)
Tet methylcytosine dioxygenase 3	TET3	The role of Tet3 DNA dioxygenase in epigenetic reprogramming by oocytes. PMID: 21892189
Three prime repair exonuclease 1	TREX1	Disease: Aicardi–Goutieres syndrome 1 (AGS1)
Three prime repair exonuclease 2	TREX2	A genome-wide function of THSC/TREX-2 at active genes prevents transcription-replication collisions. PMID: 25294824
Topoisomerase (DNA) I	TOP1	**Irinotecan** – antineoplastic agents. Note: A chromosomal aberration involving TOP1 is found in a form of therapy-related myelodysplastic syndrome. Translocation t(1120)(p15q11) with NUP98
Topoisomerase (DNA) I, mitochondrial	TOP1MT	**Topotecan** – antineoplastic agents
Topoisomerase (DNA) II alpha 170 kDa	TOP2A	**Mitoxantrone** – antineoplastic agents
Topoisomerase (DNA) II beta 180 kDa	TOP2B	TOP2β is essential for ovarian follicles that are hypersensitive to chemotherapeutic drugs. PMID: 24002654
Topoisomerase (DNA) III alpha	TOP3A	Developing T lymphocytes are uniquely sensitive to a lack of topoisomerase III alpha. PMID: 20623552

HGNC-approved name	Symbol	Target information
Topoisomerase (DNA) III beta	TOP3B	Deletion of TOP3β, a component of FMRP-containing mRNPs, contributes to neurodevelopmental disorders. PMID: 23912948
Tyrosyl-DNA phosphodiesterase 1	TDP1	Identification of a putative Tdp1 inhibitor (CD00509) by *in vitro* and cell-based assays. PMID: 25117203. Disease: spinocerebellar ataxia, autosomal recessive, with axonal neuropathy (SCAN1)
Tyrosyl-DNA phosphodiesterase 2	TDP2	Involvement of the host DNA-repair enzyme TDP2 in formation of the covalently closed circular DNA persistence reservoir of hepatitis B viruses. PMID: 25201958
X-ray repair complementing defective repair in Chinese hamster cells 2	XRCC2	Association of reduced XRCC2 expression with lymph node metastasis in breast cancer tissues. PMID: 25159888
X-ray repair complementing defective repair in Chinese hamster cells 5 (double-strand-break rejoining)	XRCC5	K14-EGFP-miR-31 transgenic mice have high susceptibility to chemical-induced squamous cell tumorigenesis that is associating with Ku80 repression. PMID: 25082302
X-ray repair complementing defective repair in Chinese hamster cells 6	XRCC6	Ku70 regulates Bax-mediated pathogenesis in laminin-alpha2-deficient human muscle cells and mouse models of congenital muscular dystrophy. PMID: 19692349
8-Oxoguanine DNA glycosylase	OGG1	Whole transcriptome analysis reveals an 8-oxoguanine DNA glycosylase-1-driven DNA repair-dependent gene expression linked to essential biological processes. PMID: 25614460. Disease: renal cell carcinoma (RCC)
Activating signal cointegrator 1 complex subunit 3	ASCC3	DNA unwinding by ASCC3 helicase is coupled to ALKBH3-dependent DNA alkylation repair and cancer cell proliferation. PMID: 22055184
Activation-induced cytidine deaminase	AICDA	Disease: immunodeficiency with hyper-IgM 2 (HIGM2)
Alpha thalassemia/mental retardation syndrome X linked	ATRX	Disease: alpha-thalassemia mental retardation syndrome, X-linked (ATRX)
Aprataxin	APTX	Aprataxin resolves adenylated RNA-DNA junctions to maintain genome integrity. PMID: 24362567. Disease: ataxia-oculomotor apraxia syndrome (AOA)
Aprataxin and PNKP-like factor	APLF	PARP-3 and APLF function together to accelerate nonhomologous end-joining. PMID: 21211721
Asteroid homolog 1 (*Drosophila*)	ASTE1	Overlapping ATP2C1 and ASTE1 Genes in Human Genome: Implications for SPCA1 Expression? PMID: 23344038
Bloom syndrome, RecQ helicase-like	BLM	Disease: Bloom syndrome (BLM)
BRCA1-interacting protein C-terminal helicase 1	BRIP1	Disease: breast cancer (BC)
BTAF1 RNA polymerase II, B-TFIID transcription factor-associated, 170 kDa	BTAF1	Underexpression of transcriptional regulators is common in metastatic breast cancer cells overexpressing Bcl-xL. PMID: 16492678
Chromatin accessibility complex 1	CHRAC1	
Claspin	CLSPN	Apoptosis-related molecular differences for response to tyrosin kinase inhibitors in drug-sensitive and drug-resistant human bladder cancer cells. PMID: 24518715
CTF18, chromosome transmission fidelity factor 18 homolog (*S. cerevisiae*)	CHTF18	Disruption of CHTF18 causes defective meiotic recombination in male mice. PMID: 23133398
DEAQ box RNA-dependent ATPase 1	DQX1	
DNA fragmentation factor, 40 kDa, beta polypeptide (caspase-activated DNase)	DFFB	GM-CSF-DFF40: A novel humanized immunotoxin induces apoptosis in acute myeloid leukemia cells. PMID: 23529188

(Continued)

HGNC-approved name	Symbol	Target information
DNA meiotic recombinase 1	DMC1	Interaction between Lim15/Dmc1 and the homologue of the large subunit of CAF-1: A molecular link between recombination and chromatin assembly during meiosis. PMID: 18355319
DNA nucleotidylexotransferase	DNTT	New nucleotide-competitive non-nucleoside inhibitors of terminal deoxynucleotidyl transferase: Discovery, characterization, and crystal structure in complex with the target. PMID: 23968551
DNA replication helicase/nuclease 2	DNA2	Disease: progressive external ophthalmoplegia with mitochondrial DNA deletions, autosomal dominant, 6 (PEOA6)
Ectonucleotide pyrophosphatase/ phosphodiesterase 3	ENPP3	**AGS-16M8F** – antineoplastic agents
Endo/exonuclease (5′-3′), endonuclease G-like	EXOG	Loss of mitochondrial exo/endonuclease EXOG affects mitochondrial respiration and induces ROS-mediated cardiomyocyte hypertrophy. PMID: 25377088
Exonuclease 1	EXO1	Exo1 phosphorylation status controls the hydroxyurea sensitivity of cells lacking the Pol32 subunit of DNA polymerases delta and zeta. PMID: 25457771
Exonuclease 5	EXO5	Human exonuclease 5 is a novel sliding exonuclease required for genome stability. PMID: 23095756
FANCD2/FANCI-associated nuclease 1	FAN1	Disease: interstitial nephritis, karyomegalic (KMIN)
Fat mass and obesity associated	FTO	The biology of FTO: From nucleic acid demethylase to amino acid sensor. PMID: 23896822. Disease: growth retardation developmental delay coarse facies early death (GDFD)
Flap structure-specific endonuclease 1	FEN1	Genomic and protein expression analysis reveals flap endonuclease 1 (FEN1) as a key biomarker in breast and ovarian cancer. PMID: 24880630
GEN1 Holliday junction 5′ flap endonuclease	GEN1	A novel role of human holliday junction resolvase GEN1 in the maintenance of centrosome integrity. PMID: 23166748
GTPase-activating protein (SH3 domain) binding protein 1	G3BP1	The stress granule protein G3BP1 recruits PKR to promote multiple innate immune antiviral responses. PMID: 25520508
HFM1, ATP-dependent DNA helicase homolog (S. cerevisiae)	HFM1	Mutations in HFM1 in recessive primary ovarian insufficiency. PMID: 24597873
Helicase-like transcription factor	HLTF	Helicase-like transcription factor: A new marker of well-differentiated thyroid cancers. PMID: 25005870
Immunoglobulin mu binding protein 2	IGHMBP2	Disease: neuronopathy, distal hereditary motor, 6 (HMN6)
INO80 complex subunit	INO80	INO80-C and SWR-C: Guardians of the genome. PMID: 25451604
Interferon induced with helicase C domain 1	IFIH1	Disease: diabetes mellitus, insulin-dependent, 19 (IDDM19)
KIAA2022	KIAA2022	Disease: mental retardation, X-linked 98 (MRX98)
Mitochondrial genome maintenance exonuclease 1	MGME1	Disease: mitochondrial DNA depletion syndrome 11 (MTDPS11)
MRE11 meiotic recombination 11 homolog A (S. cerevisiae)	MRE11A	Disease: ataxia–telangiectasia-like disorder (ATLD)
MUS81 structure-specific endonuclease subunit	MUS81	The MUS81 endonuclease is essential for telomerase negative cell proliferation. PMID: 19617716
MutY homolog	MUTYH	Disease: familial adenomatous polyposis 2 (FAP2)
NEDD4 binding protein 2	N4BP2	Identification and characterization of BCL-3-binding protein: Implications for transcription and DNA repair or recombination. PMID: 12730195
Neuron navigator 2	NAV2	14-3-3ε and NAV2 interact to regulate neurite outgrowth and axon elongation. PMID: 24161943

HGNC-approved name	Symbol	Target information
N-Methylpurine-DNA glycosylase	MPG	Enzymatic MPG DNA repair assays for two different oxidative DNA lesions reveal associations with increased lung cancer risk. PMID: 25355292
Nth endonuclease III-like 1 (*E. coli*)	NTHL1	Inhibition of DNA glycosylases via small molecule purine analogs. PMID: 24349107
Nuclear GTPase, germinal center associated	NUGGC	Altered pattern of immunoglobulin hypermutation in mice deficient in Slip-GC protein. PMID: 22833677
PIF1 5′-to-3′ DNA helicase	PIF1	Human PIF1 helicase supports DNA replication and cell growth under oncogenic-stress. PMID: 25359767
Polynucleotide kinase 3′-phosphatase	PNKP	Disease: epileptic encephalopathy, early infantile, 10 (EIEE10)
Primase and polymerase (DNA directed)	PRIMPOL	Disease: myopia 22, autosomal dominant (MYP22)
Recombination activating gene 1	RAG1	Disease: combined cellular and humoral immune defects with granulomas (CHIDG)
Regulator of telomere elongation helicase 1	RTEL1	Disease: dyskeratosis congenita, autosomal recessive, 5 (DKCB5)
Retinoblastoma binding protein 8	RBBP8	CtIP maintains stability at common fragile sites and inverted repeats by end resection-independent endonuclease activity. PMID: 24837675. Disease: Seckel syndrome 2 (SCKL2)
Senataxin	SETX	Disease: spinocerebellar ataxia, autosomal recessive, 1 (SCAR1)
Single-strand-selective monofunctional uracil-DNA glycosylase 1	SMUG1	Single-strand selective monofunctional uracil-DNA glycosylase (SMUG1) deficiency is linked to aggressive breast cancer and predicts response to adjuvant therapy. PMID: 24253812
SLX4 structure-specific endonuclease subunit	SLX4	Disease: Fanconi anemia complementation group P (FANCP)
SWIM-type zinc finger 7-associated protein 1	SWSAP1	hSWS1·SWSAP1 is an evolutionarily conserved complex required for efficient homologous recombination repair. PMID: 21965664
TCDD-inducible poly(ADP-ribose) polymerase	TIPARP	Aryl hydrocarbon receptor repressor and TiPARP (ARTD14) use similar, but also distinct mechanisms to repress aryl hydrocarbon receptor signaling. PMID: 24806346
Telomerase reverse transcriptase	TERT	**Imetelstat** – antineoplastic agents. Note: Activation of telomerase has been implicated in cell immortalization and cancer cell pathogenesis. Disease: aplastic anemia (AA)
THAP domain containing 9	THAP9	The human THAP9 gene encodes an active P element DNA transposase. PMID: 23349291
Thymine DNA glycosylase	TDG	Human DNA glycosylase enzyme TDG repairs thymine mispaired with exocyclic etheno-DNA adducts. PMID: 25151120
Translin	TSN	Rescuing dicer defects via inhibition of an anti-dicing nuclease. PMID: 25457613
Tudor domain containing 15	TDRD15	
Uracil-DNA glycosylase	UNG	Disease: immunodeficiency with hyper-IgM 5 (HIGM5)
Werner helicase-interacting protein 1	WRNIP1	Human Werner helicase-interacting protein 1 (WRNIP1) functions as a novel modulator for DNA polymerase delta. PMID: 15670210
Werner syndrome, RecQ helicase-like	WRN	Disease: Werner syndrome (WRN)
Zinc finger, RAN-binding domain containing 3	ZRANB3	ZRANB3 is a structure-specific ATP-dependent endonuclease involved in replication stress response. PMID: 22759634

5.7   RNA-processing enzymes

HGNC-approved name	Symbol	Target information
5′-3′ Exoribonuclease 1	XRN1	XRN 5′→3′ exoribonucleases: Structure, mechanisms and functions. PMID: 23517755
5′-3′ Exoribonuclease 2	XRN2	The multifunctional RNase XRN2. PMID: 23863139
Adenosine deaminase domain containing 1 (testis specific)	ADAD1	Genome-wide association scan in psoriasis: New insights into chronic inflammatory disease. PMID: 20476959
Adenosine deaminase domain containing 2	ADAD2	
Adenosine deaminase, RNA-specific	ADAR	ADAR1: A promising new biomarker for esophageal squamous cell carcinoma? PMID: 24928581. Disease: dyschromatosis symmetrica hereditaria (DSH)
Adenosine deaminase, RNA-specific, B1	ADARB1	ADAR2-editing activity inhibits glioblastoma growth through the modulation of the CDC14B/Skp2/p21/p27 axis. PMID: 22525274
Adenosine deaminase, RNA-specific, B2 (non-functional)	ADARB2	Altered adenosine-to-inosine RNA editing in human cancer. PMID: 17908822
Adenosine deaminase, tRNA-specific 1	ADAT1	RNA editing by adenosine deaminases generates RNA and protein diversity. PMID: 12457566
Adenosine deaminase, tRNA-specific 2	ADAT2	
Adenosine deaminase, tRNA-specific 3	ADAT3	Disease: mental retardation, autosomal recessive 36 (MRT36)
Alanyl-tRNA synthetase	AARS	Evolutionary and structural annotation of disease-associated mutations in human aminoacyl-tRNA synthetases. PMID: 25476837. Disease: Charcot–Marie–Tooth disease 2N (CMT2N)
Alanyl-tRNA synthetase 2, mitochondrial	AARS2	Disease: combined oxidative phosphorylation deficiency 8 (COXPD8)
AlkB, alkylation repair homolog 5 (E. coli)	ALKBH5	ALKBH5 is a mammalian RNA demethylase that impacts RNA metabolism and mouse fertility. PMID: 23177736
AlkB, alkylation repair homolog 8 (E. coli)	ALKBH8	Human AlkB homolog ABH8 Is a tRNA methyltransferase required for wobble uridine modification and DNA damage survival. PMID: 20308323
Arginyl-tRNA synthetase	RARS	Mutations in RARS cause hypomyelination. PMID: 24777941
Arginyl-tRNA synthetase 2, mitochondrial	RARS2	Disease: pontocerebellar hypoplasia 6 (PCH6)
Asparaginyl-tRNA synthetase	NARS	Fibroblast growth factor 2-induced cytoplasmic asparaginyl-tRNA synthetase promotes survival of osteoblasts by regulating anti apoptotic PI3K/Akt signaling. PMID: 19631775
Asparaginyl-tRNA synthetase 2, mitochondrial (putative)	NARS2	Two siblings with homozygous pathogenic splice-site variant in mitochondrial asparaginyl-tRNA synthetase (NARS2). PMID: 25385316
Aspartyl-tRNA synthetase	DARS	Disease: hypomyelination with brainstem and spinal cord involvement and leg spasticity (HBSL)
Aspartyl-tRNA synthetase 2, mitochondrial	DARS2	Disease: leukoencephalopathy with brainstem and spinal cord involvement and lactate elevation (LBSL)
Cap methyltransferase 1	CMTR1	Structural analysis of human 2′-O-ribose methyltransferases involved in mRNA cap structure formation. PMID: 24402442
Cap methyltransferase 2	CMTR2	
Cleavage and polyadenylation specific factor 3, 73 kDa	CPSF3	CSR1 induces cell death through inactivation of CPSF3. PMID: 18806823
Cleavage and polyadenylation specific factor 3-like	CPSF3L	snRNA 3′ end formation requires heterodimeric association of integrator subunits. PMID: 22252320

HGNC-approved name	Symbol	Target information
Cysteinyl-tRNA synthetase	CARS	Note: A chromosomal aberration involving CARS is associated with inflammatory myofibroblastic tumors (IMTs). Translocation t(2;11) (p23;p15) with ALK
Cysteinyl-tRNA synthetase 2, mitochondrial (putative)	CARS2	A homozygous splice-site mutation in CARS2 is associated with progressive myoclonic epilepsy. PMID: 25361775
Cytosolic thiouridylase subunit 1	CTU1	Biosynthesis and functions of sulfur modifications in tRNA. PMID: 24765101
Cytosolic thiouridylase subunit 2 homolog (S. pombe)	CTU2	
DEAD (Asp–Glu–Ala–Asp) box helicase 1	DDX1	The RNA-binding protein DDX1 promotes primary microRNA maturation and inhibits ovarian tumor progression. PMID: 25176654
DEAD (Asp–Glu–Ala–Asp) box helicase 3, X-linked	DDX3X	Identification of recurrent truncated DDX3X mutations in chronic lymphocytic leukaemia. PMID: 25382417
DEAD (Asp–Glu–Ala–Asp) box helicase 3, Y-linked	DDX3Y	Note: DDX3Y is located in the 'azoospermia factor a' (AZFa) region on chromosome Y which is deleted in Sertoli cell-only syndrome. This is an infertility disorder in which no germ cells are visible in seminiferous tubules leading to azoospermia
DEAD (Asp–Glu–Ala–Asp) box polypeptide 4	DDX4	DDX4 (DEAD box polypeptide 4) colocalizes with cancer stem cell marker CD133 in ovarian cancers. PMID: 24727449
DEAD (Asp–Glu–Ala–Asp) box helicase 5	DDX5	DDX5 regulates DNA replication and is required for cell proliferation in a subset of breast cancer cells. PMID: 22750847
DEAD (Asp–Glu–Ala–Asp) box helicase 6	DDX6	Note: A chromosomal aberration involving DDX6 may be a cause of hematopoietic tumors such as B-cell lymphomas. Translocation t(11;14) (q23;q32)
DEAD (Asp–Glu–Ala–Asp) box polypeptide 10	DDX10	Effects of the NUP98-DDX10 oncogene on primary human CD34+ cells: Role of a conserved helicase motif. PMID: 20339440
DEAD/H (Asp–Glu–Ala–Asp/His) box helicase 11	DDX11	Disease: Warsaw breakage syndrome (WBRS)
DEAD (Asp–Glu–Ala–Asp) box helicase 17	DDX17	Stem-loop recognition by DDX17 facilitates miRNA processing and antiviral defense. PMID: 25126784
DEAD (Asp–Glu–Ala–Asp) box polypeptide 18	DDX18	Genomic interaction between ER and HMGB2 identifies DDX18 as a novel driver of endocrine resistance in breast cancer cells. PMID: 25286587
DEAD (Asp–Glu–Ala–Asp) box polypeptide 19A	DDX19A	
DEAD (Asp–Glu–Ala–Asp) box polypeptide 19B	DDX19B	RNA export factor Ddx19 is required for nuclear import of the SRF coactivator MKL1. PMID: 25585691
DEAD (Asp–Glu–Ala–Asp) box polypeptide 20	DDX20	DEAD-box helicase DP103 defines metastatic potential of human breast cancers. PMID: 25083991
DEAD (Asp–Glu–Ala–Asp) box helicase 21	DDX21	Elevated DDX21 regulates c-Jun activity and rRNA processing in human breast cancers. PMID: 25260534
DEAD (Asp–Glu–Ala–Asp) box polypeptide 23	DDX23	Phosphorylation of human PRP28 by SRPK2 is required for integration of the U4/U6-U5 tri-snRNP into the spliceosome. PMID: 18425142
DEAD (Asp–Glu–Ala–Asp) box helicase 24	DDX24	DDX24 negatively regulates cytosolic RNA-mediated innate immune signaling. PMID: 24204270
DEAD (Asp–Glu–Ala–Asp) box helicase 25	DDX25	Role of gonadotropin regulated testicular RNA helicase (GRTH/Ddx25) on polysomal associated mRNAs in mouse testis. PMID: 22479328

(Continued)

HGNC-approved name	Symbol	Target information
DEAD (Asp–Glu–Ala–Asp) box polypeptide 27	DDX27	
DEAD (Asp–Glu–Ala–Asp) box polypeptide 28	DDX28	Regulated compartmentalization of the putative DEAD-box helicase MDDX28 within the mitochondria in COS-1 cells. PMID: 15350529
DEAD (Asp–Glu–Ala–Asp) box polypeptide 31	DDX31	DDX31 regulates the p53-HDM2 pathway and rRNA gene transcription through its interaction with NPM1 in renal cell carcinomas. PMID: 23019224
DEAD (Asp–Glu–Ala–Asp) box polypeptide 39A	DDX39A	DDX39 acts as a suppressor of invasion for bladder cancer. PMID: 22494014
DEAD (Asp–Glu–Ala–Asp) box polypeptide 39B	DDX39B	DDX39B (BAT1), TNF and IL6 gene polymorphisms and association with clinical outcomes of patients with Plasmodium vivax malaria. PMID: 25038626
DEAD (Asp–Glu–Ala–Asp) box polypeptide 41	DDX41	The helicase DDX41 recognizes the bacterial secondary messengers cyclic di-GMP and cyclic di-AMP to activate a type I interferon immune response. PMID: 23142775
DEAD (Asp–Glu–Ala–Asp) box helicase 42	DDX42	The DEAD box protein Ddx42p modulates the function of ASPP2, a stimulator of apoptosis. PMID: 19377511
DEAD (Asp–Glu–Ala–Asp) box polypeptide 43	DDX43	Overexpression of DDX43 mediates MEK inhibitor resistance through RAS Upregulation in uveal melanoma cells. PMID: 24899684
DEAD (Asp–Glu–Ala–Asp) box polypeptide 46	DDX46	A novel mechanism for Prp5 function in prespliceosome formation and proofreading the branch site sequence. PMID: 25561497
DEAD (Asp–Glu–Ala–Asp) box polypeptide 47	DDX47	Duplex destabilization by four ribosomal DEAD-box proteins. PMID: 23153376
DEAD (Asp–Glu–Ala–Asp) box polypeptide 49	DDX49	
DEAD (Asp–Glu–Ala–Asp) box polypeptide 50	DDX50	The DEXD/H-box RNA helicase RHII/Gu is a co-factor for c-Jun-activated transcription. PMID: 11823437
DEAD (Asp–Glu–Ala–Asp) box polypeptide 51	DDX51	Mammalian DEAD box protein Ddx51 acts in 3′ end maturation of 28S rRNA by promoting the release of U8 snoRNA. PMID: 20404093
DEAD (Asp–Glu–Ala–Asp) box polypeptide 52	DDX52	
DEAD (Asp–Glu–Ala–Asp) box polypeptide 53	DDX53	Cancer/testis antigen CAGE exerts negative regulation on p53 expression through HDAC2 and confers resistance to anti-cancer drugs. PMID: 20534591
DEAD (Asp–Glu–Ala–Asp) box polypeptide 54	DDX54	A DEAD-box RNA helicase Ddx54 protein in oligodendrocytes is indispensable for myelination in the central nervous system. PMID: 23239230
DEAD (Asp–Glu–Ala–Asp) box polypeptide 55	DDX55	
DEAD (Asp–Glu–Ala–Asp) box helicase 56	DDX56	The capsid-binding nucleolar helicase DDX56 is important for infectivity of West Nile virus. PMID: 21411523
DEAD (Asp–Glu–Ala–Asp) box polypeptide 58	DDX58	RIG-I suppresses the migration and invasion of hepatocellular carcinoma cells by regulating MMP9. PMID: 25626059
DEAD (Asp–Glu–Ala–Asp) box polypeptide 59	DDX59	Disease: orofaciodigital syndrome 5 (OFD5)
DEAD (Asp–Glu–Ala–Asp) box polypeptide 60	DDX60	DDX60, a DEXD/H box helicase, is a novel antiviral factor promoting RIG-I-like receptor-mediated signaling. PMID: 21791617
DEAD (Asp–Glu–Ala–Asp) box polypeptide 60-like	DDX60L	
DEAH (Asp–Glu–Ala–His) box polypeptide 8	DHX8	Identification of cellular proteins required for replication of human immunodeficiency virus type 1. PMID: 22404213

HGNC-approved name	Symbol	Target information
DEAH (Asp–Glu–Ala–His) box helicase 9	DHX9	Suppression of the DHX9 helicase induces premature senescence in human diploid fibroblasts in a p53-dependent manner. PMID: 24990949
DEAH (Asp–Glu–Ala–His) box helicase 15	DHX15	DHX15 senses double-stranded RNA in myeloid dendritic cells. PMID: 24990078
DEAH (Asp–Glu–Ala–His) box polypeptide 16	DHX16	Contribution of DEAH-box protein DHX16 in human pre-mRNA splicing. PMID: 20423332
DEAH (Asp–Glu–Ala–His) box polypeptide 29	DHX29	Helicase proteins DHX29 and RIG-I cosense cytosolic nucleic acids in the human airway system. PMID: 24821782
DEAH (Asp–Glu–Ala–His) box helicase 30	DHX30	DEXH-box protein DHX30 is required for optimal function of the zinc-finger antiviral protein. PMID: 21204022
DEAH (Asp–Glu–Ala–His) box polypeptide 32	DHX32	A role for DHX32 in regulating T-cell apoptosis. PMID: 17352256
DEAH (Asp–Glu–Ala–His) box polypeptide 33	DHX33	The DHX33 RNA helicase senses cytosolic RNA and activates the NLRP3 inflammasome. PMID: 23871209
DEAH (Asp–Glu–Ala–His) box polypeptide 34	DHX34	The RNA helicase DHX34 activates NMD by promoting a transition from the surveillance to the decay-inducing complex. PMID: 25220460
DEAH (Asp–Glu–Ala–His) box polypeptide 35	DHX35	
DEAH (Asp–Glu–Ala–His) box polypeptide 36	DHX36	DHX36 enhances RIG-I signaling by facilitating PKR-mediated antiviral stress granule formation. PMID: 24651521
DEAH (Asp–Glu–Ala–His) box polypeptide 37	DHX37	
DEAH (Asp–Glu–Ala–His) box polypeptide 38	DHX38	A missense mutation in the splicing factor gene DHX38 is associated with early-onset retinitis pigmentosa with macular coloboma. PMID: 24737827
DEAH (Asp–Glu–Ala–His) box polypeptide 40	DHX40	Identification of a novel human DDX40gene, a new member of the DEAH-box protein family. PMID: 12522690
DEAH (Asp–Glu–Ala–Asp/His) box polypeptide 57	DHX57	
Decapping mRNA 1A	DCP1A	c-Jun N-terminal kinase phosphorylates DCP1a to control formation of P bodies. PMID: 21859862
Decapping mRNA 1B	DCP1B	
Decapping mRNA 2	DCP2	The activation of the decapping enzyme DCP2 by DCP1 occurs on the EDC4 scaffold and involves a conserved loop in DCP1. PMID: 24510189
Decapping enzyme, scavenger	DCPS	Decapping scavenger (DcpS) enzyme: Advances in its structure, activity and roles in the cap-dependent mRNA metabolism. PMID: 24742626
DIS3 exosome endoribonuclease and 3′-5′ exoribonuclease	DIS3	Gene-dosage dependent overexpression at the 13q amplicon identifies DIS3 as candidate oncogene in colorectal cancer progression. PMID: 24478024
DIS3-like 3′-5′ exoribonuclease 2	DIS3L2	The RNase II/RNB family of exoribonucleases: Putting the 'Dis' in disease. PMID: 23776156. Disease: Perlman syndrome (PRLMNS)
DIS3-like exosome 3′-5′ exoribonuclease	DIS3L	The human core exosome interacts with differentially localized processive RNases: hDIS3 and hDIS3L. PMID: 20531386
D-Tyrosyl-tRNA deacylase 1	DTD1	Mechanism of chiral proofreading during translation of the genetic code. PMID: 24302572
D-Tyrosyl-tRNA deacylase 2 (putative)	DTD2	

(Continued)

HGNC-approved name	Symbol	Target information
ElaC ribonuclease Z 1	ELAC1	Localization of human RNase Z isoforms: Dual nuclear/mitochondrial targeting of the ELAC2 gene product by alternative translation initiation. PMID: 21559454
ElaC ribonuclease Z 2	ELAC2	Disease: prostate cancer, hereditary, 2 (HPC2)
Endonuclease V	ENDOV	Endonuclease V cleaves at inosines in RNA. PMID: 23912683
Endonuclease, polyU-specific	ENDOU	EndoU is a novel regulator of AICD during peripheral B cell selection. PMID: 24344237
Exoribonuclease 1	ERI1	Eri1: A conserved enzyme at the crossroads of multiple RNA-processing pathways. PMID: 24929628
ERI1 exoribonuclease family member 2	ERI2	
ERI1 exoribonuclease family member 3	ERI3	
Eukaryotic translation initiation factor 2, subunit 3 gamma, 52 kDa	EIF2S3	eIF2γ mutation that disrupts eIF2 complex integrity links intellectual disability to impaired translation initiation. PMID: 23063529
Eukaryotic translation initiation factor 4A1	EIF4A1	**Silvestrol** – antineoplastic agents
Eukaryotic translation initiation factor 4A2	EIF4A2	**Silvestrol** – antineoplastic agents
Eukaryotic translation initiation factor 4A3	EIF4A3	The diverse roles of the eIF4A family: You are the company you keep. PMID: 24450646
Eukaryotic translation initiation factor 5	EIF5	eIF5 is a dual function GAP and GDI for eukaryotic translational control. PMID: 21686265
Eukaryotic translation initiation factor 5B	EIF5B	Upregulation of eIF5B controls cell-cycle arrest and specific developmental stages. PMID: 25261552
FtsJ RNA methyltransferase homolog 1 (E. coli)	FTSJ1	Disease: mental retardation, X-linked 44 (MRX44)
FtsJ RNA methyltransferase homolog 2 (E. coli)	FTSJ2	FTSJ2, a heat shock-inducible mitochondrial protein, suppresses cell invasion and migration. PMID: 24595062
FtsJ homolog 3 (E. coli)	FTSJ3	The human nucleolar protein FTSJ3 associates with NIP7 and functions in pre-rRNA processing. PMID: 22195017
Glutamyl-tRNA(Gln) amidotransferase, subunit B	GATB	
Glutamyl-tRNA(Gln) amidotransferase, subunit C	GATC	
Histidyl-tRNA synthetase	HARS	A loss-of-function variant in the human histidyl-tRNA synthetase (HARS) gene is neurotoxic in vivo. PMID: 22930593. Disease: Usher syndrome 3B (USH3B)
Histidyl-tRNA synthetase 2, mitochondrial	HARS2	Disease: Perrault syndrome 2 (PRLTS2)
Interferon stimulated exonuclease gene 20 kDa	ISG20	ISG20, a new interferon-induced RNase specific for single-stranded RNA, defines an alternative antiviral pathway against RNA genomic viruses. PMID: 12594219
Interferon stimulated exonuclease gene 20 kDa-like 2	ISG20L2	ISG20L2, a novel vertebrate nucleolar exoribonuclease involved in ribosome biogenesis. PMID: 18065403
Isoleucyl-tRNA synthetase	IARS	Isoleucyl-tRNA synthetase levels modulate the penetrance of a homoplasmic m.4277T>C mitochondrial tRNA(Ile) mutation causing hypertrophic cardiomyopathy. PMID: 21945886
Isoleucyl-tRNA synthetase 2, mitochondrial	IARS2	Mutation in the nuclear-encoded mitochondrial isoleucyl-tRNA synthetase IARS2 in patients with cataracts, growth hormone deficiency with short stature, partial sensorineural deafness, and peripheral neuropathy or with Leigh syndrome. PMID: 25130867

HGNC-approved name	Symbol	Target information
Leucyl-tRNA synthetase	LARS	Disease: infantile liver failure syndrome 1 (ILFS1)
Leucyl-tRNA synthetase 2, mitochondrial	LARS2	Disease: Perrault syndrome 4 (PRLTS4)
Methionyl-tRNA synthetase	MARS	Disease: infantile liver failure syndrome 2 (ILFS2)
Methionyl-tRNA synthetase 2, mitochondrial	MARS2	Disease: spastic ataxia 3, autosomal recessive (SPAX3)
Methyltransferase-like 1	METTL1	tRNA modifying enzymes, NSUN2 and METTL1, determine sensitivity to 5-fluorouracil in HeLa cells. PMID: 25233213
Methyltransferase-like 2A	METTL2A	
Methyltransferase-like 2B	METTL2B	
Methyltransferase-like 3	METTL3	Methyltransferases modulate RNA stability in embryonic stem cells. PMID: 24481042
Methyltransferase-like 14	METTL14	A METTL3-METTL14 complex mediates mammalian nuclear RNA N6-adenosine methylation. PMID: 24316715
NOP2/Sun RNA methyltransferase family, member 2	NSUN2	Methylation by NSun2 represses the levels and function of microRNA 125b. PMID: 25047833. Disease: mental retardation, autosomal recessive 5 (MRT5)
NOP2/Sun domain family, member 3	NSUN3	
NOP2/Sun domain family, member 4	NSUN4	NSUN4 is a dual function mitochondrial protein required for both methylation of 12S rRNA and coordination of mitoribosomal assembly. PMID: 24516400
NOP2/Sun domain family, member 5	NSUN5	Note: NSUN5 is located in the Williams-Beuren syndrome (WBS) critical region
NOP2/Sun domain family, member 6	NSUN6	
NOP2/Sun domain family, member 7	NSUN7	Sperm motility defects and infertility in male mice with a mutation in Nsun7, a member of the Sun domain-containing family of putative RNA methyltransferases. PMID: 17442852
Nudix (nucleoside diphosphate linked moiety X)-type motif 16	NUDT16	hNUDT16: A universal decapping enzyme for small nucleolar RNA and cytoplasmic mRNA. PMID: 21337011
Nudix (nucleoside diphosphate linked moiety X)-type motif 21	NUDT21	CFIm25 links alternative polyadenylation to glioblastoma tumour suppression. PMID: 24814343
PAP-associated domain containing 4	PAPD4	Mono-uridylation of pre-microRNA as a key step in the biogenesis of group II let-7 microRNAs. PMID: 22063654
PAP-associated domain containing 5	PAPD5	PAPD5-mediated 3′ adenylation and subsequent degradation of miR-21 is disrupted in proliferative disease. PMID: 25049417
PAP-associated domain containing 7	PAPD7	Molecular cloning and characterization of a novel isoform of the non-canonical poly(A) polymerase PAPD7. PMID: 23376078
Peptidyl-tRNA hydrolase 1 homolog (S. cerevisiae)	PTRH1	
Peptidyl-tRNA hydrolase 2	PTRH2	Metastasis of tumor cells is enhanced by downregulation of Bit1. PMID: 21886829
Peptidyl-tRNA hydrolase domain containing 1	PTRHD1	
Phenylalanyl-tRNA synthetase, alpha subunit	FARSA	
Phenylalanyl-tRNA synthetase, beta subunit	FARSB	

(Continued)

HGNC-approved name	Symbol	Target information
Phenylalanyl-tRNA synthetase 2, mitochondrial	FARS2	Disease: combined oxidative phosphorylation deficiency 14 (COXPD14)
Poly(A) polymerase alpha	PAPOLA	The structure of the 5′-untranslated region of mammalian poly(A) polymerase-alpha mRNA suggests a mechanism of translational regulation. PMID: 20174964
Poly(A) polymerase beta (testis specific)	PAPOLB	Testis-specific expression of an intronless gene encoding a human poly(A) polymerase. PMID: 11459229
Poly(A) polymerase gamma	PAPOLG	Crystal structure of human poly(A) polymerase gamma reveals a conserved catalytic core for canonical poly(A) polymerases. PMID: 24076191
Polymerase (RNA) I polypeptide A, 194 kDa	POLR1A	**CX-5461** – antineoplastic agents
Polymerase (RNA) I polypeptide B, 128 kDa	POLR1B	**CX-5461** – antineoplastic agents
Polymerase (RNA) II (DNA directed) polypeptide A, 220 kDa	POLR2A	Nuclear localization of CD26 induced by a humanized monoclonal antibody inhibits tumor cell growth by modulating of POLR2A transcription. PMID: 23638030
Polymerase (RNA) II (DNA directed) polypeptide B, 140 kDa	POLR2B	
Polymerase (RNA) III (DNA directed) polypeptide A, 155 kDa	POLR3A	Association of the autoimmune disease scleroderma with an immunologic response to cancer. PMID: 24310608. Disease: leukodystrophy, hypomyelinating, 7, with or without oligodontia and/or hypogonadotropic hypogonadism (HLD7)
Polymerase (RNA) III (DNA directed) polypeptide B	POLR3B	Disease: leukodystrophy, hypomyelinating, 8, with or without oligodontia and/or hypogonadotropic hypogonadism (HLD8)
Polymerase (RNA) mitochondrial (DNA directed)	POLRMT	Human mitochondrial RNA polymerase: Structure-function, mechanism and inhibition. PMID: 22551784
Pseudouridylate synthase 1	PUS1	Disease: myopathy with lactic acidosis and sideroblastic anemia 1 (MLASA1)
Pseudouridylate synthase-like 1	PUSL1	
Pseudouridylate synthase 3	PUS3	
Pseudouridylate synthase 7 (putative)	PUS7	Transcriptome-wide mapping reveals widespread dynamic-regulated pseudouridylation of ncRNA and mRNA. PMID: 25219674
Pseudouridylate synthase 7 homolog (S. cerevisiae)-like	PUS7L	
Pseudouridylate synthase 10	PUS10	A meta-analysis of genome-wide association scans identifies IL18RAP, PTPN2, TAGAP, and PUS10 as shared risk loci for Crohn's disease and celiac disease. PMID: 21298027
Queuine tRNA-ribosyltransferase 1	QTRT1	Plant, animal, and fungal micronutrient queuosine is salvaged by members of the DUF2419 protein family. PMID: 24911101
Queuine tRNA-ribosyltransferase domain containing 1	QTRTD1	Queuosine formation in eukaryotic tRNA occurs via a mitochondria-localized heteromeric transglycosylase. PMID: 19414587
REX1, RNA exonuclease 1 homolog (S. cerevisiae)	REXO1	Identification of EloA-BP1, a novel Elongin A binding protein with an exonuclease homology domain. PMID: 12943681
RNA exonuclease 2	REXO2	REXO2 is an oligoribonuclease active in human mitochondria. PMID: 23741365
REX4, RNA exonuclease 4 homolog (S. cerevisiae)	REXO4	
Ribonuclease, RNase A family, 1 (pancreatic)	RNASE1	RNase1 prevents the damaging interplay between extracellular RNA and tumour necrosis factor-α in cardiac ischaemia/reperfusion injury. PMID: 25354936

HGNC-approved name	Symbol	Target information
Ribonuclease, RNase A family, 2 (liver, eosinophil-derived neurotoxin)	RNASE2	Antimicrobial activity of human eosinophil granule proteins: Involvement in host defence against pathogens. PMID: 22239733
Ribonuclease, RNase A family, 3	RNASE3	Protein post-translational modification in host defense: The antimicrobial mechanism of action of human eosinophil cationic protein native forms. PMID: 25271100
Ribonuclease, RNase A family, 4	RNASE4	Ribonuclease 4 protects neuron degeneration by promoting angiogenesis, neurogenesis, and neuronal survival under stress. PMID: 23143660
Ribonuclease, RNase A family, k6	RNASE6	Ribonucleases 6 and 7 have antimicrobial function in the human and murine urinary tract. PMID: 25075772
Ribonuclease, RNase A family, 7	RNASE7	An endogenous ribonuclease inhibitor regulates the antimicrobial activity of ribonuclease 7 in the human urinary tract. PMID: 24107847
Ribonuclease, RNase A family, 8	RNASE8	RNase 8, a novel RNase A superfamily ribonuclease expressed uniquely in placenta. PMID: 11861908
Ribonuclease, RNase A family, 9 (non-active)	RNASE9	Impaired sperm maturation in RNASE9 knockout mice. PMID: 24719258
Ribonuclease, RNase A family, 10 (non-active)	RNASE10	Epididymal protein Rnase10 is required for post-testicular sperm maturation and male fertility. PMID: 22750516
Ribonuclease, RNase A family, 11 (non-active)	RNASE11	
Ribonuclease, RNase A family, 12 (non-active)	RNASE12	
Ribonuclease, RNase A family, 13 (non-active)	RNASE13	
Ribonuclease H1	RNASEH1	RNaseH1 regulates TERRA-telomeric DNA hybrids and telomere maintenance in ALT tumour cells. PMID: 25330849
Ribonuclease H2, subunit A	RNASEH2A	Identification of two HIV inhibitors that also inhibit human RNaseH2. PMID: 24008364. Disease: Aicardi–Goutieres syndrome 4 (AGS4)
RNA pseudouridylate synthase domain containing 1	RPUSD1	
RNA pseudouridylate synthase domain containing 2	RPUSD2	
RNA pseudouridylate synthase domain containing 3	RPUSD3	
RNA pseudouridylate synthase domain containing 4	RPUSD4	
Seryl-tRNA synthetase	SARS	
Seryl-tRNA synthetase 2, mitochondrial	SARS2	Disease: hyperuricemia pulmonary hypertension renal failure and alkalosis (HUPRA)
Threonyl-tRNA synthetase	TARS	Threonyl-tRNA synthetase overexpression correlates with angiogenic markers and progression of human ovarian cancer. PMID: 25163878
Threonyl-tRNA synthetase 2, mitochondrial (putative)	TARS2	VARS2 and TARS2 mutations in patients with mitochondrial encephalomyopathies. PMID: 24827421
Threonyl-tRNA synthetase-like 2	TARSL2	Reinvestigation of aminoacyl-tRNA synthetase core complex by affinity purification-mass spectrometry reveals TARSL2 as a potential member of the complex. PMID: 24312579
Transcription factor B1, mitochondrial	TFB1M	Note: Variations in TFB1M may influence the clinical expression of aminoglycoside-induced deafness caused by the A1555G mutation in the mitochondrial 12S rRNA

(Continued)

HGNC-approved name	Symbol	Target information
Transcription factor B2, mitochondrial	TFB2M	Accessorizing the human mitochondrial transcription machinery. PMID: 23632312
tRNA methyltransferase 1 homolog (*S. cerevisiae*)	TRMT1	
tRNA methyltransferase 1 homolog (*S. cerevisiae*)-like	TRMT1L	The mouse Trm1-like gene is expressed in neural tissues and plays a role in motor coordination and exploratory behaviour. PMID: 17198746
tRNA methyltransferase 2 homolog A (*S. cerevisiae*)	TRMT2A	The expression of TRMT2A, a novel cell cycle regulated protein, identifies a subset of breast cancer patients with HER2 over-expression that are at an increased risk of recurrence. PMID: 20307320
tRNA methyltransferase 2 homolog B (*S. cerevisiae*)	TRMT2B	
tRNA methyltransferase 5	TRMT5	Conservation of structure and mechanism by Trm5 enzymes. PMID: 23887145
tRNA methyltransferase 6 homolog (*S. cerevisiae*)	TRMT6	The bipartite structure of the tRNA m1A58 methyltransferase from *S. cerevisiae* is conserved in humans. PMID: 16043508
tRNA methyltransferase 10 homolog A (*S. cerevisiae*)	TRMT10A	TRMT10A dysfunction is associated with abnormalities in glucose homeostasis, short stature and microcephaly. PMID: 25053765
tRNA methyltransferase 10 homolog B (*S. cerevisiae*)	TRMT10B	
tRNA methyltransferase 10 homolog C (*S. cerevisiae*)	TRMT10C	A subcomplex of human mitochondrial RNase P is a bifunctional methyltransferase – extensive moonlighting in mitochondrial tRNA biogenesis. PMID: 23042678
tRNA methyltransferase 11 homolog (*S. cerevisiae*)	TRMT11	Two-subunit enzymes involved in eukaryotic post-transcriptional tRNA modification. PMID: 25625329
tRNA methyltransferase 11-2 homolog (*S. cerevisiae*)	TRMT112	Trmt112 gene expression in mouse embryonic development. PMID: 22685353
tRNA methyltransferase 12 homolog (*S. cerevisiae*)	TRMT12	Chromosome 8 BAC array comparative genomic hybridization and expression analysis identify amplification and overexpression of TRMT12 in breast cancer. PMID: 17440925
tRNA methyltransferase 13 homolog (*S. cerevisiae*)	TRMT13	
tRNA methyltransferase 44 homolog (*S. cerevisiae*)	TRMT44	
tRNA methyltransferase 61 homolog A (*S. cerevisiae*)	TRMT61A	
tRNA methyltransferase 61 homolog B (*S. cerevisiae*)	TRMT61B	Trmt61B is a methyltransferase responsible for 1-methyladenosine at position 58 of human mitochondrial tRNAs. PMID: 23097428
tRNA-yW synthesizing protein 1 homolog (*S. cerevisiae*)	TYW1	Radical mediated ring formation in the biosynthesis of the hypermodified tRNA base wybutosine. PMID: 23856057
tRNA-yW synthesizing protein 1 homolog B (*S. cerevisiae*)	TYW1B	
tRNA-yW synthesizing protein 3 homolog (*S. cerevisiae*)	TYW3	
tRNA-yW synthesizing protein 5	TYW5	Expanding role of the jumonji C domain as an RNA hydroxylase. PMID: 20739293
TruB pseudouridine (psi) synthase family member 1	TRUB1	The human TruB family of pseudouridine synthase genes, including the Dyskeratosis Congenita 1 gene and the novel member TRUB1. PMID: 12736709
TruB pseudouridine (psi) synthase family member 2	TRUB2	

HGNC-approved name	Symbol	Target information
Tryptophanyl-tRNA synthetase	WARS	Hypoxia signature of splice forms of tryptophanyl-tRNA synthetase marks pancreatic cancer cells with distinct metastatic abilities. PMID: 21926542
Tryptophanyl tRNA synthetase 2, mitochondrial	WARS2	
TSEN2 tRNA splicing endonuclease subunit	TSEN2	Disease: pontocerebellar hypoplasia 2B (PCH2B)
TSEN34 tRNA splicing endonuclease subunit	TSEN34	Disease: pontocerebellar hypoplasia 2C (PCH2C)
Tudor domain containing 9	TDRD9	The TDRD9-MIWI2 complex is essential for piRNA-mediated retrotransposon silencing in the mouse male germline. PMID: 20059948
Tudor domain containing 12	TDRD12	Tudor domain containing 12 (TDRD12) is essential for secondary PIWI interacting RNA biogenesis in mice. PMID: 24067652
Tyrosyl-tRNA synthetase	YARS	A human tRNA synthetase is a potent PARP1-activating effector target for resveratrol. PMID: 25533949. Disease: Charcot–Marie–Tooth disease, dominant, intermediate type, C (CMTDIC)
Tyrosyl-tRNA synthetase 2, mitochondrial	YARS2	Disease: myopathy with lactic acidosis and sideroblastic anemia 2 (MLASA2)
Valyl-tRNA synthetase	VARS	
Valyl-tRNA synthetase 2, mitochondrial	VARS2	VARS2 V552V variant as prognostic marker in patients with early breast cancer. PMID: 20503108
Zinc finger CCCH-type containing 12A	ZC3H12A	Reqnase-1, a ribonuclease involved in the regulation of immune responses. PMID: 24163394
Zinc finger CCCH-type containing 12B	ZC3H12B	
Zinc finger CCCH-type containing 12C	ZC3H12C	Zc3h12c inhibits vascular inflammation by repressing NF-κB activation and pro-inflammatory gene expression in endothelial cells. PMID: 23360436
Zinc finger CCCH-type containing 12D	ZC3H12D	Note: A chromosomal aberration involving ZC3H12D may be the cause of the transformation of follicular lymphoma (FL) to diffuse large B-cell lymphoma (DLBCL). Translocation t(2;6)(p12;q25) with IGK. Resulting protein may not be expressed
Zinc finger, CCHC domain containing 6	ZCCHC6	Selective microRNA uridylation by Zcchc6 (TUT7) and Zcchc11 (TUT4). PMID: 25224700
Zinc finger, CCHC domain containing 11	ZCCHC11	Zcchc11 uridylates mature miRNAs to enhance neonatal IGF-1 expression, growth, and survival. PMID: 23209448
Angiogenin, ribonuclease, RNase A family, 5	ANG	Disease: amyotrophic lateral sclerosis 9 (ALS9)
Argonaute RISC catalytic component 2	AGO2	Down-regulation of Dicer and Ago2 is associated with cell proliferation and apoptosis in prostate cancer. PMID: 25135428
BCDIN3 domain containing	BCDIN3D	Human RNA methyltransferase BCDIN3D regulates microRNA processing. PMID: 23063121
Cleavage and polyadenylation factor I subunit 1	CLP1	Note: Neurodegeneration with progressive brain atrophy 1 (NBA1): A neurological syndrome affecting both the central and peripheral nervous systems characterized by onset of slow, progressive, neurodegenerative features and/or static encephalopathy
DALR anticodon binding domain containing 3	DALRD3	
Decapping exoribonuclease	DXO	A mammalian pre-mRNA 5′ end capping quality control mechanism and an unexpected link of capping to pre-mRNA processing. PMID: 23523372

(Continued)

HGNC-approved name	Symbol	Target information
DEXH (Asp–Glu–X–His) box polypeptide 58	DHX58	The innate immune sensor LGP2 activates antiviral signaling by regulating MDA5-RNA interaction and filament assembly. PMID: 25127512
Dicer 1, ribonuclease type III	DICER1	Identification and characterization of Dicer1e, a Dicer1 protein variant, in oral cancer cells. PMID: 25115815. Disease: pleuropulmonary blastoma (PPB)
DIM1 dimethyladenosine transferase 1 homolog (*S. cerevisiae*)	DIMT1	
Dyskeratosis congenita 1, dyskerin	DKC1	Disease: dyskeratosis congenita, X-linked (DKCX)
EMG1 *N*1-specific pseudouridine methyltransferase	EMG1	Disease: Bowen–Conradi syndrome (BWCNS)
Exosome component 10	EXOSC10	Structure of an Rrp6-RNA exosome complex bound to poly(A) RNA. PMID: 25043052
Glutaminyl-tRNA synthase (glutamine-hydrolyzing)-like 1	QRSL1	Glutamyl-tRNAGln amidotransferase is essential for mammalian mitochondrial translation in vivo. PMID: 24579914
Glutaminyl-tRNA synthetase	QARS	Expression profile of aminoacyl-tRNA synthetases in dorsal root ganglion neurons after peripheral nerve injury. PMID: 25467976
Glutamyl-prolyl-tRNA synthetase	EPRS	Halofuginone and other febrifugine derivatives inhibit prolyl-tRNA synthetase. PMID: 22327401
Glutamyl-tRNA synthetase 2, mitochondrial	EARS2	Disease: combined oxidative phosphorylation deficiency 12 (COXPD12)
Glycyl-tRNA synthetase	GARS	Disease: Charcot–Marie–Tooth disease 2D (CMT2D)
GTP binding protein 3 (mitochondrial)	GTPBP3	Characterization of human GTPBP3, a GTP-binding protein involved in mitochondrial tRNA modification. PMID: 18852288
HEN1 methyltransferase homolog 1 (*Arabidopsis*)	HENMT1	Structural insights into mechanisms of the small RNA methyltransferase HEN1. PMID: 19812675
Leucine rich repeat containing 47	LRRC47	Derangement of hypothetical proteins in fetal Down's syndrome brain. PMID: 15176487
Lysyl-tRNA synthetase	KARS	Disease: Charcot–Marie–Tooth disease, recessive, intermediate type, B (CMTRIB)
Mitochondrial methionyl-tRNA formyltransferase	MTFMT	Disease: combined oxidative phosphorylation deficiency 15 (COXPD15)
Mitochondrial poly(A) polymerase	MTPAP	Disease: spastic ataxia 4, autosomal recessive (SPAX4)
Mitochondrial rRNA methyltransferase 1 homolog (*S. cerevisiae*)	MRM1	Assignment of 2′-O-methyltransferases to modification sites on the mammalian mitochondrial large subunit 16 S ribosomal RNA (rRNA). PMID: 25074936
Mitochondrial tRNA translation optimization 1	MTO1	MTO1 mediates tissue specificity of OXPHOS defects via tRNA modification and translation optimization, which can be bypassed by dietary intervention. PMID: 25552653. Disease: combined oxidative phosphorylation deficiency 10 (COXPD10)
NOP2 nucleolar protein	NOP2	Nop2 is expressed during proliferation of neural stem cells and in adult mouse and human brain. PMID: 25481415
PAN2 poly(A)-specific ribonuclease subunit	PAN2	Structural basis for Pan3 binding to Pan2 and its function in mRNA recruitment and deadenylation. PMID: 24872509
Pelota homolog (*Drosophila*)	PELO	Pelota regulates the development of extraembryonic endoderm through activation of bone morphogenetic protein (BMP) signaling. PMID: 24835669
Phosphoseryl-tRNA kinase	PSTK	PSTK is a novel gene associated with early lung injury in Paraquat poisoning. PMID: 25592138
Polyribonucleotide nucleotidyltransferase 1	PNPT1	Disease: combined oxidative phosphorylation deficiency 13 (COXPD13)

HGNC-approved name	Symbol	Target information
Processing of precursor 1, ribonuclease P/MRP subunit (*S. cerevisiae*)	POP1	Note: Defects in POP1 may be the cause of a severe skeletal dysplasia reminiscent of anauxetic dysplasia. Affected individuals show severe growth retardation of prenatal onset, a bone dysplasia affecting the epiphyses and metaphyses of the long bones particularly in the lower limbs, and abnormalities of the spine including irregularly shaped vertebral bodies and marked cervical spine instability
Prolyl-tRNA synthetase 2, mitochondrial (putative)	PARS2	
Protein associated with topoisomerase II homolog 1 (yeast)	PATL1	RNA-related nuclear functions of human Pat1b, the P-body mRNA decay factor. PMID: 22090346
Ribonuclease L (2′,5′-oligoisoadenylate synthetase-dependent)	RNASEL	Pathologic effects of RNase-L dysregulation in immunity and proliferative control. PMID: 22202089. Disease: prostate cancer, hereditary, 1 (HPC1)
Ribonuclease T2	RNASET2	Disease: leukoencephalopathy, cystic, without megalencephaly (LCWM)
Ribonuclease, RNase K	RNASEK	Effect of cytostatic drugs on the mRNA expression levels of ribonuclease κ in breast and ovarian cancer cells. PMID: 23848202
Ribosomal RNA adenine dimethylase domain containing 1	RRNAD1	
Ribosomal RNA processing 8, methyltransferase, homolog (yeast)	RRP8	
RNA (guanine-7-) methyltransferase	RNMT	Human cap methyltransferase (RNMT) N-terminal non-catalytic domain mediates recruitment to transcription initiation sites. PMID: 23863084
RNA methyltransferase-like 1	RNMTL1	RM2 and MRM3 are involved in biogenesis of the large subunit of the mitochondrial ribosome. PMID: 25009282
RNA 3′-terminal phosphate cyclase	RTCA	HSPC111 governs breast cancer growth by regulating ribosomal biogenesis. PMID: 24425784
RNA 2′,3′-cyclic phosphate and 5′-OH ligase	RTCB	A synthetic biology approach identifies the mammalian UPR RNA ligase RtcB. PMID: 25087875
RNA guanylyltransferase and 5′-phosphatase	RNGTT	
Sep (*O*-phosphoserine) tRNA:Sec (selenocysteine) tRNA synthase	SEPSECS	Disease: pontocerebellar hypoplasia 2D (PCH2D)
Small nuclear ribonucleoprotein 200 kDa (U5)	SNRNP200	Disease: retinitis pigmentosa 33 (RP33)
Superkiller viralicidic activity 2-like (*S. cerevisiae*)	SKIV2L	The SKIV2L RNA exosome limits activation of the RIG-I-like receptors PMID: 25064072. Disease: trichohepatoenteric syndrome 2 (THES2)
Superkiller viralicidic activity 2-like 2 (*S. cerevisiae*)	SKIV2L2	Localization of a novel RNA-binding protein, SKIV2L2, to the nucleus in the round spermatids of mice. PMID: 21467735
Terminal uridylyl transferase 1, U6 snRNA-specific	TUT1	Nucleotidyl transferase TUT1 inhibits lipogenesis in osteosarcoma cells through regulation of microRNA-24 and microRNA-29a. PMID: 25142229
Transcription termination factor, RNA polymerase II	TTF2	hLodestar/HuF2 interacts with CDC5L and is involved in pre-mRNA splicing. PMID: 12927788
Trimethylguanosine synthase 1	TGS1	Structural basis for m7G-cap hypermethylation of small nuclear, small nucleolar and telomerase RNA by the dimethyltransferase TGS1. PMID: 19386620
tRNA 5-methylaminomethyl-2-thiouridylate methyltransferase	TRMU	Altered 2-thiouridylation impairs mitochondrial translation in reversible infantile respiratory chain deficiency. PMID: 23814040. Disease: liver failure, infantile, transient (LFIT)

(*Continued*)

HGNC-approved name	Symbol	Target information
tRNA aspartic acid methyltransferase 1	TRDMT1	Dnmt2 methyltransferases and immunity: An ancient overlooked connection between nucleotide modification and host defense? PMID: 24019003
tRNA isopentenyltransferase 1	TRIT1	The cytoplasmic and nuclear populations of the eukaryote tRNA-isopentenyl transferase have distinct functions with implications in human cancer. PMID: 25261850
tRNA nucleotidyl transferase, CCA-adding, 1	TRNT1	Mutations in TRNT1 cause congenital sideroblastic anemia with immunodeficiency, fevers, and developmental delay (SIFD). PMID: 25193871
tRNA phosphotransferase 1	TRPT1	An intact unfolded protein response in Trpt1 knockout mice reveals phylogenic divergence in pathways for RNA ligation. PMID: 18094117
tRNA-histidine guanylyltransferase 1-like (*S. cerevisiae*)	THG1L	Identification and characterization of a novel cytoplasm protein ICF45 that is involved in cell cycle regulation. PMID: 15459185
U6 snRNA biogenesis 1	USB1	Disease: poikiloderma with neutropenia (PN)
Williams–Beuren syndrome chromosome region 22	WBSCR22	The methyltransferase WBSCR22/Merm1 enhances glucocorticoid receptor function and is regulated in lung inflammation and cancer. PMID: 24488492. Note: WBSCR22 is located in the Williams-Beuren syndrome (WBS) critical region

5.8   Stress response and homeostasis

HGNC-approved name	Symbol	Target information
2-Oxoglutarate and iron-dependent oxygenase domain containing 1	OGFOD1	OGFOD1 catalyzes prolyl hydroxylation of RPS23 and is involved in translation control and stress granule formation. PMID: 24550447
2-Oxoglutarate and iron-dependent oxygenase domain containing 2	OGFOD2	
2-Oxoglutarate and iron-dependent oxygenase domain containing 3	OGFOD3	
Acid phosphatase 2, lysosomal	ACP2	Purkinje cell compartmentation in the cerebellum of the lysosomal Acid phosphatase 2 mutant mouse (nax – naked-ataxia mutant mouse). PMID: 24722417. Disease: acid phosphatase deficiency (ACPHD)
Acid phosphatase 5, tartrate resistant	ACP5	Disease: spondyloenchondrodysplasia with immune dysregulation (SPENCDI)
Acylphosphatase 1, erythrocyte (common) type	ACYP1	Crystallization and preliminary crystallographic analysis of human common-type acylphosphatase. PMID: 16511269
Acylphosphatase 2, muscle type	ACYP2	
ADP-ribosylhydrolase-like 1	ADPRHL1	
ADP-ribosylhydrolase-like 2	ADPRHL2	ADP-ribosyl-acceptor hydrolase 3 regulates poly (ADP-ribose) degradation and cell death during oxidative stress. PMID: 24191052
Alcohol dehydrogenase 1A (class I), alpha polypeptide	ADH1A	**Fomepizole** – alcohol poisoning
Alcohol dehydrogenase 1B (class I), beta polypeptide	ADH1B	**Fomepizole** – alcohol poisoning
Alcohol dehydrogenase 1C (class I), gamma polypeptide	ADH1C	**Fomepizole** – alcohol poisoning
Alcohol dehydrogenase 4 (class II), pi polypeptide	ADH4	

HGNC-approved name	Symbol	Target information
Alcohol dehydrogenase 5 (class III), chi polypeptide	ADH5	**N-6022** – cystic fibrosis treatments
Alcohol dehydrogenase 6 (class V)	ADH6	Identification of S-nitroso-CoA reductases that regulate protein S-nitrosylation. PMID: 25512491
Alcohol dehydrogenase 7 (class IV), mu or sigma polypeptide	ADH7	An enhancer-blocking element regulates the cell-specific expression of alcohol dehydrogenase 7. PMID: 24971505
Alcohol dehydrogenase, iron containing, 1	ADHFE1	Alcohol induces cell proliferation via hypermethylation of ADHFE1 in colorectal cancer cells. PMID: 24886599
Aldehyde dehydrogenase 1 family, member B1	ALDH1B1	ALDH1B1 is a potential stem/progenitor marker for multiple pancreas progenitor pools. PMID: 23142317
Aldehyde dehydrogenase 2 family (mitochondrial)	ALDH2	**Disulfiram** – alcohol deterrents
Aldehyde dehydrogenase 3 family, member A1	ALDH3A1	Aldehyde dehydrogenase 3A1 associates with prostate tumorigenesis. PMID: 24762960
Aldehyde dehydrogenase 3 family, member B2	ALDH3B2	Mouse aldehyde dehydrogenase ALDH3B2 is localized to lipid droplets via two C-terminal tryptophan residues and lipid modification. PMID: 25286108
Aldehyde dehydrogenase 7 family, member A1	ALDH7A1	Disease: pyridoxine-dependent epilepsy (PDE)
Aldehyde dehydrogenase 16 family, member A1	ALDH16A1	ALDH16A1 is a novel non-catalytic enzyme that may be involved in the etiology of gout via protein-protein interactions with HPRT1. PMID: 23348497
Aldo-keto reductase family 1, member A1 (aldehyde reductase)	AKR1A1	Increased aldehyde reductase expression mediates acquired radioresistance of laryngeal cancer cells via modulating p53. PMID: 22555805
Aldo-keto reductase family 1, member B1 (aldose reductase)	AKR1B1	**Ranirestat** – diabetic retinopathy
Aldo-keto reductase family 1, member B10 (aldose reductase)	AKR1B10	Selective inhibition of the tumor marker aldo-keto reductase family member 1B10 by oleanolic acid. PMID: 21561086
Aldo-keto reductase family 1, member E2	AKR1E2	
Aldo-keto reductase family 7, member A2 (aflatoxin aldehyde reductase)	AKR7A2	Elevation of AKR7A2 (succinic semialdehyde reductase) in neurodegenerative disease. PMID: 11597610
Aldo-keto reductase family 7, member A3 (aflatoxin aldehyde reductase)	AKR7A3	Protection against aflatoxin B1-induced cytotoxicity by expression of the cloned aflatoxin B1-aldehyde reductases rat AKR7A1 and human AKR7A3. PMID: 10416522
Aldo-keto reductase family 7-like	AKR7L	Aflatoxin B1 aldehyde reductase (AFAR) genes cluster at 1p35-1p36.1 in a region frequently altered in human tumour cells. PMID: 12879023
Alkaline phosphatase, intestinal	ALPI	Modulators of intestinal alkaline phosphatase. PMID: 23860652
Alkaline phosphatase, liver/bone/kidney	ALPL	Disease: hypophosphatasia (HOPS)
Alkaline phosphatase, placental	ALPP	Design and synthesis of selective inhibitors of placental alkaline phosphatase. PMID: 20031422
Alkaline phosphatase, placental-like 2	ALPPL2	**Amifostine** – radiation-protective agents
Amine oxidase, copper containing 1	AOC1	Amine oxidase copper-containing 1 (AOC1) is a downstream target gene of the Wilms tumor protein, WT1, during kidney development. PMID: 25037221
Amine oxidase, copper containing 2 (retina-specific)	AOC2	The unique substrate specificity of human AOC2, a semicarbazide-sensitive amine oxidase. PMID: 19588076

(Continued)

HGNC-approved name	Symbol	Target information
Amine oxidase, copper containing 3 (see also signaling/adhesion molecules)	AOC3	Semicarbazide-sensitive amine oxidase/vascular adhesion protein 1: Recent developments concerning substrates and inhibitors of a promising therapeutic target. PMID: 18691041
Apoptosis-inducing factor, mitochondrion-associated, 1	AIFM1	Cerulenin-induced apoptosis is mediated by disrupting the interaction between AIF and hexokinase II. PMID: 22426850. Disease: combined oxidative phosphorylation deficiency 6 (COXPD6)
Apoptosis-inducing factor, mitochondrion-associated, 2	AIFM2	AMID: New insights on its intracellular localization and expression at apoptosis. PMID: 18368494
Apoptosis-inducing factor, mitochondrion-associated, 3	AIFM3	Molecular cloning and characterization of a human AIF-like gene with ability to induce apoptosis. PMID: 15764604
Arylacetamide deacetylase	AADAC	Screening of specific inhibitors for human carboxylesterases or arylacetamide deacetylase. PMID: 24751575
Arylacetamide deacetylase-like 2	AADACL2	
Arylacetamide deacetylase-like 3	AADACL3	
Arylacetamide deacetylase-like 4	AADACL4	
Arylsulfatase A	ARSA	Correction of brain oligodendrocytes by AAVrh.10 intracerebral gene therapy in metachromatic leukodystrophy mice. PMID: 22642214. Disease: leukodystrophy metachromatic (MLD)
Arylsulfatase B	ARSB	Arylsulfatase B improves locomotor function after mouse spinal cord injury. PMID: 23520469. Disease: mucopolysaccharidosis 6 (MPS6)
Arylsulfatase C, isozyme F	ARSC2	
Arylsulfatase D	ARSD	Gene expression profiling identifies ARSD as a new marker of disease progression and the sphingolipid metabolism as a potential novel metabolism in chronic lymphocytic leukemia. PMID: 22820137
Arylsulfatase E (chondrodysplasia punctata 1)	ARSE	Disease: chondrodysplasia punctata 1, X-linked recessive (CDPX1)
Arylsulfatase F	ARSF	
Arylsulfatase G	ARSG	Arylsulfatase G inactivation causes loss of heparan sulfate 3-*O*-sulfatase activity and mucopolysaccharidosis in mice. PMID: 22689975
Arylsulfatase family, member H	ARSH	
Arylsulfatase family, member I	ARSI	
Arylsulfatase family, member J	ARSJ	
Arylsulfatase family, member K	ARSK	Arylsulfatase K, a novel lysosomal sulfatase. PMID: 23986440
Biliverdin reductase A	BLVRA	Tat-biliverdin reductase A inhibits inflammatory response by regulation of MAPK and NF-κB pathways in Raw 264.7 cells and edema mouse model. PMID: 25239864. Disease: hyperbiliverdinemia (HBLVD)
Biliverdin reductase B (flavin reductase (NADPH))	BLVRB	MALDI-MS tissue imaging identification of biliverdin reductase B overexpression in prostate cancer. PMID: 23954705
Carbonic anhydrase I	CA1	**Chlorothiazide** – diuretics
Carbonic anhydrase II	CA2	**Dorzolamide** – opthalmologicals. Disease: osteopetrosis, autosomal recessive 3 (OPTB3)
Carbonic anhydrase III, muscle specific	CA3	Mass spectrometry-based proteomic analysis of middle-aged vs. aged vastus lateralis reveals increased levels of carbonic anhydrase isoform 3 in senescent human skeletal muscle. PMID: 22797148
Carbonic anhydrase IV	CA4	**Chlorothiazide** – diuretics. Disease: retinitis pigmentosa 17 (RP17)
Carbonic anhydrase VA, mitochondrial	CA5A	Mitochondrial carbonic anhydrase VA deficiency resulting from CA5A alterations presents with hyperammonemia in early childhood. PMID: 24530203

HGNC-approved name	Symbol	Target information
Carbonic anhydrase VB, mitochondrial	CA5B	
Carbonic anhydrase VI	CA6	The role of carbonic anhydrase VI in bitter taste perception: Evidence from the Car6-/- mouse model. PMID: 25134447
Carbonic anhydrase VII	CA7	
Carbonic anhydrase VIII	CA8	Disease: cerebellar ataxia, mental retardation, and dysequilibrium syndrome 3 (CMARQ3)
Carbonic anhydrase IX	CA9	**Girentuximab** – antineoplastic agents
Carbonic anhydrase X	CA10	
Carbonic anhydrase XI	CA11	Overexpression of carbonic anhydrase-related protein XI promotes proliferation and invasion of gastrointestinal stromal tumors. PMID: 15942747
Carbonic anhydrase XII	CA12	Pharmacological inhibition of carbonic anhydrase XII interferes with cell proliferation and induces cell apoptosis in T-cell lymphomas PMID: 23348702. Disease: hyperchlorhidrosis, isolated (HCHLH)
Carbonic anhydrase XIII	CA13	Indapamide-like benzenesulfonamides as inhibitors of carbonic anhydrases I, II, VII, and XIII. PMID: 20926301
Carbonic anhydrase XIV	CA14	Carbonic anhydrases CA4 and CA14 both enhance AE3-mediated Cl–HCO3-exchange in hippocampal neurons. PMID: 19279262
Carboxylesterase 1	CES1	Carboxylesterase1/esterase-x regulates chylomicron production in mice. PMID: 23145182
Carboxylesterase 2	CES2	Down-regulation of carboxylesterases 1 and 2 plays an important role in prodrug metabolism in immunological liver injury rats. PMID: 25499727
Carboxylesterase 3	CES3	Integrated phenotypic and activity-based profiling links Ces3 to obesity and diabetes. PMID: 24362705
Carboxylesterase 4	CES4A	
Carboxylesterase 5	CES5A	An epididymis-specific carboxyl esterase CES5A is required for sperm capacitation and male fertility in the rat. PMID: 25475668
Crystallin, mu	CRYM	Loss of the thyroid hormone-binding protein Crym renders striatal neurons more vulnerable to mutant huntingtin in Huntington's disease. PMID: 25398949
Crystallin, zeta (quinone reductase)	CRYZ	Genetics of crystallins: Cataract and beyond. PMID: 19007775
Crystallin, zeta (quinone reductase)-like 1	CRYZL1	
Cytochrome b5 type A (microsomal)	CYB5A	Role of CYB5A in pancreatic cancer prognosis and autophagy modulation. PMID: 24301457. Disease: methemoglobinemia CYB5A-related (METHB-CYB5A)
Cytochrome b5 type B (outer mitochondrial membrane)	CYB5B	Constitutively overexpressed 21 kDa protein in Hodgkin lymphoma and aggressive non-Hodgkin lymphomas identified as cytochrome B5b (CYB5B). PMID: 20100355
Cytochrome b5 reductase 1	CYB5R1	
Cytochrome b5 reductase 2	CYB5R2	Cytochrome b5 reductase 2 is a novel candidate tumor suppressor gene frequently inactivated by promoter hypermethylation in human nasopharyngeal carcinoma. PMID: 24338690
Cytochrome b5 reductase 3	CYB5R3	Disease: methemoglobinemia CYB5R3-related (METHB-CYB5R3)
Cytochrome b5 reductase 4	CYB5R4	Natural mutations lead to enhanced proteasomal degradation of human Ncb5or, a novel flavoheme reductase. PMID: 23523930
Cytochrome b5 reductase-like	CYB5RL	

(Continued)

HGNC-approved name	Symbol	Target information
Cytochrome P450, family 1, subfamily A, polypeptide 1	CYP1A1	Cytochrome P450 1 family and cancers. PMID: 25448748
Cytochrome P450, family 1, subfamily A, polypeptide 2	CYP1A2	
Cytochrome P450, family 1, subfamily B, polypeptide 1	CYP1B1	Disease: Peters anomaly (PETAN)
Cytochrome P450, family 2, subfamily A	CYP2A	Extra-hepatic isozymes from the CYP1 and CYP2 families as potential chemotherapeutic targets. PMID: 23688134
Cytochrome P450, family 2, subfamily A, polypeptide 6	CYP2A6	Rational design of novel CYP2A6 inhibitors. PMID: 25458499
Cytochrome P450, family 2, subfamily A, polypeptide 7	CYP2A7	
Cytochrome P450, family 2, subfamily A, polypeptide 13	CYP2A13	Benzylmorpholine analogs as selective inhibitors of lung cytochrome P450 2A13 for the chemoprevention of lung cancer in tobacco users. PMID: 23756756
Cytochrome P450, family 2, subfamily B, polypeptide 6	CYP2B6	
Cytochrome P450, family 2, subfamily C, polypeptide 8	CYP2C8	
Cytochrome P450, family 2, subfamily C, polypeptide 9	CYP2C9	
Cytochrome P450, family 2, subfamily C, polypeptide 18	CYP2C18	
Cytochrome P450, family 2, subfamily C, polypeptide 19	CYP2C19	Decreased hippocampal volume and increased anxiety in a transgenic mouse model expressing the human CYP2C19 gene. PMID: 23877834
Cytochrome P450, family 2, subfamily D, polypeptide 6	CYP2D6	
Cytochrome P450, family 2, subfamily E, polypeptide 1	CYP2E1	
Cytochrome P450, family 2, subfamily F, polypeptide 1	CYP2F1	
Cytochrome P450, family 2, subfamily J, polypeptide 2	CYP2J2	CYP epoxygenase derived EETs: From cardiovascular protection to human cancer therapy. PMID: 23688135
Cytochrome P450, family 2, subfamily R, polypeptide 1	CYP2R1	Disease: rickets vitamin D-dependent 1B (VDDR1B)
Cytochrome P450, family 2, subfamily S, polypeptide 1	CYP2S1	CYP2S1 depletion enhances colorectal cell proliferation is associated with PGE2-mediated activation of β-catenin signaling. PMID: 25557876
Cytochrome P450, family 2, subfamily U, polypeptide 1	CYP2U1	Disease: spastic paraplegia 56, autosomal recessive (SPG56)
Cytochrome P450, family 2, subfamily W, polypeptide 1	CYP2W1	CYP2W1 is highly expressed in adrenal glands and is positively associated with the response to mitotane in adrenocortical carcinoma. PMID: 25144458
Cytochrome P450, family 3, subfamily A	CYP3A	Modulation of CYP3a expression and activity in mice models of type 1 and type 2 diabetes. PMID: 25505621
Cytochrome P450, family 3, subfamily A, polypeptide 4	CYP3A4	Clinical significance of steroid and xenobiotic receptor and its targeted gene CYP3A4 in human prostate cancer. PMID: 22050110
Cytochrome P450, family 3, subfamily A, polypeptide 5	CYP3A5	CYP3A5 regulates prostate cancer cell growth by facilitating nuclear translocation of AR. PMID: 25586052
Cytochrome P450, family 3, subfamily A, polypeptide 7	CYP3A7	

HGNC-approved name	Symbol	Target information
Cytochrome P450, family 3, subfamily A, polypeptide 43	CYP3A43	
Cytochrome P450, family 4, subfamily A, polypeptide 11	CYP4A11	CYP4 enzymes as potential drug targets: Focus on enzyme multiplicity, inducers and inhibitors, and therapeutic modulation of 20-hydroxyeicosatetraenoic acid (20-HETE) synthase and fatty acid ω-hydroxylase activities. PMID: 23688133
Cytochrome P450, family 4, subfamily A, polypeptide 22	CYP4A22	
Cytochrome P450, family 4, subfamily B, polypeptide 1	CYP4B1	Synthesis and evaluation of a 18F-labeled 4-ipomeanol as an imaging agent for CYP4B1 gene prodrug activation therapy. PMID: 23682585
Cytochrome P450, family 4, subfamily F, polypeptide 2	CYP4F2	Disease: coumarin resistance (CMRES)
Cytochrome P450, family 4, subfamily F, polypeptide 3	CYP4F3	Human cytochrome P450 4F3: Structure, functions, and prospects. PMID: 22706230
Cytochrome P450, family 4, subfamily F, polypeptide 8	CYP4F8	Arachidonic acid pathway members PLA2G7, HPGD, EPHX2, and CYP4F8 identified as putative novel therapeutic targets in prostate cancer. PMID: 21281786
Cytochrome P450, family 4, subfamily F, polypeptide 11	CYP4F11	
Cytochrome P450, family 4, subfamily F, polypeptide 12	CYP4F12	
Cytochrome P450, family 4, subfamily F, polypeptide 22	CYP4F22	Disease: ichthyosis, congenital, autosomal recessive 5 (ARCI5)
Cytochrome P450, family 4, subfamily V, polypeptide 2	CYP4V2	Disease: Bietti crystalline corneoretinal dystrophy (BCD)
Cytochrome P450, family 4, subfamily X, polypeptide 1	CYP4X1	
Cytochrome P450, family 4, subfamily Z, polypeptide 1	CYP4Z1	Increased expression of CYP4Z1 promotes tumor angiogenesis and growth in human breast cancer. PMID: 22841774
Cytochrome P450, family 7, subfamily A, polypeptide 1	CYP7A1	
Cytochrome P450, family 7, subfamily B, polypeptide 1	CYP7B1	Disease: spastic paraplegia 5A, autosomal recessive (SPG5A)
Cytochrome P450, family 8, subfamily B, polypeptide 1	CYP8B1	Loss of Cyp8b1 improves glucose homeostasis by increasing GLP-1. PMID: 25338812
Cytochrome P450, family 11, subfamily A, polypeptide 1	CYP11A1	**Aminoglutethimide** – Cushing's syndrome treatments. Disease: adrenal insufficiency, congenital, with 46,XY sex reversal (AICSR)
Cytochrome P450, family 11, subfamily B, polypeptide 1	CYP11B1	**Metyrapone** – diagnostics. Disease: adrenal hyperplasia 4 (AH4)
Cytochrome P450, family 11, subfamily B, polypeptide 2	CYP11B2	Disease: corticosterone methyloxidase 1 deficiency (CMO-1 deficiency)
Cytochrome P450, family 17, subfamily A, polypeptide 1	CYP17A1	**Abiraterone** – antineoplastic agents. Disease: adrenal hyperplasia 5 (AH5)
Cytochrome P450, family 19, subfamily A, polypeptide 1	CYP19A1	**Anastrozole** – antineoplastic agents. Disease: aromatase excess syndrome (AEXS)
Cytochrome P450, family 20, subfamily A, polypeptide 1	CYP20A1	
Cytochrome P450, family 21, subfamily A, polypeptide 2	CYP21A2	Disease: adrenal hyperplasia 3 (AH3)

(Continued)

HGNC-approved name	Symbol	Target information
Cytochrome P450, family 24, subfamily A, polypeptide 1	CYP24A1	Disease: hypercalcemia infantile (HCAI)
Cytochrome P450, family 26, subfamily A, polypeptide 1	CYP26A1	Molecular recognition of CYP26A1 binding pockets and structure-activity relationship studies for design of potent and selective retinoic acid metabolism blocking agents. PMID: 25541526
Cytochrome P450, family 26, subfamily B, polypeptide 1	CYP26B1	Disease: radiohumeral fusions with other skeletal and craniofacial anomalies (RHFCA)
Cytochrome P450, family 26, subfamily C, polypeptide 1	CYP26C1	Disease: focal facial dermal dysplasia 4 (FFDD4)
Cytochrome P450, family 27, subfamily A, polypeptide 1	CYP27A1	27-Hydroxycholesterol links hypercholesterolemia and breast cancer pathophysiology. PMID: 24288332. Disease: cerebrotendinous xanthomatosis (CTX)
Cytochrome P450, family 27, subfamily B, polypeptide 1	CYP27B1	Disease: rickets vitamin D-dependent 1A (VDDR1A)
Cytochrome P450, family 27, subfamily C, polypeptide 1	CYP27C1	
Cytochrome P450, family 39, subfamily A, polypeptide 1	CYP39A1	
Cytochrome P450, family 46, subfamily A, polypeptide 1	CYP46A1	Pharmacologic stimulation of cytochrome P450 46A1 and cerebral cholesterol turnover in mice. PMID: 24352658
Cytochrome P450, family 51, subfamily A, polypeptide 1	CYP51A1	The effects of rosuvastatin and the CYP51A1 inhibitor LEK-935 on the proteome of primary human hepatocytes. PMID: 22180046
Dehydrogenase/reductase (SDR family) member 3	DHRS3	The retinaldehyde reductase activity of DHRS3 is reciprocally activated by retinol dehydrogenase 10 to control retinoid homeostasis. PMID: 24733397
Dehydrogenase/reductase (SDR family) member 4	DHRS4	Characterization of human DHRS4: An inducible short-chain dehydrogenase/reductase enzyme with 3beta-hydroxysteroid dehydrogenase activity. PMID: 18571493
Dehydrogenase/reductase (SDR family) member 7	DHRS7	Biochemical properties of human dehydrogenase/reductase (SDR family) member 7. PMID: 24246760
Dehydrogenase/reductase (SDR family) member 7B	DHRS7B	
Dehydrogenase/reductase (SDR family) member 7C	DHRS7C	DHRS7c, a novel cardiomyocyte-expressed gene that is down-regulated by adrenergic stimulation and in heart failure. PMID: 22143074
Dehydrogenase/reductase (SDR family) member 9	DHRS9	Multiple retinol and retinal dehydrogenases catalyze all-trans-retinoic acid biosynthesis in astrocytes. PMID: 21138835
Dual oxidase 1	DUOX1	Duox1-derived H2O2 modulates Cxcl8 expression and neutrophil recruitment via JNK/c-JUN/AP-1 signaling and chromatin modifications. PMID: 25582859
Dual oxidase 2	DUOX2	Dual oxidase 2 is essential for the toll-like receptor 5-mediated inflammatory response in airway mucosa. PMID: 21714724. Disease: thyroid dyshormonogenesis 6 (TDH6)
Ecto-NOX disulfide-thiol exchanger 1	ENOX1	The novel antiangiogenic VJ115 inhibits the NADH oxidase ENOX1 and cytoskeleton-remodeling proteins. PMID: 23054211
Ecto-NOX disulfide-thiol exchanger 2	ENOX2	Metabolite modulation of HeLa cell response to ENOX2 inhibitors EGCG and phenoxodiol. PMID: 21571040
Egl-9 family hypoxia-inducible factor 1	EGLN1	(R)-2-Hydroxyglutarate is sufficient to promote leukemogenesis and its effects are reversible. PMID: 23393090. Disease: erythrocytosis, familial, 3 (ECYT3)

HGNC-approved name	Symbol	Target information
Egl-9 family hypoxia-inducible factor 2	EGLN2	**TRC160334** – renal disease treatments
Egl-9 family hypoxia-inducible factor 3	EGLN3	Prolyl hydroxylases 2 and 3 act in gliomas as protective negative feedback regulators of hypoxia-inducible factors. PMID: 20028863
Epoxide hydrolase 1, microsomal (xenobiotic)	EPHX1	Note: In some populations, the high activity haplotype tyr113/his139 is overrepresented among women suffering from pregnancy-induced hypertension (pre-eclampsia) when compared with healthy controls. Disease: familial hypercholanemia (FHCA)
Epoxide hydrolase 2, cytoplasmic	EPHX2	**AR9281** – antihypertensives
Epoxide hydrolase 3	FPHX3	EH3 (ABHD9): The first member of a new epoxide hydrolase family with high activity for fatty acid epoxides. PMID: 22798687
Epoxide hydrolase 4	FPHX4	
Flavin-containing monooxygenase 1	FMO1	The phenotype of a flavin-containing monooyxgenase knockout mouse implicates the drug-metabolizing enzyme FMO1 as a novel regulator of energy balance. PMID: 24792439
Flavin-containing monooxygenase 2 (non-functional)	FMO2	
Flavin-containing monooxygenase 3	FMO3	Flavin monooxygenase metabolism: Why medicinal chemists should matter. PMID: 25003501. Disease: trimethylaminuria (TMAU)
Flavin-containing monooxygenase 4	FMO4	
Flavin-containing monooxygenase 5	FMO5	
Glutaredoxin (thioltransferase)	GLRX	Alteration of thioredoxin and glutaredoxin in the progression of Alzheimer's disease. PMID: 24270206
Glutaredoxin 2	GLRX2	Glutaredoxin-2 is required to control oxidative phosphorylation in cardiac muscle by mediating deglutathionylation reactions. PMID: 24727547
Glutaredoxin 3	GLRX3	Caspase-3-mediated cleavage of PICOT in apoptosis. PMID: 23415866
Glutaredoxin 5	GLRX5	Disease: anemia, sideroblastic, pyridoxine-refractory, autosomal recessive (PRARSA)
Glutathione peroxidase 1	GPX1	Glutathione peroxidase-1 as a novel therapeutic target for COPD. PMID: 23849338
Glutathione peroxidase 2 (gastrointestinal)	GPX2	GPx2 suppression of H2O2 stress links the formation of differentiated tumor mass to metastatic capacity in colorectal cancer. PMID: 25261240
Glutathione peroxidase 3 (plasma)	GPX3	Clinical significance and therapeutic value of glutathione peroxidase 3 (GPx3) in hepatocellular carcinoma. PMID: 25333265
Glutathione peroxidase 4	GPX4	Glutathione peroxidase 4 is reversibly induced by HCV to control lipid peroxidation and to increase virion infectivity. PMID: 25516417
Glutathione peroxidase 5 (epididymal androgen-related protein)	GPX5	Epididymal specific, selenium-independent GPX5 protects cells from oxidative stress-induced lipid peroxidation and DNA mutation. PMID: 23696541
Glutathione peroxidase 6 (olfactory)	GPX6	Synthetic lethal screening in the mammalian central nervous system identifies Gpx6 as a modulator of Huntington's disease. PMID: 25535386
Glutathione peroxidase 7	GPX7	Disease: Barrett esophagus (BE)
Glutathione peroxidase 8 (putative)	GPX8	Glutathione peroxidase 8 is transcriptionally regulated by HIFα and modulates growth factor signaling in HeLa cells. PMID: 25557012
Glutathione S-transferase alpha 1	GSTA1	2,2′-Dihydroxybenzophenones and their carbonyl N-analogues as inhibitor scaffolds for MDR-involved human glutathione transferase isoenzyme A1-1. PMID: 25002233
Glutathione S-transferase alpha 2	GSTA2	

(Continued)

HGNC-approved name	Symbol	Target information
Glutathione S-transferase alpha 3	GSTA3	Genetic or pharmacologic activation of Nrf2 signaling fails to protect against aflatoxin genotoxicity in hypersensitive GSTA3 knockout mice. PMID: 24675090
Glutathione S-transferase alpha 4	GSTA4	Protection from oxidative and electrophilic stress in the Gsta4-null mouse heart. PMID: 23690225
Glutathione S-transferase alpha 5	GSTA5	
Glutathione S-transferase kappa 1	GSTK1	Glutathione transferase kappa deficiency causes glomerular nephropathy without overt oxidative stress. PMID: 21826057
Glutathione S-transferase mu 1	GSTM1	Glutathione S-transferase gene GSTM1, gene-gene interaction, and gastric cancer susceptibility: Evidence from an updated meta-analysis. PMID: 25477765
Glutathione S-transferase mu 2 (muscle)	GSTM2	Glutathione transferase mu 2 protects glioblastoma cells against aminochrome toxicity by preventing autophagy and lysosome dysfunction. PMID: 24434817
Glutathione S-transferase mu 3 (brain)	GSTM3	GSTM3 reverses the resistance of hepatoma cells to radiation by regulating the expression of cell cycle/apoptosis-related molecules. PMID: 25202346
Glutathione S-transferase mu 4	GSTM4	Targeting glutathione S-transferase M4 in Ewing sarcoma. PMID: 25147782
Glutathione S-transferase mu 5	GSTM5	
Glutathione S-transferase omega 1	GSTO1	Identification of glutathione S-transferase omega 1 (GSTO1) protein as a novel tumor-associated antigen and its autoantibody in human esophageal squamous cell carcinoma. PMID: 25085586
Glutathione S-transferase omega 2	GSTO2	Structural insights into omega-class glutathione transferases: A snapshot of enzyme reduction and identification of a non-catalytic ligandin site. PMID: 23593192
Glutathione S-transferase pi 1	GSTP1	Targeting GSTP1-1 induces JNK activation and leads to apoptosis in cisplatin-sensitive and -resistant human osteosarcoma cell lines. PMID: 22068640
Glutathione S-transferase theta 1	GSTT1	Dual glutathione-S-transferase-θ1 and -μ1 gene deletions determine imatinib failure in chronic myeloid leukemia. PMID: 25188725
Glutathione S-transferase theta 2	GSTT2	GSTT2, a phase II gene induced by apple polyphenols, protects colon epithelial cells against genotoxic damage. PMID: 19753610
Glutathione S-transferase theta 2B (gene/pseudogene)	GSTT2B	
Glutathione S-transferase zeta 1	GSTZ1	Glutathione transferase zeta: Discovery, polymorphic variants, catalysis, inactivation, and properties of Gstz1-/- mice. PMID: 21303221
Heme oxygenase (decycling) 1	HMOX1	Heme oxygenase 1 as a target for the design of gene and pharmaceutical therapies for autoimmune diseases. PMID: 24766133. Disease: heme oxygenase 1 deficiency (HMOX1D)
Heme oxygenase (decycling) 2	HMOX2	Neuroprotective effect of heme oxygenase-2 knockout in the blood injection model of intracerebral hemorrhage. PMID: 25149897
Hephaestin	HEPH	The ceruloplasmin homolog hephaestin and the control of intestinal iron absorption. PMID: 12547227
Hephaestin-like 1	HEPHL1	Dietary iron, iron homeostatic gene polymorphisms and the risk of advanced colorectal adenoma and cancer. PMID: 24536049
Hydroxyacylglutathione hydrolase	HAGH	Fructose compared with glucose is more a potent glycoxidation agent in vitro, but not under carbohydrate-induced stress in vivo: Potential role of antioxidant and antiglycation enzymes. PMID: 24361593
Hydroxyacylglutathione hydrolase-like	HAGHL	

HGNC-approved name	Symbol	Target information
Microsomal glutathione S-transferase 1	MGST1	Characterization of new potential anticancer drugs designed to overcome glutathione transferase mediated resistance. PMID: 21851097
Microsomal glutathione S-transferase 2	MGST2	
Microsomal glutathione S-transferase 3	MGST3	Joint genetic analysis of hippocampal size in mouse and human identifies a novel gene linked to neurodegenerative disease. PMID: 25280473
NAD(P)H dehydrogenase, quinone 1	NQO1	An NQO1-initiated and p53-independent apoptotic pathway determines the anti-tumor effect of tanshinone IIA against non-small cell lung cancer. PMID: 22848731
NAD(P)H dehydrogenase, quinone 2	NQO2	Mechanism-based inhibition of quinone reductase 2 (NQO2): Selectivity for NQO2 over NQO1 and structural basis for flavoprotein inhibition. PMID: 21506232
NADPH oxidase 1	NOX1	Redox unbalance: NADPH oxidase as therapeutic target in blood pressure control. PMID: 20549031
NADPH oxidase 3	NOX3	siRNA-mediated knock-down of NOX3: Therapy for hearing loss? PMID: 22562580
NADPH oxidase 4	NOX4	Nox4 as a potential therapeutic target for treatment of uremic toxicity associated to chronic kidney disease. PMID: 23538692
NADPH oxidase, EF-hand calcium binding domain 5	NOX5	NADPH oxidase 5 and renal disease. PMID: 25415612
N-Acetyltransferase 1 (arylamine N-acetyltransferase)	NAT1	Arylamine N-acetyltransferases: From drug metabolism and pharmacogenetics to drug discovery. PMID: 24667436
N-Acetyltransferase 2 (arylamine N-acetyltransferase)	NAT2	Inflammation-induced phenoconversion of polymorphic drug metabolizing enzymes: A hypothesis with implications for personalized medicine. PMID: 25519488
Paraoxonase 1	PON1	Paraoxonase1 192 (PON1 192) gene polymorphism and serum paraoxonase activity in panic disorder patients. PMID: 25600530. Disease: microvascular complications of diabetes 5 (MVCD5)
Paraoxonase 2	PON2	DJ-1 interacts with and regulates paraoxonase-2, an enzyme critical for neuronal survival in response to oxidative stress. PMID: 25210784
Paraoxonase 3	PON3	PON3 knockout mice are susceptible to obesity, gallstone formation, and atherosclerosis. PMID: 25477283
Peroxidasin homolog (Drosophila)	PXDN	Peroxidasin: Tying the collagen-sulfilimine knot. PMID: 22907088
Peroxidasin homolog (Drosophila) like	PXDNL	Peroxidasin-like protein: A novel peroxidase homologue in the human heart. PMID: 24253521
Peroxiredoxin 1	PRDX1	Proteomic analysis of bladder cancer indicates Prx-I as a key molecule in BI-TK/GCV treatment system. PMID: 24904997
Peroxiredoxin 2	PRDX2	Peroxiredoxin 2 is involved in vasculogenic mimicry formation by targeting VEGFR2 activation in colorectal cancer. PMID: 25471788
Peroxiredoxin 3	PRDX3	Heart mitochondrial proteome study elucidates changes in cardiac energy metabolism and antioxidant PRDX3 in human dilated cardiomyopathy. PMID: 25397948
Peroxiredoxin 4	PRDX4	Investigation of peroxiredoxin IV as a calpain-regulated pathway in cancer. PMID: 21187494
Peroxiredoxin 5	PRDX5	**Auranofin** – anti-inflammatory and antirheumatic products

(Continued)

HGNC-approved name	Symbol	Target information
Peroxiredoxin 6	PRDX6	PRDX6 exacerbates dopaminergic neurodegeneration in a MPTP mouse model of Parkinson's disease. PMID: 25193021
Prolyl 4-hydroxylase, alpha polypeptide I	P4HA1	**GSK 1278863** – antianaemic preparations
Prolyl 4-hydroxylase, alpha polypeptide II	P4HA2	Prolyl-4-hydroxylase α subunit 2 promotes breast cancer progression and metastasis by regulating collagen deposition. PMID: 24383403
Prolyl 4-hydroxylase, alpha polypeptide III	P4HA3	Cloning of a novel prolyl 4-hydroxylase subunit expressed in the fibrous cap of human atherosclerotic plaque. PMID: 12874193
Prolyl 4-hydroxylase, beta polypeptide	P4HB	
Prolyl 4-hydroxylase, transmembrane (endoplasmic reticulum)	P4HTM	Transmembrane prolyl 4-hydroxylase is a fourth prolyl 4-hydroxylase regulating EPO production and erythropoiesis. PMID: 22955912
Prune exopolyphosphatase	PRUNE	H-Prune through GSK-3β interaction sustains canonical WNT/β-catenin signaling enhancing cancer progression in NSCLC. PMID: 25026278
Prune homolog 2 (*Drosophila*)	PRUNE2	BMCC1 is an AP-2 associated endosomal protein in prostate cancer cells. PMID: 24040105
Pyrophosphatase (inorganic) 1	PPA1	Regulation of neurite growth by inorganic pyrophosphatase 1 via JNK dephosphorylation. PMID: 23626709
Pyrophosphatase (inorganic) 2	PPA2	
Quiescin Q6 sulfhydryl oxidase 1	QSOX1	Elevated transcription of the gene QSOX1 encoding quiescin Q6 sulfhydryl oxidase 1 in breast cancer. PMID: 23460839
Quiescin Q6 sulfhydryl oxidase 2	QSOX2	Neuroblastoma-derived sulfhydryl oxidase, a new member of the sulfhydryl oxidase/Quiescin6 family, regulates sensitization to interferon gamma-induced cell death in human neuroblastoma cells. PMID: 14633699
Six-transmembrane epithelial antigen of the prostate 1	STEAP1	**RG7450** – antineoplastic agents
STEAP family member 1B	STEAP1B	Expression of STEAP1 and STEAP1B in prostate cell lines, and the putative regulation of STEAP1 by post-transcriptional and post-translational mechanisms. PMID: 25053991
STEAP family member 2, metalloreductase	STEAP2	A role for STEAP2 in prostate cancer progression. PMID: 25248617
STEAP family member 3, metalloreductase	STEAP3	Disease: anemia, hypochromic microcytic, with iron overload 2 (AHMIO2)
STEAP family member 4	STEAP4	STEAP4 and insulin resistance. PMID: 24627165
Sulfatase-modifying factor 1	SUMF1	A non-conserved miRNA regulates lysosomal function and impacts on a human lysosomal storage disorder. PMID: 25524633. Disease: multiple sulfatase deficiency (MSD)
Sulfatase-modifying factor 2	SUMF2	Sos recruitment system for the analysis of the interaction between sulfatase-modifying factor 2 subtypes and interleukin-13. PMID: 24301935
Sulfotransferase family, cytosolic, 1A, phenol-preferring, member 1	SULT1A1	Expression of sulfotransferase SULT1A1 in cancer cells predicts susceptibility to the novel anticancer agent NSC-743380. PMID: 25514600
Sulfotransferase family, cytosolic, 1A, phenol-preferring, member 2	SULT1A2	Crystal structures of human sulfotransferases: Insights into the mechanisms of action and substrate selectivity. PMID: 22512672
Sulfotransferase family, cytosolic, 1A, phenol-preferring, member 3	SULT1A3	Cytosolic sulfotransferase 1A3 is induced by dopamine and protects neuronal cells from dopamine toxicity: Role of D1 receptor-*N*-methyl-D-aspartate receptor coupling. PMID: 24136195
Sulfotransferase family, cytosolic, 1A, phenol-preferring, member 4	SULT1A4	
Sulfotransferase family, cytosolic, 1B, member 1	SULT1B1	

HGNC-approved name	Symbol	Target information
Sulfotransferase family, cytosolic, 1C, member 2	SULT1C2	
Sulfotransferase family, cytosolic, 1C, member 3	SULT1C3	SULT1C3, an orphan sequence of the human genome, encodes an enzyme activating various promutagens. PMID: 17936463
Sulfotransferase family, cytosolic, 1C, member 4	SULT1C4	
Sulfotransferase family 1E, estrogen-preferring, member 1	SULT1E1	Estrogen sulfotransferase/SULT1E1 promotes human adipogenesis. PMID: 24567372
Sulfotransferase family, cytosolic, 2A, dehydroepiandrosterone (DHEA)-preferring, member 1	SULT2A1	Paradigms of sulfotransferase catalysis: The mechanism of SULT2A1. PMID: 25056952
Sulfotransferase family, cytosolic, 2B, member 1	SULT2B1	SULT2B1: Unique properties and characteristics of a hydroxysteroid sulfotransferase family. PMID: 24020383
Sulfotransferase family 4A, member 1	SULT4A1	Expression of the orphan cytosolic sulfotransferase SULT4A1 and its major splice variant in human tissues and cells: Dimerization, degradation and polyubiquitination. PMID: 24988429
Sulfotransferase family, cytosolic, 6B, member 1	SULT6B1	Molecular cloning, expression and characterization of a novel mouse SULT6 cytosolic sulfotransferase. PMID: 19505954
Superoxide dismutase 1, soluble	SOD1	**ATN-224** – antineoplastic agents. Disease: amyotrophic lateral sclerosis 1 (ALS1)
Superoxide dismutase 2, mitochondrial	SOD2	Epigenetic attenuation of mitochondrial superoxide dismutase 2 in pulmonary arterial hypertension: A basis for excessive cell proliferation and a new therapeutic target. PMID: 20529999. Disease: microvascular complications of diabetes 6 (MVCD6)
Superoxide dismutase 3, extracellular	SOD3	Overexpression of superoxide dismutase 3 gene blocks high-fat diet-induced obesity, fatty liver and insulin resistance. PMID: 25030609
Thioredoxin	TXN	Changes in the mitochondrial antioxidant systems in neurodegenerative diseases and acute brain disorders. PMID: 25576182
Thioredoxin 2	TXN2	Adiponectin induces apoptosis in hepatocellular carcinoma through differential modulation of thioredoxin proteins. PMID: 25514170
Thioredoxin domain containing 5 (endoplasmic reticulum)	TXNDC5	TXNDC5, a newly discovered disulfide isomerase with a key role in cell physiology and pathology. PMID: 25526565
Thioredoxin domain containing 12 (endoplasmic reticulum)	TXNDC12	Solution structure and dynamics of ERp18, a small endoplasmic reticulum resident oxidoreductase. PMID: 19361220
Thioredoxin domain containing 17	TXNDC17	Thioredoxin-related protein of 14 kDa is an efficient ʟ-cystine reductase and S-denitrosylase. PMID: 24778250
Thioredoxin reductase 1	TXNRD1	**Arsenic trioxide** – antineoplastic agents
Thioredoxin reductase 2	TXNRD2	Novel role for thioredoxin reductase-2 in mitochondrial redox adaptations to obesogenic diet and exercise in heart and skeletal muscle. PMID: 23613536
Thioredoxin reductase 3	TXNRD3	The selenoproteins GPx2, TrxR2 and TrxR3 are regulated by Wnt signalling in the intestinal epithelium. PMID: 22683372
Thioredoxin reductase 3 neighbor	TXNRD3NB	
Thioredoxin-like 1	TXNL1	TXNL1 induces apoptosis in cisplatin resistant human gastric cancer cell lines. PMID: 25348020
Thioredoxin-related transmembrane protein 1	TMX1	The protective role of the transmembrane thioredoxin-related protein TMX in inflammatory liver injury. PMID: 22924822

(Continued)

HGNC-approved name	Symbol	Target information
Thioredoxin-related transmembrane protein 3	TMX3	Structure-function analysis of the endoplasmic reticulum oxidoreductase TMX3 reveals interdomain stabilization of the N-terminal redox-active domain. PMID: 17881353
Thioredoxin-related transmembrane protein 4	TMX4	Novel thioredoxin-related transmembrane protein TMX4 has reductase activity. PMID: 20056998
UDP glucuronosyltransferase 1 family, polypeptide A complex locus	UGT1A	UDP glucuronosyltransferase 1A expression levels determine the response of colorectal cancer cells to the heat shock protein 90 inhibitor ganetespib. PMID: 25210794
UDP glucuronosyltransferase 2 family, polypeptide A1, complex locus	UGT2A1	Identification and functional characterization of a novel UDP-glucuronosyltransferase 2A1 splice variant: Potential importance in tobacco-related cancer susceptibility. PMID: 22984225
UDP glucuronosyltransferase 2 family, polypeptide A2	UGT2A2	Importance of UDP-glucuronosyltransferases 2A2 and 2A3 in tobacco carcinogen metabolism. PMID: 23086198
UDP glucuronosyltransferase 2 family, polypeptide A3	UGT2A3	Importance of UDP-glucuronosyltransferases 2A2 and 2A3 in tobacco carcinogen metabolism. PMID: 23086198
UDP glucuronosyltransferase 2 family, polypeptide B4	UGT2B4	High enzyme activity UGT1A1 or low activity UGT1A8 and UGT2B4 genotypes increase esophageal cancer risk. PMID: 22367021
UDP glucuronosyltransferase 2 family, polypeptide B7	UGT2B7	
UDP glucuronosyltransferase 2 family, polypeptide B10	UGT2B10	N-Glucuronidation of drugs and other xenobiotics by human and animal UDP-glucuronosyltransferases. PMID: 21434773
UDP glucuronosyltransferase 2 family, polypeptide B11	UGT2B11	
UDP glucuronosyltransferase 2 family, polypeptide B15	UGT2B15	Human hepatic UGT2B15 developmental expression. PMID: 24980262
UDP glucuronosyltransferase 2 family, polypeptide B17	UGT2B17	Androgen glucuronidation: An unexpected target for androgen deprivation therapy, with prognosis and diagnostic implications. PMID: 24121496
UDP glucuronosyltransferase 2 family, polypeptide B28	UGT2B28	
UDP glycosyltransferase 3 family, polypeptide A1	UGT3A1	Identification of UDP glycosyltransferase 3A1 as a UDP N-acetylglucosaminyltransferase. PMID: 18981171
UDP glycosyltransferase 3 family, polypeptide A2	UGT3A2	The novel UDP glycosyltransferase 3A2: Cloning, catalytic properties, and tissue distribution. PMID: 21088224
UDP glycosyltransferase 8	UGT8	A novel function for UDP glycosyltransferase 8 (UGT8): Galactosidation of bile acids. PMID: 25519837
Vitamin K epoxide reductase complex, subunit 1	VKORC1	Warfarin – antithrombotic agents. Disease: combined deficiency of vitamin K-dependent clotting factors 2 (VKCFD2)
Vitamin K epoxide reductase complex, subunit 1-like 1	VKORC1L1	VKORC1L1, an enzyme rescuing the vitamin K 2,3 epoxide reductase activity in some extrahepatic tissues during anticoagulation therapy. PMID: 23928358
5-Oxoprolinase (ATP-hydrolysing)	OPLAH	Disease: 5-oxoprolinase deficiency (OPLAHD)
ADP-ribosylarginine hydrolase	ADPRH	ADP-ribosylarginine hydrolase regulates cell proliferation and tumorigenesis. PMID: 21697277
Aldehyde oxidase 1	AOX1	The role of aldehyde oxidase in drug metabolism. PMID: 22335465
Apolipoprotein A-I binding protein	APOA1BP	Extremely conserved ATP- or ADP-dependent enzymatic system for nicotinamide nucleotide repair. PMID: 21994945
Arsenic (+3 oxidation state) methyltransferase	AS3MT	Human arsenic methyltransferase (AS3MT) pharmacogenetics: Gene resequencing and functional genomics studies. PMID: 16407288
ATPase family, AAA domain containing 1	ATAD1	Msp1/ATAD1 maintains mitochondrial function by facilitating the degradation of mislocalized tail-anchored proteins. PMID: 24843043

HGNC-approved name	Symbol	Target information
Biphenyl hydrolase-like (serine hydrolase)	BPHL	Human valacyclovir hydrolase/biphenyl hydrolase-like protein is a highly efficient homocysteine thiolactonase. PMID: 25333274
Carboxymethylenebutenolidase homolog (*Pseudomonas*)	CMBL	Interindividual variability of carboxymethylenebutenolidase homolog, a novel olmesartan medoxomil hydrolase, in the human liver and intestine. PMID: 23471504
Catalase	CAT	**Fomepizole** – alcohol poisoning antidotes. Disease: acatalasemia (ACATLAS)
Ceruloplasmin (ferroxidase)	CP	Disease: aceruloplasminemia (ACERULOP)
Cytochrome b reductase 1	CYBRD1	Functional characterization of human duodenal cytochrome b (Cybrd1): Redox properties in relation to iron and ascorbate metabolism. PMID: 18194661
D-Dopachrome tautomerase	DCT	DCT protects human melanocytic cells from UVR and ROS damage and increases cell viability. PMID: 25346513
Eosinophil peroxidase	EPX	Disease: eosinophil peroxidase deficiency (EPXD)
Esterase D	ESD	Activity-based proteomics: Identification of ABHD11 and ESD activities as potential biomarkers for human lung adenocarcinoma. PMID: 21596165
FAD dependent oxidoreductase domain containing 2	FOXRED2	The flavoprotein FOXRED2 reductively activates nitro-chloromethylbenzindolines and other hypoxia-targeting prodrugs. PMID: 24632291
Ferric-chelate reductase 1	FRRS1	Acquisition of biologically relevant gene expression data by Affymetrix microarray analysis of archival formalin-fixed paraffin-embedded tumours. PMID: 18382428
Ferrochelatase	FECH	**Deferoxamine** – photodynamic therapy adjuvant. Disease: erythropoietic protoporphyria (EPP)
Gamma-glutamylcyclotransferase	GGCT	γ-Glutamylcyclotransferase as a novel immunohistochemical biomarker for the malignancy of esophageal squamous tumors. PMID: 24342434
Glutathione reductase	GSR	**Carmustine** – antineoplastic agents
Glutathione synthetase	GSS	Disease: glutathione synthetase deficiency (GSS deficiency)
HIV-1 Tat interactive protein 2, 30 kDa	HTATIP2	Factors that retard remyelination in multiple sclerosis with a focus on TIP30: A novel therapeutic target. PMID: 19839715
Holocytochrome c synthase	HCCS	Disease: microphthalmia, syndromic, 7 (MCOPS7)
Indolethylamine *N*-methyltransferase	INMT	Noncompetitive inhibition of indolethylamine-*N*-methyltransferase by *N,N*-dimethyltryptamine and *N,N*-dimethylaminopropyltryptamine. PMID: 24730580
Lactoperoxidase	LPO	Mode of action of lactoperoxidase as related to its antimicrobial activity: A review. PMID: 25309750
Mercaptopyruvate sulfurtransferase	MPST	Note: Aberrant MPST activity is found in a few cases of mercaptolactate-cysteine disulfiduria (MCDU)
Multiple inositol-polyphosphate phosphatase 1	MINPP1	Note: Defects in MINPP1 may be involved in follicular thyroid tumors development
Myeloperoxidase	MPO	**AZD3241** – anti-Parkinson drugs. Disease: myeloperoxidase deficiency (MPOD)
Nicotinamide *N*-methyltransferase	NNMT	Nicotinamide *N*-methyltransferase knockdown protects against diet-induced obesity. PMID: 24717514
Nucleoredoxin	NXN	Nucleoredoxin regulates glucose metabolism via phosphofructokinase 1. PMID: 24120946
Obg-like ATPase 1	OLA1	OLA1, an Obg-like ATPase, suppresses antioxidant response via nontranscriptional mechanisms. PMD: 19706404
Oxidation resistance 1	OXR1	Human OXR1 maintains mitochondrial DNA integrity and counteracts hydrogen peroxide-induced oxidative stress by regulating antioxidant pathways involving p21. PMID: 25236744
P450 (cytochrome) oxidoreductase	POR	Disease: Antley–Bixler syndrome, with genital anomalies and disordered steroidogenesis (ABS1)

(Continued)

HGNC-approved name	Symbol	Target information
Paroxysmal nonkinesigenic dyskinesia	PNKD	Myofibrillogenesis regulator 1 (MR-1) is a novel biomarker and potential therapeutic target for human ovarian cancer. PMID: 21702971. Disease: dystonia 8 (DYT8)
Phospholysine phosphohistidine inorganic pyrophosphate phosphatase	LHPP	Evidence for HTR1A and LHPP as interacting genetic risk factors in major depression. PMID: 18268499
Pirin (iron-binding nuclear protein)	PIR	Pirin is an iron-dependent redox regulator of NF-κB. PMID: 23716661
Renalase, FAD-dependent amine oxidase	RNLS	Metabolic function for human renalase: Oxidation of isomeric forms of β-NAD(P)H that are inhibitory to primary metabolism. PMID: 25531177
Sepiapterin reductase (7,8-dihydrobiopterin:NADP+ oxidoreductase)	SPR	Disease: dystonia, DOPA-responsive, due to sepiapterin reductase deficiency (DRDSPRD)
SH3 domain binding glutamate-rich protein-like 3	SH3BGRL3	Crystal structure of the glutaredoxin-like protein SH3BGRL3 at 1.6 Angstrom resolution. PMID: 15120624
Sulfiredoxin 1	SRXN1	Sulfiredoxin-1 protects PC12 cells against oxidative stress induced by hydrogen peroxide. PMID: 23553940
Thyroid peroxidase	TPO	**Methimazole** – antithyroid agents. Note: An alternative splicing in the thyroperoxidase mRNA can cause Graves' disease. Disease: thyroid dyshormonogenesis 2A (TDH2A)
Tumor protein p53-inducible protein 3	TP53I3	PIG3 functions in DNA damage response through regulating DNA-PKcs homeostasis. PMID: 23678292
UbiA prenyltransferase domain containing 1	UBIAD1	Disease: corneal dystrophy, Schnyder type (SCCD)
WW domain-containing oxidoreductase	WWOX	Down-regulation of WWOX is associated with poor prognosis in patients with intrahepatic cholangiocarcinoma after Curative Resection. PMID: 25168293. Note: Defects in WWOX may be involved in several cancer types. The gene spans the second most common chromosomal fragile site (FRA16D) which is frequently altered in cancers
Thiosulfate sulfurtransferase (rhodanese)	TST	Decreased mucosal sulfide detoxification is related to an impaired butyrate oxidation in ulcerative colitis. PMID: 22434643

5.9   Miscellaneous enzymes

HGNC-approved name	Symbol	Target information
5′-Nucleotidase domain containing 1	NT5DC1	Single-nucleotide polymorphisms in the TSPYL-4 and NT5DC1 genes are associated with susceptibility to chronic obstructive pulmonary disease. PMID: 22736055
5′-Nucleotidase domain containing 2	NT5DC2	
5′-Nucleotidase domain containing 3	NT5DC3	
5′-Nucleotidase domain containing 4	NT5DC4	
Dehydrogenase/reductase (SDR family) member 1	DHRS1	Mutant IDH1 confers an in vivo growth in a melanoma cell line with BRAF mutation. PMID: 21356389
Dehydrogenase/reductase (SDR family) member 2	DHRS2	Mitochondrial Hep27 is a c-Myb target gene that inhibits Mdm2 and stabilizes p53. PMID: 20547751
Dehydrogenase/reductase (SDR family) member 4-like 2	DHRS4L2	
Dehydrogenase/reductase (SDR family) member 11	DHRS11	

HGNC-approved name	Symbol	Target information
Dehydrogenase/reductase (SDR family) member 12	DHRS12	
Dehydrogenase/reductase (SDR family) member 13	DHRS13	
Dehydrogenase/reductase (SDR family) X-linked	DHRSX	DHRSX, a novel non-classical secretory protein associated with starvation induced autophagy. PMID: 25076851
Fumarylacetoacetate hydrolase domain containing 1	FAHD1	Identification of FAH domain containing protein 1 (FAHD1) as oxaloacetate decarboxylase. PMID: 25575590
Fumarylacetoacetate hydrolase domain containing 2A	FAHD2A	Differential expression of murine CGI-105 gene in 3T3-L1 cells by adrenocorticotropic hormones. PMID: 15774318
Fumarylacetoacetate hydrolase domain containing 2B	FAHD2B	
Glucose–fructose oxidoreductase domain containing 1	GFOD1	
Glucose–fructose oxidoreductase domain containing 2	GFOD2	Effect of a GFOD2 variant on responses in total and LDL cholesterol in Mexican subjects with hypercholesterolemia after soy protein and soluble fiber supplementation. PMID: 24064143
Immunity-related GTPase family, cinema	IRGC	
Immunity-related GTPase family, M	IRGM	Disease: inflammatory bowel disease 19 (IBD19)
Lactamase, beta	LACTB	LACTB is a filament-forming protein localized in mitochondria. PMID: 19858488
Lactamase, beta-like 1	LACTBL1	
Lactamase, beta 2	LACTB2	Protein profilings in mouse liver regeneration after partial hepatectomy using iTRAQ technology. PMID: 19099420
Metallo-beta-lactamase domain containing 1	MBLAC1	
Metallo-beta-lactamase domain containing 2	MBLAC2	
Metallophosphoesterase domain containing 1	MPPED1	Expression analysis of an evolutionarily conserved metallophosphodiesterase gene, Mpped1, in the normal and beta-catenin-deficient malformed dorsal telencephalon. PMID: 20503375
Metallophosphoesterase domain containing 2	MPPED2	The metallophosphodiesterase Mpped2 impairs tumorigenesis in neuroblastoma. PMID: 22262177
Methyltransferase-like 4	METTL4	
Methyltransferase-like 5	METTL5	
Methyltransferase-like 6	METTL6	An integrated genomics approach identifies drivers of proliferation in luminal-subtype human breast cancer. PMID: 25151356
Methyltransferase-like 7A	METTL7A	
Methyltransferase-like 12	METTL12	
Methyltransferase-like 13	METTL13	
Methyltransferase-like 15	METTL15	
Methyltransferase-like 16	METTL16	
Methyltransferase-like 24	METTL24	
Methyltransferase-like 25	METTL25	

(*Continued*)

HGNC-approved name	Symbol	Target information
MX dynamin-like GTPase 1	MX1	Mx GTPases: Dynamin-like antiviral machines of innate immunity. PMID: 25572883
MX dynamin-like GTPase 2	MX2	Structural insight into the assembly of human anti-HIV dynamin-like protein MxB/Mx2. PMID: 25446123
Nudix (nucleoside diphosphate linked moiety X)-type motif 13	NUDT13	
Nudix (nucleoside diphosphate linked moiety X)-type motif 22	NUDT22	
Pyridine nucleotide-disulphide oxidoreductase domain 1	PYROXD1	
Pyridine nucleotide-disulphide oxidoreductase domain 2	PYROXD2	A genome-wide assessment of variability in human serum metabolism. PMID: 23281178
Adipocyte plasma membrane-associated protein	APMAP	The adipocyte differentiation protein APMAP is an endogenous suppressor of Aβ production in the brain. PMID: 25180020
ArsA arsenite transporter, ATP-binding, homolog 1 (bacterial)	ASNA1	ASNA-1 activity modulates sensitivity to cisplatin. PMID: 20966125
ATPase family, AAA domain containing 2	ATAD2	Fragment-based screening of the bromodomain of ATAD2. PMID: 25314628
Chromosome 9 open reading frame 156	C9orf156	
Chromosome 11 open reading frame 54	C11orf54	Crystal structure of Homo sapiens PTD012 reveals a zinc-containing hydrolase fold. PMID: 16522806
Cytidine and dCMP deaminase domain containing 1	CDADC1	NYD-SP15: A novel gene potentially involved in regulating testicular development and spermatogenesis. PMID: 16955368
Cytochrome b561 family, member A3	CYB561A3	An ascorbate-reducible cytochrome b561 is localized in macrophage lysosomes. PMID: 16996694
FAD-dependent oxidoreductase domain containing 1	FOXRED1	Disease: mitochondrial complex I deficiency (MT-C1D)
IlvB (bacterial acetolactate synthase)-like	ILVBL	
KIAA1191	KIAA1191	Characterizing the novel protein p33MONOX. PMID: 21153684
Kinesin family member 20B	KIF20B	MPHOSPH1: A potential therapeutic target for hepatocellular carcinoma. PMID: 25269478
Monooxygenase, DBH-like 1	MOXD1	Novel markers reveal subpopulations of subplate neurons in the murine cerebral cortex. PMID: 19008461
N-Acetyltransferase 14 (GCN5-related, putative)	NAT14	Genomic structure and regulation of a novel human gene, Klp1. PMID: 11779635
NADH dehydrogenase (ubiquinone) complex I, assembly factor 5	NDUFAF5	Disease: mitochondrial complex I deficiency (MT-C1D)
NADPH-dependent diflavin oxidoreductase 1	NDOR1	Molecular view of an electron transfer process essential for iron-sulfur protein biogenesis. PMID: 23596212
Oxidoreductase NAD-binding domain containing 1	OXNAD1	New gene targets of PGC-1α and ERRα co-regulation in C2C12 myotubes. PMID: 25192891
Phenazine biosynthesis-like protein domain containing	PBLD	MAWBP and MAWD inhibit proliferation and invasion in gastric cancer. PMID: 23687415
Phosphotriesterase related	PTER	Phosphotriesterase-related protein sensed albuminuria and conferred renal tubular cell activation in membranous nephropathy. PMID: 24750591
Pyridoxal-dependent decarboxylase domain containing 1	PDXDC1	Exome sequencing and genome-wide copy number variant mapping reveal novel associations with sensorineural hereditary hearing loss. PMID: 25528277

HGNC-approved name	Symbol	Target information
Reticulon 4-interacting protein 1	RTN4IP1	RTN4IP1 is down-regulated in thyroid cancer and has tumor-suppressive function. PMID: 23393170
Retinoblastoma binding protein 9	RBBP9	Rapid development of a potent photo-triggered inhibitor of the serine hydrolase RBBP9. PMID: 22907802
Saccharopine dehydrogenase (putative)	SCCPDH	Proteomic analysis of proteins associated with lipid droplets of basal and lipolytically stimulated 3T3-L1 adipocytes. PMID: 15337753
Serine hydrolase-like 2	SERHL2	Early gene expression in human lymphocytes after gamma-irradiation-a genetic pattern with potential for biodosimetry. PMID: 18464067
Short-chain dehydrogenase/reductase family 39U, member 1	SDR39U1	
Transmembrane protein with metallophosphoesterase domain	TMPPE	
Valosin-containing protein	VCP	Demethylation-mediated miR-129-5p up-regulation inhibits malignant phenotype of osteogenic osteosarcoma by targeting Homo sapiens valosin-containing protein (VCP). PMID: 25566966
WD-repeat domain 93	WDR93	
Zinc binding alcohol dehydrogenase domain containing 2	ZADH2	Targets and candidate agents for type 2 diabetes treatment with computational bioinformatics approach. PMID: 25401107

Chapter 6
Remaining annotated entries grouped by subcellular location

The compendium entries in the previous chapters have been assigned into groups of proteins associated with known drug targets. These entries, which cover receptors, transporters and enzymes, have been annotated with drug names, literature references and links to disease using the processes described in Chapter 2. However, many existing and potential drug targets do not fit into these categories; in order to simplify the layout, they are instead grouped according to their subcellular locations, as listed in the UniProt Knowledgebase. Unlike those in previous chapters, only the entries with annotations are included so as to constrain the book to a reasonable size. Those HGNC genes which have not been included in the book will be the subject of a future volume.

6.1   Cell surface and secreted proteins

HGNC-approved name	Symbol	Target information
ADP-ribosylation factor-like 6	ARL6	Disease: Bardet–Biedl syndrome 3 (BBS3)
ADP ribosylation factor like 13B	ARL13B	Disease: Joubert syndrome 8 (JBTS8)
Anthrax toxin receptor 1	ANTXR1	Disease: hemangioma, capillary infantile (HCI)
Anthrax toxin receptor 2	ANTXR2	Disease: hyaline fibromatosis syndrome (HFS)
Bardet–Biedl syndrome 1	BBS1	Note: Ciliary dysfunction leads to a broad spectrum of disorders, collectively termed ciliopathies. Overlapping clinical features include retinal degeneration, renal cystic disease, skeletal abnormalities, fibrosis of various organ, and a complex
Bardet–Biedl syndrome 2	BBS2	Disease: Bardet–Biedl syndrome 2 (BBS2)
Bardet–Biedl syndrome 5	BBS5	Disease: Bardet–Biedl syndrome 5 (BBS5)
Bardet–Biedl syndrome 7	BBS7	Note: Ciliary dysfunction leads to a broad spectrum of disorders, collectively termed ciliopathies. Overlapping clinical features include retinal degeneration, renal cystic disease, skeletal abnormalities, fibrosis of various organ, and a complex

(Continued)

Human Drug Targets: A Compendium for Pharmaceutical Discovery, First Edition. Edward D. Zanders.
© 2016 John Wiley & Sons, Ltd. Published 2016 by John Wiley & Sons, Ltd.

HGNC-approved name	Symbol	Target information
Bardet–Biedl syndrome 10	BBS10	Disease: Bardet–Biedl syndrome 10 (BBS10)
Bardet–Biedl syndrome 12	BBS12	Disease: Bardet–Biedl syndrome 12 (BBS12)
B-cell CLL/lymphoma 2	BCL2	**Oblimersen** – antineoplastic agents. Note: A chromosomal aberration involving BCL2 has been found in chronic lymphatic leukemia. Translocation t(1418) (q32q21) with immunoglobulin gene regions
BCL2-like 2	BCL2L2	**ABT-263** – antineoplastic agents
Cellular repressor of E1A-stimulated genes 1	CREG1	CREG mediated adventitial fibroblast phenotype modulation: A possible therapeutic target for proliferative vascular disease. PMID: 22543074
Cellular repressor of E1A-stimulated genes 2	CREG2	Identification and characterization of novel members of the CREG family, putative secreted glycoproteins expressed specifically in brain. PMID: 12408961
Chromosome 2 open reading frame 71	C2orf71	Disease: retinitis pigmentosa 54 (RP54)
Chromosome 3 open reading frame 33	C3orf33	AC3-33, a novel secretory protein, inhibits Elk1 transcriptional activity via ERK pathway. PMID: 20680465
Chromosome 4 open reading frame 26	C4orf26	Disease: amelogenesis imperfecta, hypomaturation type, 2A4 (AI2A4)
Chromosome 5 open reading frame 42	C5orf42	Disease: Joubert syndrome 17 (JBTS17)
Chromosome X open reading frame 36	CXorf36	Note: Genetic variations in CXorf36 may be associated with susceptibility to autism
Coiled-coil domain containing 88A	CCDC88A	GIV/Girdin is a central hub for profibrogenic signalling networks during liver fibrosis. PMID: 25043713
Coiled-coil domain containing 114	CCDC114	Disease: ciliary dyskinesia, primary, 20 (CILD20)
Collagen, type I, alpha 1	COL1A1	Disease: Caffey disease (CAFFD)
Collagen, type I, alpha 2	COL1A2	Disease: Ehlers–Danlos syndrome 7B (EDS7B)
Collagen, type II, alpha 1	COL2A1	Disease: spondyloepiphyseal dysplasia, congenital type (SEDC)
Collagen, type III, alpha 1	COL3A1	Disease: Ehlers–Danlos syndrome 3 (EDS3)
Collagen, type IV, alpha 1	COL4A1	Disease: brain small vessel disease with hemorrhage (BSVDH)
Collagen, type IV, alpha 2	COL4A2	Notch3 overexpression promotes anoikis resistance in epithelial ovarian cancer via upregulation of COL4A2. PMID: 25169943. Disease: porencephaly 2 (POREN2)
Collagen, type IV, alpha 3 (goodpasture antigen)	COL4A3	Note: Autoantibodies against the NC1 domain of alpha 3(IV) are found in Goodpasture syndrome, an autoimmune disease of lung and kidney. Disease: Alport syndrome, autosomal recessive (APSAR)
Collagen, type IV, alpha 4	COL4A4	Disease: Alport syndrome, autosomal recessive (APSAR)
Collagen, type IV, alpha 5	COL4A5	Disease: Alport syndrome, X-linked (APSX)
Collagen, type IV, alpha 6	COL4A6	Note: Deletions covering the N-terminal regions of COL4A5 and COL4A6, which are localized in a head-to-head manner, are found in the chromosome Xq22.3 centromeric deletion syndrome
Collagen, type V, alpha 1	COL5A1	Disease: Ehlers–Danlos syndrome 1 (EDS1)
Collagen, type V, alpha 2	COL5A2	Disease: Ehlers–Danlos syndrome 1 (EDS1)
Collagen, type VI, alpha 1	COL6A1	Disease: Bethlem myopathy (BM)
Collagen, type VI, alpha 2	COL6A2	Disease: Bethlem myopathy (BM)
Collagen, type VI, alpha 3	COL6A3	Disease: Bethlem myopathy (BM)
Collagen, type VI, alpha 5	COL6A5	Note: Patients affected by atopic dermatitis display an abnormal distribution of COL29A1 mRNA and protein in skin suggesting that COL29A1 may be involved in the pathogenesis of the disease

HGNC-approved name	Symbol	Target information
Collagen, type VII, alpha 1	COL7A1	Note: Epidermolysis bullosa acquisita (EBA) is an autoimmune acquired blistering skin disease resulting from autoantibodies to type VII collagen. Disease: epidermolysis bullosa dystrophica, autosomal dominant (DDEB)
Collagen, type VIII, alpha 2	COL8A2	Disease: corneal dystrophy, Fuchs endothelial, 1 (FECD1)
Collagen, type IX, alpha 1	COL9A1	Disease: multiple epiphyseal dysplasia 6 (EDM6)
Collagen, type IX, alpha 2	COL9A2	Disease: multiple epiphyseal dysplasia 2 (EDM2)
Collagen, type IX, alpha 3	COL9A3	Disease: multiple epiphyseal dysplasia 3 (EDM3)
Collagen, type X, alpha 1	COL10A1	Disease: Schmid-type metaphyseal chondrodysplasia (SMCD)
Collagen, type XI, alpha 1	COL11A1	Disease: Stickler syndrome 2 (STL2)
Collagen, type XI, alpha 2	COL11A2	Disease: Stickler syndrome 3 (STL3)
Collagen, type XVIII, alpha 1	COL18A1	The effect of intracellular protein delivery on the anti-tumor activity of recombinant human endostatin. PMID: 23714245. Disease: Knobloch syndrome 1 (KNO1)
Contactin 1	CNTN1	Disease: Compton–North congenital myopathy (CNCM)
Contactin 2 (axonal)	CNTN2	Disease: epilepsy, familial adult myoclonic, 5 (FAME5)
Contactin 4	CNTN4	Note: A chromosomal aberration involving CNTN4 has been found in a boy with characteristic physical features of 3p deletion syndrome (3PDS)
Contactin-associated protein-like 2	CNTNAP2	Disease: cortical dysplasia-focal epilepsy syndrome (CDFES)
Contactin-associated protein-like 4	CNTNAP4	Cntnap4 differentially contributes to GABAergic and dopaminergic synaptic transmission. PMID: 24870235
Cyclin M2	CNNM2	Disease: hypomagnesemia 6 (HOMG6)
Cyclin M4	CNNM4	Disease: Jalili syndrome (JALIS)
Cyclin Y	CCNY	Cell cycle protein cyclin Y is associated with human non-small-cell lung cancer proliferation and tumorigenesis. PMID: 21273179
Cytochrome b-245, alpha polypeptide	CYBA	Disease: granulomatous disease, chronic, cytochrome-b-negative, autosomal recessive (ARCGD)
Cytochrome b-245, beta polypeptide	CYBB	Disease: granulomatous disease, chronic, X-linked (CGD)
Delta-like 1 homolog (*Drosophila*)	DLK1	DLK1: A novel target for immunotherapeutic remodeling of the tumor blood vasculature. PMID: 23896726
Delta like 3 (*Drosophila*)	DLL3	Disease: spondylocostal dysostosis 1, autosomal recessive (SCDO1)
Delta like 1 (*Drosophila*)	DLL1	Demcizumab – antineoplastic agents
Dickkopf WNT signaling pathway inhibitor 1	DKK1	**BHQ880** – drugs affecting bone structure and mineralization
Dickkopf WNT signaling pathway inhibitor 2	DKK2	The WNT antagonist Dickkopf2 promotes angiogenesis in rodent and human endothelial cells. PMID: 21540552
Dickkopf WNT signaling pathway inhibitor 3	DKK3	Dickkopf-related protein 3 promotes pathogenic stromal remodeling in benign prostatic hyperplasia and prostate cancer. PMID: 23765731
Discs, large homolog 1 (*Drosophila*)	DLG1	Selective phosphorylation of the Dlg1AB variant is critical for TCR-induced p38 activation and induction of proinflammatory cytokines in CD8+ T cells. PMID: 25098293
Discs, large homolog 4 (*Drosophila*)	DLG4	**Tat-NR2B9c** – vasoprotectives
EGF-containing fibulin-like extracellular matrix protein 1	EFEMP1	Disease: Doyne honeycomb retinal dystrophy (DHRD)

(*Continued*)

HGNC-approved name	Symbol	Target information
EGF-containing fibulin-like extracellular matrix protein 2	EFEMP2	Overexpression of fibulin 4 is associated with tumor progression and poor prognosis in patients with cervical carcinoma. PMID: 24737201. Disease: cutis laxa, autosomal recessive, 1B (ARCL1B)
EGF-like-domain, multiple 6	EGFL6	EGFL6 promotes endothelial cell migration and angiogenesis through the activation of extracellular signal-regulated kinase. PMID: 21531721
EGF-like-domain, multiple 7	EGFL7	**Parsatuzumab** – antineoplastic agents
Ellis–van Creveld syndrome	EVC	Disease: Ellis–van Creveld syndrome (EVC)
Ellis–van Creveld syndrome 2	EVC2	Disease: Ellis–van Creveld syndrome (EVC)
Epithelial membrane protein 2	EMP2	Epithelial membrane protein-2 is a novel therapeutic target in ovarian cancer. PMID: 20670949
Epithelial membrane protein 3	EMP3	EMP3 as a candidate tumor suppressor gene for solid tumors. PMID: 19466912
ER degradation enhancer, mannosidase alpha-like 1	EDEM1	Endoplasmic reticulum degradation-enhancing α-mannosidase-like protein 1 targets misfolded HLA-B27 dimers for endoplasmic reticulum-associated degradation. PMID: 25132672
ER degradation enhancer, mannosidase alpha-like 2	EDEM2	EDEM2 initiates mammalian glycoprotein ERAD by catalyzing the first mannose trimming step. PMID: 25092655
Family with sequence similarity 3, member A	FAM3A	FAM3A promotes vascular smooth muscle cell proliferation and migration and exacerbates neointima formation in rat artery after balloon injury. PMID: 24857820
Family with sequence similarity 3, member B	FAM3B	PANDER transgenic mice display fasting hyperglycemia and hepatic insulin resistance. PMID: 24468680
Family with sequence similarity 3, member C	FAM3C	ILEI: A novel target for epithelial-mesenchymal transition and poor prognosis in colorectal cancer. PMID: 24738665
Family with sequence similarity 3, member D	FAM3D	Oit1/Fam3D, a gut-secreted protein displaying nutritional status-dependent regulation. PMID: 22226334
Family with sequence similarity 20, member A	FAM20A	Disease: amelogenesis imperfecta and gingival fibromatosis syndrome (AIGFS)
Family with sequence similarity 20, member C	FAM20C	Disease: Raine syndrome (RNS)
Fibrillin 1	FBN1	Disease: Marfan syndrome (MFS)
Fibrillin 2	FBN2	Disease: arthrogryposis, distal, 9 (DA9)
Fibrinogen alpha chain	FGA	Molecular genetics of quantitative fibrinogen disorders. PMID: 17430139
Fibrinogen beta chain	FGB	Molecular genetics of quantitative fibrinogen disorders. PMID: 17430139
Fibrinogen gamma chain	FGG	Molecular genetics of quantitative fibrinogen disorders. PMID: 17430139
Fibrinogen-like 2	FGL2	The duality of Fgl2-secreted immune checkpoint regulator versus membrane-associated procoagulant: Therapeutic potential and implications. PMID: 25259408
Flotillin 1	FLOT1	Knockdown of FLOT1 impairs cell proliferation and tumorigenicity in breast cancer through upregulation of FOXO3a. PMID: 21447726
Flotillin 2	FLOT2	Up-regulation of flotillin-2 is associated with renal cell carcinoma progression. PMID: 25053596
Follistatin-like 1	FSTL1	Targeting FSTL1 prevents tumor bone metastasis and consequent immune dysfunction. PMID: 23966294
Follistatin-like 3 (secreted glycoprotein)	FSTL3	Note: A chromosomal aberration involving FSTL3 is found in a case of B-cell chronic lymphocytic leukemia. Translocation t(1119)(q13p13) with CCDN1
FRAS1-related extracellular matrix 1	FREM1	Disease: bifid nose, with or without anorectal and renal anomalies (BNAR)

HGNC-approved name	Symbol	Target information
FRAS1-related extracellular matrix protein 2	FREM2	Disease: Fraser syndrome (FRASS)
Follicular dendritic cell-secreted protein	FDCSP	Transfection with follicular dendritic cell secreted protein to affect phenotype expression of human periodontal ligament cells. PMID: 24357406
FXYD domain containing ion transport regulator 2	FXYD2	**Cyclothiazide** – diuretics. Disease: hypomagnesemia 2 (HOMG2)
FXYD domain containing ion transport regulator 6	FXYD6	FXYD6: A novel therapeutic target toward hepatocellular carcinoma. PMID: 24715268. Disease: schizophrenia 2 (SCZD2)
Growth arrest-specific 1	GAS1	A soluble form of GAS1 inhibits tumor growth and angiogenesis in a triple negative breast cancer model. PMID: 24992044
Growth arrest-specific 2	GAS2	Growth arrest specific 2 is up-regulated in chronic myeloid leukemia cells and required for their growth. PMID: 24465953
Growth arrest-specific 6	GAS6	Pleiotropic role of growth arrest-specific gene 6 in atherosclerosis. PMID: 19644365
Growth arrest-specific 7	GAS7	Note: A chromosomal aberration involving GAS7 is found in acute myeloid leukemia. Translocation t(1117)(q23p13) with KMT2A/MLL1
Growth arrest-specific 8	GAS8	Growth arrest specific 8 (Gas8) and G protein-coupled receptor kinase 2 (GRK2) cooperate in the control of Smoothened signaling. PMID: 21659505
Guanine nucleotide binding protein (G protein), alpha activating activity polypeptide O	GNAO1	Disease: epileptic encephalopathy, early infantile, 17 (EIEE17)
Guanine nucleotide binding protein (G protein), beta polypeptide 2-like 1	GNB2L1	Downregulation of receptor for activated C-kinase 1 (RACK1) suppresses tumor growth by inhibiting tumor cell proliferation and tumor-associated angiogenesis. PMID: 21848913
Hepatitis A virus cellular receptor 1	HAVCR1	Kidney injury molecule-1. PMID: 20930626
Hepatitis A virus cellular receptor 2	HAVCR2	TIM-3 as a therapeutic target for malignant stem cells in acute myelogenous leukemia. PMID: 22901263
Hyaluronan and proteoglycan link protein 3	HAPLN3	Significant elevation of CLDN16 and HAPLN3 gene expression in human breast cancer. PMID: 20664984
Hyaluronan and proteoglycan link protein 4	HAPLN4	Reduced expression of the hyaluronan and proteoglycan link proteins in malignant gliomas. PMID: 19633295
Growth regulation by estrogen in breast cancer 1	GREB1	Consideration of GREB1 as a potential therapeutic target for hormone-responsive or endocrine resistant cancers. PMID: 24998469
Guanylate cyclase activator 1B (retina)	GUCA1B	Disease: retinitis pigmentosa 48 (RP48)
Haptoglobin	HP	Disease: anhaptoglobinemia (AHP)
Hemicentin 1	HMCN1	Expression of fibulin-6 in failing hearts and its role for cardiac fibroblast migration. PMID: 24951538. Disease: macular degeneration, age-related, 1 (ARMD1)
Hemochromatosis type 2 (juvenile)	HFE2	Disease: hemochromatosis 2A (HFE2A)
Heparan sulfate proteoglycan 2	HSPG2	Disease: Schwartz–Jampel syndrome (SJS1)
Interferon-induced transmembrane protein 1	IFITM1	Knockdown of interferon-induced transmembrane protein 1 (IFITM1) inhibits proliferation, migration, and invasion of glioma cells. PMID: 20838853
Interferon-induced transmembrane protein 3	IFITM3	Mechanism and biological significance of the overexpression of IFITM3 in gastric cancer. PMID: 25270246

(Continued)

HGNC-approved name	Symbol	Target information
Interferon-induced transmembrane protein 5	IFITM5	Disease: osteogenesis imperfecta 5 (OI5)
Interphotoreceptor matrix proteoglycan 1	IMPG1	Mutations in IMPG1 cause vitelliform macular dystrophies. PMID: 23993198
Interphotoreceptor matrix proteoglycan 2	IMPG2	Disease: retinitis pigmentosa 56 (RP56)
Jagged 1	JAG1	Disease: Alagille syndrome 1 (ALGS1)
Jagged 2	JAG2	Hypoxia-induced Jagged2 promotes breast cancer metastasis and self-renewal of cancer stem-like cells. PMID: 21499308
Junctophilin 2	JPH2	Disease: cardiomyopathy, familial hypertrophic 17 (CMH17)
Junctophilin 3	JPH3	Disease: Huntington disease-like 2 (HDL2)
KIAA0319	KIAA0319	Disease: dyslexia 2 (DYX2)
KIAA1524	KIAA1524	CIP2A is a predictor of survival and a novel therapeutic target in bladder urothelial cell carcinoma. PMID: 23275123
KIAA1549	KIAA1549	Note: A chromosomal aberration involving KIAA1549 is found in pilocytic astrocytoma
Kinesin family member 7	KIF7	Note: Ciliary dysfunction leads to a broad spectrum of disorders, collectively termed ciliopathies. The ciliopathy range of diseases includes Meckel–Gruber syndrome, Bardet–Biedl syndrome, Joubert syndrome, and hydrolethalus syndrome among others
Kinesin family member 18A	KIF18A	Kif18A is involved in human breast carcinogenesis. PMID: 20595236
Klotho	KL	The role of Klotho in energy metabolism. PMID: 22641000. Disease: tumoral calcinosis, hyperphosphatemic, familial (HFTC)
Klotho beta	KLB	**PEG-FGF21** – drugs used in diabetes
Latent transforming growth factor beta binding protein 2	LTBP2	Disease: glaucoma 3, primary congenital, D (GLC3D)
Latent transforming growth factor beta binding protein 3	LTBP3	Disease: tooth agenesis selective 6 (STHAG6)
Latent transforming growth factor beta binding protein 4	LTBP4	Disease: Urban–Rifkin–Davis syndrome (URDS)
Leucine-rich repeats and immunoglobulin-like domains 1	LRIG1	Upregulation of LRIG1 suppresses malignant glioma cell growth by attenuating EGFR activity. PMID: 19300910
Leucine-rich repeats and immunoglobulin-like domains 2	LRIG2	Disease: urofacial syndrome 2 (UFS2)
Lipocalin 2	LCN2	Lipocalin-2 negatively modulates the epithelial-to-mesenchymal transition in hepatocellular carcinoma through the epidermal growth factor (TGF-beta1)/Lcn2/Twist1 pathway. PMID: 23696034
Lipocalin 6	LCN6	LCN6, a novel human epididymal lipocalin. PMID: 14617364
Lipoma HMGIC fusion partner	LHFP	Note: A chromosomal aberration involving LHFP is associated with a subclass of benign mesenchymal tumors known as lipomas. Translocation t(1213)(q13-q15q12) with HMGA2 is shown in lipomas
Lipoma HMGIC fusion partner-like 5	LHFPL5	Disease: deafness, autosomal recessive, 67 (DFNB67)
Low-density lipoprotein receptor-related protein 5	LRP5	Norrin, frizzled-4, and Lrp5 signaling in endothelial cells controls a genetic program for retinal vascularization. PMID: 19837032. Disease: vitreoretinopathy, exudative 4 (EVR4)

HGNC-approved name	Symbol	Target information
Low-density lipoprotein receptor-related protein 6	LRP6	Niclosamide suppresses cancer cell growth by inducing Wnt co-receptor LRP6 degradation and inhibiting the Wnt/β-catenin pathway. PMID: 22195040. Disease: coronary artery disease, autosomal dominant, 2 (ADCAD2)
Lymphocyte antigen 6 complex, locus D	LY6D	Human Ly-6 antigen E48 (Ly-6D) regulates important interaction parameters between endothelial cells and head-and-neck squamous carcinoma cells. PMID: 11948455
Lymphocyte antigen 6 complex, locus K	LY6K	The regulatory mechanism of the LY6K gene expression in human breast cancer cells. PMID: 22988241
Lymphocyte antigen 9	LY9	Identification of SLAMF3 (CD229) as an inhibitor of hepatocellular carcinoma cell proliferation and tumour progression. PMID: 24376606
Lymphocyte antigen 96	LY96	Myeloid differentiation 2 as a therapeutic target of inflammatory disorders. PMID: 22119168
LPS-responsive vesicle trafficking, beach and anchor containing	LRBA	Disease: immunodeficiency, common variable, 8, with autoimmunity (CVID8)
Lysosomal-associated membrane protein 2	LAMP2	Disease: Danon disease (DAND)
Major facilitator superfamily domain containing 2A	MFSD2A	Mfsd2a is a transporter for the essential omega-3 fatty acid docosahexaenoic acid. PMID: 24828044
Major facilitator superfamily domain containing 2B	MFSD2B	Major facilitator superfamily domain-containing protein 2a (MFSD2A) has roles in body growth, motor function, and lipid metabolism. PMID: 23209793
Major facilitator superfamily domain containing 8	MFSD8	Disease: ceroid lipofuscinosis, neuronal, 7 (CLN7)
Major facilitator superfamily domain containing 10	MFSD10	Selective suppression of Th2 cell mediated lung eosinophilic inflammation by anti-major facilitator super family domain containing 10 monoclonal antibody. PMID: 24809373
Matrilin 1, cartilage matrix protein	MATN1	Matrilin-1 is an inhibitor of neovascularization. PMID: 24692560
Matrilin 3	MATN3	The matrilin-3 VWA1 domain modulates interleukin-6 release from primary human chondrocytes. PMID: 23523902. Disease: multiple epiphyseal dysplasia 5 (EDM5)
Melanocortin 2 receptor accessory protein	MRAP	Disease: glucocorticoid deficiency 2 (GCCD2)
Melanocortin 2 receptor accessory protein 2	MRAP2	Disease: obesity (OBESITY)
Melanoma antigen family A, 1 (directs expression of antigen MZ2-E)	MAGEA1	Mage-A cancer/testis antigens inhibit p53 function by blocking its interaction with chromatin. PMID: 21056992
Melanoma antigen family A, 4	MAGEA4	Proteomic profiling of triple-negative breast carcinomas in combination with a three-tier orthogonal technology approach identifies Mage-A4 as potential therapeutic target in estrogen receptor negative breast cancer. PMID: 23172894
Melanoma antigen family D, 4B	MAGED4B	MAGE-D4B is a novel marker of poor prognosis and potential therapeutic target involved in breast cancer tumorigenesis. PMID: 21618523
Membrane protein, palmitoylated 1, 55 kDa	MPP1	Erythrocyte scaffolding protein p55/MPP1 functions as an essential regulator of neutrophil polarity. PMID: 19897731
Membrane protein, palmitoylated 3 (MAGUK p55 subfamily member 3)	MPP3	Membrane palmitoylated protein 3 promotes hepatocellular carcinoma cell migration and invasion via up-regulating matrix metalloproteinase 1. PMID: 24513266

(Continued)

HGNC approved name	Symbol	Target information
Nidogen 1	NID1	Tetanus toxin entry. Nidogens are therapeutic targets for the prevention of tetanus. PMID: 25430769
Nidogen 2 (osteonidogen)	NID2	Tetanus toxin entry. Nidogens are therapeutic targets for the prevention of tetanus. PMID: 25430769
Nephrosis 1, congenital, Finnish type (nephrin)	NPHS1	Disease: nephrotic syndrome 1 (NPHS1)
Nephrosis 2, idiopathic, steroid-resistant (podocin)	NPHS2	Disease: nephrotic syndrome 2 (NPHS2)
Neuroligin 1	NLGN1	Autism-related deficits via dysregulated eIF4E-dependent translational control. PMID: 23172145
Neuroligin 3	NLGN3	Disease: autism, X-linked 1 (AUTSX1)
Neuroligin 4, X-linked	NLGN4X	Disease: autism, X-linked 2 (AUTSX2)
Notch 1	NOTCH1	**OMP-52M51** – antineoplastic agents. Disease: aortic valve disease 1 (AOVD1)
Notch 2	NOTCH2	**Tarextumab** – antineoplastic agents. Disease: Alagille syndrome 2 (ALGS2)
Notch 3	NOTCH3	Silencing of Notch3 Using shRNA driven by survivin promoter inhibits growth and promotes apoptosis of human T-cell acute lymphoblastic leukemia cells. PMID: 21940234. Disease: cerebral arteriopathy with subcortical infarcts and leukoencephalopathy, autosomal dominant (CASIL)
Notch 4	NOTCH4	NOTCH4 is a potential therapeutic target for triple-negative breast cancer. PMID: 24403446
Patched 1	PTCH1	PTCH 1 staining of pancreatic neuroendocrine tumor (PNET) samples from patients with and without multiple endocrine neoplasia (MEN-1) syndrome reveals a potential therapeutic target. PMID: 25482929. Disease: basal cell nevus syndrome (BCNS)
Patched 2	PTCH2	Disease: medulloblastoma (MDB)
Podocalyxin-like	PODXL	Podocalyxin is a marker of poor prognosis in colorectal cancer. PMID: 25004863
Podocalyxin-like 2	PODXL2	Endoglycan, a member of the CD34 family, functions as an L-selectin ligand through modification with tyrosine sulfation and sialyl Lewis x. PMID: 12889478
Protein phosphatase 1, regulatory subunit 3A	PPP1R3A	Disease: diabetes mellitus, non-insulin-dependent (NIDDM)
Protein phosphatase 1, regulatory subunit 3F	PPP1R3F	R3F, a novel membrane-associated glycogen targeting subunit of protein phosphatase 1 regulates glycogen synthase in astrocytoma cells in response to glucose and extracellular signals. PMID: 21668450
Protein phosphatase 1, regulatory subunit 16A	PPP1R16A	PPP1R16A, the membrane subunit of protein phosphatase 1beta, signals nuclear translocation of the nuclear receptor constitutive active/androstane receptor. PMID: 18202305
Protein phosphatase 1, regulatory subunit 16B	PPP1R16B	TIMAP promotes angiogenesis by suppressing PTEN-mediated Akt inhibition in human glomerular endothelial cells. PMID: 25007873
Proteoglycan 2, bone marrow (natural killer cell activator, eosinophil granule major basic protein)	PRG2	Eosinophil major basic protein activates human cord blood mast cells primed with fibroblast membranes by integrin-β1. PMID: 24112102
Proteoglycan 4	PRG4	Disease: camptodactyly–arthropathy–coxa vara–pericarditis syndrome (CACP)
RAB25, member RAS oncogene family	RAB25	Rab25 is a tumor suppressor gene with antiangiogenic and anti-invasive activities in esophageal squamous cell carcinoma. PMID: 22991305
RAB39B, member RAS oncogene family	RAB39B	Disease: mental retardation, X-linked 72 (MRX72)

HGNC-approved name	Symbol	Target information
RAB40A, member RAS oncogene family-like	RAB40AL	Disease: mental retardation, X-linked, syndromic, Martin–Probst type (MRXSMP)
Receptor accessory protein 2	REEP2	Disease: spastic paraplegia 72 (SPG72)
Receptor accessory protein 6	REEP6	Polymorphisms of the apoptosis-associated gene DP1L1 (deleted in polyposis 1-like 1) in colon cancer and inflammatory bowel disease. PMID: 19924442
Regulator of G-protein signaling 2, 24 kDa	RGS2	Regulator of G protein signaling 2 is a key modulator of airway hyperresponsiveness. PMID: 22704538
Regulator of G-protein signaling 9	RGS9	Disease: prolonged electroretinal response suppression (PERRS)
Regulator of G-protein signaling 9 binding protein	RGS9BP	Disease: prolonged electroretinal response suppression (PERRS)
Roundabout, axon guidance receptor, homolog 1 (*Drosophila*)	ROBO1	A (90)Y-labelled anti-ROBO1 monoclonal antibody exhibits antitumour activity against hepatocellular carcinoma xenografts during ROBO1-targeted radioimmunotherapy. PMID: 25006547
Roundabout, axon guidance receptor, homolog 2 (*Drosophila*)	ROBO2	Disease: vesicoureteral reflux 2 (VUR2)
Roundabout, axon guidance receptor, homolog 3 (*Drosophila*)	ROBO3	Disease: familial horizontal gaze palsy with progressive scoliosis (HGPPS)
Roundabout, axon guidance receptor, homolog 4 (*Drosophila*)	ROBO4	Targeting Robo4-dependent Slit signaling to survive the cytokine storm in sepsis and influenza. PMID: 20375003
S100 calcium binding protein A7	S100A7	Identification and characterization of binding sites on S100A7, a participant in cancer and inflammation pathways. PMID: 19810752
S100 calcium binding protein A8	S100A8	Impact of S100A8 expression on kidney cancer progression and molecular docking studies for kidney cancer therapeutics. PMID: 24692722
S100 calcium binding protein A9	S100A9	Platelet-derived S100 family member myeloid-related protein-14 regulates thrombosis. PMID: 24691441
Secreted frizzled-related protein 1	SFRP1	Expression of secreted frizzled related protein 1, β-catenin and E-cadherin in colorectal carcinoma and its clinicopathological significances. PMID: 25187453
Secreted frizzled related protein 2	SFRP2	A novel monoclonal antibody to secreted frizzled-related protein 2 inhibits tumor growth. PMID: 23604067
SLIT and NTRK like family member 1	SLITRK1	Disease: Gilles de la Tourette syndrome (GTS)
SLIT- and NTRK-like family, member 3	SLITRK3	Selective control of inhibitory synapse development by Slitrk3-PTPδ trans-synaptic interaction. PMID: 22286174
SLIT- and NTRK-like family, member 5	SLITRK5	Slitrk5 deficiency impairs corticostriatal circuitry and leads to obsessive-compulsive-like behaviors in mice. PMID: 20418887
SLIT- and NTRK-like family, member 6	SLITRK6	**ASG-15ME** – antineoplastic agents. Disease: deafness and myopia (DFNMYP)
Slit homolog 1 (*Drosophila*)	SLIT1	Slit1 promotes regenerative neurite outgrowth of adult dorsal root ganglion neurons *in vitro* via binding to the Robo receptor. PMID: 20172023
Slit homolog 2 (*Drosophila*)	SLIT2	SLIT2 attenuation during lung cancer progression deregulates beta-catenin and E-cadherin and associates with poor prognosis. PMID: 20068157
Slit homolog 3 (*Drosophila*)	SLIT3	Control of human hematopoietic stem/progenitor cell migration by the extracellular matrix protein Slit3. PMID: 22614124

(Continued)

HGNC-approved name	Symbol	Target information
SPARC-related modular calcium binding 1	SMOC1	Disease: ophthalmoacromelic syndrome (OAS)
SPARC-related modular calcium binding 2	SMOC2	Disease: dentin dysplasia 1 (DTDP1)
Spinster homolog 1 (*Drosophila*)	SPNS1	HSpin1, a transmembrane protein interacting with Bcl-2/Bcl-xL, induces a caspase-independent autophagic cell death. PMID: 12815463
Spinster homolog 2 (*Drosophila*)	SPNS2	Spinster homolog 2 (spns2) deficiency causes early onset progressive hearing loss. PMID: 25356849
Spinster homolog 3 (*Drosophila*)	SPNS3	Identification of six putative human transporters with structural similarity to the drug transporter SLC22 family. PMID: 17714910
Sprouty-related, EVH1 domain containing 1	SPRED1	Disease: neurofibromatosis 1-like syndrome (NFLS)
Sprouty-related, EVH1 domain containing 2	SPRED2	Regulation of human hepatocellular carcinoma cells by Spred2 and correlative studies on its mechanism. PMID: 21703232
Stromal interaction molecule 1	STIM1	Suppression of STIM1 inhibits human glioblastoma cell proliferation and induces G0/G1 phase arrest. PMID: 23578185. Disease: immunodeficiency 10 (IMD10)
Stromal interaction molecule 2	STIM2	A reciprocal shift in transient receptor potential channel 1 (TRPC1) and stromal interaction molecule 2 (STIM2) contributes to Ca2+ remodeling and cancer hallmarks in colorectal carcinoma cells. PMID: 25143380
Surfactant protein A1	SFTPA1	Disease: pulmonary fibrosis, idiopathic (IPF)
Surfactant protein A2	SFTPA2	Disease: pulmonary fibrosis, idiopathic (IPF)
Surfactant protein B	SFTPB	Disease: pulmonary surfactant metabolism dysfunction 1 (SMDP1)
Surfactant protein C	SFTPC	Disease: pulmonary surfactant metabolism dysfunction 2 (SMDP2)
Surfactant associated 2	SFTA2	SFTA2 – a novel secretory peptide highly expressed in the lung – is modulated by lipopolysaccharide but not hyperoxia. PMID: 22768197
Tectonic family member 2	TCTN2	Disease: Meckel syndrome 8 (MKS8)
Tectonic family member 3	TCTN3	Disease: orofaciodigital syndrome 4 (OFD4)
Tenascin C	TNC	**Neuradiab** – antineoplastic agents. Disease: deafness, autosomal dominant, 56 (DFNA56)
Tenascin XB	TNXB	Disease: tenascin-X deficiency (TNXD)
Tetraspanin 7	TSPAN7	Disease: mental retardation, X-linked 58 (MRX58)
Tetraspanin 12	TSPAN12	Disease: vitreoretinopathy, exudative 5 (EVR5)
Transmembrane 4 L six family member 1	TM4SF1	TM4SF1: A new vascular therapeutic target in cancer. PMID: 24986520
Transmembrane 4 L six family member 5	TM4SF5	Monoclonal antibody targeting of the cell surface molecule TM4SF5 inhibits the growth of hepatocellular carcinoma. PMID: 24802189
Transmembrane 4 L six family member 20	TM4SF20	Disease: specific language impairment 5 (SLI5)
Transmembrane protein 5	TMEM5	Disease: muscular dystrophy–dystroglycanopathy congenital with brain and eye anomalies A10 (MDDGA10)
Transmembrane protein 8B	TMEM8B	NGX6 gene mediated by promoter methylation as a potential molecular marker in colorectal cancer. PMID: 20423473
Transmembrane protein 67	TMEM67	Note: TMEM67 mutations result in ciliary dysfunction leading to a broad spectrum of disorders, collectively termed ciliopathies
Transmembrane protein 114	TMEM114	Note: Chromosomal aberrations involving TMEM114 may be a cause of congenital and juvenile cataracts. Translocation t(1622) (p13.3q11.2)

HGNC-approved name	Symbol	Target information
Transmembrane protein 127	TMEM127	Disease: pheochromocytoma (PCC)
Transmembrane protein 132E	TMEM132E	Note: TMEM132E is located in a region involved in a heterozygous deletion of approximately 4.7 Mb; this deletion, involving the NF1 gene and contiguous genes lying in its flanking regions, is observed in a patient 17q11.2 microdeletion syndrome
Transmembrane protein 216	TMEM216	Disease: Joubert syndrome 2 (JBTS2)
Transmembrane protein 231	TMEM231	Disease: Joubert syndrome 20 (JBTS20)
Transmembrane protein 237	TMEM237	Disease: Joubert syndrome 14 (JBTS14)
Transmembrane protein with EGF-like and two follistatin-like domains 1	TMEFF1	Identification of tomoregulin-1 as a novel addicsin-associated factor. PMID: 24680664
Transmembrane protein with EGF-like and two follistatin-like domains 2	TMEFF2	TMEFF2 deregulation contributes to gastric carcinogenesis and indicates poor survival outcome. PMID: 24987055
Trefoil factor 1	TFF1	The trefoil factor 1 (TFF1) protein involved in doxorubicin-induced apoptosis resistance is upregulated by estrogen in breast cancer cells. PMID: 23836323
Trefoil factor 2	TFF2	Trefoil factor family member 2 (Tff2) KO mice are protected from high-fat diet-induced obesity. PMID: 23754443
Trefoil factor 3 (intestinal)	TFF3	Upregulation of trefoil factor 3 (TFF3) after rectal cancer chemoradiotherapy is an adverse prognostic factor and a potential therapeutic target. PMID: 22516806
TRPM8 channel-associated factor 1	TCAF1	TRP channel-associated factors are a novel protein family that regulates TRPM8 trafficking and activity. PMID: 25559186
TRPM8 channel-associated factor 2	TCAF2	TRP channel-associated factors are a novel protein family that regulates TRPM8 trafficking and activity. PMID: 25559186
Tubulointerstitial nephritis antigen	TINAG	Role of extracellular matrix renal tubulo-interstitial nephritis antigen (TINag) in cell survival utilizing integrin (alpha)vbeta3/focal adhesion kinase (FAK)/ phosphatidylinositol 3-kinase (PI3K)/protein kinase B-serine/threonine kinase (AKT) signaling pathway. PMID: 21795690
Tubulointerstitial nephritis antigen-like 1	TINAGL1	TINAGL1 and B3GALNT1 are potential therapy target genes to suppress metastasis in non-small cell lung cancer. PMID: 25521548
VANGL planar cell polarity protein 1	VANGL1	Disease: neural tube defects (NTD)
VANGL planar cell polarity protein 2	VANGL2	Disease: neural tube defects (NTD)
WD repeat domain 11	WDR11	Note: A chromosomal aberration involving WDR11 is found in a form of glioblastoma. Translocation t(1019)(q26q13.3) with ZNF320. Note: A chromosomal aberration involving WDR11 is found in a form of Kallmann syndrome
WD repeat domain 19	WDR19	Disease: cranioectodermal dysplasia 4 (CED4)
WD repeat domain 60	WDR60	Disease: short-rib thoracic dysplasia 8 with or without polydactyly (SRTD8)
Wingless-type MMTV integration site family, member 1	WNT1	Disease: osteoporosis (OSTEOP)
Wingless-type MMTV integration site family, member 2B	WNT2B	Silencing Wnt2B by siRNA interference inhibits metastasis and enhances chemotherapy sensitivity in ovarian cancer. PMID: 22635028
Wingless-type MMTV integration site family, member 3	WNT3	Disease: tetraamelia, autosomal recessive (ARTTRA)

(Continued)

HGNC-approved name	Symbol	Target information
Wingless-type MMTV integration site family, member 4	WNT4	Disease: Rokitansky–Kuster–Hauser syndrome (RKH syndrome)
Wingless-type MMTV integration site family, member 5A	WNT5A	WNT5A has anti-prostate cancer effects *in vitro* and reduces tumor growth in the skeleton in vivo. PMID: 25224731. Disease: Robinow syndrome, autosomal dominant (DRS)
Wingless-type MMTV integration site family, member 7A	WNT7A	Disease: limb pelvis hypoplasia aplasia syndrome (LPHAS)
Wingless-type MMTV integration site family, member 10A	WNT10A	Note: Defects in WNT10A may be a cause of hypohidrotic/anhidrotic ectodermal dysplasia, a disorder characterized by sparse hair (atrichosis or hypotrichosis), abnormal or missing teeth and the inability to sweat due to the absence of sweat glands
Wingless-type MMTV integration site family, member 10B	WNT10B	Disease: split-hand/foot malformation 6 (SHFM6)
ABI family, member 3 (NESH) binding protein	ABI3BP	A novel p53-dependent apoptosis function of TARSH in tumor development. PMID: 19994723
Adaptor-related protein complex 2, sigma 1 subunit	AP2S1	Disease: hypocalciuric hypercalcemia, familial 3 (HHC3)
Adenomatosis polyposis coli down-regulated 1	APCDD1	Disease: hypotrichosis 1 (HYPT1)
Aggrecan	ACAN	Disease: spondyloepiphyseal dysplasia type Kimberley (SEDK)
Agrin	AGRN	Injection of a soluble fragment of neural agrin (NT-1654) considerably improves the muscle pathology caused by the disassembly of the neuromuscular junction. PMID: 24520420. Disease: myasthenia, limb-girdle, familial (LGM)
Albumin	ALB	Disease: dysalbuminemic hyperthyroxinemia (DH)
Alpha-2-glycoprotein 1, zinc-binding	AZGP1	Alpha-2-glycoprotein 1(AZGP1) regulates biological behaviors of LoVo cells by down-regulating mTOR signaling pathway and endogenous fatty acid synthesis. PMID: 24918753
Alpha-fetoprotein	AFP	Homology modeling, molecular docking, and molecular dynamics simulations elucidated α-fetoprotein binding modes. PMID: 24266910
Amelogenin, X-linked	AMELX	Disease: amelogenesis imperfecta 1E (AI1E)
Amyloid beta (A4) precursor protein	APP	**Ponezumab** – anti-dementia drugs. Disease: Alzheimer disease 1 (AD1)
Amyloid P component, serum	APCS	Note: SAP is a precursor of amyloid component P which is found in basement membrane and associated with amyloid deposits
Angiogenic factor with G patch and FHA domains 1	AGGF1	AGGF1 protects from myocardial ischemia/reperfusion injury by regulating myocardial apoptosis and angiogenesis. PMID: 24893993. Disease: Klippel–Trenaunay syndrome (KTS)
Ankyrin repeat and sterile alpha motif domain containing 6	ANKS6	Disease: nephronophthisis 16 (NPHP16)
Annexin A2	ANXA2	Proteomics identification of annexin A2 as a key mediator in the metastasis and proangiogenesis of endometrial cells in human adenomyosis. PMID: 22493182
Anterior gradient 2	AGR2	AGR2 is induced in asthma and promotes allergen-induced mucin overproduction. PMID: 22403803
Anti-Mullerian hormone	AMH	Disease: persistent Mullerian duct syndrome 1 (PMDS1)
ARFGEF family member 3	ARFGEF3	Targeting BIG3-PHB2 interaction to overcome tamoxifen resistance in breast cancer cells. PMID: 24051437
Asporin	ASPN	Disease: osteoarthritis 3 (OS3)
Ataxia, cerebellar, Cayman type	ATCAY	Disease: cerebellar ataxia, Cayman type (ATCAY)
Attractin	ATRN	Does human attractin have DP4 activity? PMID: 17261078

HGNC-approved name	Symbol	Target information
BCL2-associated agonist of cell death	BAD	BAD, a proapoptotic member of the BCL2 family, is a potential therapeutic target in hepatocellular carcinoma. PMID: 20647330
Biglycan	BGN	Biglycan enhances gastric cancer invasion by activating FAK signaling pathway. PMID: 24681892
BMP-binding endothelial regulator	BMPER	Disease: diaphanospondylodysostosis (DSD)
bolA family member 3	BOLA3	Disease: multiple mitochondrial dysfunctions syndrome 2 (MMDS2)
Bone gamma-carboxyglutamate (gla) protein	BGLAP	The preventive effect of uncarboxylated osteocalcin against free fatty acid-induced endothelial apoptosis through the activation of phosphatidylinositol 3-kinase/Akt signaling pathway. PMID: 23639572
Brain expressed, associated with NEDD4, 1	BEAN1	Disease: spinocerebellar ataxia 31 (SCA31)
Brevican	BCAN	Brevican knockdown reduces late-stage glioma tumor aggressiveness. PMID: 25052349
Calcium modulating ligand	CAMLG	**Cyclosporine A** – immunomodulating agents
Calsequestrin 2 (cardiac muscle)	CASQ2	Disease: ventricular tachycardia, catecholaminergic polymorphic, 2 (CPVT2)
Cancer/testis antigen 1B	CTAG1B	A novel human-derived antibody against NY-ESO-1 improves the efficacy of chemotherapy. PMID: 23390374
Carcinoembryonic antigen-related cell adhesion molecule 16	CEACAM16	Disease: deafness, autosomal dominant, 4B (DFNA4B)
Cartilage-associated protein	CRTAP	Disease: osteogenesis imperfecta 7 (OI7)
Cartilage intermediate layer protein, nucleotide pyrophosphohydrolase	CILP	Disease: intervertebral disc disease (IDD)
Cartilage oligomeric matrix protein	COMP	Disease: multiple epiphyseal dysplasia 1 (EDM1)
CD69 molecule	CD69	
Cell adhesion associated, oncogene regulated	CDON	Disease: holoprosencephaly 11 (HPE11)
Chondroadherin	CHAD	The C-terminal domain of chondroadherin: A new regulator of osteoclast motility counteracting bone loss. PMID: 24616121
Chimerin-like 1	CHRDL1	Disease: megalocornea 1, X-linked (MGC1)
CKLF like MARVEL transmembrane domain containing 1	CMTM1	CMTM1_v17 is a novel potential therapeutic target in breast cancer. PMID: 23177800
Clarin 1	CLRN1	Disease: Usher syndrome 3A (USH3A)
Cochlin	COCH	Disease: deafness, autosomal dominant, 9 (DFNA9)
Collagen and calcium binding EGF domains 1	CCBE1	Disease: Hennekam lymphangiectasia–lymphedema syndrome (HLLS)
Collagen triple helix repeat containing 1	CTHRC1	Disease: Barrett esophagus (BE)
Complement component 4B (Chido blood group)	C4B	Disease: systemic lupus erythematosus (SLE)
Corneodesmosin	CDSN	Disease: hypotrichosis 2 (HYPT2)
Cripto, FRL-1, cryptic family 1	CFC1	Targeting the embryonic gene Cripto-1 in cancer and beyond. PMID: 21073352. Disease: heterotaxy, visceral, 2, autosomal (HTX2)

(Continued)

HGNC-approved name	Symbol	Target information
Crumbs family member 1, photoreceptor morphogenesis associated	CRB1	Note: CRB1 mutations have been found in various retinal dystrophies, chronic and disabling disorders of visual function
CTAGE family, member 5	CTAGE5	Note: Autoantibodies against CTAGE5 are present in several cancer types, including benign meningioma and cutaneous T-cell lymphoma (CTCL)
CUB domain containing protein 1	CDCP1	Proteolysis-induced N-terminal ectodomain shedding of the integral membrane glycoprotein CUB domain-containing protein 1 (CDCP1) is accompanied by tyrosine phosphorylation of its C-terminal domain and recruitment of Src and PKCdelta. PMID: 20551327
Cubilin (intrinsic factor–cobalamin receptor)	CUBN	Cubilin maintains blood levels of HDL and albumin. PMID: 24357674. Disease: recessive hereditary megaloblastic anemia 1 (RH-MGA1)
Cysteine-rich hydrophobic domain 2	CHIC2	Note: A chromosomal aberration involving CHIC2 is found in a form of acute myeloid leukemia (AML). Translocation t(412)(q12p13) with ETV6
Cysteine-rich with EGF-like domains 1	CRELD1	Disease: atrioventricular septal defect 2 (AVSD2)
Cysteine-rich, angiogenic inducer, 61	CYR61	Blockade of cysteine-rich protein 61 attenuates renal inflammation and fibrosis after ischemic kidney injury. PMID: 24920753
Cytochrome c oxidase assembly homolog 14 (*Saccharomyces cerevisiae*)	COX14	Note: Defects in COX14 may be a cause of a mitochondrial disorder presenting with severe congenital lactic acidosis and dysmorphic features associated with a COX assembly defect
DDB1- and CUL4-associated factor 17	DCAF17	Disease: Woodhouse–Sakati syndrome (WoSaS)
Decorin	DCN	Disease: corneal dystrophy, congenital stromal (CSCD)
Deleted in colorectal carcinoma	DCC	Disease: mirror movements 1 (MRMV1)
Deleted in malignant brain tumors 1	DMBT1	Disease: glioma (GLM)
Dentin sialophosphoprotein	DSPP	Disease: deafness, autosomal dominant, 39, with dentinogenesis imperfecta 1 (DFNA39/DGI1)
Dermatopontin	DPT	Identification of fibronectin binding sites in dermatopontin and their biological function. PMID: 25092262
Dermokine	DMKN	Mice deficient for the epidermal dermokine β and γ isoforms display transient cornification defects. PMID: 24794495
Diaphanous-related formin 1	DIAPH1	Disease: deafness, autosomal dominant, 1 (DFNA1)
Discoidin, CUB and LCCL domain containing 2	DCBLD2	Transmembrane protein ESDN promotes endothelial VEGF signaling and regulates angiogenesis. PMID: 24177422
Dishevelled segment polarity protein 1	DVL1	**NSC 668036** – antineoplastic agents
Docking protein 7	DOK7	Neuromuscular disease. DOK7 gene therapy benefits mouse models of diseases characterized by defects in the neuromuscular junction. PMID: 25237101. Disease: myasthenia, limb-girdle, familial (LGM)
Dolichyl-diphosphooligosaccharide–protein glycosyltransferase subunit (non-catalytic)	DDOST	Disease: congenital disorder of glycosylation 1R (CDG1R)
Dynein, axonemal, assembly factor 1	DNAAF1	Disease: ciliary dyskinesia, primary, 13 (CILD13)
Dysferlin	DYSF	Dysferlin stabilizes stress-induced Ca2+ signaling in the transverse tubule membrane. PMID: 24302765. Disease: limb-girdle muscular dystrophy 2B (LGMD2B)
Dystroglycan 1 (dystrophin-associated glycoprotein 1)	DAG1	Phosphorylation within the cysteine-rich region of dystrophin enhances its association with β-dystroglycan and identifies a potential novel therapeutic target for skeletal muscle wasting. PMID: 25082828. Disease: muscular dystrophy–dystroglycanopathy limb-girdle C9 (MDDGC9)

HGNC-approved name	Symbol	Target information
Dystrophin	DMD	Antisense oligonucleotide-mediated exon skipping for Duchenne muscular dystrophy: Progress and challenges. PMID: 22533380. Disease: Duchenne muscular dystrophy (DMD)
EGF-like repeats and discoidin I-like domains 3	EDIL3	Elevated autocrine EDIL3 protects hepatocellular carcinoma from anoikis through RGD-mediated integrin activation. PMID: 25273699
Elastin	ELN	Disease: cutis laxa, autosomal dominant, 1 (ADCL1)
ELMO/CED-12 domain containing 3	ELMOD3	Disease: deafness, autosomal recessive, 88 (DFNB88)
Enamelin	ENAM	Disease: amelogenesis imperfecta 1B (AI1B)
Endoglin	ENG	A phase I study of TRC105 anti-CD105 (endoglin) antibody in metastatic castration-resistant prostate cancer. PMID: 25407442. Disease: telangiectasia, hereditary hemorrhagic, 1 (HHT1)
Endothelial cell-specific molecule 1	ESM1	ESM-1 regulates cell growth and metastatic process through activation of NF-κB in colorectal cancer. PMID: 22735811
Eva-1 homolog C (*Caenorhabditis elegans*)	EVA1C	The expression pattern of EVA1C, a novel Slit receptor, is consistent with an axon guidance role in the mouse nervous system. PMID: 24040182
Extracellular matrix protein 1	ECM1	Disease: lipoid proteinosis (LiP)
Eyes shut homolog (*Drosophila*)	EYS	Association of common variants in the human eyes shut ortholog (EYS) with statin-induced myopathy: Evidence for additional functions of EYS. PMID: 21826682. Disease: retinitis pigmentosa 25 (RP25)
FERM domain containing 7	FRMD7	Disease: Nystagmus congenital X-linked 1 (NYS1)
Fermitin family member 3	FERMT3	Disease: leukocyte adhesion deficiency 3 (LAD3)
Fibromodulin	FMOD	Fibromodulin, an oxidative stress-sensitive proteoglycan, regulates the fibrogenic response to liver injury in mice. PMID: 22138190
Fibronectin 1	FN1	**Radretumab** – antineoplastic agents. Disease: glomerulopathy with fibronectin deposits 2 (GFND2)
Fibronectin leucine-rich transmembrane protein 3	FLRT3	Disease: hypogonadotropic hypogonadism 21 with or without anosmia (HH21)
Fibronectin type III domain containing 5	FNDC5	Maternal inheritance of circulating irisin in humans. PMID: 24397868
Folate receptor 1 (adult)	FOLR1	**Farletuzumab** – antineoplastic agents. Disease: neurodegeneration due to cerebral folate transport deficiency (NCFTD)
Fraser extracellular matrix complex subunit 1	FRAS1	Disease: Fraser syndrome (FRASS)
Frizzled-related protein	FRZB	Disease: osteoarthritis 1 (OS1)
Ganglioside-induced differentiation associated protein 1	GDAP1	Disease: Charcot–Marie–Tooth disease 4A (CMT4A)
Gastric intrinsic factor (vitamin B synthesis)	GIF	Disease: hereditary intrinsic factor deficiency (IFD)
Glutaredoxin, cysteine rich 1	GRXCR1	Disease: deafness, autosomal recessive, 25 (DFNB25)
Golgi glycoprotein 1	GLG1	Lipopolysaccharide-induced early response genes in bovine peripheral blood mononuclear cells implicate GLG1/E-selectin as a key ligand-receptor interaction. PMID: 19263101
GRB2-binding adaptor protein, transmembrane	GAPT	Identification of a new transmembrane adaptor protein that constitutively binds Grb2 in B cells. PMID: 18559951
Gremlin 1, DAN family BMP antagonist	GREM1	Gremlin is a downstream profibrotic mediator of transforming growth factor-beta in cultured renal cells. PMID: 23548835. Disease: polyposis syndrome, mixed hereditary 1 (HMPS1)

(Continued)

HGNC-approved name	Symbol	Target information
Hepcidin antimicrobial peptide	HAMP	**XEN70**1 – antianaemic preparations. Disease: hemochromatosis 2B (HFE2B)
Hyaluronan-mediated motility receptor (RHAMM)	HMMR	RHAMM (CD168) is overexpressed at the protein level and may constitute an immunogenic antigen in advanced prostate cancer disease. PMID: 19724689
HYDIN, axonemal central pair apparatus protein	HYDIN	Disease: ciliary dyskinesia, primary, 5 (CILD5)
Immunoglobulin superfamily, member 1	IGSF1	Disease: hypothyroidism, central, and testicular enlargement (CHTE)
Immunoglobulin-like domain containing receptor 1	ILDR1	Disease: deafness, autosomal recessive, 42 (DFNB42)
Indian hedgehog	IHH	Indian Hedgehog, a critical modulator in osteoarthritis, could be a potential therapeutic target for attenuating cartilage degeneration disease. PMID: 24844414. Disease: brachydactyly A1 (BDA1)
Inositol 1,4,5-trisphosphate receptor-interacting protein	ITPRIP	Death-associated protein kinase-mediated cell death modulated by interaction with DANGER. PMID: 20053891
Integrin-binding sialoprotein	IBSP	Bone sialoprotein deficiency impairs osteoclastogenesis and mineral resorption *in vitro*. PMID: 20812227
Intraflagellar transport 172 homolog (*Chlamydomonas*)	IFT172	Disease: short-rib thoracic dysplasia 10 with or without polydactyly (SRTD10)
IQ motif containing GTPase activating protein 1	IQGAP1	IQGAP1 plays an important role in the invasiveness of thyroid cancer. PMID: 20959410
Izumo sperm-egg fusion 1	IZUMO1	Molecular and cellular mechanisms of sperm-oocyte interactions opinions relative to *in vitro* fertilization (IVF). PMID: 25054321
IZUMO1 receptor, JUNO	IZUMO1R	Juno is the egg Izumo receptor and is essential for mammalian fertilization. PMID: 24739963
Jagunal homolog 1 (*Drosophila*)	JAGN1	Note: Neutropenia, severe congenital, autosomal recessive (SCNN). A disorder of hematopoiesis characterized by maturation arrest of granulopoiesis at the level of promyelocyte/myelocyte stage
Keratin 1	KRT1	Disease: epidermolytic hyperkeratosis (EHK)
Keratocan	KERA	Disease: autosomal recessive cornea plana 2 (CNA2)
Kin of IRRE like 3 (*Drosophila*)	KIRREL3	Note: A chromosomal aberration involving KIRREL3 and CDH15 is found in a patient with severe mental retardation and dysmorphic facial features. Translocation t(11;16)(q24.2;q24). Disease: mental retardation, autosomal dominant 4 (MRD4)
KN motif and ankyrin repeat domains 1	KANK1	Disease: cerebral palsy, spastic quadriplegic 2 (CPSQ2)
Layilin	LAYN	Secretion of inflammatory factors from chondrocytes by layilin signaling. PMID: 25150153
Left–right determination factor 2	LEFTY2	Lefty A attenuates the TGF-beta1-induced epithelial to mesenchymal transition of human renal proximal epithelial tubular cells. PMID: 20157767. Disease: left–right axis malformations (LRAM)
Lens intrinsic membrane protein 2, 19 kDa	LIM2	Disease: cataract 19 (CTRCT19)
Leucine-rich repeat and Ig domain containing 1	LINGO1	**ANTI-LINGO** – selective immunosuppressants
Leucine-rich repeat containing 8 family, member A	LRRC8A	Disease: agammaglobulinemia 5, autosomal dominant (AGM5)
Leucine-rich repeat neuronal 1	LRRN1	NLRR1 enhances EGF-mediated MYCN induction in neuroblastoma and accelerates tumor growth in vivo. PMID: 22815527
Leucine-rich alpha-2-glycoprotein 1	LRG1	LRG1 promotes angiogenesis by modulating endothelial TGF-β signalling. PMID: 23868260

HGNC-approved name	Symbol	Target information
Leucine-rich, glioma inactivated 1	LGI1	Disease: epilepsy, familial temporal lobe, 1 (ETL1)
Leukocyte cell-derived chemotaxin 2	LECT2	The tumor suppressor function of LECT2 in human hepatocellular carcinoma makes it a potential therapeutic target. PMID: 21394108
Limb development membrane protein 1	LMBR1	Disease: preaxial polydactyly 2 (PPD2)
Lipoxygenase homology domains 1	LOXHD1	Disease: deafness, autosomal recessive, 77 (DFNB77)
LMBR1 domain containing 1	LMBRD1	LMBD1 protein serves as a specific adaptor for insulin receptor internalization. PMID: 24078630. Disease: methylmalonic aciduria and homocystinuria type cblF (MMAHCF)
Low-density lipoprotein receptor class A domain containing 4	LDLRAD4	C18 ORF1: A novel negative regulator of TGF-β signaling. PMID: 24627487
Mannose-P-dolichol utilization defect 1	MPDU1	Disease: congenital disorder of glycosylation 1F (CDG1F)
MARVEL domain containing 2	MARVELD2	Disease: deafness, autosomal recessive, 49 (DFNB49)
Matrix Gla protein	MGP	Disease: Keutel syndrome (KS)
Megalencephalic leukoencephalopathy with subcortical cysts 1	MLC1	Disease: leukoencephalopathy, megalencephalic, with subcortical cysts, 1 (MLC1)
Melanoma inhibitory activity	MIA	Modulation of cartilage differentiation by melanoma inhibiting activity/ cartilage-derived retinoic acid-sensitive protein (MIA/CD RAP). PMID: 20164682
Membrane-associated guanylate kinase, WW and PDZ domain containing 3	MAGI3	MAGI-3 regulates LPA-induced activation of Erk and RhoA. PMID: 16904209
Membrane frizzled-related protein	MFRP	Disease: nanophthalmos 2 (NNO2)
Mesencephalic astrocyte-derived neurotrophic factor	MANF	Functions for the cardiomyokine, MANF, in cardioprotection, hypertrophy and heart failure. PMID: 20970425
Mesothelin	MSLN	**Amatuximab** – antineoplastic agents. Note: Antibodies against MSLN are detected in patients with mesothelioma and ovarian cancer
Metadherin	MTDH	Astrocyte elevated gene-1: A novel target for human glioma therapy. PMID: 20059777
Microseminoprotein, beta	MSMB	The potential value of microseminoprotein beta as a prostate cancer biomarker and therapeutic target. PMID: 10700236. Disease: prostate cancer, hereditary, 13 (HPC13)
Milk fat globule-EGF factor 8 protein	MFGE8	Recombinant milk fat globule-EGF factor-8 reduces oxidative stress via integrin β3/nuclear factor erythroid 2-related factor 2/heme oxygenase pathway in subarachnoid hemorrhage rats. PMID: 25342030
Mitofusin 2	MFN2	Anti-tumour efficacy of mitofusin-2 in urinary bladder carcinoma. PMID: 20803103. Disease: Charcot–Marie–Tooth disease 2A2 (CMT2A2)
Mitochondrially encoded 16S RNA	MT-RNR2	S14G-humanin inhibits Aβ1-42 fibril formation, disaggregates preformed fibrils, and protects against Aβ- induced cytotoxicity in vitro. PMID: 23349038
Multimerin 1	MMRN1	Note: Deficiency in multimerin-1 due to proteolytic degradation within the platelet alpha granules is associated with an autosomal dominant bleeding disorder (factor V Quebec)
Multiple PDZ domain protein	MPDZ	Disease: hydrocephalus, non-syndromic, autosomal recessive 2 (HYC2)

(Continued)

HGNC-approved name	Symbol	Target information
Mutated in colorectal cancers	MCC	Mutated in colorectal cancer (MCC) is a novel oncogene in B lymphocytes. PMID: 25200342
Myelin-associated glycoprotein	MAG	**GSK249320** – vasoprotectives
Myelin basic protein	MBP	Note: The reduction in the surface charge of citrullinated and/or methylated MBP could result in a weakened attachment to the myelin membrane
Myelin oligodendrocyte glycoprotein	MOG	Disease: narcolepsy 7 (NRCLP7)
Myelin protein zero	MPZ	Disease: Charcot–Marie–Tooth disease 1B (CMT1B)
Myeloid cell leukemia 1	MCL1	Targeting Mcl-1 for multiple myeloma (MM) therapy: Drug-induced generation of Mcl-1 fragment Mcl-1(128-350) triggers MM cell death via c-Jun upregulation. PMID: 24120758
Myocilin, trabecular meshwork-inducible glucocorticoid response	MYOC	Disease: glaucoma 1, open angle, A (GLC1A)
Myosin XVA	MYO15A	Disease: deafness, autosomal recessive, 3 (DFNB3)
Myotilin	MYOT	Disease: limb-girdle muscular dystrophy 1A (LGMD1A)
Na+/K+ transporting ATPase interacting 2	NKAIN2	Note: A chromosomal aberration involving NKAIN2 is a cause of lymphoma. Deletion del(6)(q13q21) within NKAIN2 and involving SUSP1 generates the SUSP1-NKAIN2 product
NADPH oxidase organizer 1	NOXO1	TNF-α/TNFR1 signaling promotes gastric tumorigenesis through induction of Noxo1 and Gna14 in tumor cells. PMID: 23975421
Nance–Horan syndrome (congenital cataracts and dental anomalies)	NHS	Disease: Nance–Horan syndrome (NHS)
NEDD8-activating enzyme E1 subunit 1	NAE1	**MLN4924** – antineoplastic agents
Negative regulator of reactive oxygen species	NRROS	NRROS negatively regulates reactive oxygen species during host defence and autoimmunity. PMID: 24739962
Neogenin 1	NEO1	Neogenin1 is a Sonic Hedgehog target in medulloblastoma and is necessary for cell cycle progression. PMID: 23775842
Nephroblastoma overexpressed	NOV	CCN3 promotes prostate cancer bone metastasis by modulating the tumor-bone microenvironment through RANKL-dependent pathway. PMID: 23536580
Nephronophthisis 3 (adolescent)	NPHP3	Disease: nephronophthisis 3 (NPHP3)
Netrin 1	NTN1	The neuroimmune guidance cue netrin-1: A new therapeutic target in cardiovascular disease. PMID: 23991347
Neudesin neurotrophic factor	NENF	Identification of hypothalamic neuron-derived neurotrophic factor as a novel factor modulating appetite. PMID: 23576617
Neurofibromin 2 (merlin)	NF2	Disease: neurofibromatosis 2 (NF2)
Neuron-derived neurotrophic factor	NDNF	Neuron-derived neurotrophic factor functions as a novel modulator that enhances endothelial cell function and revascularization processes. PMID: 24706764
Nicalin	NCLN	The Nicastrin-like protein Nicalin regulates assembly and stability of the Nicalin-nodal modulator (NOMO) membrane protein complex. PMID: 17261586
Nischarin	NISCH	Imidazoline antihypertensive drugs: Selective i(1) -imidazoline receptors activation. PMID: 21884004
Noggin	NOG	Noggin is novel inducer of mesenchymal stem cell adipogenesis: Implications for bone health and obesity. PMID: 22351751. Disease: symphalangism, proximal 1A (SYM1A)
Norrie disease (pseudoglioma)	NDP	The Norrin/Frizzled4 signaling pathway in retinal vascular development and disease. PMID: 20688566. Disease: Norrie disease (ND)

HGNC-approved name	Symbol	Target information
Numb homolog (*Drosophila*)	NUMB	NUMB inhibition of NOTCH signalling as a therapeutic target in prostate cancer. PMID: 25134838
Nyctalopin	NYX	Disease: night blindness, congenital stationary, 1A (CSNB1A)
Occludin	OCLN	Disease: band-like calcification with simplified gyration and polymicrogyria (BLCPMG)
Opioid binding protein/cell adhesion molecule-like	OPCML	Disease: ovarian cancer (OC)
Osteoglycin	OGN	Circulating osteoglycin and NGAL/MMP9 complex concentrations predict 1-year major adverse cardiovascular events after coronary angiography. PMID: 24651681
Osteomodulin	OMD	The biology of small leucine-rich proteoglycans in bone pathophysiology. PMID: 22879588
Otoancorin	OTOA	Disease: deafness, autosomal recessive, 22 (DFNB22)
Oviductal glycoprotein 1, 120 kDa	OVGP1	Oviductal glycoprotein (OVGP1, MUC9): A differentiation-based mucin present in serum of women with ovarian cancer. PMID: 20130498
Paired immunoglobulin-like type 2 receptor beta	PILRB	PirB is a novel potential therapeutic target for enhancing axonal regeneration and synaptic plasticity following CNS injury in mammals. PMID: 24405091
PDZ domain containing 7	PDZD7	Note: A chromosomal aberration disrupting PDZD7 has been found in patients with non-syndromic sensorineural deafness. Translocation t(10;11),t(10;11). Disease: Usher syndrome 2C (USH2C)
Periostin	POSTN	Periostin: A novel prognostic and therapeutic target for genitourinary cancer? PMID: 24656869
Peripheral myelin protein 22	PMP22	Disease: Charcot–Marie–Tooth disease 1A (CMT1A)
Peripherin 2 (retinal degeneration, slow)	PRPH2	Disease: retinitis pigmentosa 7 (RP7)
Phospholamban	PLN	Disease: cardiomyopathy, dilated 1P (CMD1P)
Phospholipase A2 receptor 1, 180 kDa	PLA2R1	New physiopathological roles for the PLA2R1 receptor in cancer and membranous nephropathy. PMID: 24939538
Piezo-type mechanosensitive ion channel component 2	PIEZO2	Disease: arthrogryposis, distal, 5 (DA5)
Placenta-specific 1	PLAC1	PLAC1 (Placenta-specific 1): A novel, X-linked gene with roles in reproductive and cancer biology. PMID: 20509147
Platelet-derived growth factor receptor like	PDGFRL	Disease: colorectal cancer (CRC)
Pleckstrin homology domain containing, family O member 1	PLEKHO1	CKIP-1: A scaffold protein and potential therapeutic target integrating multiple signaling pathways and physiological functions. PMID: 22878216
Podocan	PODN	Novel small leucine-rich repeat protein podocan is a negative regulator of migration and proliferation of smooth muscle cells, modulates neointima formation, and is expressed in human atheroma. PMID: 24043300
Polycystic kidney and hepatic disease 1 (autosomal recessive)	PKHD1	Disease: polycystic kidney disease, autosomal recessive (ARPKD)
Polycystic kidney disease 1 (autosomal dominant)	PKD1	Disease: polycystic kidney disease 1 (PKD1)
Post-GPI attachment to proteins 1	PGAP1	Disease: mental retardation, autosomal recessive 42 (MRT42)
Potassium channel tetramerization domain containing 7	KCTD7	Disease: epilepsy, progressive myoclonic 3, with or without intracellular inclusions (EPM3)

(*Continued*)

HGNC-approved name	Symbol	Target information
Prion protein	PRNP	Peripheral administration of a humanized anti-PrP antibody blocks Alzheimer's disease Aβ synaptotoxicity. PMID: 24790184
Progressive rod-cone degeneration	PRCD	Disease: retinitis pigmentosa 36 (RP36)
Proline/arginine-rich end leucine-rich repeat protein	PRELP	AAV-mediated expression of human PRELP inhibits complement activation, choroidal neovascularization and deposition of membrane attack complex in mice. PMID: 24670995
Proline-rich transmembrane protein 2	PRRT2	Disease: episodic kinesigenic dyskinesia 1 (EKD1)
Prominin 1	PROM1	CD133+ cells as a therapeutic target for kidney diseases. PMID: 22324879. Disease: retinitis pigmentosa 41 (RP41)
Prostate stem cell antigen	PSCA	The therapeutic efficacy of I131-PSCA-mAb in orthotopic mouse models of prostate cancer. PMID: 24330823
Prostate transmembrane protein, androgen induced 1	PMEPA1	TMEPAI, a transmembrane TGF-beta-inducible protein, sequesters Smad proteins from active participation in TGF-beta signaling. PMID: 20129061
Protein kinase, cAMP-dependent, regulatory, type I, alpha	PRKAR1A	PRKAR1A is overexpressed and represents a possible therapeutic target in human cholangiocarcinoma. PMID: 20824711. Disease: Carney complex 1 (CNC1)
Proteolipid protein 1	PLP1	Disease: leukodystrophy, hypomyelinating, 1 (HLD1)
Rac/Cdc42 guanine nucleotide exchange factor (GEF) 6	ARHGEF6	Disease: mental retardation, X-linked 46 (MRX46)
Radixin	RDX	Disease: deafness, autosomal recessive, 24 (DFNB24)
Receptor-associated protein of the synapse	RAPSN	Disease: myasthenic syndrome, congenital, associated with acetylcholine receptor deficiency (CMS-ACHRD)
Regulating synaptic membrane exocytosis 1	RIMS1	Disease: cone-rod dystrophy 7 (CORD7)
Repulsive guidance molecule family member a	RGMA	RGMa modulates T cell responses and is involved in autoimmune encephalomyelitis. PMID: 21423182
Retinal outer segment membrane protein 1	ROM1	MiR-22 suppresses the proliferation and invasion of gastric cancer cells by inhibiting CD151. PMID: 24495805. Disease: retinitis pigmentosa 7 (RP7)
Retinitis pigmentosa 2 (X-linked recessive)	RP2	Disease: retinitis pigmentosa 2 (RP2)
Retinitis pigmentosa GTPase regulator-interacting protein 1	RPGRIP1	Disease: Leber congenital amaurosis 6 (LCA6)
Retinoschisin 1	RS1	Disease: retinoschisis juvenile X-linked 1 (XLRS1)
Retrotransposon-like 1	RTL1	Identification of rtl1, a retrotransposon-derived imprinted gene, as a novel driver of hepatocarcinogenesis. PMID: 23593033
RFT1 homolog (S. cerevisiae)	RFT1	Disease: congenital disorder of glycosylation 1N (CDG1N)
Rho GTPase-activating protein 31	ARHGAP31	Disease: Adams–Oliver syndrome 1 (AOS1)
Rho guanine nucleotide exchange factor (GEF) 26	ARHGEF26	The guanine-nucleotide exchange factor SGEF plays a crucial role in the formation of atherosclerosis. PMID: 23372835
Ribosomal protein SA	RPSA	Expression of 67-kDa laminin receptor was associated with tumor progression and poor prognosis in epithelial ovarian cancer. PMID: 22285843
Sclerostin	SOST	**Romosozumab** – drugs affecting bone structure and mineralization. Disease: sclerosteosis 1 (SOST1)
Scribbled planar cell polarity protein	SCRIB	Disease: neural tube defects (NTD)
Secreted LY6/PLAUR domain containing 1	SLURP1	Disease: mal de Meleda (MDM)

HGNC-approved name	Symbol	Target information
Secreted phosphoprotein 2, 24 kDa	SPP1	Osteopontin as potential biomarker and therapeutic target in gastric and liver cancers. PMID: 22912540
Selenoprotein P, plasma, 1	SEPP1	A liver-derived secretory protein, selenoprotein P, causes insulin resistance. PMID: 21035759
SH3 and multiple ankyrin repeat domains 2	SHANK2	Disease: autism 17 (AUTS17)
Signal peptide, CUB domain, EGF-like 2	SCUBE2	Tumor suppressor SCUBE2 inhibits breast-cancer cell migration and invasion through the reversal of epithelial-mesenchymal transition. PMID: 24213532
Signal regulatory protein alpha	SIRPA	The CD47-SIRPα signalling system: Its physiological roles and therapeutic application. PMID: 24627525
SLIT-ROBO Rho GTPase-activating protein 2	SRGAP2	Note: A chromosomal aberration disrupting SRGAP2 has been found in a patient with early infantile epileptic encephalopathy. Balanced translocation t(1:9)(q32:q13)
Sonic hedgehog	SHH	Disease: microphthalmia, isolated, with coloboma, 5 (MCOPCB5)
Sortilin 1	SORT1	Note: A common polymorphism located in a non-coding region between CELSR2 and PSRC1 alters a CEBP transcription factor binding site and is responsible for changes in hepatic expression of SORT1
Sortilin-related receptor, L(DLR class) A repeats containing	SORL1	Disease: Alzheimer disease (AD)
Sortilin-related VPS10 domain containing receptor 3	SORCS3	Sortilin-related receptor SORCS3 is a postsynaptic modulator of synaptic depression and fear extinction. PMID: 24069373
Spastic paraplegia 11 (autosomal recessive)	SPG11	Disease: spastic paraplegia 11, autosomal recessive (SPG11)
Spastin	SPAST	Disease: spastic paraplegia 4, autosomal dominant (SPG4)
Stanniocalcin 1	STC1	Overexpression of stanniocalcin-1 inhibits reactive oxygen species and renal ischemia/reperfusion injury in mice. PMID: 22695329
Stereocilin	STRC	Disease: deafness, autosomal recessive, 16 (DFNB16)
Stimulated by retinoic acid 6	STRA6	Disease: microphthalmia, syndromic, 9 (MCOPS9)
Sushi-repeat containing protein, X-linked 2	SRPX2	Disease: rolandic epilepsy with speech dyspraxia and mental retardation X-linked (RESDX)
Synaptotagmin XIV	SYT14	Disease: spinocerebellar ataxia, autosomal recessive, 11 (SCAR11)
Syntaxin 11	STX11	Disease: familial hemophagocytic lymphohistiocytosis 4 (FHL4)
Syntrophin, alpha 1	SNTA1	Disease: long QT syndrome 12 (LQT12)
Tapenn	TPRN	Disease: deafness, autosomal recessive, 79 (DFNB79)
TBC1 domain family, member 20	TBC1D20	Disease: Warburg micro syndrome 4 (WARBM4)
T-cell immunoglobulin and mucin domain containing 4	TIMD4	T-cell immunoglobulin and mucin domain 4 (TIM-4) signaling in innate immune-mediated liver ischemia-reperfusion injury. PMID: 25066922
T-cell leukemia translocation altered	TCTA	Note: A chromosomal aberration involving TCTA is associated with T-cell acute lymphoblastic leukemia (T-ALL). Translocation t(13)(p34p21)
Tectorin alpha	TECTA	Disease: deafness, autosomal dominant, 12 (DFNA12)
Teneurin transmembrane protein 3	TENM3	Disease: microphthalmia, isolated, with coloboma, 9 (MCOPCB9)
Teratocarcinoma-derived growth factor 1	TDGF1	Cripto-1 expression in glioblastoma multiforme. PMID: 24521322
TGF-beta-activated kinase 1/ MAP3K7 binding protein 2	TAB2	Disease: congenital heart defects, multiple types, 2 (CHTD2)
Thrombospondin-type laminin G domain and EAR repeats	TSPEAR	Disease: deafness, autosomal recessive, 98 (DFNB98)

(Continued)

HGNC-approved name	Symbol	Target information
Thyroglobulin	TG	Disease: thyroid dyshormonogenesis 3 (TDH3)
Transforming growth factor, beta-induced, 68 kDa	TGFBI	Disease: corneal dystrophy, epithelial basement membrane (EBMD)
Transmembrane channel-like 1	TMC1	Disease: deafness, autosomal dominant, 36 (DFNA36)
Transmembrane inner ear	TMIE	Disease: deafness, autosomal recessive, 6 (DFNB6)
Triadin	TRDN	Disease: ventricular tachycardia, catecholaminergic polymorphic, 5, with or without muscle weakness (CPVT5)
Trophoblast glycoprotein	TPBG	**PF-06263507** – antineoplastic agents
Tsukushi, small leucine-rich proteoglycan	TSKU	Tsukushi is involved in the wound healing by regulating the expression of cytokines and growth factors. PMID: 25159578
Tumor necrosis factor, alpha-induced protein 6	TNFAIP6	TSG-6 inhibits neutrophil migration via direct interaction with the chemokine CXCL8. PMID: 24501198
Tumor-associated calcium signal transducer 2	TACSTD2	A novel human Fab antibody for Trop2 inhibits breast cancer growth *in vitro* and in vivo. PMID: 23982827. Disease: corneal dystrophy, gelatinous drop-like (GDLD)
Twisted gastrulation BMP signaling modulator 1	TWSG1	Twisted gastrulation, a BMP antagonist, exacerbates podocyte injury. PMID: 24586548
Ubiquinol–cytochrome c reductase complex assembly factor 3	UQCC3	Note: UQCC3 mutations are a cause of mitochondrial complex III deficiency, a disorder of the mitochondrial respiratory chain resulting in a highly variable phenotype depending on which tissues are affected
Uromodulin	UMOD	Disease: familial juvenile hyperuricemic nephropathy 1 (HNFJ1)
Usher syndrome 2A (autosomal recessive, mild)	USH2A	Disease: Usher syndrome 2A (USH2A)
Versican	VCAN	Disease: Wagner vitreoretinopathy (WGVRP)
Vitronectin	VTN	Vitronectin: A migration and wound healing factor for human corneal epithelial cells. PMID: 25237160
V-set and immunoglobulin domain containing 1	VSIG1	Decreased expression of V-set and immunoglobulin domain containing 1 (VSIG1) is associated with poor prognosis in primary gastric cancer. PMID: 22095633
WD repeat containing planar cell polarity effector	WDPCP	Disease: Bardet–Biedl syndrome 15 (BBS15)
X-linked Kx blood group	XK	Disease: McLeod syndrome (MLS)
Zona pellucida glycoprotein 1 (sperm receptor)	ZP1	Disease: oocyte maturation defect (OOMD)
Zymogen granule protein 16B	ZG16B	Efficient targeting and tumor retardation effect of pancreatic adenocarcinoma up-regulated factor (PAUF)-specific RNA replacement in pancreatic cancer mouse model. PMID: 24189457

6.2 Cytoskeleton

HGNC-approved name	Symbol	Target information
Actin, alpha 1, skeletal muscle	ACTA1	Disease: nemaline myopathy 3 (NEM3)
Actin, alpha 2, smooth muscle, aorta	ACTA2	Note: ACTA2 mutations predispose patients to a variety of diffuse and diverse vascular diseases, premature onset coronary artery disease (CAD), premature ischemic strokes and Moyamoya disease. Disease: aortic aneurysm, familial thoracic 6 (AAT6)
Actin, alpha, cardiac muscle 1	ACTC1	Disease: cardiomyopathy, dilated 1R (CMD1R)
Actin, beta	ACTB	Disease: dystonia, juvenile-onset (DYTJ)

HGNC-approved name	Symbol	Target information
Actin, gamma 1	ACTG1	Disease: deafness, autosomal dominant, 20 (DFNA20)
Actin filament-associated protein 1-like 1	AFAP1L1	AFAP1L1, a novel associating partner with vinculin, modulates cellular morphology and motility, and promotes the progression of colorectal cancers. PMID: 24723436
Actin filament-associated protein 1-like 2	AFAP1L2	XB130, a new adaptor protein, regulates expression of tumor suppressive microRNAs in cancer cells. PMID: 23527086
Ankyrin 1, erythrocytic	ANK1	Disease: spherocytosis 1 (SPH1)
Ankyrin 2, neuronal	ANK2	Disease: long QT syndrome 4 (LQT4)
Ankyrin 3, node of Ranvier (ankyrin G)	ANK3	Note: Genetic variations in ANK3 may be associated with autism spectrum disorders susceptibility. Disease: mental retardation, autosomal recessive 37 (MRT37)
B9 protein domain 1	B9D1	Disease: Meckel syndrome 9 (MKS9)
B9 protein domain 2	B9D2	Disease: Meckel syndrome 10 (MKS10)
Bardet–Biedl syndrome 4	BBS4	Disease: Bardet–Biedl syndrome 4 (BBS4)
Bardet–Biedl syndrome 9	BBS9	Note: A chromosomal aberration involving PTHB1 has been found in Wilms tumor. Translocation t(1;7)(q42;p15) with OBSCN. Disease: Bardet–Biedl syndrome 9 (BBS9)
Calmodulin 1 (phosphorylase kinase, delta)	CALM1	Disease: ventricular tachycardia, catecholaminergic polymorphic, 4 (CPVT4)
Calmodulin 2 (phosphorylase kinase, delta)	CALM2	Disease: ventricular tachycardia, catecholaminergic polymorphic, 4 (CPVT4)
Calmodulin 3 (phosphorylase kinase, delta)	CALM3	Disease: ventricular tachycardia, catecholaminergic polymorphic, 4 (CPVT4)
Catenin (cadherin-associated protein), alpha-like 1	CTNNAL1	α-Catulin marks the invasion front of squamous cell carcinoma and is important for tumor cell metastasis. PMID: 22648798
Catenin (cadherin-associated protein), alpha 3	CTNNA3	Disease: arrhythmogenic right ventricular dysplasia, familial, 13 (ARVD13)
Centrosomal protein 19 kDa	CEP19	Disease: morbid obesity and spermatogenic failure (MOSPGF)
Centrosomal protein 41 kDa	CEP41	Disease: Joubert syndrome 15 (JBTS15)
Centrosomal protein 55 kDa	CEP55	Expression of FLJ10540 is correlated with aggressiveness of oral cavity squamous cell carcinoma by stimulating cell migration and invasion through increased FOXM1 and MMP-2 activity. PMID: 19525975
Centrosomal protein 63 kDa	CEP63	Disease: Seckel syndrome 6 (SCKL6)
Centrosomal protein 83 kDa	CEP83	Disease: nephronophthisis 18 (NPHP18)
Centrosomal protein 85 kDa-like	CEP85L	Note: A chromosomal aberration involving CEP85L is found in a patient with T-lymphoblastic lymphoma (T-ALL) and an associated myeloproliferative neoplasm (MPN) with eosinophilia. Translocation t(56)(q33-34q23) with PDGFRB
Centrosomal protein 135 kDa	CEP135	Disease: microcephaly 8, primary, autosomal recessive (MCPH8)
Centrosomal protein 152 kDa	CEP152	Disease: microcephaly 9, primary, autosomal recessive (MCPH9)
Centrosomal protein 164 kDa	CEP164	Disease: nephronophthisis 15 (NPHP15)
Centrosomal protein 290 kDa	CEP290	Disease: Joubert syndrome 5 (JBTS5)
Coiled-coil domain containing 28B	CCDC28B	Disease: Bardet–Biedl syndrome (BBS)
Coiled-coil domain containing 39	CCDC39	Disease: ciliary dyskinesia, primary, 14 (CILD14)
Coiled-coil domain containing 78	CCDC78	Disease: myopathy, centronuclear, 4 (CNM4)

(Continued)

HGNC-approved name	Symbol	Target information
Dynamin 1	DNM1	Suppression of dynamin GTPase decreases α-synuclein uptake by neuronal and oligodendroglial cells: A potent therapeutic target for synucleinopathy. PMID: 22892036
Dynamin 2	DNM2	Disease: myopathy, centronuclear, 1 (CNM1)
Dynein, axonemal, heavy chain 5	DNAH5	Disease: ciliary dyskinesia, primary, 3 (CILD3)
Dynein, axonemal, heavy chain 11	DNAH11	Disease: Kartagener syndrome (KTGS)
Dynein, axonemal, intermediate chain 1	DNAI1	Disease: ciliary dyskinesia, primary, 1 (CILD1)
Dynein, axonemal, intermediate chain 2	DNAI2	Disease: ciliary dyskinesia, primary, 9 (CILD9)
Dynein, cytoplasmic 1, heavy chain 1	DYNC1H1	Disease: Charcot–Marie–Tooth disease 20 (CMT20)
Dynein, cytoplasmic 2, heavy chain 1	DYNC2H1	Small-molecule inhibitors of the AAA+ ATPase motor cytoplasmic dynein. PMID: 22425997. Disease: short-rib thoracic dysplasia 3 with or without polydactyly (SRTD3)
Erythrocyte membrane protein band 4.1 (elliptocytosis 1, RH-linked)	EPB41	Disease: elliptocytosis 1 (EL1)
Erythrocyte membrane protein band 4.1-like 1	EPB41L1	Disease: mental retardation, autosomal dominant 11 (MRD11)
Family with sequence similarity 110, member B	FAM110B	Integrative genomic, transcriptomic, and RNAi analysis indicates a potential oncogenic role for FAM110B in castration-resistant prostate cancer. PMID: 21919029
Family with sequence similarity 161, member A	FAM161A	Disease: retinitis pigmentosa 28 (RP28)
Fascin actin-bundling protein 1	FSCN1	Association of fascin-1 with mortality, disease progression and metastasis in carcinomas: A systematic review and meta-analysis. PMID: 23442983
Fascin actin-bundling protein 2, retinal	FSCN2	Disease: retinitis pigmentosa 30 (RP30)
Formin binding protein 1	FNBP1	Note: A chromosomal aberration involving FNBP1 is found in acute leukemias. Translocation t(911)(q34q23) with KMT2A/MLL1
Formin binding protein 1-like	FNBP1L	Transducer of Cdc42-dependent actin assembly promotes breast cancer invasion and metastasis. PMID: 22824798
Growth arrest-specific 2-like 1	GAS2L1	GAR22: A novel target gene of thyroid hormone receptor causes growth inhibition in human erythroid cells. PMID: 19375645
Growth arrest-specific 2-like 2	GAS2L2	GAS2-like proteins mediate communication between microtubules and actin through interactions with end-binding proteins. PMID: 24706950
Growth arrest-specific 2-like 3	GAS2L3	Gas2l3 is essential for brain morphogenesis and development. PMID: 25131197
Intraflagellar transport 43 homolog (*Chlamydomonas*)	IFT43	Disease: cranioectodermal dysplasia 3 (CED3)
Intraflagellar transport 80 homolog (*Chlamydomonas*)	IFT80	Disease: short-rib thoracic dysplasia 2 with or without polydactyly (SRTD2)
Intraflagellar transport 88 homolog (*Chlamydomonas*)	IFT88	Tg737 signaling is required for hypoxia-enhanced invasion and migration of hepatoma cells. PMID: 22974282
Intraflagellar transport 140 homolog (*Chlamydomonas*)	IFT140	Disease: short-rib thoracic dysplasia 9 with or without polydactyly (SRTD9)
Kelch-like family member 3	KLHL3	Disease: pseudohypoaldosteronism 2D (PHA2D)
Kelch-like family member 41	KLHL41	Disease: nemaline myopathy 9 (NEM9)

HGNC-approved name	Symbol	Target information
Kinesin family member 1A	KIF1A	Disease: spastic paraplegia 30, autosomal recessive (SPG30)
Kinesin family member 1C	KIF1C	Disease: spastic ataxia 2, autosomal recessive (SPAX2)
Kinesin heavy chain member 2A	KIF2A	Disease: cortical dysplasia, complex, with other brain malformations 3 (CDCBM3)
Kinesin family member 2C	KIF2C	Mitosis phase enrichment with identification of mitotic centromere-associated kinesin as a therapeutic target in castration-resistant prostate cancer. PMID: 22363599
Kinesin family member 5C	KIF5C	Disease: cortical dysplasia, complex, with other brain malformations 2 (CDCBM2)
Kinesin family member 11	KIF11	**Monastrol** – antineoplastic agents. Disease: microcephaly with or without chorioretinopathy, lymphedema, or mental retardation (MCLMR)
Kinesin family member 21A	KIF21A	Disease: congenital fibrosis of extraocular muscles 1 (CFEOM1)
Microtubule-associated protein 1 light chain 3 alpha	MAP1LC3A	An activation of LC3A-mediated autophagy contributes to de novo and acquired resistance to EGFR tyrosine kinase inhibitors in lung adenocarcinoma. PMID: 24687913
Microtubule-associated protein 1A	MAP1A	**Estramustine** – antineoplastic agents
Microtubule-associated protein 2	MAP2	**Estramustine** – antineoplastic agents
Microtubule-associated protein 4	MAP4	**Paclitaxel** – antineoplastic agents
Microtubule-associated protein, RP/ EB family, member 1	MAPRE1	Proteomic-based identification of the APC-binding protein EB1 as a candidate of novel tissue biomarker and therapeutic target for colorectal cancer. PMID: 22735596
Myosin IE	MYO1E	Disease: focal segmental glomerulosclerosis 6 (FSGS6)
Myosin IIIA	MYO3A	Disease: deafness, autosomal recessive, 30 (DFNB30)
Ninein (GSK3B-interacting protein)	NIN	Disease: Seckel syndrome 7 (SCKL7)
Ninein-like	NINL	Centrosomal Nlp is an oncogenic protein that is gene-amplified in human tumors and causes spontaneous tumorigenesis in transgenic mice. PMID: 20093778
Radial spoke head 4 homolog A (*Chlamydomonas*)	RSPH4A	Disease: ciliary dyskinesia, primary, 11 (CILD11)
Radial spoke head 9 homolog (*Chlamydomonas*)	RSPH9	Disease: ciliary dyskinesia, primary, 12 (CILD12)
Retinitis pigmentosa 1 (autosomal dominant)	RP1	Disease: retinitis pigmentosa 1 (RP1)
Retinitis pigmentosa 1-like 1	RP1L1	Disease: occult macular dystrophy (OCMD)
Ribosomal protein S2	RPS2	RPS2: A novel therapeutic target in prostate cancer. PMID: 19138403
Ribosomal protein S7	RPS7	Disease: Diamond–Blackfan anemia 8 (DBA8)
Septin 9	SEPT9	The influence of methylated septin 9 gene on RNA and protein level in colorectal cancer. PMID: 21267688. Note: A chromosomal aberration involving SEPT9/MSF is found in therapy-related acute myeloid leukemia (t-AML)
Septin 11	SEPT11	Note: A chromosomal aberration involving SEPT11 may be a cause of chronic neutrophilic leukemia. Translocation t(411)(q21q23) with KMT2A/MLL1
Septin 12	SEPT12	Disease: spermatogenic failure 10 (SPGF10)
Spectrin, alpha, erythrocytic 1 (elliptocytosis 2)	SPTA1	Disease: elliptocytosis 2 (EL2)

(*Continued*)

HGNC-approved name	Symbol	Target information
Spectrin, alpha, non-erythrocytic 1	SPTAN1	Disease: epileptic encephalopathy, early infantile, 5 (EIEE5)
Spectrin, beta, erythrocytic	SPTB	Disease: elliptocytosis 3 (EL3)
Spectrin, beta, non-erythrocytic 2	SPTBN2	Disease: spinocerebellar ataxia 5 (SCA5)
Tetratricopeptide repeat domain 8	TTC8	Disease: retinitis pigmentosa 51 (RP51)
Tetratricopeptide repeat domain 21B	TTC21B	Note: Ciliary dysfunction leads to a broad spectrum of disorders, collectively termed ciliopathies
Tropomyosin 1 (alpha)	TPM1	Disease: cardiomyopathy, familial hypertrophic 3 (CMH3)
Tropomyosin 2 (beta)	TPM2	Disease: nemaline myopathy 4 (NEM4)
Tropomyosin 3	TPM3	Disease: nemaline myopathy 1 (NEM1)
Tubulin, alpha 1a	TUBA1A	Disease: lissencephaly 3 (LIS3)
Tubulin, alpha 8	TUBA8	Disease: polymicrogyria, with optic nerve hypoplasia (PMGONH)
Tubulin, beta 1 class VI	TUBB1	Disease: macrothrombocytopenia, autosomal dominant, TUBB1-related (MAD-TUBB1)
Tubulin, beta 2A class IIa	TUBB2A	Disease: cortical dysplasia, complex, with other brain malformations 5 (CDCBM5)
Tubulin, beta class I	TUBB	Disease: cortical dysplasia, complex, with other brain malformations 6 (CDCBM6)
Tubulin, beta 2B class IIb	TUBB2B	Disease: polymicrogyria, symmetric or asymmetric (PMGYSA)
Tubulin, beta 3 class III	TUBB3	The seco-taxane IDN5390 is able to target class III beta-tubulin and to overcome paclitaxel resistance. PMID: 15781655. Disease: congenital fibrosis of extraocular muscles 3A (CFEOM3A)
Tubulin, beta 4A class IVa	TUBB4A	Disease: dystonia 4, torsion, autosomal dominant (DYT4)
Tubulin, gamma 1	TUBG1	Disease: cortical dysplasia, complex, with other brain malformations 4 (CDCBM4)
Tubulin, gamma complex-associated protein 6	TUBGCP6	Disease: microcephaly and chorioretinopathy with or without mental retardation, autosomal recessive (MCPHCR)
WD repeat domain 34	WDR34	Disease: short-rib thoracic dysplasia 11 with or without polydactyly (SRTD11)
WD repeat domain 35	WDR35	Disease: cranioectodermal dysplasia 2 (CED2)
Wiskott–Aldrich syndrome	WAS	A peptide derived from the Wiskott–Aldrich syndrome (WAS) protein-interacting protein (WIP) restores WAS protein level and actin cytoskeleton reorganization in lymphocytes from patients with WAS mutations that disrupt WIP binding. PMID: 21376381. Disease: Wiskott–Aldrich syndrome 1 (WAS1)
Wiskott–Aldrich syndrome-like	WASL	N-WASP is highly expressed in hepatocellular carcinoma and associated with poor prognosis. PMID: 23218879
Zinc finger, FYVE domain containing 19	ZFYVE19	Note: A chromosomal aberration involving ZFYVE19 is associated with acute myeloblastic leukemia (AML). Translocation t(1115)(q23q14) with KMT2A/MLL1. PMID: 12618766
Zinc finger, FYVE domain containing 26	ZFYVE26	Disease: spastic paraplegia 15, autosomal recessive (SPG15)
A kinase (PRKA) anchor protein 9	AKAP9	Disease: long QT syndrome 11 (LQT11)
Abelson helper integration site 1	AHI1	AHI-1: A novel signaling protein and potential therapeutic target in human leukemia and brain disorders. PMID: 22248740. Disease: Joubert syndrome 3 (JBTS3)
Actin, gamma 2, smooth muscle, enteric	ACTG2	Disease: visceral myopathy (VSCM)

HGNC-approved name	Symbol	Target information
Actinin, alpha 1	ACTN1	Disease: bleeding disorder, platelet-type 15 (BDPLT15)
Alstrom syndrome 1	ALMS1	Disease: Alstrom syndrome (ALMS)
Armadillo repeat containing 4	ARMC4	Disease: ciliary dyskinesia, primary, 23 (CILD23)
Asp (abnormal spindle) homolog, microcephaly associated (*Drosophila*)	ASPM	Disease: microcephaly 5, primary, autosomal recessive (MCPH5)
C2 calcium-dependent domain containing 3	C2CD3	Note: Orofaciodigital syndrome (OFDS), a heterogeneous disease characterized by abnormalities in the oral cavity, face and digits, and associated phenotypic abnormalities that lead to the delineation of various subtypes
Calmodulin-regulated spectrin-associated protein family, member 2	CAMSAP2	Note: Defects in CAMSAP2 may be a cause of susceptibility to epilepsy in the Chinese population
CAP-GLY domain containing linker protein 2	CLIP2	Note: CLIP2 is located in the Williams–Beuren syndrome (WBS) critical region
Catenin (cadherin-associated protein), alpha 1, 102 kDa	CTNNA1	Disease: hereditary diffuse gastric cancer (HDGC)
CDK5 regulatory subunit-associated protein 2	CDK5RAP2	Disease: microcephaly 3, primary, autosomal recessive (MCPH3)
Cell division cycle 20	CDC20	Synergistic blockade of mitotic exit by two chemical inhibitors of the APC/C. PMID: 25156254
Centriolin	CNTRL	Note: A chromosomal aberration involving CEP110 may be a cause of stem cell myeloproliferative disorder (MPD)
Centromere protein J	CENPJ	Disease: microcephaly 6, primary, autosomal recessive (MCPH6)
Centrosome and spindle pole-associated protein 1	CSPP1	Disease: Joubert syndrome 21 (JBTS21)
Coiled-coil and C2 domain containing 2A	CC2D2A	Disease: Meckel syndrome 6 (MKS6)
Coiled-coil domain containing 6	CCDC6	Disease: thyroid papillary carcinoma (TPC)
Coronin, actin binding protein, 1A	CORO1A	Coronin-1a is a potential therapeutic target for activated T cell-related immune disorders. PMID: 24862375. Disease: immunodeficiency 8 (IMD8)
Disrupted in schizophrenia 1	DISC1	DISC1 as a therapeutic target for mental illnesses. PMID: 23130881. Note: A chromosomal aberration involving DISC1 segregates with schizophrenia and related psychiatric disorders in a large Scottish family
Dynactin 1	DCTN1	dnc-1/dynactin 1 knockdown disrupts transport of autophagosomes and induces motor neuron degeneration. PMID: 23408943. Disease: neuronopathy, distal hereditary motor, 7B (HMN7B)
Dynein regulatory complex subunit 1 homolog (*Chlamydomonas*)	DRC1	Disease: ciliary dyskinesia, primary, 21 (CILD21)
Dystonin	DST	Disease: neuropathy, hereditary sensory and autonomic, 6 (HSAN6)
Echinoderm microtubule-associated protein-like 1	EML1	Note: Mutations in this gene are associated with atypical heterotopia, epilepsy and mental retardation. Patients present giant bilateral periventricular and ribbon-like subcortical heterotopia with polymicrogyria and agenesis of the corpus callosum
Espin	ESPN	Disease: deafness, autosomal recessive, 36, with or without vestibular involvement (DFNB36)
FERM domain containing 4A	FRMD4A	FRMD4A upregulation in human squamous cell carcinoma promotes tumor growth and metastasis and is associated with poor prognosis. PMID: 22564525
Fermitin family member 1	FERMT1	Disease: Kindler syndrome (KINDS)

(Continued)

HGNC-approved name	Symbol	Target information
FGFR1 oncogene partner	FGFR1OP	Note: A chromosomal aberration involving FGFR1OP may be a cause of stem cell myeloproliferative disorder (MPD). Translocation t(68)(q27p11) with FGFR1
FYVE, RhoGEF and PH domain containing 4	FGD4	Disease: Charcot–Marie–Tooth disease 4H (CMT4H)
Gelsolin	GSN	Gelsolin as therapeutic target in Alzheimer's disease. PMID: 20433353. Disease: amyloidosis 5 (AMYL5)
Gigaxonin	GAN	Disease: giant axonal neuropathy (GAN)
Inversin	INVS	Disease: nephronophthisis 2 (NPHP2)
IQ motif containing B1	IQCB1	Disease: Senior–Loken syndrome 5 (SLSN5)
Kaptin (actin binding protein)	KPTN	Mutations in KPTN cause macrocephaly, neurodevelopmental delay, and seizures. PMID: 24239382
KIAA1279	KIAA1279	Disease: Goldberg–Shprintzen megacolon syndrome (GOSHS)
Kizuna centrosomal protein	KIZ	Disease: retinitis pigmentosa 69 (RP69)
KRIT1, ankyrin repeat containing	KRIT1	Disease: cerebral cavernous malformations 1 (CCM1)
Leber congenital amaurosis 5	LCA5	Disease: Leber congenital amaurosis 5 (LCA5)
Lymphocyte cytosolic protein 1 (L-plastin)	LCP1	Note: Chromosomal aberrations involving LCP1 is a cause of B-cell non-Hodgkin lymphomas (B-cell NHL). Translocation t(313)(q27q14), with BCL6
McKusick–Kaufman syndrome	MKKS	Disease: McKusick–Kaufman syndrome (MKKS)
Meckel syndrome, type 1	MKS1	Disease: Meckel syndrome 1 (MKS1)
Mediterranean fever	MEFV	Disease: familial Mediterranean fever, autosomal recessive (ARFMF)
Metastasis suppressor 1	MTSS1	Expression of metastasis suppressor 1 in cervical carcinoma and the clinical significance. PMID: 25295101
Microcephalin 1	MCPH1	The overexpression of MCPH1 inhibits cell growth through regulating cell cycle-related proteins and activating cytochrome c-caspase 3 signaling in cervical cancer. PMID: 24633962. Disease: microcephaly 1, primary, autosomal recessive (MCPH1)
Myosin, heavy chain 9, non-muscle	MYH9	Disease: May–Hegglin anomaly (MHA)
Myristoylated alanine-rich protein kinase C substrate	MARCKS	Myristoylated alanine-rich protein kinase C substrate (MARCKS) expression modulates the metastatic phenotype in human and murine colon carcinoma *in vitro* and in vivo. PMID: 23376641
Nephronophthisis 4	NPHP4	Disease: nephronophthisis 4 (NPHP4)
Neuron navigator 1	NAV1	Dynamic microtubules catalyze formation of navigator-TRIO complexes to regulate neurite extension. PMID: 25065758
Nexilin (F actin binding protein)	NEXN	Disease: cardiomyopathy, dilated 1CC (CMD1CC)
nudE neurodevelopment protein 1	NDE1	Disease: lissencephaly 4 (LIS4)
Obscurin-like 1	OBSL1	The genetics of 3-M syndrome: Unravelling a potential new regulatory growth pathway. PMID: 22156540
Oral–facial–digital syndrome 1	OFD1	Disease: orofaciodigital syndrome 1 (OFD1)
Palladin, cytoskeletal-associated protein	PALLD	Disease: pancreatic cancer 1 (PNCA1)
Pericentrin	PCNT	Disease: microcephalic osteodysplastic primordial dwarfism 2 (MOPD2)
Pericentriolar material 1	PCM1	Disease: thyroid papillary carcinoma (TPC)
Phosphoprotein enriched in astrocytes 15	PEA15	Quantitative proteomics reveals that PEA15 regulates astroglial $A\beta$ phagocytosis in an Alzheimer's disease mouse model. PMID: 25108202
Platelet-activating factor acetylhydrolase 1b, regulatory subunit 1 (45 kDa)	PAFAH1B1	Disease: lissencephaly 1 (LIS1)

HGNC-approved name	Symbol	Target information
Plectin	PLEC	Disease: epidermolysis bullosa simplex with pyloric atresia (EBS-PA)
POC1 centriolar protein A	POC1A	Disease: short stature, onychodysplasia, facial dysmorphism, and hypotrichosis (SOFT)
Pre-B-cell leukemia homeobox-interacting protein 1	PBXIP1	Pre-B-cell leukemia homeobox interacting protein 1 is overexpressed in astrocytoma and promotes tumor cell growth and migration. PMID: 24470547
Profilin 1	PFN1	Protein phosphatase 1 dephosphorylates profilin-1 at Ser-137. PMID: 22479341. Disease: amyotrophic lateral sclerosis 18 (ALS18)
Ras association (RalGDS/AF-6) domain family member 1	RASSF1	Ras-association domain family 1C protein promotes breast cancer cell migration and attenuates apoptosis. PMID: 20955597
Retinitis pigmentosa GTPase regulator	RPGR	Disease: retinitis pigmentosa 3 (RP3)
Rotatin	RTTN	Disease: polymicrogyria with seizures (PMGYS)
RPGRIP1-like	RPGRIP1L	Note. Ciliary dysfunction leads to a broad spectrum of disorders, collectively termed ciliopathies. Overlapping clinical features include retinal degeneration, renal cystic disease, skeletal abnormalities, fibrosis of various organ, and a complex
Septin 10	SEPT10	Disease: spermatogenic failure 10 (SPGF10)
Serologically defined colon cancer antigen 8	SDCCAG8	Disease: Senior–Loken syndrome 7 (SLSN7)
Shroom family member 4	SHROOM4	Disease: mental retardation, X-linked, syndromic, Stocco dos Santos type (SDSX)
Sperm antigen with calponin homology and coiled-coil domains 1-like	SPECC1L	Disease: facial clefting, oblique, 1 (OBLFC1)
StAR-related lipid transfer (START) domain containing 9	STARD9	STARD9/Kif16a is a novel mitotic kinesin and antimitotic target. PMID: 22754624
Stathmin 1	STMN1	Stathmin 1 in normal and malignant hematopoiesis. PMID: 24667172
Tectonic family member 1	TCTN1	Expression and prognostic significance of TCTN1 in human glioblastoma. PMID: 25304031. Disease: Joubert syndrome 13 (JBTS13)
Tensin 4	TNS4	C-terminal tensin-like (CTEN): A promising biomarker and target for cancer. PMID: 24735711
Testis-specific serine kinase substrate	TSKS	Validation of a testis specific serine/threonine kinase (TSSK) family and the substrate of TSSK1 & 2, TSKS as contraceptive targets PMID: 17568264
Thymosin beta 4, X-linked	TMSB4X	**RGN-352** – cardiac therapy
Thyroid hormone receptor interactor 6	TRIP6	TRIP6 regulates p27 KIP1 to promote tumorigenesis. PMID: 23339869
Tubulin-folding cofactor E	TBCE	Disease: hypoparathyroidism–retardation–dysmorphism syndrome (HRD)
unc-119 homolog (*C. elegans*)	UNC119	Note: Defects in UNC119 may be a cause of cone-rod dystrophy
Villin 1	VIL1	Note: Biliary atresia is a chronic and progressive cholestatic liver disease of childhood characterized by an abnormal villin gene expression and severe malformation of canalicular microvillus structure
Vinculin	VCL	Disease: cardiomyopathy, dilated 1W (CMD1W)
WAS protein family, member 3	WASF3	Increased expression levels of WAVE3 are associated with the progression and metastasis of triple negative breast cancer. PMID: 22952619
Zinc finger, MYND-type containing 10	ZMYND10	Disease: ciliary dyskinesia, primary, 22 (CILD22)
Zyxin	ZYX	Zyxin is a novel target for β-amyloid peptide: characterization of its role in Alzheimer's pathogenesis. PMID: 23330981

(Continued)

6.3   Cytoplasm to nucleus

HGNC-approved name	Symbol	Target information
Aminoacyl tRNA synthetase complex-interacting multifunctional protein 1	AIMP1	Disease: leukodystrophy, hypomyelinating, 3 (HLD3)
Aminoacyl tRNA synthetase complex-interacting multifunctional protein 2	AIMP2	Splicing variant of AIMP2 as an effective target against chemoresistant ovarian cancer. PMID: 22532625
Ataxin 1	ATXN1	Disease: spinocerebellar ataxia 1 (SCA1)
Ataxin 10	ATXN10	Disease: spinocerebellar ataxia 10 (SCA10)
B-cell CLL/lymphoma 3	BCL3	Note: A chromosomal aberration involving BCL3 may be a cause of B-cell chronic lymphocytic leukemia (B-CLL). Translocation t(1419)(q32q13.1) with immunoglobulin gene regions
B-cell CLL/lymphoma 10	BCL10	Note: A chromosomal aberration involving BCL10 is recurrent in low-grade mucosa-associated lymphoid tissue (MALT lymphoma). Translocation t(114)(p22q32)
B-cell CLL/lymphoma 11A (zinc finger protein)	BCL11A	Note: Chromosomal aberrations involving BCL11A may be a cause of lymphoid malignancies. Translocation t(214)(p13q32.3) causes BCL11A deregulation and amplification
Cell death-inducing DFFA-like effector a	CIDEA	Note: In omental and subcutaneous adipose tissue of obese patients matched for BMI, expression levels correlate with insulin sensitivity
Cell death-inducing DFFA-like effector c	CIDEC	Note: In omental adipose tissue of obese patients matched for BMI, expression levels tend to correlate with insulin sensitivity
Cryptochrome circadian clock 1	CRY1	Cryptochrome 1 overexpression correlates with tumor progression and poor prognosis in patients with colorectal cancer. PMID: 23626715
Cryptochrome circadian clock 2	CRY2	CRY2 genetic variants associate with dysthymia. PMID: 23951166
Crystallin, alpha A	CRYAA	Identification of the HSPB4/TLR2/NF-κB axis in macrophage as a therapeutic target for sterile inflammation of the cornea. PMID: 22359280. Note: Alpha-crystallin A 1-172 is found at nearly twofold higher levels in diabetic lenses than in age-matched controls
Crystallin, alpha B	CRYAB	Disease: myopathy, myofibrillar, 2 (MFM2)
Cyclin-dependent kinase inhibitor 1A (p21, Cip1)	CDKN1A	Less understood issues: p21(Cip1) in mitosis and its therapeutic potential. PMID: 24858045
Cyclin-dependent kinase inhibitor 1B (p27, Kip1)	CDKN1B	Cyclin-dependent kinase inhibitor p27 as a prognostic biomarker and potential cancer therapeutic target. PMID: 21142858. Disease: multiple endocrine neoplasia 4 (MEN4)
Cyclin-dependent kinase inhibitor 2A	CDKN2A	Geriatric muscle stem cells switch reversible quiescence into senescence. PMID: 24622534. Note: The association between cutaneous and uveal melanomas in some families suggests that mutations in CDKN2A may account for a proportion of uveal melanomas
Deleted in azoospermia 1	DAZ1	Disease: spermatogenic failure Y-linked 2 (SPGFY2)
Deleted in azoospermia 2	DAZ2	Disease: spermatogenic failure Y-linked 2 (SPGFY2)
Deleted in azoospermia 3	DAZ3	Disease: spermatogenic failure Y-linked 2 (SPGFY2)
Deleted in azoospermia 4	DAZ4	Disease: spermatogenic failure Y-linked 2 (SPGFY2)
Eukaryotic translation initiation factor 4H	EIF4H	An alternative splicing isoform of eukaryotic initiation factor 4H promotes tumorigenesis in vivo and is a potential therapeutic target for human cancer. PMID: 20473909. Note: EIF4H is located in the Williams–Beuren syndrome (WBS) critical region

HGNC-approved name	Symbol	Target information
Eukaryotic translation initiation factor 5A2	EIF5A2	Overexpression of EIF5A2 promotes colorectal carcinoma cell aggressiveness by upregulating MTA1 through C-myc to induce epithelial-mesenchymal transition. PMID: 21813470
Fanconi anemia, complementation group A	FANCA	Disease: Fanconi anemia (FA)
Fanconi anemia, complementation group C	FANCC	Disease: Fanconi anemia complementation group C (FANCC)
Fanconi anemia, complementation group G	FANCG	Disease: Fanconi anemia complementation group G (FANCG)
Forkhead box A2	FOXA2	Inhibition of the liver enriched protein FOXA2 recovers HNF6 activity in human colon carcinoma and liver hepatoma cells. PMID: 20967225
Forkhead box O1	FOXO1	FOXO1 binds to the TAU5 motif and inhibits constitutively active androgen receptor splice variants. PMID: 23389878. Disease: rhabdomyosarcoma 2 (RMS2)
Forkhead box O3	FOXO3	Deregulation of FoxO3a accelerates prostate cancer progression in TRAMP mice. PMID: 23765843. Note: A chromosomal aberration involving FOXO3 is found in secondary acute leukemias. Translocation t(611)(q21q23) with KMT2A/MLL1
Forkhead box O4	FOXO4	The transcription factor FOXO4 is down-regulated and inhibits tumor proliferation and metastasis in gastric cancer. PMID: 24886657. Note: A chromosomal aberration involving FOXO4 is found in acute leukemias. Translocation t(X11)(q13q23) with KMT2A/MLL1. The result is a rogue activator protein
Forkhead box O6	FOXO6	FoxO6 in glucose metabolism (FoxO6). PMID: 23324123
GLI family zinc finger 1	GLI1	Synthesis and biological evaluation of SANT-2 and analogues as inhibitors of the hedgehog signaling pathway. PMID: 19541490
GLI family zinc finger 3	GLI3	Disease: Greig cephalopolysyndactyly syndrome (GCPS)
Heat shock transcription factor 1	HSF1	The heat shock transcription factor 1 as a potential new therapeutic target in multiple myeloma. PMID: 23252346
Heat shock transcription factor 2	HSF2	Heat shock factor 2 is required for maintaining proteostasis against febrile-range thermal stress and polyglutamine aggregation. PMID: 21813737
IKAROS family zinc finger 1 (Ikaros)	IKZF1	Lenalidomide causes selective degradation of IKZF1 and IKZF3 in multiple myeloma cells. PMID: 24292625. Note: Defects in IKZF1 are frequent occurrences (28.6%) in acute lymphoblastic leukemia (ALL)
IKAROS family zinc finger 3 (Aiolos)	IKZF3	Lenalidomide causes selective degradation of IKZF1 and IKZF3 in multiple myeloma cells. PMID: 24292625
Inhibitor of DNA binding 1, dominant-negative helix-loop-helix protein	ID1	Id-1 is a key transcriptional regulator of glioblastoma aggressiveness and a novel therapeutic target. PMID: 23243024
Inhibitor of DNA binding 2, dominant-negative helix-loop-helix protein	ID2	Essential role of Id2 in negative regulation of IgE class switching. PMID: 12483209
Interferon regulatory factor 1	IRF1	Blocking of interferon regulatory factor 1 reduces tumor necrosis factor α-induced interleukin-18 bioactivity in rheumatoid arthritis synovial fibroblasts by induction of interleukin-18 binding protein a: Role of the nuclear interferon regulatory factor 1. PMID: 21834067

(Continued)

HGNC-approved name	Symbol	Target information
Interferon regulatory factor 5	IRF5	Interferon regulatory factor (IRF)-5: A potential therapeutic target for ankylosing spondylitis. PMID: 21928121. Disease: inflammatory bowel disease 14 (IBD14)
Interferon regulatory factor 6	IRF6	Disease: van der Woude syndrome 1 (VWS1)
Interferon regulatory factor 9	IRF9	Interferon regulatory factor 9 is an essential mediator of heart dysfunction and cell death following myocardial ischemia/reperfusion injury. PMID: 25150882
Keratin 8	KRT8	Disruption of cytokeratin-8 interaction with F508del-CFTR corrects its functional defect. PMID: 22038833. Disease: cirrhosis (CIRRH)
Keratin 14	KRT14	Disease: epidermolysis bullosa simplex, Dowling–Meara type (DM-EBS)
Keratin 18	KRT18	Disease: cirrhosis (CIRRH)
Kinesin family member 5A	KIF5A	Disease: spastic paraplegia 10, autosomal dominant (SPG10)
Kinesin family member 14	KIF14	Kinesin family member 14: An independent prognostic marker and potential therapeutic target for ovarian cancer. PMID: 21618518
Kinesin family member 22	KIF22	Disease: spondyloepimetaphyseal dysplasia with joint laxity, 2 (SEMDJL2)
Lin-28 homolog A (*C. elegans*)	LIN28A	Lin28 regulates HER2 and promotes malignancy through multiple mechanisms. PMID: 22713243
Lin-28 homolog B (*C. elegans*)	LIN28B	Lin28B is a novel prognostic marker in gastric adenocarcinoma. PMID: 25197381
Nuclear factor of kappa light polypeptide gene enhancer in B cells 2 (p49/p100)	NFKB2	Note: A chromosomal aberration involving NFKB2 is found in a case of B-cell non Hodgkin lymphoma (B-NHL). Translocation t(1014)(q24q32) with IGHA1. The resulting oncogene is also called Lyt-10C alpha variant
Nuclear factor of kappa light polypeptide gene enhancer in B-cells inhibitor, alpha	NFKBIA	Disease: ectodermal dysplasia, anhidrotic, with T-cell immunodeficiency autosomal dominant (ADEDAID)
POU class 2 homeobox 2	POU2F2	Expression and prognostic significance of Oct2 and Bob1 in multiple myeloma: Implications for targeted therapeutics. PMID: 21438833
POU class 5 homeobox 1	POU5F1	Implications of transcriptional factor, OCT-4, in human bladder malignancy and tumor recurrence. PMID: 21533858
Protein phosphatase 1, regulatory subunit 13-like	PPP1R13L	RNA interference-mediated silencing of iASPP induces cell proliferation inhibition and G0/G1 cell cycle arrest in U251 human glioblastoma cells. PMID: 21184255
Protein phosphatase 1, regulatory subunit 13B	PPP1R13B	ASPP1 and ASPP2 bind active RAS, potentiate RAS signalling and enhance p53 activity in cancer cells. PMID: 23392125
Ribosomal protein L5	RPL5	Disease: Diamond–Blackfan anemia 6 (DBA6)
Ribosomal protein S10	RPS10	Disease: Diamond–Blackfan anemia 9 (DBA9)
S100 calcium binding protein B	S100B	Identification of small-molecule inhibitors of the human S100B-p53 interaction and evaluation of their activity in human melanoma cells. PMID: 23375094
S100 calcium binding protein P	S100P	S100P: A novel therapeutic target for cancer. PMID: 20509035
Signal transducer and activator of transcription 1, 91 kDa	STAT1	Pravastatin attenuates interferon-gamma action via modulation of STAT1 to prevent aortic atherosclerosis in apolipoprotein E-knockout mice. PMID: 19018808. Disease: STAT1 deficiency complete (STAT1D)
Signal transducer and activator of transcription 3 (acute-phase response factor)	STAT3	**AZD9150** – antineoplastic agents. Disease: hyperimmunoglobulin E recurrent infection syndrome, autosomal dominant (AD-HIES)
Signal transducer and activator of transcription 4	STAT4	Disease: systemic lupus erythematosus 11 (SLEB11)

HGNC-approved name	Symbol	Target information
Signal transducer and activator of transcription 5A	STAT5A	The STAT5 inhibitor pimozide decreases survival of chronic myelogenous leukemia cells resistant to kinase inhibitors. PMID: 21233313
Signal transducer and activator of transcription 5B	STAT5B	The STAT5 inhibitor pimozide decreases survival of chronic myelogenous leukemia cells resistant to kinase inhibitors. PMID: 21233313. Disease: growth hormone insensitivity with immunodeficiency (GHII)
Signal transducer and activator of transcription 6, interleukin-4 induced	STAT6	Novel 7H-pyrrolo[2,3-d]pyrimidine derivatives as potent and orally active STAT6 inhibitors. PMID: 19747833
SIX homeobox 1	SIX1	The SIX1-EYA transcriptional complex as a therapeutic target in cancer. PMID: 25555392. Disease: deafness, autosomal dominant, 23 (DFNA23)
SIX homeobox 5	SIX5	Disease: branchiootorenal syndrome 2 (BOR2)
SMAD family member 1	SMAD1	Note: SMAD1 variants may be associated with susceptibility to pulmonary hypertension, a disorder characterized by plexiform lesions of proliferating endothelial cells in pulmonary arterioles
SMAD family member 3	SMAD3	PTTG1 inhibits SMAD3 in prostate cancer cells to promote their proliferation. PMID: 24627133. Disease: colorectal cancer (CRC)
SMAD family member 4	SMAD4	A novel human Smad4 mutation is involved in papillary thyroid carcinoma progression. PMID: 22109972. Disease: pancreatic cancer (PNCA)
SMAD family member 7	SMAD7	Inhibitory Smad7: Emerging roles in health and disease. PMID: 21222648. Disease: colorectal cancer 3 (CRCS3)
SMAD family member 9	SMAD9	Disease: pulmonary hypertension, primary, 2 (PPH2)
Snail family zinc finger 1	SNAI1	Snail1 mediates hypoxia-induced melanoma progression. PMID: 21996677
Snail family zinc finger 2	SNAI2	SLUG promotes prostate cancer cell migration and invasion via CXCR4/CXCL12 axis. PMID: 22074556. Disease: Waardenburg syndrome 2D (WS2D)
Survival of motor neuron 1, telomeric	SMN1	**RG7800** – spinal muscular atrophy treatments. Disease: spinal muscular atrophy 1 (SMA1)
Survival of motor neuron 2, centromeric	SMN2	**ISIS 396443** – spinal muscular atrophy treatments. Disease: spinal muscular atrophy 1 (SMA1)
Tripartite motif containing 24	TRIM24	Trim24 targets endogenous p53 for degradation. PMID: 19556538. Disease: thyroid papillary carcinoma (TPC)
Tripartite motif containing 55	TRIM55	Cooperative control of striated muscle mass and metabolism by MuRF1 and MuRF2. PMID: 18157088
Tumor necrosis factor, alpha-induced protein 1 (endothelial)	TNFAIP1	Knockdown of TNFAIP1 inhibits growth and induces apoptosis in osteosarcoma cells through inhibition of the nuclear factor-κB pathway. PMID: 24969828
Tumor necrosis factor, alpha-induced protein 2	TNFAIP2	A novel role for TNFAIP2: Its correlation with invasion and metastasis in nasopharyngeal carcinoma. PMID: 21057457
Zic family member 1	ZIC1	ZIC1 modulates cell-cycle distributions and cell migration through regulation of sonic hedgehog, PI(3)K and MAPK signaling pathways in gastric cancer. PMID: 22799764
Zic family member 2	ZIC2	Disease: holoprosencephaly 5 (HPE5)
Zic family member 3	ZIC3	Disease: heterotaxy, visceral, 1, X-linked (HTX1)
Abl interactor 1	ABI1	Abl interactor 1 regulates Src-Id1-matrix metalloproteinase 9 axis and is required for invadopodia formation, extracellular matrix degradation and tumor growth of human breast cancer cells. PMID: 19843640
Actinin, alpha 4	ACTN4	Disease: focal segmental glomerulosclerosis 1 (FSGS1)

(Continued)

HGNC-approved name	Symbol	Target information
Activating transcription factor 5	ATF5	Interference with ATF5 function enhances the sensitivity of human pancreatic cancer cells to paclitaxel-induced apoptosis. PMID: 23060563
Adaptor-related protein complex 5, zeta 1 subunit	AP5Z1	Disease: spastic paraplegia 48, autosomal recessive (SPG48)
Amyloid beta (A4) precursor protein-binding, family A, member 3	APBA3	Mint3 enhances the activity of hypoxia-inducible factor-1 (HIF-1) in macrophages by suppressing the activity of factor inhibiting HIF-1. PMID: 19726677
Ankyrin repeat and SOCS box containing 10	ASB10	Disease: glaucoma 1, open angle, F (GLC1F)
Annexin A1	ANXA1	**Beclomethasone** – anti-inflammatory and antirheumatic products
Arrestin, beta 1	ARRB1	B-arrestin: A signaling molecule and potential therapeutic target for heart failure. PMID: 21074538
Aryl hydrocarbon receptor	AHR	Inhibition of constitutive aryl hydrocarbon receptor (AhR) signaling attenuates androgen independent signaling and growth in (C4-2) prostate cancer cells. PMID: 23266674
Aryl hydrocarbon receptor-interacting protein-like 1	AIPL1	Disease: Leber congenital amaurosis 4 (LCA4)
Aryl hydrocarbon receptor nuclear translocator-like	ARNTL	The circadian clock gene BMAL1 is a novel therapeutic target for malignant pleural mesothelioma. PMID: 22510946
Atrophin 1	ATN1	Disease: dentatorubral–pallidoluysian atrophy (DRPLA)
Autoimmune regulator	AIRE	Disease: autoimmune polyendocrine syndrome 1, with or without reversible metaphyseal dysplasia (APS1)
Axin1	AXIN1	Disease: hepatocellular carcinoma (HCC)
Barrier to autointegration factor 1	BANF1	Disease: Nestor–Guillermo progeria syndrome (NGPS)
Basic helix-loop-helix domain containing, class B, 9	BHLHB9	P60TRP interferes with the GPCR/secretase pathway to mediate neuronal survival and synaptogenesis. PMID: 21199326
Basic helix-loop-helix family, member e40	BHLHE40	The transcription factor DEC1 (BHLHE40/STRA13/SHARP-2) is negatively associated with TNM stage in non-small-cell lung cancer and inhibits the proliferation through cyclin D1 in A549 and BE1 cells. PMID: 23423709
BCL2-associated athanogene 6	BAG6	Bat3 promotes T cell responses and autoimmunity by repressing Tim-3–mediated cell death and exhaustion. PMID: 22863785
BMI1 proto-oncogene, polycomb ring finger	BMI1	Bmi1 knockdown inhibits hepatocarcinogenesis. PMID: 23138990
Bridging integrator 1	BIN1	Disease: myopathy, centronuclear, 2 (CNM2)
Calcium binding protein 2	CABP2	Disease: deafness, autosomal recessive, 93 (DFNB93)
Calcyclin binding protein	CACYBP	CacyBP/SIP expression is involved in the clinical progression of breast cancer. PMID: 20585948
Calmodulin-binding transcription activator 1	CAMTA1	Disease: cerebellar ataxia, non-progressive, with mental retardation (CANPMR)
CASP2 and RIPK1 domain containing adaptor with death domain	CRADD	Disease: mental retardation, autosomal recessive 34 (MRT34)
Catenin (cadherin-associated protein), beta 1, 88 kDa	CTNNB1	**CEQ508** – antineoplastic agents. Disease: colorectal cancer (CRC)
Cell division cycle 6	CDC6	Disease: Meier–Gorlin syndrome 5 (MGORS5)
Cell division cycle-associated 7-like	CDCA7L	The reduction of R1, a novel repressor protein for monoamine oxidase A, in major depressive disorder. PMID: 21654740
Centrosomal protein 57 kDa	CEP57	Disease: mosaic variegated aneuploidy syndrome 2 (MVA2)

HGNC-approved name	Symbol	Target information
Chromosome 9 open reading frame 72	C9orf72	c9RAN translation: A potential therapeutic target for the treatment of amyotrophic lateral sclerosis and frontotemporal dementia. PMID: 23844663. Disease: frontotemporal dementia and/or amyotrophic lateral sclerosis (FTDALS)
Codanin 1	CDAN1	Disease: anemia, congenital dyserythropoietic, 1A (CDAN1A)
Coiled-coil and C2 domain containing 1A	CC2D1A	Disease: mental retardation, autosomal recessive 3 (MRT3)
COP9 signalosome subunit 3	COPS3	Silencing of the COPS3 gene by siRNA reduces proliferation of lung cancer cells most likely via induction of cell cycle arrest and apoptosis. PMID: 22631635
CREB binding protein	CREBBP	**PRI-724** – antineoplastic agents. Note: Chromosomal aberrations involving CREBBP may be a cause of acute myeloid leukemias. Translocation t(816)(p11p13) with KAT6A translocation t(1116)(q23p13.3) with KMT2A/MLL1 translocation t(1016)(q22p13) with KAT6B
CREB-regulated transcription coactivator 1	CRTC1	Note: A chromosomal aberration involving CRTC1 is found in mucoepidermoid carcinomas, benign Warthin tumors and clear cell hidradenomas. Translocation t(1119)(q21p13) with MAML2
C-terminal binding protein 1	CTBP1	CtBP1 is involved in epithelial-mesenchymal transition and is a potential therapeutic target for hepatocellular carcinoma. PMID: 23756565
CUE domain containing 2	CUEDC2	Note: May predict the clinical outcome of tamoxifen therapy of breast cancer patients. Patients with tumors that highly express CUEDC2 do not respond to tamoxifen treatment as effectively as those with tumors with low expression
Cyclin D1	CCND1	RNA inhibition highlights cyclin D1 as a potential therapeutic target for mantle cell lymphoma. PMID: 22905260. Note: A chromosomal aberration involving CCND1 may be a cause of B-lymphocytic malignancy, particularly mantle-cell lymphoma (MCL)
Cysteine- and glycine-rich protein 3 (cardiac LIM protein)	CSRP3	Disease: cardiomyopathy, dilated 1M (CMD1M)
Cytokine-induced apoptosis inhibitor 1	CIAPIN1	CIAPIN1 as a therapeutic target in cancer. PMID: 20367238
DEAF1 transcription factor	DEAF1	Interactome mapping of the phosphatidylinositol 3-kinase-mammalian target of rapamycin pathway identifies deformed epidermal autoregulatory factor-1 as a new glycogen synthase kinase-3 interactor. PMID: 20368287
Death effector domain containing	DEDD	DEDD, a novel tumor repressor, reverses epithelial-mesenchymal transition by activating selective autophagy. PMID: 22874565
DnaJ (Hsp40) homolog, subfamily B, member 1	DNAJB1	Detection of a recurrent DNAJB1-PRKACA chimeric transcript in fibrolamellar hepatocellular carcinoma. PMID: 24578576
DnaJ (Hsp40) homolog, subfamily B, member 6	DNAJB6	Disease: limb-girdle muscular dystrophy 1E (LGMD1E)
Dentin matrix acidic phosphoprotein 1	DMP1	Disease: hypophosphatemic rickets, autosomal recessive, 1 (ARHR1)
Dishevelled-binding antagonist of beta-catenin 1	DACT1	Downregulation of HDPR1 is associated with poor prognosis and affects expression levels of p120-catenin and beta-catenin in nonsmall cell lung cancer. PMID: 20232357. Disease: neural tube defects (NTD)
DNA-damage-induced apoptosis suppressor	DDIAS	Human Noxin is an anti-apoptotic protein in response to DNA damage of A549 non-small cell lung carcinoma. PMID: 24214091
DNA-damage-inducible transcript 3	DDIT3	Involvement of endoplasmic reticulum stress-mediated CHOP (GADD153) induction in the cytotoxicity of 2-aminophenoxazine-3-one in cancer cells. PMID: 21667023. Disease: myxoid liposarcoma (MXLIPO)

(Continued)

HGNC-approved name	Symbol	Target information
Dyslexia susceptibility 1 candidate 1	DYX1C1	Disease: dyslexia 1 (DYX1)
E1A binding protein p300	EP300	Small-molecule inhibitors of acetyltransferase p300 identified by high-throughput screening are potent anticancer agents. PMID: 23625935. Note: Defects in EP300 may play a role in epithelial cancer
E74-like factor 1 (ets domain transcription factor, epithelial-specific)	ELF3	ELF3 is a repressor of androgen receptor action in prostate cancer cells. PMID: 23435425
Ecotropic viral integration site 5	EVI5	Note: A chromosomal aberration involving EVI5 is found is a patient with stage 4S neuroblastoma. Translocation t(110)(p22q21) that forms a EVI5-TRNG10 fusion protein. TRNG10 is a probable structural transcript which is normally not translated
ELAV-like RNA binding protein 1	ELAVL1	HuR, a key post-transcriptional regulator, and its implication in progression of breast cancer. PMID: 20712017
EWS RNA binding protein 1	EWSR1	Small-molecule screen identifies modulators of EWS/FLI1 target gene expression and cell survival in Ewing's sarcoma. PMID: 22323082. Disease: Ewing sarcoma (ES)
Exportin 1	XPO1	**Leptomycin B** – antineoplastic agents
Family with sequence similarity 111, member A	FAM111A	Disease: Kenny–Caffey syndrome 2 (KCS2)
F-box protein 32	FBXO32	Inhibition of atrogin-1/MAFbx expression by adenovirus-delivered small hairpin RNAs attenuates muscle atrophy in fasting mice. PMID: 21126200
Folliculin	FLCN	Disease: Birt–Hogg–Dube syndrome (BHD)
Fragile X mental retardation 1	FMR1	Disease: fragile X syndrome (FRAX)
General transcription factor IIi	GTF2I	Note: GTF2I is located in the Williams–Beuren syndrome (WBS) critical region
GLE1 RNA export mediator	GLE1	Disease: lethal congenital contracture syndrome 1 (LCCS1)
Growth factor receptor-bound protein 2	GRB2	SUMOylation of Grb2 enhances the ERK activity by increasing its binding with Sos1. PMID: 24775912
Heterogeneous nuclear ribonucleoprotein A1	HNRNPA1	IL-6-induced enhancement of c-Myc translation in multiple myeloma cells: critical role of cytoplasmic localization of the rna-binding protein hnRNP A1. PMID: 20974848. Disease: inclusion body myopathy with early-onset Paget disease with or without frontotemporal dementia 3 (MSP3)
Host cell factor C1 (VP16 accessory protein)	HCFC1	Disease: mental retardation, X-linked 3 (MRX3)
Human immunodeficiency virus type I enhancer binding protein 3	HIVEP3	ZAS3 promotes TNFα-induced apoptosis by blocking NFκB-activated expression of the anti-apoptotic genes TRAF1 and TRAF2. PMID: 21524353
Huntingtin	HTT	**PRO289** – anti-dementia drugs. Disease: Huntington disease (HD)
Huntingtin-interacting protein 1	HTP1	Note: A chromosomal aberration involving HIP1 is found in a form of chronic myelomonocytic leukemia (CMML). Translocation t(57)(q33q11.2) with PDGFRB. The chimeric HIP1-PDGFRB transcript results from an in-frame fusion of the two genes
Hypoxia-inducible factor 1, alpha subunit (basic helix-loop-helix transcription factor)	HIF1A	A new anti-angiogenic small molecule, G0811, inhibits angiogenesis via targeting hypoxia inducible factor (HIF)-1α signal transduction. PMID: 24161739
Inhibitor of kappa light polypeptide gene enhancer in B cells, kinase complex-associated protein	IKBKAP	Disease: neuropathy, hereditary sensory and autonomic, 3 (HSAN3)
Inhibitor of kappa light polypeptide gene enhancer in B cells, kinase gamma	IKBKG	Disease: ectodermal dysplasia, anhidrotic, with immunodeficiency X-linked (EDAID)

HGNC-approved name	Symbol	Target information
Insulin-like growth factor 2 mRNA binding protein 3	IGF2BP3	Up-regulation of Imp3 confers in vivo tumorigenicity on murine osteosarcoma cells. PMID: 23226335
Inverted formin, FH2 and WH2 domain containing	INF2	Disease: focal segmental glomerulosclerosis 5 (FSGS5)
Jade family PHD finger 1	JADE1	Polycystin-1 regulates the stability and ubiquitination of transcription factor Jade-1. PMID: 23001567
Kelch-like ECH-associated protein 1	KEAP1	NRF2 and KEAP1 mutations: Permanent activation of an adaptive response in cancer. PMID: 19321346
KIN, antigenic determinant of recA protein homolog (mouse)	KIN	Up-regulation of kin17 is essential for proliferation of breast cancer. PMID: 21980430
LIM domain binding 3	LDB3	Disease: cardiomyopathy, dilated 1C (CMD1C)
LIM domain containing preferred translocation partner in lipoma	LPP	Note: A chromosomal aberration involving LPP is associated with a subclass of benign mesenchymal tumors known as lipomas. Translocation t(3;12)(q27;q20q13 q15) with HMGA2 is shown in lipomas
Loricrin	LOR	Disease: Vohwinkel syndrome with ichthyosis (VSI)
Metastasis associated in colon cancer 1	MACC1	MACC1: A novel target for solid cancers. PMID: 23815185
Muscleblind-like splicing regulator 1	MBNL1	Disease: dystrophia myotonica 1 (DM1)
Myeloid leukemia factor 1	MLF1	Note: A chromosomal aberration involving MLF1 is a cause of myelodysplastic syndrome (MDS). Translocation t(3;5)(q25.1;q34) with NPM1/NPM
Myopalladin	MYPN	Disease: cardiomyopathy, dilated 1KK (CMD1KK)
Nanos homolog 1 (*Drosophila*)	NANOS1	Disease: spermatogenic failure 12 (SPGF12)
Neuronal differentiation 1	NEUROD1	Disease: maturity-onset diabetes of the young 6 (MODY6)
Nuclear factor of activated T cells, cytoplasmic, calcineurin-dependent 2	NFATC2	NFATc2 is a potential therapeutic target in human melanoma. PMID: 22718120
Nuclear factor, erythroid 2-like 2	NFE2L2	Nuclear erythroid 2-related factor 2: A novel potential therapeutic target for liver fibrosis. PMID: 23793039
Nuclear RNA export factor 5	NXF5	Note: A chromosomal aberration involving NXF5 has been observed in one patient with a syndromic form of mental retardation and short stature. Pericentric inversion inv(X)(p21.1q22) that interrupts NXF5
Nucleus accumbens associated 1, BEN and BTB (POZ) domain containing	NACC1	Fatty acid synthase expression associated with NAC1 is a potential therapeutic target in ovarian clear cell carcinomas. PMID: 22653145
NUT midline carcinoma, family member 1	NUTM1	Note: A chromosomal aberration involving NUT is found in a rare, aggressive, and lethal carcinoma arising in midline organs of young people. Translocation t(15;19)(q14;p13) with BRD4 which produces a BRD4-NUT fusion protein
Oligodendrocyte lineage transcription factor 2	OLIG2	Note: A chromosomal aberration involving OLIG2 may be a cause of a form of T-cell acute lymphoblastic leukemia (T-ALL). Translocation t(14;21)(q11.2;q22) with TCRA
Optineurin	OPTN	Disease: glaucoma 1, open angle, E (GLC1E)
Pancreas-specific transcription factor, 1a	PTF1A	Disease: pancreatic and cerebellar agenesis (PACA)
Pancreatic and duodenal homeobox 1	PDX1	PDX-1 is a therapeutic target for pancreatic cancer, insulinoma and islet neoplasia using a novel RNA interference platform. PMID: 22905092. Disease: pancreatic agenesis, congenital (PAGEN)
PARP1 binding protein	PARPBP	PARI overexpression promotes genomic instability and pancreatic tumorigenesis. PMID: 23436799

(Continued)

HGNC-approved name	Symbol	Target information
PDZ and LIM domain 2 (mystique)	PDLIM2	DNA methylation-dependent repression of PDZ-LIM domain-containing protein 2 in colon cancer and its role as a potential therapeutic target. PMID: 20145149
Periaxin	PRX	Disease: Dejerine–Sottas syndrome (DSS)
Period circadian clock 2	PER2	Disease: advanced sleep phase syndrome, familial, 1 (FASPS1)
PHD finger protein 1	PHF1	Note: A chromosomal aberration involving PHF1 may be a cause of endometrial stromal tumors. Translocation t(67)(p21p22) with JAZF1. Translocation t(16)(p34p21) with MEAF6
Phosphoinositide-3-kinase, regulatory subunit 5	PIK3R5	Disease: ataxia-oculomotor apraxia 3 (AOA3)
Pituitary tumor-transforming 1	PTTG1	PTTG1 overexpression in adrenocortical cancer is associated with poor survival and represents a potential therapeutic target. PMID: 24238056
Piwi-like RNA-mediated gene silencing 4	PIWIL4	PIWIL4 regulates cervical cancer cell line growth and is involved in down-regulating the expression of p14ARF and p53. PMID: 22483988
Poly(A) binding protein, nuclear 1	PABPN1	PABPN1: Molecular function and muscle disease. PMID: 23601051. Disease: oculopharyngeal muscular dystrophy (OPMD)
PRKC, apoptosis, WT1, regulator	PAWR	PAR-4 as a possible new target for pancreatic cancer therapy. PMID: 20426700
Programmed cell death 4 (neoplastic transformation inhibitor)	PDCD4	PDCD4 inhibits the malignant phenotype of ovarian cancer cells. PMID: 19493270
Proliferation-associated 2G4, 38 kDa	PA2G4	The downregulation of ErbB3 binding protein 1 (EBP1) is associated with poor prognosis and enhanced cell proliferation in hepatocellular carcinoma. PMID: 25081333
Proteasome maturation protein	POMP	Disease: keratosis linearis with ichthyosis congenita and sclerosing keratoderma (KLICK)
RAN binding protein 9	RANBP9	A fragment of the scaffolding protein RanBP9 is increased in Alzheimer's disease brains and strongly potentiates amyloid-beta peptide generation. PMID: 19729516
RAS and EF-hand domain containing	RASEF	RASEF is a novel diagnostic biomarker and a therapeutic target for lung cancer. PMID: 23686708
Ras association (RalGDS/AF-6) domain family member 2	RASSF2	Loss of RASSF2 enhances tumorigenicity of lung cancer cells and confers resistance to chemotherapy. PMID: 22693671
RCD1 required for cell differentiation1 homolog (*Schizosaccharomyces pombe*)	RQCD1	Involvement of RQCD1 overexpression, a novel cancer-testis antigen, in the Akt pathway in breast cancer cells. PMID: 19724902
Recombination signal binding protein for immunoglobulin kappa J region	RBPJ	Disease: Adams–Oliver syndrome 3 (AOS3)
Regulator of cell cycle	RGCC	Knockdown of response gene to complement 32 (RGC32) induces apoptosis and inhibits cell growth, migration, and invasion in human lung cancer cells. PMID: 24833469
Retinoic acid induced 1	RAI1	Disease: Smith–Magenis syndrome (SMS)
RNA binding motif protein 38	RBM38	Radiation sensitivity of esophageal adenocarcinoma: The contribution of the RNA-binding protein RNPC1 and p21-mediated cell cycle arrest to radioresistance. PMID: 22214381
Runt-related transcription factor 3	RUNX3	Lactam-based HDAC inhibitors for anticancer chemotherapy: Restoration of RUNX3 by posttranslational modification and epigenetic control. PMID: 24376239
SEC14-like 2 (*S. cerevisiae*)	SEC14L2	Tocopherol-associated protein suppresses prostate cancer cell growth by inhibition of the phosphoinositide 3-kinase pathway. PMID: 16267002
Serine/arginine-rich splicing factor 3	SRSF3	SRp20: An overview of its role in human diseases. PMID: 23685143

HGNC-approved name	Symbol	Target information
SERTA domain containing 2	SERTAD2	Ablation of TRIP-Br2, a regulator of fat lipolysis, thermogenesis and oxidative metabolism, prevents diet-induced obesity and insulin resistance. PMID: 23291629
Shwachman–Bodian–Diamond syndrome	SBDS	Disease: Shwachman–Diamond syndrome (SDS)
Small nuclear ribonucleoprotein polypeptide E	SNRPE	Disease: hypotrichosis 11 (HYPT11)
soc-2 suppressor of clear homolog (*C. elegans*)	SHOC2	Disease: Noonan syndrome-like disorder with loose anagen hair (NSLH)
Spalt-like transcription factor 4	SALL4	Sal-like protein 4 (SALL4), a stem cell biomarker in liver cancers. PMID: 23175232. Disease: Duane-radial ray syndrome (DRRS)
Spermatogenesis- and oogenesis-specific basic helix-loop-helix 1	SOHLH1	Note: Genetic variations in SOHLH1 may be associated with non-obstructive azoospermia
Squamous cell carcinoma antigen recognized by T cells 3	SART3	Disease: disseminated superficial actinic porokeratosis 1 (DSAP1)
SRY (sex-determining region Y)-box 10	SOX10	Disease: Waardenburg syndrome 2E (WS2E)
STE20-related kinase adaptor alpha	STRADA	Note: A homozygous 7-kb deletion involving STRADA is a cause of a syndrome characterized by polyhydramnios, megalencephaly and symptomatic epilepsy
Storkhead box 1	STOX1	Disease: pre-eclampsia/eclampsia 4 (PEE4)
Stress-induced phosphoprotein 1	STIP1	Tumor stress-induced phosphoprotein1 (STIP1) as a prognostic biomarker in ovarian cancer. PMID: 23468915
Structural maintenance of chromosomes 2	SMC2	Human SMC2 protein, a core subunit of human condensin complex, is a novel transcriptional target of the WNT signaling pathway and a new therapeutic target. PMID: 23095742
Suppressor of fused homolog (*Drosophila*)	SUFU	Disease: medulloblastoma (MDB)
Synaptosomal-associated protein, 25 kDa	SNAP25	**Botulinum toxin type A** – antispasmodics
Synuclein, gamma (breast cancer-specific protein 1)	SNCG	Synuclein-gamma (SNCG) may be a novel prognostic biomarker in uterine papillary serous carcinoma. PMID: 19476987
Tax1 (human T-cell leukemia virus type I) binding protein 3	TAX1BP3	The PDZ protein TIP-1 facilitates cell migration and pulmonary metastasis of human invasive breast cancer cells in athymic mice. PMID: 22564736
T-cell leukemia/lymphoma 1A	TCL1A	Note: Chromosomal aberrations activating TCL1A are found in chronic T-cell leukemias (T-CLL). Translocation t(1414)(q11q32) translocation t(714)(q35q32) inversion inv(14)(q11q32) that involves the T-cell receptor alpha/delta loci
Testis-specific protein, Y-linked 1	TSPY1	Note: TSPY is located in the gonadoblastoma critical region and is preferentially expressed in tumor germ cells of gonadoblastoma specimens. Expression also correlates with testicular seminoma and tumorigenesis of the prostate gland
Tetratricopeptide repeat domain 37	TTC37	Disease: trichohepatoenteric syndrome 1 (THES1)
Topoisomerase (DNA) II binding protein 1	TOPBP1	TopBP1 mediates mutant p53 gain of function through NF-Y and p63/p73. PMID: 21930790
TPX2, microtubule-associated	TPX2	The TPX2 gene is a promising diagnostic and therapeutic target for cervical cancer. PMID: 22307108
Trafficking protein particle complex 2	TRAPPC2	Disease: spondyloepiphyseal dysplasia tarda (SEDT)

(Continued)

HGNC-approved name	Symbol	Target information
Transcription factor EB	TFEB	TFEB regulates lysosomal proteostasis. PMID: 23393155
Transducer of ERBB2, 1	TOB1	The role of antiproliferative gene Tob1 in the immune system. PMID: 25071870
Transforming, acidic coiled-coil-containing protein 2	TACC2	Transforming acidic coiled-coil-containing protein 2 (TACC2) in human breast cancer, expression pattern and clinical/prognostic relevance. PMID: 20335520
TRIO and F-actin binding protein	TRIOBP	Disease: deafness, autosomal recessive, 28 (DFNB28)
Tumor protein p53	TP53	**Cenersen** – antineoplastic agents. Note: TP53 is found in increased amounts in a wide variety of transformed cells. TP53 is frequently mutated or inactivated in about 60% of cancers
Tumor protein p53-inducible nuclear protein 1	TP53INP1	TP53INP1 as new therapeutic target in castration-resistant prostate cancer. PMID: 22213058
Twist family bHLH transcription factor 2	TWIST2	Disease: focal facial dermal dysplasia 3, Setleis type (FFDD3)
Ubiquilin 2	UBQLN2	Disease: amyotrophic lateral sclerosis 15, with or without frontotemporal dementia (ALS15)
UPF3 regulator of nonsense transcripts homolog B (yeast)	UPF3B	Disease: mental retardation, X-linked, syndromic, 14 (MRXS14)
Upregulator of cell proliferation	URGCP	RNA interference-mediated URG4 gene silencing diminishes cyclin D1 mRNA expression in HepG2 cells. PMID: 20714998
v-ets avian erythroblastosis virus E26 oncogene homolog	ERG	Disease: Ewing sarcoma (ES)
v-rel avian reticuloendotheliosis viral oncogene homolog A	RELA	C11orf95-RELA fusions drive oncogenic NF-κB signalling in ependymoma. PMID: 24553141
WD repeat domain 62	WDR62	Disease: microcephaly 2, primary, autosomal recessive, with or without cortical malformations (MCPH2)
WW domain containing transcription regulator 1	WWTR1	TAZ expression as a prognostic indicator in colorectal cancer. PMID: 23372686
Xeroderma pigmentosum, complementation group C	XPC	Disease: xeroderma pigmentosum complementation group C (XP-C)
XIAP-associated factor 1	XAF1	XAF1 as a prognostic biomarker and therapeutic target in squamous cell lung cancer. PMID: 22088514
Y-box binding protein 1	YBX1	YB-1: Oncoprotein, prognostic marker and therapeutic target? PMID: 23216250
Yes-associated protein 1	YAP1	The role and clinical significance of YES-associated protein 1 in human osteosarcoma. PMID: 23527718
YY1-associated protein 1	YY1AP1	Genetic and pharmacological disruption of the TEAD-YAP complex suppresses the oncogenic activity of YAP. PMID: 22677547
ZFP36 ring finger protein-like 2	ZFP36L2	Note: Defects in ZFP36L2 may be a cause of leukemias. Frameshifts mutations disrupting ZFP36L2 have been found in a patient with acute myeloid leukemia. PMID: 21109922
Zinc finger protein 703	ZNF703	ZNF703 promotes tumor cell proliferation and invasion and predicts poor prognosis in patients with colorectal cancer. PMID: 25017610. Note: Luminal B breast cancers are the clinically more aggressive estrogen receptor-positive tumors. Amplification of a distal 8p12 locus occurs in around one third of the cases and ZNF703 is the single gene within the minimal amplicon
Zinc finger, C4H2 domain containing	ZC4H2	Disease: Wieacker–Wolff syndrome (WRWF)

6.4   Nucleus

HGNC-approved name	Symbol	Target information
Activating transcription factor 1	ATF1	Disease: angiomatoid fibrous histiocytoma (AFH)
Activating transcription factor 3	ATF3	Targeting activating transcription factor 3 by galectin-9 induces apoptosis and overcomes various types of treatment resistance in chronic myelogenous leukemia. PMID: 20571063
Additional sex combs-like 1 (*Drosophila*)	ASXL1	Disease: Bohring–Opitz syndrome (BOPS)
Additional sex combs-like 2 (*Drosophila*)	ASXL2	Note: A chromosomal aberration involving ASXL2 is a cause of therapy-related myelodysplastic syndrome. Translocation t(28) (p23p11.2) with KAT6A generates a KAT6A-ASXL2 fusion protein
Additional sex combs-like 3 (*Drosophila*)	ASXL3	Disease: Bainbridge–Ropers syndrome (BRPS)
AF4/FMR2 family, member 1	AFF1	Note: A chromosomal aberration involving AFF1 is associated with acute leukemias. Translocation t(411)(q21q23) with KMT2A/MLL1. The result is a rogue activator protein
AF4/FMR2 family, member 2	AFF2	Disease: mental retardation, X-linked, associated with fragile site FRAXE (MRFRAXE)
AF4/FMR2 family, member 4	AFF4	Note: A chromosomal aberration involving AFF4 is found in acute lymphoblastic leukemia (ALL). Insertion ins(511)(q31q13q23) that forms a KMT2A/MLL1-AFF4 fusion protein
ALX homeobox 1	ALX1	Disease: frontonasal dysplasia 3 (FND3)
ALX homeobox 3	ALX3	Disease: frontonasal dysplasia 1 (FND1)
ALX homeobox 4	ALX4	Disease: parietal foramina 2 (PFM2)
Ankyrin repeat domain 1 (cardiac muscle)	ANKRD1	Disease: total anomalous pulmonary venous return (TAPVR)
Ankyrin repeat domain 11	ANKRD11	Disease: KBG syndrome (KBGS)
AT-rich interactive domain 1A (SWI-like)	ARID1A	Disease: mental retardation, autosomal dominant 14 (MRD14)
AT-rich interactive domain 1B (SWI1-like)	ARID1B	ARID1B is a specific vulnerability in ARID1A-mutant cancers. PMID: 24562383. Disease: mental retardation, autosomal dominant 12 (MRD12)
AT-rich interactive domain 5B (MRF1-like)	ARID5B	Note: Defects in ARID5B may be a cause of susceptibility to coronary atherosclerosis in the Japanese population. Disease: leukemia, acute lymphoblastic (ALL)
Ataxin 7	ATXN7	Disease: spinocerebellar ataxia 7 (SCA7)
Ataxin 8	ATXN8	Disease: spinocerebellar ataxia 8 (SCA8)
Atonal homolog 1 (*Drosophila*)	ATOH1	Atonal homolog 1 expression in lung cancer correlates with inhibitors of the Wnt pathway as well as the differentiation and primary tumor stage. PMID: 23030416
Atonal homolog 7 (*Drosophila*)	ATOH7	Disease: retinal non-attachment, congenital, non-syndromic (RNANC)
B-cell CLL/lymphoma 6	BCL6	Note: Chromosomal aberrations involving BCL6 are a cause of B-cell non-Hodgkin lymphomas (B-cell NHL), including diffuse large B-cell lymphoma and follicular lymphoma
B-cell CLL/lymphoma 9	BCL9	Note: A chromosomal aberration involving BCL9 is found in a patient with precursor B-cell acute lymphoblastic leukemia (ALL). Translocation t(114) (q21q32)
Bromodomain adjacent to zinc finger domain, 1B	BAZ1B	Note: BAZ1B is located in the Williams–Beuren syndrome (WBS) critical region

(Continued)

HGNC-approved name	Symbol	Target information
Bromodomain adjacent to zinc finger domain, 2B	BAZ2B	Targeting low-druggability bromodomains: Fragment based screening and inhibitor design against the BAZ2B bromodomain. PMID: 24304323
Bromodomain containing 2	BRD2	**GSK525762** – antineoplastic agents
Bromodomain containing 3	BRD3	**GSK525762** – antineoplastic agents. Note: A chromosomal aberration involving BRD3 is found in a rare, aggressive, and lethal carcinoma arising in midline organs of young people
Bromodomain containing 4	BRD4	RNAi screen identifies Brd4 as a therapeutic target in acute myeloid leukaemia. PMID: 21814200. Note: A chromosomal aberration involving BRD4 is found in a rare, aggressive, and lethal carcinoma arising in midline organs of young people
Bromodomain containing 7	BRD7	Targeting the anaphase-promoting complex/cyclosome (APC/C)-bromodomain containing 7 (BRD7) pathway for human osteosarcoma. PMID: 24840027
Centromere protein A	CENPA	Short hairpin RNA-mediated down-regulation of CENP-A attenuates the aggressive phenotype of lung adenocarcinoma cells. PMID: 25228009
Centromere protein E, 312 kDa	CENPE	**GSK923295** – antineoplastic agents
Centromere protein K	CENPK	Note: Chromosomal aberrations involving CENPK are a cause of acute leukemias. Translocation t(511)(q12q23) with KMT2A/MLL1
Chromobox homolog 1	CBX1	HP1β suppresses metastasis of human cancer cells by decreasing the expression and activation of MMP2. PMID: 25201136
Chromobox homolog 2	CBX2	Disease: 46,XY sex reversal 5 (SRXY5)
Chromobox homolog 3	CBX3	Overexpression of HP1γ is associated with poor prognosis in non-small cell lung cancer cell through promoting cell survival. PMID: 24981246
Cullin 1	CUL1	Cullin1 is a novel marker of poor prognosis and a potential therapeutic target in human breast cancer. PMID: 23592700
Cullin 3	CUL3	Disease: pseudohypoaldosteronism 2E (PHA2E)
Cullin 4A	CUL4A	Oncogenic CUL4A determines the response to thalidomide treatment in prostate cancer. PMID: 22422151
Cullin 4B	CUL4B	Disease: mental retardation, X-linked, syndromic, 15 (MRXS15)
Damage-specific DNA binding protein 1, 127 kDa	DDB1	Structure of the DDB1-CRBN E3 ubiquitin ligase in complex with thalidomide. PMID: 25043012
Damage-specific DNA binding protein 2, 48 kDa	DDB2	DDB2: A novel regulator of NF-κB and breast tumor invasion. PMID: 23774208. Disease: xeroderma pigmentosum complementation group E (XP-E)
Deoxynucleotidyltransferase, terminal, interacting protein 1	DNTTIP1	TdIF1 recognizes a specific DNA sequence through its helix-turn-helix and AT-hook motifs to regulate gene transcription. PMID: 23874396
Deoxynucleotidyltransferase, terminal, interacting protein 2	DNTTIP2	TdIF2 is a nucleolar protein that promotes rRNA gene promoter activity. PMID: 21668587
Distal-less homeobox 3	DLX3	Disease: trichodontoosseous syndrome (TDO)
Distal-less homeobox 5	DLX5	DLX5 (distal-less homeobox 5) promotes tumor cell proliferation by transcriptionally regulating MYC. PMID: 19497851. Disease: split-hand/foot malformation 1 with sensorineural hearing loss, autosomal recessive (SHFM1D)
E2F transcription factor 1	E2F1	E2F1: A potential therapeutic target for systematic lupus erythematosus. PMID: 24071937
E2F transcription factor 3	E2F3	Curtailing overexpression of E2F3 in breast cancer using siRNA (E2F3)-based gene silencing. PMID: 22960857

HGNC-approved name	Symbol	Target information
Early growth response 1	EGR1	An early response transcription factor, Egr-1, enhances insulin resistance in type 2 diabetes with chronic hyperinsulinism. PMID: 21321112
Early growth response 2	EGR2	Disease: neuropathy, congenital hypomyelinating or amyelinating (CHN)
ets variant 1	ETV1	Disease: Ewing sarcoma (ES)
ets variant 6	ETV6	Note: A chromosomal aberration involving ETV6 is found in a form of chronic myelomonocytic leukemia (CMML)
Fanconi anemia, complementation group B	FANCB	Disease: Fanconi anemia complementation group B (FANCB)
Fanconi anemia, complementation group D2	FANCD2	FANCD2 is a potential therapeutic target and biomarker in alveolar rhabdomyosarcoma harboring the PAX3-FOXO1 fusion gene. PMID: 24787670. Disease: Fanconi anemia complementation group D2 (FANCD2)
Fanconi anemia, complementation group E	FANCE	Disease: Fanconi anemia complementation group E (FANCE)
Fanconi anemia, complementation group F	FANCF	RNA interference-mediated FANCF silencing sensitizes OVCAR3 ovarian cancer cells to adriamycin through increased adriamycin-induced apoptosis dependent on JNK activation. PMID: 23440494. Disease: Fanconi anemia complementation group F (FANCF)
Fanconi anemia, complementation group I	FANCI	Disease: Fanconi anemia complementation group I (FANCI)
F-box protein 5	FBXO5	Expression and clinicopathological significance of Emi1 in breast carcinoma. PMID: 25312659
F-box protein 11	FBXO11	Note: Defects in FBXO11 may be a cause of diffuse large B-cell lymphoma by allowing the accumulation of BCL6, an oncoprotein that has a critical role in lymphomas (PMID: 22113614)
F-box protein 25	FBXO25	Note: A chromosomal aberration involving FBXO25 is a cause of X-linked mental retardation (XLMR). Translocation t(X8)(p11.22p23.3) with SHROOM4
Forkhead box A1	FOXA1	FOXA1 is a potential oncogene in anaplastic thyroid carcinoma. PMID: 19470727
Forkhead box C1	FOXC1	FOXC1 contributes to microvascular invasion in primary hepatocellular carcinoma via regulating epithelial-mesenchymal transition. PMID: 22991501. Disease: Axenfeld–Rieger syndrome 3 (RIEG3)
Forkhead box C2 (MFH-1, mesenchyme forkhead 1)	FOXC2	Disease: lymphedema, hereditary, 2 (LMPH2)
Forkhead box D3	FOXD3	Disease: autoimmune disease 1 (AIS1)
Forkhead box E1 (thyroid transcription factor 2)	FOXE1	Expression and clinical significance of FOXE1 in papillary thyroid carcinoma. PMID: 23715628. Disease: Bamforth-Lazarus syndrome (BLS)
Forkhead box E3	FOXE3	Disease: anterior segment mesenchymal dysgenesis (ASMD)
Forkhead box F1	FOXF1	Disease: alveolar capillary dysplasia with misalignment of pulmonary veins (ACDMPV)
Forkhead box G1	FOXG1	Disease: Rett syndrome, congenital variant (RTTCV)
Forkhead box L1	FOXL1	Forkhead box L1 is frequently downregulated in gallbladder cancer and inhibits cell growth through apoptosis induction by mitochondrial dysfunction. PMID: 25010679
Forkhead box L2	FOXL2	Disease: blepharophimosis, ptosis, and epicanthus inversus syndrome (BPES)
Forkhead box M1	FOXM1	Targeting FOXM1 in cancer. PMID: 23103567
Forkhead box N1	FOXN1	Disease: T-cell immunodeficiency, congenital alopecia, and nail dystrophy (TIDAND)

(Continued)

HGNC-approved name	Symbol	Target information
Forkhead box P1	FOXP1	FOXP1 acts through a negative feedback loop to suppress FOXO-induced apoptosis. PMID: 23832113. Note: A chromosomal aberration involving FOXP1 is found in acute lymphoblastic leukemia. Disease: mental retardation (MRLIAF)
Forkhead box P2	FOXP2	Disease: speech–language disorder 1 (SPCH1)
Forkhead box P3	FOXP3	Disease: immunodeficiency polyendocrinopathy, enteropathy, X-linked syndrome (IPEX)
GATA binding protein 1 (globin transcription factor 1)	GATA1	Disease: X-linked dyserythropoietic anemia and thrombocytopenia (XDAT)
GATA binding protein 2	GATA2	Disease: dendritic cell monocyte lymphocyte B and natural killer lymphocyte deficiency (DCML)
GATA binding protein 3	GATA3	Linking GATA-3 and interleukin-13: implications in asthma. PMID: 24363163. Disease: hypoparathyroidism, sensorineural deafness, and renal disease (HDR)
GATA binding protein 4	GATA4	Disease: atrial septal defect 2 (ASD2)
GATA binding protein 5	GATA5	Note: Rare variants in GATA5 may be a cause of susceptibility to atrial fibrillation, a common sustained cardiac rhythm disturbance
GATA binding protein 6	GATA6	Disease: conotruncal heart malformations (CTHM)
GATA zinc finger domain containing 1	GATAD1	Disease: cardiomyopathy, dilated 2B (CMD2B)
GATA zinc finger domain containing 2B	GATAD2B	Disease: mental retardation, autosomal dominant 18 (MRD18)
GLIS family zinc finger 2	GLIS2	Disease: nephronophthisis 7 (NPHP7)
GLIS family zinc finger 3	GLIS3	Disease: diabetes mellitus, neonatal, with congenital hypothyroidism (NDH)
Growth factor-independent 1 transcription repressor	GFI1	Enhancer hijacking activates GFI1 family oncogenes in medulloblastoma. PMID: 25043047. Disease: neutropenia, severe congenital 2, autosomal dominant (SCN2)
Growth factor-independent 1B transcription repressor	GFI1B	Enhancer hijacking activates GFI1 family oncogenes in medulloblastoma. PMID: 25043047. Disease: neutropenia, severe congenital 2, autosomal dominant (SCN2)
GTF2I repeat domain containing 1	GTF2IRD1	Note: GTF2IRD1 is located in the Williams–Beuren syndrome (WBS) critical region
GTF2I repeat domain containing 2	GTF2IRD2	Note: GTF2IRD2 is located in the Williams–Beuren syndrome (WBS) critical region
GTF2I repeat domain containing 2B	GTF2IRD2B	Note: GTF2IRD2B is located in the Williams–Beuren syndrome (WBS) critical region
H3 histone, family 3A	H3F3A	Use of human embryonic stem cells to model pediatric gliomas with H3.3K27M histone mutation. PMID: 25525250
H3 histone, family 3B (H3.3B)	H3F3B	Citrullinated histone H3: A novel target for the treatment of sepsis. PMID: 24957671
Heterogeneous nuclear ribonucleoprotein A2/B1	HNRNPA2B1	Disease: inclusion body myopathy with early-onset Paget disease with or without frontotemporal dementia 2 (IBMPFD2)
Heterogeneous nuclear ribonucleoprotein D (AU-rich element RNA binding protein 1, 37 kDa)	HNRNPD	Post-transcriptional regulation of CD83 expression by AUF1 proteins. PMID: 23161671
High mobility group box 1	HMGB1	Orientin inhibits HMGB1-induced inflammatory responses in HUVECs and in murine polymicrobial sepsis. PMID: 24771074
High mobility group box 3	HMGB3	Disease: microphthalmia, syndromic, 13 (MCOPS13)

HGNC-approved name	Symbol	Target information
Histone cluster 4, H4	HIST4H4	Note: Chromosomal aberrations involving HISTONE H4 is a cause of B-cell non-Hodgkin lymphomas (B-cell NHL). Translocation t(36)(q27p21), with BCL6
Histone cluster 1, H4a	HIST1H4A	Note: Chromosomal aberrations involving HISTONE H4 is a cause of B-cell non-Hodgkin lymphomas (B-cell NHL). Translocation t(36)(q27p21), with BCL6
Histone cluster 1, H4b	HIST1H4B	Note: Chromosomal aberrations involving HISTONE H4 is a cause of B-cell non-Hodgkin lymphomas (B-cell NHL). Translocation t(36)(q27p21), with BCL6
Histone cluster 1, H4c	HIST1H4C	Note: Chromosomal aberrations involving HISTONE H4 is a cause of B-cell non-Hodgkin lymphomas (B-cell NHL). Translocation t(36)(q27p21), with BCL6
Histone cluster 1, H4d	HIST1H4D	Note: Chromosomal aberrations involving HISTONE H4 is a cause of B-cell non-Hodgkin lymphomas (B-cell NHL). Translocation t(36)(q27p21), with BCL6
Histone cluster 1, H4e	HIST1H4E	Note: Chromosomal aberrations involving HISTONE H4 is a cause of B-cell non-Hodgkin lymphomas (B-cell NHL). Translocation t(36)(q27p21), with BCL6
Histone cluster 1, H4f	HIST1H4F	Note: Chromosomal aberrations involving HISTONE H4 is a cause of B-cell non-Hodgkin lymphomas (B-cell NHL). Translocation t(36)(q27p21), with BCL6
Histone cluster 1, H4h	HIST1H4H	Note: Chromosomal aberrations involving HISTONE H4 is a cause of B-cell non-Hodgkin lymphomas (B-cell NHL). Translocation t(36)(q27p21), with BCL6
Histone cluster 1, H4i	HIST1H4I	Note: Chromosomal aberrations involving HISTONE H4 is a cause of B-cell non-Hodgkin lymphomas (B-cell NHL). Translocation t(36)(q27p21), with BCL6
Histone cluster 1, H4j	HIST1H4J	Note: Chromosomal aberrations involving HISTONE H4 is a cause of B-cell non-Hodgkin lymphomas (B-cell NHL). Translocation t(36)(q27p21), with BCL6
Histone cluster 1, H4k	HIST1H4K	Note: Chromosomal aberrations involving HISTONE H4 is a cause of B-cell non-Hodgkin lymphomas (B-cell NHL). Translocation t(36)(q27p21), with BCL6
Histone cluster 1, H4l	HIST1H4L	Note: Chromosomal aberrations involving HISTONE H4 is a cause of B-cell non-Hodgkin lymphomas (B-cell NHL). Translocation t(36)(q27p21), with BCL6
Histone cluster 2, H4a	HIST2H4A	Note: Chromosomal aberrations involving HISTONE H4 is a cause of B-cell non-Hodgkin lymphomas (B-cell NHL). Translocation t(36)(q27p21), with BCL6
Histone cluster 2, H4b	HIST2H4B	Note: Chromosomal aberrations involving HISTONE H4 is a cause of B-cell non-Hodgkin lymphomas (B-cell NHL). Translocation t(36)(q27p21), with BCL6
HNF1 homeobox A	HNF1A	Disease: hepatic adenomas familial (HEPAF)
HNF1 homeobox B	HNF1B	Downregulation of HNF-1B in renal cell carcinoma is associated with tumor progression and poor prognosis. PMID: 20538322. Disease: renal cysts and diabetes syndrome (RCAD)
Homeobox A1	HOXA1	Disease: Athabaskan brainstem dysgenesis syndrome (ABDS)
Homeobox A2	HOXA2	Disease: microtia, hearing impairment, and cleft palate (MHICP)

(Continued)

HGNC-approved name	Symbol	Target information
Homeobox A9	HOXA9	Peptide-based inhibition of the HOXA9/PBX interaction retards the growth of human meningioma. PMID: 24141373. Note: A chromosomal aberration involving HOXA9 is found in a form of acute myeloid leukemia. Translocation t(711)(p15p15) with NUP98
Homeobox A11	HOXA11	Disease: radioulnar synostosis with amegakaryocytic thrombocytopenia (RSAT)
Homeobox A13	HOXA13	Disease: hand–foot–genital syndrome (HFG)
Homeobox B1	HOXB1	Disease: facial paresis, hereditary congenital, 3 (HCFP3)
Homeobox B3	HOXB3	HoxB3 promotes prostate cancer cell progression by transactivating CDCA3. PMID: 23219899
Homeobox C5	HOXC5	Homeobox C5 expression is associated with the progression of 4-nitroquinoline 1-oxide-induced rat tongue carcinogenesis. PMID: 22385119
Homeobox C13	HOXC13	Disease: ectodermal dysplasia 9, hair/nail type (ECTD9)
Homeobox D9	HOXD9	Functional analysis of HOXD9 in human gliomas and glioma cancer stem cells. PMID: 21600039
Homeobox D10	HOXD10	Disease: vertical talus, congenital (CVT)
Homeobox D13	HOXD13	Disease: synpolydactyly 1 (SPD1)
Importin 4	IPO4	CCAAT/enhancer binding protein delta (C/EBPdelta, CEBPD)-mediated nuclear import of FANCD2 by IPO4 augments cellular response to DNA damage. PMID: 20805509
Importin 5	IPO5	Karyopherin beta3: A new cellular target for the HPV-16 E5 oncoprotein. PMID: 18455505
Importin 7	IPO7	Importin-7 mediates glucocorticoid receptor nuclear import and is impaired by oxidative stress, leading to glucocorticoid insensitivity. PMID: 23934279
Importin 8	IPO8	Importin 8 regulates the transport of mature microRNAs into the cell nucleus. PMID: 24596094
Importin 9	IPO9	Novel role for molecular transporter importin 9 in posttranscriptional regulation of IFN-ε expression. PMID: 23851686
Importin 11	IPO11	The ubiquitin conjugating enzyme, UBE2E3 and its import receptor, importin-11 regulate the localization and activity of the anti-oxidant transcription factor NRF2. PMID: 25378586
Importin 13	IPO13	Importin 13 regulates nuclear import of the glucocorticoid receptor in airway epithelial cells. PMID: 16809634
Inhibitor of DNA binding 3, dominant negative helix-loop-helix protein	ID3	Burkitt lymphoma pathogenesis and therapeutic targets from structural and functional genomics. PMID: 22885699
Inhibitor of DNA binding 4, dominant negative helix-loop-helix protein	ID4	ID4 methylation predicts high risk of leukemic transformation in patients with myelodysplastic syndrome. PMID: 19853913
Inhibitor of growth family, member 1	ING1	Disease: squamous cell carcinoma of the head and neck (HNSCC)
Inhibitor of growth family, member 3	ING3	Expression and prognostic value of ING3 in human primary hepatocellular carcinoma. PMID: 22550337. Disease: squamous cell carcinoma of the head and neck (HNSCC)
Inhibitor of growth family, member 5	ING5	Decreased nuclear expression and increased cytoplasmic expression of ING5 may be linked to tumorigenesis and progression in human head and neck squamous cell carcinoma. PMID: 20182888
Interferon regulatory factor 4	IRF4	IRF4 and its regulators: Evolving insights into the pathogenesis of inflammatory arthritis? PMID: 20192994. Disease: multiple myeloma (MM)
Interferon regulatory factor 8	IRF8	Constitutive IRF8 expression inhibits AML by activation of repressed immune response signaling. PMID: 24957708. Disease: IRF8 deficiency, autosomal dominant (IRF8DD)
Interleukin enhancer-binding factor 2	ILF2	Expression of NF45 correlates with malignant grade in gliomas and plays a pivotal role in tumor growth. PMID: 25023405

HGNC-approved name	Symbol	Target information
Iroquois homeobox 3	IRX3	Obesity-associated variants within FTO form long-range functional connections with IRX3. PMID: 24646999
Iroquois homeobox 5	IRX5	Disease: Hamamy syndrome (HMMS)
Karyopherin alpha 1 (importin alpha 5)	KPNA1	Targeting nuclear import shuttles, importins/karyopherins alpha by a peptide mimicking the NFκB1/p50 nuclear localization sequence. PMID: 24042087
Karyopherin alpha 2 (RAG cohort 1, importin alpha 1)	KPNA2	Nuclear karyopherin-α2 expression in primary lesions and metastatic lymph nodes was associated with poor prognosis and progression in gastric cancer. PMID: 23749771
Karyopherin alpha 3 (importin alpha 4)	KPNA3	The nuclear import factor importin α4 can protect against oxidative stress. PMID: 23773962
Karyopherin alpha 4 (importin alpha 3)	KPNA4	Identification of critical motifs within HIV-1 integrase required for importin α3 interaction and viral cDNA nuclear import. PMID: 21763491
Karyopherin alpha 5 (importin alpha 6)	KPNA5	Ebola virus VP24 targets a unique NLS binding site on karyopherin alpha 5 to selectively compete with nuclear import of phosphorylated STAT1. PMID: 25121748
Karyopherin alpha 6 (importin alpha 7)	KPNA6	Importin-α7 is required for enhanced influenza A virus replication in the alveolar epithelium and severe lung damage in mice. PMID: 24829333
Karyopherin alpha 7 (importin alpha 8)	KPNA7	KPNA7, a nuclear transport receptor, promotes malignant properties of pancreatic cancer cells *in vitro*. PMID: 24275456
Karyopherin (importin) beta 1	KPNB1	Inhibition of the nuclear transporter, Kpnβ1, results in prolonged mitotic arrest and activation of the intrinsic apoptotic pathway in cervical cancer cells. PMID: 24390070
Kruppel-like factor 1 (erythroid)	KLF1	Disease: anemia, congenital dyserythropoietic, 4 (CDAN4)
Kruppel-like factor 2	KLF2	KLF2 is a rate-limiting transcription factor that can be targeted to enhance regulatory T-cell production. PMID: 24979767
Kruppel-like factor 4 (gut)	KLF4	KLF4 regulates abdominal aortic aneurysm morphology and deletion attenuates aneurysm formation. PMID: 24030402
Kruppel-like factor 5 (intestinal)	KLF5	Essential role of KLF5 transcription factor in cell proliferation and differentiation and its implications for human diseases. PMID: 19448973
Kruppel-like factor 6	KLF6	Expression and significance of Kruppel-like factor 6 gene in osteosarcoma. PMID: 22855058. Disease: gastric cancer (GASC)
Kruppel-like factor 8	KLF8	Krüppel-like factor 8 overexpression is correlated with angiogenesis and poor prognosis in gastric cancer. PMID: 23805141
Kruppel-like factor 11	KLF11	Disease: maturity-onset diabetes of the young 7 (MODY7)
Kruppel-like factor 15	KLF15	Note: KLF15 deficiency results in loss of rhythmic QT variation and abnormal heart repolarization (PMID: 22367544)
Kruppel-like factor 17	KLF17	KLF17 is a negative regulator of epithelial-mesenchymal transition and metastasis in breast cancer. PMID: 19801974
l(3)mbt-like 1 (*Drosophila*)	L3MBTL1	Small-molecule ligands of methyl-lysine binding proteins. PMID: 21417280
l(3)mbt-like 3 (*Drosophila*)	L3MBTL3	Small-molecule ligands of methyl-lysine binding proteins: Optimization of selectivity for L3MBTL3. PMID: 24040942
Lamin A/C	LMNA	Disease: Emery–Dreifuss muscular dystrophy 2, autosomal dominant (EDMD2)
Lamin B receptor	LBR	Disease: Pelger–Huet anomaly (PHA)
Lamin B1	LMNB1	Disease: leukodystrophy, demyelinating, autosomal dominant, adult-onset (ADLD)
Lamin B2	LMNB2	Disease: partial acquired lipodystrophy (APLD)

(Continued)

HGNC-approved name	Symbol	Target information
LIM domain only 1 (rhombotin 1)	LMO1	Note: A chromosomal aberration involving LMO1 may be a cause of a form of T-cell acute lymphoblastic leukemia (T-ALL). Translocation t(11,14) (p15q11) with TCRD
LIM domain only 2 (rhombotin like 1)	LMO2	Note: A chromosomal aberration involving LMO2 may be a cause of a form of T-cell acute lymphoblastic leukemia (T-ALL). Translocation t(11,14) (p13q11) with TCRD
LIM homeobox 3	LHX3	Disease: pituitary hormone deficiency, combined, 3 (CPHD3)
LIM homeobox 4	LHX4	Disease: pituitary hormone deficiency, combined, 4 (CPHD4)
Mastermind-like domain containing 1	MAMLD1	Disease: hypospadias 2, X-linked (HYSP2)
Mastermind-like 2 (Drosophila)	MAML2	Note: A chromosomal aberration involving MAML2 is found in mucoepidermoid carcinomas, benign Warthin tumors and clear cell hidradenomas. Translocation t(1119)(q21p13) with CRTC1/MECT1
Mediator complex subunit 12	MED12	MED12 controls the response to multiple cancer drugs through regulation of TGF-β receptor signaling. PMID: 23178117. Disease: Opitz–Kaveggia syndrome (OKS)
Mediator complex subunit 13-like	MED13L	Disease: transposition of the great arteries, dextro-looped 1 (DTGA1)
Mediator complex subunit 17	MED17	Disease: microcephaly, postnatal progressive, with seizures and brain atrophy (MCPHSBA)
Mediator complex subunit 19	MED19	Knockdown of MED19 by lentivirus-mediated shRNA in human osteosarcoma cells inhibits cell proliferation by inducing cell cycle arrest in the G0/G1 phase. PMID: 21542455
Mediator complex subunit 23	MED23	Downregulation of MED23 promoted the tumorigenecity of esophageal squamous cell carcinoma. PMID: 23625751. Disease: mental retardation, autosomal recessive 18 (MRT18)
Mediator complex subunit 25	MED25	Disease: Charcot–Marie–Tooth disease 2B2 (CMT2B2)
Methyl-CpG binding domain protein 1	MBD1	Up-regulation of MBD1 promotes pancreatic cancer cell epithelial-mesenchymal transition and invasion by epigenetic down-regulation of E-cadherin. PMID: 23331011
Methyl CpG binding protein 2 (Rett syndrome)	MECP2	Disease: Angelman syndrome (AS)
Methyl-CpG binding domain protein 5	MBD5	Disease: Mental retardation, autosomal dominant 1 (MRD1)
Msh homeobox 1	MSX1	Disease: Tooth agenesis selective 1 (STHAG1)
Msh homeobox 2	MSX2	Disease: Parietal foramina 1 (PFM1)
MutL homolog 1	MLH1	Disease: Hereditary non-polyposis colorectal cancer 2 (HNPCC2)
MutL homolog 3	MLH3	Disease: Hereditary non-polyposis colorectal cancer 7 (HNPCC7)
MutS homolog 2	MSH2	Disease: hereditary non-polyposis colorectal cancer 1 (HNPCC1)
MutS homolog 3	MSH3	DNA mismatch repair complex MutSβ promotes GAA·TTC repeat expansion in human cells. PMID: 22787155. Disease: endometrial cancer (ENDMC)
MutS homolog 4	MSH4	MutS homologue hMSH4: Interaction with eIF3f and a role in NHEJ-mediated DSB repair. PMID: 23725059
MutS homolog 5	MSH5	MutS homologues hMSH4 and hMSH5: Genetic variations, functions, and implications in human diseases. PMID: 24082819
MutS homolog 6	MSH6	Disease: hereditary non-polyposis colorectal cancer 5 (HNPCC5)
Myeloid/lymphoid or mixed-lineage leukemia (trithorax homolog, Drosophila); translocated to, 3	MLLT3	Note: A chromosomal aberration involving MLLT3 is associated with acute leukemias. Translocation t(911)(p22q23) with KMT2A/MLL1. The result is a rogue activator protein

HGNC-approved name	Symbol	Target information
Myeloid/lymphoid or mixed-lineage leukemia (trithorax homolog, *Drosophila*); translocated to, 6	MLLT6	Note: A chromosomal aberration involving MLLT6 is associated with acute leukemias. Translocation t(1117)(q23q21) with KMT2A/MLL1. The result is a rogue activator protein
Myeloid/lymphoid or mixed-lineage leukemia (trithorax homolog, *Drosophila*); translocated to, 10	MLLT10	Note: A chromosomal aberration involving MLLT10 is associated with acute leukemias. Translocation t(1011)(p12q23) with KMT2A/MLL1. The result is a rogue activator protein
Myocyte enhancer factor 2A	MEF2A	Disease: coronary artery disease, autosomal dominant, 1 (ADCAD1)
Myocyte enhancer factor 2C	MEF2C	Disease: mental retardation, autosomal dominant 20 (MRD20)
Neuronal PAS domain protein 3	NPAS3	Note: A chromosomal aberration involving NPAS3 is found in a family with schizophrenia. Translocation t(914)(q34q13)
Neuronal PAS domain protein 4	NPAS4	Npas4 is a novel activity-regulated cytoprotective factor in pancreatic β-cells. PMID: 23656887
NK2 homeobox 1	NKX2-1	Disease: chorea, hereditary benign (BHC)
NK2 homeobox 5	NKX2-5	Disease: atrial septal defect 7, with or without atrioventricular conduction defects (ASD7)
NK2 homeobox 6	NKX2-6	Disease: conotruncal heart malformations (CTHM)
NK3 homeobox 2	NKX3-2	Disease: spondylo-megaepiphyseal-metaphyseal dysplasia (SMMD)
NOP10 ribonucleoprotein	NOP10	Disease: dyskeratosis congenita, autosomal recessive, 1 (DKCB1)
NOP56 ribonucleoprotein	NOP56	Disease: spinocerebellar ataxia 36 (SCA36)
Nucleoporin 62 kDa	NUP62	Disease: infantile striatonigral degeneration (SNDI)
Nucleoporin 98 kDa	NUP98	Nucleoporin Nup98 mediates galectin-3 nuclear-cytoplasmic trafficking. PMID: 23541576. Note: A chromosomal aberration involving NUP98 is found in a form of acute myeloid leukemia. Translocation t(711)(p15p15) with HOXA9
Nucleoporin 153 kDa	NUP153	The nucleoporin 153, a novel factor in double-strand break repair and DNA damage response. PMID: 22249246
Nucleoporin 188 kDa	NUP188	Note: Copy number variations of NUP188 gene may be a cause of heterotaxy, a congenital heart disease resulting from abnormalities in left-right (LR) body patterning. PMID: 21282601
Nucleoporin 214 kDa	NUP214	Note: A chromosomal aberration involving NUP214 is found in a subset of acute myeloid leukemia (AML) also known as acute non-lymphocytic leukemia. Translocation t(69)(p23q34) with DEK. It will i in the formation of a DEK-CAN fusion gene
Origin recognition complex, subunit 1	ORC1	Disease: Meier–Gorlin syndrome 1 (MGORS1)
Origin recognition complex, subunit 4	ORC4	Disease: Meier–Gorlin syndrome 2 (MGORS2)
Origin recognition complex, subunit 6	ORC6	Disease: Meier–Gorlin syndrome 3 (MGORS3)
Ornithine decarboxylase antizyme 2	OAZ2	The polyamine metabolism genes ornithine decarboxylase and antizyme 2 predict aggressive behavior in neuroblastomas with and without MYCN amplification. PMID: 19960435
Ornithine decarboxylase antizyme 3	OAZ3	Ornithine decarboxylase antizyme Oaz3 modulates protein phosphatase activity. PMID: 21712390
Orthodenticle homeobox 1	OTX1	OTX1 promotes colorectal cancer progression through epithelial-mesenchymal transition. PMID: 24388989
Orthodenticle homeobox 2	OTX2	Disease: microphthalmia, syndromic, 5 (MCOPS5)
Paired box 1	PAX1	Disease: otofaciocervical syndrome 2 (OFC2)

(*Continued*)

HGNC-approved name	Symbol	Target information
Paired box 2	PAX2	PAX2 is an antiapoptotic molecule with deregulated expression in medulloblastoma. PMID: 22552444. Disease: renal-coloboma syndrome (RCS)
Paired box 3	PAX3	Disease: Waardenburg syndrome 1 (WS1)
Paired box 4	PAX4	Disease: diabetes mellitus, non-insulin-dependent (NIDDM)
Paired box 5	PAX5	Note: A chromosomal aberration involving PAX5 is a cause of acute lymphoblastic leukemia. Translocation t(9;18)(p13;q11.2) with ZNF521. Translocation t(9;3)(p13;p14.1) with FOXP1. Translocation t(9;12) (p13;p13) with ETV6. Disease: leukemia, acute lymphoblastic, 3 (ALL3)
Paired box 6	PAX6	Disease: aniridia (AN)
Paired box 7	PAX7	Disease: rhabdomyosarcoma 2 (RMS2)
Paired box 8	PAX8	Disease: hypothyroidism, congenital, non-goitrous, 2 (CHNG2)
Paired box 9	PAX9	Disease: tooth agenesis selective 3 (STHAG3)
Paired-related homeobox 1	PRRX1	Disease: agnathia-otocephaly complex (AGOTC)
Paired-like homeobox 2a	PHOX2A	Disease: congenital fibrosis of extraocular muscles 2 (CFEOM2)
Paired-like homeobox 2b	PHOX2B	Disease: congenital central hypoventilation syndrome (CCHS)
Paired-like homeodomain 1	PITX1	Disease: clubfoot, congenital, with or without deficiency of long bones and/or mirror-image polydactyly (CCF)
Paired-like homeodomain 2	PITX2	Disease: Axenfeld–Rieger syndrome 1 (RIEG1)
Paired-like homeodomain 3	PITX3	Disease: anterior segment mesenchymal dysgenesis (ASMD)
Polymerase (DNA directed), delta 2, accessory subunit	POLD2	POLD2 and KSP37 (FGFBP2) correlate strongly with histology, stage and outcome in ovarian carcinomas. PMID: 21079801
Polymerase (DNA directed), delta-interacting protein 2	POLDIP2	Poldip2, a novel regulator of Nox4 and cytoskeletal integrity in vascular smooth muscle cells. PMID: 19574552
Polymerase (RNA) I polypeptide C, 30 kDa	POLR1C	Disease: Treacher Collins syndrome 3 (TCS3)
Polymerase (RNA) I polypeptide D, 16 kDa	POLR1D	Disease: Treacher Collins syndrome 2 (TCS2)
Polymerase (RNA) II (DNA directed) polypeptide F	POLR2F	POLR2F, ATP6V0A1 and PRNP expression in colorectal cancer: New molecules with prognostic significance? PMID: 18505059
POU class 1 homeobox 1	POU1F1	Deregulation of the Pit-1 transcription factor in human breast cancer cells promotes tumor growth and metastasis. PMID: 21060149. Disease: pituitary hormone deficiency, combined, 1 (CPHD1)
POU class 2-associating factor 1	POU2AF1	Note: A chromosomal aberration involving POU2AF1/OBF1 may be a cause of a form of B-cell leukemia. Translocation t(311)(q27q23) with BCL6
POU class 3 homeobox 4	POU3F4	Disease: deafness, X-linked, 2 (DFNX2)
POU class 4 homeobox 3	POU4F3	Disease: deafness, autosomal dominant, 15 (DFNA15)
POU class 6 homeobox 2	POU6F2	Disease: hereditary susceptibility to Wilms tumor 5 (WT5)
PR domain containing 1, with ZNF domain	PRDM1	Role of Blimp-1 in programing Th effector cells into IL-10 producers. PMID: 25073792
PR domain containing 5	PRDM5	Disease: brittle cornea syndrome 2 (BCS2)
PR domain containing 16	PRDM16	Disease: left ventricular non-compaction 8 (LVNC8)
Pre-B-cell leukemia homeobox 1	PBX1	Note: A chromosomal aberration involving PBX1 is a cause of pre-B-cell acute lymphoblastic leukemia (B-ALL)
Pre-B-cell leukemia homeobox 3	PBX3	PBX3 is an important cofactor of HOXA9 in leukemogenesis. PMID: 23264595
Pre-mRNA processing factor 3	PRPF3	Disease: retinitis pigmentosa 18 (RP18)
Pre-mRNA processing factor 6	PRPF6	Disease: retinitis pigmentosa 60 (RP60)

HGNC-approved name	Symbol	Target information
Pre-mRNA processing factor 8	PRPF8	Disease: retinitis pigmentosa 13 (RP13)
Pre-mRNA processing factor 31	PRPF31	Disease: retinitis pigmentosa 11 (RP11)
Prickle homolog 1 (*Drosophila*)	PRICKLE1	Disease: epilepsy, progressive myoclonic 1B (EPM1B)
Prickle homolog 2 (*Drosophila*)	PRICKLE2	Disease: epilepsy, progressive myoclonic 5 (EPM5)
Protein phosphatase 1, regulatory subunit 10	PPP1R10	PNUTS functions as a proto-oncogene by sequestering PTEN. PMID: 23117887
Protein phosphatase 2, regulatory subunit B′, gamma	PPP2R5C	The protein phosphatase 2A regulatory subunit B56γ mediates suppression of T cell receptor (TCR)-induced nuclear factor-κB (NF-κB) activity. PMID: 24719332
Protein phosphatase 2, regulatory subunit B″, gamma	PPP2R3C	Protein phosphatase complex PP5/PPP2R3C dephosphorylates P-glycoprotein/ABCB1 and down-regulates the expression and function. PMID: 24333728
RAD9 homolog B (*S. pombe*)	RAD9B	Rad9B responds to nucleolar stress through ATR and JNK signalling, and delays the G1-S transition. PMID: 22399810
RAD21 homolog (*S. pombe*)	RAD21	Enhanced RAD21 cohesin expression confers poor prognosis and resistance to chemotherapy in high grade luminal, basal and HER2 breast cancers. PMID: 21255398. Disease: Cornelia de Lange syndrome 4 (CDLS4)
RAD21-like 1 (*S. pombe*)	RAD21L1	Cohesin complexes with a potential to link mammalian meiosis to cancer. PMID: 25408876
RAD50 homolog (*S. cerevisiae*)	RAD50	RAD50 targeting impairs DNA damage response and sensitizes human breast cancer cells to cisplatin therapy. PMID: 24642965. Disease: Nijmegen breakage syndrome like disorder (NBSLD)
RAD51-associated protein 1	RAD51AP1	RAD51AP1-deficiency in vertebrate cells impairs DNA replication. PMID: 25288561
RAD51-associated protein 2	RAD51AP2	RAD51AP2, a novel vertebrate- and meiotic-specific protein, shares a conserved RAD51-interacting C-terminal domain with RAD51AP1/PIR51. PMID: 16990250
Regulatory factor X, 5 (influences HLA class II expression)	RFX5	Disease: bare lymphocyte syndrome 2 (BLS2)
Regulatory factor X, 6	RFX6	Disease: Mitchell–Riley syndrome (MIRIS)
Regulatory factor X-associated ankyrin-containing protein	RFXANK	Disease: bare lymphocyte syndrome 2 (BLS2)
Regulatory factor X-associated protein	RFXAP	Disease: bare lymphocyte syndrome 2 (BLS2)
Replication factor C (activator 1) 1, 145 kDa	RFC1	Down-regulation of replication factor C-40 (RFC40) causes chromosomal missegregation in neonatal and hypertrophic adult rat cardiac myocytes. PMID: 22720015
Replication factor C (activator 1) 2, 40 kDa	RFC2	Note: RFC2 is located in the Williams–Beuren syndrome (WBS) critical region
Replication factor C (activator 1) 3, 38 kDa	RFC3	Overexpression of RFC3 is correlated with ovarian tumor development and poor prognosis. PMID: 25030735
Replication factor C (activator 1) 4, 37 kDa	RFC4	Levels of human replication factor C4, a clamp loader, correlate with tumor progression and predict the prognosis for colorectal cancer. PMID: 25407051
Retina and anterior neural fold homeobox	RAX	Disease: microphthalmia, isolated, 3 (MCOP3)
Retina and anterior neural fold homeobox 2	RAX2	Disease: macular degeneration, age-related, 6 (ARMD6)
Ribonuclease H2, subunit B	RNASEH2B	Disease: Aicardi–Goutieres syndrome 2 (AGS2)
Ribonuclease H2, subunit C	RNASEH2C	Disease: Aicardi–Goutieres syndrome 3 (AGS3)

(Continued)

HGNC-approved name	Symbol	Target information
Ribosomal protein L11	RPL11	Disease: Diamond–Blackfan anemia 7 (DBA7)
Ribosomal protein S19	RPS19	Disease: Diamond–Blackfan anemia 1 (DBA1)
RNA binding motif (RNP1, RRM) protein 3	RBM3	RBM3 mediates structural plasticity and protective effects of cooling in neurodegeneration. PMID: 25607368
RNA binding motif protein 10	RBM10	Disease: TARP syndrome (TARPS)
RNA binding motif protein 15	RBM15	Note: A chromosomal aberration involving RBM15 may be a cause of acute megakaryoblastic leukemia. Translocation t(122)(p13q13) with MKL1
RNA binding motif protein 20	RBM20	Disease: cardiomyopathy, dilated 1DD (CMD1DD)
RNA binding motif protein 28	RBM28	Disease: alopecia, neurologic defects, and endocrinopathy syndrome (ANES)
Runt-related transcription factor 1	RUNX1	A stable transcription factor complex nucleated by oligomeric AML1-ETO controls leukaemogenesis. PMID: 23812588. Note: A chromosomal aberration involving RUNX1/AML1 is a cause of M2 type acute myeloid leukemia (AML-M2)
Runt-related transcription factor 1; translocated to, 1 (cyclin D-related)	RUNX1T1	Note: A chromosomal aberration involving RUNX1T1 is a cause of acute myeloid leukemia (AML-M2). Translocation t(8;21)(q22;q22) with RUNX1/AML1. Disease: colorectal cancer (CRC)
Runt-related transcription factor 2	RUNX2	Runx2 mediates epigenetic silencing of the bone morphogenetic protein-3B (BMP-3B/GDF10) in lung cancer cells. PMID: 22537242. Disease: cleidocranial dysplasia (CLCD)
SATB homeobox 1	SATB1	Special AT-rich sequence-binding protein 1 promotes cell growth and metastasis in colorectal cancer. PMID: 23613626
SATB homeobox 2	SATB2	Note: Chromosomal aberrations involving SATB2 are found in isolated cleft palate. Translocation t(2;7); translocation t(2;11). Disease: cleft palate isolated (CPI)
SIX homeobox 3	SIX3	Disease: holoprosencephaly 2 (HPE2)
SIX homeobox 6	SIX6	Disease: microphthalmia, isolated, with cataract, 2 (MCOPCT2)
Spectrin repeat containing, nuclear envelope 1	SYNE1	Disease: spinocerebellar ataxia, autosomal recessive, 8 (SCAR8)
Spectrin repeat containing, nuclear envelope 2	SYNE2	Disease: Emery–Dreifuss muscular dystrophy 5, autosomal dominant (EDMD5)
Spectrin repeat containing, nuclear envelope family member 4	SYNE4	Disease: deafness, autosomal recessive, 76 (DFNB76)
Splicing factor 3b, subunit 1, 155 kDa	SF3B1	Spliceostatin A targets SF3b and inhibits both splicing and nuclear retention of pre-mRNA. PMID: 17643111
Splicing factor 3b, subunit 4, 49 kDa	SF3B4	Disease: acrofacial dysostosis 1, Nager type (AFD1)
SRY (sex-determining region Y)-box 2	SOX2	SOX2 overexpression correlates with poor prognosis in laryngeal squamous cell carcinoma. PMID: 23462687. Disease: microphthalmia, syndromic, 3 (MCOPS3)
SRY (sex-determining region Y)-box 3	SOX3	Disease: panhypopituitarism X-linked (PHPX)
SRY (sex-determining region Y)-box 4	SOX4	Prognostic significance of Sox4 expression in human cutaneous melanoma and its role in cell migration and invasion. PMID: 20952589
SRY (sex-determining region Y)-box 9	SOX9	Disease: campomelic dysplasia (CMD1)
SRY (sex-determining region Y)-box 11	SOX11	SOX11 promotes tumor angiogenesis through transcriptional regulation of PDGFA in mantle cell lymphoma. PMID: 25092176
SRY (sex-determining region Y)-box 17	SOX17	Sox17 regulates proliferation and cell cycle during gastric cancer progression. PMID: 21514720. Disease: vesicoureteral reflux 3 (VUR3)
SRY (sex-determining region Y)-box 18	SOX18	The lymphangiogenic factor SOX 18: A key indicator to stage gastric tumor progression. PMID: 21796627. Disease: hypotrichosis–lymphedema–telangiectasia syndrome (HLTS)

HGNC-approved name	Symbol	Target information
Structural maintenance of chromosomes 1A	SMC1A	Disease: Cornelia de Lange syndrome 2 (CDLS2)
Structural maintenance of chromosomes 3	SMC3	Disease: Cornelia de Lange syndrome 3 (CDLS3)
SWI/SNF-related, matrix-associated, actin-dependent regulator of chromatin, subfamily b, member 1	SMARCB1	Loss of SNF5 expression correlates with poor patient survival in melanoma. PMID: 19808872. Disease: rhabdoid tumor predisposition syndrome 1 (RTPS1)
SWI/SNF-related, matrix-associated, actin-dependent regulator of chromatin, subfamily e, member 1	SMARCE1	Note: Defects in SMARCE1 may be a cause of Coffin–Siris syndrome, a highly variable disease characterized by mental retardation associated with a broad spectrum of different clinical features
Synovial sarcoma translocation, chromosome 18	SS18	Note: A chromosomal aberration involving SS18 may be a cause of synovial sarcoma. Translocation t(X18)(p11.2q11.2). The translocation is specifically found in more than 80% of synovial sarcoma
Synovial sarcoma, X breakpoint 2	SSX2	Note: A chromosomal aberration involving SSX2 may be a cause of synovial sarcoma. Translocation t(X18)(p11.2q11.2). The translocation is specifically found in more than 80% of synovial sarcoma
Synovial sarcoma, X breakpoint 2B	SSX2B	Note: A chromosomal aberration involving SSX2 may be a cause of synovial sarcoma. Translocation t(X18)(p11.2q11.2). The translocation is specifically found in more than 80% of synovial sarcoma
TAF1 RNA polymerase II, TATA box binding protein (TBP)-associated factor, 250 kDa	TAF1	Biased multicomponent reactions to develop novel bromodomain inhibitors. PMID: 25314271. Disease: dystonia 3, torsion, X-linked (DYT3)
TAF2 RNA polymerase II, TATA box binding protein (TBP)-associated factor, 150 kDa	TAF2	Disease: mental retardation, autosomal recessive 40 (MRT40)
TAF15 RNA polymerase II, TATA box binding protein (TBP)-associated factor, 68 kDa	TAF15	Note: A chromosomal aberration involving TAF15/TAF2N is found in a form of extraskeletal myxoid chondrosarcomas (EMC). Translocation t(917)(q22q11) with NR4A3
T-box 1	TBX1	Note: Haploinsufficiency of the TBX1 gene is responsible for most of the physical malformations present in DiGeorge syndrome (DGS) and velocardiofacial syndrome (VCFS)
T-box 3	TBX3	Disease: ulnar–mammary syndrome (UMS)
T-box 4	TBX4	Disease: small patella syndrome (SPS)
T-box 5	TBX5	Disease: Holt–Oram syndrome (HOS)
T-box 6	TBX6	Disease: spondylocostal dysostosis 5, autosomal dominant (SCDO5)
T-box 15	TBX15	Disease: Cousin syndrome (COUSS)
T-box 19	TBX19	Disease: ACTH deficiency, isolated (IAD)
T-box 20	TBX20	Disease: atrial septal defect 4 (ASD4)
T-box 21	TBX21	The transcription factor T-bet regulates mucosal T cell activation in experimental colitis and Crohn's disease. PMID: 11994418. Disease: asthma, with nasal polyps and aspirin intolerance (ANPAI)
T-box 22	TBX22	Disease: cleft palate with or without ankyloglossia, X-linked (CPX)
TEA domain family member 1 (SV40 transcriptional enhancer factor)	TEAD1	Genetic and pharmacological disruption of the TEAD-YAP complex suppresses the oncogenic activity of YAP. PMID: 22677548. Disease: Sveinsson chorioretinal atrophy (SCRA)
TEA domain family member 2	TEAD2	Genetic and pharmacological disruption of the TEAD-YAP complex suppresses the oncogenic activity of YAP. PMID: 22677549
TEA domain family member 3	TEAD3	Genetic and pharmacological disruption of the TEAD-YAP complex suppresses the oncogenic activity of YAP. PMID: 22677550
TEA domain family member 4	TEAD4	Genetic and pharmacological disruption of the TEAD-YAP complex suppresses the oncogenic activity of YAP. PMID: 22677551

(Continued)

HGNC-approved name	Symbol	Target information
THO complex 1	THOC1	Thoc1 inhibits cell growth via induction of cell cycle arrest and apoptosis in lung cancer cells. PMID: 24682263
THO complex 6 homolog (*Drosophila*)	THOC6	Disease: Beaulieu–Boycott–Innes syndrome (BBIS)
Transcription factor 3	TCF3	Burkitt lymphoma pathogenesis and therapeutic targets from structural and functional genomics. PMID: 22885699. Note: Chromosomal aberrations involving TCF3 are cause of forms of pre-B-cell acute lymphoblastic leukemia (B-ALL). Translocation t(119)(q23p13.3).
Transcription factor 4	TCF4	Inhibition of Tcf-4 induces apoptosis and enhances chemosensitivity of colon cancer cells. PMID: 23029137. Disease: Pitt–Hopkins syndrome (PTHS)
Transcription factor 12	TCF12	Disease: craniosynostosis 3 (CRS3)
Transcription factor AP-2 alpha (activating enhancer binding protein 2 alpha)	TFAP2A	Disease: branchiooculofacial syndrome (BOFS)
Transcription factor AP-2 beta (activating enhancer binding protein 2 beta)	TFAP2B	Disease: Char syndrome (CHAR)
Transcription factor AP-4 (activating enhancer binding protein 4)	TFAP4	Molecular mechanism of activating protein-4 regulated growth of hepatocellular carcinoma. PMID: 25195135
Transportin 1	TNPO1	The nuclear import receptor Kpnβ1 and its potential as an anticancer therapeutic target. PMID: 23557333
Transportin 2	TNPO2	Transportin 2 regulates apoptosis through the RNA-binding protein HuR. PMID: 21646354
Transportin 3	TNPO3	Structural basis for nuclear import of splicing factors by human transportin 3. PMID: 24449914
Visual system homeobox 1	VSX1	Disease: corneal dystrophy, posterior polymorphous, 1 (PPCD1)
Visual system homeobox 2	VSX2	Disease: microphthalmia, isolated, 2 (MCOP2)
v-maf avian musculoaponeurotic fibrosarcoma oncogene homolog	MAF	Note: A chromosomal aberration involving MAF is found in some forms of multiple myeloma (MM). Translocation t(14;16)(q32.3;q23) with an IgH locus. Disease: cataract 21, multiple types (CTRCT21)
v-maf avian musculoaponeurotic fibrosarcoma oncogene homolog A	MAFA	Role of MafA in pancreatic beta-cells. PMID: 19393272
v-maf avian musculoaponeurotic fibrosarcoma oncogene homolog B	MAFB	Disease: multicentric carpotarsal osteolysis syndrome (MCTO)
v-myb avian myeloblastosis viral oncogene homolog	MYB	MYB suppresses differentiation and apoptosis of human breast cancer cells. PMID: 20659323
v-myc avian myelocytomatosis viral oncogene homolog	MYC	Note: Overexpression of MYC is implicated in the etiology of a variety of hematopoietic tumors. Note: A chromosomal aberration involving MYC may be a cause of a form of B-cell chronic lymphocytic leukemia
v-myc avian myelocytomatosis viral oncogene neuroblastoma derived homolog	MYCN	Note: Amplification of the N-MYC gene is associated with a variety of human tumors, most frequently neuroblastoma, where the level of amplification appears to increase as the tumor progresses. Disease: Feingold syndrome 1 (FGLDS1)
Zinc finger and BTB domain containing 7A	ZBTB7A	The oncogene LRF is a survival factor in chondrosarcoma and contributes to tumor malignancy and drug resistance. PMID: 22847180
Zinc finger and BTB domain containing 16	ZBTB16	Disease: skeletal defects, genital hypoplasia, and mental retardation (SGYMR)
Zinc finger and BTB domain containing 18	ZBTB18	Disease: mental retardation, autosomal dominant 22 (MRD22)
Zinc finger and BTB domain containing 24	ZBTB24	Disease: immunodeficiency–centromeric instability–facial anomalies syndrome 2 (ICF2)

HGNC-approved name	Symbol	Target information
Zinc finger homeobox 3	ZFHX3	ATBF1 is a novel amyloid-β protein precursor (AβPP) binding protein that affects AβPP expression. PMID: 25079792
Zinc finger homeobox 4	ZFHX4	Note: A chromosomal aberration involving ZFHX4 is found in one patient with ptosis. Translocation t(18)(p34.3q21.12)
Zinc finger protein 41	ZNF41	Note: A chromosomal aberration involving ZNF41 has been found in a patient with severe mental retardation. Translocation t(X7)(p11.3q11.21)
Zinc finger protein 81	ZNF81	Disease: mental retardation, X-linked 45 (MRX45)
Zinc finger protein 141	ZNF141	Disease: polydactyly, postaxial A6 (PAPA6)
Zinc finger protein 143	ZNF143	The combination of strong expression of ZNF143 and high MIB-1 labelling index independently predicts shorter disease-specific survival in lung adenocarcinoma. PMID: 24736586
Zinc finger protein 217	ZNF217	The transcription factor ZNF217 is a prognostic biomarker and therapeutic target during breast cancer progression. PMID: 22728437
Zinc finger protein 267	ZNF267	Increased expression of zinc finger protein 267 in non-alcoholic fatty liver disease. PMID: 22076166
Zinc finger protein 320	ZNF320	Note: A chromosomal aberration involving ZNF320 is found in a form of glioblastoma. Translocation t(1019)(q26q13.3) with BRWD2/WDR11
Zinc finger protein 335	ZNF335	Disease: microcephaly 10, primary, autosomal recessive (MCPH10)
Zinc finger protein 423	ZNF423	Restored immunosuppressive effect of mesenchymal stem cells on B cells after OAZ downregulation in patients with systemic lupus erythematosus. PMID: 25218347. Disease: nephronophthisis 14 (NPHP14)
Zinc finger protein 469	ZNF469	Disease: brittle cornea syndrome 1 (BCS1)
Zinc finger protein 513	ZNF513	Disease: retinitis pigmentosa 58 (RP58)
Zinc finger protein 521	ZNF521	Note: A chromosomal aberration involving ZNF521 is found in acute lymphoblastic leukemia. Translocation t(918)(p13q11.2) with PAX5
Zinc finger protein 592	ZNF592	Disease: spinocerebellar ataxia, autosomal recessive, 5 (SCAR5)
Zinc finger protein 644	ZNF644	Disease: myopia 21, autosomal dominant (MYP21)
Zinc finger protein 711	ZNF711	Disease: mental retardation, X-linked, ZNF711-related (MRXZ)
Zinc finger protein 750	ZNF750	Disease: seborrhea-like dermatitis with psoriasiform elements (SLDP)
Zinc finger, MYM-type 2	ZMYM2	Note: A chromosomal aberration involving ZMYM2 may be a cause of stem cell leukemia lymphoma syndrome (SCLL). Translocation t(813)(p11q12) with FGFR1
Zinc finger, MYM-type 3	ZMYM3	Note: A chromosomal aberration involving ZMYM3 may be a cause of X-linked mental retardation in Xq13.1. Translocation t(X13)(q13.13)
Zinc finger, MYND-type containing 11	ZMYND11	ZMYND11 links histone H3.3K36me3 to transcription elongation and tumour suppression. PMID: 24590075. Note: A chromosomal aberration involving ZMYND11 is a cause of acute poorly differentiated myeloid leukemia. Translocation (1017)(p15q21) with MBTD1
Zinc finger, MYND-type containing 15	ZMYND15	Disease: spermatogenic failure 14 (SPGF14)
Achaete-scute family bHLH transcription factor 2	ASCL2	Transcription factor achaete-scute homologue 2 initiates follicular T-helper-cell development. PMID: 24463518
Achalasia, adrenocortical insufficiency, alacrima	AAAS	Disease: achalasia–addisonianism–alacrima syndrome (AAAS)
Activity-dependent neuroprotector homeobox	ADNP	Protection against tauopathy by the drug candidates NAP (davunetide) and D-SAL: Biochemical, cellular and behavioral aspects. PMID: 21728979
Apoptotic chromatin condensation inducer 1	ACIN1	NGF inhibits human leukemia proliferation by downregulating cyclin A1 expression through promoting acinus/CtBP2 association. PMID: 19668232

(Continued)

HGNC-approved name	Symbol	Target information
Arginine-glutamic acid dipeptide (RE) repeats	RERE	Note: A chromosomal aberration involving RERE is found in the neuroblastoma cell line NGP. Translocation t(115)(p36.2q24)
Aristaless-related homeobox	ARX	Disease: lissencephaly, X-linked 2 (LISX2)
Aryl hydrocarbon receptor nuclear translocator 2	ARNT2	Disease: Webb–Dattani syndrome (WEDAS)
ATPase family, AAA domain containing 2B	ATAD2B	ATAD2B is a phylogenetically conserved nuclear protein expressed during neuronal differentiation and tumorigenesis. PMID: 21158754
Basic helix-loop-helix family, member a9	BHLHA9	Disease: split-hand/foot malformation with long bone deficiency 3 (SHFLD3)
Basic leucine zipper transcription factor, ATF-like 2	BATF2	The function of SARI in modulating epithelial-mesenchymal transition and lung adenocarcinoma metastasis. PMID: 23049725
BCL2-associated athanogene	BAG1	BAG1 is neuroprotective in in vivo and *in vitro* models of Parkinson's disease. PMID: 25106480
BCL6 corepressor	BCOR	Disease: microphthalmia, syndromic, 2 (MCOPS2)
BRCA1-associated ATM activator 1	BRAT1	Disease: rigidity and multifocal seizure syndrome, lethal neonatal (RMFSL)
Breast cancer 2, early onset	BRCA2	Disease: breast cancer (BC)
Breast carcinoma-amplified sequence 3	BCAS3	Note: A chromosomal aberration involving BCAS3 has been found in some breast carcinoma cell lines. Translocation t(1720)(q23q13) with BCAS4
BRF2, RNA polymerase III transcription initiation factor 50 kDa subunit	BRF2	RNA polymerase III transcription in cancer: The BRF2 connection. PMID: 21518452
BTB (POZ) domain containing 7	BTBD7	BTB/POZ domain-containing protein 7: Epithelial-mesenchymal transition promoter and prognostic biomarker of hepatocellular carcinoma. PMID: 23325674
cAMP-responsive element binding protein 1	CREB1	Active CREB1 promotes a malignant TGFβ2 autocrine loop in glioblastoma. PMID: 25084773. Disease: angiomatoid fibrous histiocytoma (AFH)
Cancer susceptibility candidate 5	CASC5	Note: A chromosomal aberration involving CASC5 is associated with acute myeloblastic leukemia (AML). Translocation t(11;15)(q23;q14) with KMT2A/MLL1. May give rise to a KMT2A/MLL1-CASC5 fusion protein. Disease: microcephaly 4, primary, autosomal recessive (MCPH4)
Caudal-type homeobox 2	CDX2	Expression and functional role of Cdx2 in intestinal metaplasia of cystitis glandularis. PMID: 23545100
Cbp/p300-interacting transactivator, with Glu/Asp-rich carboxy-terminal domain, 2	CITED2	Disease: ventricular septal defect 2 (VSD2)
CCCTC-binding factor (zinc finger protein)	CTCF	Disease: mental retardation, autosomal dominant 21 (MRD21)
CDKN1A-interacting zinc finger protein 1	CIZ1	Note: Defects in CIZ1 may be a cause of adult onset primary cervical dystonia. Dystonia is defined by the presence of sustained involuntary muscle contractions, often leading to abnormal postures
CXXC finger protein 5	CXXC5	Expression of the potential therapeutic target CXXC5 in primary acute myeloid leukemia cells – high expression is associated with adverse prognosis as well as altered intracellular signaling and transcriptional regulation. PMID: 25605239
Cell division cycle 5-like	CDC5L	Note: A chromosomal aberration involving CDC5L is found in multicystic renal dysplasia. Translocation t(619)(p21q13.1) with USF2
Cell division cycle 73	CDC73	Disease: familial isolated hyperparathyroidism (FIHP)
Cell division cycle associated 8	CDCA8	Borealin/Dasra B is overexpressed in colorectal cancers and contributes to proliferation of cancer cells. PMID: 25259650
Cell migration-inducing protein, hyaluronan binding	CEMIP	KIAA1199 interacts with glycogen phosphorylase kinase β-subunit (PHKB) to promote glycogen breakdown and cancer cell survival. PMID: 25051373
Chromatin licensing and DNA replication factor 1	CDT1	Disease: Meier–Gorlin syndrome 4 (MGORS4)

HGNC-approved name	Symbol	Target information
Cirrhosis, autosomal recessive 1A (cirhin)	CIRH1A	Disease: North American Indian childhood cirrhosis (NAIC)
Class II, major histocompatibility complex, transactivator	CIITA	Disease: bare lymphocyte syndrome 2 (BLS2)
Cleavage stimulation factor, 3′ pre-RNA, subunit 2, 64 kDa	CSTF2	Characterization of a cleavage stimulation factor, 3′ pre-RNA, subunit 2, 64 kDa (CSTF2) as a therapeutic target for lung cancer. PMID: 21813631
Cofilin 2 (muscle)	CFL2	Disease: nemaline myopathy 7 (NEM7)
Coilin	COIL	The coilin interactome identifies hundreds of small noncoding RNAs that traffic through Cajal bodies. PMID: 25514182
Cone-rod homeobox	CRX	Disease: Leber congenital amaurosis 7 (LCA7)
Core-binding factor, beta subunit	CBFB	Note: A chromosomal aberration involving CBFB is associated with acute myeloid leukemia of M4EO subtype. Pericentric inversion inv(16)(p13q22)
Core-binding factor, runt domain, alpha subunit 2; translocated to, 3	CBFA2T3	Note: A chromosomal aberration involving CBFA2T3 is found in therapy-related myeloid malignancies. Translocation t(1621)(q24q22) that forms a RUNX1 CBFA2T3 fusion protein
C-terminal binding protein 2	CTBP2	Changes in the numbers of ribbon synapses and expression of RIBEYE in salicylate-induced tinnitus. PMID: 25170565
CTS telomere maintenance complex component 1	CTC1	Disease: cerebroretinal microangiopathy with calcifications and cysts (CRMCC)
Cut-like homeobox 1	CUX1	Transcription factor CUTL1 is a negative regulator of drug resistance in gastric cancer. PMID: 23255599
Cyclin E1	CCNE1	Gene amplification CCNE1 is related to poor survival and potential therapeutic target in ovarian cancer. PMID: 20336784
Cyclin-dependent kinase inhibitor 1C (p57, Kip2)	CDKN1C	Disease: Beckwith–Wiedemann syndrome (BWS)
DEK oncogene	DEK	DEK overexpression is correlated with the clinical features of breast cancer. PMID: 22360505. Note: A chromosomal aberration involving DEK is found in a subset of acute myeloid leukemia (AML) also known as acute non-lymphocytic leukemia
DNA methyltransferase 1-associated protein 1	DMAP1	Novel 1p tumour suppressor Dnmt1-associated protein 1 regulates MYCN/ataxia telangiectasia mutated/p53 pathway. PMID: 24559687
Doublesex- and mab-3-related transcription factor 1	DMRT1	Disease: testicular germ cell tumor (TGCT)
E1A binding protein p400	EP400	The E1A-associated p400 protein modulates cell fate decisions by the regulation of ROS homeostasis. PMID: 20540951
E74-like factor 1 (ets domain transcription factor)	ELF4	Note: A chromosomal aberration involving ELF4 has been found in a case of acute myeloid leukemia (AML). Translocation t(X21)(q25-26q22) with ERG
ELK4, ETS-domain protein (SRF accessory protein 1)	ELK4	ELK4 neutralization sensitizes glioblastoma to apoptosis through downregulation of the anti-apoptotic protein Mcl-1. PMID: 21846680
Elongation factor RNA polymerase II	ELL	Note: A chromosomal aberration involving ELL is found in acute leukemias. Translocation t(1119)(q23p13.1) with KMT2A/MLL1. The result is a rogue activator protein
Elongation factor Tu GTP-binding domain containing 2	EFTUD2	Disease: mandibulofacial dysostosis with microcephaly (MFDM)
Emerin	EMD	Disease: Emery–Dreifuss muscular dystrophy 1, X-linked (EDMD1)
Empty spiracles homeobox 2	EMX2	EMX2 is downregulated in endometrial cancer and correlated with tumor progression. PMID: 23370654. Disease: schizencephaly (SCHZC)
Endothelial PAS domain protein 1	EPAS1	HIF-2α as a possible therapeutic target of osteoarthritis. PMID: 20950696. Disease: erythrocytosis, familial, 4 (ECYT4)

(Continued)

HGNC-approved name	Symbol	Target information
Engrailed homeobox 2	EN2	Note: Genetic variations in EN2 may be associated with susceptibility to autism
Eomesodermin	EOMES	Note: A translocation t(310)(p24q23) located 215 kb 3′ to the EOMES gene but leading to loss of its expression was identified in a large consanguineous family. Homozygous silencing produces microcephaly associated with corpus callosum agenesis
Epilepsy, progressive myoclonus type 2A, Lafora disease (laforin)	EPM2A	Protein degradation and quality control in cells from laforin and malin knockout mice. PMID: 24914213
Erythroid differentiation regulatory factor 1	EDRF1	Erythroid differentiation regulator 1 (Erdr1) is a proapoptotic factor in human keratinocytes. PMID: 21995813
Etoposide induced 2.4	EI24	Ei24, a novel E2F target gene, affects p53-independent cell death upon ultraviolet C irradiation. PMID: 24014029. Note: EI24 is on a chromosomal region frequently deleted in solid tumors, and it is thought to play a role in breast and cervical cancer
Ets2 repressor factor	ERF	Disease: craniosynostosis 4 (CRS4)
Excision repair cross-complementation group 8	ERCC8	Disease: Cockayne syndrome A (CSA)
Factor interacting with PAPOLA and CPSF1	FIP1L1	Note: A chromosomal aberration involving FIP1L1 is found in some cases of hypereosinophilic syndrome. Interstitial chromosomal deletion del(4)(q12q12) causes the fusion of FIP1L1 and PDGFRA (FIP1L1-PDGFRA)
Family with sequence similarity 175, member A	FAM175A	Disease: breast cancer (BC)
F-box and WD repeat domain containing 7, E3 ubiquitin protein ligase	FBXW7	The two faces of FBW7 in cancer drug resistance. PMID: 22006825
FEV (ETS oncogene family)	FEV	Disease: sudden infant death syndrome (SIDS)
Fli-1 proto-oncogene, ETS transcription factor	FLI1	Small-molecule screen identifies modulators of EWS/FLI1 target gene expression and cell survival in Ewing's sarcoma. PMID: 22323082. Disease: Ewing sarcoma (ES)
Folliculogenesis specific basic helix-loop-helix	FIGLA	Disease: premature ovarian failure 6 (POF6)
FSHD region gene 1	FRG1	Disease: facioscapulohumeral muscular dystrophy 1 (FSHD1)
Fused in sarcoma	FUS	Note: A chromosomal aberration involving FUS is found in a patient with malignant myxoid liposarcoma. Translocation t(1216)(q13p11) with DDIT3. Note: A chromosomal aberration involving FUS is a cause of acute myeloid leukemia (AML)
General transcription factor IIII, polypeptide 5	GTF2I5	Disease: trichothiodystrophy photosensitive (TTDP)
GLI family zinc finger 2	GLI2	GLI2 is a potential therapeutic target in pediatric medulloblastoma. PMID: 21572341. Disease: holoprosencephaly 9 (HPE9)
Glial cells missing homolog 2 (*Drosophila*)	GCM2	Disease: hypoparathyroidism, familial isolated (FIII)
Glioma tumor suppressor candidate region gene 2	GLTSCR2	Nucleolar protein GLTSCR2 stabilizes p53 in response to ribosomal stresses. PMID: 22522597
Goosecoid homeobox	GSC	Disease: short stature, auditory canal atresia, mandibular hypoplasia, skeletal abnormalities (SAMS)
Grainyhead-like 2 (*Drosophila*)	GRHL2	Disease: deafness, autosomal dominant, 28 (DFNA28)
Guanine nucleotide binding protein (G protein), q polypeptide	GNAQ	Activating mutations of the GNAQ gene: A frequent event in primary melanocytic neoplasms of the central nervous system. PMID: 19936769. Disease: capillary malformations, congenital (CMC)
Guanine nucleotide binding protein-like 3 (nucleolar)	GNL3	*In vitro* study of nucleostemin gene as a potential therapeutic target for human lung carcinoma. PMID: 22424632

HGNC-approved name	Symbol	Target information
H2A histone family, member Z	H2AFZ	Histone H2A.Z subunit exchange controls consolidation of recent and remote memory. PMID: 25218986
H6 family homeobox 1	HMX1	Disease: oculoauricular syndrome (OCLAUS)
Heat shock transcription factor 4	HSF4	Disease: cataract 5, multiple types (CTRCT5)
Hepatic leukemia factor	HLF	Note: A chromosomal aberration involving HLF is a cause of pre-B-cell acute lymphoblastic leukemia (B-ALL). Translocation t(1719)(q22p13.3) with TCF3
Hes family bHLH transcription factor 7	HES7	Disease: spondylocostal dysostosis 4, autosomal recessive (SCDO4)
Hes-related family bHLH transcription factor with YRPW motif 1	HEY1	Expression of the transcription factor HEY1 in glioblastoma: A preliminary clinical study. PMID: 20437865
HESX homeobox 1	HESX1	Disease: septooptic dysplasia (SOD)
Heterochromatin protein 1, binding protein 3	HP1BP3	Quantitative profiling of chromatome dynamics reveals a novel role for HP1BP3 in hypoxia-induced oncogenesis. PMID: 25100860
Holliday junction recognition protein	HJURP	Modulation of HJURP (Holliday junction-recognizing protein) levels is correlated with glioblastoma cells survival. PMID: 23638004
Homeobox B13	HOXB13	Disease: prostate cancer (PC)
HORMA domain containing 1	HORMAD1	Biological significance of HORMA domain containing protein 1 (HORMAD1) in epithelial ovarian carcinoma. PMID: 22776561
Intestine-specific homeobox	ISX	Proinflammatory homeobox gene, ISX, regulates tumor growth and survival in hepatocellular carcinoma. PMID: 23221382
JAZF zinc finger 1	JAZF1	Note: A chromosomal aberration involving JAZF1 may be a cause of endometrial stromal tumors. Translocation t(717)(p15q21) with SUZ12
Kelch-like family member 7	KLHL7	Disease: retinitis pigmentosa 42 (RP42)
KH domain containing, RNA binding, signal transduction associated 1	KHDRBS1	**CWP232291** – antineoplastic agents
Killin, p53-regulated DNA replication inhibitor	KLLN	Disease: Cowden syndrome 4 (CWS4)
Kinetochore-localized astrin/SPAG5 binding protein	KNSTRN	Note: Cutaneous squamous cell carcinomas (SCC). A malignancy of the skin. The hallmark of cutaneous SCC is malignant transformation of normal epidermal keratinocytes
La ribonucleoprotein domain family, member 7	LARP7	Disease: Alazami syndrome (ALAZS)
LanC lantibiotic synthetase component C-like 2 (bacterial)	LANCL2	Lanthionine synthetase component C-like protein 2: A new drug target for inflammatory diseases and diabetes. PMID: 24628287
LEM domain containing 3	LEMD3	Disease: Buschke–Ollendorff syndrome (BOS)
LIM homeobox transcription factor 1, beta	LMX1B	Disease: nail–patella syndrome (NPS)
Lymphoblastic leukemia-associated hematopoiesis regulator 1	LYL1	Note: A chromosomal aberration involving LYL1 may be a cause of a form of T-cell acute lymphoblastic leukemia (T-ALL). Translocation t(719)(q35p13) with TCRB
Lymphoid enhancer-binding factor 1	LEF1	LEF1 regulates glioblastoma cell proliferation, migration, invasion, and cancer stem-like cell self-renewal. PMID: 25128061
LYR motif containing 1	LYRM1	Note: When overexpressed, may be involved in obesity-associated insulin resistance, possibly by causing mitochondrial dysfunction in adipocytes
mab-21-like 2 (*C. elegans*)	MAB21L2	Disease: microphthalmia, syndromic, 14 (MCOPS14)
MAD1 mitotic arrest deficient-like 1 (yeast)	MAD1L1	Note: Defects in MAD1L1 are involved in the development and/or progression of various types of cancer
Matrin 3	MATR3	Disease: myopathy, distal, 2 (MPD2)

(Continued)

HGNC-approved name	Symbol	Target information
MAX dimerization protein 3	MXD3	Downregulation of Max dimerization protein 3 is involved in decreased visceral adipose tissue by inhibiting adipocyte differentiation in zebrafish and mice. PMID: 24254064
MAX interactor 1, dimerization protein	MXI1	Disease: prostate cancer (PC)
mbt domain containing 1	MBTD1	Note: A chromosomal aberration involving MBTD1 is a cause of acute poorly differentiated myeloid leukemia. Translocation (1017)(p15q21) with ZMYND11
Mdm4 p53 binding protein homolog (mouse)	MDM4	MDM4 is a key therapeutic target in cutaneous melanoma. PMID: 22820643
MDS1 and EVI1 complex locus	MECOM	Evaluation of EVI1 and EVI1s (Delta324) as potential therapeutic targets in ovarian cancer. PMID: 20462630. Note: A chromosomal aberration involving EVI1 is a cause of chronic myelogenous leukemia (CML). Translocation t(321)(q26q22) with RUNX1/AML1
Mediator of DNA-damage checkpoint 1	MDC1	NFBD1/MDC1 is a protein of oncogenic potential in human cervical cancer. PMID: 21853275
Meis homeobox 1	MEIS1	Meis1 regulates postnatal cardiomyocyte cell cycle arrest. PMID: 23594737. Disease: restless legs syndrome 7 (RLS7)
Mesenchyme homeobox 1	MEOX1	Disease: Klippel–Feil syndrome 2, autosomal recessive (KFS2)
Mesoderm posterior 2 homolog (mouse)	MESP2	Disease: spondylocostal dysostosis 2, autosomal recessive (SCDO2)
Metal regulatory transcription factor 1	MTF1	Regulation of the catabolic cascade in osteoarthritis by the zinc-ZIP8-MTF1 axis. PMID: 24529376
Microphthalmia-associated transcription factor	MITF	Disease: Waardenburg syndrome 2A (WS2A)
Microspherule protein 1	MCRS1	MSP58 knockdown inhibits the proliferation of esophageal squamous cell carcinoma in vitro and in vivo. PMID: 22994740
MKL/myocardin-like 2	MKL2	Note: A chromosomal aberration involving C11orf95 is found in 3 chondroid lipomas. Translocation t(11;16)(q13;p13) with C11orf95 produces a C11orf95-MKL2 fusion protein
MLX-interacting protein-like	MLXIPL	Note: WBSCR14 is located in the Williams–Beuren syndrome (WBS) critical region
Motor neuron and pancreas homeobox 1	MNX1	Disease: Currarino syndrome (CURRAS)
M-phase-specific PLK1-interacting protein	MPLKIP	Disease: trichothiodystrophy non-photosensitive 1 (TTDN1)
MRG/MORF4L binding protein	MRGBP	C20orf20 (MRG-binding protein) as a potential therapeutic target for colorectal cancer. PMID: 20051959
Multiciliate differentiation and DNA synthesis-associated cell cycle protein	MCIDAS	Note: Ciliary dyskinesia, primary (CILDN). A disorder characterized by abnormalities of motile cilia
Multiple endocrine neoplasia I	MEN1	Menin-MLL inhibitors reverse oncogenic activity of MLL fusion proteins in leukemia. PMID: 22286128. Disease: familial multiple endocrine neoplasia type 1 (MEN1)
MYC-associated zinc finger protein (purine-binding transcription factor)	MAZ	The prostate cancer-up-regulated Myc-associated zinc-finger protein (MAZ) modulates proliferation and metastasis through reciprocal regulation of androgen receptor. PMID: 23609189
Myogenic factor 6 (herculin)	MYF6	Disease: myopathy, centronuclear, 3 (CNM3)
MYST/Esa1-associated factor 6	MEAF6	Note: A chromosomal aberration involving MEAF6 may be a cause of endometrial stromal tumors. Translocation t(16)(p34p21) with PHF1
Nanog homeobox	NANOG	Increased Nanog expression promotes tumor development and cisplatin resistance in human esophageal cancer cells. PMID: 23221432
NCK-interacting protein with SH3 domain	NCKIPSD	Note: A chromosomal aberration involving NCKIPSD/AF3p21 is found in therapy-related leukemia. Translocation t(311)(p21q23) with KMT2A/MLL1

HGNC-approved name	Symbol	Target information
NDC80 kinetochore complex component	NDC80	Characterization of the biological activity of a potent small molecule Hec1 inhibitor TAI-1. PMID: 24401611
Necdin, melanoma antigen (MAGE) family member	NDN	Necdin: A multi functional protein with potential tumor suppressor role? PMID: 19626646
NEDD4 binding protein 1	N4BP1	N4BP1 is a newly identified nucleolar protein that undergoes SUMO-regulated polyubiquitylation and proteasomal turnover at promyelocytic leukemia nuclear bodies. PMID: 20233849
Negative regulator of ubiquitin-like proteins 1	NUB1	Identification of NUB1 as a suppressor of mutant Huntington toxicity via enhanced protein clearance. PMID: 23525043
Neural retina leucine zipper	NRL	Disease: retinitis pigmentosa 27 (RP27)
Neurofibromin 1	NF1	Disease: neurofibromatosis 1 (NF1)
Neurogenin 3	NEUROG3	Disease: diarrhea 4, malabsorptive, congenital (DIAR4)
Neuron navigator 3	NAV3	Neuron navigator 3 alterations in nervous system tumors associate with tumor malignancy grade and prognosis. PMID: 23097141. Note: A chromosomal aberration disrupting NAV3 has been found in patients with Sezary syndrome. Translocation t(1218)(q21q21.2)
NHP2 ribonucleoprotein	NHP2	Disease: dyskeratosis congenita, autosomal recessive, 2 (DKCB2)
Nibrin	NBN	Disease: Nijmegen breakage syndrome (NBS)
NIN1/RPN12 binding protein 1 homolog (S. cerevisiae)	NOB1	Downregulation of NIN/RPN12 binding protein inhibit the growth of human hepatocellular carcinoma cells. PMID: 21573803
Nipped-B homolog (*Drosophila*)	NIPBL	Disease: Cornelia de Lange syndrome 1 (CDLS1)
NMDA receptor synaptonuclear signaling and neuronal migration factor	NSMF	Disease: hypogonadotropic hypogonadism 9 with or without anosmia (HH9)
NOBOX oogenesis homeobox	NOBOX	Disease: premature ovarian failure 5 (POF5)
Nonhomologous end-joining factor 1	NHEJ1	Disease: severe combined immunodeficiency due to NHEJ1 deficiency (NHEJ1-SCID)
Non-POU domain containing, octamer-binding	NONO	Note: A chromosomal aberration involving NONO may be a cause of papillary renal cell carcinoma (PRCC). Translocation t(XX)(p11.2q13.1) with TFE3
Nuclear factor I/X (CCAAT-binding transcription factor)	NFIX	Disease: Sotos syndrome 2 (SOTOS2)
Nuclear factor of kappa light polypeptide gene enhancer in B-cells inhibitor-like 1	NFKBIL1	Disease: rheumatoid arthritis (RA)
Nuclear transcription factor Y, alpha	NFYA	NFYA underlies EZH2 upregulation and is essential for proliferation of human epithelial ovarian cancer cells. PMID: 23360797
Nucleolar protein 3 (apoptosis repressor with CARD domain)	NOL3	Disease: myoclonus, familial cortical (FCM)
Nucleolin	NCL	Nucleolin promotes TGF-β signaling initiation via TGF-β receptor I in glioblastoma. PMID: 24682943
Nucleophosmin (nucleolar phosphoprotein B23, numatrin)	NPM1	Nucleophosmin and its complex network: A possible therapeutic target in hematological diseases. PMID: 21278791. Note: A chromosomal aberration involving NPM1 is found in a form of non-Hodgkin lymphoma. Translocation t(25)(p23q35) with ALK
Nucleoporin 155 kDa	NUP155	Disease: atrial fibrillation, familial, 15 (ATFB15)
Opa-interacting protein 5	OIP5	OIP5 is a highly expressed potential therapeutic target for colorectal and gastric cancers. PMID: 20510019
Opioid growth factor receptor	OGFR	Targeting the opioid growth factor: Opioid growth factor receptor axis for treatment of human ovarian cancer. PMID: 23856908

(Continued)

HGNC-approved name	Symbol	Target information
Papillary renal cell carcinoma (translocation-associated)	PRCC	Note: A chromosomal aberration involving PRCC is found in patients with papillary renal cell carcinoma. Translocation t(X1)(p11.2q21.2) with TFE3
Parathymosin	PTMS	Macromolecular translocation inhibitor II (Zn(2+)-binding protein, parathymosin) interacts with the glucocorticoid receptor and enhances transcription in vivo. PMID: 16150697
Partner and localizer of BRCA2	PALB2	Disease: breast cancer (BC)
PC4- and SFRS1-interacting protein 1	PSIP1	The same site on LEDGF IBD domain represents therapeutic target for MLL leukemia and HIV. PMID: 25305204. Note: A chromosomal aberration involving PSIP1 is associated with pediatric acute myeloid leukemia (AML) with intermediate characteristics between M2 and M3 French-American-British (FAB) subtypes
Peroxisome proliferator-activated receptor gamma, coactivator 1 alpha	PPARGC1A	The PGC-1 cascade as a therapeutic target for heart failure. PMID: 20888832
PHD finger protein 6	PHF6	Disease: Borjeson–Forssman–Lehmann syndrome (BFLS)
Pleiomorphic adenoma gene 1	PLAG1	Note: A chromosomal aberration involving PLAG1 is found in salivary gland pleiomorphic adenomas, the most common benign epithelial tumors of the salivary gland. Translocation t(38)(p21q12) with constitutively expressed beta-catenin/CTNNB1
Polybromo 1	PBRM1	Disease: renal cell carcinoma (RCC)
Polycomb group ring finger 2	PCGF2	The novel tumor-suppressor Mel-18 in prostate cancer: Its functional polymorphism, expression and clinical significance. PMID: 19585577
Polyglutamine binding protein 1	PQBP1	Disease: Renpenning syndrome 1 (RENS1)
Polyhomeotic homolog 1 (Drosophila)	PHC1	Disease: microcephaly 11, primary, autosomal recessive (MCPH11)
Poly-U-binding splicing factor 60 kDa	PUF60	Disease: Verheij syndrome (VRJS)
Potassium channel tetramerization domain containing 1	KCTD1	KCTD1 suppresses canonical Wnt signaling pathway by enhancing β-catenin degradation. PMID: 24736394. Disease: scalp–ear–nipple syndrome (SENS)
POZ (BTB) and AT hook containing zinc finger 1	PATZ1	Note: A chromosomal aberration involving PATZ1 is associated with small round cell sarcoma. Translocation t(122)(p36.1q12) with EWSR1
Pre-mRNA processing factor 4	PRPF4	Disease: retinitis pigmentosa 70 (RP70)
Proliferating cell nuclear antigen	PCNA	PCNA: A silent housekeeper or a potential therapeutic target? PMID: 24655521
Proline-, glutamate- and leucine-rich protein 1	PELP1	PELP1: A novel therapeutic target for hormonal cancers. PMID: 20014005
PROP paired-like homeobox 1	PROP1	Disease: pituitary hormone deficiency, combined, 2 (CPHD2)
Protection of telomeres 1	POT1	Telomere 1 (POT1) gene expression and its association with telomerase activity in colorectal tumor samples with different pathological features. PMID: 25194444
Protein inhibitor of activated STAT, 4	PIAS4	PIAS4 is an activator of hypoxia signalling via VHL suppression during growth of pancreatic cancer cells. PMID: 24002598
Prothymosin, alpha	PTMA	Novel molecular targets regulated by tumor suppressors microRNA-1 and microRNA-133a in bladder cancer. PMID: 22378464
PSMC3-interacting protein	PSMC3IP	Disease: ovarian dysgenesis 3 (ODG3)
Purine-rich element binding protein A	PURA	Purine-rich element binding protein (PUR) alpha induces endoplasmic reticulum stress response, and cell differentiation pathways in prostate cancer cells. PMID: 19267365
Rac GTPase-activating protein 1	RACGAP1	Discovery of MINC1, a GTPase-activating protein small molecule inhibitor, targeting MgcRacGAP. PMID: 25479424

HGNC-approved name	Symbol	Target information
RAN binding protein 2	RANBP2	Selective impairment of a subset of Ran-GTP-binding domains of Ran-binding protein 2 (Ranbp2) suffices to recapitulate the degeneration of the retinal pigment epithelium (RPE) triggered by Ranbp2 ablation. PMID: 25187515. Disease: encephalopathy, acute, infection-induced, 3 (IIAE3)
RE1-silencing transcription factor	REST	REST is a novel prognostic factor and therapeutic target for medulloblastoma. PMID: 22848092
Recombination activating gene 2	RAG2	Disease: combined cellular and humoral immune defects with granulomas (CHIDG)
Remodeling and spacing factor 1	RSF1	Rsf-1 is overexpressed in non-small cell lung cancers and regulates cyclinD1 expression and ERK activity. PMID: 22387541
Replication initiator 1	REPIN1	Replication initiator 1 in adipose tissue function and human obesity. PMID: 23374714
Retinitis pigmentosa 9 (autosomal dominant)	RP9	Disease: retinitis pigmentosa 9 (RP9)
Retinoblastoma 1	RB1	Disease: childhood cancer retinoblastoma (RB)
Ring finger protein 1	RING1	Snail recruits Ring1B to mediate transcriptional repression and cell migration in pancreatic cancer cells. PMID: 24903147
Rogdi homolog (*Drosophila*)	ROGDI	Disease: Kohlschutter–Tonz syndrome (KTZS)
S100 calcium binding protein A6	S100A6	S100A6 as a potential serum prognostic biomarker and therapeutic target in gastric cancer. PMID: 24705642
SAM pointed domain containing ETS transcription factor	SPDEF	SPDEF functions as a colorectal tumor suppressor by inhibiting β-catenin activity. PMID: 23376423
SECIS binding protein 2	SECISBP2	Disease: abnormal thyroid hormone metabolism (ATHYHM)
SET binding protein 1	SETBP1	Disease: Schinzel–Giedion midface retraction syndrome (SGMFS)
Sex-determining region Y	SRY	Disease: 46,XY sex reversal 1 (SRXY1)
Short stature homeobox	SHOX	Disease: Leri–Weill dyschondrosteosis (LWD)
Shugoshin-like 1 (*S. pombe*)	SGOL1	Lentivirus-mediated siRNA interference targeting SGO-1 inhibits human NSCLC cell growth. PMID: 22161216
Sigma non-opioid intracellular receptor 1	SIGMAR1	**Cutamesine** – neuropsychiatric agents. Disease: amyotrophic lateral sclerosis 16, juvenile (ALS16)
SMAD family member 6	SMAD6	Disease: aortic valve disease 2 (AOVD2)
Smad nuclear-interacting protein 1	SNIP1	Disease: psychomotor retardation, epilepsy, and craniofacial dysmorphism (PMRED)
Small nuclear ribonucleoprotein 70 kDa (U1)	SNRNP70	U1 small nuclear ribonucleoprotein complex and RNA splicing alterations in Alzheimer's disease. PMID: 24023061
Sp1 transcription factor	SP1	Down-regulation of Sp1 suppresses cell proliferation, clonogenicity and the expressions of stem cell markers in nasopharyngeal carcinoma. PMID: 25099028
SP110 nuclear body protein	SP110	Disease: hepatic venoocclusive disease with immunodeficiency (VODI)
Sp7 transcription factor	SP7	Disease: osteogenesis imperfecta 12 (OI12)
Spalt-like transcription factor 1	SALL1	Disease: Townes–Brocks syndrome (TBS)
Speckle-type POZ protein	SPOP	The emerging role of speckle-type POZ protein (SPOP) in cancer development. PMID: 25058385
Sperm antigen with calponin homology and coiled-coil domains 1	SPECC1	NSP 5a3a: A potential novel cancer target in head and neck carcinoma. PMID: 21311098. Note: A chromosomal aberration involving CYTSB may be a cause of juvenile myelomonocytic leukemia. Translocation t(517)(q33p11.2) with PDGFRB

(Continued)

HGNC-approved name	Symbol	Target information
Splicing factor proline/glutamine-rich	SFPQ	Note: A chromosomal aberration involving SFPQ may be a cause of papillary renal cell carcinoma (PRCC). Translocation t(X1)(p11.2p34) with TFE3
Stromal antigen 3	STAG3	Disease: premature ovarian failure 8 (POF8)
Structural maintenance of chromosomes flexible hinge domain containing 1	SMCHD1	Disease: facioscapulohumeral muscular dystrophy 2 (FSHD2)
SUB1 homolog (*S. cerevisiae*)	SUB1	Human positive coactivator 4 is a potential novel therapeutic target in non-small cell lung cancer. PMID: 22918472
Suppressor of Ty 6 homolog (*S. cerevisiae*)	SUPT6H	SUPT6H controls estrogen receptor activity and cellular differentiation by multiple epigenomic mechanisms. PMID: 24441044
SUZ12 polycomb repressive complex 2 subunit	SUZ12	SUZ12 depletion suppresses the proliferation of gastric cancer cells. PMID: 23735840. Note: A chromosomal aberration involving SUZ12 may be a cause of endometrial stromal tumors. Translocation t(717)(p15q21) with JAZF1
Synaptonemal complex protein 3	SYCP3	Disease: spermatogenic failure 4 (SPGF4)
T, brachyury homolog (mouse)	T	From notochord formation to hereditary chordoma: The many roles of Brachyury. PMID: 23662285. Disease: neural tube defects (NTD)
TAF4b RNA polymerase II, TATA box binding protein (TBP)-associated factor, 105 kDa	TAF4B	Disease: spermatogenic failure 13 (SPGF13)
TAR DNA binding protein	TARDBP	Cellular model of TAR DNA-binding protein 43 (TDP-43) aggregation based on its C-terminal Gln/Asn-rich region. PMID: 22235134. Disease: amyotrophic lateral sclerosis 10 (ALS10)
TATA box binding protein	TBP	Disease: spinocerebellar ataxia 17 (SCA17)
TATA box binding protein (TRP)-associated factor, RNA polymerase I, A, 48 kDa	TAF1A	Targeting RNA polymerase I with an oral small molecule CX-5461 inhibits ribosomal RNA synthesis and solid tumor growth. PMID: 21159662
T-cell acute lymphocytic leukemia 1	TAL1	Note: A chromosomal aberration involving TAL1 may be a cause of some T-cell acute lymphoblastic leukemias (T-ALL). Translocation t(114)(p32q11) with T-cell receptor alpha chain (TCRA) genes
T-cell leukemia homeobox 1	TLX1	Note: A chromosomal aberration involving TLX1 may be a cause of a form of T-cell acute lymphoblastic leukemia (T-ALL). Translocation t(1014)(q24q11) with TCRD
TCF3 (E2A) fusion partner (in childhood leukemia)	TFPT	Note: A chromosomal aberration involving TFPT is a cause of pre-B-cell acute lymphoblastic leukemia (B-ALL). Inversion inv(19)(p13q13) with TCF3
Teashirt zinc finger homeobox 1	TSHZ1	Disease: aural atresia, congenital (CAA)
telomeric repeat-binding factor 2	TERF2	Molecular targeting of TRF2 suppresses the growth and tumorigenesis of glioblastoma stem cells. PMID: 24909307
TERF1 (TRF1)-interacting nuclear factor 2	TINF2	Disease: dyskeratosis congenita, autosomal dominant, 3 (DKCA3)
TGFB-induced factor homeobox 1	TGIF1	Disease: holoprosencephaly 4 (HPE4)
THAP domain containing, apoptosis-associated protein 1	THAP1	Disease: dystonia 6, torsion (DYT6)
Thymopoietin	TMPO	Disease: cardiomyopathy, dilated 1T (CMD1T)
Thyroid hormone receptor interactor 13	TRIP13	Thyroid hormone receptor interacting protein 13 (TRIP13) AAA-ATPase is a novel mitotic checkpoint-silencing protein. PMID: 25012665
Transcription elongation factor A (SII), 1	TCEA1	Note: A chromosomal aberration involving TCEA1 may be a cause of salivary gland pleiomorphic adenomas (PA) [181030]. Pleiomorphic adenomas are the most common benign epithelial tumors of the salivary gland. Translocation t(38)(p21q12) with PLAG1
Transcription factor 7-like 2 (T-cell specific, HMG-box)	TCF7L2	Note: Constitutive activation and subsequent transactivation of target genes may lead to the maintenance of stem-cell characteristics (cycling and longevity) in cells that should normally undergo terminal differentiation

HGNC-approved name	Symbol	Target information
Transcription factor binding to IGHM enhancer 3	TFE3	Note: A chromosomal aberration involving TFE3 is found in patients with alveolar soft part sarcoma. Translocation t(X17)(p11q25) with ASPSCR1 forms a ASPSCR1-TFE3 fusion protein
Transformation/transcription domain-associated protein	TRRAP	Note: TRRAP mutation Phe-722 has been frequently found in cutaneous malignant melanoma, suggesting that TRRAP may play a role in the pathogenesis of melanoma (PMID: 21499247)
Transformer 2 beta homolog (*Drosophila*)	TRA2B	Transformer 2β (Tra2β/SFRS10) positively regulates the progression of NSCLC via promoting cell proliferation. PMID: 24952301
Translocated promoter region, nuclear basket protein	TPR	Disease: thyroid papillary carcinoma (TPC)
Treacher Collins–Franceschetti syndrome 1	TCOF1	Disease: Treacher Collins syndrome 1 (TCS1)
Tribbles pseudokinase 3	TRIB3	High throughput kinase inhibitor screens reveal TRB3 and MAPK-ERK/TGFβ pathways as fundamental Notch regulators in breast cancer. PMID: 23319603
Trichorhinophalangeal syndrome I	TRPS1	Disease: tricho–rhino–phalangeal syndrome 1 (TRPS1)
TSEN54 tRNA splicing endonuclease subunit	TSEN54	Disease: pontocerebellar hypoplasia 4 (PCH4)
TSPY-like 1	TSPYL1	Disease: sudden infant death with dysgenesis of the testes syndrome (SIDDT)
Tumor protein p53 binding protein 1	TP53BP1	Note: A chromosomal aberration involving TP53BP1 is found in a form of myeloproliferative disorder chronic with eosinophilia. Translocation t(515)(q33q22) with PDGFRB creating a TP53BP1-PDGFRB fusion protein
Tumor protein p63	TP63	Disease: acro–dermato–ungual–lacrimal–tooth syndrome (ADULT syndrome)
Twist family bHLH transcription factor 1	TWIST1	Twist, a master regulator of morphogenesis, plays an essential role in tumor metastasis. PMID: 15210113. Disease: Saethre–Chotzen syndrome (SCS)
Ubiquitin interaction motif containing 1	UIMC1	Molecular basis for impaired DNA damage response function associated with the RAP80 ΔE81 defect. PMID: 24627472
Ubiquitin-like with PHD and ring finger domains 1	UHRF1	Increasing role of UHRF1 in the reading and inheritance of the epigenetic code as well as in tumorogenesis. PMID: 24134914. Note: Defects in UHRF1 may be a cause of cancers. Overexpressed in many different forms of human cancers, including bladder, breast, cervical, colorectal and prostate cancers, as well as pancreatic adenocarcinomas, rhabdomyosarcomas and gliomas
Upstream transcription factor 1	USF1	Disease: hyperlipidemia combined 1 (HYPLIP1)
UV-stimulated scaffold protein A	UVSSA	Disease: UV-sensitive syndrome 3 (UVSS3)
Ventral anterior homeobox 1	VAX1	Disease: microphthalmia, syndromic, 11 (MCOPS11)
v-ski avian sarcoma viral oncogene homolog	SKI	Ski acts as therapeutic target of qingyihuaji formula in the treatment of SW1990 pancreatic cancer. PMID: 20308085
WD repeat containing, antisense to TP53	WRAP53	Disease: dyskeratosis congenita, autosomal recessive, 3 (DKCB3)
WD repeat domain 36	WDR36	Disease: glaucoma 1, open angle, G (GLC1G)
Wilms tumor 1	WT1	Targeting the intracellular WT1 oncogene product with a therapeutic human antibody. PMID: 23486779. Disease: Frasier syndrome (FS)
X-box binding protein 1	XBP1	XBP1 promotes triple-negative breast cancer by controlling the HIF1α pathway. PMID: 24670461. Disease: major affective disorder 7 (MAFD7)
Xeroderma pigmentosum, complementation group A	XPA	Disease: xeroderma pigmentosum complementation group A (XP-A)
X-ray repair complementing defective repair in Chinese hamster cells 3	XRCC3	Disease: breast cancer (BC)
YY1 transcription factor	YY1	Yin Yang 1 promotes hepatic gluconeogenesis through upregulation of glucocorticoid receptor. PMID: 23193188

(Continued)

HGNC-approved name	Symbol	Target information
ZFP57 zinc finger protein	ZFP57	Disease: transient neonatal diabetes mellitus 1 (TNDM1)
Zinc finger E-box binding homeobox 1	ZEB1	Zinc finger E-box binding factor 1 plays a central role in regulating Epstein-Barr virus (EBV) latent-lytic switch and acts as a therapeutic target in EBV-associated gastric cancer. PMID: 21717425. Disease: corneal dystrophy, posterior polymorphous, 3 (PPCD3)
Zinc finger protein, FOG family member 2	ZFPM2	Disease: tetralogy of Fallot (TOF)
Zinc finger protein, X-linked	ZFX	Knockdown of zinc finger protein X-linked inhibits prostate cancer cell proliferation and induces apoptosis by activating caspase-3 and caspase-9. PMID: 22898899
Zinc finger, BED-type containing 1	ZBED1	Adenovirus E1A targets the DREF nuclear factor to regulate virus gene expression, DNA replication, and growth. PMID: 25210186

6.5   Internal membranes and organelles

HGNC-approved name	Symbol	Target information
Activating transcription factor 4	ATF4	ATF4: A novel potential therapeutic target for Alzheimer's disease. PMID: 25381575
Activating transcription factor 6	ATF6	Role of disulfide bridges formed in the luminal domain of ATF6 in sensing endoplasmic reticulum stress. PMID: 17101776
Adaptor related protein complex 1, sigma 1 subunit	AP1S1	Disease: mental retardation, enteropathy, deafness, peripheral neuropathy, ichthyosis, and keratoderma (MEDNIK)
Adaptor-related protein complex 1, sigma 2 subunit	AP1S2	Disease: mental retardation, X-linked 59 (MRX59)
Adaptor-related protein complex 3, beta 1 subunit	AP3B1	Disease: Hermansky–Pudlak syndrome 2 (HPS2)
Adaptor-related protein complex 4, beta 1 subunit	AP4B1	Disease: cerebral palsy, spastic quadriplegic 5 (CPSQ5)
Adaptor-related protein complex 4, epsilon 1 subunit	AP4E1	Disease: cerebral palsy, spastic quadriplegic 4 (CPSQ4)
Adaptor-related protein complex 4, mu 1 subunit	AP4M1	Disease: cerebral palsy, spastic quadriplegic 3 (CPSQ3)
Adaptor-related protein complex 4, sigma 1 subunit	AP4S1	Disease: cerebral palsy, spastic quadriplegic 6 (CPSQ6)
ADP-ribosylation factor-like 4C	ARL4C	Arl4c expression in colorectal and lung cancers promotes tumorigenesis and may represent a novel therapeutic target. PMID: 25486429
ADP-ribosylation factor-like 6-interacting protein 1	ARL6IP1	Disease: spastic paraplegia 61, autosomal recessive (SPG61)
ADP-ribosylation factor like 6-interacting protein 5	ARL6IP5	Downregulation of JWA promotes tumor invasion and predicts poor prognosis in human hepatocellular carcinoma. PMID: 23169062
ATPase family, AAA domain containing 3A	ATAD3A	Mitochondrial protein ATPase family, AAA domain containing 3A correlates with radioresistance in glioblastoma. PMID: 24057885
ATPase family, AAA domain containing 3B	ATAD3B	ATAD3B is a human embryonic stem cell specific mitochondrial protein, re-expressed in cancer cells, that functions as dominant negative for the ubiquitous ATAD3A. PMID: 22664726
ATPase family, AAA domain containing 5	ATAD5	ATAD5 deficiency decreases B cell division and Igh recombination. PMID: 25404367

HGNC-approved name	Symbol	Target information
Beaded filament structural protein 1, filensin	BFSP1	Disease: cataract 33 (CTRCT33)
Beaded filament structural protein 2, phakinin	BFSP2	Disease: cataract 12, multiple types (CTRCT12)
cAMP-responsive element binding protein 3	CREB3	Small leucine zipper protein (sLZIP) negatively regulates skeletal muscle differentiation via interaction with α-actinin-4. PMID: 24375477
cAMP-responsive element binding protein 3-like 1	CREB3L1	Doxorubicin blocks proliferation of cancer cells through proteolytic activation of CREB3L1. PMID: 23256041
cAMP-responsive element binding protein 3-like 2	CREB3L2	Note: A chromosomal aberration involving CREB3L2 is found in low grade fibromyxoid sarcoma (LGFMS). Translocation t(716)(q33p11) with FUS
cAMP-responsive element binding protein 3-like 3	CREB3L3	Hepatic CREB3L3 controls whole-body energy homeostasis and improves obesity and diabetes. PMID: 25231913
Caveolin 1, caveolae protein, 22 kDa	CAV1	Serum caveolin-1, a biomarker of drug response and therapeutic target in prostate cancer models. PMID: 23114714. Disease: congenital generalized lipodystrophy 3 (CGL3)
Caveolin 3	CAV3	Disease: limb-girdle muscular dystrophy 1C (LGMD1C)
Ceroid lipofuscinosis, neuronal 3	CLN3	Disease: ceroid lipofuscinosis, neuronal, 3 (CLN3)
Ceroid lipofuscinosis, neuronal 5	CLN5	Disease: ceroid lipofuscinosis, neuronal, 5 (CLN5)
Ceroid lipofuscinosis, neuronal 6, late infantile, variant	CLN6	Disease: ceroid lipofuscinosis, neuronal, 6 (CLN6)
Ceroid lipofuscinosis, neuronal 8 (epilepsy, progressive with mental retardation)	CLN8	Disease: ceroid lipofuscinosis, neuronal, 8 (CLN8)
Chromogranin A (parathyroid secretory protein 1)	CHGA	The surging role of chromogranin A in cardiovascular homeostasis. PMID: 25177680
Chromogranin B (secretogranin 1)	CHGB	Chromogranin B: Intra- and extra-cellular mechanisms to regulate catecholamine storage and release, in catecholaminergic cells and organisms. PMID: 24266713
Chromosome 3 open reading frame 58	C3orf58	Note: Genetic variations in C3orf58 may be associated with susceptibility to autism
Chromosome 10 open reading frame 2	C10orf2	Disease: progressive external ophthalmoplegia with mitochondrial DNA deletions, autosomal dominant, 3 (PEOA3)
Chromosome 12 open reading frame 65	C12orf65	Disease: combined oxidative phosphorylation deficiency 7 (COXPD7)
Chromosome 19 open reading frame 12	C19orf12	Disease: neurodegeneration with brain iron accumulation 4 (NBIA4)
Claudin 1	CLDN1	Disease: neonatal ichthyosis–sclerosing cholangitis syndrome (NISCH)
Claudin 3	CLDN3	Development and characterization of a human single-chain antibody fragment against claudin-3: A novel therapeutic target in ovarian and uterine carcinomas. PMID: 19426958. Note: CLDN3 is located in the Williams–Beuren syndrome (WBS) critical region
Claudin 4	CLDN4	Claudin-4 as therapeutic target in cancer. PMID: 22286027. Note: CLDN4 is located in the Williams–Beuren syndrome (WBS) critical region
Claudin 11	CLDN11	Claudin-11 decreases the invasiveness of bladder cancer cells. PMID: 21468549
Claudin 14	CLDN14	Disease: deafness, autosomal recessive, 29 (DFNB29)
Claudin 16	CLDN16	Significant elevation of CLDN16 and HAPLN3 gene expression in human breast cancer. PMID: 20664984. Disease: hypomagnesemia 3 (HOMG3)
Claudin 19	CLDN19	Disease: hypomagnesemia 5 (HOMG5)

(Continued)

HGNC-approved name	Symbol	Target information
Coiled-coil domain containing 40	CCDC40	Disease: ciliary dyskinesia, primary, 15 (CILD15)
Coiled-coil domain containing 103	CCDC103	Disease: ciliary dyskinesia, primary, 17 (CILD17)
Coiled-coil-helix-coiled-coil-helix domain containing 7	CHCHD7	Note: A chromosomal aberration involving CHCHD7 is found in salivary gland pleiomorphic adenomas, the most common benign epithelial tumors of the salivary gland. Translocation t(68) (p21.3-22q13) with PLAG1
Coiled-coil-helix-coiled-coil-helix domain containing 10	CHCHD10	Disease: frontotemporal dementia and/or amyotrophic lateral sclerosis 2 (FTDALS2)
Component of oligomeric golgi complex 1	COG1	Disease: congenital disorder of glycosylation 2G (CDG2G)
Component of oligomeric golgi complex 4	COG4	Disease: congenital disorder of glycosylation 2J (CDG2J)
Component of oligomeric golgi complex 6	COG6	Disease: congenital disorder of glycosylation 2L (CDG2L)
Component of oligomeric golgi complex 7	COG7	Disease: congenital disorder of glycosylation 2E (CDG2E)
Component of oligomeric golgi complex 8	COG8	Disease: congenital disorder of glycosylation 2H (CDG2H)
Cytochrome c, somatic	CYCS	Disease: thrombocytopenia 4 (THC4)
Cytochrome c-1	CYC1	CYC1 silencing sensitizes osteosarcoma cells to TRAIL-induced apoptosis. PMID: 25562155
Cytochrome c oxidase subunit IV isoform 2 (lung)	COX4I2	Disease: exocrine pancreatic insufficiency dyserythropoietic anemia and calvarial hyperostosis (EPIDACH)
Cytochrome c oxidase subunit Va	COX5A	The role of cytochrome c oxidase subunit Va in non-small cell lung carcinoma cells: Association with migration, invasion and prediction of distant metastasis. PMID: 22748147
Cytochrome c oxidase subunit Vb	COX5B	COX5B regulates MAVS-mediated antiviral signaling through interaction with ATG5 and repressing ROS production. PMID: 23308066
Cytochrome c oxidase subunit VIa polypeptide 1	COX6A1	A mutation of COX6A1 causes a recessive axonal or mixed form of Charcot–Marie–Tooth disease. PMID: 25152455
Cytochrome c oxidase subunit VIa polypeptide 2	COX6A2	Mice deficient in the respiratory chain gene Cox6a2 are protected against high-fat diet-induced obesity and insulin resistance. PMID: 23460811
Cytochrome c oxidase subunit VIb polypeptide 1 (ubiquitous)	COX6B1	Disease: mitochondrial complex IV deficiency (MT-C4D)
Cytochrome c oxidase subunit VIc	COX6C	Lung epithelial cells resist influenza A infection by inducing the expression of cytochrome c oxidase VIc which is modulated by miRNA 4276. PMID: 25203353
Cytochrome c oxidase subunit VIIa polypeptide 1 (muscle)	COX7A1	Deletion of heart-type cytochrome c oxidase subunit 7a1 impairs skeletal muscle angiogenesis and oxidative phosphorylation. PMID: 22869013
Cytochrome c oxidase subunit VIIa polypeptide 2 (liver)	COX7A2	Cox7a2 mediates steroidogenesis in TM3 mouse Leydig cells. PMID: 16752004
Cytochrome c oxidase subunit VIIa polypeptide 2-like	COX7A2L	The respiratory chain supercomplex organization is independent of COX7a2l isoforms. PMID: 25470551
Cytochrome c oxidase subunit VIIb	COX7B	Disease: aplasia cutis congenita, reticulolinear, with microcephaly, facial dysmorphism and other congenital anomalies (APLCC)
Dedicator of cytokinesis 1	DOCK1	Overexpression of dedicator of cytokinesis I (Dock180) in ovarian cancer correlated with aggressive phenotype and poor patient survival. PMID: 22175896
Dedicator of cytokinesis 7	DOCK7	Disease: epileptic encephalopathy, early infantile, 23 (EIEE23)
DnaJ (Hsp40) homolog, subfamily C, member 5	DNAJC5	Disease: ceroid lipofuscinosis, neuronal, 4B (CLN4B)

HGNC-approved name	Symbol	Target information
DnaJ (Hsp40) homolog, subfamily C, member 15	DNAJC15	Note: Absent or down-regulated in many advanced cases of ovarian adenocarcinoma, due to hypermethylation and allelic loss. Loss of expression correlates with increased resistance to antineoplastic drugs, such as cisplatin
DnaJ (Hsp40) homolog, subfamily C, member 19	DNAJC19	Disease: 3-methylglutaconic aciduria 5 (MGA5)
Electron-transfer-flavoprotein, alpha polypeptide	ETFA	Disease: glutaric aciduria 2A (GA2A)
Electron-transfer-flavoprotein, beta polypeptide	ETFB	Disease: glutaric aciduria 2B (GA2B)
Electron-transferring-flavoprotein dehydrogenase	ETFDH	Disease: glutaric aciduria 2C (GA2C)
Family with sequence similarity 126, member A	FAM126A	Disease: leukodystrophy, hypomyelinating, 5 (HLD5)
Family with sequence similarity 134, member B	FAM134B	Disease: neuropathy, hereditary sensory and autonomic, 2B (HSAN2B)
FAST kinase domains 2	FASTKD2	Fas activated serine-threonine kinase domains 2 (FASTKD2) mediates apoptosis of breast and prostate cancer cells through its novel FAST2 domain. PMID: 25409762
FAST kinase domains 3	FASTKD3	Fast kinase domain-containing protein 3 is a mitochondrial protein essential for cellular respiration. PMID: 20869947
Growth factor receptor-bound protein 7	GRB7	Targeting the calmodulin-regulated ErbB/Grb7 signaling axis in cancer therapy. PMID: 23958188
Growth factor receptor-bound protein 14	GRB14	Structural basis for the interaction of the adaptor protein grb14 with activated ras. PMID: 23967305
Guanine nucleotide binding protein (G protein), alpha transducing activity polypeptide 1	GNAT1	Disease: night blindness, congenital stationary, autosomal dominant 3 (CSNBAD3)
Guanine nucleotide binding protein (G protein), alpha transducing activity polypeptide 2	GNAT2	Disease: achromatopsia 4 (ACHM4)
Hydroxysteroid dehydrogenase-like 1	HSDL1	Human and zebrafish hydroxysteroid dehydrogenase like 1 (HSDL1) proteins are inactive enzymes but conserved among species. PMID: 19026618
Hydroxysteroid dehydrogenase like 2	HSDL2	In search for function of two human orphan SDR enzymes: hydroxysteroid dehydrogenase like 2 (HSDL2) and short-chain dehydrogenase/reductase-orphan (SDR-O). PMID: 19703561
Inositol 1,4,5-trisphosphate receptor, type 1	ITPR1	Targeting Bcl-2 based on the interaction of its BH4 domain with the inositol 1,4,5-trisphosphate receptor. PMID: 19056433. Disease: spinocerebellar ataxia 15 (SCA15)
Inositol 1,4,5-trisphosphate receptor, type 2	ITPR2	Targeting Bcl-2 based on the interaction of its BH4 domain with the inositol 1,4,5-trisphosphate receptor. PMID: 19056433
Inositol 1,4,5-trisphosphate receptor, type 3	ITPR3	Caffeine-mediated inhibition of calcium release channel inositol 1,4,5-trisphosphate receptor subtype 3 blocks glioblastoma invasion and extends survival. PMID: 20103623
Isochorismatase domain containing 1	ISOC1	Systemic identification of estrogen-regulated genes in breast cancer cells through cap analysis of gene expression mapping. PMID: 24746470
Isochorismatase domain containing 2	ISOC2	Identification and characterization of a novel protein ISOC2 that interacts with p16INK4a. PMID: 17658461

(Continued)

HGNC-approved name	Symbol	Target information
Kinesin family member 1B	KIF1B	Disease: Charcot–Marie–Tooth disease 2A1 (CMT2A1)
Kinesin family member 20A	KIF20A	KIF20A-mediated RNA granule transport system promotes the invasiveness of pancreatic cancer cells. PMID: 25499221
LYR motif containing 4	LYRM4	Disease: combined oxidative phosphorylation deficiency 19 (COXPD19)
LYR motif containing 7	LYRM7	Disease: mitochondrial complex III deficiency, nuclear 8 (MC3DN8)
Lysosomal protein transmembrane 4 beta	LAPTM4B	Overexpression of LAPTM4B promotes growth of gallbladder carcinoma cells in vitro. PMID: 19954766
Lysosomal protein transmembrane 5	LAPTM5	LAPTM5 protein is a positive regulator of proinflammatory signaling pathways in macrophages. PMID: 22733818
Mitochondrial ribosomal protein L3	MRPL3	Disease: combined oxidative phosphorylation deficiency 9 (COXPD9)
Mitochondrial ribosomal protein L44	MRPL44	Disease: combined oxidative phosphorylation deficiency 16 (COXPD16)
Mitochondrial ribosomal protein S16	MRPS16	Disease: combined oxidative phosphorylation deficiency 2 (COXPD2)
Mitochondrial ribosomal protein S22	MRPS22	Disease: combined oxidative phosphorylation deficiency 5 (COXPD5)
Mitochondrially encoded cytochrome c oxidase II	MT-CO2	Disease: mitochondrial complex IV deficiency (MT-C4D)
Mitochondrially encoded cytochrome c oxidase III	MT-CO3	Disease: Leber hereditary optic neuropathy (LHON)
Mitochondrially encoded NADH dehydrogenase 1	MT-ND1	Disease: Leber hereditary optic neuropathy (LHON)
Mitochondrially encoded NADH dehydrogenase 2	MT-ND2	Disease: Leber hereditary optic neuropathy (LHON)
Mitochondrially encoded NADH dehydrogenase 3	MT-ND3	Disease: Leigh syndrome (LS)
Mitochondrially encoded NADH dehydrogenase 4	MT-ND4	Disease: Leber hereditary optic neuropathy (LHON)
Mitochondrially encoded NADH dehydrogenase 4L	MT-ND4L	Disease: Leber hereditary optic neuropathy (LHON)
Mitochondrially encoded NADH dehydrogenase 5	MT-ND5	Disease: Leber hereditary optic neuropathy (LHON)
Mitochondrially encoded NADH dehydrogenase 6	MT-ND6	Disease: Leber hereditary optic neuropathy (LHON)
NADH dehydrogenase (ubiquinone) 1 alpha subcomplex, 1, 7.5 kDa	NDUFA1	Disease: mitochondrial complex I deficiency (MT-C1D)
NADH dehydrogenase (ubiquinone) 1 alpha subcomplex, 10, 42 kDa	NDUFA10	PINK1 loss-of-function mutations affect mitochondrial complex I activity via NdufA10 ubiquinone uncoupling. PMID: 24652937
NADH dehydrogenase (ubiquinone) 1 alpha subcomplex, 11, 14.7 kDa	NDUFA11	Disease: mitochondrial complex I deficiency (MT-C1D)
NADH dehydrogenase (ubiquinone) 1 alpha subcomplex, 12	NDUFA12	Disease: Leigh syndrome (LS)
NADH dehydrogenase (ubiquinone) 1 alpha subcomplex, 13	NDUFA13	Disease: Hurthle cell thyroid carcinoma (HCTC)
NADH dehydrogenase (ubiquinone) complex I, assembly factor 1	NDUFAF1	Leukodystrophy associated with mitochondrial complex I deficiency due to a novel mutation in the NDUFAF1 gene. PMID: 24963768
NADH dehydrogenase (ubiquinone) complex I, assembly factor 2	NDUFAF2	Disease: mitochondrial complex I deficiency (MT-C1D)
NADH dehydrogenase (ubiquinone) complex I, assembly factor 3	NDUFAF3	Disease: mitochondrial complex I deficiency (MT-C1D)

HGNC-approved name	Symbol	Target information
NADH dehydrogenase (ubiquinone) complex I, assembly factor 4	NDUFAF4	Disease: mitochondrial complex I deficiency (MT-C1D)
NADH dehydrogenase (ubiquinone) complex I, assembly factor 6	NDUFAF6	Disease: mitochondrial complex I deficiency (MT-C1D)
Peroxisomal biogenesis factor 1	PEX1	Disease: peroxisome biogenesis disorder complementation group 1 (PBD-CG1)
Peroxisomal biogenesis factor 2	PEX2	Disease: peroxisome biogenesis disorder complementation group 5 (PBD-CG5)
Peroxisomal biogenesis factor 3	PEX3	Disease: peroxisome biogenesis disorder complementation group 12 (PBD-CG12)
Peroxisomal biogenesis factor 5	PEX5	Disease: peroxisome biogenesis disorder 2A (PBD2A)
Peroxisomal biogenesis factor 6	PEX6	Disease: peroxisome biogenesis disorder complementation group 4 (PBD-CG4)
Peroxisomal biogenesis factor 7	PEX7	Disease: peroxisome biogenesis disorder complementation group 11 (PBD-CG11)
Peroxisomal biogenesis factor 10	PEX10	Disease: peroxisome biogenesis disorder complementation group 7 (PBD-CG7)
Peroxisomal biogenesis factor 11 beta	PEX11B	Disease: peroxisome biogenesis disorder 14B (PBD14B)
Peroxisomal biogenesis factor 12	PEX12	Disease: peroxisome biogenesis disorder complementation group 3 (PBD-CG3)
Peroxisomal biogenesis factor 13	PEX13	Disease: peroxisome biogenesis disorder complementation group 13 (PBD-CG13)
Peroxisomal biogenesis factor 14	PEX14	Disease: peroxisome biogenesis disorder complementation group K (PBD-CGK)
Peroxisomal biogenesis factor 16	PEX16	Disease: peroxisome biogenesis disorder complementation group 9 (PBD-CG9)
Peroxisomal biogenesis factor 19	PEX19	Disease: peroxisome biogenesis disorder complementation group 14 (PBD-CG14)
Peroxisomal biogenesis factor 26	PEX26	Disease: peroxisome biogenesis disorder complementation group 8 (PBD-CG8)
Post-GPI attachment to proteins 2	PGAP2	Disease: hyperphosphatasia with mental retardation syndrome 3 (HPMRS3)
Post-GPI attachment to proteins 3	PGAP3	Disease: hyperphosphatasia with mental retardation syndrome 4 (HPMRS4)
Prohibitin	PHB	The natural anticancer compounds rocaglamides inhibit the Raf-MEK-ERK pathway by targeting prohibitin 1 and 2. PMID: 22999878
Prohibitin 2	PHB2	The natural anticancer compounds rocaglamides inhibit the Raf-MEK-ERK pathway by targeting prohibitin 1 and 2. PMID: 22999878
Regulator of G protein signaling 5	RGS5	Regulator of G-protein signaling 5 (RGS5) is a novel repressor of hedgehog signaling. PMID: 23037832
Regulator of G-protein signaling 6	RGS6	Decreased RGS6 expression is associated with poor prognosis in pancreatic cancer patients. PMID: 25120791
Regulator of G-protein signaling 17	RGS17	RGS17: An emerging therapeutic target for lung and prostate cancers. PMID: 23734683
Reticulon 2	RTN2	Disease: spastic paraplegia 12, autosomal dominant (SPG12)
Reticulon 4	RTN4	**Atinumab** – immunomodulating agents
Rho/Rac guanine nucleotide exchange factor (GEF) 2	ARHGEF2	The RhoA activator GEF-H1/Lfc is a transforming growth factor-beta target gene and effector that regulates alpha-smooth muscle actin expression and cell migration. PMID: 20089843
Rho guanine nucleotide exchange factor (GEF) 7	ARHGEF7	The role of Pak-interacting exchange factor-β phosphorylation at serines 340 and 583 by PKCλ in dopamine release. PMID: 25009260
Rho guanine nucleotide exchange factor (GEF) 12	ARHGEF12	Note: A chromosomal aberration involving ARHGEF12 may be a cause of acute leukemia. Translocation t(1111)(q2323) with KMT2A/MLL1

(Continued)

HGNC-approved name	Symbol	Target information
SCO1 cytochrome c oxidase assembly protein	SCO1	Disease: mitochondrial complex IV deficiency (MT-C4D)
SCO2 cytochrome c oxidase assembly protein	SCO2	Disease: cardioencephalomyopathy, fatal infantile, due to cytochrome c oxidase deficiency 1 (CEMCOX1)
Sec23 homolog A (*S. cerevisiae*)	SEC23A	Disease: craniolenticulosutural dysplasia (CLSD)
Sec23 homolog B (*S. cerevisiae*)	SEC23B	Disease: anemia, congenital dyserythropoietic, 2 (CDAN2)
SEC31 homolog A (*S. cerevisiae*)	SEC31A	Note: A chromosomal aberration involving SEC31A is associated with inflammatory myofibroblastic tumors (IMTs). Translocation t(24)(p23q21) with ALK
SEC63 homolog (*S. cerevisiae*)	SEC63	Disease: polycystic liver disease (PCLD)
Secretion associated, Ras-related GTPase 1A	SAR1A	Hydroxyurea-inducible SAR1 gene acts through the $Gi\alpha\pm$/JNK/Jun pathway to regulate γ-globin expression. PMID: 24914133
Secretion associated, Ras-related GTPase 1B	SAR1B	Disease: chylomicron retention disease (CMRD)
Secretory carrier membrane protein 1	SCAMP1	Inhibition of SCAMP1 suppresses cell migration and invasion in human pancreatic and gallbladder cancer cells. PMID: 23653380
Secretory carrier membrane protein 2	SCAMP2	Modulation of the dopamine transporter by interaction with secretory carrier membrane protein 2. PMID: 21295544
Secretory carrier membrane protein 3	SCAMP3	SCAMP3 negatively regulates epidermal growth factor receptor degradation and promotes receptor recycling. PMID: 19158374
Secretory carrier membrane protein 5	SCAMP5	SCAMP5 plays a critical role in synaptic vesicle endocytosis during high neuronal activity. PMID: 25057210
SH3 and PX domains 2A	SH3PXD2A	Tks5 activation in mesothelial cells creates invasion front of peritoneal carcinomatosis. PMID: 25088196
SH3 and PX domains 2B	SH3PXD2B	Disease: Frank–ter Haar syndrome (FTHS)
Sorting nexin 3	SNX3	Disease: microphthalmia, syndromic, 8 (MCOPS8)
Sorting nexin 10	SNX10	Disease: osteopetrosis, autosomal recessive 8 (OPTB8)
Succinate dehydrogenase complex assembly factor 1	SDHAF1	Disease: mitochondrial complex II deficiency (MT-C2D)
Succinate dehydrogenase complex assembly factor 2	SDHAF2	Succinate dehydrogenase 5 (SDH5) regulates glycogen synthase kinase 3β-β-catenin-mediated lung cancer metastasis. PMID: 23983127. Disease: paragangliomas 2 (PGL2)
Succinate dehydrogenase complex, subunit A, flavoprotein (Fp)	SDHA	Disease: mitochondrial complex II deficiency (MT-C2D)
Succinate dehydrogenase complex, subunit B, iron sulfur (Ip)	SDHB	Disease: pheochromocytoma (PCC)
Succinate dehydrogenase complex, subunit C, integral membrane protein, 15 kDa	SDHC	Disease: paragangliomas 3 (PGL3)
Succinate dehydrogenase complex, subunit D, integral membrane protein	SDHD	Disease: paragangliomas 1 (PGL1)
Synaptic vesicle glycoprotein 2A	SV2A	**Levetiracetam** – antiepileptics
Synaptic vesicle glycoprotein 2B	SV2B	Lack of synaptic vesicle protein SV2B protects against amyloid-β_{25-35}-induced oxidative stress, cholinergic deficit and cognitive impairment in mice. PMID: 24937053
Synaptic vesicle glycoprotein 2C	SV2C	A role for Sv2c in basal ganglia functions. PMID: 23458503
Syntaxin 1A (brain)	STX1A	Note: STX1A is located in the Williams–Beuren syndrome (WBS) critical region
Syntaxin 16	STX16	Disease: pseudohypoparathyroidism 1B (PHP1B)

HGNC-approved name	Symbol	Target information
TNF receptor-associated protein 1	TRAP1	Expression of TRAP1 predicts poor survival of malignant glioma patients. PMID: 25189320
TNF receptor-associated factor 3	TRAF3	Disease: herpes simplex encephalitis 3 (HSE3)
Trafficking protein particle complex 4	TRAPPC4	Synbindin in extracellular signal-regulated protein kinase spatial regulation and gastric cancer aggressiveness. PMID: 24104608
Trafficking protein particle complex 9	TRAPPC9	Disease: mental retardation, autosomal recessive 13 (MRT13)
Trafficking protein particle complex 11	TRAPPC11	Disease: limb-girdle muscular dystrophy 2S (LGMD2S)
Translocase of inner mitochondrial membrane 8 homolog A (yeast)	TIMM8A	Disease: Mohr–Tranebjaerg syndrome (MTS)
Translocase of inner mitochondrial membrane 17 homolog A (yeast)	TIMM17A	High TIMM17A expression is associated with adverse pathological and clinical outcomes in human breast cancer. PMID: 20972741
Translocase of inner mitochondrial membrane 23 homolog (yeast)	TIMM23	Intra-mitochondrial degradation of Tim23 curtails the survival of cells rescued from apoptosis by caspase inhibitors. PMID: 18174902
Translocase of inner mitochondrial membrane 44 homolog (yeast)	TIMM44	Therapeutic approach for diabetic nephropathy using gene delivery of translocase of inner mitochondrial membrane 44 by reducing mitochondrial superoxide production. PMID: 16510762
Translocase of inner mitochondrial membrane 50 homolog (S. cerevisiae)	TIMM50	Upregulation of the mitochondrial transport protein, Tim50, by mutant p53 contributes to cell growth and chemoresistance. PMID: 21621504
Translocase of inner mitochondrial membrane domain containing 1	TIMMDC1	Depletion of C3orf1/TIMMDC1 inhibits migration and proliferation in 95D lung carcinoma cells. PMID: 25391042
Translocase of outer mitochondrial membrane 22 homolog (yeast)	TOMM22	The mitochondrial import gene tomm22 is specifically required for hepatocyte survival and provides a liver regeneration model. PMID: 20483998
Translocase of outer mitochondrial membrane 34	TOMM34	TOMM34 expression in early invasive breast cancer: A biomarker associated with poor outcome. PMID: 23053644
Translocase of outer mitochondrial membrane 40 homolog (yeast)	TOMM40	TOMM40 alterations in Alzheimer's disease over a 2-year follow-up period. PMID: 25201778
Translocase of outer mitochondrial membrane 70 homolog A (S. cerevisiae)	TOMM70A	Translocase of outer mitochondrial membrane 70 expression is induced by hepatitis C virus and is related to the apoptotic response. PMID: 21412788
Transmembrane channel-like 6	TMC6	Disease: epidermodysplasia verruciformis (EV)
Transmembrane channel-like 8	TMC8	Disease: epidermodysplasia verruciformis (EV)
Transmembrane protein 38B	TMEM38B	Disease: osteogenesis imperfecta 14 (OI14)
Transmembrane protein 43	TMEM43	Disease: arrhythmogenic right ventricular dysplasia, familial, 5 (ARVD5)
Transmembrane protein 70	TMEM70	Disease: mitochondrial complex V deficiency, nuclear 2 (MC5DN2)
Transmembrane protein 79	TMEM79	Note: Defects in TMEM79 may be associated with susceptibility to atopic dermatitis
Transmembrane protein 98	TMEM98	Disease: nanophthalmos 4 (NNO4)
Transmembrane protein 106B	TMEM106B	Disease: ubiquitin-positive frontotemporal dementia (UP-FTD)
Transmembrane protein 126A	TMEM126A	Disease: optic atrophy 7 (OPA7)
Transmembrane protein 138	TMEM138	Disease: Joubert syndrome 16 (JBTS16)
Transmembrane protein 165	TMEM165	Disease: congenital disorder of glycosylation 2K (CDG2K)
Transmembrane protein 173	TMEM173	STING manifests self DNA-dependent inflammatory disease. PMID: 23132945
Transmembrane protein 240	TMEM240	Disease: spinocerebellar ataxia 21 (SCA21)

(Continued)

HGNC-approved name	Symbol	Target information
Tripartite motif containing 3	TRIM3	Decreased expression of TRIM3 is associated with poor prognosis in patients with primary hepatocellular carcinoma. PMID: 24994609
Tripartite motif containing 72, E3 ubiquitin protein ligase	TRIM72	Central role of E3 ubiquitin ligase MG53 in insulin resistance and metabolic disorders. PMID: 23354051
Tuberous sclerosis 1	TSC1	Disease: tuberous sclerosis 1 (TSC1)
Tuberous sclerosis 2	TSC2	Disease: tuberous sclerosis 2 (TSC2)
Tyrosine 3 monooxygenase/tryptophan 5-monooxygenase activation protein, epsilon	YWHAE	Overexpression of 14-3-3ε predicts tumour metastasis and poor survival in hepatocellular carcinoma. PMID: 21401702
Tyrosine 3-monooxygenase/tryptophan 5-monooxygenase activation protein, zeta	YWHAZ	14-3-3ζ as a prognostic marker and therapeutic target for cancer. PMID: 21058923
Ubiquinol–cytochrome c reductase binding protein	UQCRB	Development of a novel class of mitochondrial ubiquinol-cytochrome c reductase binding protein (UQCRB) modulators as promising antiangiogenic leads. PMID: 25242095. Disease: mitochondrial complex III deficiency, nuclear 3 (MC3DN3)
Ubiquinol–cytochrome c reductase core protein II	UQCRC2	Disease: mitochondrial complex III deficiency, nuclear 5 (MC3DN5)
Ubiquinol–cytochrome c reductase hinge protein	UQCRH	UQCRH gene encoding mitochondrial Hinge protein is interrupted by a translocation in a soft-tissue sarcoma and epigenetically inactivated in some cancer cell lines. PMID: 12881716
Ubiquinol–cytochrome c reductase, complex III subunit VII, 9.5 kDa	UQCRQ	Disease: mitochondrial complex III deficiency, nuclear 4 (MC3DN4)
Ubiquinol–cytochrome c reductase, Rieske iron-sulfur polypeptide 1	UQCRFS1	Impaired OXPHOS complex III in breast cancer. PMID: 21901141
unc-13 homolog D (*C. elegans*)	UNC13D	Disease: familial hemophagocytic lymphohistiocytosis 3 (FHL3)
unc-93 homolog B1 (*C. elegans*)	UNC93B1	Disease: herpes simplex encephalitis 1 (HSE1)
Vacuolar protein sorting 33 homolog B (yeast)	VPS33B	Disease: arthrogryposis, renal dysfunction and cholestasis syndrome 1 (ARCS1)
Vacuolar protein sorting 35 homolog (*S. cerevisiae*)	VPS35	Disease: Parkinson disease 17 (PARK17)
Vacuolar protein sorting 37 homolog A (*S. cerevisiae*)	VPS37A	Disease: spastic paraplegia 53, autosomal recessive (SPG53)
Vacuolar protein sorting 37 homolog D (*S. cerevisiae*)	VPS37D	Note: VPS37D is located in the Williams–Beuren syndrome (WBS) critical region
Vacuolar protein sorting 41 homolog (*S. cerevisiae*)	VPS41	VPS41, a protein involved in lysosomal trafficking, is protective in Caenorhabditis elegans and mammalian cellular models of Parkinson's disease. PMID: 19850127
Vacuolar protein sorting 45 homolog (*S. cerevisiae*)	VPS45	Disease: neutropenia, severe congenital 5, autosomal recessive (SCN5)
Vesicle-associated membrane protein 1 (synaptobrevin 1)	VAMP1	**Senrebotase** – urologicals
Vesicle-associated membrane protein 8	VAMP8	VAMP8 is a vesicle SNARE that regulates mucin secretion in airway goblet cells. PMID: 22144578
A kinase (PRKA) anchor protein 10	AKAP10	Disease: sudden cardiac death (SCD)
Adaptor protein, phosphotyrosine interaction, PH domain and leucine zipper containing 2	APPL2	Note: A chromosomal aberration involving APPL2/DIP13B is found in patients with chromosome 22q13.3 deletion syndrome. Translocation t(1222)(q24.1q13.3) with SHANK3/PSAP2
Adenomatous polyposis coli	APC	Disease: familial adenomatous polyposis (FAP)

HGNC-approved name	Symbol	Target information
ADP-ribosylation factor GTPase-activating protein 1	ARFGAP1	Small-molecule synergist of the Wnt/beta-catenin signaling pathway. PMID: 17460038
ADP-ribosylation factor guanine nucleotide exchange factor 2 (brefeldin A-inhibited)	ARFGEF2	Disease: periventricular nodular heterotopia 2 (PVNH2)
ADP-ribosylation factor-like 2 binding protein	ARL2BP	Disease: retinitis pigmentosa with or without situs inversus (RPSI)
Alveolar soft part sarcoma chromosome region, candidate 1	ASPSCR1	Note: A chromosomal aberration involving ASPSCR1 is found in patients with alveolar soft part sarcoma. Translocation t(X17)(p11q25) with TFE3 forms a ASPSCR1-TFE3 fusion protein
ATP synthase mitochondrial F1 complex assembly factor 2	ATPAF2	Disease: mitochondrial complex V deficiency, nuclear 1 (MC5DN1)
Autophagy-related 16-like 1 (*S. cerevisiae*)	ATG16L1	A Crohn's disease variant in Atg16l1 enhances its degradation by caspase 3. PMID: 24553140. Disease: inflammatory bowel disease 10 (IBD10)
BBSome-interacting protein 1	BBIP1	Disease: Bardet–Biedl syndrome 18 (BBS18)
BC1 (ubiquinol–cytochrome c reductase) synthesis-like	BCS1L	Disease: GRACILE syndrome (GRACILE)
B-cell linker	BLNK	Disease: agammaglobulinemia 4, autosomal recessive (AGM4)
B-cell receptor-associated protein 31	BCAP31	Disease: deafness, dystonia, and cerebral hypomyelination (DDCH)
BCL2-interacting killer (apoptosis-inducing)	BIK	BIK/NBK gene as potential marker of prognostic and therapeutic target in breast cancer patients. PMID: 22855140
BCL2-like 1	BCL2L1	**ABT-263** – antineoplastic agents
Beclin 1, autophagy related	BECN1	Identification of a candidate therapeutic autophagy-inducing peptide. PMID: 23364696
Berardinelli–Seip congenital lipodystrophy 2 (seipin)	BSCL2	Disease: congenital generalized lipodystrophy 2 (CGL2)
BH3-like motif containing, cell death inducer	BLID	BRCC2 inhibits breast cancer cell growth and metastasis *in vitro* and in vivo via downregulating AKT pathway. PMID: 23928696
Bicaudal D homolog 2 (*Drosophila*)	BICD2	Disease: spinal muscular atrophy, lower extremity-predominant 2, autosomal dominant (SMALED2)
Biogenesis of lysosomal organelles complex-1, subunit 6, pallidin	BLOC1S6	Disease: Hermansky–Pudlak syndrome 9 (HPS9)
Breast cancer anti-estrogen resistance 1	BCAR1	p130Cas is an essential transducer element in ErbB2 transformation. PMID: 20505116
Calcium-activated nucleotidase 1	CANT1	Disease: Desbuquois dysplasia (DBQD)
Calcium/calmodulin-dependent protein kinase II inhibitor 1	CAMK2N1	The tumor suppressive role of CAMK2N1 in castration-resistant prostate cancer. PMID: 25003983
Calreticulin 3	CALR3	Disease: cardiomyopathy, familial hypertrophic 19 (CMH19)
Caspase recruitment domain family, member 11	CARD11	Disease: persistent polyclonal B-cell lymphocytosis (PPBL)
CCHC-type zinc finger, nucleic acid binding protein	CNBP	Disease: dystrophia myotonica 2 (DM2)
CDGSH iron-sulfur domain 2	CISD2	Disease: Wolfram syndrome 2 (WFS2)
CDK5 regulatory subunit-associated protein 1-like 1	CDKAL1	Disease: diabetes mellitus, non-insulin-dependent (NIDDM)
Cingulin-like 1	CGNL1	Disease: aromatase excess syndrome (AEXS)
ClpB caseinolytic peptidase B homolog (*Escherichia coli*)	CLPB	CLPB variants associated with autosomal-recessive mitochondrial disorder with cataract, neutropenia, epilepsy, and methylglutaconic aciduria. PMID: 25597511

(Continued)

HGNC-approved name	Symbol	Target information
Coenzyme Q9	COQ9	Disease: coenzyme Q10 deficiency, primary, 5 (COQ10D5)
Collagen, type XVII, alpha 1	COL17A1	Disease: generalized atrophic benign epidermolysis bullosa (GABEB)
Collagen-like tail subunit (single strand of homotrimer) of asymmetric acetylcholinesterase	COLQ	Disease: myasthenic syndrome, congenital, Engel type (CMSE)
COX17 cytochrome c oxidase copper chaperone	COX17	Cisplatin binds to human copper chaperone Cox17: The mechanistic implication of drug delivery to mitochondria. PMID: 24473407
C-x(9)-C motif containing 4	CMC4	Note: Overexpressed in T-cell leukemia bearing a t(X14) translocation
CXADR-like membrane protein	CLMP	Disease: congenital short bowel syndrome (CSBS)
Cytochrome b561	CYB561	The trans-membrane cytochrome b561 proteins: Structural information and biological function. PMID: 25163754
DAB2-interacting protein	DAB2IP	Role of DAB2IP in modulating epithelial-to-mesenchymal transition and prostate cancer metastasis. PMID: 20080667. Note: A chromosomal aberration involving DAB2IP is found in a patient with acute myeloid leukemia (AML)
Deafness, autosomal recessive 31	DFNB31	Disease: deafness, autosomal recessive, 31 (DFNB31)
Derlin 1	DERL1	Derlin-1 is overexpressed in non-small cell lung cancer and promotes cancer cell invasion via EGFR-ERK-mediated up-regulation of MMP-2 and MMP-9. PMID: 23306155
Desmoplakin	DSP	Disease: keratoderma, palmoplantar, striate 2 (SPPK2)
Diablo, IAP-binding mitochondrial protein	DIABLO	Solid phase synthesis of Smac/DIABLO-derived peptides using a "Safety-Catch" resin: Identification of potent XIAP BIR3 antagonists. PMID: 23886811. Disease: deafness, autosomal dominant, 64 (DFNA64)
Diazepam-binding inhibitor (GABA receptor modulator, acyl-CoA binding protein)	DBI	Endozepines. PMID: 25600369
Differentially expressed in FDCP 6 homolog (mouse)	DEF6	Overexpression of the Interferon regulatory factor 4-binding protein in human colorectal cancer and its clinical significance. PMID: 19679060
Discs, large homolog 2 (*Drosophila*)	DLG2	PSD93 regulates synaptic stability at neuronal cholinergic synapses. PMID: 14724236
Disrupted in renal carcinoma 2	DIRC2	Disrupted in renal carcinoma 2 (DIRC2), a novel transporter of the lysosomal membrane, is proteolytically processed by cathepsin L. PMID: 21692750
DLC1 Rho GTPase activating protein	DLC1	DLC 1, a candidate tumor suppressor gene, inhibits the proliferation, migration and tumorigenicity of human nasopharyngeal carcinoma cells. PMID: 23588806
DNA-damage-inducible transcript 4	DDIT4	**PF-04523655** – opthalmologicals
Doublecortin	DCX	Disease: lissencephaly, X-linked 1 (LISX1)
Dual oxidase maturation factor 2	DUOXA2	Disease: thyroid dyshormonogenesis 5 (TDH5)
Dymeclin	DYM	Disease: Dyggve–Melchior–Clausen syndrome (DMC)
Dystrobrevin binding protein 1	DTNBP1	Disease: Hermansky–Pudlak syndrome 7 (HPS7)
Dystrobrevin, alpha	DTNA	Disease: left ventricular non-compaction 1 (LVNC1)
EH domain binding protein 1	EHBP1	Disease: prostate cancer, hereditary, 12 (HPC12)
EH domain containing 1	EHD1	Structured cyclic peptides that bind the EH domain of EHD1. PMID: 25014215
ELKS/RAB6-interacting/CAST family member 1	ERC1	Disease: thyroid papillary carcinoma (TPC)
Epidermal growth factor receptor pathway substrate 15	EPS15	Note: A chromosomal aberration involving EPS15 is found in acute leukemias. Translocation t(111)(p32q23) with KMT2A/MLL1. The result is a rogue activator protein

HGNC-approved name	Symbol	Target information
Epsin 1	EPN1	Are epsins a therapeutic target for tumor angiogenesis? PMID: 23187137
ER lipid raft associated 2	ERLIN2	Erlin-2 is associated with active γ-secretase in brain and affects amyloid β-peptide production. PMID: 22771797. Disease: spastic paraplegia 18, autosomal recessive (SPG18)
ER membrane-associated RNA degradation	ERMARD	Disease: periventricular nodular heterotopia 6 (PVNH6)
Family with sequence similarity 65, member B	FAM65B	Note: FAM65B mutations may be a cause of non-syndromic deafness. A splice site mutation causing in-frame skipping of exon 3 has been found in a large consanguineous kindred with recessive non-syndromic, prelingual, profound hearing loss
FIG4 homolog, SAC1 lipid phosphatase domain containing (*S. cerevisiae*)	FIG4	Disease: Charcot–Marie–Tooth disease 4J (CMT4J)
Filamin A-interacting protein 1-like	FILIP1L	Filamin A interacting protein 1-like as a therapeutic target in cancer. PMID: 25200207
Filamin C, gamma	FLNC	Disease: myopathy, myofibrillar, 5 (MFM5)
Frataxin	FXN	Disease: Friedreich ataxia (FRDA)
Fukutin	FKTN	Disease: muscular dystrophy–dystroglycanopathy congenital with brain and eye anomalies A4 (MDDGA4)
FYVE and coiled-coil domain containing 1	FYCO1	Disease: cataract 18 (CTRCT18)
FYVE, RhoGEF and PH domain containing 1	FGD1	Disease: Aarskog–Scott syndrome (AAS)
Gamma-secretase-activating protein	GSAP	Gamma-secretase activating protein is a therapeutic target for Alzheimer's disease. PMID: 20811458
GDP dissociation inhibitor 1	GDI1	Disease: mental retardation, X-linked 41 (MRX41)
Gephyrin	GPHN	Disease: molybdenum cofactor deficiency, complementation group C (MOCODC)
Glutamate receptor-interacting protein 1	GRIP1	Disease: Fraser syndrome (FRASS)
Glycosylphosphatidylinositol-anchored high-density lipoprotein binding protein 1	GPIHBP1	Disease: hyperlipoproteinemia 1D (HLPP1D)
GM2 ganglioside activator	GM2A	Disease: GM2-gangliosidosis AB (GM2GAB)
GNAS complex locus	GNAS	Disease: ACTH-independent macronodular adrenal hyperplasia (AIMAH)
Golgi membrane protein 1	GOLM1	Expression of GOLPH2 is associated with the progression of and poor prognosis in gastric cancer. PMID: 25119897
Golgi phosphoprotein 3 (coat-protein)	GOLPH3	GOLPH3 regulates the migration and invasion of glioma cells though RhoA. PMID: 23500462
Golgi SNAP receptor complex member 2	GOSR2	Disease: epilepsy, progressive myoclonic 6 (EPM6)
Golgi-associated PDZ and coiled-coil motif containing	GOPC	Note: A chromosomal aberration involving GOPC is found in a glioblastoma multiforme sample
Golgin A5	GOLGA5	Disease: thyroid papillary carcinoma (TPC)
Golgin, RAB6-interacting	GORAB	Disease: geroderma osteodysplasticum (GO)
GRB2-associated binding protein 2	GAB2	Gab2 signaling in chronic myeloid leukemia cells confers resistance to multiple Bcr-Abl inhibitors. PMID: 22858987
GTP binding protein 2	GTPBP2	Ribosome stalling induced by mutation of a CNS-specific tRNA causes neurodegeneration. PMID: 25061210
Guanine nucleotide binding protein (G protein), alpha activating activity polypeptide, olfactory type	GNAL	Disease: dystonia 25 (DYT25)
Guanine nucleotide binding protein (G protein), beta polypeptide 4	GNB4	Disease: Charcot–Marie–Tooth disease, dominant, intermediate type, F (CMTDIF)
Guanylate cyclase activator 1A (retina)	GUCA1A	Disease: cone dystrophy 3 (COD3)

(*Continued*)

HGNC-approved name	Symbol	Target information
HCLS1-associated protein X-1	HAX1	Expression of HAX-1 in human colorectal cancer and its clinical significance. PMID: 24057929. Disease: neutropenia, severe congenital 3, autosomal recessive (SCN3)
Hermansky–Pudlak syndrome 6	HPS6	Disease: Hermansky–Pudlak syndrome 6 (HPS6)
Homocysteine-inducible, endoplasmic reticulum stress-inducible, ubiquitin-like domain member 1	HERPUD1	A deficiency of Herp, an endoplasmic reticulum stress protein, suppresses atherosclerosis in ApoE knockout mice by attenuating inflammatory responses. PMID: 24204574
Immediate early response 3-interacting protein 1	IER3IP1	Disease: microcephaly, epilepsy, and diabetes syndrome (MEDS)
Integral membrane protein 2B	ITM2B	Disease: cerebral amyloid angiopathy, ITM2B-related 1 (CAA-ITM2B1)
Intraflagellar transport 122 homolog (*Chlamydomonas*)	IFT122	Disease: cranioectodermal dysplasia 1 (CED1)
Iron-sulfur cluster assembly enzyme	ISCU	Disease: myopathy with exercise intolerance Swedish type (MEIS)
Junction plakoglobin	JUP	Disease: Naxos disease (NXD)
KIAA0226	KIAA0226	Disease: spinocerebellar ataxia, autosomal recessive, 15 (SCAR15)
Late endosomal/lysosomal adaptor, MAPK and MTOR activator 2	LAMTOR2	Disease: immunodeficiency due to defect in MAPBP-interacting protein (ID-MAPBPIP)
Leucine-rich repeat containing 6	LRRC6	Disease: ciliary dyskinesia, primary, 19 (CILD19)
Leucine zipper, putative tumor suppressor 1	LZTS1	Disease: esophageal cancer (ESCR)
Leucine-rich pentatricopeptide repeat containing	LRPPRC	Disease: Leigh syndrome French-Canadian type (LSFC)
Leucine-rich repeat, immunoglobulin-like and transmembrane domains 3	LRIT3	Disease: night blindness, congenital stationary, 1F (CSNB1F)
Lipase maturation factor 1	LMF1	Disease: combined lipase deficiency (CLD)
Lipopolysaccharide-induced TNF factor	LITAF	Disease: Charcot–Marie–Tooth disease 1C (CMT1C)
Low-density lipoprotein receptor-related protein-associated protein 1	LRPAP1	Disease: myopia 23, autosomal recessive (MYP23)
MAGE-like 2	MAGEL2	Note: May play a role in Prader–Willi syndrome (PWS) which is a contiguous gene syndrome resulting from inactivity of the paternal copies of a number of genes on 15q11
Mannose-6-phosphate receptor (cation dependent)	M6PR	**Alglucosidase alfa** – enzyme replacement agents
Mannosidase, alpha, class 1B, member 1	MAN1B1	Disease: mental retardation, autosomal recessive 15 (MRT15)
MAP-kinase-activating death domain	MADD	MADD promotes the survival of human lung adenocarcinoma cells by inhibiting apoptosis. PMID: 23443411
MCF.2 cell line-derived transforming sequence	MCF2	Note: MCF2 and DBL represent two activated versions of the same proto oncogene
Membrane-associated guanylate kinase, WW and PDZ domain containing 2	MAGI2	Expression profile of MAGI2 gene as a novel biomarker in combination with major deregulated genes in prostate cancer. PMID: 24985972
Metaxin 1	MTX1	Metaxin deficiency alters mitochondrial membrane permeability and leads to resistance to TNF-induced cell killing. PMID: 21088703
Microtubule-associated tumor suppressor 1	MTUS1	ATIP3, a novel prognostic marker of breast cancer patient survival, limits cancer cell migration and slows metastatic progression by regulating microtubule dynamics. PMID: 23396587. Disease: hepatocellular carcinoma (HCC)
Mitochondrially encoded cytochrome b	MT-CYB	Note: Defects in MT-CYB are a rare cause of mitochondrial dysfunction underlying different myopathies. They include mitochondrial encephalomyopathy, hypertrophic cardiomyopathy (HCM), and sporadic mitochondrial myopathy (MM)

HGNC-approved name	Symbol	Target information
MpV17 mitochondrial inner membrane protein	MPV17	Disease: mitochondrial DNA depletion syndrome 6 (MTDPS6)
Multiple coagulation factor deficiency 2	MCFD2	Disease: factor V and factor VIII combined deficiency 2 (F5F8D2)
Myeloid/lymphoid or mixed-lineage leukemia (trithorax homolog, *Drosophila*); translocated to, 4	MLLT4	Note: A chromosomal aberration involving MLLT4 is associated with acute leukemias. Translocation t(611)(q27q23) with KMT2A/MLL1. The result is a rogue activator protein
Myosin regulatory light chain interacting protein	MYLIP	The E3 ubiquitin ligase IDOL induces the degradation of the low density lipoprotein receptor family members VLDLR and ApoER2. PMID: 20427281
Myosin VI	MYO6	Myosin VI is a modulator of androgen-dependent gene expression. PMID: 19787211. Disease: deafness, autosomal dominant, 22 (DFNA22)
Myosin, heavy chain 11, smooth muscle	MYH11	Note: A chromosomal aberration involving MYH11 is found in acute myeloid leukemia of M4EO subtype
NADH dehydrogenase (ubiquinone) Fe–S protein 4, 18 kDa (NADH–coenzyme Q reductase)	NDUFS4	Disease: mitochondrial complex I deficiency (MT-C1D)
NECAP endocytosis associated 1	NECAP1	Disease: epileptic encephalopathy, early infantile, 21 (EIEE21)
NEDD4 binding protein 3	N4BP3	The Nedd4-binding protein 3 (N4BP3) is crucial for axonal and dendritic branching in developing neurons. PMID: 24044555
Nephronophthisis 1 (juvenile)	NPHP1	Disease: nephronophthisis 1 (NPHP1)
Neurobeachin-like 2	NBEAL2	Disease: gray platelet syndrome (GPS)
NFU1 iron-sulfur cluster scaffold homolog (*S. cerevisiae*)	NFU1	Disease: multiple mitochondrial dysfunctions syndrome 1 (MMDS1)
Nicastrin	NCSTN	Disease: acne inversa, familial, 1 (ACNINV1)
Nitric oxide associated 1	NOA1	hNOA1 interacts with complex I and DAP3 and regulates mitochondrial respiration and apoptosis. PMID: 19103604
Nucleotide binding protein-like	NUBPL	Disease: mitochondrial complex I deficiency (MT-C1D)
Oculocerebrorenal syndrome of Lowe	OCRL	Disease: Lowe oculocerebrorenal syndrome (OCRL)
Oligophrenin 1	OPHN1	Disease: mental retardation, X-linked, syndromic, OPHN1-related (MRXSO)
Optic atrophy 3 (autosomal recessive, with chorea and spastic paraplegia)	OPA3	Disease: 3-methylglutaconic aciduria 3 (MGA3)
ORM1-like 3 (*S. cerevisiae*)	ORMDL3	Disease: asthma (ASTHMA)
Osteopetrosis-associated transmembrane protein 1	OSTM1	Disease: osteopetrosis, autosomal recessive 5 (OPTB5)
Otoferlin	OTOF	Disease: deafness, autosomal recessive, 9 (DFNB9)
PDZ and LIM domain 7 (enigma)	PDLIM7	Loss of the cytoskeletal protein Pdlim7 predisposes mice to heart defects and hemostatic dysfunction. PMID: 24278323
Perilipin 1	PLIN1	Disease: lipodystrophy, familial partial, 4 (FPLD4)
Phorbol-12-myristate-13-acetate-induced protein 1	PMAIP1	Noxa in rheumatic diseases: Present understanding and future impact. PMID: 24352336
Phosphatidylethanolamine binding protein 4	PEBP4	Phosphatidylethanolamine-binding protein 4 is associated with breast cancer metastasis through Src-mediated Akt tyrosine phosphorylation. PMID: 24276246
Phosphatidylinositol-binding clathrin assembly protein	PICALM	Note: A chromosomal aberration involving PICALM is found in diffuse histiocytic lymphomas. Translocation t(1011)(p13q14) with MLLT10
Phosphodiesterase 4D-interacting protein	PDE4DIP	Note: A chromosomal aberration involving PDE4DIP may be the cause of a myeloproliferative disorder (MBD) associated with eosinophilia. Translocation t(15)(q23q33) that forms a PDE4DIP-PDGFRB fusion protein
Phosphofurin acidic cluster sorting protein 1	PACS1	Disease: mental retardation, autosomal dominant 17 (MRD17)

(Continued)

HGNC-approved name	Symbol	Target information
Phospholipase A2 activating protein	PLAA	Membrane-mediated actions of 1,25-dihydroxy vitamin D3: A review of the roles of phospholipase A2 activating protein and Ca2+/calmodulin-dependent protein kinase II. PMID: 25448737
Piezo-type mechanosensitive ion channel component 1	PIEZO1	Piezo1 is as a novel trefoil factor family 1 binding protein that promotes gastric cancer cell mobility *in vitro*. PMID: 24798994. Disease: dehydrated hereditary stomatocytosis with or without pseudohyperkalemia and/or perinatal edema (DHS)
PITPNM family member 3	PITPNM3	Disease: cone-rod dystrophy 5 (CORD5)
Plasmalemma vesicle-associated protein	PLVAP	Plasmalemmal vesicle associated protein (PLVAP) as a therapeutic target for treatment of hepatocellular carcinoma. PMID: 25376302
Polymerase (DNA directed), gamma 2, accessory subunit	POLG2	Disease: progressive external ophthalmoplegia with mitochondrial DNA deletions, autosomal dominant, 4 (PEOA4)
Polymerase I and transcript release factor	PTRF	Disease: congenital generalized lipodystrophy 4 (CGL4)
Porcupine homolog (*Drosophila*)	PORCN	**LGK974** – antineoplastic agents. Disease: focal dermal hypoplasia (FODH)
Potassium channel tetramerization domain containing 12	KCTD12	Pfetin as a prognostic biomarker in gastrointestinal stromal tumor: Novel monoclonal antibody and external validation study in multiple clinical facilities. PMID: 19815537
Premature ovarian failure, 1B	POF1B	Disease: premature ovarian failure 2B (POF2B)
Presenilin enhancer gamma secretase subunit	PSENEN	Disease: acne inversa, familial, 2 (ACNINV2)
Programmed cell death 10	PDCD10	Exceptional aggressiveness of cerebral cavernous malformation disease associated with PDCD10 mutations. PMID: 25122144. Disease: cerebral cavernous malformations 3 (CCM3)
Prosaposin	PSAP	Disease: combined saposin deficiency (CSAPD)
Protein kinase C substrate 80K-H	PRKCSH	Disease: polycystic liver disease (PCLD)
Protein phosphatase 1, regulatory subunit 15A	PPP1R15A	Selective inhibition of a regulatory subunit of protein phosphatase 1 restores proteostasis. PMID: 21385720
Rab-interacting lysosomal protein	RILP	RILP regulates vacuolar ATPase through interaction with the V1G1 subunit. PMID: 24762812
RAB11 family-interacting protein 1 (class I)	RAB11FIP1	RCP is a human breast cancer-promoting gene with Ras-activating function. PMID: 19620787
RAB33B, member RAS oncogene family	RAB33B	Disease: Smith–McCort dysplasia 2 (SMC2)
Radial spoke head 1 homolog (*Chlamydomonas*)	RSPH1	Disease: ciliary dyskinesia, primary, 24 (CILD24)
Ral GEF with PH domain and SH3 binding motif 2	RALGPS2	Note: RALGPS2 is a potential candidate gene for susceptibility to Alzheimer disease linked to 1q24
ralA binding protein 1	RALBP1	Structure and function of RLIP76 (RalBP1): An intersection point between Ras and Rho signalling. PMID: 24450627
Rap guanine nucleotide exchange factor (GEF) 1	RAPGEF1	Signalling to actin: role of C3G, a multitasking guanine-nucleotide-exchange factor. PMID: 21366540
Reactive oxygen species modulator 1	ROMO1	Romo1 is associated with ROS production and cellular growth in human gliomas. PMID: 25193023
Receptor accessory protein 1	REEP1	Disease: spastic paraplegia 31, autosomal dominant (SPG31)
Required for meiotic nuclear division 1 homolog (*S. cerevisiae*)	RMND1	Disease: combined oxidative phosphorylation deficiency 11 (COXPD11)
Retinal pigment epithelium-specific protein 65 kDa	RPE65	Disease: Leber congenital amaurosis 2 (LCA2)
Rho GTPase-activating protein 26	ARHGAP26	Disease: leukemia, juvenile myelomonocytic (JMML)
Seizure threshold 2 homolog (mouse)	SZT2	Disease: epileptic encephalopathy, early infantile, 18 (EIEE18)

HGNC-approved name	Symbol	Target information
Selenoprotein N, 1	SEPN1	Disease: rigid spine muscular dystrophy 1 (RSMD1)
Sequestosome 1	SQSTM1	Increased signaling through p62 in the marrow microenvironment increases myeloma cell growth and osteoclast formation. PMID: 19282458. Disease: Paget disease of bone (PDB)
SH3 and multiple ankyrin repeat domains 3	SHANK3	Note: A chromosomal aberration involving SHANK3 is found in patients with chromosome 22q13.3 deletion syndrome. Translocation t(1222) (q24.1q13.3) with APPL2/DIP13B. Note: Defects in SHANK3 are associated with neuropsychiatric disorders
SH3-domain GRB2-like 1	SH3GL1	Note: In some cases of acute leukemia, a translocation results in the formation of a KMT2A/MLL1-EEN fusion gene
SHC (Src homology 2 domain containing) transforming protein 1	SHC1	Novel role of p66Shc in ROS-dependent VEGF signaling and angiogenesis in endothelial cells. PMID: 22101521
Signal recognition particle 72 kDa	SRP72	Disease: bone marrow failure syndrome 1 (BMFS1)
SIL1 nucleotide exchange factor	SIL1	Disease: Marinesco–Sjogren syndrome (MSS)
Single-pass membrane protein with aspartate-rich tail 1	SMDT1	EMRE is an essential component of the mitochondrial calcium uniporter complex. PMID: 24231807
Solute carrier family 9, subfamily A (NHE3, cation proton antiporter 3), member 3 regulator 1	SLC9A3R1	Disease: nephrolithiasis/osteoporosis, hypophosphatemic, 2 (NPHLOP2)
Spermatogenesis associated 16	SPATA16	Disease: spermatogenic failure 6 (SPGF6)
S-phase kinase-associated protein 1	SKP1	Targeting SKP1, an ubiquitin E3 ligase component found decreased in sporadic Parkinson's disease. PMID: 22205206
Sprouty homolog 4 (*Drosophila*)	SPRY4	Disease: hypogonadotropic hypogonadism 17 with or without anosmia (HH17)
StAR related lipid transfer (START) domain containing 13	STARD13	StarD13 is a tumor suppressor in breast cancer that regulates cell motility and invasion. PMID: 24627003
Staufen double-stranded RNA binding protein 1	STAU1	The RNA-binding protein Staufen1 is increased in DM1 skeletal muscle and promotes alternative pre-mRNA splicing. PMID: 22431750
Steroidogenic acute regulatory protein	STAR	Upregulation of steroidogenic acute regulatory protein by hypoxia stimulates aldosterone synthesis in pulmonary artery endothelial cells to promote pulmonary vascular fibrosis. PMID: 25001622. Disease: adrenal hyperplasia 1 (AH1)
Sterol regulatory element binding transcription factor 2	SREBF2	Sterol regulatory element binding protein 2 activation of NLRP3 inflammasome in endothelium mediates hemodynamic-induced atherosclerosis susceptibility. PMID: 23838163
Surfeit 1	SURF1	Disease: Leigh syndrome (LS)
Synapsin I	SYN1	Disease: epilepsy X-linked, with variable learning disabilities and behavior disorders (XELBD)
Synaptophysin	SYP	Disease: mental retardation, X-linked, SYP-related (MRXSYP)
Synaptosomal-associated protein, 29 kDa	SNAP29	Disease: cerebral dysgenesis, neuropathy, ichthyosis, and palmoplantar keratoderma syndrome (CEDNIK)
Synaptotagmin II	SYT2	**Botulinum toxin type B** – antispasmodics
Syntaxin binding protein 1	STXBP1	Disease: epileptic encephalopathy, early infantile, 4 (EIEE4)
Synuclein, alpha (non-A4 component of amyloid precursor)	SNCA	Gallic acid interacts with α-synuclein to prevent the structural collapse necessary for its aggregation. PMID: 24769497. Note: Genetic alterations of SNCA resulting in aberrant polymerization into fibrils, are associated with several neurodegenerative diseases
Talin 1	TLN1	A directional switch of integrin signalling and a new anti-thrombotic strategy. PMID: 24162846

(Continued)

HGNC-approved name	Symbol	Target information
TBC1 domain family, member 7	TBC1D7	Disease: megalencephaly, autosomal recessive (MGCPH)
T-cell lymphoma invasion and metastasis 1	TIAM1	Elevated expression of T-lymphoma invasion and metastasis inducing factor 1 in squamous-cell carcinoma of the head and neck and its clinical significance. PMID: 24189000
Testis-derived transcript (3 LIM domains)	TES	TESTIN suppresses tumor growth and invasion via manipulating cell cycle progression in endometrial carcinoma. PMID: 24929083
Tetraspanin 1	TSPAN1	The effect of NET-1 on the proliferation, migration and endocytosis of the SMMC-7721 HCC cell line. PMID: 22378020
Tetratricopeptide repeat domain 19	TTC19	Disease: mitochondrial complex III deficiency, nuclear 2 (MC3DN2)
Thyroid hormone receptor interactor 11	TRIP11	Note: A chromosomal aberration involving TRIP11 may be a cause of acute myelogenous leukemia. Translocation t(5;14)(q33;q32) with PDGFRB. The fusion protein may be involved in clonal evolution of leukemia and eosinophilia. Disease: achondrogenesis 1A (ACG-IA)
TIA1 cytotoxic granule-associated RNA binding protein	TIA1	Disease: Welander distal myopathy (WDM)
Tight junction protein 2	TJP2	Disease: familial hypercholanemia (FHCA)
Transforming growth factor beta 1-induced transcript 1	TGFB1I1	Hydrogen peroxide-inducible clone 5 (Hic-5) as a potential therapeutic target for vascular and other disorders. PMID: 22472216
Translational activator of mitochondrially encoded cytochrome c oxidase I	TACO1	Disease: Leigh syndrome (LS)
Transmembrane and coiled-coil domains 1	TMCO1	Disease: craniofacial dysmorphism, skeletal anomalies and mental retardation syndrome (CFSMR)
Transmembrane BAX inhibitor motif containing 6	TMBIM6	Lentivirus-mediated RNA interference targeting Bax inhibitor-1 suppresses ex vivo cell proliferation and in vivo tumor growth of nasopharyngeal carcinoma. PMID: 21545297
Ts translation elongation factor, mitochondrial	TSFM	Disease: combined oxidative phosphorylation deficiency 3 (COXPD3)
Tubby-like protein 1	TULP1	Disease: retinitis pigmentosa 14 (RP14)
Tumor suppressor candidate 3	TUSC3	Disease: mental retardation, autosomal recessive 7 (MRT7)
Ubiquinol–cytochrome c reductase complex assembly factor 2	UQCC2	Disease: mitochondrial complex III deficiency, nuclear 7 (MC3DN7)
Up-regulated during skeletal muscle growth 5 homolog (mouse)	USMG5	Knockdown of DAPIT (diabetes-associated protein in insulin-sensitive tissue) results in loss of ATP synthase in mitochondria. PMID: 21345788
Uroplakin 3A	UPK3A	Disease: renal adysplasia (RADYS)
Utrophin	UTRN	Daily treatment with SMTC1100, a novel small molecule utrophin upregulator, dramatically reduces the dystrophic symptoms in the mdx mouse. PMID: 21573153
UV radiation resistance associated	UVRAG	Note: A chromosomal aberration involving UVRAG has been observed in a patient with heterotaxy (left-right axis malformation). Inversion Inv(11)(q13.5q25)
Vacuolar protein sorting 53 homolog (S. cerevisiae)	VPS53	Disease: pontocerebellar hypoplasia 2E (PCH2E)
Vacuole membrane protein 1	VMP1	Novel role of VMP1 as modifier of the pancreatic tumor cell response to chemotherapeutic drugs. PMID: 23460482
VAMP (vesicle-associated membrane protein)-associated protein B and C	VAPB	Disease: amyotrophic lateral sclerosis 8 (ALS8)
VMA21 vacuolar H+-ATPase homolog (S. cerevisiae)	VMA21	Disease: myopathy, X-linked, with excessive autophagy (MEAX)

HGNC-approved name	Symbol	Target information
VPS33B-interacting protein, apical–basolateral polarity regulator, spe-39 homolog	VIPAS39	Disease: arthrogryposis, renal dysfunction and cholestasis syndrome 2 (ARCS2)
WAS/WASL-interacting protein family, member 1	WIPF1	Disease: Wiskott–Aldrich syndrome 2 (WAS2)
Wolfram syndrome 1 (wolframin)	WFS1	Disease: Wolfram syndrome 1 (WFS1)
yrdC *N*(6)-threonylcarbamoyltransferase domain containing	YRDC	Ischemia-reperfusion-inducible protein modulates cell sensitivity to anticancer drugs by regulating activity of efflux transporter. PMID: 19279227
Zinc finger, FYVE domain containing 27	ZFYVE27	Disease: spastic paraplegia 33, autosomal dominant (SPG33)

6.6 Cytoplasmic proteins

HGNC-approved name	Symbol	Target information
Baculoviral IAP repeat containing 2	BIRC2	**Birinapant** – antineoplastic agents
Baculoviral IAP repeat containing 3	BIRC3	Note: A chromosomal aberration involving BIRC3 is recurrent in low-grade mucosa-associated lymphoid tissue (MALT lymphoma). Translocation t(11;18)(q21;q21) with MALT1
Baculoviral IAP repeat containing 8	BIRC8	Differential expression of BIRC family genes in chronic myeloid leukaemia—BIRC3 and BIRC8 as potential new candidates to identify disease progression. PMID: 24266799
Caspase recruitment domain family, member 9	CARD9	Disease: candidiasis, familial, 2 (CANDF2)
Caspase recruitment domain family, member 10	CARD10	CARMA3 is crucial for EGFR-induced activation of NF-κB and tumor progression. PMID: 21406399
Caspase recruitment domain family, member 14	CARD14	Disease: psoriasis 2 (PSORS2)
Charged multivesicular body protein 2B	CHMP2B	Disease: frontotemporal dementia, chromosome 3-linked (FTD3)
Charged multivesicular body protein 4B	CHMP4B	Disease: cataract 31, multiple types (CTRCT31)
Chromosome 8 open reading frame 37	C8orf37	Disease: cone-rod dystrophy 16 (CORD16)
Chromosome 12 open reading frame 57	C12orf57	Disease: Temtamy syndrome (TEMTYS)
Chromosome 21 open reading frame 59	C21orf59	Disease: ciliary dyskinesia, primary, 26 (CILD26)
Choroideremia (Rab escort protein 1)	CHM	Disease: choroideremia (CHM)
Diaphanous-related formin 2	DIAPH2	Disease: premature ovarian failure 2A (POF2A)
Diaphanous-related formin 3	DIAPH3	Disease: auditory neuropathy, autosomal dominant, 1 (AUNA1)
DnaJ (Hsp40) homolog, subfamily B, member 4	DNAJB4	Tumour suppressor HLJ1: A potential diagnostic, preventive and therapeutic target in non-small cell lung cancer. PMID: 25493224
DnaJ (Hsp40) homolog, subfamily C, member 6	DNAJC6	DNAJC6 promotes hepatocellular carcinoma progression through induction of epithelial-mesenchymal transition. PMID: 25446072. Disease: Parkinson disease 19, juvenile-onset (PARK19)
Dynein, axonemal, assembly factor 2	DNAAF2	Disease: ciliary dyskinesia, primary, 10 (CILD10)
Dynein, axonemal, assembly factor 3	DNAAF3	Disease: ciliary dyskinesia, primary, 2 (CILD2)
Dynein, axonemal, assembly factor 5	DNAAF5	Disease: ciliary dyskinesia, primary, 18 (CILD18)

(Continued)

HGNC-approved name	Symbol	Target information
Eukaryotic translation initiation factor 3, subunit B	EIF3B	Translation initiation factor eIF3b expression in human cancer and its role in tumor growth and lung colonization. PMID: 23575475
Eukaryotic translation initiation factor 3, subunit C	EIF3C	eIF3c: A potential therapeutic target for cancer. PMID: 23623922
Eukaryotic translation initiation factor 4E	EIF4E	**ISIS-EIF4ERx** – antineoplastic agents. Disease: autism 19 (AUTS19)
F-box and leucine-rich repeat protein 4	FBXL4	Disease: mitochondrial DNA depletion syndrome 13 (MTDPS13)
F-box and leucine-rich repeat protein 5	FBXL5	FBXL5-mediated degradation of single-stranded DNA-binding protein hSSB1 controls DNA damage response. PMID: 25249620
F-box protein 7	FBXO7	Disease: Parkinson disease 15 (PARK15)
F-box protein 10	FBXO10	Note: Defects in FBXO10 may be a cause of diffuse large B-cell lymphoma by allowing the accumulation of BCL2, an oncoprotein that has a critical role in lymphomas. PMID: 23431138
F-box protein 38	FBXO38	Disease: neuronopathy, distal hereditary motor, 2D (HMN2D)
Filamin A, alpha	FLNA	Reducing amyloid-related Alzheimer's disease pathogenesis by a small molecule targeting filamin A. PMID: 22815492. Disease: periventricular nodular heterotopia 1 (PVNH1)
Filamin B, beta	FLNB	Note: Interaction with FLNA may compensate for dysfunctional FLNA homodimer in the periventricular nodular heterotopia (PVNH) disorder. Disease: atelosteogenesis 1 (AO1)
GPN-loop GTPase 1	GPN1	Human GTPases associate with RNA polymerase II to mediate its nuclear import. PMID: 21768307
GPN-loop GTPase 3	GPN3	Human GTPases associate with RNA polymerase II to mediate its nuclear import. PMID: 21768307
Hermansky–Pudlak syndrome 3	HPS3	Disease: Hermansky–Pudlak syndrome 3 (HPS3)
Hermansky–Pudlak syndrome 5	HPS5	Disease: Hermansky–Pudlak syndrome 5 (HPS5)
Kinase suppressor of ras 1	KSR1	The dual function of KSR1: A pseudokinase and beyond. PMID: 23863182
Kinase suppressor of ras 2	KSR2	KSR2 mutations are associated with obesity, insulin resistance, and impaired cellular fuel oxidation. PMID: 24209692
Methylmalonic aciduria (cobalamin deficiency) cblC type, with homocystinuria	MMACHC	Disease: methylmalonic aciduria and homocystinuria type cblC (MMAHCC)
Methylmalonic aciduria (cobalamin deficiency) cblD type, with homocystinuria	MMADHC	Disease: methylmalonic aciduria and homocystinuria type cblD (MMAHCD)
Musashi RNA binding protein 1	MSI1	Musashi1 as a potential therapeutic target and diagnostic marker for lung cancer. PMID: 23715514
Musashi RNA binding protein 2	MSI2	Musashi-2 controls cell fate, lineage bias, and TGF-β signaling in HSCs. PMID: 24395885
Neutrophil cytosolic factor 1	NCF1	Mice lacking NCF1 exhibit reduced growth of implanted melanoma and carcinoma tumors. PMID: 24358335
Neutrophil cytosolic factor 2	NCF2	NCF2/p67phox: A novel player in the anti-apoptotic functions of p53. PMID: 23255096
Neutrophil cytosolic factor 4, 40 kDa	NCF4	Activation of NADPH oxidase subunit NCF4 induces ROS-mediated EMT signaling in HeLa cells. PMID: 24378533
NLR family, pyrin domain containing 1	NLRP1	Amyloid-β induces NLRP1-dependent neuronal pyroptosis in models of Alzheimer's disease. PMID: 25144717. Disease: vitiligo (VTLG)
NLR family, pyrin domain containing 2	NLRP2	Human astrocytes express a novel NLRP2 inflammasome. PMID: 23625868

HGNC-approved name	Symbol	Target information
NLR family, pyrin domain containing 3	NLRP3	NLRP3 inflammasome: From a danger signal sensor to a regulatory node of oxidative stress and inflammatory diseases. PMID: 25625584
NLR family, pyrin domain containing 4	NLRP4	NLRP4 negatively regulates autophagic processes through an association with beclin1. PMID: 21209283
NLR family, pyrin domain containing 5	NLRP5	Potential role for MATER in cytoplasmic lattice formation in murine oocytes. PMID: 20830304
NLR family, pyrin domain containing 6	NLRP6	NLRP6 inflammasome orchestrates the colonic host-microbial interface by regulating goblet cell mucus secretion. PMID: 24581500
NLR family, pyrin domain containing 7	NLRP7	Disease: hydatidiform mole, recurrent, 1 (HYDM1)
NLR family, pyrin domain containing 10	NLRP10	Roles of NLRP10 in innate and adaptive immunity. PMID: 23562614
NLR family, pyrin domain containing 12	NLRP12	The multifaceted nature of NLRP12. PMID: 25249449
NLR family, pyrin domain containing 14	NLRP14	Mutations in the testis-specific NALP14 gene in men suffering from spermatogenic failure. PMID: 16931801
Phospholipase C-like 1	PLCL1	Tumor suppressor role of phospholipase C epsilon in Ras-triggered cancers. PMID: 24591640
Phospholipase C-like 2	PLCL2	Role of phospholipase C-L2, a novel phospholipase C-like protein that lacks lipase activity, in B-cell receptor signaling. PMID: 14517301
Protein phosphatase 1, regulatory (inhibitor) subunit 1B	PPP1R1B	Darpp-32 and t-Darpp are differentially expressed in normal and malignant mouse mammary tissue. PMID: 25128420
Protein phosphatase 1, regulatory subunit 8	PPP1R8	A role for PP1/NIPP1 in steering migration of human cancer cells. PMID: 22815811
Protein phosphatase 1, regulatory subunit 9B	PPP1R9B	Spinophilin loss correlates with poor patient prognosis in advanced stages of colon carcinoma. PMID: 23729363
Protein phosphatase 1, regulatory subunit 12A	PPP1R12A	Myosin phosphatase target subunit: Many roles in cell function. PMID: 18155661
Protein phosphatase 1, regulatory subunit 12B	PPP1R12B	Characterization and function of MYPT2, a target subunit of myosin phosphatase in heart. PMID: 16431080
Protein phosphatase 1, regulatory (inhibitor) subunit 14A	PPP1R14A	Role of CPI-17 in restoring skin homoeostasis in cutaneous field of cancerization: Effects of topical application of a film-forming medical device containing photolyase and UV filters. PMID: 23800065
Protein phosphatase 1, regulatory (inhibitor) subunit 14C	PPP1R14C	Expression of the protein phosphatase 1 inhibitor KEPI is downregulated in breast cancer cell lines and tissues and involved in the regulation of the tumor suppressor EGR1 via the MEK-ERK pathway. PMID: 17516844
Protein phosphatase 2, regulatory subunit A, alpha	PPP2R1A	PR65A phosphorylation regulates PP2A complex signaling. PMID: 24465463
Protein phosphatase 2, regulatory subunit B', alpha	PPP2R5A	Loss of PPP2R2A inhibits homologous recombination DNA repair and predicts tumor sensitivity to PARP inhibition. PMID: 23087057
Protein phosphatase 2, regulatory subunit B, beta	PPP2R2B	Disease: spinocerebellar ataxia 12 (SCA12)
Protein phosphatase 2, regulatory subunit B', beta	PPP2R5B	B56beta, a regulatory subunit of protein phosphatase 2A, interacts with CALEB/NGC and inhibits CALEB/NGC-mediated dendritic branching. PMID: 18385213
Protein phosphatase 2, regulatory subunit B, delta	PPP2R2D	In vivo discovery of immunotherapy targets in the tumour microenvironment. PMID: 24476824
Protein phosphatase 2, regulatory subunit B', delta	PPP2R5D	Mice lacking phosphatase PP2A subunit PR61/B'delta (Ppp2r5d) develop spatially restricted tauopathy by deregulation of CDK5 and GSK3beta. PMID: 21482799
Protein phosphatase 2, regulatory subunit B', epsilon isoform	PPP2R5E	Downregulation of PPP2R5E is a common event in acute myeloid leukemia that affects the oncogenic potential of leukemic cells. PMID: 23812941

(Continued)

HGNC-approved name	Symbol	Target information
Protein phosphatase 3, regulatory subunit B, alpha	PPP3R1	Variants in PPP3R1 and MAPT are associated with more rapid functional decline in Alzheimer's disease: The Cache County Dementia Progression Study. PMID: 23727081
Protein phosphatase 4, regulatory subunit 2	PPP4R2	PPP4R2 regulates neuronal cell differentiation and survival, functionally cooperating with SMN. PMID: 22559936
RAS guanyl releasing protein 1 (calcium and DAG-regulated)	RASGRP1	Targeted deletion of RasGRP1 impairs skin tumorigenesis. PMID: 24464785
RAS guanyl releasing protein 2 (calcium and DAG-regulated)	RASGRP2	Disease: bleeding disorder, platelet-type 18 (BDPLT18)
Rho GDP dissociation inhibitor (GDI) alpha	ARHGDIA	Knockdown of RhoGDIα induces apoptosis and increases lung cancer cell chemosensitivity to paclitaxel. PMID: 22668020. Disease: nephrotic syndrome 8 (NPHS8)
Rho GDP dissociation inhibitor (GDI) beta	ARHGDIB	RhoGDI2 as a therapeutic target in cancer. PMID: 20001211
RIC8 guanine nucleotide exchange factor A	RIC8A	The G protein α chaperone Ric-8 as a potential therapeutic target. PMID: 25319541
RIC8 guanine nucleotide exchange factor B	RIC8B	The G protein α chaperone Ric-8 as a potential therapeutic target. PMID: 25319541
Ring finger and CCCH-type domains 1	RC3H1	Cleavage of roquin and regnase-1 by the paracaspase MALT1 releases their cooperatively repressed targets to promote T(H)17 differentiation. PMID: 25282160
Ring finger and CCCH-type domains 2	RC3H2	Roquin-2 promotes ubiquitin-mediated degradation of ASK1 to regulate stress responses. PMID: 24448648
S100 calcium binding protein A1	S100A1	S100A1: A major player in cardiovascular performance. PMID: 25157660
S100 calcium binding protein A14	S100A14	The role of S100A14 in epithelial ovarian tumors. PMID: 24939856
Spastic paraplegia 20 (Troyer syndrome)	SPG20	The role of spartin and its novel ubiquitin binding region in DALIS occurrence. PMID: 24523286
Spastic paraplegia 21 (autosomal recessive, Mast syndrome)	SPG21	Targeted disruption of the Mast syndrome gene SPG21 in mice impairs hind limb function and alters axon branching in cultured cortical neurons. PMID: 20661613
Actinin, alpha 2	ACTN2	Disease: cardiomyopathy, dilated 1AA (CMD1AA)
Activating signal cointegrator 1 complex subunit 1	ASCC1	Disease: Barrett esophagus (BE)
Age-related maculopathy susceptibility 2	ARMS2	Disease: macular degeneration, age-related, 8 (ARMD8)
AKT-interacting protein	AKTIP	EGCG suppresses Fused Toes Homolog protein through p53 in cervical cancer cells.
Alpha- and gamma-adaptin binding protein	AAGAB	Disease: keratoderma, palmoplantar, punctate 1A (PPKP1A)
Amyloid beta (A4) precursor protein-binding, family A, member 1	APBA1	Expression of the neuronal adaptor protein X11alpha protects against memory dysfunction in a transgenic mouse model of Alzheimer's disease. PMID: 20378958
Arrestin domain containing 3	ARRDC3	ARRDC3 suppresses breast cancer progression by negatively regulating integrin beta4. PMID: 20603614
Aryl hydrocarbon receptor interacting protein	AIP	Disease: growth hormone-secreting pituitary adenoma (GHSPA)
Ataxin 2	ATXN2	Disease: spinocerebellar ataxia 2 (SCA2)
Axin2	AXIN2	Axin2 as regulatory and therapeutic target in newborn brain injury and remyelination. PMID: 21706018. Disease: colorectal cancer (CRC)

HGNC-approved name	Symbol	Target information
B-cell scaffold protein with ankyrin repeats 1	BANK1	Inactivation of BANK1 in a novel IGH-associated translocation t(4;14)(q24;q32) suggests a tumor suppressor role in B-cell lymphoma. PMID: 24879116. Disease: systemic lupus erythematosus (SLE)
BCL2-related protein A1	BCL2A1	Characterization of peptide aptamers targeting Bfl-1 anti-apoptotic protein. PMID: 21563784
BicC family RNA binding protein 1	BICC1	Disease: renal dysplasia, cystic (CYSRD)
Biogenesis of lysosomal organelles complex-1, subunit 3	BLOC1S3	Disease: Hermansky–Pudlak syndrome 8 (HPS8)
Breast carcinoma-amplified sequence 4	BCAS4	Note: A chromosomal aberration involving BCAS4 has been found in some breast carcinoma cell lines. Translocation t(1720)(q23q13) with BCAS3
Calcium binding protein 4	CABP4	Disease: night blindness, congenital stationary, 2B (CSNB2B)
CASP8- and FADD-like apoptosis regulator	CFLAR	Targeting c-FLIP in cancer. PMID: 21071136
Cdc42 guanine nucleotide exchange factor (GEF) 9	ARHGEF9	Disease: epileptic encephalopathy, early infantile, 8 (EIEE8)
Cell cycle-associated protein 1	CAPRIN1	Caprin-1, a novel Cyr61-interacting protein, promotes osteosarcoma tumor growth and lung metastasis in mice. PMID: 23528710
Cell division cycle 37	CDC37	Suppressing the CDC37 cochaperone in hepatocellular carcinoma cells inhibits cell cycle progression and cell growth. PMID: 25098386
Centrosomal protein 89 kDa	CEP89	Note: Homozygous deletion comprising CEP89 and SLC7A9 genes has been reported in a patient with isolated complex IV deficiency, intellectual disability and multisystemic problems.
Cereblon	CRBN	Structure of the human Cereblon-DDB1-lenalidomide complex reveals basis for responsiveness to thalidomide analogs. PMID: 25108355
Cerebral cavernous malformation 2	CCM2	Disease: cerebral cavernous malformations 2 (CCM2)
Coiled-coil domain containing 50	CCDC50	Disease: deafness, autosomal dominant, 44 (DFNA44)
Component of oligomeric golgi complex 5	COG5	Disease: Congenital disorder of glycosylation 2I (CDG2I)
Cullin 7	CUL7	Overexpressed ubiquitin ligase Cullin7 in breast cancer promotes cell proliferation and invasion via down-regulating p53. PMID: 25003318. Disease: 3M syndrome 1 (3M1)
Cysteine-rich PDZ binding protein	CRIPT	Disease: short stature with microcephaly and distinctive facies (SSMF)
Cytoglobin	CYGB	Cytoglobin in tumor hypoxia: Novel insights into cancer suppression. PMID: 24816917
Dedicator of cytokinesis 3	DOCK3	Dock3 protects myelin in the cuprizone model for demyelination. PMID: 25165881. Note: A chromosomal aberration involving DOCK3 has been found in a family with early-onset behavioral/developmental disorder with features of attention deficit-hyperactivity disorder and intellectual disability. Inversion inv(3)(p14:q21). The inversion disrupts DOCK3 and SLC9A9
Dedicator of cytokinesis 6	DOCK6	Disease: Adams–Oliver syndrome 2 (AOS2)
Deleted in lung and esophageal cancer 1	DLEC1	Note: DLEC1 silencing due to promoter methylation and aberrant transcription are implicated in the development of different cancers, including esophageal (ESCR), renal and lung cancers (LNCR). Disease: lung cancer (LNCR)
DEP domain containing 5	DEPDC5	Disease: epilepsy, familial focal, with variable foci (FFEVF)
Desmin	DES	Disease: myopathy, myofibrillar, 1 (MFM1)
Discs, large (*Drosophila*) homolog-associated protein 5	DLGAP5	Silencing of DLGAP5 by siRNA significantly inhibits the proliferation and invasion of hepatocellular carcinoma cells. PMID: 24324629
ECSIT signalling integrator	ECSIT	Hepatitis B virus X protein increases the IL-1β-induced NF-κB activation via interaction with evolutionarily conserved signaling intermediate in Toll pathways (ECSIT). PMID: 25449573

(*Continued*)

HGNC-approved name	Symbol	Target information
EDAR-associated death domain	EDARADD	Disease: ectodermal dysplasia 11A, hypohidrotic/hair/nail type, autosomal dominant (ECTD11A)
Epidermal growth factor receptor pathway substrate 8	EPS8	Note: Defects in EPS8 are associated with some cancers, such as pancreatic, oral squamous cell carcinomas or pituitary cancers
Eukaryotic elongation factor, selenocysteine-tRNA-specific	EEFSEC	The selenocysteine-specific elongation factor contains a novel and multi-functional domain. PMID: 22992746
Exosome component 3	EXOSC3	Disease: pontocerebellar hypoplasia 1B (PCH1B)
Fas apoptotic inhibitory molecule	FAIM	FAIM-L is an IAP-binding protein that inhibits XIAP ubiquitinylation and protects from Fas-induced apoptosis. PMID: 24305822
F-box and WD repeat domain containing 4	FBXW4	Disease: split-hand/foot malformation 3 (SHFM3)
Fermitin family member 2	FERMT2	Kindlin-2 regulates renal tubular cell plasticity by activation of Ras and its downstream signaling. PMID: 24226523
FGFR1 oncogene partner 2	FGFR1OP2	Note: A chromosomal aberration involving FGFR1OP2 may be a cause of stem cell myeloproliferative disorder (MPD). Insertion ins(128)(p11p11p22) with FGFR1
Finkel–Biskis–Reilly murine sarcoma virus (FBR-MuSV) ubiquitously expressed	FAU	Candidate tumour suppressor Fau regulates apoptosis in human cells: An essential role for Bcl-G. PMID: 21550398
Four and a half LIM domains 1	FHL1	Disease: scapuloperoneal myopathy, X-linked dominant (SPM)
G-2 and S-phase expressed 1	GTSE1	GTSE1 is a microtubule plus-end tracking protein that regulates EB1-dependent cell migration. PMID: 23236459
Glial fibrillary acidic protein	GFAP	Disease: Alexander disease (ALEXD)
G-protein signaling modulator 2	GPSM2	Disease: Chudley–McCullough syndrome (CMCS)
GRB2-associated binding protein 1	GAB1	Novel inhibitors induce large conformational changes of GAB1 pleckstrin homology domain and kill breast cancer cells. PMID: 25569504
Growth factor receptor-bound protein 10	GRB10	Lentivirus shRNA Grb10 targeting the pancreas induces apoptosis and improved glucose tolerance due to decreased plasma glucagon levels. PMID: 22222503
GTP cyclohydrolase I feedback regulator	GCHFR	A polymorphism of the GTP-cyclohydrolase I feedback regulator gene alters transcriptional activity and may affect response to SSRI antidepressants. PMID: 20351752
Guanine nucleotide binding protein (G protein), alpha inhibiting activity polypeptide 3	GNAI3	Disease: auriculocondylar syndrome 1 (ARCND1)
HEAT repeat containing 2	HEATR2	Disease: ciliary dyskinesia, primary, 18 (CILD18)
HSPA (heat shock 70 kDa) binding protein, cytoplasmic cochaperone 1	HSPBP1	HSP70-binding protein HSPBP1 regulates chaperone expression at a posttranslational level and is essential for spermatogenesis. PMID: 24899640
HSPB (heat shock 27 kDa) associated protein 1	HSPBAP1	Androgen receptor-interacting protein HSPBAP1 facilitates growth of prostate cancer cells in androgen-deficient conditions. PMID: 25359680. Note: A chromosomal aberration involving HSPBAP1 has been found in a family with renal carcinoma. Translocation t(23)(q35q21) with the putative pseudogene DIRC3.
Hydrolethalus syndrome 1	HYLS1	Hydrolethalus syndrome is caused by a missense mutation in a novel gene HYLS1. PMID: 15843405
Immunoglobulin (CD79A) binding protein 1	IGBP1	Alpha4 is a ubiquitin-binding protein that regulates protein serine/threonine phosphatase 2A ubiquitination. PMID: 20092282
Interaction protein for cytohesin exchange factors 1	IPCEF1	CNK3 and IPCEF1 produce a single protein that is required for HGF dependent Arf6 activation and migration. PMID: 22085542
Low-density lipoprotein receptor adaptor protein 1	LDLRAP1	Mining the genome for lipid genes. PMID: 24798233

HGNC-approved name	Symbol	Target information
Melanophilin	MLPH	Identification of novel rab27a/melanophilin blockers by pharmacophore-based virtual screening. PMID: 24293275
Membrane-associated guanylate kinase, WW and PDZ domain containing 1	MAGI1	Regulation and involvement in cancer and pathological conditions of MAGI1, a tight junction protein. PMID: 24982328
Migration and invasion enhancer 1	MIEN1	MicroRNA-940 suppresses prostate cancer migration and invasion by regulating MIEN1. PMID: 25406943
Mitogen-activated protein kinase 8-interacting protein 1	MAPK8IP1	Identification of small-molecule inhibitors of the JIP–JNK interaction. PMID: 19243309. Disease: diabetes mellitus, non-insulin-dependent (NIDDM)
Myeloid differentiation primary response 88	MYD88	Targeting the Toll-like receptor/interleukin 1 receptor pathway in human diseases: Rational design of MyD88 inhibitors. PMID: 23490990. Disease: MYD88 deficiency (MYD88D)
NEDD4 binding protein 2-like 2	N4BP2L2	Contributions to neutropenia from PFAAP5 (N4BP2L2), a novel protein mediating transcriptional repressor cooperation between Gfi1 and neutrophil elastase. PMID: 19506020
Neural precursor cell expressed, developmentally down-regulated 9	NEDD9	Involvement of NEDD9 in the invasion and migration of gastric cancer. PMID: 25577245
Nuclear receptor binding protein 2	NRBP2	Nuclear receptor binding protein 2 is induced during neural progenitor differentiation and affects cell survival. PMID: 18619852
Nudix (nucleoside diphosphate linked moiety X)-type motif 6	NUDT6	Alternative splicing and differential subcellular localization of the rat FGF antisense gene product. PMID: 18215310
Ornithine decarboxylase antizyme 1	OAZ1	Effects of ornithine decarboxylase antizyme 1 on the proliferation and differentiation of human oral cancer cells. PMID: 25318549
Piwi-like RNA-mediated gene silencing 1	PIWIL1	Hiwi facilitates chemoresistance as a cancer stem cell marker in cervical cancer. PMID: 25119492
Protein inhibitor of activated STAT, 3	PIAS3	PIAS3 activates the intrinsic apoptotic pathway in non-small cell lung cancer cells independent of p53 status. PMID: 23959540
Protein interacting with PRKCA 1	PICK1	Identification of a small-molecule inhibitor of the PICK1 PDZ domain that inhibits hippocampal LTP and LTD. PMID: 20018661
Protein kinase, interferon-inducible double-stranded RNA-dependent activator	PRKRA	Disease: dystonia 16 (DYT16)
Purkinje cell protein 4	PCP4	Anti-apoptotic effects of PCP4/PEP19 in human breast cancer cell lines: a novel oncotarget. PMID: 25153723
Rho/Rac guanine nucleotide exchange factor (GEF) 18	ARHGEF18	Expression of p114RhoGEF predicts lymph node metastasis and poor survival of squamous-cell lung carcinoma patients. PMID: 23512329
Ribosomal protein L23a	RPL23A	Detection of T cell responses to a ubiquitous cellular protein in autoimmune disease. PMID: 25324392
Ring finger protein 17	RNF17	Mouse piwi interactome identifies binding mechanism of Tdrkh tudor domain to arginine methylated miwi. PMID: 19918066
Serine/threonine/tyrosine interacting-like 1	STYXL1	MK-STYX, a catalytically inactive phosphatase regulating mitochondrially dependent apoptosis. PMID: 21262771
SET nuclear proto-oncogene	SET	Stable SET knockdown in head and neck squamous cell carcinoma promotes cell invasion and the mesenchymal-like phenotype *in vitro*, as well as necrosis, cisplatin sensitivity and lymph node metastasis in xenograft tumor models. PMID: 24555657
SH3-domain GRB2-like endophilin B1	SH3GLB1	Sh3glb1/Bif-1 and mitophagy: Acquisition of apoptosis resistance during Myc-driven lymphomagenesis. PMID: 23680845
S-phase kinase-associated protein 2, E3 ubiquitin protein ligase	SKP2	SKP2 inactivation suppresses prostate tumorigenesis by mediating JARID1B ubiquitination. PMID: 25596733

(Continued)

HGNC-approved name	Symbol	Target information
Sperm-associated antigen 1	SPAG1	Sperm-associated antigens as targets for cancer immunotherapy: Expression pattern and humoral immune response in cancer patients. PMID: 21150711. Disease: ciliary dyskinesia, primary, 28 (CILD28)
Spermatogenesis associated 5	SPATA5	Genome-wide pooling approach identifies SPATA5 as a new susceptibility locus for alopecia areata. PMID: 22027810
Staphylococcal nuclease and tudor domain containing 1	SND1	SND1 acts downstream of TGFβ1 and upstream of Smurf1 to promote breast cancer metastasis. PMID: 25596283
Synovial sarcoma, X breakpoint 2-interacting protein	SSX2IP	SSX2IP promotes metastasis and chemotherapeutic resistance of hepatocellular carcinoma. PMID: 23452395
Trio Rho guanine nucleotide exchange factor	TRIO	Identification of a mitotic Rac-GEF, Trio, that counteracts MgcRacGAP function during cytokinesis. PMID: 25355950
Tumor protein p53 binding protein 2	TP53BP2	ASPP2 controls epithelial plasticity and inhibits metastasis through β-catenin-dependent regulation of ZEB1. PMID: 25344754
Tumor protein, translationally controlled 1	TPT1	TCTP as therapeutic target in cancers. PMID: 24650927
Ubiquitin domain containing 2	UBTD2	Cloning and identification of a novel human ubiquitin-like protein, DC-UbP, from dendritic cells. PMID: 12507522
Ubiquitin-like 5	UBL5	UBL5 is essential for pre-mRNA splicing and sister chromatid cohesion in human cells. PMID: 25092792
Z-DNA binding protein 1	ZBP1	A role for Z-DNA binding in vaccinia virus pathogenesis. PMID: 12777633

6.7 Subcellular location not annotated

HGNC-approved name	Symbol	Target information
Annexin A3	ANXA3	Annexin A3 as a potential target for immunotherapy of liver cancer stem-like cells. PMID: 25264174
Annexin A4	ANXA4	Enhanced expression of Annexin A4 in clear cell carcinoma of the ovary and its association with chemoresistance to carboplatin. PMID: 19598262
Annexin A5	ANXA5	Recombinant human annexin A5: A novel drug candidate for treatment of sepsis? PMID: 24346537. Disease: pregnancy loss, recurrent, 3 (RPRGL3)
Annexin A8	ANXA8	Annexin A8 is a prognostic marker and potential therapeutic target for pancreatic cancer. PMID: 25267197
B-cell CLL/lymphoma 7A	BCL7A	Note: Chromosomal aberrations involving BCL7A may be a cause of B-cell non-Hodgkin lymphoma. Three-way translocation t(81412) (q24.1q32.3q24.1) with MYC and with immunoglobulin gene regions
B-cell CLL/lymphoma 7B	BCL7B	Note: BCL7B is located in the Williams–Beuren syndrome (WBS) critical region
Chromosome 12 open reading frame 5	C12orf5	A TIGAR-regulated metabolic pathway is critical for protection of brain ischemia. PMID: 24872551
Chromosome 10 open reading frame 11	C10orf11	Disease: albinism, oculocutaneous, 7 (OCA7)
Chromosome 15 open reading frame 41	C15orf41	Disease: anemia, congenital dyserythropoietic, 1B (CDAN1B)
Chromosome 11 open reading frame 95	C11orf95	C11orf95-RELA fusions drive oncogenic NF-κB signalling in ependymoma. PMID: 24553141
Cilia- and flagella-associated protein 53	CFAP53	Disease: heterotaxy, visceral, 6, autosomal (HTX6)
Cilia- and flagella-associated protein 57	CFAP57	Disease: van der Woude syndrome 2 (VWS2)

HGNC-approved name	Symbol	Target information
Coiled-coil domain containing 8	CCDC8	Disease: 3M syndrome 3 (3M3)
Coiled-coil domain containing 22	CCDC22	Note: May be involved in X-linked syndromic mental retardation (PMID: 21826058)
Coiled-coil domain containing 28A	CCDC28A	Note: A chromosomal aberration involving CCDC28A has been identified in acute leukemias. Translocation t(611)(q24.1p15.5) with NUP98. The chimeric transcript is an in-frame fusion of NUP98 exon 13 to CCDC28A exon 2. Ectopic expression of NUP98-C
Coiled-coil domain containing 34	CCDC34	Note: A chromosomal aberration involving CCDC34 is found in a patient with hamartoma of the retinal pigment epithelium and retina. Translocation t(1118) (p13p11.2)
Coiled-coil domain containing 65	CCDC65	Disease: ciliary dyskinesia, primary, 27 (CILD27)
Coiled-coil domain containing 88C	CCDC88C	Disease: hydrocephalus, non-syndromic, autosomal recessive 1 (HYC1)
Crystallin, beta A1	CRYBA1	Disease: Cataract 10, multiple types (CTRCT10)
Crystallin, beta A2	CRYBA2	Note: Defects in CRYBA2 may be a cause of congenital cataract. Cataract is an opacification of the crystalline lens of the eye that frequently resulting in visual impairment or blindness
Crystallin, beta A4	CRYBA4	Disease: cataract 23 (CTRCT23)
Crystallin, beta B1	CRYBB1	Disease: cataract 17, multiple types (CTRCT17)
Crystallin, beta B2	CRYBB2	Disease: cataract 3, multiple types (CTRCT3)
Crystallin, beta B3	CRYBB3	Disease: cataract 22 (CTRCT22)
Crystallin, gamma B	CRYGB	Disease: cataract 39, multiple types (CTRCT39)
Crystallin, gamma C	CRYGC	Disease: cataract 2, multiple types (CTRCT2)
Crystallin, gamma D	CRYGD	Disease: cataract 4, multiple types (CTRCT4)
Crystallin, gamma S	CRYGS	Disease: cataract 20, multiple types (CTRCT20)
DnaJ (Hsp40) homolog, subfamily B, member 2	DNAJB2	Disease: distal spinal muscular atrophy, autosomal recessive, 5 (DSMA5)
DnaJ (Hsp40) homolog, subfamily C, member 30	DNAJC30	Note: DNAJC30 is located in the Williams–Beuren syndrome (WBS) critical region
Eukaryotic translation initiation factor 2B, subunit 1 alpha, 26 kDa	EIF2B1	Disease: leukodystrophy with vanishing white matter (VWM)
Eukaryotic translation initiation factor 2B, subunit 2 beta, 39 kDa	EIF2B2	Disease: leukodystrophy with vanishing white matter (VWM)
Eukaryotic translation initiation factor 2B, subunit 3 gamma, 58 kDa	EIF2B3	Disease: leukodystrophy with vanishing white matter (VWM)
Eukaryotic translation initiation factor 2B, subunit 4 delta, 67 kDa	EIF2B4	Disease: leukodystrophy with vanishing white matter (VWM)
Eukaryotic translation initiation factor 2B, subunit 5 epsilon, 82 kDa	EIF2B5	Disease: leukodystrophy with vanishing white matter (VWM)
Eukaryotic translation initiation factor 4 gamma, 1	EIF4G1	Small-molecule inhibition of the interaction between the translation initiation factors eIF4E and eIF4G. PMID: 17254965. Disease: Parkinson disease 18 (PARK18)
Eukaryotic translation initiation factor 4E binding protein 1	EIF4EBP1	Anti-oncogenic potential of the eIF4E-binding proteins. PMID: 22508483
Eukaryotic translation initiation factor 4E binding protein 2	EIF4EBP2	Anti-oncogenic potential of the eIF4E-binding proteins. PMID: 22508483
Eukaryotic translation initiation factor 4E binding protein 3	EIF4EBP3	Anti-oncogenic potential of the eIF4E-binding proteins. PMID: 22508483

(Continued)

HGNC-approved name	Symbol	Target information
Family with sequence similarity 58, member A	FAM58A	Disease: toe syndactyly, telecanthus, and anogenital and renal malformations (STAR)
Family with sequence similarity 83, member H	FAM83H	Disease: amelogenesis imperfecta 3 (AI3)
Family with sequence similarity 111, member B	FAM111B	Disease: poikiloderma, hereditary fibrosing, with tendon contractures, myopathy, and pulmonary fibrosis (POIKTMP)
Hermansky–Pudlak syndrome 1	HPS1	Disease: Hermansky–Pudlak syndrome 1 (HPS1)
Hermansky–Pudlak syndrome 4	HPS4	Disease: Hermansky–Pudlak syndrome 4 (HPS4)
Kelch-like family member 10	KLHL10	Disease: spermatogenic failure 11 (SPGF11)
Kelch-like family member 40	KLHL40	Disease: nemaline myopathy 8 (NEM8)
Keratin 2	KRT2	Disease: ichthyosis bullosa of Siemens (IBS)
Keratin 3	KRT3	Disease: corneal dystrophy, Meesmann (MECD)
Keratin 4	KRT4	Disease: white sponge nevus of Cannon (WSN)
Keratin 5	KRT5	Disease: epidermolysis bullosa simplex, Dowling–Meara type (DM-EBS)
Keratin 6A	KRT6A	Disease: pachyonychia congenita 1 (PC1)
Keratin 6B	KRT6B	Disease: pachyonychia congenita 2 (PC2)
Keratin 6C	KRT6C	Disease: palmoplantar keratoderma, non-epidermolytic, focal or diffuse (PPKNEFD)
Keratin 9	KRT9	Disease: keratoderma, palmoplantar, epidermolytic (EPPK)
Keratin 10	KRT10	Disease: epidermolytic hyperkeratosis (EHK)
Keratin 12	KRT12	Disease: corneal dystrophy, Meesmann (MECD)
Keratin 13	KRT13	Disease: white sponge nevus of Cannon (WSN)
Keratin 16	KRT16	Disease: pachyonychia congenita 1 (PC1)
Keratin 17	KRT17	Disease: pachyonychia congenita 2 (PC2)
Keratin 71	KRT71	Disease: hypotrichosis 13 (HYPT13)
Keratin 74	KRT74	Disease: woolly hair autosomal dominant (ADWH)
Keratin 75	KRT75	Disease: loose anagen hair syndrome (LAHS)
Keratin 81	KRT81	Disease: monilethrix (MLTRX)
Keratin 83	KRT83	Disease: monilethrix (MLTRX)
Keratin 85	KRT85	Disease: ectodermal dysplasia 4, hair/nail type (ECTD4)
Keratin 86	KRT86	Disease: monilethrix (MLTRX)
KIAA0196	KIAA0196	Disease: spastic paraplegia 8, autosomal dominant (SPG8)
KIAA1033	KIAA1033	Disease: mental retardation, autosomal recessive 43 (MRT43)
Metallothionein 1A	MT1A	Metallothioneins, ageing and cellular senescence: A future therapeutic target. PMID: 23061732
Metallothionein 3	MT3	Characterization of the role of metallothionein-3 in an animal model of Alzheimer's disease. PMID: 22722772
Myeloid/lymphoid or mixed-lineage leukemia (trithorax homolog, *Drosophila*) translocated to, 1	MLLT1	Note: A chromosomal aberration involving MLLT1 is associated with acute leukemias. Translocation t(1119)(q23p13.3) with KMT2A/MLL1. The result is a rogue activator protein
Myeloid/lymphoid or mixed-lineage leukemia (trithorax homolog, *Drosophila*); translocated to, 11	MLLT11	Note: A chromosomal aberration involving MLLT11 is found in acute leukemias. Translocation t(111)(q21q23) with KMT2A/MLL1
Myosin binding protein C, cardiac	MYBPC3	Disease: cardiomyopathy, familial hypertrophic 4 (CMH4)
Myosin binding protein C, slow type	MYBPC1	Disease: arthrogryposis, distal, 1B (DA1B)

HGNC-approved name	Symbol	Target information
Myosin IA	MYO1A	Disease: deafness, autosomal dominant, 48 (DFNA48)
Myosin VA (heavy chain 12, myoxin)	MYO5A	Disease: Griscelli syndrome 1 (GS1)
Myosin VB	MYO5B	Disease: diarrhea 2, with microvillus atrophy (DIAR2)
MYO7A: myosin VIIA	MYO7A	Disease: Usher syndrome 1B (USH1B)
Myosin IXB	MYO9B	Disease: celiac disease 4 (CELIAC4)
Myosin, heavy chain 2, skeletal muscle, adult	MYH2	Disease: inclusion body myopathy 3 (IBM3)
Myosin, heavy chain 3, skeletal muscle, embryonic	MYH3	Disease: arthrogryposis, distal, 2A (DA2A)
Myosin, heavy chain 6, cardiac muscle, alpha	MYH6	Disease: atrial septal defect 3 (ASD3)
Myosin, heavy chain 7, cardiac muscle, beta	MYH7	Disease: cardiomyopathy, familial hypertrophic 1 (CMH1)
Myosin, heavy chain 8, skeletal muscle, perinatal	MYH8	Disease: Carney complex variant (CACOV)
Neurofilament, heavy polypeptide	NEFH	Disease: amyotrophic lateral sclerosis (ALS)
Neurofilament, light polypeptide	NEFL	Disease: Charcot–Marie–Tooth disease 1F (CMT1F)
Nitrogen permease regulator-like 2 (S. cerevisiae)	NPRL2	Note: Inactivating mutations and truncating deletions in the genes encoding GATOR1 proteins, including NPRL2, are detected in glioblastoma and ovarian tumors and are associated with loss of heterozygosity events
Nitrogen permease regulator like 3 (S. cerevisiae)	NPRL3	Note: Inactivating mutations and truncating deletions in the genes encoding GATOR1 proteins are detected in glioblastoma and ovarian tumors and are associated with loss of heterozygosity events
Phosphoinositide-3-kinase, regulatory subunit 1 (alpha)	PIK3R1	Abrogation of PIK3CA or PIK3R1 reduces proliferation, migration, and invasion in glioblastoma multiforme cells. PMID: 22064833. Disease: agammaglobulinemia 7, autosomal recessive (AGM7)
Phosphoinositide-3-kinase, regulatory subunit 2 (beta)	PIK3R2	Disease: megalencephaly–polymicrogyria–polydactyly–hydrocephalus syndrome (MPPH)
Potassium channel tetramerization domain containing 6	KCTD6	Identification and characterization of KCASH2 and KCASH3, 2 novel Cullin3 adaptors suppressing histone deacetylase and Hedgehog activity in medulloblastoma. PMID: 21472142
Potassium channel tetramerization domain containing 21	KCTD21	Identification and characterization of KCASH2 and KCASH3, 2 novel Cullin3 adaptors suppressing histone deacetylase and Hedgehog activity in medulloblastoma. PMID: 21472142
Protein kinase, AMP-activated, beta 1 non-catalytic subunit	PRKAB1	**Metformin** – drugs used in diabetes
Protein kinase, AMP-activated, gamma 2 non-catalytic subunit	PRKAG2	Disease: Wolff–Parkinson–White syndrome (WPWS)
Protein phosphatase 1, regulatory (inhibitor) subunit 2	PPP1R2	Identification and characterization of two distinct PPP1R2 isoforms in human spermatozoa. PMID: 23506001
Protein phosphatase 1, regulatory subunit 3B	PPP1R3B	Mutated PPP1R3B is recognized by T cells used to treat a melanoma patient who experienced a durable complete tumor regression. PMID: 23690473
Protein phosphatase 1, regulatory subunit 3C	PPP1R3C	PTG depletion removes Lafora bodies and rescues the fatal epilepsy of Lafora disease. PMID: 21552327
Protein phosphatase 1, regulatory subunit 3D	PPP1R3D	Glycogenic activity of R6, a protein phosphatase 1 regulatory subunit, is modulated by the laforin-malin complex. PMID: 23624058

(Continued)

HGNC-approved name	Symbol	Target information
Protein phosphatase 1, regulatory subunit 3G	PPP1R3G	Regulation of glucose homeostasis and lipid metabolism by PPP1R3G-mediated hepatic glycogenesis. PMID: 24264575
Protein phosphatase 1, regulatory subunit 7	PPP1R7	Sds22 and Repo-Man stabilize chromosome segregation by counteracting Aurora B on anaphase kinetochores. PMID: 22801782
Protein phosphatase 1, regulatory subunit 15B	PPP1R15B	Protein phosphatase 1, regulatory subunit 15B is a survival factor for ERα-positive breast cancer. PMID: 23169272
Protein phosphatase 2, regulatory subunit B, alpha	PPP2R2A	The protein phosphatase 2A regulatory subunit B55α is a modulator of signaling and microRNA expression in acute myeloid leukemia cells. PMID: 24858343
Protein phosphatase 2, regulatory subunit B, gamma	PPP2R2C	Over expression of PPP2R2C inhibits human glioma cells growth through the suppression of mTOR pathway. PMID: 24126060
Protein phosphatase 3, regulatory subunit B, beta	PPP3R2	**Cyclosporine** – immunomodulating agents
Protein phosphatase 4, regulatory subunit 1	PPP4R1	The PP4R1 subunit of protein phosphatase PP4 targets TRAF2 and TRAF6 to mediate inhibition of NF-κB activation. PMID: 25134449
Ras association (RalGDS/AF-6) domain family member 6	RASSF6	The RASSF6 tumor suppressor protein regulates apoptosis and the cell cycle via MDM2 protein and p53 protein. PMID: 24003224
Ras association (RalGDS/AF-6) domain family member 8	RASSF8	Note: A chromosomal aberration involving RASSF8 is found in a complex type of synpolydactyly referred to as 3/3-prime/4 synpolydactyly associated with metacarpal and metatarsal synostoses. Reciprocal translocation t(1222)(p11.2q13.3) with FBLN1
Ribosomal protein L7a	RPL7A	Note: Chromosomal recombination involving RPL7A activates the receptor kinase domain of the TRK oncogene
Ribosomal protein L10	RPL10	Disease: autism, X-linked 5 (AUTSX5)
Ribosomal protein L15	RPL15	Disease: Diamond–Blackfan anemia 12 (DBA12)
Ribosomal protein L21	RPL21	Note: Defects in RPL21 are a cause of generalized hypotrichosis simplex (HTS). A rare form of non-syndromic hereditary hypotrichosis without characteristic hair shaft anomalies
Ribosomal protein L24	RPL24	RPL24: A potential therapeutic target whose depletion or acetylation inhibits polysome assembly and cancer cell growth. PMID: 24970821
Ribosomal protein L26	RPL26	Disease: Diamond–Blackfan anemia 11 (DBA11)
Ribosomal protein L35a	RPL35A	Disease: Diamond–Blackfan anemia 5 (DBA5)
Ribosomal protein S15	RPS15	Down regulation of ribosomal protein S15A mRNA with a short hairpin RNA inhibits human hepatic cancer cell growth *in vitro*. PMID: 24334120
Ribosomal protein S17	RPS17	Disease: Diamond–Blackfan anemia 4 (DBA4)
Ribosomal protein S24	RPS24	Disease: Diamond–Blackfan anemia 3 (DBA3)
Ribosomal protein S26	RPS26	Disease: Diamond–Blackfan anemia 10 (DBA10)
S100 calcium binding protein A4	S100A4	RNA interference targeting against S100A4 suppresses cell growth and motility and induces apoptosis in human pancreatic cancer cells. PMID: 19799859
S100 calcium binding protein A10	S100A10	p11 and its role in depression and therapeutic responses to antidepressants. PMID: 24002251
Troponin C type 1 (slow)	TNNC1	**Levosimendan** – cardiac therapy. Disease: cardiomyopathy, dilated 1Z (CMD1Z)
Troponin C type 2 (fast)	TNNC2	**Tirasemtiv** – amyotrophic lateral sclerosis treatments
Troponin I type 2 (skeletal, fast)	TNNI2	Disease: arthrogryposis, distal, 2B (DA2B)
Troponin I type 3 (cardiac)	TNNI3	Disease: cardiomyopathy, familial hypertrophic 7 (CMH7)

HGNC-approved name	Symbol	Target information
Troponin T type 1 (skeletal, slow)	TNNT1	Disease: nemaline myopathy 5 (NEM5)
Troponin T type 2 (cardiac)	TNNT2	Disease: cardiomyopathy, familial hypertrophic 2 (CMH2)
Troponin T type 3 (skeletal, fast)	TNNT3	Disease: arthrogryposis, distal, 2B (DA2B)
TRPM8 channel-associated factor 1	TCAF1	TRP channel-associated factors are a novel protein family that regulates TRPM8 trafficking and activity. PMID: 25559186
TRPM8 channel-associated factor 2	TCAF2	TRP channel-associated factors are a novel protein family that regulates TRPM8 trafficking and activity. PMID: 25559186
Tumor protein D52	TPD52	PrLZ protects prostate cancer cells from apoptosis induced by androgen deprivation via the activation of Stat3/Bcl-2 pathway. PMID: 21385902
Tumor protein D52-like 2	TPD52L2	Knockdown of tumor protein D52-like 2 induces cell growth inhibition and apoptosis in oral squamous cell carcinoma. PMID: 25260534
Usher syndrome 1C (autosomal recessive, severe)	USH1C	Disease: Usher syndrome 1C (USH1C)
Usher syndrome 1G (autosomal recessive)	USH1G	Disease: Usher syndrome 1G (USH1G)
Vacuolar protein sorting 13 homolog A (*S. cerevisiae*)	VPS13A	Disease: choreoacanthocytosis (CHAC)
Vacuolar protein sorting 13 homolog B (yeast)	VPS13B	Disease: Cohen syndrome (COH1)
WD repeat domain 45	WDR45	Disease: neurodegeneration with brain iron accumulation 5 (NBIA5)
WD repeat domain 72	WDR72	Disease: amelogenesis imperfecta, hypomaturation type, 2A3 (AI2A3)
WD repeat domain 81	WDR81	Disease: cerebellar ataxia, mental retardation, and dysequilibrium syndrome 2 (CAMRQ2)
Williams–Beuren syndrome chromosome region 16	WBSCR16	Note: WBSCR16 is located in the Williams–Beuren syndrome (WBS) critical region
Williams–Beuren syndrome chromosome region 27	WBSCR27	Note: WBSCR27 is located in the Williams–Beuren syndrome (WBS) critical region
ADP-ribosylation factor-like 11	ARL11	Disease: leukemia, chronic lymphocytic (CLL)
Alport syndrome, mental retardation, midface hypoplasia and elliptocytosis chromosomal region gene 1	AMMECR1	Disease: Alport syndrome with mental retardation, midface hypoplasia and elliptocytosis (ATS-MR)
Amyotrophic lateral sclerosis 2 (juvenile)	ALS2	Disease: amyotrophic lateral sclerosis 2 (ALS2)
Anaphase-promoting complex subunit 5	ANAPC5	The anaphase promoting complex protein 5 (AnapC5) associates with A20 and inhibits IL-17-mediated signal transduction. PMID: 23922952
Antioxidant 1 copper chaperone	ATOX1	Novel role of copper transport protein antioxidant-1 in neointimal formation after vascular injury. PMID: 23349186
APC membrane recruitment protein 1	AMER1	Disease: osteopathia striata with cranial sclerosis (OSCS)
Armadillo repeat containing 8	ARMC8	ARMC8α promotes proliferation and invasion of non-small cell lung cancer cells by activating the canonical Wnt signaling pathway. PMID: 24894675
AT hook, DNA binding motif, containing 1	AHDC1	Disease: mental retardation, autosomal dominant 25 (MRD25)
Autism susceptibility candidate 2	AUTS2	Disease: mental retardation, autosomal dominant 26 (MRD26)
B-cell translocation gene 1, anti-proliferative	BTG1	Note: A chromosomal aberration involving BTG1 may be a cause of a form of B-cell chronic lymphocytic leukemia. Translocation t(812)(q24q22) with MYC
BCL2-associated athanogene 3	BAG3	Overexpressed BAG3 is a potential therapeutic target in chronic lymphocytic leukemia. PMID: 23978946. Disease: myopathy, myofibrillar, 6 (MFM6)

(Continued)

HGNC-approved name	Symbol	Target information
Breakpoint cluster region	BCR	Disease: leukemia, chronic myeloid (CML)
Bromodomain and WD repeat domain containing 3	BRWD3	Note: A chromosomal aberration involving BRWD3 can be found in patients with B-cell chronic lymphocytic leukemia (B-CLL). Translocation t(X;11) (q21;q23) with ARHGAP20 does not result in fusion transcripts but disrupts both genes. Disease: mental retardation, X-linked 93 (MRX93)
BTB (POZ) domain containing 9	BTBD9	Disease: restless legs syndrome 6 (RLS6)
BTG family, member 2	BTG2	B-cell translocation gene 2 regulates hepatic glucose homeostasis via induction of orphan nuclear receptor Nur77 in diabetic mouse model. PMID: 24647738
Calbindin 2	CALB2	Calretinin interacts with huntingtin and reduces mutant huntingtin-caused cytotoxicity. PMID: 22891683
Calpain, small subunit 1	CAPNS1	Capn4 is a marker of poor clinical outcomes and promotes nasopharyngeal carcinoma metastasis via nuclear factor-κB-induced matrix metalloproteinase 2 expression. PMID: 24703594
Caspase recruitment domain family, member 16	CARD16	Definition of ubiquitination modulator COP1 as a novel therapeutic target in human hepatocellular carcinoma. PMID: 20959491
Chimerin 1	CHN1	Disease: Duane retraction syndrome 2 (DURS2)
Coiled-coil domain containing 151	CCDC151	Disease: ciliary dyskinesia, primary, 30 (CILD30)
Cyclin O	CCNO	Disease: ciliary dyskinesia, primary, 29 (CILD29)
Cytochrome c oxidase assembly factor 5	COA5	Disease: mitochondrial complex IV deficiency (MT-C4D)
DDB1- and CUL4-associated factor 8	DCAF8	Disease: giant axonal neuropathy 2, autosomal dominant (GAN2)
Deafness, autosomal dominant 5	DFNA5	Disease: deafness, autosomal dominant, 5 (DFNA5)
Deafness, autosomal recessive 59	DFNB59	Disease: deafness, autosomal recessive, 59 (DFNB59)
Dedicator of cytokinesis 8	DOCK8	Disease: hyperimmunoglobulin E recurrent infection syndrome, autosomal recessive (AR-HIES)
DEP domain containing MTOR-interacting protein	DEPTOR	Targeting the mTOR-DEPTOR pathway by CRL E3 ubiquitin ligases: Therapeutic application. PMID: 22745582
Discs, large homolog 3 (*Drosophila*)	DLG3	Disease: mental retardation, X-linked 90 (MRX90)
Disrupted in renal carcinoma 1	DIRC1	Note: A chromosomal aberration involving DIRC1 is associated with familial clear cell renal carcinoma. Translocation t(23)(q33q21)
Double homeobox 4	DUX4	Disease: facioscapulohumeral muscular dystrophy 1 (FSHD1)
Doublecortin domain containing 2	DCDC2	Disease: dyslexia 2 (DYX2)
Dynein, axonemal, light chain 1	DNAL1	Disease: ciliary dyskinesia, primary, 16 (CILD16)
Ectopic P-granules autophagy protein 5 homolog (*C. elegans*)	EPG5	Disease: Vici syndrome (VICIS)
EF-hand domain (C-terminal) containing 1	EFHC1	Disease: juvenile myoclonic epilepsy 1 (EJM1)
Elongation factor Tu GTP-binding domain containing 1	EFTUD1	Functional analysis of a novel glioma antigen, EFTUD1. PMID: 25015090
Exophilin 5	EXPH5	Disease: epidermolysis bullosa, non-specific, autosomal recessive (EBNS)
Fas (TNFRSF6)-associated via death domain	FADD	Increased expression of phosphorylated FADD in anaplastic large cell and other T-cell lymphomas. PMID: 25232277. Disease: infections, recurrent, associated with encephalopathy, hepatic dysfunction and cardiovascular malformations (IEHDCM)
Filaggrin	FLG	Possible new therapeutic strategy to regulate atopic dermatitis through upregulating filaggrin expression. PMID: 24055295. Disease: ichthyosis vulgaris (VI)
Forkhead box J1	FOXJ1	Decreased expression of FOXJ1 is a potential prognostic predictor for progression and poor survival of gastric cancer. PMID: 24809300. Disease: allergic rhinitis (ALRH)

HGNC-approved name	Symbol	Target information
FRY-like	FRYL	Note: A chromosomal aberration involving FRYL is found in treatment-related acute lymphoblastic leukemia (ALL). Translocation t(411)(p12q23) that forms a KMT2A/MLL1-FRYL fusion protein
GIPC PDZ domain containing family, member 3	GIPC3	Disease: deafness, autosomal recessive, 15 (DFNB15)
Glia maturation factor, beta	GMFB	Glia maturation factor deficiency suppresses 1-methyl-4-phenylpyridinium-induced oxidative stress in astrocytes. PMID: 24430624
Glomulin, FKBP-associated protein	GLMN	Disease: glomuvenous malformations (GVMs)
Glutaredoxin, cysteine rich 2	GRXCR2	Disease: deafness, autosomal recessive, 101 (DFNB101)
GRB10-interacting GYF protein 2	GIGYF2	Disease: Parkinson disease 11 (PARK11)
Growth arrest and DNA-damage-inducible, beta	GADD45B	Abnormal expression of GADD45B in human colorectal carcinoma. PMID: 23110778
Hexamethylene bis-acetamide inducible 1	HEXIM1	Cardiomyocyte-specific overexpression of HEXIM1 prevents right ventricular hypertrophy in hypoxia-induced pulmonary hypertension in mice. PMID: 23300697
Huntingtin-associated protein 1	HAP1	Huntingtin-associated protein-1 interacts with pro-brain-derived neurotrophic factor and mediates its transport and release. PMID: 19996106
Insulin receptor substrate 1	IRS1	**Aganirsen** – antineovascularization agents. Disease: diabetes mellitus, non-insulin-dependent (NIDDM)
Interferon, gamma inducible protein 16	IFI16	IFI16 DNA sensor is required for death of lymphoid CD4 T cells abortively infected with HIV. PMID: 24356113
Intraflagellar transport 27	IFT27	Disease: Bardet–Biedl syndrome 19 (BBS19)
IQ motif and Sec7 domain 2	IQSEC2	Disease: mental retardation, X-linked 1 (MRX1)
Kelch domain containing 8B	KLHDC8B	Disease: Hodgkin lymphoma (HL)
Kelch repeat and BTB (POZ) domain containing 13	KBTBD13	Disease: nemaline myopathy 6 (NEM6)
KH domain containing 3-like, subcortical maternal complex member	KHDC3L	Disease: hydatidiform mole, recurrent, 2 (HYDM2)
Kinase non-catalytic C-lobe domain (KIND) containing 1	KNDC1	KNDC1 knockdown protects human umbilical vein endothelial cells from senescence. PMID: 24788352
Leucine-rich repeat and sterile alpha motif containing 1	LRSAM1	Disease: Charcot–Marie–Tooth disease 2P (CMT2P)
Leucine zipper transcription factor-like 1	LZTFL1	Disease: Bardet–Biedl syndrome 17 (BBS17)
Leucine-zipper-like transcription regulator 1	LZTR1	Disease: schwannomatosis 2 (SWNTS2)
Leukemia NUP98 fusion partner 1	LNP1	Note: A chromosomal aberration involving LNP1 is found in a form of T-cell acute lymphoblastic leukemia (T-ALL). Translocation t(311) (q12.2p15.4) with NUP98
Lines homolog (*Drosophila*)	LINS	Disease: mental retardation, autosomal recessive 27 (MRT27)
Lysosomal trafficking regulator	LYST	Disease: Chediak–Higashi syndrome (CHS)
Malignant fibrous histiocytoma-amplified sequence 1	MFHAS1	Note: A chromosomal aberration involving MFHAS1 may be a cause of B-cell lymphoma. Translocation t(814)(p23.1q21) with a cryptic exon named '14q21 element'. The resulting fusion protein named 'chimeric MASL1' is tumorigenic in nude mice
Mature T-cell proliferation 1	MTCP1	Note: Detected in T-cell leukemia bearing a t(X14) translocation. Plays a key role in T-cell prolymphocytic leukemia
MCF.2 cell line-derived transforming sequence-like 2	MCF2L2	Disease: diabetes mellitus, non-insulin-dependent (NIDDM)

(Continued)

HGNC-approved name	Symbol	Target information
MDM2 binding protein	MTBP	MTBP is overexpressed in triple-negative breast cancer and contributes to its growth and survival. PMID: 24866769
Mediator of cell motility 1	MEMO1	Memo is a copper-dependent redox protein with an essential role in migration and metastasis. PMID: 24917593
Megakaryoblastic leukemia (translocation) 1	MKL1	Note: A chromosomal aberration involving MKL1 may be a cause of acute megakaryoblastic leukemia. Translocation t(122)(p13q13) with RBM15
Meningioma (disrupted in balanced translocation) 1	MN1	Note: A chromosomal aberration involving MN1 may be a cause of acute myeloid leukemia (AML). Translocation t(1222)(p13q11) with TEL
Microtubule-associated protein tau	MAPT	Note: In Alzheimer disease, the neuronal cytoskeleton in the brain is progressively disrupted and replaced by tangles of paired helical filaments (PHF) and straight filaments, mainly composed of hyperphosphorylated forms of TAU (PHF-TAU or AD P-TAU)
Mirror-image polydactyly 1	MIPOL1	Note: A chromosomal aberration involving MIPOL1 is found in a patient with mirror-image polydactyly of hands and feet without other anomalies (MIP). Translocation t(214)(p23.3q13)
Myelodysplastic syndrome 2 translocation associated	MDS2	Note: A chromosomal aberration involving MDS2 is a cause of myelodysplastic syndrome (MDS). Translocation t(112)(p36.1p13) with ETV6
Myosin, heavy chain 14, non-muscle	MYH14	Disease: deafness, autosomal dominant, 4A (DFNA4A)
Myosin, light chain 2, regulatory, cardiac, slow	MYL2	Disease: cardiomyopathy, familial hypertrophic 10 (CMH10)
Myosin, light chain 3, alkali; ventricular, skeletal, slow	MYL3	Disease: cardiomyopathy, familial hypertrophic 8 (CMH8)
Myozenin 2	MYOZ2	Disease: cardiomyopathy, familial hypertrophic 16 (CMH16)
Nebulin	NEB	Disease: nemaline myopathy 2 (NEM2)
Nestin	NES	Nestin as a novel therapeutic target for pancreatic cancer via tumor angiogenesis. PMID: 22246533
Neuroblastoma-amplified sequence	NBAS	Disease: short stature, optic nerve atrophy, and Pelger–Huet anomaly (SOPH)
N-myc downstream regulated 1	NDRG1	Disease: Charcot–Marie–Tooth disease 4D (CMT4D)
Nuclear receptor coactivator 4	NCOA4	Quantitative proteomics identifies NCOA4 as the cargo receptor mediating ferritinophagy. PMID: 24695223. Disease: thyroid papillary carcinoma (TPC)
Nucleotide-binding oligomerization domain containing 2	NOD2	Disease: Blau syndrome (BS)
Oral cancer overexpressed 1	ORAOV1	Oral cancer overexpressed 1 (ORAOV1) regulates cell cycle and apoptosis in cervical cancer HeLa cells. PMID: 20105337
Orofacial cleft 1 candidate 1	OFCC1	Note: A chromosomal aberration involving OFCC1 is found in patients with orofacial cleft
PHD finger protein 23	PHF23	Note: A chromosomal aberration involving PHF23 is found in a patient with acute myeloid leukemia (AML). Translocation t(1117)(p15p13) with NUP98
Phospholipase C, gamma 2 (phosphatidylinositol-specific)	PLCG2	Disease: familial cold autoinflammatory syndrome 3 (FCAS3)
Plastin 3	PLS3	Disease: osteoporosis (OSTEOP)
Pleckstrin homology domain containing, family G (with RhoGef domain) member 5	PLEKHG5	Disease: distal spinal muscular atrophy, autosomal recessive, 4 (DSMA4)
Pleckstrin homology domain containing, family M (with RUN domain) member 1	PLEKHM1	Disease: osteopetrosis, autosomal recessive 6 (OPTB6)

HGNC-approved name	Symbol	Target information
Programmed cell death 5	PDCD5	Programmed cell death 5 factor enhances triptolide-induced fibroblast-like synoviocyte apoptosis of rheumatoid arthritis. PMID: 20047520
Proline–serine–threonine phosphatase-interacting protein 1	PSTPIP1	Disease: PAPA syndrome (PAPAS)
Promyelocytic leukemia	PML	Note: A chromosomal aberration involving PML may be a cause of acute promyelocytic leukemia (APL). Translocation t(1517)(q21q21) with RARA. The PML breakpoints (type A and type B) lie on either side of an alternatively spliced exon
Ral guanine nucleotide dissociation stimulator-like 2	RGL2	The ral exchange factor rgl2 promotes cardiomyocyte survival and inhibits cardiac fibrosis. PMID: 24069211
RALY RNA binding protein-like	RALYL	RALY RNA binding protein-like reduced expression is associated with poor prognosis in clear cell renal cell carcinoma. PMID: 22994768
RanBP-type and C3HC4-type zinc finger containing 1	RBCK1	Disease: polyglucosan body myopathy, early-onset, with or without immunodeficiency (PBMEI)
Ras and Rab interactor 2	RIN2	Disease: MACS syndrome (MACS)
RAS guanyl releasing protein 3 (calcium and DAG-regulated)	RASGRP3	RasGRP3, a Ras activator, contributes to signaling and the tumorigenic phenotype in human melanoma. PMID: 21602881
RAS p21 protein activator (GTPase-activating protein) 1	RASA1	Note: Mutations in the SH2 domain of RASA seem to be oncogenic and cause basal cell carcinomas. Disease: capillary malformation–arteriovenous malformation (CMAVM)
Ras protein-specific guanine nucleotide-releasing factor 1	RASGRF1	RasGRF1 regulates proliferation and metastatic behavior of human alveolar rhabdomyosarcomas. PMID: 22752028
Regulator of calcineurin 1	RCAN1	Calcineurin and its regulator, RCAN1, confer time-of-day changes in susceptibility of the heart to ischemia/reperfusion. PMID: 24838101
Regulator of G-protein signaling 4	RGS4	Disease: schizophrenia 9 (SCZD9)
Retinal degeneration 3	RD3	Disease: Leber congenital amaurosis 12 (LCA12)
Rho GTPase-activating protein 20	ARHGAP20	Note: A chromosomal aberration involving ARHGAP20 may be a cause of B-cell chronic lymphocytic leukemia (B-CLL). Translocation t(X11) (q21q23) with BRWD3 does not result in fusion transcripts but disrupts both genes
Rho guanine nucleotide exchange factor (GEF) 10	ARHGEF10	Disease: slowed nerve conduction velocity (SNCV)
Ribosomal protein S29	RPS29	Disease: Diamond–Blackfan anemia 13 (DBA13)
Sacsin molecular chaperone	SACS	Disease: spastic ataxia Charlevoix–Saguenay type (SACS)
S-antigen; retina and pineal gland (arrestin)	SAG	Disease: night blindness, congenital stationary, Oguchi type 1 (CSNBO1)
SCL/TAL1-interrupting locus	STIL	Note: A chromosomal aberration involving STIL may be a cause of some T-cell acute lymphoblastic leukemias (T-ALL). A deletion at 1p32 between STIL and TAL1 genes leads to STIL/TAL1 fusion mRNA with STIL exon 1 slicing to TAL1 exon 3
SH2 domain containing 1A	SH2D1A	Disease: lymphoproliferative syndrome, X-linked, 1 (XLP1)
SH2B adaptor protein 3	SH2B3	The adaptor Lnk (SH2B3): An emerging regulator in vascular cells and a link between immune and inflammatory signaling. PMID: 21723852. Disease: celiac disease 13 (CELIAC13)
SH3 and cysteine-rich domain 3	STAC3	Disease: Native American myopathy (NAM)
SH3 domain and tetratricopeptide repeats 2	SH3TC2	Disease: Charcot–Marie–Tooth disease 4C (CMT4C)
SH3 domain binding protein 2	SH3BP2	The adaptor 3BP2 is required for early and late events in FcεRI signaling in human mast cells. PMID: 22896635. Disease: cherubism (CRBM)

(Continued)

HGNC-approved name	Symbol	Target information
Sine oculis binding protein homolog (*Drosophila*)	SOBP	Disease: mental retardation, anterior maxillary protrusion, and strabismus (MRAMS)
SKI-like oncogene	SKIL	SnoN/SkiL, a TGFβ signaling mediator: A participant in autophagy induced by arsenic trioxide. PMID: 20699661
SLIT-ROBO Rho GTPase-activating protein 3	SRGAP3	Note: A chromosomal aberration involving SRGAP3 is found in a patient with severe idiopathic mental retardation. Translocation t(X3)(p11.2p25)
SLX4-interacting protein	SLX4IP	Note: Chromosomal aberrations involving SLX4IP are found in acute lymphoblastic leukemia
Small glutamine-rich tetratricopeptide repeat (TPR)-containing, alpha	SGTA	Expression and clinical role of small glutamine-rich tetratricopeptide repeat (TPR)-containing protein alpha (SGTA) as a novel cell cycle protein in NSCLC. PMID: 23857189
Small muscle protein, X-linked	SMPX	Disease: deafness, X-linked, 4 (DFNX4)
Son of sevenless homolog 1 (*Drosophila*)	SOS1	Disease: gingival fibromatosis 1 (GGF1)
Sorcin	SRI	Sorcin, a calcium binding protein involved in the multidrug resistance mechanisms in cancer cells. PMID: 25197934
Speedy/RINGO cell cycle regulator family member E1	SPDYE1	Note: SPDYE1 is located in the Williams–Beuren syndrome (WBS) critical region
Spermatogenesis associated 7	SPATA7	Disease: Leber congenital amaurosis 3 (LCA3)
Sterile alpha motif domain containing 9	SAMD9	Disease: tumoral calcinosis, normophosphatemic, familial (NFTC)
Suppressor of cytokine signaling 3	SOCS3	SOCS3: A potential therapeutic target for many human diseases. PMID: 22010342. Note: There is some evidence that SOCS3 may be a susceptibility gene for atopic dermatitis linked to 17q25
Synaptic Ras GTPase-activating protein 1	SYNGAP1	Disease: mental retardation, autosomal dominant 5 (MRD5)
Synovial sarcoma, X breakpoint 1	SSX1	Note: A chromosomal aberration involving SSX1 may be a cause of synovial sarcoma. Translocation t(X18)(p11.2q11.2). The translocation is specifically found in more than 80% of synovial sarcoma
Syntaxin binding protein 2	STXBP2	Disease: familial hemophagocytic lymphohistiocytosis 5 (FHL5)
Synuclein, alpha interacting protein	SNCAIP	Disease: Parkinson disease (PARK)
TBC1 domain family, member 24	TBC1D24	Disease: familial infantile myoclonic epilepsy (FIME)
T-cell acute lymphocytic leukemia 2	TAL2	Note: A chromosomal aberration involving TAL2 may be a cause of some T-cell acute lymphoblastic leukemia (T-ALL). Translocation t(79)(q34q32) with TCRB
Tectonin beta-propeller repeat containing 2	TECPR2	Disease: spastic paraplegia 49, autosomal recessive (SPG49)
TELO2-interacting protein 2	TTI2	Disease: mental retardation, autosomal recessive 39 (MRT39)
Tetratricopeptide repeat domain 7A	TTC7A	Disease: intestinal atresia, multiple (MINAT)
Thyroid adenoma associated	THADA	Note: Chromosomal aberrations involving THADA have been observed in benign thyroid adenomas
Titin-cap	TCAP	Disease: cardiomyopathy, familial hypertrophic (CMH)
Toll-like receptor adaptor molecule 1	TICAM1	Disease: herpes simplex encephalitis 4 (HSE4)
TRAF3-interacting protein 2	TRAF3IP2	Disease: psoriasis 13 (PSORS13)
Transducin (beta)-like 2	TBL2	Note: TBL2 is located in the Williams–Beuren syndrome (WBS) critical region
Translocator protein (18 kDa)	TSPO	**SSR-180575** – neuroprotective agents
TRK-fused gene	TFG	Disease: thyroid papillary carcinoma (TPC)
Tudor domain containing 7	TDRD7	Disease: cataract 36 (CTRCT36)
Tumor suppressor candidate 2	TUSC2	Phase I clinical trial of systemically administered TUSC2(FUS1)-nanoparticles mediating functional gene transfer in humans. PMID: 22558101

HGNC-approved name	Symbol	Target information
Tyrosine 3-monooxygenase/tryptophan 5-monooxygenase activation protein, eta	YWHAH	Roles of 14-3-3η in mitotic progression and its potential use as a therapeutic target for cancers. PMID: 22562251
Ubiquitin domain containing 1	UBTD1	UBTD1 induces cellular senescence through an UBTD1-Mdm2/p53 positive feedback loop. PMID: 25382750
Ubiquitin-like 7	UBL7	Solution structure of the ubiquitin-associated domain of human BMSC-UbP and its complex with ubiquitin. PMID: 16731964
vav 1 guanine nucleotide exchange factor	VAV1	VAV1 represses E-cadherin expression through the transactivation of Snail and Slug: a potential mechanism for aberrant epithelial to mesenchymal transition in human epithelial ovarian cancer. PMID: 23856093
v-crk avian sarcoma virus CT10 oncogene homolog-like	CRKL	The CRKL gene encoding an adaptor protein is amplified, overexpressed, and a possible therapeutic target in gastric cancer. PMID: 22591714
Vimentin	VIM	Disease: cataract 30 (CTRCT30)
Zinc finger E-box binding homeobox 2	ZEB2	Smad interacting protein 1 (SIP1) is associated with peritoneal carcinomatosis in intestinal type gastric cancer. PMID: 23143680. Disease: Mowat–Wilson syndrome (MOWS)
Zinc finger protein 365	ZNF365	Disease: uric acid nephrolithiasis (UAN)
Zinc finger, SWIM-type containing 6	ZSWIM6	Disease: acromelic frontonasal dysostosis (AFND)

<h1 style="text-align:center">Chapter 7</h1>

Non-coding RNAs

Most of the entries in this book relate to proteins translated from nuclear and mitochondrial messenger RNA. The bulk of nuclear RNA however is non-coding ('dark matter RNA', as described in one publication [1]). This may just have been an interesting aspect of genetic regulation with little consequence for pharmaceutical research were it not for the fact that some non-coding RNAs are involved in disease pathogenesis and are potential drug targets (e.g. Refs [2–4]). This chapter lists annotated entries taken from the long non-coding RNA (lncRNA) and microRNAs (miRNAs). Both entries represent a small proportion of total lnc and miRNAs listed by the HGNC, 0.7 and 8%, respectively. MRX34 is a mimic of microRNA 34a undergoing phase I clinical trials at the time of writing [5]. It is the only drug candidate listed in this chapter as others at a more advanced stage of clinical development are directed against viral targets.

References

1. Kapranov P *et al.* (2010) The majority of total nuclear-encoded non-ribosomal RNA in a human cell is 'dark matter' un-annotated RNA. *BMC Biology* **8**, 149.
2. Alexander R, Lodish H, Sun L. (2011) MicroRNAs in adipogenesis and as therapeutic targets for obesity. *Expert Opinion on Therapeutic Targets* **15**, 623–36.
3. He Y *et al.* (2014) Long noncoding RNAs: novel insights into hepatocellular carcinoma. *Cancer Letters* **344**, 20–7.
4. Li Z, Rana TM. (2014) Therapeutic targeting of microRNAs: current status and future challenges. *Nature Reviews Drug Discovery* **13**, 622–38.
5. ClinicalTrials.gov (2014) *A Multicenter Phase I Study of MRX34, MicroRNA miR-RX34 Liposomal Injection*. Available at https://clinicaltrials.gov/ct2/show/NCT01829971 (accessed 11 May 2015).

HGNC approved name	Symbol	Target information
Long non-coding RNAs		
ADAMTS9 antisense RNA 2	ADAMTS9-AS2	A new tumor suppressor LncRNA ADAMTS9-AS2 is regulated by DNMT1 and inhibits migration of glioma cells. PMID: 24833086
CADM2 antisense RNA 1	CADM2-AS1	LncRNA TSLC1-AS1 is a novel tumor suppressor in glioma. PMID: 25031725
CBR3 antisense RNA 1	CBR3-AS1	Upregulation of the long non-coding RNA PlncRNA-1 promotes esophageal squamous carcinoma cell proliferation and correlates with advanced clinical stage. PMID: 24337686
Colon cancer-associated transcript 1 (non-protein coding)	CCAT1	Long non-coding RNA CARLo-5 is a negative prognostic factor and exhibits tumor pro-oncogenic activity in non-small cell lung cancer. PMID: 25129441
FOXC1 upstream transcript (non-protein coding)	FOXCUT	A novel long non-coding RNA FOXCUT and mRNA FOXC1 pair promote progression and predict poor prognosis in esophageal squamous cell carcinoma. PMID: 25031703
GATA6 antisense RNA 1 (head to head)	GATA6-AS1	A known expressed sequence tag, BM742401, is a potent lincRNA inhibiting cancer metastasis. PMID: 23846333
HOX transcript antisense RNA	HOTAIR	Overexpression of long non-coding RNA HOTAIR predicts tumor recurrence in hepatocellular carcinoma patients following liver transplantation. PMID: 21327457
Long intergenic non-protein-coding RNA, regulator of reprogramming	LINC-ROR	LincRNA-ROR induces epithelial-to-mesenchymal transition and contributes to breast cancer tumorigenesis and metastasis. PMID: 24922071
Maternally expressed 3 (non-protein coding)	MEG3	Inhibitory effects of long noncoding RNA MEG3 on hepatic stellate cells activation and liver fibrogenesis. PMID: 25201080
Metastasis-associated lung adenocarcinoma transcript 1 (non-protein coding)	MALAT1	Long noncoding RNA MALAT-1 is a new potential therapeutic target for castration resistant prostate cancer.
Myosin heavy chain-associated RNA transcript	MHRT	A long noncoding RNA protects the heart from pathological hypertrophy. PMID: 25119045
NPTN intronic transcript 1 (non-protein coding)	NPTN-IT1	Long non-coding RNA-LET is a positive prognostic factor and exhibits tumor-suppressive activity in gallbladder cancer. PMID: 25213660
PCNA antisense RNA 1	PCNA-AS1	Antisense long non-coding RNA PCNA-AS1 promotes tumor growth by regulating proliferating cell nuclear antigen in hepatocellular carcinoma. PMID: 24704293
Pvt1 oncogene (non-protein coding)	PVT1	PVT1 dependence in cancer with MYC copy-number increase. PMID: 25043044
Small nucleolar RNA, C/D box 113-1	SNORD113-1	Small nucleolar RNA 113-1 suppresses tumorigenesis in hepatocellular carcinoma. PMID: 25216612
Small nucleolar RNA host gene 14 (non-protein coding)	SNHG14	Towards a therapy for Angelman syndrome by targeting a long non-coding RNA. PMID: 26470045
Taurine up-regulated 1 (non-protein coding)	TUG1	Long intergenic non-coding RNA TUG1 is overexpressed in urothelial carcinoma of the bladder. PMID: 22961206
Telomerase RNA component	TERC	Expression of targeted ribozyme against telomerase RNA causes altered expression of several other genes in tumor cells. PMID: 24664581
Urothelial cancer-associated 1 (non-protein coding)	UCA1	Long non-coding RNA UCA1a(CUDR) promotes proliferation and tumorigenesis of bladder cancer. PMID: 22576688
MicroRNAs		
MicroRNA 1-1	MIR1-1	MicroRNA-1 (miR-1) inhibits chordoma cell migration and invasion by targeting slug. PMID: 24760686
MicroRNA 7-1	MIR7-1	EGFR promotes lung tumorigenesis by activating miR-7 through a Ras/ERK/Myc pathway that targets the Ets2 transcriptional repressor ERF. PMID: 20978205

HGNC approved name	Symbol	Target information
MicroRNA 7-2	MIR7-2	MiR-7b directly targets DC-STAMP causing suppression of NFATc1 and c-Fos signaling during osteoclast fusion and differentiation. PMID: 25123438
MicroRNA 9-1	MIR9-1	miR-9 is an essential oncogenic microRNA specifically overexpressed in mixed lineage leukemia-rearranged leukemia. PMID: 23798388
MicroRNA 10a	MIR10A	Transcriptional regulation of miR-10a/b by TWIST-1 in myelodysplastic syndromes. PMID: 22983574
MicroRNA 10b	MIR10B	Human glioma growth is controlled by microRNA-10b. PMID: 21471404
MicroRNA 15a	MIR15A	MicroRNA 15a, inversely correlated to PKCα, is a potential marker to differentiate between benign and malignant renal tumors in biopsy and urine samples. PMID: 22429968
MicroRNA 15b	MIR15B	miR-15b/16-2 regulates factors that promote p53 phosphorylation and augments the DNA damage response following radiation in the lung. PMID: 25092292
MicroRNA 16-1	MIR16-1	MicroRNA-16 inhibits glioma cell growth and invasion through suppression of BCL2 and the nuclear factor-κB1/MMP9 signaling pathway. PMID: 24418124
MicroRNA 17	MIR17	Increased expression of microRNA-17 predicts poor prognosis in human glioma. PMID: 23226946
MicroRNA 18a	MIR18A	miR-18a promotes malignant progression by impairing microRNA biogenesis in nasopharyngeal carcinoma. PMID: 23097559
MicroRNA 19b-1	MIR19B1	Effects of miR-19b overexpression on proliferation, differentiation, apoptosis and Wnt/β-catenin signaling pathway in P19 cell model of cardiac differentiation *in vitro*. PMID: 23443808
MicroRNA 20a	MIR20A	miR-20a targets BNIP2 and contributes chemotherapeutic resistance in colorectal adenocarcinoma SW480 and SW620 cell lines. PMID: 21242194
MicroRNA 21	MIR21	MicroRNA-21: From cancer to cardiovascular disease. PMID: 20415649
MicroRNA 22	MIR22	miR-22 as a prognostic factor targets glucose transporter protein type 1 in breast cancer. PMID: 25304371
MicroRNA 23a	MIR23A	Effect of miR-23 on oxidant-induced injury in human retinal pigment epithelial cells. PMID: 21693609
MicroRNA 23b	MIR23B	MicroRNA-23b promotes tolerogenic properties of dendritic cells *in vivo* through inhibiting Notch1/NF κB signalling pathways. PMID: 22229716
MicroRNA 24-1	MIR24-1	Tumor suppressor miR-24 restrains gastric cancer progression by downregulating RegIV. PMID: 24886316
MicroRNA 25	MIR25	Inhibition of miR-25 improves cardiac contractility in the failing heart. PMID: 24670661
MicroRNA 26a-1	MIR26A1	MicroRNA-26 was decreased in rat cardiac hypertrophy model and may be a promising therapeutic target. PMID: 23719092
MicroRNA 26b	MIR26B	Human embryonic stem cells and metastatic colorectal cancer cells shared the common endogenous human microRNA-26b. PMID: 20831567
MicroRNA 27a	MIR27A	MicroRNA-27a inhibitors alone or in combination with perifosine suppress the growth of gastric cancer cells. PMID: 23175237
MicroRNA 27b	MIR27B	MicroRNA-27b contributes to lipopolysaccharide-mediated peroxisome proliferator-activated receptor gamma (PPARgamma) mRNA destabilization. PMID: 20164187
MicroRNA 29a	MIR29A	MicroRNA-29 family, a crucial therapeutic target for fibrosis diseases. PMID: 23542596

(Continued)

HGNC approved name	Symbol	Target information
MicroRNA-29b-1	MIR29B1	Identification of collagen 1 as a post-transcriptional target of miR-29b in skin fibroblasts: Therapeutic implication for scar reduction. PMID: 23221517
MicroRNA 30a	MIR30A	MicroRNA-30a inhibits cell migration and invasion by downregulating vimentin expression and is a potential prognostic marker in breast cancer. PMID: 22476851
MicroRNA 30b	MIR30B	MicroRNA-30b functions as a tumour suppressor in human colorectal cancer by targeting KRAS, PIK3CD and BCL2. PMID: 24293274
MicroRNA 30d	MIR30D	MicroRNA-30d promotes tumor invasion and metastasis by targeting Galphai2 in hepatocellular carcinoma. PMID: 20054866
MicroRNA 30e	MIR30E	MicroRNA-30e* promotes human glioma cell invasiveness in an orthotopic xenotransplantation model by disrupting the NF-κB/IκBα negative feedback loop. PMID: 22156201
MicroRNA 31	MIR31	MiR-31 is an independent prognostic factor and functions as an oncomir in cervical cancer via targeting ARID1A. PMID: 24793973
MicroRNA 33a	MIR33A	MicroRNA-33 and the SREBP host genes cooperate to control cholesterol homeostasis. PMID: 20466882
MicroRNA 33b	MIR33B	MicroRNA-33b knock-in mice for an intron of sterol regulatory element-binding factor 1 (Srebf1) exhibit reduced HDL-C in vivo. PMID: 24931346
MicroRNA 34a	**MIR34A**	**MRX34** – antineoplastic agents
MicroRNA 34b	MIR34B	Frequent methylation and oncogenic role of microRNA-34b/c in small-cell lung cancer. PMID: 22047961
MicroRNA 34c	MIR34C	Frequent methylation and oncogenic role of microRNA-34b/c in small-cell lung cancer. PMID: 22047961
MicroRNA 92a-1	MIR92A1	MicroRNA-92a functions as an oncogene in colorectal cancer by targeting PTEN. PMID: 24026406
MicroRNA 95	MIR95	MiR-95 induces proliferation and chemo- or radioresistance through directly targeting sorting nexin1 (SNX1) in non-small cell lung cancer. PMID: 24835695
MicroRNA 96	MIR96	Suppression of microRNA-96 expression inhibits the invasion of hepatocellular carcinoma cells. PMID: 22160187
MicroRNA 98	MIR98	Identification of microRNA-98 as a therapeutic target inhibiting prostate cancer growth and a biomarker induced by vitamin D. PMID: 23188821
MicroRNA 99a	MIR99A	Downregulation of microRNA 99a in oral squamous cell carcinomas contributes to the growth and survival of oral cancer cells. PMID: 22751686
MicroRNA 100	MIR100	Downregulation of microRNA-100 correlates with tumor progression and poor prognosis in hepatocellular carcinoma. PMID: 23842624
MicroRNA 101-1	MIR101-1	Carcinogenesis of intraductal papillary mucinous neoplasm of the pancreas: loss of microRNA-101 promotes overexpression of histone methyltransferase EZH2. PMID: 21932133
MicroRNA 103a-1	MIR103A1	MicroRNA-103 promotes colorectal cancer by targeting tumor suppressor DICER and PTEN. PMID: 24828205
MicroRNA 105-1	MIR105-1	MicroRNA-105 suppresses cell proliferation and inhibits PI3K/AKT signaling in human hepatocellular carcinoma. PMID: 25280563
MicroRNA 106a	MIR106A	Regulation of STAT3 by miR-106a is linked to cognitive impairment in ovariectomized mice. PMID: 23399684
MicroRNA 106b	MIR106B	Down-regulation of microRNA 106b is involved in p21-mediated cell cycle arrest in response to radiation in prostate cancer cells. PMID: 20878953

HGNC approved name	Symbol	Target information
MicroRNA 107	MIR107	MicroRNA-107, an oncogene microRNA that regulates tumour invasion and metastasis by targeting DICER1 in gastric cancer. PMID: 21029372
MicroRNA 122	MIR122	MicroRNA miR-122 as a therapeutic target for oligonucleotides and small molecules. PMID: 23745562
MicroRNA 124-1	MIR124-1	MicroRNA-124 regulates the proliferation of colorectal cancer cells by targeting iASPP. PMID: 23691514
MicroRNA 125b-1	MIR125B1	Oncomir miR-125b suppresses p14(ARF) to modulate p53-dependent and p53-independent apoptosis in prostate cancer. PMID: 23585871
MicroRNA 126	MIR126	Downregulation of miR-126 induces angiogenesis and lymphangiogenesis by activation of VEGF-A in oral cancer. PMID: 22836510
MicroRNA 127	MIR127	High expression of microRNA-127 is involved in cell cycle arrest in MC-3 mucoepidermoid carcinoma cells. PMID: 23232714
MicroRNA 128-1	MIR128-1	MiR-128 inhibits tumor growth and angiogenesis by targeting p70S6K1. PMID: 22442669
MicroRNA 129-1	MIR129-1	miR-129 as a novel therapeutic target and biomarker in gastrointestinal cancer. PMID: 25187728
MicroRNA 132	MIR132	MicroRNA-132 suppresses autoimmune encephalomyelitis by inducing cholinergic anti-inflammation: A new Ahr-based exploration. PMID: 23780851
MicroRNA 133a-1	MIR133A1	MicroRNA-133a regulates the cell cycle and proliferation of breast cancer cells by targeting epidermal growth factor receptor through the EGFR/Akt signaling pathway. PMID: 23786162
MicroRNA 134	MIR134	miR-134 regulates ischemia/reperfusion injury-induced neuronal cell death by regulating CREB signaling. PMID: 25316150
MicroRNA 135a-1	MIR135A1	miR-135a inhibition protects A549 cells from LPS-induced apoptosis by targeting Bcl-2. PMID: 25230140
MicroRNA 135b	MIR135B	MicroRNA-135b promotes lung cancer metastasis by regulating multiple targets in the Hippo pathway and LZTS1. PMID: 23695671
MicroRNA 138-1	MIR138-1	Downregulation of the Rho GTPase signaling pathway is involved in the microRNA-138-mediated inhibition of cell migration and invasion in tongue squamous cell carcinoma. PMID: 20232393
MicroRNA 141	MIR141	miR-141 targets ZEB2 to suppress HCC progression. PMID: 25008569
MicroRNA 142	MIR142	An oncogenic role of miR-142-3p in human T-cell acute lymphoblastic leukemia (T-ALL) by targeting glucocorticoid receptor-α and cAMP/PKA pathways. PMID: 21979877
MicroRNA 143	MIR143	MicroRNA-143 targets syndecan-1 to repress cell growth in melanoma. PMID: 24722758
MicroRNA 145	MIR145	MicroRNA-145 targeted therapy reduces atherosclerosis. PMID: 22965997
MicroRNA 146a	MIR146A	MicroRNA-146a acts as a metastasis suppressor in gastric cancer by targeting WASF2. PMID: 23435376
MicroRNA 146b	MIR146B	MiR-146b is a regulator of human visceral preadipocyte proliferation and differentiation and its expression is altered in human obesity. PMID: 24931160
MicroRNA 148a	MIR148A	MicroRNA-148a: A potential therapeutic target for cancer. PMID: 24084367
MicroRNA 150	MIR150	The role of miR-150 in normal and malignant hematopoiesis. PMID: 23955084
MicroRNA 152	MIR152	MiR-152 reduces human umbilical vein endothelial cell proliferation and migration by targeting ADAM17. PMID: 24813629

(Continued)

HGNC approved name	Symbol	Target information
MicroRNA 155	MIR155	miR-155 inhibitor reduces the proliferation and migration in osteosarcoma MG-63 cells. PMID: 25289062
MicroRNA 181a-1	MIR181A1	Downregulation of miR-181a upregulates sirtuin-1 (SIRT1) and improves hepatic insulin sensitivity. PMID: 22476949
MicroRNA 181b-1	MIR181B1	MicroRNA-181b regulates articular chondrocytes differentiation and cartilage integrity. PMID: 23313477
MicroRNA 181c	MIR181C	Association of microRNA-181c expression with the progression and prognosis of human gastric carcinoma. PMID: 23425811
MicroRNA 182	MIR182	Aberrant microRNA-182 expression is associated with glucocorticoid resistance in lymphoblastic malignancies. PMID: 22582938
MicroRNA 185	MIR185	MicroRNA-185 and 342 inhibit tumorigenicity and induce apoptosis through blockade of the SREBP metabolic pathway in prostate cancer cells. PMID: 23951060
MicroRNA 191	MIR191	MicroRNA-191 correlates with poor prognosis of colorectal carcinoma and plays multiple roles by targeting tissue inhibitor of metalloprotease 3. PMID: 24195505
MicroRNA 192	MIR192	BMP-6 inhibits cell proliferation by targeting microRNA-192 in breast cancer. PMID: 24012720
MicroRNA 193a	MIR193A	Focal segmental glomerulosclerosis is induced by microRNA-193a and its downregulation of WT1. PMID: 23502960
MicroRNA 195	MIR195	Genome-wide screening reveals that miR-195 targets the TNF-α/NF-κB pathway by down-regulating IκB kinase alpha and TAB3 in hepatocellular carcinoma. PMID: 23487264
MicroRNA 196a-1	MIR196A1	MicroRNA-196a is a putative diagnostic biomarker and therapeutic target for laryngeal cancer. PMID: 23967217
MicroRNA 197	MIR197	miR-197 induces epithelial-mesenchymal transition in pancreatic cancer cells by targeting p120 catenin. PMID: 23139153
MicroRNA 199a-1	MIR199A1	miR-199a regulates the tumor suppressor mitogen-activated protein kinase kinase kinase 11 in gastric cancer. PMID: 21048306
MicroRNA 199b	MIR199B	MicroRNA-199b targets the nuclear kinase Dyrk1a in an auto-amplification loop promoting calcineurin/NFAT signalling. PMID: 21102440
MicroRNA 200a	MIR200A	DCAMKL-1 regulates epithelial-mesenchymal transition in human pancreatic cells through a miR-200a dependent mechanism. PMID: 21285251
MicroRNA 200c	MIR200C	Down-regulation of microRNA-200c is associated with drug resistance in human breast cancer. PMID: 22101791
MicroRNA 203a	MIR203A	MiR-203 controls proliferation, migration and invasive potential of prostate cancer cell lines. PMID: 21368580
MicroRNA 204	MIR204	miR-204 suppresses cochlear spiral ganglion neuron survival *in vitro* by targeting TMPRSS3. PMID: 24924414
MicroRNA 205	MIR205	Downregulation of miR-205 modulates cell susceptibility to oxidative and endoplasmic reticulum stresses in renal tubular cells. PMID: 22859986
MicroRNA 206	MIR206	Downregulation of microRNA-206 is a potent prognostic marker for patients with gastric cancer. PMID: 23751352
MicroRNA 210	MIR210	Downregulation of miR-210 expression inhibits proliferation, induces apoptosis and enhances radiosensitivity in hypoxic human hepatoma cells *in vitro*. PMID: 22387901
MicroRNA 214	MIR214	MiR-214 inhibits cell growth in hepatocellular carcinoma through suppression of β-catenin. PMID: 23068095

HGNC approved name	Symbol	Target information
MicroRNA 215	MIR215	MicroRNA-215 regulates fibroblast function: Insights from a human fibrotic disease. PMID: 25565137
MicroRNA 216b	MIR216B	Regulation of the P2X7R by microRNA-216b in human breast cancer. PMID: 25078617
MicroRNA 217	MIR217	MicroRNA-217 promotes ethanol-induced fat accumulation in hepatocytes by down-regulating SIRT1. PMID: 22308024
MicroRNA 218-1	MIR219	Paxillin predicts survival and relapse in non-small cell lung cancer by microRNA-218 targeting. PMID: 21159652
MicroRNA 219	MIR218-2	miR-219 inhibits the proliferation, migration and invasion of medulloblastoma cells by targeting CD164. PMID: 24756834
MicroRNA 221	MIR221	Clinical significance of miR-221 and its inverse correlation with p27Kip[1] in hepatocellular carcinoma. PMID: 20146005
MicroRNA 222	MIR222	Down-regulation of miR-221 and miR-222 correlates with pronounced Kit expression in gastrointestinal stromal tumors. PMID: 21132270
MicroRNA 223	MIR223	Stathmin1 plays oncogenic role and is a target of microRNA-223 in gastric cancer. PMID: 22470493
MicroRNA 224	MIR224	Role of miR-224 in hepatocellular carcinoma: A tool for possible therapeutic intervention? PMID: 22122284
MicroRNA 296	MIR296	Regulation of HMGA1 expression by microRNA-296 affects prostate cancer growth and invasion. PMID: 21138859
MicroRNA 320a	MIR320A	MicroRNA 320a functions as a novel endogenous modulator of aquaporins 1 and 4 as well as a potential therapeutic target in cerebral ischemia. PMID: 20628061
MicroRNA 320c-1	MIR320C1	miR-320c regulates gemcitabine-resistance in pancreatic cancer via SMARCC1. PMID: 23799850
MicroRNA 323a	MIR323A	Identification of microRNA-221/222 and microRNA-323-3p association with rheumatoid arthritis via predictions using the human tumour necrosis factor transgenic mouse model. PMID:
MicroRNA 328	MIR328	MicroRNA-328 contributes to adverse electrical remodeling in atrial fibrillation. PMID: 21098446
MicroRNA 330	MIR330	MicroRNA-330 is an oncogenic factor in glioblastoma cells by regulating SH3GL2 gene. PMID: 23029364
MicroRNA 331	MIR331	MicroRNA-331-3p promotes proliferation and metastasis of hepatocellular carcinoma by targeting PH domain and leucine rich repeat protein phosphatase. PMID: 24825302
MicroRNA 335	MIR335	Effect of miR-335 upregulation on the apoptosis and invasion of lung cancer cell A549 and H1299. PMID: 23740614
MicroRNA 339	MIR339	MiR-339-5p regulates the growth, colony formation and metastasis of colorectal cancer cells by targeting PRL-1. PMID: 23696794
MicroRNA 340	MIR340	MicroRNA-340 suppresses osteosarcoma tumor growth and metastasis by directly targeting ROCK1. PMID: 23872151
MicroRNA 342	MIR342	MicroRNA-185 and 342 inhibit tumorigenicity and induce apoptosis through blockade of the SREBP metabolic pathway in prostate cancer cells. PMID: 23951060
MicroRNA 365a	MIR365A	miR-365 promotes cutaneous squamous cell carcinoma (CSCC) through targeting nuclear factor I/B (NFIB). PMID: 24949940
MicroRNA 372	MIR372	Upregulation of microRNA-372 associates with tumor progression and prognosis in hepatocellular carcinoma. PMID: 23291979
MicroRNA 374a	MIR374A	Identification of miR-374a as a prognostic marker for survival in patients with early-stage nonsmall cell lung cancer. PMID: 21748820

(Continued)

HGNC approved name	Symbol	Target information
MicroRNA 375	MIR375	miR-375 is upregulated in acquired paclitaxel resistance in cervical cancer. PMID: 23778521
MicroRNA 376a-1	MIR376A1	Attenuated adenosine-to-inosine editing of microRNA-376a* promotes invasiveness of glioblastoma cells. PMID: 23093778
MicroRNA 376c	MIR376C	MicroRNA-376c inhibits cell proliferation and invasion in osteosarcoma by targeting to transforming growth factor-alpha. PMID: 23631646
MicroRNA 377	MIR377	MicroRNA-377 inhibited proliferation and invasion of human glioblastoma cells by directly targeting specificity protein 1. PMID: 24951112
MicroRNA 378a	MIR378A	Interplay between heme oxygenase-1 and miR-378 affects non-small cell lung carcinoma growth, vascularization, and metastasis. PMID: 23617628
MicroRNA 380	MIR380	miR-380-5p represses p53 to control cellular survival and is associated with poor outcome in MYCN-amplified neuroblastoma. PMID: 20871609
MicroRNA 382	MIR382	RP5-833A20.1/miR-382-5p/NFIA-dependent signal transduction pathway contributes to the regulation of cholesterol homeostasis and inflammatory reaction. PMID: 25265644
MicroRNA 383	MIR383	Downregulation of miR-383 promotes glioma cell invasion by targeting insulin-like growth factor 1 receptor. PMID: 23564324
MicroRNA 409	MIR409	miR-409-3p/-5p promotes tumorigenesis, epithelial-to-mesenchymal transition, and bone metastasis of human prostate cancer. PMID: 24963047
MicroRNA 421	MIR421	Angiotensin-converting enzyme 2 is subject to post-transcriptional regulation by miR-421. PMID: 24564768
MicroRNA 423	MIR423	miRNA423-5p regulates cell proliferation and invasion by targeting trefoil factor 1 in gastric cancer cells. PMID: 24486742
MicroRNA 424	MIR424	Suppressed miR-424 expression via upregulation of target gene Chk1 contributes to the progression of cervical cancer. PMID: 22469983
MicroRNA 451a	MIR451A	MiR-451 is decreased in hypertrophic cardiomyopathy and regulates autophagy by targeting TSC1. PMID: 25209900
MicroRNA 452	MIR452	Downregulation of miR-452 promotes stem-like traits and tumorigenicity of gliomas. PMID: 23695168
MicroRNA 492	MIR492	MicroRNA-492 expression promotes the progression of hepatic cancer by targeting PTEN. PMID: 25253996
MicroRNA 493	MIR493	MicroRNA-493 suppresses tumor growth, invasion and metastasis of lung cancer by regulating E2F1. PMID: 25105419
MicroRNA 494	MIR494	MicroRNA-494 within an oncogenic microRNA megacluster regulates G1/S transition in liver tumorigenesis through suppression of mutated in colorectal cancer. PMID: 23913442
MicroRNA 495	MIR495	MicroRNA-495 induces breast cancer cell migration by targeting JAM-A. PMID: 25070379
MicroRNA 497	MIR497	MicroRNA-497 suppresses proliferation and induces apoptosis in prostate cancer cells. PMID: 23886135
MicroRNA 499a	MIR499A	MicroRNA-499-5p promotes cellular invasion and tumor metastasis in colorectal cancer by targeting FOXO4 and PDCD4. PMID: 21934092
MicroRNA 501	MIR501	MicroRNA-501 promotes HBV replication by targeting HBXIP. PMID: 23266610
MicroRNA 503	MIR503	Deregulation of microRNA-503 contributes to diabetes mellitus-induced impairment of endothelial function and reparative angiogenesis after limb ischemia. PMID: 21220732

HGNC approved name	Symbol	Target information
MicroRNA 506	MIR506	Up-regulation of microRNA 506 leads to decreased Cl-/HCO3- anion exchanger 2 expression in biliary epithelium of patients with primary biliary cirrhosis. PMID: 22383162
MicroRNA 516	MIR516	The metastasis-associated microRNA miR-516a-3p is a novel therapeutic target for inhibiting peritoneal dissemination of human scirrhous gastric cancer. PMID: 21169410
MicroRNA 526b	MIR526B	By downregulating Ku80, hsa-miR-526b suppresses non-small cell lung cancer. PMID: 25596743
MicroRNA 574	MIR574	Concerted functions of HDAC1 and microRNA-574-5p repress alternatively spliced ceramide synthase 1 expression in human cancer cells. PMID: 22180294
MicroRNA 708	MIR1228	miR-1228 prevents cellular apoptosis through targeting of MOAP1 protein. PMID: 22434376
MicroRNA 708	MIR708	miRNA-708 control of CD44(+) prostate cancer-initiating cells. PMID: 22552290
MicroRNA 940	MIR940	MicroRNA-940 suppresses prostate cancer migration and invasion by regulating MIEN1. PMID: 25406943
MicroRNA 1202	MIR1202	miR-1202 is a primate-specific and brain-enriched microRNA involved in major depression and antidepressant treatment. PMID: 24908571
MicroRNA 1228	MIR1258	MicroRNA-1258 suppresses breast cancer brain metastasis by targeting heparanase. PMID: 21266359
MicroRNA 1258	MIR1275	Contribution of microRNA-1275 to Claudin11 protein suppression via a polycomb-mediated silencing mechanism in human glioma stem-like cells. PMID: 22736761
MicroRNA 1275	MIR548D	MicroRNA miR-548d is a superior regulator in pancreatic cancer. PMID: 21946813
MicroRNA let-7a-1	MIRLET7A1	Let-7a is a direct EWS-FLI-1 target implicated in Ewing's sarcoma development. PMID: 21853155
MicroRNA let-7b	MIRLET7B	Let-7b-5p regulates proliferation and apoptosis in multiple myeloma by targeting IGF1R. PMID: 25271256
MicroRNA let-7c	MIRLET7C	MicroRNA let-7c is downregulated in prostate cancer and suppresses prostate cancer growth. PMID: 22479342
MicroRNA let-7i	MIRLET7I	Decreased expression of microRNA let-7i and its association with chemotherapeutic response in human gastric cancer. PMID: 23107361